PERGAMON INTERNATIONAL LIBRARY
of Science, Technology, Engineering and Social Studies
The 1000-volume original paperback library in aid of education, industrial training and the enjoyment of leisure
Publisher: Robert Maxwell, M.C.

STRUCTURE OF METALS

Third Revised Edition

International Series on
MATERIALS SCIENCE AND TECHNOLOGY

Volume 35 — *Editor:* H. G. HOPKINS, D.Sc.†

STRUCTURE OF METALS

Crystallographic Methods, Principles and Data

Third Revised Edition

C. S. BARRETT, Ph.D.
Emeritus Professor, University of Chicago
Adjunct Professor and Senior Research Scientist
University of Denver
Colorado, USA

T. B. MASSALSKI, Ph.D., D.Sc.
Professor of Metallurgy, Materials Science and Physics
Carnegie-Mellon University
Pittsburgh, Pennsylvania, USA

PERGAMON PRESS
OXFORD · NEW YORK · BEIJING · FRANKFURT
SÃO PAULO · SYDNEY · TOKYO · TORONTO

U.K.	Pergamon Press, Headington Hill Hall, Oxford OX3 0BW, England
U.S.A.	Pergamon Press, Maxwell House, Fairview Park, Elmsford, New York 10523, U.S.A.
PEOPLE'S REPUBLIC OF CHINA	Pergamon Press, Room 4037, Qianmen Hotel, Beijing, People's Republic of China
FEDERAL REPUBLIC OF GERMANY	Pergamon Press, Hammerweg 6, D-6242 Kronberg, Federal Republic of Germany
BRAZIL	Pergamon Editora, Rua Eça de Queiros, 346, CEP 04011, Paraiso, São Paulo, Brazil
AUSTRALIA	Pergamon Press Australia, P.O. Box 544, Potts Point, N.S.W. 2011, Australia
JAPAN	Pergamon Press, 8th Floor, Matsuoka Central Building, 1-7-1 Nishishinjuku, Shinjuku-ku, Tokyo 160, Japan
CANADA	Pergamon Press Canada, Suite No 271, 253 College Street, Toronto, Ontario, Canada M5T 1R5

First edition 1943 (McGraw-Hill Inc.)
Second edition 1952 (McGraw-Hill Inc.)
Third edition 1966 (McGraw-Hill Inc.)
Third (revised) edition 1980
Reprinted 1982, 1986, 1987

British Library Cataloguing in Publication Data

Barrett, Charles Sanborn
Structure of metals.—3rd revised ed.
(Pergamon international library: international series on materials science and technology; vol. 35).
1. Metallography
I. Title II. Massalski, Tadeusz Bronislaw
669′.95 TN690 80-49878

ISBN 0-08-026171-X (Hardcover)
ISBN 0-08-026172-8 (Flexicover)

Printed in Great Britain by A. Wheaton & Co. Ltd., Exeter

Preface

This book presents the methods most used in determining the structures of crystalline and noncrystalline materials, introduces the nomenclature with which they and their symmetry properties are described, and summarizes the nature of many of the most important structures and theories regarding them. Defects in crystals, the means of observing these, and transformations from one crystal structure to another are also treated.

The book is intended to be used as a convenient reference volume for those concerned with research in metallurgy, materials science, chemistry, and solid state physics, but many of the chapters are also intended to serve students at both undergraduate and graduate levels. The authors feel that a single volume of moderate size can be useful in these diverse ways; this optimistic view has arisen from the wide acceptance of the preceding editions of the book and from the diverse uses that have been made of these editions. Discussions with many readers led to the conclusion that although extensive revision was desirable, the book should be neither greatly expanded in size nor severely reduced in breadth of coverage.

The first four chapters present the fundamentals of crystal lattices, crystal symmetry, projections and their use, and the basic principles of the reciprocal lattice and of diffraction by crystals. Chapters 5 to 9, 15 to 17, 22, and 23 present experimental methods; the remaining chapters discuss the results of research. Although metals and alloys are given more attention than nonmetallic substances, there has been no intention to limit discussions solely to the metallic. In fact, sections dealing with methods are almost equally applicable to the wider range of materials; some examples are the sections on structure determination, defects and defect registration by modern microscopic methods, single-crystal orientation and polycrystalline preferred orientation determination, phase-diagram determination, identification of materials, and stress analysis.

The treatments of many subjects have been extensively revised in this edition, especially the following: crystal structures of alloys; superlattices and their transformations; short-range order in superlattices; theories of the structures of metals and alloys; textures, texture determination, and texture theories; phase transformations and precipitation; stacking faults and crystal defects of many kinds and methods of studying these; cold-worked metals; liquids; chemical analysis by diffraction; grain-size and particle-size determination; diffractometer operation; x-ray topography; electron microscopy; electron diffraction; and neutron diffraction. Some of the most useful computer programs for crystallographic work have been listed. Many tables have been revised and brought up to date; new tables have been added, including data on structures of metals, superlattices, alloys, and structures of metals and compounds at high pressures; and data on structures at high and low temperatures have been expanded.

In attempting to retain approximately the same broad coverage as in earlier editions without expanding the book unduly, we faced a difficult problem, and we hope that our method of handling it will prove satisfactory. We have shortened (but not eliminated) the most elementary sections; this material is available in many books and is already familiar to most of our readers. We have given only fundamental principles, examples, and critical discussions of the most advanced research topics, avoiding elaborate details, since anyone undertaking research in these areas would have need to consult the original sources in any event—but we have attempted to cite important references where details, further results, and extensive bibliographies will be found. Finally, we have saved valuable space by using smaller print for sections we judge will be referred to relatively infrequently by most readers.

We are grateful to many friends for suggestions and advice and for permissions to reproduce figures and data. Particular thanks are due the following for reading portions of the manuscript and suggesting improvements: J. A. Beardon, Paul A. Beck, M. B. Bever, J. B. Cohen, Mrs. V. B. Compton, R. M. J. Cotterill, A. English, L. M. Falicov, John Goldak, Peter Haasen, Karl Haefner, W. H. Hu, R. W. Johnson, J. K. Mackenzie, R. E. Marburger, M. Marezio, M. Meshii, M. H. Mueller, M. V. Nevitt, H. M. Otte, E. Parthé, Horace Pops, W. Pfann, R. E. Reed-Hill, Hiroshi Sato, C. G. Shull, J. W. Stout, Gareth Thomsa, L. F. Vassamillet, Ray Waddoups, C. B. Walker, and Charles Wert.

Our work on this edition was facilitated by several arrangements that were very much appreciated by us. In the early stages, we were visiting professors together in the summer term of 1963 at the Department of Materials Science of Stanford University; in the final stages (1965–1966) we were both in Professor Hume-Rothery's Department of Metallurgy at Oxford University, as a result of an appointment of a George Eastman Visiting Professorship to C.S.B., and of a John Simon Guggenheim Fellowship to T.B.M. The support and encouragement of our own institutions, the University of Chicago and the Mellon Institute, throughout the whole period are gratefully acknowledged. We are indebted to Professor R. F. Mehl for advice and criticism in connection with earlier editions, to various readers who have offered suggestions over the years, and to Mrs. Barrett for editorial assistance again in this edition.

Charles S. Barrett
T. B. Massalski

Contents

11 Superlattices 270

12 Electrons in metallic crystals 306

13 Theories of metallic phases 338

1

The fundamentals of crystallography

Introduction Atoms in a crystal are arrayed in a pattern that repeats itself in three dimensions throughout the interior of the crystal. Various early crystallographers (particularly those of the 1700s) had proposed that a crystal is made up of numerous identical units stacked tightly together in a three-dimensional array. Spherical and spheroidal units had been suggested in previous centuries by Kepler, Hooke, and Huygens, and polyhedral parts had been proposed by P. C. Grignon, Tobern Bergman, and Romé de l'Isle. Following these qualitative ideas, Haüy initiated the quantitative science of crystallography by working out mathematical expressions for the precisely measurable and unvarying angles between crystal faces of a given crystalline substance in terms of the stacking together of identical units of structure.[1] The proof of the general correctness of these ideas was not obtained until fairly recently—in 1912, to be exact, when x-rays were first used to reveal the actual internal structure of a crystal. In the meantime, these angular relationships and their symmetries were found to provide a good basis for classifying crystals and proved to be particularly useful over a period of many years for identifying minerals.

A real understanding of the crystalline state in terms of the distribution of atoms began with Max von Laue's discovery in 1912 of the diffraction of x rays by crystals and with W. L. (Sir Laurence) Bragg's first analysis of the structure of a crystal (rock salt) and with the determination of the wavelength of monochromatic x-ray beams by his father, W. H. (Sir William) Bragg.

Suddenly the crystallographer had available a means by which he could deduce the exact size and shape of the fundamental unit of structure of a crystal and the spatial grouping of atoms that made up this unit. For the first time the origin of the observable properties of a crystal could be sought in terms of an actual, not merely a postulated, internal structure. The impact of this development and its ramifications on pure and applied science has been immense and has seldom been equaled.

The crystallography that was initiated by x-ray diffraction in 1912 and was later extended to electron and neutron diffraction has been of such widespread value that half a century later it is a more active field than

[1] René Just Haüy, *Phil. Mag.*, vol. 1, pp. 35, 153, 287, 376, 1798. His ideas also appeared in book form in "Essai d'une théorie sur la Structure de Crystaux," 1784, and in his other books of 1801 and 1822.

ever before.[1] It has made countless contributions in the fields of minerals, metals, foods, drugs, living tissues, textiles, fibers, plastics, and organic and inorganic materials of all kinds. Therefore it has become increasingly important for physicists, chemists, mineralogists, metallurgists, and engineers to be aware of the fundamentals of crystallography, its methods, its major successes, and its potentialities for the future.

To grasp the fundamentals and to understand the developments in any of these fields requires an understanding of the crystallographic language with which they are reported—a kind of crystallographer's "shorthand." With this nomenclature, which is used throughout the world, one neatly and accurately expresses the entire symmetry of a crystal, the symmetry not only of external faces but also of the internal structure, the atom positions in a crystal, and the diffraction characteristics of a crystal. We therefore begin with a condensed presentation of this nomenclature, with an explanation of symmetry and of the periodicity of the inner structure and with references to the standard tabulations that are so valuable in crystallographic work. Graphical methods that aid in visualizing the crystallographic relationships are explained (Chap. 2), and then a number of chapters take up the diffraction methods and their use. In later chapters the structures of actual crystals and some theories of these, imperfections in crystals and their origins, orientations of crystals in polycrystalline materials, and the nature of solid-state transformations are presented. Throughout the book attention is given to the means of observing these diverse features with modern techniques employing x-rays, electrons, and neutrons.

Noncrystalline and semicrystalline states Crystallinity is completely absent in a *gas*, owing to the random motion and spacing of the constituent atoms or molecules. It is also absent in liquids and glasses; in these there is a tendency for atoms to group themselves at close, uniform distances from each other, but without building up long-range periodicity in the array. Some of the groupings presumably resemble tiny crystals having the atomic configuration that the material would have if crystallized, but other local groupings differ from these, some having arrangements not found in any three-dimensional crystals. Each specific local arrangement has a lifetime of very short duration, and is constantly undergoing change as the atoms migrate in thermally induced motion.

A *liquid* has a definite volume but no definite shape; at lower temperatures a liquid becomes a *solid* and has a definite shape as well as a definite volume. The solid may be crystalline or noncrystalline; if noncrystalline (amorphous, vitreous, glassy) it is actually a *supercooled liquid* and differs from ordinary liquids only in its physical properties. By convention, the property of viscosity is normally used to distinguish a glass from a liquid; if its viscosity is less than 10^{15} poises (cgs units), it is called a liquid and if

[1] A recent historical treatment deserves mention here: P. P. Ewald (ed.), "Fifty Years of X-ray Diffraction," N. V. A. Oosterhoek's Uitgeversmaatschappij, Utrecht, The Netherlands, 1962.

greater than this, a solid. The transition between the liquid and glassy states on heating or cooling is a gradual one, whereas the more fundamental transition between the liquid and crystalline states is more abrupt. Crystals may form and grow in a noncrystalline solid if the solid is held at a temperature high enough to permit atoms to interchange positions. This is the process of devitrification of a glass. A crystalline solid does not deform in a viscous manner; there is practically no permanent deformation if the applied stresses are less than a certain critical value, and at higher stresses the rates of plastic deformation are very different from those of the liquid or glassy states.

Glasses have been formed not only with substances having complex structures in their crystalline state, but also with some having simple structures, and in substances representing every type of interatomic bond—metallic, covalent, ionic, and van der Waals. A critical review of theories of the liquid and glassy states has been published by Turnbull.[1] Metals do not ordinarily supercool to the glassy state, but it is possible to produce an amorphous or nearly amorphous condition in some metals and alloys by various methods. Some metals deposited from the vapor state at very low temperatures are amorphous, and some alloys have been supercooled to the glassy state by shooting liquid drops at high velocity against a cold metal surface.[2] Bridgman showed that the crystallinity of some metals could be disrupted almost completely by extreme amounts of plastic shearing between rotating anvils under compressive stress.[3] The plastic deformation in the surface layers of a solid during mechanical polishing also results in a near destruction of crystallinity in some instances.

An amorphous substance usually is isotropic, i.e., it exhibits the same physical and chemical properties when tested in any direction (provided it is homogeneous and free from internal strains), but a crystal has many properties that are directional. Directionality (anisotropy) in a crystal is found to a greater or lesser degree, for example, in electrical and thermal conductivity, thermal expansion, elastic constants, optical constants, and chemical reactivity of exposed surfaces. Directionality is particularly evident when a crystal is growing, for if it is allowed to grow without being inhibited by containing walls or other disturbing influences, it will develop growth faces, which are intimately related to the internal structure. The symmetry of the growth faces has served, in fact, as an important means of identifying minerals and in determining the orientation of crystals (see Chap. 2).

Substances may be polycrystalline with a grain size so small that long-range periodicity is practically absent. Various other semicrystalline arrangements are also found. Long-range periodicity may exist in one direction only, as when layers are stacked upon each other with constant interlayer distances but with random orientation of each layer. Crystallinity in two

[1] D. Turnbull, *Trans. AIME*, vol. 221, p. 422, 1961.

[2] W. Klement, R. H. Willens, and P. Duwez, *Nature*, vol. 187, p. 869, 1960. Additional references to this type of experiment are included later (p. 161).

[3] P. W. Bridgman, *Phys. Rev.*, vol. 48. p. 825, 1935.

dimensions can exist within a single layer (for example, in monomolecular layers of one substance on the surface of another). High polymers composed of long-chain molecules can exist with varying amounts of short-range order, determined by the degree of cross-linking of the molecules. The degree of alignment of the long-chain molecules parallel to each other varies over wide limits in different substances and in some substances under different conditions. All real crystals have occasional atomic positions vacant, i.e., they contain "vacancies," and most crystals also have atomic layers or segments of layers out of place. Impurities or alloying elements in crystals may introduce atoms interstitially between the host atoms, or may replace the host atoms at random or in a way that shows some degree of regularity. All of these deviations from perfection have important effects on some of the physical, chemical, and mechanical properties (the "structure-sensitive" properties). But before giving consideration to deviations from crystal perfection, it is logical to discuss the nature of ideally perfect crystals and to introduce important facts, concepts, and nomenclature that should become familiar if one is to comprehend current thought and research dealing with crystals.

Lattices Since the essential characteristic of a crystal is the periodic nature of its internal structure, it is natural to relate the atomic arrangement to a network of points in space, called a *space lattice* or simply a *lattice*, the surroundings of each point being identical with the surroundings of each of the others. An example is sketched in Fig. 1–1. Consider, first, any *row* of atoms in a crystal. The atomic arrangement along the row will repeat itself at regular intervals so that the repetitive scheme can be represented by a row of equally spaced points. These points would be points of a space lattice. Similarly, the periodic arrangement on an atomic *plane* will have a periodicity that can be represented by a two-dimensional array of points, each point having associated with it an identical cluster of atoms or perhaps only a single atom. If a vector **a** specifies the direction and magnitude of the translation from one point to the next along one row of points in the plane, and if the vector **b** does the same for another row in a different direction, as indicated in Fig. 1–1, then the vector relation $\mathbf{r}_{uv} = u\mathbf{a} + v\mathbf{b}$, where u and v are integers, describes the position of other lattice points on the plane. It is possible to choose the set of vectors **a** and **b** so as to have the numbers u and v be integral for *all* these points. On the other hand, it might be convenient to choose them otherwise, in order, for

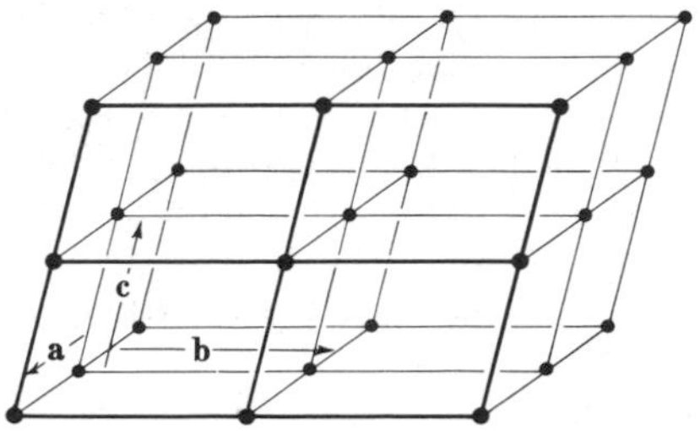

Fig. 1-1 A space lattice.

example, to have **a** and **b** at right angles to each other; u and v might then have fractional values for some of the points. Finally, consider the periodicity throughout a three-dimensional crystal. Three fundamental translations, **a**, **b**, **c**, are needed, and not all these can lie in a plane. Any point in the three-dimensional space lattice can be specified by $\mathbf{r}_{uvw} = u\mathbf{a} + v\mathbf{b} + w\mathbf{c}$, and as before, the vectors **a**, **b**, **c** can be chosen, if desired, so that u, v, and w are integers. The translation **c** superimposes the lattice plane containing **a** and **b** upon another lattice plane that lies parallel to it.

The points of the space lattice of a crystal are the positions occupied by *single* atoms, in many of the common metals and other elements of the periodic table; in most crystals, however, they are points at each of which there is centered an identical *group* or *basis* of atoms. In a perfect crystal each such group is identical both in composition and in orientation with every other, so that a tiny observer hopping from one lattice point to another would be unable to distinguish one from another, the outlook from any one being the same as from any other.

The fundamental translation vectors serve as reference axes for the crystal. These *crystal axes* may or may not be at right angles and may or may not be of equal length, depending on the crystal considered. If all combinations of equality and inequality in interaxial angles and in lengths are counted, it is found that a total of 14 kinds of lattices are possible; no more than 14 types of arrangements of points in space are possible without having some of the points distinguishable from others. Since the first correct formulation of the lattice concept was by Bravais,[1] they are known as the *Bravais lattices*. Each of these 14 is also called a *translation group* since the totality of all translations of the type $\mathbf{r}_{uvw}$ in a crystal constitutes a group and these 14 (as well as the symmetry characteristics of crystals mentioned in the following pages) can be derived on the basis of group theory.

Frequently the term "lattice" is loosely used as a synonym for *crystal structure*, a practice that is incorrect and confusing. The term "lattice" should be used to refer to the scheme of repetition in the crystal, as mentioned above, and not to the actual arrangement of atoms in a crystal. The *atomic* arrangement is properly called the *structure* of a crystal. An unlimited number of different crystal structures can be conceived, but only 14 lattices are possible.

Crystal systems To specify a given point in a lattice or atom in a structure, its coordinates are referred to the crystal axes. The most symmetrical crystals, for example, are referred to axes at right angles to each other that form three edges of a cube, a *cubic system* of axes. Each space lattice has some convenient set of axes that is conventionally used with it; a total of seven different *systems of axes*, each with specified equality or inequality of lengths and angles, is used in crystallography. These are the basis of the seven *crystal systems* employed, for example, in the classification of minerals. In Fig. 1–2 the vectors **a**, **b**, **c**, are shown in the right-hand orientation that

[1] A. Bravais, *J. École Polytech. Paris*, vol. 19, p. 1, 1850.

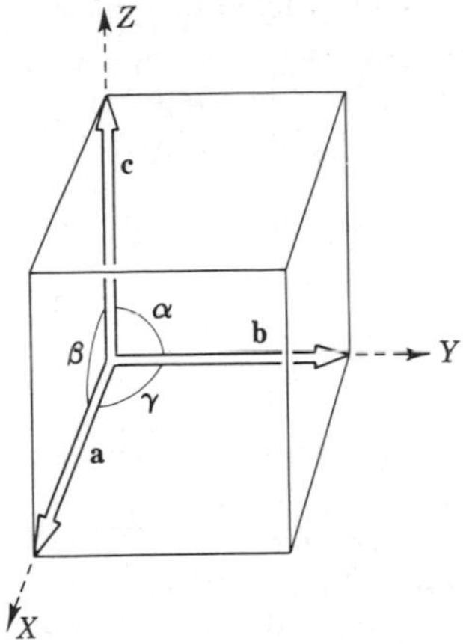

Fig. 1-2 Lattice axes, interaxial angles, and unit cell.

Table 1–1 The crystal systems

System	Axes and interaxial angles	Examples
Triclinic	Three axes not at right angles, of any lengths $a \neq b \neq c$* $\quad \alpha \neq \beta \neq \gamma \neq 90°$	K_2CrO_7
Monoclinic	Three axes, one pair not at right angles, of any lengths $a \neq b \neq c$ $\quad \alpha = \gamma = 90° \neq \beta$	β-S $CaSO_4 \cdot 2H_2O$ (gypsum)
Orthorhombic (rhombic)	Three axes at right angles; all unequal $a \neq b \neq c$ $\quad \alpha = \beta = \gamma = 90°$	α-S Ga Fe_3C (cementite)
Tetragonal	Three axes at right angles; two equal $a = b \neq c$ $\quad \alpha = \beta = \gamma = 90°$	β-Sn (white) TiO_2
Cubic	Three axes at right angles; all equal $a = b = c$ $\quad \alpha = \beta = \gamma = 90°$	Cu, Ag, Au Fe NaCl
Hexagonal	Three axes coplanar at 120°, equal Fourth axis at right angles to these $a_1 = a_2 = a_3 \neq c$ (or $a_1 = b \neq c$) $\quad \alpha = \beta = 90°, \gamma = 120°$	Zn, Cd NiAs
Rhombohedral (trigonal)	Three axes equally inclined, not at right angles; all equal $a = b = c$ $\quad \alpha = \beta = \gamma \neq 90°$	As, Sb, Bi Calcite

* (In this table ≠ means "not necessarily equal to, and generally different from."

is always used in crystallography. The angle opposite the **a** axis is called α, etc. When a crystal structure is determined, definite values in terms of angstrom units are found for the lengths a, b, c of the axes, and definite angles for α, β, γ. The crystal systems are listed in Table 1–1, together with a few examples of crystals belonging to each.

Unit cells The crystal axes form the edges of a parallelepiped called a *unit cell*. Each crystal is built up of a repetitive stacking of unit cells, each identical in size, shape, and orientation with every other one in the crystal. The unit cell is thus the fundamental building block of the crystal, with an atomic arrangement within it which, when repeated in three dimensions, gives the total structure of the crystal. A unit cell is always drawn with lattice points at each corner, but there are also lattice points at the center of certain faces or at the center of volume in the cells of some lattices. Unit cells with lattice points at corners only, called *primitive cells*, could, in fact, be used for all the 14 lattices, but some of these would then exhibit less symmetry than the lattices they represent, and unnecessary complications would result.[1]

The 14 space lattices are pictured in Fig. 1–3, with the conventional unit cells indicated, and are listed in Table 1–2. It will be noted that one space lattice in each system is primitive, in that it has lattice points only at the corners; all of these are given the symbol P in the universally adopted *Hermann-Mauguin* notation. In the monoclinic system there is a cell with a lattice point in the center of the face opposite the c axis (the face containing the a and b axes), i.e., in addition to the corner points, the c face also has a point; this cell is given the symbol C, as is also the base-centered orthorhombic cell. Cells with body-centered points are labeled I (for inner-centered), and those with all faces centered are labeled F. The rhombohedral cell, although primitive, has R as a symbol.

It might be supposed that a face-centered tetragonal (f.c.t.) lattice should be added to this list, since such an arrangement fulfills the requirements for a space lattice. If one makes a sketch of this arrangement, however, it will become clear that with a different choice of axes and origin, the lattice may be described as a body-centered tetragonal (b.c.t.) lattice, and thus it is not different from one of those listed. Similarly, it is unnecessary to list a base-centered tetragonal lattice, for it is equivalent to a P lattice; other examples of equivalence are not difficult to find.

The hexagonal unit cell is drawn as a parallelepiped having edges parallel to a_1, a_2, and c, the heavier lines in Fig. 1–4. It is thus constructed like all the others, and it is not immediately apparent from the unit cell why it is hexagonal. But if a number of these cells are packed tightly together with the axes parallel in all, the hexagonal character of the array becomes obvious, as in Fig. 1–4. The hexagonal prism outlined in Fig. 1–4 contains two whole unit cells and two halves.

Indices in *rhombohedral crystals* can be based on hexagonal axes if desired, or alternatively they can be based on rhombohedral axes, which are of equal lengths and at equal angles to each other. The rhombohedral axes lie at equal angles from the c axis of the equivalent hexagonal axis system. A source of confusion, however, lies in the fact that there are two possible choices of rhombohedral axes; one rhombohedral cell has edges that are half the face diagonals of the other.[2]

[1] One is illustrated on p. 226, Chap. 10.

[2] "International Tables for X-ray Crystallography," vol. 1, p. 20, Kynoch, Birmingham. England, 1952.

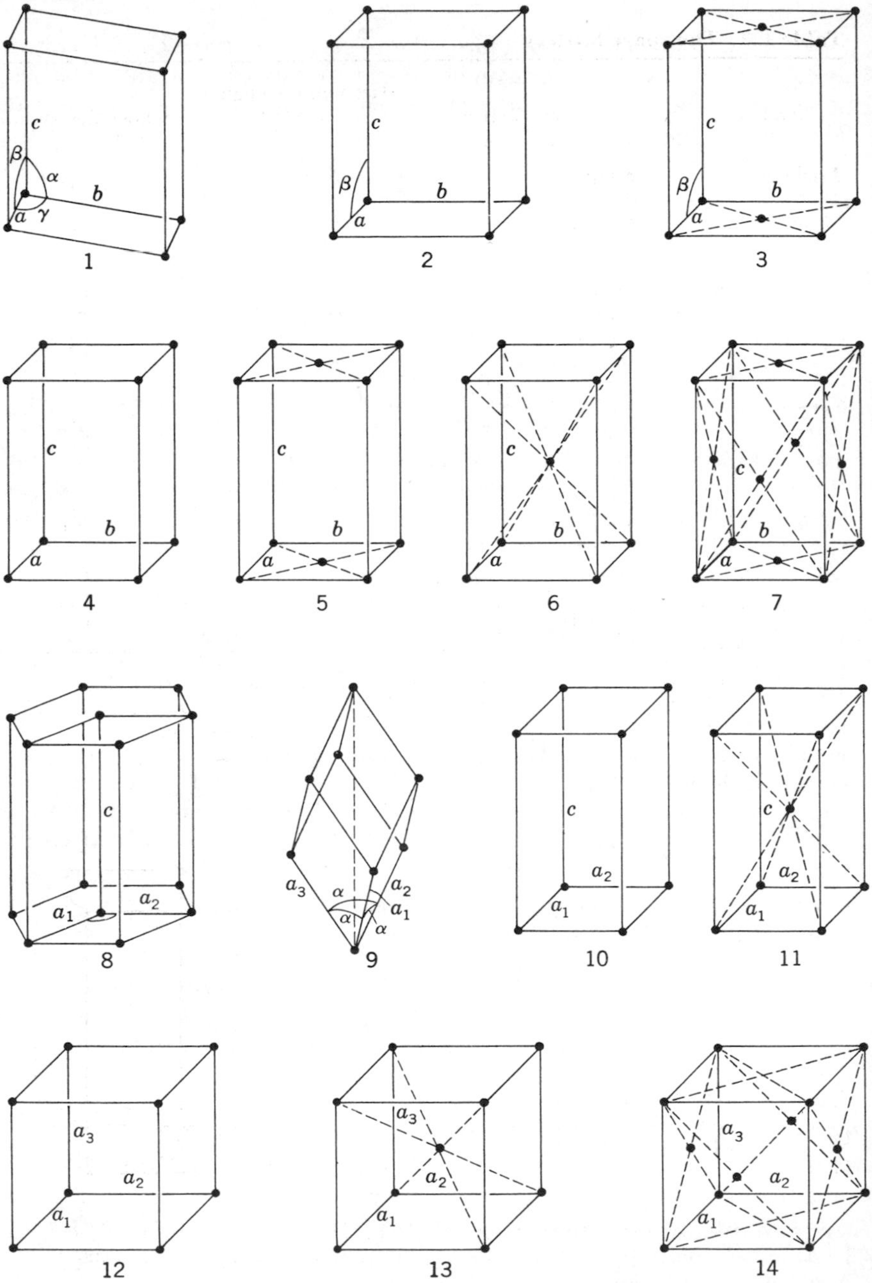

Fig. 1-3 The 14 space lattices illustrated by a unit cell of each: (*1*) triclinic, simple, (*2*) monoclinic, simple, (*3*) monoclinic, base-centered, (*4*) orthorhombic, simple, (*5*) orthorhombic, base-centered, (*6*) orthorhombic, body-centered, (*7*) orthorhombic, face-centered, (*8*) hexagonal, (*9*) rhombohedral, (*10*) tetragonal, simple, (*11*) tetragonal, body-centered, (*12*) cubic, simple, (*13*) cubic, body-centered, (*14*) cubic, face-centered.

Table 1–2 The space lattices

System	Space lattice	Hermann-Mauguin symbol	Schoenflies symbol
Triclinic	Simple	P	Γ_{tr}
Monoclinic	Simple	P	Γ_m
	Base-centered*	C	$\Gamma_{m'}$
Orthorhombic	Simple	P	Γ_o
	Base-centered*	C	$\Gamma_{o'}$
	Face-centered	F	$\Gamma_{o''}$
	Body-centered	I	$\Gamma_{o'''}$
Tetragonal	Simple	P	Γ_t
	Body-centered	I	$\Gamma_{t'}$
Hexagonal	Simple	P (or C)†	Γ_h
Rhombohedral	Simple	R	Γ_{rh}
Cubic	Simple	P	Γ_c
	Face-centered	F	$\Gamma_{c'}$
	Body-centered	I	$\Gamma_{c''}$

* The face that has a lattice point at its center may be chosen as the c face (the XY plane) denoted by the symbol C, or the a or b face denoted by A or B, since the choice of axes is arbitrary and does not alter the actual translations of the lattice.

† The symbol C may be used for hexagonal crystals since they may be regarded as base-centered orthorhombic.

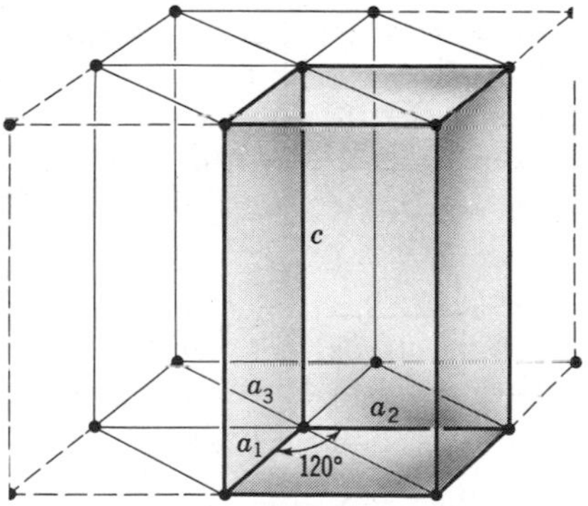

Fig. 1-4 Relation of the unit cell in the hexagonal system (shaded) to a six-sided prism with hexagonal symmetry.

Coordinates of position in the unit cell A position in a space lattice or in a unit cell is specified in terms of its coordinates. If the vector from the origin to the point x, y, z, is $\mathbf{r}_{xyz} = x\mathbf{a} + y\mathbf{b} + z\mathbf{c}$, then the coordinates of the point are x, y, z. In other words, the coordinates are expressed in terms of the unit cell edges, not in terms of centimeters or angstroms. Thus, with an origin chosen at a lattice point, the point at position 3, 2, 1 lies at the end of the vector $\mathbf{r}_{321}$ and is reached by moving along the $\mathbf{a}$ axis a distance of three times the length of the vector $\mathbf{a}$, then parallel to $\mathbf{b}$ a distance

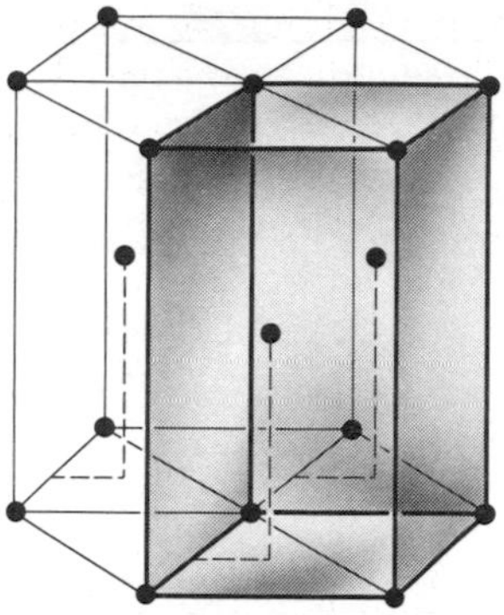

Fig. 1-5 The c.p.h. structure.

twice the length of **b**, and finally parallel to **c** a distance equal to the length of **c**. This point will be at a unit cell corner, and other corner points will have coordinates that are all integers. On the other hand, points not at cell corners will have some nonintegral coordinates; for example, $\frac{1}{2}, \frac{1}{2}, \frac{1}{2}$ will be a body-centered point, and $\frac{1}{2}, \frac{1}{2}, 0$ will be a face-centered point.

The position of an atom in a unit cell is very often reported in terms of one or more *parameters* (variables). It may be given, say, as $x + \frac{1}{2}, y, 0$, with the parameters x and y having certain values to be determined; thus, if it were found that $x = 0.333$ and $y = 0.250$, the coordinates giving the atom position would be 0.333, 0.250, 0 in terms of the cell edge lengths a, b, and c, respectively.

A common crystal structure among the metals is the *close-packed hexagonal (c.p.h.) structure* possessed by zinc, cadmium, magnesium, and others. In this structure the atoms are centered at the positions shown in Fig. 1–5, which have coordinates 0, 0, 0 and $\frac{2}{3}, \frac{1}{3}, \frac{1}{2}$ in each unit cell. A study of the drawing will reveal that the atom positions in this structure do not constitute a space lattice, since the surroundings of the interior atom are not identical with those of the atoms at the cell corners. The actual space lattice is *primitive*, with points at cell corners only, and the structure can be viewed as *pairs* of atoms associated with each lattice point.

Indices of lattice directions and planes A notation is needed to specify directions in a crystal. The notation $[uvw]$ is used to indicate the direction of a line from the origin to a point whose coordinates are uvw. It is customary to use square brackets, to avoid fractional coordinates, and to use the smallest integers that will locate a point on the line. Thus, since a line through 0, 0, 0 and 2, 2, 0 also passes through 1, 1, 0, the indices used for this line and any one parallel to it are [110]. Any negative indices are written with a bar above them. Thus, $[u\bar{v}0]$ signifies the direction through 0, 0, 0 and $u, \bar{v}, 0$. Because of symmetry, various directions in a crystal are equivalent. A full set of equivalent directions (directions of a *form*) is indicated by carets: $\langle uvw \rangle$; the cube edges of a unit cell of a cubic crystal are indicated, for example, by $\langle 100 \rangle$.

Miller indices are universally used as a system of notation for faces of a crystal or planes within a crystal or a space lattice. They specify the orientation of planes relative to the crystal axes without giving the position of the plane in space with respect to the origin. These indices are based on the intercepts of a plane with the three crystal axes, each intercept with an axis being measured in terms of the unit cell dimension (a, b, or c) along that axis (not centimeters or angstrom units). To determine the Miller indices of a plane, the following procedure is used:

1. Find the intercepts on the three axes.
2. Take the reciprocals of these numbers.
3. Reduce to the three smallest integers having the same ratio.
4. Enclose in parentheses: (hkl).

The indices of some important planes are illustrated in Fig. 1–6. A plane with intercepts 2, ∞, 1, has reciprocal intercepts $\frac{1}{2}$, 0, 1 and Miller indices (102). All parallel planes have the same indices. Negative intercepts result in indices indicated with a bar above: $(\bar{h}kl)$, for example, indicates that the intercept along the a axis is negative.

The use of parentheses (hkl) signifies a single plane or a set of parallel planes (all parallel planes have the same indices); curly brackets {braces} signify *planes of a form*—those which are equivalent in the crystal—such as the cube faces of a cubic crystal:

$$\{100\} = (100) + (010) + (001) + (\bar{1}00) + (0\bar{1}0) + (00\bar{1})$$

Planes of a form all have the same atomic configurations.

Crystal forms are also referred to by names that signify the number of reference axes cut by each of their planes: (1) *Pinacoids* are forms in which each plane intersects a single axis and is parallel to the other two. (2) *Prisms* and *domes* have each plane intersecting two axes and lying parallel to the third; if the planes are parallel to the c axis, thus having indices $\{hk0\}$, the form is a prism; if the planes have indices $\{0kl\}$ or $\{h0l\}$ the form is a dome. (3) *Pyramids* have general indices of the type $\{hkl\}$ and

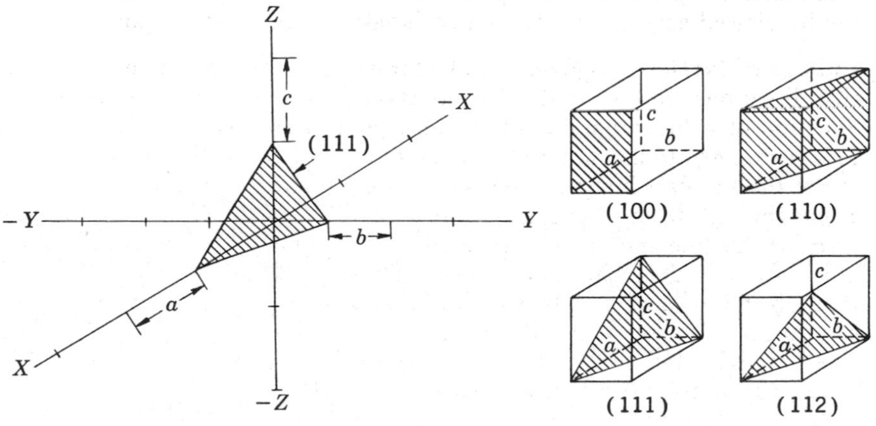

Fig. 1-6 Miller indices of some important planes.

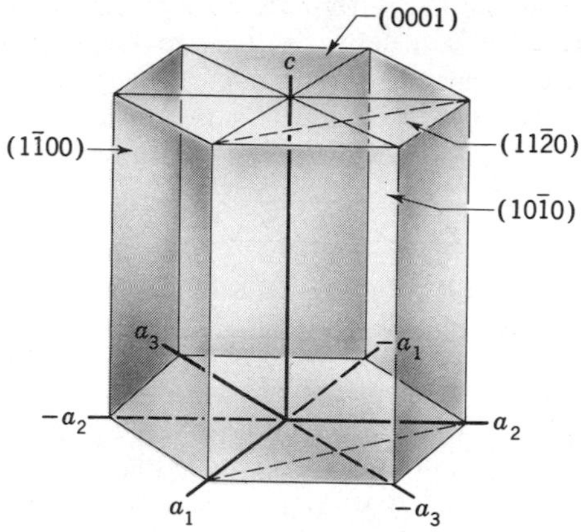

Fig. 1-7 Indices of some planes in a hexagonal crystal.

have planes intersecting all three axes. Additional terms in this nomenclature can be found in the older crystallographic books.

A natural consequence of the periodic internal structure of crystals is that the indices of natural growth faces and cleavage faces of a crystal are small integers. This fact is known as the *law of rational indices.*[1]

Hexagonal indices Indices based on the three axes a_1, a_2, and c are frequently used with hexagonal crystals but are open to the objection that equivalent planes do not have similar indices. For instance, the planes (100) and ($\bar{1}$10) are equivalent. The same objection applies to the indices of directions. For this reason many crystallographers prefer to use four indices, based on the four axes indicated in Fig. 1–4, namely, a_1, a_2, a_3, c. Equivalent planes then are indicated by permutations of the first three indices, since all three a axes are equivalent. For example, the equivalent planes which in former years were known as *prism planes of type I* have indices (1$\bar{1}$00), (10$\bar{1}$0), (01$\bar{1}$0), ($\bar{1}$100), ($\bar{1}$010), and (0$\bar{1}$10). Some hexagonal indices are indicated in Fig. 1–7.

When reciprocal intercepts of a plane on all four axes are found and reduced to smallest integers, the indices will be of the type $(hkil)$, where the first three indices will always be related by the equation

$$i = -(h + k)$$

Since the third index is always completely determined by the first two, it is not necessary to write it down, and the abbreviated style $(hk \cdot l)$ may be used in which the third index is replaced by a dot. The dot immediately discloses that the crystal is being referred to hexagonal indices and also that the four-index system is being used.

Directions in the hexagonal system can be expressed in various ways. In terms of the *three* axes $\mathbf{a}_1$, $\mathbf{a}_2$, and $\mathbf{c}$, a direction may be specified by the three vector components U, V, W of the vector $\mathbf{r}_{UVW} = U\mathbf{a} + V\mathbf{b} + W\mathbf{c}$. But with this system, equivalent directions may not appear to be equiva-

[1] The law was first stated by Abbe René Just Haüy (1743–1822).

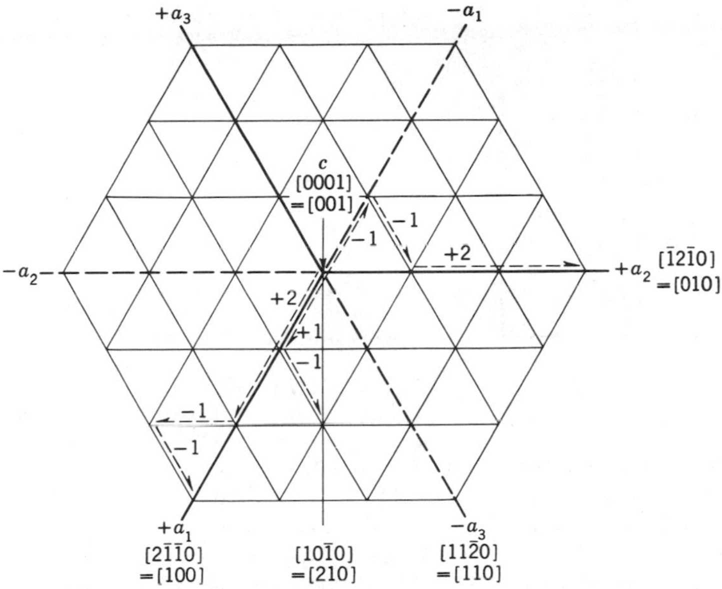

Fig. 1-8 Indices of directions in the hexagonal system with both three- and four-digit indices. The "primary" axis is c (normal to the plane of the drawing); a_1, a_2, and a_3 are "secondary" axes and in c.p.h. crystals are the close-packed rows of atoms.

lent. In terms of the *four* hexagonal axes $\mathbf{a}_1$, $\mathbf{a}_2$, $\mathbf{a}_3$, $\mathbf{c}$ the vector components u, v, t, w must be used, and then by imposing the condition $u + v + t = 0$, one obtains the four-index notation in which all equivalent directions do appear to be equivalent; the vector concerned is then $\mathbf{r}_{uvtw} = u\mathbf{a}_1 + v\mathbf{a}_2 + t\mathbf{a}_3 + w\mathbf{c}$, and the direction is written $[uvtw]$. In Fig. 1-8 the translations leading from the origin to several points are indicated, and both kinds of indices are given.

In the following section still another set of axes for the hexagonal system is introduced, namely, axes that are at right angles to each other, *orthohexagonal axes*. The use of these simplifies several formulas.

Transformations of indices It is frequently desirable to change from one set of axes to another and to transform the indices of a crystallographic plane accordingly. This presents no problem when changing the indices of *planes* of hexagonal crystals from three to four, for the index i in the notation $(hkil)$ is always given by the equation $i = -(h + k)$. To transform from four to three requires dropping the third index i.

A *direction* in the hexagonal system may be written with either three or four indices, as explained in the previous section, and $[UVW]$ in the one case will have the indices $[uvtw]$ in the other if

$$U = u - t \qquad V = v - t \qquad W = w$$

$$u = \tfrac{1}{3}(2U - V) \qquad v = \tfrac{1}{3}(2V - U) \qquad t = -(u + v) \qquad w = W$$

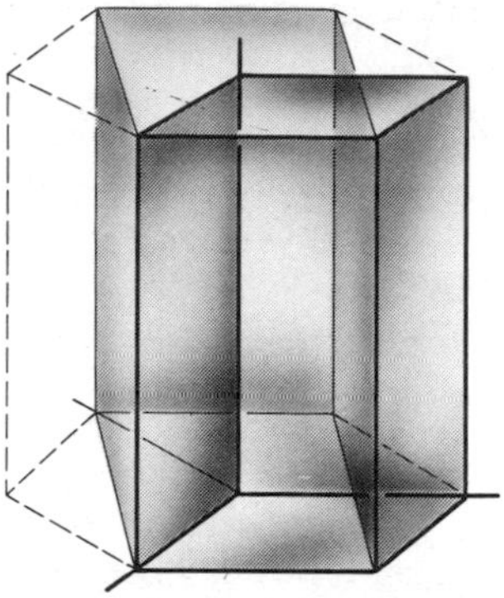

Fig. 1-9 Relation of the simple hexagonal cell (heavy lines) to the orthohexagonal cell (light full lines) and the hexagonal prism.

For crystals that can be described on the basis of either hexagonal or rhombohedral axes, if the $(10\bar{1}1)$ plane in the hexagonal cell is made the (100) face of the rhombohedral cell, then any plane $(hkil)$ in the hexagonal system will have indices in the rhombohedral system that are given by the equations $H = \frac{1}{3}(2h + k + l)$, $K = \frac{1}{3}(h - k + l)$, and $L = \frac{1}{3}(-2k - h + l)$. Thus, for example, the hexagonal $(11\bar{2}0)$ plane has rhombohedral indices $(10\bar{1})$. Transformation of indices from rhombohedral to hexagonal is done by the relations $h = (H - K)$, $k = (K - L)$, $i = (L - H)$, and $l = (H + K + L)$. Both rhombohedral and hexagonal crystals can be treated on the basis of hexagonal axes; therefore a criterion is needed to decide whether such a crystal really belongs to the hexagonal system or is actually rhombohedral. If the indices of reflecting planes (deduced from the diffraction patterns discussed later) have *only* the hexagonal indices that meet the criterion that $(-h + k + l)$ is an integral multiple of 3, then the crystal is rhombohedral. The rhombohedral cell has axial lengths a_r related to the hexagonal by the equation $a_r = \frac{1}{3}(3a_{hex}^2 + c_{hex}^2)^{1/2}$ and with the rhombohedral interaxial angle given by $\sin(\alpha/2) = \frac{3}{2}[3 + (c/a)_{hex}^2]^{-1/2}$. A crystal is also rhombohedral if $(h + k + l)$ is an integral multiple of 3. Different transformation equations then apply.

Orthohexagonal axes provide a simple basis for working with hexagonal crystal problems. These axes define a unit cell illustrated in Fig. 1–9, which has axes normal to each other, and has the same **c** axis as the hexagonal cell. The vectors **a** and **b** of the orthohexagonal cell are given by $\mathbf{a} = \mathbf{a}_1$, $\mathbf{b} = \mathbf{a}_1 + 2\mathbf{a}_2$ where $\mathbf{a}_1$ and $\mathbf{a}_2$ are the edges of the hexagonal cell. A direction $[pqr]$ in the orthohexagonal system has Miller indices in the three-axis hexagonal system $[UVW]$ where

$$[UVW] = [p + q, 2q, r]$$

and the plane (efg) in the orthohexagonal system has three-axis hexagonal indices (HKL) where

$$(HKL) = (e, [f - e]/2, g)$$

The fact that the equivalent planes do not have obviously similar indices in the orthohexagonal system is not a serious disadvantage, whereas the simple relationship between the lattice of the crystal and the *reciprocal lattice*, which is introduced in Chap. 4, is a very definite advantage when orthohexagonal axes are used.[1]

In general, innumerable different sets of axes can be chosen for *any* crystal, and the one chosen for some purpose may not be the one listed in crystallographic tables. It may then be necessary to transform indices of planes, directions, or coordinates from one set to another. This is best

[1] This has been emphasized and formulas have been listed in detail in H. M. Otte and A. G. Crocker, *Phys. Stat. Solidi*, vol. 9, p. 441, 1965. The reciprocal lattice for orthohexagonal crystal axes is also orthogonal.

done by setting up the matrix of the transformation and using matrix methods.[1]

Matrices in crystallography Any change of indices accompanying a change of axes can conveniently be handled by using a matrix of the transformation; any homogeneous deformation of a lattice can likewise be treated in matrix fashion, which has been done throughout the development of the theory of phase transformations referred to in Chap. 18. The types of specific problems that are treated graphically in Chap. 2 can also be solved by the methods of matrix algebra. It is therefore appropriate to mention a few fundamentals here.

Transformation of axes from an old set **a**, **b**, **c** to a new set **A**, **B**, **C** requires expressing the new vectors in terms of the old. This is done by the vector sums

$$\mathbf{A} = a_{11}\mathbf{a} + a_{12}\mathbf{b} + a_{13}\mathbf{c}$$

$$\mathbf{B} = a_{21}\mathbf{a} + a_{22}\mathbf{b} + a_{23}\mathbf{c}$$

$$\mathbf{C} = a_{31}\mathbf{a} + a_{32}\mathbf{b} + a_{33}\mathbf{c}$$

where the constants a_{ij} characterize the transformation. These coefficients can be written as a matrix

$$A = [a_{ij}] = \begin{bmatrix} a_{11} & a_{12} & a_{13} \\ a_{21} & a_{22} & a_{23} \\ a_{31} & a_{32} & a_{33} \end{bmatrix}$$

which is the matrix of the axis transformation. After the matrix for the axis transformation for any particular change of axes or rotation of axes has been written down, the new indices of any plane (hkl) or direction $[uvw]$ can immediately be obtained from the old indices because the coefficients a_{ij} for the change of indices are the same as for the change of axes. For example, if three-axis hexagonal axes are transformed to orthohexagonal as specified on page 14, with $\mathbf{a} = 1\mathbf{a}_1 + 0\mathbf{a}_2 + 0\mathbf{c}$, $\mathbf{b} = 1\mathbf{a}_1 + 2\mathbf{a}_2 + 0\mathbf{c}$, $\mathbf{c} = 0\mathbf{a}_1 + 0\mathbf{a}_2 + 1\mathbf{c}$, the matrix is

$$\begin{bmatrix} 1 & 0 & 0 \\ 1 & 2 & 0 \\ 0 & 0 & 1 \end{bmatrix}$$

and the plane (HKL) of the hexagonal crystal would have transformed indices (efg) in the orthohexagonal system with $e = 1H + 0K + 0L = H$, $f = 1H + 2K + 0L = H + 2K$, $g = L$.

[1] An elementary explanation is given in M. J. Buerger, "X-ray Crystallography," Wiley, New York, 1942, and in texts listed in "International Tables for X-ray Crystallography," vol. 2, Kynoch, Birmingham, England, 1959, which also has a review of the subject in advanced, condensed form.

Suppose a second set of axes is changed to a third set that is related to the second set by the transformation matrix $\boldsymbol{B} = [b_{ij}]$. It is then possible to transform directly from the first to the third by using matrix multiplication; matrix $\boldsymbol{B}$ is multiplied by matrix $\boldsymbol{A}$. The product, which is the required overall transformation matrix, is written $\boldsymbol{BA}$ or $[b_{ij}][a_{ij}]$. (Since matrix multiplication is not necessarily commutative, this is not equal, in general, to $\boldsymbol{AB}$.) In the product matrix $\boldsymbol{BA}$, the ith row of $\boldsymbol{B}$ is multiplied term by term by the jth column of $\boldsymbol{A}$ and the sum of these products is entered in the ij position of the $\boldsymbol{BA}$ matrix.

The coefficients a_{ij} of a matrix need not be integers. It is often convenient to make them the direction cosines of a line, for the line may be a direction that has irrational indices. In discussions of the crystallography of phase transformations, the coefficients may represent the components of a homogeneous strain, such as a pure shear, an isotropic expansion of the lattice, or a pure rotation. Two successive deformations of a lattice can then be handled through matrix multiplication.

Zones and zone axes; crystal geometry Certain sets of crystal planes meet along a line or along parallel lines. For example, the vertical sides of a hexagonal prism intersect along lines that are parallel to the c axis. Such planes are known as planes of a zone, and the direction of their intersection is the zone axis. Any two nonparallel planes will intersect and will thus be planes of a zone, for which their line of intersection is the zone axis; but the important zones in a crystal will be those to which many different sets of planes belong. On the surface of a crystal the faces of a zone form a belt around the crystal; and by placing the crystal on the graduated circle of a goniometer with the zone axis parallel to the goniometer axis, the angles between all faces of the zone can be measured directly on the circle. Zones are also useful in interpreting x-ray diffraction patterns and in making projections of a crystal. Some particularly useful formulas of crystal geometry, including those related to zones, are given in the Appendix.

Symmetry classes and point groups A crystal possesses definite symmetry in the arrangement of its external faces, if the faces are developed, and also in the value of its physical properties in different directions, such as its thermal expansion, elastic moduli, and optical constants. The nature of the symmetry revealed by measurements of these features is the basis of the classification of the crystals into 32 symmetry classes. These can be understood best if one separates the total symmetry of a crystal into simple fundamental symmetry elements, which, when grouped together at a point within the crystal, combine to yield the total symmetry of the crystal. These 32 symmetry classes can be derived from point-group theory, for they constitute the possible ways to arrange equivalent points symmetrically around a given point in space. Accordingly they are also called point groups.

A symmetry element describes an operation by which equivalent points are brought into coincidence. For example, if a rotation of half a turn about

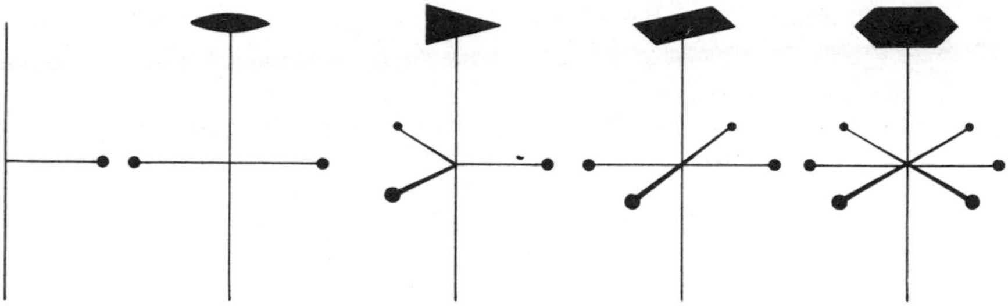

Fig. 1-10 Rotation axes of symmetry: one-, two-, three-, four-, and sixfold.

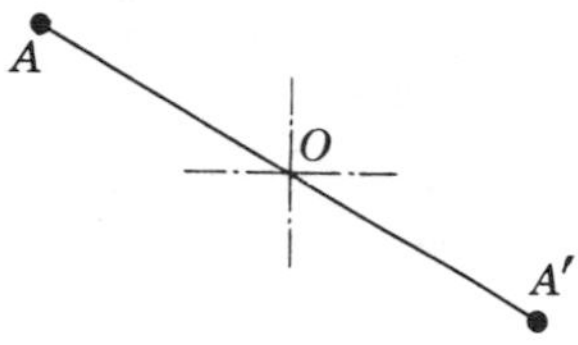

Fig. 1-11 Center of symmetry.

a certain axis brings a lattice into a position indistinguishable from its former one—brings it into coincidence with itself—the axis is a symmetry axis; specifically, it is a twofold rotation axis. If a rotation of a third of a turn (120°) brings coincidence, the axis is threefold. The list of symmetry axes that can exist in lattices is limited, and does not include, for example, a fivefold axis even though an individual object standing alone may have such symmetry. The *symmetry axes* in crystals are illustrated in Fig. 1–10 by vertical lines. The symmetry of each is shown by the geometrical figure at the top and by the arrangement of equivalent points on the arms extending out from each of the axes.

A *symmetry plane* exists if equivalent points are brought into coincidence by reflection across the plane. This is the spatial relationship between an object and its mirror image, or between one's right hand and left hand.

A *center of inversion* exists if every point on one side of the center is matched by an equivalent point on the other side, located on the same line through the center and at the same distance from the center. In Fig. 1–11, point A is related to A' by a center of inversion at O. In vector terms, any vector $\mathbf{r}$ is converted into vector $-\mathbf{r}$ by the inversion operation.

A *rotation-inversion axis* exists if equivalent points are brought into self-coincidence by a combined rotation and inversion. The operation is indicated in Fig. 1–12, which shows a twofold rotation-inversion axis standing vertically operating on the point A to rotate it to the intermediate position A', together with an inversion center at O taking A' to the point A''. Points A and A'' are related by the twofold rotation-inversion axis. The orientations of the external faces of a crystal of calcopyrite ($CuFeS_2$) exhibit fourfold rotation-inversion symmetry, as may be seen from the drawing in Fig. 1–13. Crystals can possess one-, two-, three-, four-, and sixfold rotation-inversion axes. Some of these elements, however, are equivalent to other symmetries: the twofold is equivalent to a reflection

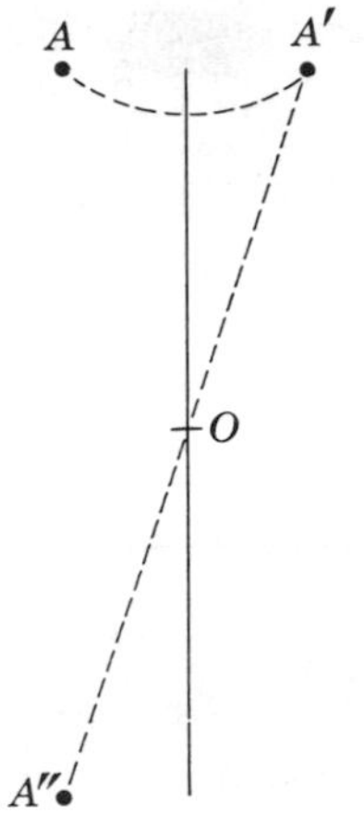

Fig. 1-12 Twofold rotation-inversion axis.

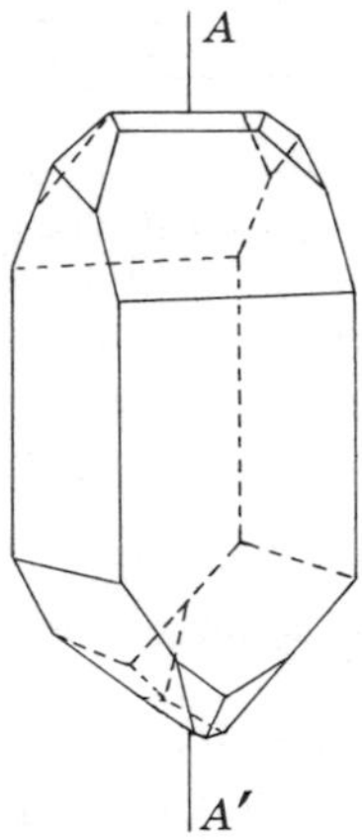

Fig. 1-13 Crystal of calcopyrite, having a vertical fourfold rotation-inversion axis. (**W. L. Bragg.**)

plane normal to the axis and the onefold is equivalent to a center of symmetry.

The symmetry elements enumerated above, singly and in combinations, specify the symmetries shown by the macroscopic properties of crystals and by the growth forms of crystals. They are called the *macroscopic symmetry elements* to distinguish them from others having to do only with the internal structure on a fine scale.

Several notations have been devised for the 32 symmetry classes, but crystallographers have agreed to adopt the Hermann-Mauguin symbols. (In the Schoenflies system, widely used in the past, the symbols are C, S, D, V, T, and O, with certain subscripts: C_2, for example, standing for the "cyclic" class having only a twofold rotation axis of symmetry, C_{3h} denoting the class having a threefold rotation axis and a reflection plane, and D_2 or V denoting the "dihedral" class having three twofold axes at right angles.) In the Hermann-Mauguin notation, the point group is merely a brief list of symmetry elements associated with three important crystallographic directions. The macroscopic symmetry elements in this notation are as follows:

Element	Hermann-Mauguin symbol
1-, 2-, 3-, 4-, and 6-fold rotation axes	1, 2, 3, 4, and 6
Plane of symmetry	m (for "mirror" plane)
Axes of rotation-inversion*	$\bar{1}$, $\bar{2}$, $\bar{3}$, $\bar{4}$, and $\bar{6}$
Center of symmetry	$\bar{1}$ (equivalent to 1-fold rotation-inversion)

* Several inversion axes are equivalent to other elements: $\bar{1}$ is equivalent to a center, $\bar{2}$ to m, $\bar{3}$ to 3 together with a center, $\bar{6}$ to 3 normal to m.

The 32 crystal classes are divided among the seven crystal systems in such a way that each system has a certain minimum of symmetry elements, as follows:

Triclinic	None
Monoclinic	A single 2-fold rotation axis or a single plane
Orthorhombic	Two perpendicular planes or three mutually perpendicular 2-fold axes of rotation
Tetragonal	A single 4-fold axis of rotation or of rotation-inversion
Rhombohedral	A single 3-fold axis of rotation or of rotation-inversion
Hexagonal	A single 6-fold axis of rotation or of rotation-inversion
Cubic	Four 3-fold rotation axes (along cube diagonals)

These elements may coexist with others, but they are sufficient to identify the system to which any crystal belongs. In each system there are several classes differing from one another; thus, in the triclinic system there is a class having no symmetry and a second class having only a center of symmetry. Some classes of crystals contain axes of symmetry that are different at their two ends. The opposite ends of such *polar axes* show different physical properties—for example, they may develop electric charges of opposite sign when heated (*pyroelectricity*) or when mechanically stressed (*piezoelectricity*).

When a crystal is built up of atoms located only at the corners of unit cells, the crystal will have the highest symmetry possible in its system, and this will also be true if the crystal has highly symmetrical groups of atoms at the lattice points. But if a low-symmetry group of atoms surrounds each lattice point, the symmetry of the crystal will be reduced. Thus, it is possible for several classes of symmetry to exist in a single system.

The importance of the point groups is not limited solely to morphology and other macroscopic studies, for the point groups describe the symmetry of the group of atoms (or of molecules) that surrounds each lattice point in a crystal. Clearly, the symmetry of the atomic grouping around an individual lattice point must be consistent with the symmetry of the lattice points as seen from any one of the points. For example, one would not

Table 1–3 The 32 symmetry classes and their symbols*

System	Hermann-Mauguin symbol: Full	Hermann-Mauguin symbol: Abbreviated	Schoenflies symbol
Triclinic	1	1	C_1
	$\bar{1}$	$\bar{1}$	C_i, (S_2)
Monoclinic	m	m	C_s, (C_{1h})
	2	2	C_2
	$2/m$	$2/m$	C_{2h}
Orthorhombic	$2mm$	mm	C_{2v}
	222	222	D_2, (V)
	$2/m\ 2/m\ 2/m$	mmm	D_{2h}, (V_h)
Tetragonal	$\bar{4}$	$\bar{4}$	S_4
	4	4	C_4
	$4/m$	$4/m$	C_{4h}
	$\bar{4}2m$	$\bar{4}2m$	D_{2d}, (V_d)
	$4mm$	$4mm$	C_{4v}
	422	42	D_4
	$4/m\ 2/m\ 2/m$	$4/mmm$	D_{4h}
Rhombohedral†	3	3	C_3
	$\bar{3}$	$\bar{3}$	C_{3i}, (S_6)
	$3m$	$3m$	C_{3v}
	32	32	D_3
	$\bar{3}\ 2/m$	$\bar{3}m$	D_{3d}
Hexagonal	$\bar{6}$	$\bar{6}$	C_{3h}
	6	6	C_6
	$6/m$	$6/m$	C_{6h}
	$\bar{6}2m$	$\bar{6}2m$	D_{3h}
	$6mm$	$6mm$	C_{6v}
	622	62	D_6
	$6/m\ 2/m\ 2/m$	$6/mmm$	D_{6h}
Cubic	23	23	T
	$2/m\ \bar{3}$	$m3$	T_h
	$\bar{4}3m$	$\bar{4}3m$	T_d
	432	43	O
	$4/m\ \bar{3}\ 2/m$	$m3m$	O_h

* In the Hermann-Mauguin notation the symmetry axes parallel to and the symmetry planes perpendicular to each of the "principal" directions in the crystal are named in order. When there is both an axis parallel to and a plane normal to a given direction, these are indicated as a fraction; thus $6/m$ means a sixfold rotation axis standing perpendicular to a plane of symmetry, while $\bar{4}$ means only a fourfold rotary inversion axis.

† Rhombohedral ("trigonal") is sometimes taken as a subgroup of the hexagonal.

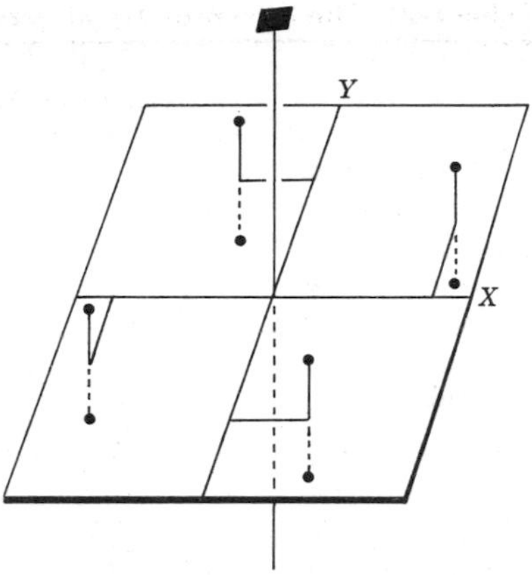

Fig. 1-14 Symmetry elements of class C_{4h}—$4/m$ (fourfold axis normal to a plane of symmetry).

expect to find a molecule having a threefold axis of symmetry to be located at lattice points of a tetragonal crystal where the lattice symmetry always involves axes of fourfold symmetry (4 or $\bar{4}$); similarly, if a molecule normally having sixfold symmetry were placed at each lattice point of a space lattice having only threefold symmetry, one would expect to find that the molecules were distorted into lower symmetry, perhaps threefold, by the less symmetrical interatomic forces.

There is no need to discuss in detail all the 32 individual classes. These are listed briefly in Table 1–3, together with the notations commonly used to designate them and to indicate their symmetry elements.[1]

As an introduction to these tabulations one of the classes of lower symmetry will be mentioned, the one whose elements are sketched in Fig. 1–14, possessing a vertical fourfold axis and a horizontal plane of symmetry. The symmetry in the older notation is C_{4h}, and in the newer notation it is $4/m$. The figure shows a set of equivalent points which come into coincidence by the operation of the symmetry elements. If the origin of coordinates in this figure is placed at one of the points of a space lattice, the coordinates of these equivalent points could give the positions of equivalent atoms in a crystal.

It will be seen from Fig. 1–14 that if one of these equivalent points is given the coordinates x, y, z, the coordinates of all points equivalent to this one in this point group will be

$$x, y, z; \quad \bar{y}, x, z; \quad \bar{x}, \bar{y}, z; \quad y, \bar{x}, z;$$

$$\bar{x}, \bar{y}, \bar{z}; \quad y, \bar{x}, \bar{z}; \quad x, y, \bar{z}; \quad \bar{y}, x, \bar{z}.$$

Crystallographic tables list not only the equivalent points of each point group but also the equivalent faces or planes—the forms. Some forms have

[1] "International Tables for X-ray Crystallography," vol. 1, Kynoch, Birmingham, England, 1952, includes not only these, but also the two-dimensional point groups and other related information.

no restrictions on the Miller indices involved and thus are called "general forms," while others include only certain types of indices and thus are "special forms." For example, a special form might have $l = 0$ for all members. In the example given in Fig. 1–14, one special form consists of (001) and $(00\bar{1})$.

Space groups The preceding section discusses the symmetry of the external faces of crystals and of their anisotropic physical properties; this permits a classification of all crystals into 32 *crystal classes*, which are divided among the 7 *systems*. The symmetry of each of these classes is described by macroscopic symmetry elements grouped at a point, termed a *point group*. It may also be described by a group of *equivalent points* that can be written in the form $x, y, z;\ \ \bar{x}, \bar{y}, \bar{z}$, etc. The operation of all the symmetry elements of the point group upon any one of the equivalent points will produce all the others. We now consider another classification of crystals that has become of primary importance since the advent of x-ray analysis of crystals, a classification that specifies the total symmetry of the arrangement of *atoms* in a crystal.

A *space group* is an array of symmetry elements in three dimensions on a space lattice. Just as a point group is a group of symmetry elements at a *point*, a space group is a group in *space*. Each element of symmetry has a specific location in a unit cell as well as a specific direction with respect to the axes of the cell, and each unit cell in the crystal has an identical array of symmetry elements within it. The elements are always arranged so that the operation of any one of them brings all others into self-coincidence, and thus they may be said to be self-consistent. The group of symmetry elements at and around every lattice point is identical throughout a crystal.

A large number of the possible space groups consists simply of point groups placed at the points of the 14 space lattices. This procedure, however, does not produce all possible space groups, for there are certain symmetry elements possible in a space group that are not possible in a point group; these are commonly termed the *microscopic symmetry elements* since they involve translations through distances of the order of a few angstrom units. All possible self-consistent arrangements of all macroscopic and microscopic symmetry elements in space constitute a total of 230 space groups, to one or more of which every crystal must belong.

The symmetry elements of a space group, operating on a point located at random in a unit cell of the lattice, will produce a set of *equivalent points* in the cell. In an actual crystal, if an atom is located at one of these equivalent points, identical atoms should be found at each of the other equivalent points. The complete tabulation of coordinates of equivalent points in all the space groups is thus of great convenience to crystallographers when they are determining complex crystal structures, for it is a *description of all possible atomic groupings in (perfect) crystals.*

Consider as an example the space group derived by placing a center of symmetry at the lattice points of the simple triclinic space lattice. The lattice is represented by a unit cell in Fig. 1–15*a*; a center of symmetry is

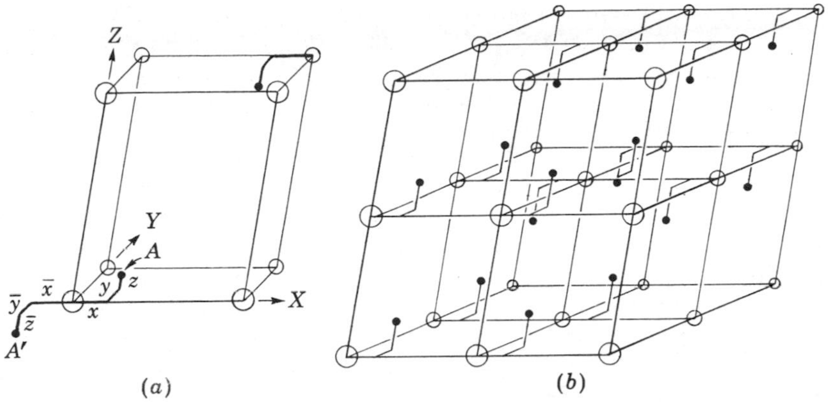

Fig. 1-15 The space group C_i^1—$P\bar{1}$. (*a*) Unit cell with center of symmetry at 000 indicated by open circles. Points A and A' at xyz and $\bar{x}\bar{y}\bar{z}$, respectively, are equivalent. (*b*) Eight unit cells, showing repetition in the crystal. Additional symmetry elements are present but are not indicated.

indicated by a small circle at the origin which turns a point A into a point A', the former having coordinates x, y, z, the latter $\bar{x}$, $\bar{y}$, $\bar{z}$, where x, y, and z may be any fractions of the axial lengths, a, b, and c. The translations of the lattice produce symmetry centers and pairs of equivalent points identical with A and A' at all other points of the space lattice, as will be seen from Fig. 1–15*b*.

It will be noticed from studying this figure that additional symmetry elements are to be found in this arrangement of equivalent points, for the full symmetry includes symmetry centers at the midpoints of each edge and of each face of the unit cell, as well as one at the center of the cell. The coordinates xyz of an equivalent point are always expressed as fractions of the axial lengths a, b, and c; these fractions may have any value whatever without altering the symmetry properties of the group of points derived from it by the operation of the symmetry elements of the space groups. The value of the coordinates does alter the *number* of equivalent points in special cases, however, for if a point is located on an axis of symmetry it is obvious that no new equivalent points will be produced by the operation of rotation around the axis. Similarly, in the example of Fig. 1–15*a*, if the point A lies at the special position 000, where a center of symmetry is located, all the equivalent points will be at the corners of the unit cells, and there will be only half as many as for the general position.

Glide planes and screw axes In the point groups the operation of a symmetry element located at the origin on a point x, y, z will always produce equivalent points that are equidistant from the origin; but in the space groups it is possible to have symmetry elements in which a translation is involved, and equivalent points will consequently be at different distances from the origin.

A *glide plane* combines a reflection plane with a translation parallel

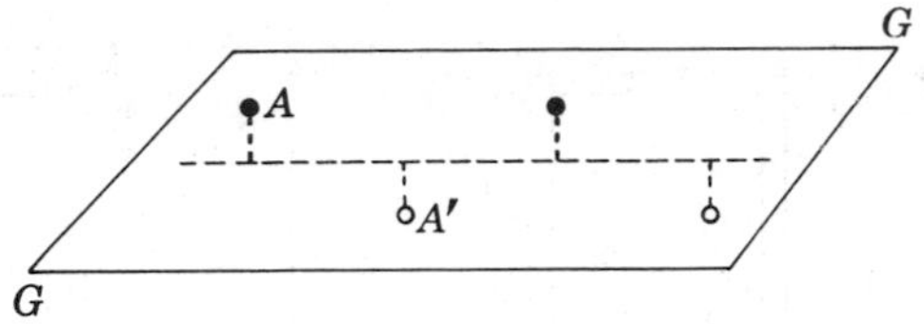

Fig. 1-16 Glide plane.

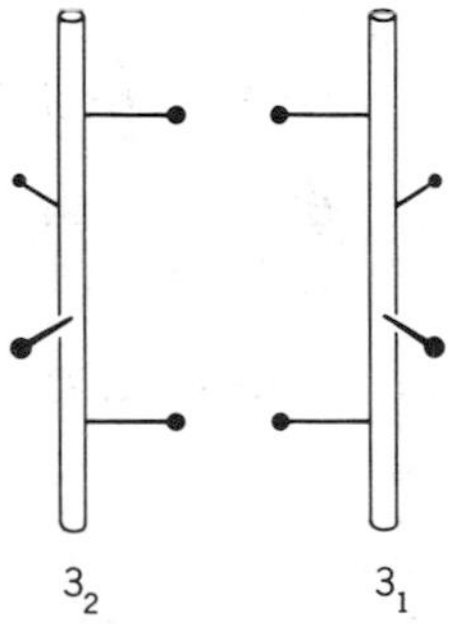

Fig. 1-17 Threefold screw axes: 3_2 (left-handed) and 3_1 (right-handed).

to the plane so that the structure is brought into coincidence by reflection across the plane and simultaneous movement along the plane a specified distance. This is illustrated in Fig. 1–16, where the point A' is produced from point A by the action of glide plane GG. In the different kinds of glide planes the translations are half the axial lengths, half the face diagonals, or one-fourth the face diagonals.

In the Hermann-Mauguin notation glide planes with a glide of $a/2$, $b/2$, and $c/2$ are represented by the symbols a, b, and c, respectively, one with a glide of half a face diagonal by n, and one with a glide of one-fourth a face diagonal by d; in each of these the translation is parallel to the axis or the diagonal concerned.

A *screw axis* combines rotation with translation parallel to the axis. A threefold screw axis parallel to the Z axis, for instance, involves a rotation of one-third turn around Z and a translation of one-third the axial length c, as indicated in Fig. 1–17. An n-fold screw axis combines a rotation of $2\pi/n$ with a translation parallel to the axis amounting to a certain fraction of the distance between lattice points in the direction of the axis; the translation is one-half this distance for a twofold screw axis, one-third in the case of a threefold, one-fourth or one-half in the case of a fourfold, and one-sixth, one-third, or one-half in the case of a sixfold. Several of these axes can be either right-handed or left-handed, since the equivalent points are on spirals that advance as either a right-handed or a left-handed screw, as shown for threefold axes in Fig. 1–17 and for sixfold in Fig. 1–18 (together with common symbols). All types of screw axes are listed in Table 1–4. The subscripts in the notation for screw axes refer to the translation accompanying a rotation of $360°/n$ that would produce coincidence *if* the axis were treated as an n-fold right-handed axis.

As far as the *external* symmetry of crystals is concerned, glide planes cannot be distinguished from reflection planes, nor can screw axes be

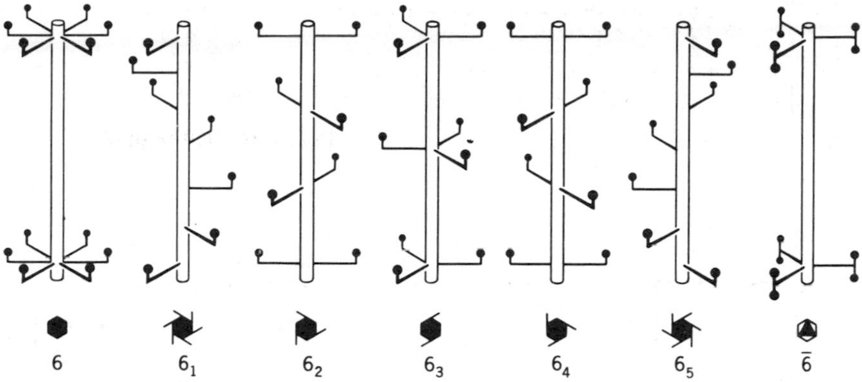

Fig. 1-18 The sixfold axes of all types. (**After C. W. Bunn, "Chemical Crystallography," Clarendon, Oxford, 1945.**)

Table 1–4 The screw axes

Symbol	Multiplicity	Translation	Nature
2_1	2-fold	1/2	
3_1	3-fold	1/3	Right-handed
3_2	3-fold	1/3	Left-handed
4_1	4-fold	1/4	Right-handed
4_2	4-fold	1/2	Includes rotation axis 2
4_3	4-fold	1/4	Left-handed
6_1	6-fold	1/6	Right-handed
6_2	6-fold	1/3	Right-handed
6_3	6-fold	1/2	Includes rotation axis 3
6_4	6-fold	1/3	Left-handed
6_5	6-fold	1/6	Left-handed

distinguished from rotation axes of the same multiplicity. For example, the orthorhombic crystal class having three mutually perpendicular twofold axes, *V*—222, has a number of possible internal arrangements of twofold rotation and screw axes, leading to different space groups pictured in Fig. 1–19. While the internal structures of the crystals belonging to these space groups are all different, their macroscopic symmetry properties are identical.

Space-group notation The symmetry elements of each of the 230 space groups have been tabulated many times.[1,2] The content of space-group

[1] R. W. G. Wyckoff, The Analytical Expression of the Results of the Theory of Space Groups, *Carnegie Inst. Wash. Publ.* 318, 1922, 1930.

[2] "International Tables for X-ray Crystallography," vol. 1, Kynoch, Birmingham, England, 1952.

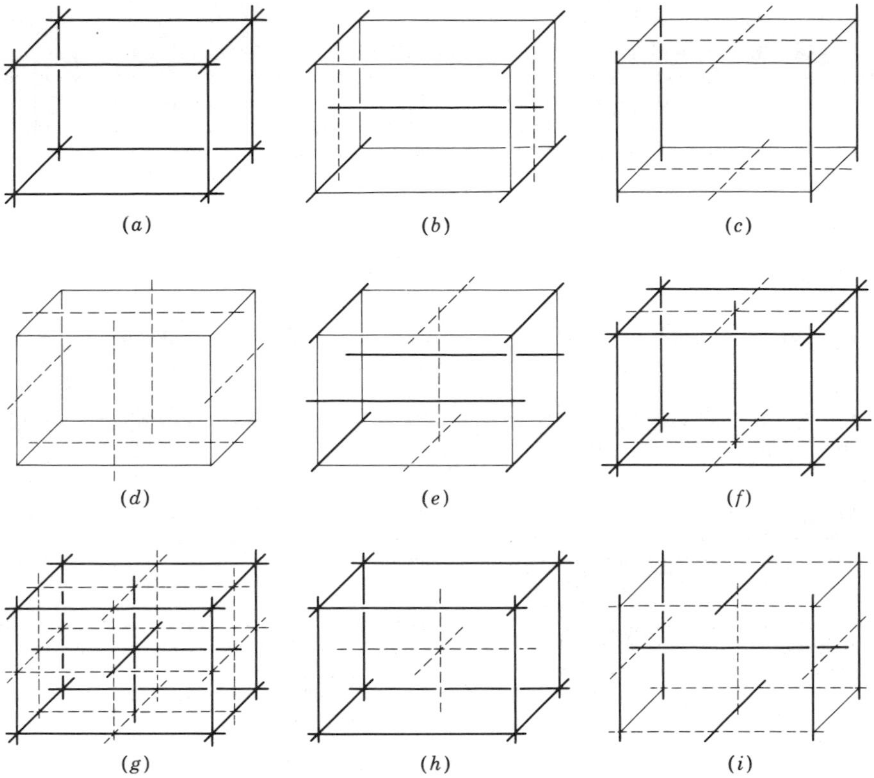

Fig. 1-19 Space groups of class D_2—222, of the orthorhombic system, showing $\frac{1}{8}$ of unit cell of each. Dashed and heavy lines are 2_1 and 2 axes, respectively. Remaining portions of unit cell are identical to these. (**Courtesy of R. W. James, "X-ray Crystallography," Dutton, New York, 1948.**)

tables will be illustrated with reference to Fig. 1–19*c*. The notation for this space group is D_2^3—$P2_12_12$, the symbol in front of the dash being the Schoenflies and the second the newer Hermann-Mauguin symbol, which has now been universally adopted by international agreement. In the older system the symbol for the point group is retained as the symbol for all space groups belonging thereto, and a superscript is added to designate the serial number of the space group belonging to that particular point group; the one under discussion is thus the third space group of the point group D_2. The symbols in the newer notation indicate the type of lattice (in this case simple, since the letter P is used) and the symmetry elements associated with the principal crystallographic directions (here twofold screw axes parallel to two of the principal axes and a twofold rotation axis parallel to the third). The new notation is, in effect, a shorthand for all the symmetry properties of a space group.

Rules that must be kept in mind for an understanding of this notation are as follows[1]:

The designation of the lattice is given by a capital letter. P denotes a simple lattice, A, B, or C a lattice centered on the a, b, or c face (i.e., a cell of a C lattice has a point $\frac{1}{2}\,\frac{1}{2}\,0$ equivalent to the corner point 000), and F a lattice centered on all faces.

In this system of notation, cells are chosen so as to be as nearly rectangular as possible, the hexagonal lattice being denoted by the letter C (an orthohexagonal cell centered on the c face with $a/b = 1:\sqrt{3}$) or by the letter H if a cell is chosen having an axial ratio $a/b = \sqrt{3}:1$. The simple rhombohedral cell having equal axes at equal angles is given a special symbol R, since it is not conveniently drawn using rectangular axes.

The remaining symbols in the space-group notation give the symmetry elements associated with special directions in the crystals. These are: in the monoclinic system the axis normal to the others; in the orthorhombic system the three mutually perpendicular axes; in the tetragonal system the "principal" axis (the one parallel to the fourfold axis), the "secondary" axis, and the "tertiary" axis 90° from the principal and 45° from the secondary; in the rhombohedral and hexagonal systems the principal axis (the one parallel to the three- or sixfold axis), the secondary axis, and the tertiary axis, which lies 90° from the principal and 30° from the secondary; and in the cubic system the directions [001], [111], and [110]. The symbols for rotation axes, screw axes, or axes of rotary inversion along these directions are written in the symbol, as are also the symbols for the reflection planes and glide planes that stand perpendicular to the directions. If two symmetry elements belong to one direction, their two symbols may be combined as a fraction; $2/m$ thus denotes a twofold axis normal to a reflection plane. Only the necessary minima of symmetry elements are given, for the remainder follow as a consequence of those given.

EXAMPLE The space group O_h^5—$Fm3m$ is based on a face-centered lattice and has reflection planes normal to [100] and [110] with threefold axes along the secondary axes [111], an arrangement possible only in the cubic system. It should be mentioned that the crystal axes sometimes can be chosen in different ways with respect to the symmetry elements; these different orientations are distinguished from one another in the notation, and a normal, or standard, orientation is chosen from the various possible ones and is used in space-group tables.

Tables of equivalent points The equivalent points in each space group are listed in tables in such a way as to show clearly the number of equivalent points belonging to each set (the "multiplicity") and thus the number of equivalent atoms that could be located at these points in a crystal. For the example we are considering (Fig. 1–19c), the points that lie on no symmetry elements form a set of four equivalent points known as the "general" set, while two sets of "special" point positions are possible in which the points lie on rotation axes; the sets are labeled a, b, and c for convenience in working with the tables. The information is tabulated as follows:

Equivalent point positions of space group D_2^3—$P2_12_12$

Multiplicity	Set	Coordinates			
2:	(a)	$0, 0, z;$	$\frac{1}{2}, \frac{1}{2}, \bar{z}$		
2:	(b)	$0, \frac{1}{2}, z;$	$\frac{1}{2}, 0, \bar{z}$		
4:	(c)	$x, y, z;$	$\bar{x}, \bar{y}, z;$	$\frac{1}{2}+x, \frac{1}{2}-y, \bar{z};$	$\frac{1}{2}-x, \frac{1}{2}+y, \bar{z}$

[1] *Ibid.*

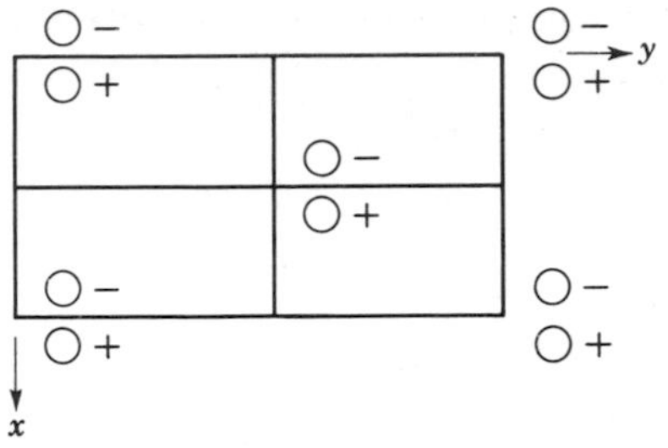

Fig. 1-20 Unit cell and equivalent points for space group C_2^3—$C121$ with notation used in "International Tables for X-ray Crystallography," vol. 1, Kynoch, Birmingham, England, 1952, 1965.

The more complete tables, such as those in the "International Tables for X-ray Crystallography," also give the symmetry around each point in each set; thus the points of set *a* in the above example each have symmetry 2 (i.e., a twofold rotation axis). To save space in tabulating the equivalent points in body-centered or face-centered lattices, the coordinates of the centering points may be given at the top of the list and must be added to each of the coordinates in turn that are listed below them.

EXAMPLE Space group C_2^3—$C121$ has a *C*-face-centered lattice; the general equivalent positions form a set of four, listed in the "International Tables" as

$$(0, 0, 0;\ \ \tfrac{1}{2}, \tfrac{1}{2}, 0) + x, y, z;\ \ \bar{x}, y, \bar{z};$$

therefore the full list, obtained by addition of coordinates, is

$$x, y, z;\ \ \bar{x}, y, \bar{z};\ \ \tfrac{1}{2} + x, \tfrac{1}{2} + y, z;\ \ \tfrac{1}{2} + \bar{x}, \tfrac{1}{2} + y, \bar{z}.$$

The drawing accompanying this space group (Fig. 1–20) shows a unit cell with general positions as circles accompanied by a + sign for the points with positive *z* coordinate, and a − sign for negative *z* coordinate. The *x* and *y* axes lie in the plane of the paper with the positive direction of *x* downward and positive *y* to the right; positive *z* extends toward the reader from the paper. It will be seen that the group of points near 0, 0, 0 is repeated near the *C*-face-centered position. From the arrangement of the equivalent points it can also be seen that there are twofold screw axes parallel to *y* and cutting the *x* axis at $\frac{1}{4}$ and at $\frac{3}{4}$; in other words, 2_1 axes lie along $\frac{1}{4}, y, 0$ and $\frac{3}{4}, y, 0$. There are also twofold rotation axes along $0, y, 0$ and $\frac{1}{2}, y, 0$. This space group is monoclinic and has no other symmetry elements.

Common structure types—The Strukturbericht designation In abstracting crystal-structure determinations that had been made, for publication in *Strukturbericht* each year, and later in *Structure Reports*, the editors found it convenient to use a simple nomenclature for the commonly encountered types of structures. The frequently encountered f.c.c. structures were called *A*1, for example; b.c.c., *A*2; c.p.h., *A*3; diamond-type structures, *A*4; etc. More is said about the common structures in Chap. 10, where use is made of these simple designations, and examples are listed. An alphabetical listing of the elements and their structures appears in the

Table 1–5 Strukturbericht–Structure Reports symbols for crystal-structure types

A types	Elements
B types	AB compounds
C types	AB_2 compounds
D types	A_mB_n compounds
E· · ·*K* types	More complex compounds
L types	Alloys
O types	Organic compounds
S types	Silicates

Appendix, together with the Strukturbericht-type number. The scheme in general is summarized in Table 1–5.

The *A*1 type (which is f.c.c.) has the symmetry of space group $Fm3m$, which is mentioned on page 27. The *A*2 type (b.c.c.) has space group $Im3m$; here again the threefold axis indicated by the 3 in the space-group symbol at the place for the second-named symmetry element immediately indicates that the lattice is cubic, and the *I* indicates that it is body-centered. The *A*3 type (c.p.h.) has space group $P6_3mmc$ and therefore has a primitive space lattice with a sixfold screw axis parallel to the principal crystal axis (the *c* axis) with the screw translation distance $c/2$, and mirror planes normal to the secondary and tertiary axes. Elements with rhombohedral structure, such as As, Sb, and Bi, are called *A*7 type; their space group $R\bar{3}m$ has a threefold rotary-inversion axis (equivalent, as we have seen, to a threefold axis plus an inversion center) and a mirror plane parallel to $\bar{3}$.

2

The stereographic projection

The angular relationships among crystal faces, crystal edges, atomic planes, zones, and crystallographic symmetry elements cannot be accurately displayed by perspective drawings, and if they are stated precisely in mathematical terms, they are often difficult to comprehend and to manipulate, but if they are displayed in a stereographic or gnomonic projection they are easily understood, conveniently manipulated, and recorded accurately enough for a great many purposes. Problems in angular relationships can be solved in a matter of a few minutes with plots on sheets of paper of ordinary size, and it is not difficult to reach a precision of about $\frac{1}{2}°$ in all angles plotted. Although a few problems are more elegantly handled with the gnomonic than with the stereographic projection, for most problems the stereographic is the more convenient.

The stereographic projection is much used for the following purposes: determining crystal orientation; reorienting crystals preparatory to cutting specified crystal faces; determining the crystallographic indices of surface markings such as slip lines, twins, deformation bands, cracks, etch pits, patterns made by magnetic powders; and solving crystallographic problems involved in precipitation, transformations in the solid state, and overgrowths. Preferred orientations of polycrystalline materials are almost invariably treated with the aid of stereographic projections (pole figures). Directional properties of a crystal or a polycrystalline material can be mapped on a stereographic projection; examples are the modulus of elasticity, the yield point, and the electrical conductivity.

Reference sphere and its stereographic projection The nature of the stereographic projection of a crystal is easily understood if the crystal is assumed to be very small and to be located exactly at the center of a sphere, the reference sphere. Crystal planes and atomic planes within the crystal can then be represented by extending them until they intersect the sphere, as in Fig. 2–1, where the plane F intersects the sphere along the circle M. The crystal is assumed to be so small that all crystal planes pass through the center of the sphere; the circle M is therefore a great circle on the sphere. On this "spherical projection" the great circles representing various planes in the crystal intersect at the same angles as do the planes and so exhibit without distortion the angular relationships of the crystal.

Crystal planes can also be represented on the reference sphere by erecting perpendiculars to the planes. These plane normals are made to pass through the center of the sphere and to pierce the spherical surface at points known

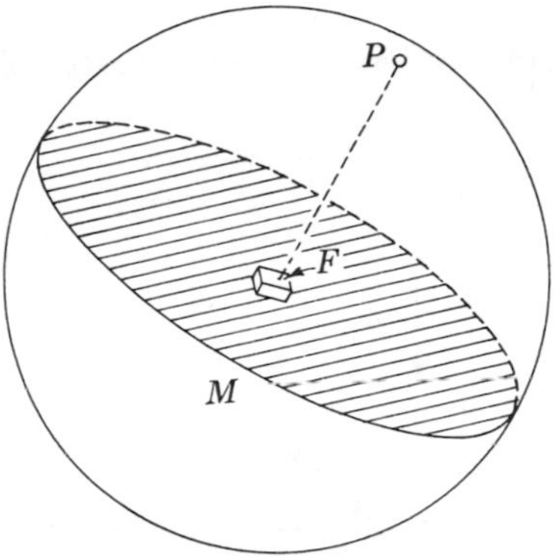

Fig. 2-1 Projection of crystal plane upon reference sphere. Plane F represented on sphere by great circle M or pole P.

as the poles of the planes. This is illustrated in Fig. 2–1, where the plane F and its pole P are shown. The array of poles on the sphere, forming a *pole figure*, represents the orientation of the crystal planes without, of course, indicating the size and shape of the crystal planes. The angle between any two planes is equal to the angle between their poles and is the number of degrees between the poles measured on a great circle through them, as indicated in Fig. 2–2.

The applications discussed in this chapter can be carried through by using the spherical projection just described, but in practice it is usually more convenient to use a map of the sphere, so that all the work can be done on flat sheets of paper. The stereographic projection is one of the methods—and generally the most satisfactory one—by which the sphere may be mapped without distorting the angular relations between planes or poles.

In Fig. 2–3 it will be seen that there is a simple relation between the sphere and its stereographic projection. If the sphere is transparent and a source of light is located *at a point on its surface*, the markings on the surface of the sphere will be projected as shadows upon a plane erected as shown. The plane is perpendicular to the diameter of the sphere that passes through the light source. The pattern made by the shadows is a stereographic projection of the sphere; the point P' is the stereographic projection of the pole P. The distance of the plane ("projection plane") from the sphere is immaterial, for changing the distance will merely change the magnification of the map and will not alter the geometrical relations (in fact, the plane is frequently considered as passing through the center of the sphere).

Obviously, only the hemisphere opposite the source of light will project within the *basic circle* shown in the figure. The hemisphere containing the source of light will project outside the basic circle and extend to infinity. It is possible, however, to represent the whole sphere within the basic circle if two projections are superimposed, the one for the left-hand hemisphere constructed as in Fig. 2–3 and the one for the right-hand hemisphere constructed by having the light source on the left and the screen on the right. The same basic circle is used for both projections, and the points on one hemisphere are distinguished from those on the other by some notation such as plus and minus signs.

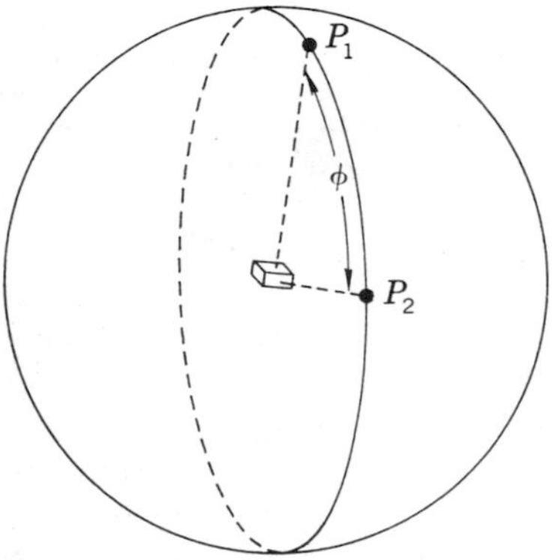

Fig. 2-2 Angle ϕ between poles P_1 and P_2 is measured on great circle through poles.

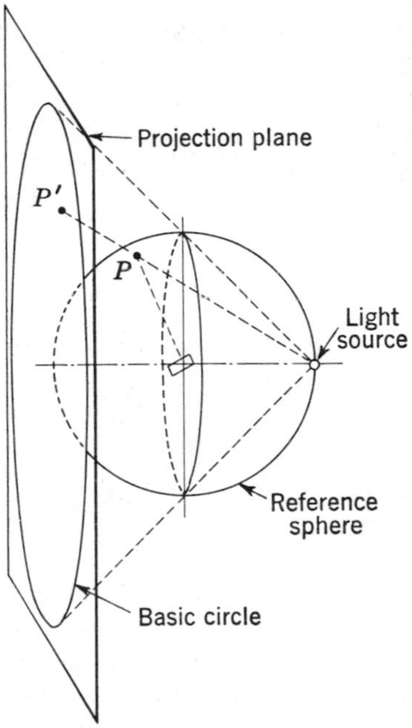

Fig. 2-3 Stereographic projection. Pole P of crystallographic plane projects to P' on projection plane.

Projection of great and small circles Let us consider how great circles and small circles inscribed on the sphere will appear on the projection (Fig. 2–4*a*). Any great circle that passes through the point N will project to form a straight line passing diametrically through the basic circle on the projection; thus, $SPNS$ projects to EE. That this is true will be seen from the fact that the great circle $SPNS$ and its projection EE are in fact lines of intersection of a plane with the sphere and projection plane, respectively. If the great circle is graduated in degrees, its projection EE will be a scale of stereographically projected degree points[1] and will be useful for reading angular distances on the projection; it is shown with 5° gradations in Fig. 2–4*b*.

A small circle inscribed about a point such as P (Fig. 2–4*a*) that lies on the great circle $SPNS$ will cut the great circle at two points, each of which is $\phi°$ from P. The point P will project to P'. The bundle of projection lines for the small circle will form an elliptical cone with its apex at S, and the cone will intersect the plane in a true circle of which the center is on the line EE, either inside or outside the basic circle. The point P' will not be at the center of area of this projected circle but will lie on line EE at a point distant an

[1] S. L. Penfield, *Am. J. Sci.*, vol. 11, pp. 1, 115, 1901.

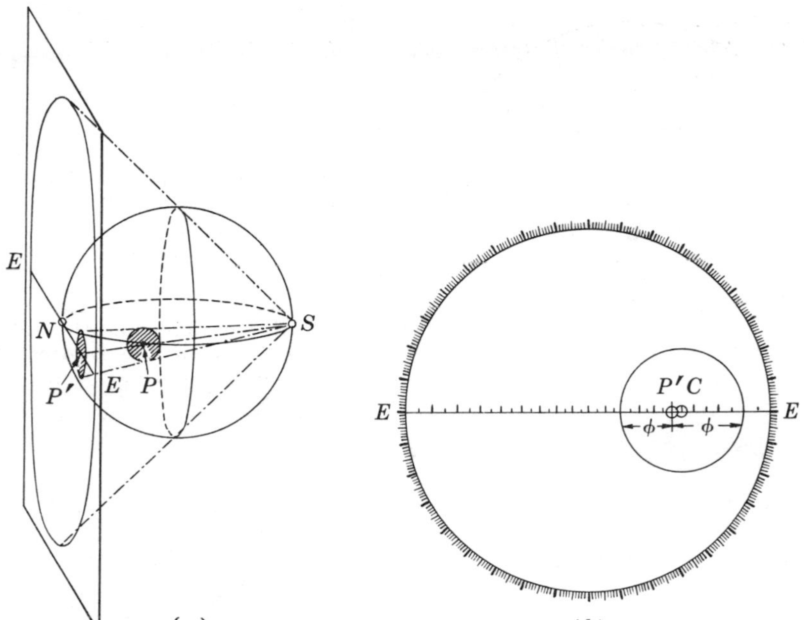

(a) (b)

Fig. 2-4 Stereographic projection of circles. (a) Projection of a great circle through N is EE; (b) projection of a small circle (shaded) is circle having center displaced from P', which is the projection of P.

equal number of *stereographically projected* degrees from all points of the projected circle. The scale of projected degrees enables the size of the projected circle to be determined quickly, as indicated in Fig. 2–4*b*: the scale EE is laid diametrically across the basic circle so as to pass through P'; then two points are laid down at a distance of $\phi°$ from P' in each direction, and a circle, centered on EE, is drawn through the two points thus located. (In Fig. 2–4*b*, $\phi = 30°$ and the center of area on the projected circle is at C.)

If the radius of the small circle about P is increased, it finally becomes a great circle. Since this great circle does not pass through the point N, its projection will not be a straight line, but will be a circle having its center on an extension of EE; it will cut the line EE at the point 90 stereographic degrees from P'.

The plotting of points, small circles, and great circles can be very simply done with the graphical aids mentioned in the following section, but before introducing these it should be noted that all plotting can be done by trigonometric relationships, directly on plain paper or on graph paper. If the radius of the basic circle of the projection is r, a pole at an angle ϕ from the normal to the projection plane will be plotted as a point at $(\phi/2)$ distance r tan from the center of the projection. The locus of poles at an angle α from a pole A on the basic circle will appear on the projection as a small

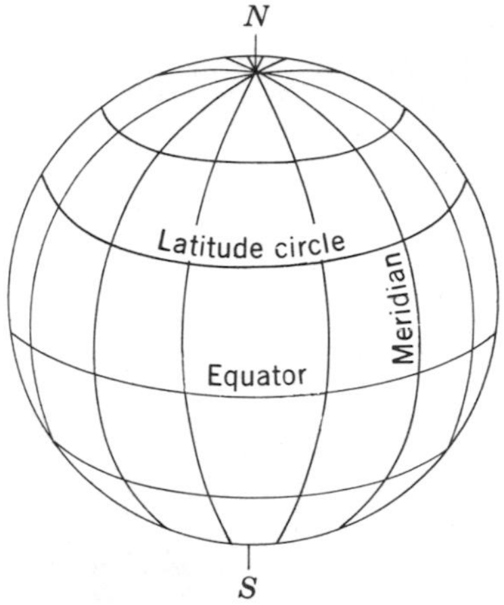

Fig. 2-5 Ruled globe. Projections of this form the stereographic nets of Figs. 2-6 and 2-7.

circle of radius $r \tan \alpha$ centered at A. A great circle inclined at angle β to the normal to the projection will appear as a circle of radius $r/\sin \beta$. Simple relationships exist between direction cosines and the stereographic projection, which are particularly convenient for cubic crystals.[1]

Ruled globe and stereographic nets A globe can be used for crystallographic problems if it is ruled as in geography with great circles through the north and south poles to form meridians connecting points of equal longitude, and with small circles concentric with the north and south poles to form lines of equal latitude (Fig. 2–5).

If the net of latitude and longitude lines on the reference sphere is projected upon a plane, it will form a stereographic net much resembling the rulings of the globe in appearance. When the north-south axis of the sphere is *parallel to the projection plane,* the latitude and longitude lines form the stereographic net of Fig. 2–6, frequently referred to as a *Wulff net.* The meridians extend from top to bottom, the latitude lines from side to side (compare with Fig. 2–5). If, on the other hand, the north-south axis is perpendicular to the projection plane, the net of Fig. 2–7 will be formed, which is known as the *polar net* or *equatorial net.* In this case the meridians radiate from the pole in the center, and the latitude lines are concentric circles.

The nets reproduced here are graduated in intervals of 2°. Larger nets of

[1] J. K. Mackenzie and J. S. Bowles, *Acta Met.,* vol. 5, p. 137, 1957, in plotting crystallographic relationships for martensitic transformations, used the following principles: For a projection with basic circle of unit radius, the rectangular coordinates of the point on the projection corresponding to a unit vector with direction cosines l, m, n with respect to suitably oriented rectangular coordinate axes in the crystal are $x = l/(1 + n)$ and $y = m/(1 + n)$. A small circle of radius ψ about the end point of this vector projects into a circle with radius $\sin \psi/(n + \cos \psi)$ and center at $l/(n + \cos \psi)$, $m/(n + \cos \psi)$. For reading the plots the inverse relations are computed from $1 + n = 2/(1 + x^2 + y^2)$, $l = x\,(1 + n)$, $m = y(1 + n)$. If the basic circle is not shown on a plot, the scale of the plot can be computed from the known direction cosines of poles that are shown.

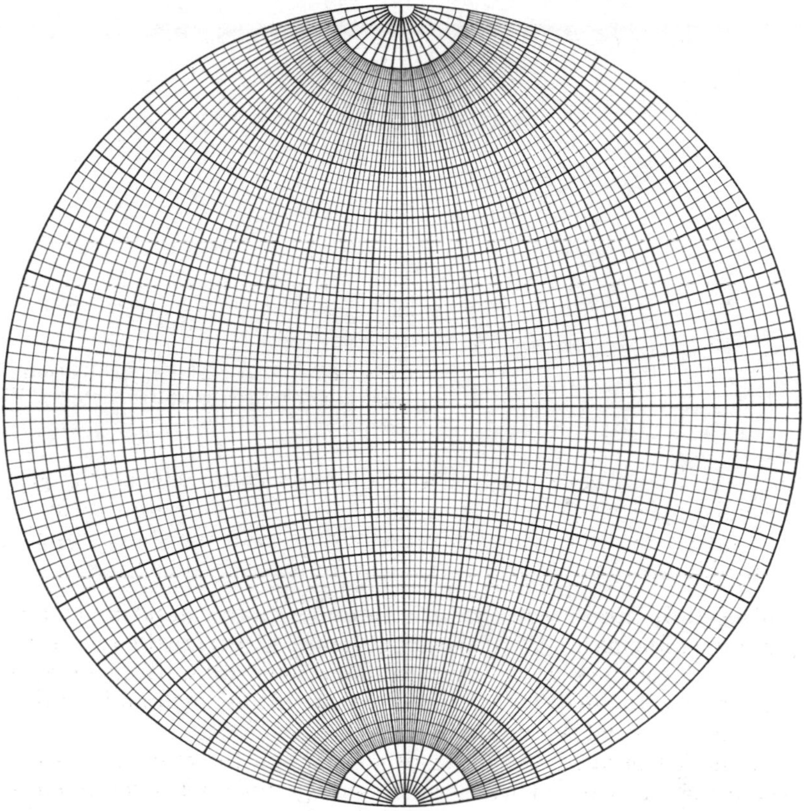

Fig. 2-6 Stereographic net, Wulff or meridional type, with 2° graduations.

greater precision have been published repeatedly.[1] Nets of reasonable size will enable problems to be solved with an error of a degree or at best a few tenths of a degree; for greater precision it is necessary to resort to mathematical analysis.

[1] An accurate stereographic net, 15¾ in. in diameter, of the type reproduced in Fig. 2–6' was engraved by Admiral Sigsbee for the Hydrographic Office of the United States Navy, and is known as *H. O. Miscellaneous* 7736–1. Wulff nets 18 cm in diameter on a transparent plastic material, together with standard projections of crystals, have been marketed by N. P. Nies, Laguna Beach, Calif., and nets of 25-cm radius by Richard Seifert and Co., Röntgenwerk, Hamburg, Germany. Various stereographic and x-ray charts are sold by the Institute of Physics, London, England (in the United States by Polycrystal Book Service, Pittsburgh, Pa. 15238); these are listed in *Acta Cryst.*, vol. 5, p. 294, 1952. A 6¼-in. diameter net and a cubic standard projection are reproduced in Elizabeth A. Wood, "Crystal Orientation Manual," Columbia, New York, 1963. E. Leitz, New York, New York, marketed a net mounted with a device for turning a tracing sheet concentrically on it.

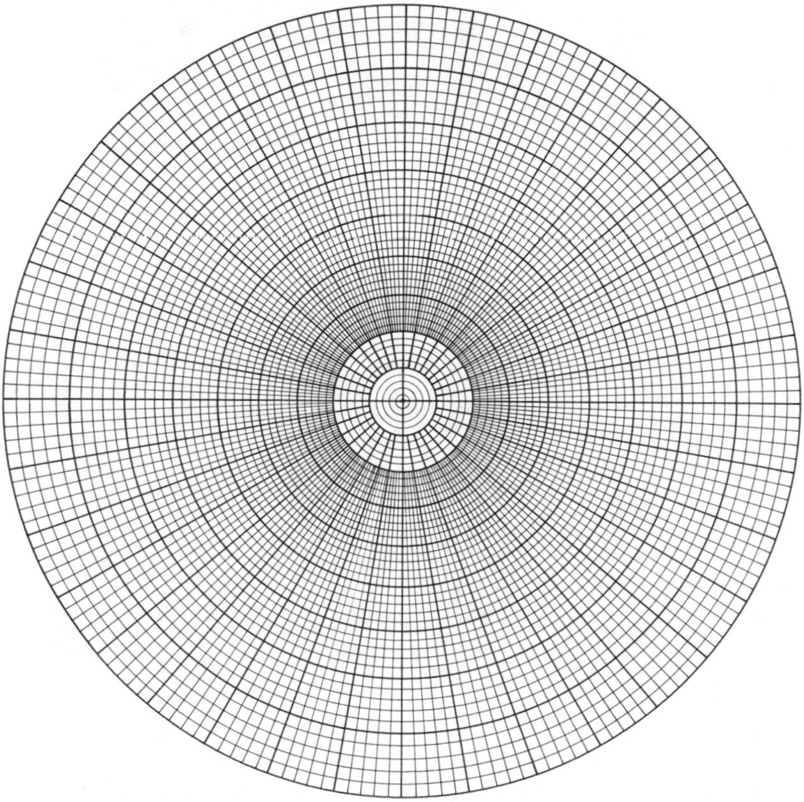

Fig. 2-7 Polar stereographic net with 2° graduations.

Rotation with the nets For the solution of crystallographic problems on a ruled globe, it is necessary to use a device similar or equivalent to the one sketched in Fig. 2–8, consisting of a transparent cap fitting accurately over the globe but free to rotate with respect to it. Poles marked on the cap, such as P_1 and P_2, may be studied with reference to the underlying net of latitude and longitude lines. Rotating this cap about the north-south axis of the globe will cause each point on the cap to move along a circle of constant latitude on the globe, as shown; in so doing, each point will cross the same number of meridians, i.e., each point will retain its latitude and will alter its longitude equally.

An exactly analogous rotation may be carried out with stereographic nets. A transparent sheet of tracing paper replaces the transparent spherical cap, and the stereographic net laid under the paper replaces the ruled globe. An array of poles on the tracing paper is rotated with respect to the net by moving each point along the latitude line that passes through it, counting off along that line the required difference in longitude. With the Wulff net of Fig. 2–6 the poles shift to the right or left, whereas with the polar net of Fig. 2–7 they rotate about the center.

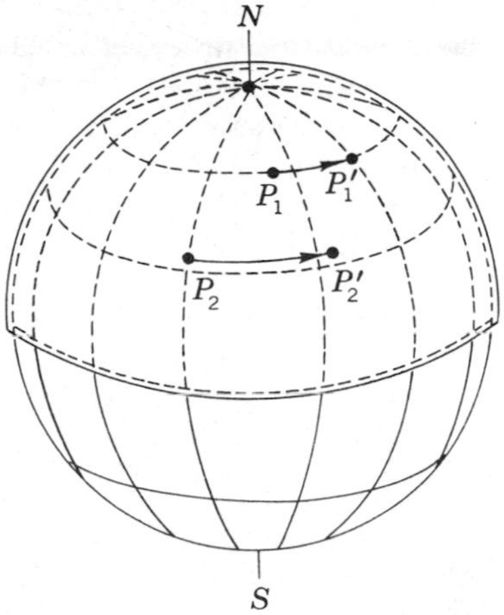

Fig. 2-8 Ruled globe with transparent cap. Rotation of cap about ***NS*** **axis moves** P_1 **to** P_1' **and** P_2 **to** P_2'**.**

A greater freedom of rotation is possible with the cap-and-globe device than with the nets, for the axis of rotation in the former case can be chosen at random, while in the latter case rotation must always be done about the north-south axis of a net. It is possible, however, to rotate first about the axis of one net and then about the axis of the other and by thus combining rotations to effect a rotation about an axis inclined to both. In this way, rotations of any amount about any axis, whatever its inclination to the projection, can be made. The method amounts to resolving the rotation into components; one component, rotation about the axis parallel to the plane of the paper, is carried out by the Wulff net, and the other component, rotation about the axis normal to the paper, is accomplished by using the polar net. There is therefore no need for the bulky cap-and-globe device; it is introduced here to aid the beginner in visualizing the operations with nets.

In practice, rotation about an inclined axis can be accomplished without transferring the tracing paper from one net to the other, for obviously the circular rotation with the polar net can be performed simply by rotating the tracing paper about a pin at the center of its basic circle. Rotations of both types can be carried out conveniently with the tracing paper lying on the Wulff net and free to swing about a central pin.

Angle measurement Angles are measured with a stereographic net by bringing the points to the same meridian of the stereographic net (i.e., on a great circle) and counting their difference in latitude. Any two points can be brought to the same meridian merely by rotating them a certain amount about the center (swinging the tracing paper about the central pin of a Wulff net).

The most frequent source of error in students' work with the projection comes from misunderstanding or forgetting the principle that *the angle between two points is equal to their difference in latitude only when they lie on the same meridian.*

When planes appear in a stereographic projection as great circles, such as the circle M of Fig. 2–1, it is easy to plot the poles of the planes and then to measure the angle between the poles. To plot the projection of pole P (Fig. 2–1), it is necessary merely to turn the tracing paper about the central pin in a Wulff net until the projection of great circle M falls on a meridian of the Wulff net; then the point on the equator 90° from that meridian is the pole P of the plane.

Properties of stereographic projection Elaborate treatises have been written on the properties and uses of stereographic projection for crystallographic work,[1] and the reader is referred to these for details not mentioned in the present discussion and for mathematical proofs. We may summarize as follows some very useful properties of the projection:

1. The reference sphere is projected as it would appear to the eye at a point on the spherical surface; hence, it is a "perspective projection." It is also the "shadow projection" when a source of light is on the sphere, as discussed above.
2. Small circles on the sphere appear as circles on the projection; however, the centers of these circles on the sphere will not project to the center of the area of the projected circles but will be displaced radially an amount corresponding to equal *angular* distances from the center to all points on the circumference.
3. Great circles on the sphere appear on the projection as circles cutting the basic circle at two diametrically opposite points; a great circle lying in a plane perpendicular to the projection plane becomes a diameter on the projection, while great circles in inclined positions on the sphere may be made to coincide with one of the meridians of a Wulff net.
4. Angles between points are measurable and may be read as a difference of latitude on a net so rotated as to give the points the same longitude.
5. Angular relations between points on the projection remain unchanged by rotation of the points about the axis of a stereographic net, as described earlier (page 36).
6. Half a sphere is projected within a basic circle; the other half is projected on a plane of infinite extent but is more conveniently projected within the basic circle and distinguished from the first by some notation such as + and −.

Standard projections of crystals A stereographic projection of all the important planes in a crystal is very useful in solving problems, since it contains in graphical form all the angles between the planes. Such a projection plotted with a plane of low indices as the plane of projection is a *stand-*

[1] S. L. Penfield, *Am. J. Sci.*, vol 11, pp. 1, 115, 1901; vol. 14, p. 249, 1902; *Z. Krist.*, vol. 35, p. 1, 1902. E. Boeke, "Die Anwendung der stereographische Projektion bei kristallographischen Untersuchungen," Bornträger, Berlin, 1911. F. E. Wright, *J. Opt. Soc. Am.*, vol. 20, p. 529, 1930; *Am. Mineralogist*, vol. 14, p. 251, 1929. A. Hutchinson, *Z. Krist.*, vol. 46, p. 225, 1909. D. Jerome Fisher, A New Projection Protractor, *J. Geol.*, vol. 49, pp. 292, 419, 1941. P. Terpstra and L. W. Codd, "Crystallometry," Longmans, New York, 1961.

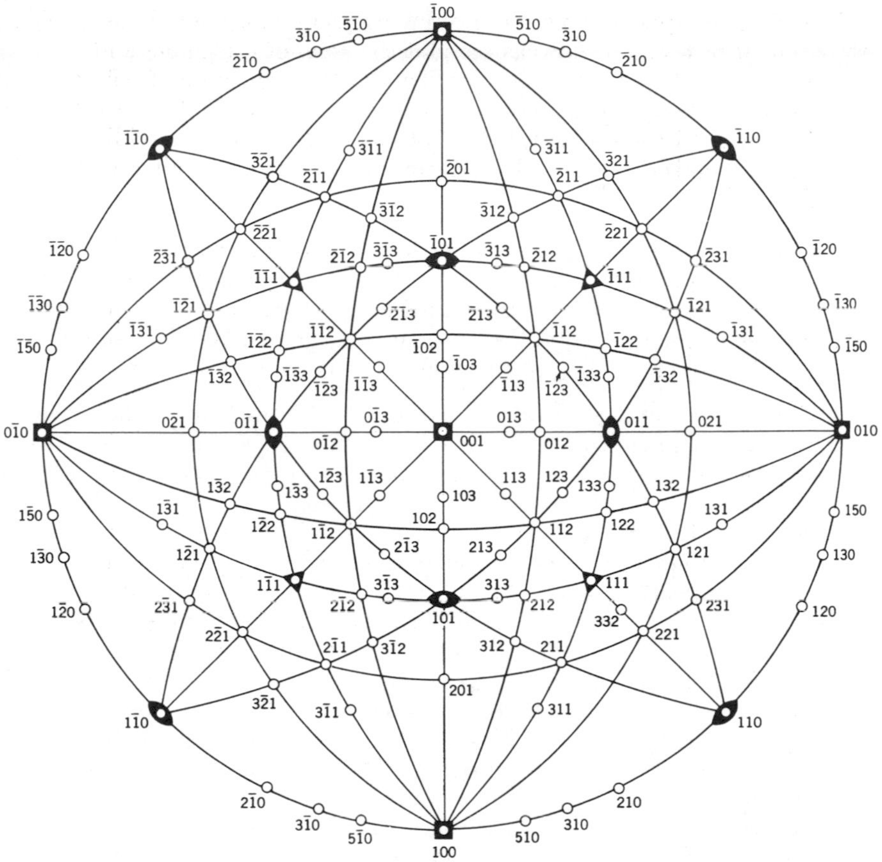

Fig. 2-9 Standard (001) stereographic projection of poles and zone circles for cubic crystals. (After E. A. Wood, "Crystal Orientation Manual," Columbia, New York, 1963.)

ard projection. An example for a cubic crystal is shown in Fig. 2–9; the projection plane is parallel to the (001) plane and the poles of (100) and (010) planes lie on the basic circle. Cubic standard projections apply to all cubic crystals since all cubic crystals have identical interplanar angles, but this is not true for noncubic crystals. Given the axial ratios or unit cell dimensions of a crystal, the interplanar angles can be computed by the use of formulas given in the Appendix, or by a computing machine, and the poles of a standard projection can then easily be plotted with the aid of a stereographic net. The process is shortened by using the symmetry properties of the crystal and by using zonal relations: the poles of planes belonging to a given zone all lie normal to the zone axis and on a great circle of the projection 90° from the zone axis. Low-index poles lie at the intersections of zone circles, as will be seen in Fig. 2–9. A number of the most important zone circles are shown on the cubic (001) standard projection of Fig. 2–9.

Table 2–1 Angles between planes in cubic crystals

HKL	*hkl*						
100	100	0.00	90.00				
	110	45.00	90.00				
	111	54.74					
	210	26.56	63.43	90.00			
	211	35.26	65.90				
	221	48.19	70.53				
	310	18.43	71.56	90.00			
	311	25.24	72.45				
	320	33.69	56.31	90.00			
	321	36.70	57.69	74.50			
	322	43.31	60.98				
	331	46.51	76.74				
	332	50.24	64.76				
	410	14.04	75.96	90.00			
	411	19.47	76.37				
110	110	0.00	60.00	90.00			
	111	35.26	90.00				
	210	18.43	50.77	71.56			
	211	30.00	54.74	73.22	90.00		
	221	19.47	45.00	76.37	90.00		
	310	26.56	47.87	63.43	77.08		
	311	31.48	64.76	90.00			
	320	11.31	53.96	66.91	78.69		
	321	19.11	40.89	55.46	67.79	79.11	
	322	30.96	46.69	80.12	90.00		
	331	13.26	49.54	71.07	90.00		
	332	25.24	41.08	81.33	90.00		
	410	30.96	46.69	59.04	80.12		
	411	33.56	60.00	70.53	90.00		
111	111	0.00	70.53				
	210	39.23	75.04				
	211	19.47	61.87	90.00			
	221	15.79	54.74	78.90			
	310	43.09	68.58				
	311	29.50	58.52	79.98			
	320	36.81	80.78				
	321	22.21	51.89	72.02	90.00		
	322	11.42	65.16	81.95			
	331	22.00	48.53	82.39			
	332	10.02	60.50	75.75			
	410	45.56	65.16				
	411	35.26	57.02	74.21			
210	210	0.00	36.87	53.13	66.42	78.46	90.00
	211	24.09	43.09	56.79	79.48	90.00	
	221	26.56	41.81	53.40	63.43	72.65	90.00
	310	8.13	31.95	45.00	64.90	73.57	81.87
	311	19.29	47.61	66.14	82.25		
	320	7.12	29.74	41.91	60.25	68.15	75.64
		82.87					
	321	17.02	33.21	53.30	61.44	68.99	83.14
		90.00					

Table 2–1 Angles between planes in cubic crystals (Cont.)

HKL	*hkl*								
	322	29.80	40.60	49.40	64.29	77.47	83.77		
	331	22.57	44.10	59.14	72.07	84.11			
	332	30.89	40.29	48.13	67.58	73.38	84.53		
	410	12.53	29.80	40.60	49.40	64.29	77.47		
		83.77							
	411	18.43	42.45	50.77	71.56	77.83	83.95		
211	211	0.00	33.56	48.19	60.00	70.53	80.40		
	221	17.72	35.26	47.12	65.90	74.21	82.18		
	310	25.35	**49.80**	58.91	75.04	82.58			
	311	10.02	42.39	60.50	75.75	90.00			
	320	25.06	37.57	55.52	63.07	83.50			
	321	10.89	29.20	40.20	49.11	56.94	70.89	77.40	83.74
		90.00							
	322	8.05	26.98	53.55	60.32	72.72	78.58	84.32	
	331	20.51	41.47	68.00	79.20				
	332	16.78	29.50	52.46	64.20	69.62	79.98	85.01	
	410	26.98	46.12	53.55	60.32	72.72	78.58		
	411	15.79	39.66	47.66	54.74	61.24	73.22	84.48	
221	221	0.00	27.27	38.94	63.61	83.62	90.00		
	310	32.51	42.45	58.19	65.06	83.95			
	311	25.24	45.29	59.83	72.45	84.23			
	320	22.41	42.30	49.67	68.30	79.34	84.70		
	321	11.49	27.02	36.70	57.69	63.55	74.50	79.74	84.89
	322	14.04	27.21	49.70	66.16	71.13	75.96	90.00	
	331	6.21	32.73	57.64	67.52	85.61			
	332	5.77	22.50	44.71	60.17	69.19	81.83	85.92	
	410	36.06	43.31	55.53	60.98	80.69			
	411	30.20	45.00	51.06	56.63	66.87	71.68	90.00	
310	310	0.00	25.84	36.87	53.13	72.54	84.26		
	311	17.55	40.29	55.10	67.58	79.01	90.00		
	320	15.26	37.87	52.12	58.25	74.74	79.90		
	321	21.62	32.31	40.48	47.46	53.73	59.53	65.00	75.31
		85.15	90.00						
	322	32.47	46.35	52.15	57.53	72.13	76.70		
	331	29.47	43.49	54.52	64.20	90.00			
	332	36.00	42.13	52.64	61.84	66.14	78.33		
	410	4.40	23.02	32.47	57.53	72.13	76.70	85.60	
	411	14.31	34.93	58.55	72.65	81.43	85.72		
311	311	0.00	35.10	50.48	62.96	84.78			
	320	23.09	41.18	54.17	65.28	75.47	85.20		
	321	14.76	36.31	49.86	61.09	71.20	80.72		
	322	18.07	36.45	48.84	59.21	68.55	85.81		
	331	25.94	40.46	51.50	61.04	69.76	78.02		
	332	25.85	39.52	50.00	59.05	67.31	75.10	90.00	
	410	18.07	36.45	59.21	68.55	77.33	85.81		
	411	5.77	31.48	44.71	55.35	64.76	81.83	90.00	
320	320	0.00	22.62	46.19	62.51	67.38	72.08		
	321	15.50	27.19	35.38	48.15	53.63	58.74	68.24	72.75
		77.15	85.75	90.00					

Table 2–1 Angles between planes in cubic crystals (Cont.)

HKL	*hkl*								
	322	29.02	36.18	47.73	70.35	82.27	90.00		
	331	17.36	45.58	55.06	63.55	79.00			
	332	27.50	39.76	44.80	72.80	79.78	90.00		
	410	19.65	36.18	42.27	47.73	57.44	70.35	78.36	82.27
	411	23.76	44.02	49.18	70.92	86.25			
321	321	0.00	21.79	31.00	38.21	44.41	49.99	64.62	69.07
		73.40	85.90						
	322	13.51	24.84	32.57	44.52	49.59	63.01	71.09	78.79
		82.55	86.28						
	331	11.19	30.85	42.63	52.18	60.63	68.42	75.80	82.96
		90.00							
	332	14.38	24.26	31.27	42.20	55.26	59.15	62.88	73.45
		80.16	83.46	86.73					
	410	24.84	32.57	44.52	49.59	54.31	63.01	67.11	71.09
		82.55	86.28						
	411	19.10	36.02	40.89	46.14	50.95	55.46	67.79	71.64
		75.40	79.11	86.39					
322	322	0.00	19.75	58.03	61.93	76.39	86.63		
	331	18.93	33.42	43.67	59.95	73.85	80.97	86.81	
	332	10.74	21.45	55.33	68.78	71.92	87.04		
	410	34.57	49.68	53.97	69.33	72.90			
	411	23.84	42.00	46.69	59.04	62.79	66.41	80.12	
331	331	0.00	26.52	37.86	61.73	80.91	86.98		
	332	11.98	28.31	38.50	54.06	72.93	84.39	90.00	
	410	33.42	43.67	52.26	59.95	67.08	86.81		
	411	30.09	40.80	57.27	64.37	77.51	83.79		
332	332	0.00	17.34	50.48	65.85	79.52	82.16		
	410	39.14	43.62	55.33	58.86	62.26	75.02		
	411	31.32	45.29	49.21	55.75	66.30	69.40	84.23	
410	410	0.00	19.75	28.07	61.93	76.39	86.63	90.00	
	411	13.63	30.96	62.78	73.39	80.12	90.00		
411	411	0.00	27.27	38.94	60.00	67.12	86.82		

Tables of angles between planes of some of the more common crystals have been published.[1] Some of the most important angles for the cubic

[1] Extensive tables for cubic crystals are given in "International Tables for X-ray Crystallography," vol. 2, p. 120, Kynoch, Birmingham, England, 1959, and in R. J. Peavler and J. N. Lenusky, "Angles between Planes in Cubic Crystals," American Institute for Mining, Metallurgical, and Petroleum Engineers, New York. Angles in bismuth, antimony, and hexagonal crystals are given in: Salkovitz, *J. Metals* (*AIME*), vol 8, p. 176, 1956; W. Vickers, *Trans. AIME.*, vol. 209, p. 827, 1957. Angles in gallium are given in C. G. Wilson, *Trans. AIME*, vol. 224, p. 1293, 1962. Angles for tetragonal crystals with c/a from 0.500 to 2.000 are given in a series of tables by R. E. Frounfelker and W. M. Hirthe, *Trans. AIME.*, vol. 224, p. 196, 1962. For uranium angles, see R. B. Russel, *Trans. AIME*, vol. 197, p. 1190, 1953.

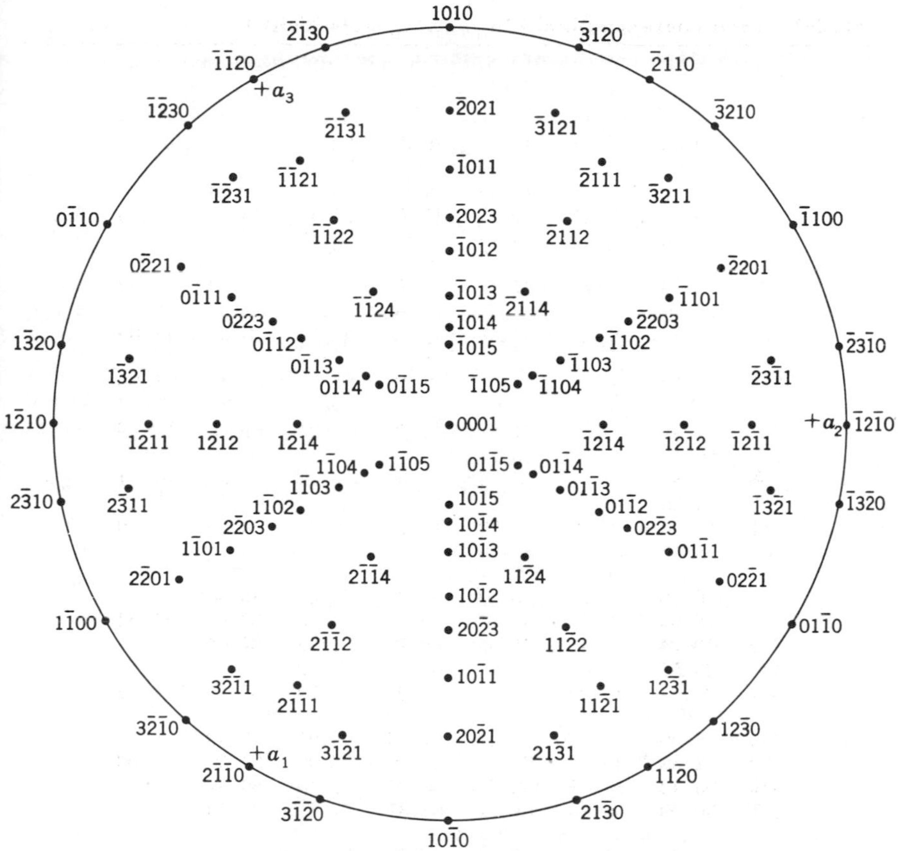

Fig. 2-10 Standard (0001) projection for zinc (hexagonal, $c/a = 1.86$).

system are given in Table 2–1. A typical standard projection for a hexagonal crystal is given in Fig. 2–10, which is for a crystal with an axial ratio of 1.86 (zinc and to a fair approximation also cadmium of $c/a = 1.89$).

Orientation of single-crystal wires and disks The orientation of the axis of a single-crystal wire, rod, or disk, or a crystal surface, can be represented in a simple way on a standard projection by merely placing a point at the appropriate place on the projection. Because of the symmetry properties of crystals, it is usually unnecessary to show the entire standard projection in order to indicate the orientation; for example, in the cubic system it is merely necessary to show the nearest poles of three low-index planes, and for this purpose it is customary to use (001), (101), and (111), which form the corner of a stereographic triangle that can be seen in Fig. 2–9. (Two adjacent triangles of this type are sometimes needed to show how a crystal changes its orientation during deformation.) Showing the crystal orientation as a single point leaves the azimuthal orientation unspecified, but the latter is frequently of no significance; if the azimuth is also important it can be specified by a second point plotted on the projection at the proper place to show the orientation of a reference mark, edge, or crystal face. To show the variation of physical properties with orientation of a rod-shaped crystal, the magnitude of the property can be written beside the point representing the crystal, and points of equal magnitude can be joined by contours.

Applications The combination of standard projection and stereographic net is particularly convenient for analyzing the crystallographic features of the deformation of crystals by slip, twinning, and cleavage or the growth habits of crystals precipitated within a crystal, forming a Widmanstätten pattern. To aid the beginner, it is desirable to list the more common problems and to give the operations by which they are graphically solved, but it should be borne in mind that as soon as one is accustomed to think clearly of the sphere and its "picture," the stereographic net, many of these operations become self-evident.

Obviously, the solutions are independent of the choice of projection plane and are applicable to problems in pure spherical trigonometry; but to make the operations more easily understood, we shall present some typical applications in considerable detail. We shall speak of polished surfaces of specimens and of traces of crystallographic planes in these surfaces (lines of intersection), and we shall generally consider the projection plane to lie in one of the surfaces.

1. Orientation of planes causing a given trace in a surface Let us consider the stereographic projection of a polished surface containing the trace of a crystal plane, the projection being made on a sheet of paper laid parallel to the polished surface. The surface will then be represented on the paper by the basic circle, and markings on the surface will be plotted as points on the circumference of this basic circle. A trace in the surface that runs lengthwise of the page ("vertically") will be plotted in the projection as the diametrically opposite points T and T' at the top and bottom of the basic circle (Fig. 2-11). A trace in any other direction in the surface would be plotted similarly, as the end points of a diameter parallel to the given direction.

To find the planes that would intersect the surface in the direction TT', the points

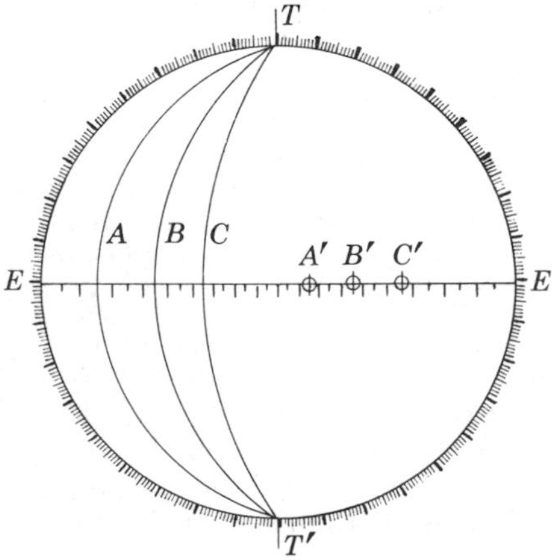

Fig. 2-11 Planes A, B, and C, with poles A', B', and C', and all other poles along EE' intersect projection plane in trace TT'.

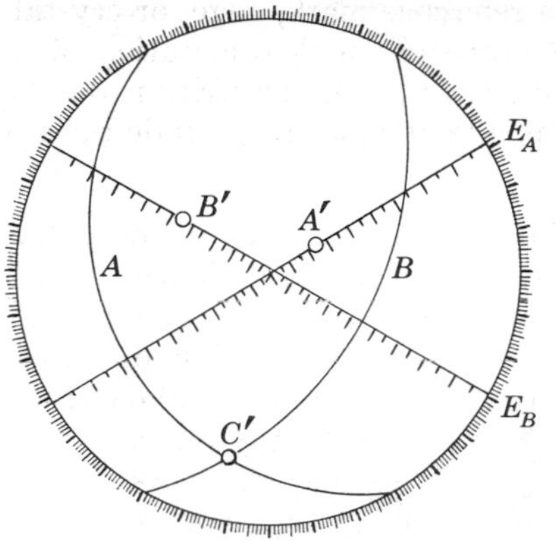

Fig. 2-12 Planes A and B, with poles A' and B', intersect along direction C'.

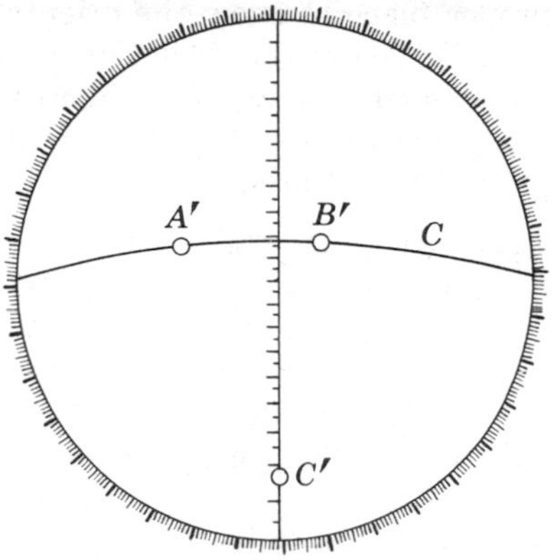

Fig. 2-13 Alternative method of locating C', which is normal to A' and B'.

TT' are superimposed on the N and S poles of a Wulff net. It will then be seen that all the meridians of the net—such as the meridians A, B, and C of Fig. 2–11—are projections of the required planes, since they intersect the basic circle at T and T'. Similarly, any other plane whose pole lies on the equator of the Wulff net will intersect the surface in the direction of the NS axis.

Conversely, if the pole of a plane is given, such as A', its trace in the projection plane is readily found. The transparent sheet on which the pole is plotted is laid on a Wulff net and turned until the pole falls on the equator, in which position the required trace will be parallel to the NS axis of the net.

2. Trace of one plane in another when both are inclined to the protection plane Given two poles A' and B' (Fig. 2–12), the planes A and B are first plotted. This is accomplished for A by rotating the projection over a Wulff net so that the pole A' lies on the equator E_A and then tracing on the projection the meridian lying at 90° to the pole A'. The operation is repeated for the second pole B', with the net turned so that its equator is in the position E_B. The point of intersection, C', of the two planes thus plotted is the projection of the required line of intersection of the planes.

3. Direction normal to two given directions (or zone axis of two planes whose poles are given) Referring again to Fig. 2–12, let us assume that directions A' and B' are given and that the direction normal to both is required. The projection is rotated over a Wulff net until both A' and B' lie on the same meridian, as in Fig. 2–13; then the point C' on the equator and 90° from this meridian is the projection of the required direction. If A' and B' are poles of planes, C is the zone circle and C' is their zone axis.

4. Determination of orientation of a plane from its traces in two surfaces The surfaces are first plotted on the projection as in Fig. 2–14*a* and *b*, one surface lying in the plane of projection and forming the basic circle A of Fig. 2–14*b*, the other surface B coinciding with the meridian of the stereographic net that lies ϕ from the first about the axis NS. (To draw this meridian, the net is rotated so that the direction NS is parallel to the line of intersection of the two surfaces.) On the planes A and B thus plotted in Fig. 2–14*b* are then located the points T_A and T_B, which represent the directions of the traces in the two surfaces, respectively. They will lie at angles laid off from the edge NS to correspond with the angles on the specimen; the angles are measured as differences of latitude, ψ_A and ψ_B, on the stereographic net. After the traces T_A and T_B have been plotted, the plane that causes them can be drawn by rotating the net so that some single meridian of the net will pass through both points; this meridian (the dashed circle C in the figure) is then the projection of the required plane.

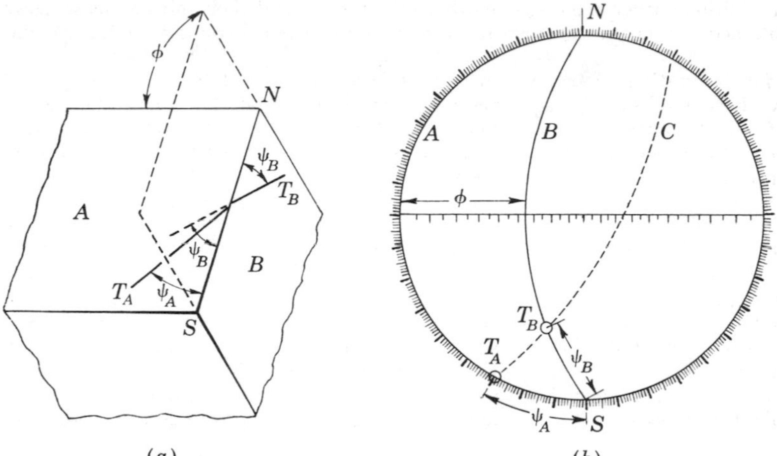

(a) (b)

Fig. 2-14 Determination of orientation of plane from its traces in two surfaces. Traces are T_A and T_B in surfaces A and B, respectively. Plane causing these is C.

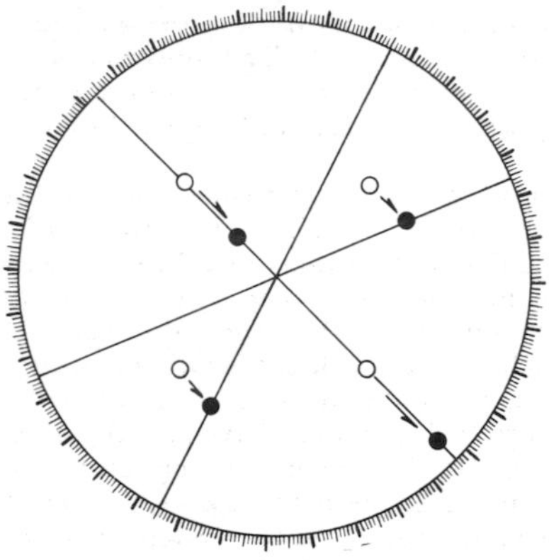

Fig. 2-15 Orientation of cubic crystal determined from traces of {111} planes. Indicated on projection are normals to traces (diameters); {111} poles of standard projection (○); {111} poles in orientation explaining traces (●).

5. Determination of crystal orientation from traces of {hkl} planes when h, k, and l are known

(*a*) *Traces in one surface only* On tracing paper a basic circle is drawn representing the specimen surface; through this circle are drawn diameters perpendicular to the directions of traces seen on the specimen surface. These diameters are then the loci of all poles capable of forming the traces. A standard projection, on a transparent sheet, of all poles of the given form {*hkl*} is then superimposed on this plot and on a Wulff net, and a pin is put centrally through all three sheets. By trial the relative position of the three sheets is found in which a pole of the standard projection may be rotated into coincidence with each of the diameters by the same amount of rotation about the axis of the net. This position of the sheets is illustrated in Fig. 2–15, in which the {111} trace normals (diameters), the {111} poles of the standard projection (○), and the position of these poles on the trace normals after the rotation with the net (●) appear. This final array of poles (●) describes an orientation of the crystal consistent with the observed traces; it is not, however, the only consistent orientation. In fact, if traces on only one surface are studied, the crystal may have the orientation shown or a mirror image of this orientation in the plane of projection, the poles lying in either of the hemispheres. In addition, either one or three other orientations and their mirror images may also be obtained, as has been shown by a complete analytical solution of the general problem.[1]

Charts for orientation determination from measurements of {100} and from {111} traces have been published,[2] and tables have been worked out for orientation determination from {111} traces.[3] When there is need to reduce errors below a few tenths of a degree and the observations justify it, a least-squares refinement of an approximate orientation can be carried out analytically.[4]

(*b*) *Traces in two surfaces* The solution is more direct and rigorous if traces can be followed from one surface around the edge to the other surface, thus eliminating any uncertainty as to the proper pairing of traces on the two surfaces. If this is possible, the first operation is to plot the orientation of each plane by method 4 above. The poles thus plotted give the crystal orientation. If the orientations of other poles of the same crystal are required, they may be obtained by rotation of the standard projection, as in the previous method—the plotted poles on one sheet and the standard projection on another being rotated with respect to the net until a difference of $\phi°$ of longitude and no difference of latitude exist between each $\{hkl\}$ pole of the standard and a corresponding plotted pole. Rotation of $\phi°$ then puts any pole of the standard into its proper position in the plot.

When the pairing of traces on the two surfaces is uncertain, it is necessary to make a plot of poles for all possible pairings. Among this array of poles there will be one or more groupings having the angular relations appropriate for $\{hkl\}$ planes, and these may be singled out from the whole number by trial rotations of the $\{hkl\}$ poles of the standard, possible solutions being those for which $\phi°$ rotation about the net axis brings the standard poles into coincidence with the sets of plotted poles.

6. Determining indices of set of planes causing traces on one or more surfaces

(*a*) *Crystal orientation unknown* The poles of the planes (or of all possible planes in cases of uncertain pairing) are plotted as in the preceding problem. Trial rotations are then performed, using different sets of standard projection poles, until a set is brought into coincidence with the plotted poles.[5]

The procedure in this problem is laborious and leads to uncertain results unless traces are measured on two surfaces and unless the planes are of low indices. It is frequently possible to save labor by noting the number of different directions of traces on each surface, for in this way certain planes may be eliminated from further consideration. If, for example, a single crystal of a cubic metal exhibits more than three directions of traces, the {100} planes alone could not be responsible; if more than four directions are found, neither {100} nor {111} planes alone could have produced them.

(*b*) *Crystal orientation known* If x-ray data or other observations have already given the orientation of a crystal, a standard projection of all likely planes can be rotated to their positions for this crystal orientation. By the methods presented above, the traces that each set of planes will make in the plane of polish may then be plotted. The coincidence of predicted and observed traces will then single out the sets of planes best able to explain the data.[6]

[1] M. P. Drazin and H. M. Otte, *Phys. Stat. Solidi*, vol. 3, p. 814, 1963.

[2] S. Takeuchi, T. Honma, and S. Ikeda, *Sci. Rept. Res. Inst. Tohoku Univ.*, vol. A11, p. 81, 1959.

[3] M. P. Drazin and H. M. Otte, "Tables for Determining Cubic Crystal Orientations from Surface Traces of Octahedral Planes," P. M. Harrod Co., 26 W. Chesapeake Ave., Baltimore, Md., 1964.

[4] J. K. Mackenzie, *Acta Cryst.*, vol. 15, p. 979, 1962. The normal equations are expressed in matrix notation.

[5] R. F. Mehl and C. S. Barrett, *Trans. AIME*, vol. 93, p. 78, 1931.

[6] When traces are available on a single surface, it is best to rotate into one stereographic triangle all arcs of the great circles representing trace normals. The arcs then intersect at the standard projection pole that accounts for all the traces. J. S. Bowles, *Trans. AIME*, vol. 191, p. 44, 1951.

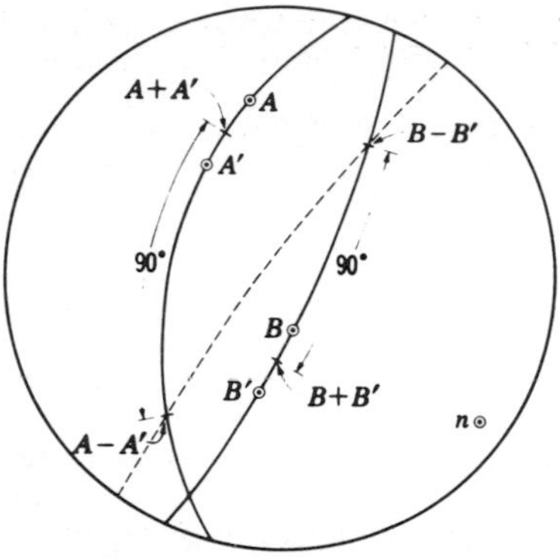

Fig. 2-16 Direct determination of rotation axis n that rotates A into A′, B into B′. Dashed circle is normal to n.

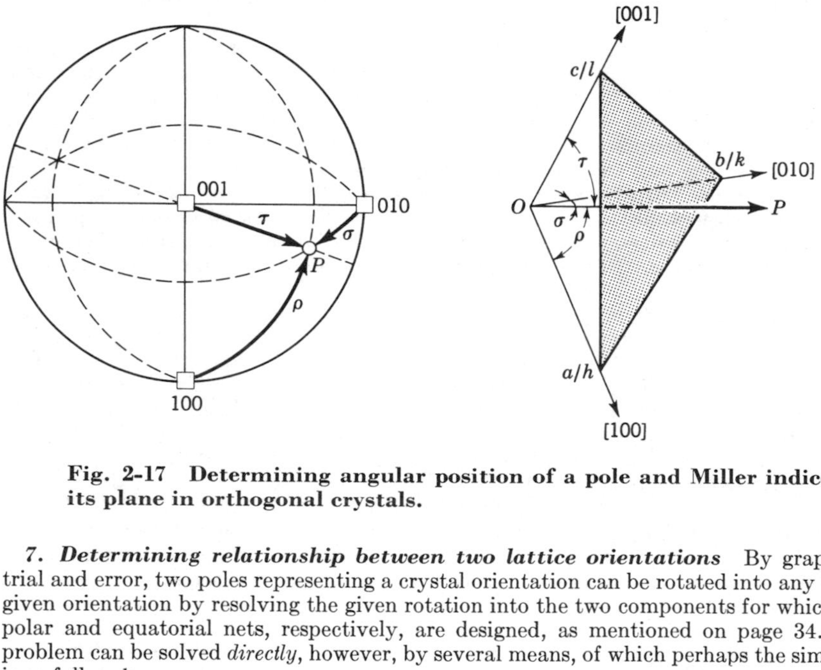

Fig. 2-17 Determining angular position of a pole and Miller indices of its plane in orthogonal crystals.

*7. **Determining relationship between two lattice orientations*** By graphical trial and error, two poles representing a crystal orientation can be rotated into any other given orientation by resolving the given rotation into the two components for which the polar and equatorial nets, respectively, are designed, as mentioned on page 34. The problem can be solved *directly*, however, by several means, of which perhaps the simplest is as follows[1]:

Let **A** and **B** be poles representing the one orientation, and **A′** and **B′** be the corresponding poles for the other orientation. Then the axis **n** and the angle θ of rotation about this axis that will rotate the one orientation into the other can be found directly. The solution involves plotting a point (**A′** + **A**) (Fig. 2–16) midway between **A′** and **A**, and the point (**A′** − **A**) on the same great circle 90° from (**A′** + **A**); it also involves plotting similar points (**B′** + **B**) and (**B′** − **B**) on the zone circle through **B′** and **B**, as indicated in the figure. The zone through (**A′** − **A**) and (**B′** − **B**), indicated by the

[1] This principle has been used in studies of twinning and of martensite transformations. G. Friedel, "Lecons de Cristallographie," Berger-Levrault, Paris, 1926. J. D. H. Donnay and G. Donnay in "International Tables for X-ray Crystallography," vol. 2, p. 104, Kynoch, Birmingham, England, 1959. H. M. Otte, *Acta Cryst.*, vol. 14, p. 360, 1961.

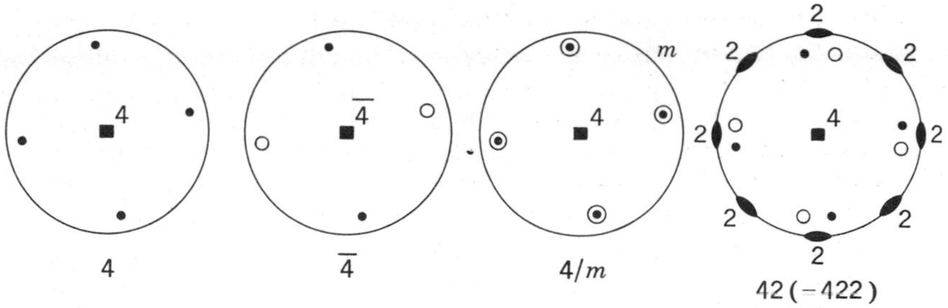

Fig. 2-18 Stereographic projections of four crystal classes of the tetragonal system, showing symmetry elements (squares and ellipses) and poles of a general plane {*hkl*} in one hemisphere (○) and the other (●), respectively.

dashed great circle, has as its pole the required axis **n**. The amount of rotation, θ, about this axis is then determined by the angular distance between **A** and **A′** (or between **B** and **B′**). (In the special case when (**A′** − **A**) and (**B′** − **B**) coincide, **n** is determined, instead, by the intersection of the great circle through **A** and **B** with the great circle through **A′** and **B′**.) When greater accuracy is required, a procedure may be used that is entirely numerical and that locates the axis and rotation as precisely as possible by a least-squares method in which weighting factors can be used with the experimental data.[1]

8. Relation among indices, direction cosines, and position on a standard projection The simple relation between direction cosines of a pole and its position on a projection were mentioned for cubic crystals on page 34; analogous relations hold with other crystals. The following statements apply to any crystal with *orthogonal* axes (cubic, tetragonal, orthorhombic, or orthohexagonal).

If the pole is, for example, pole A, Fig. 2–17a, the angles ρ, σ, τ between the pole and the a, b, c directions in the crystal are measured along the great circles indicated, by turning the plot to the proper orientations on a stereographic net. Let the direction cosines determined by these angles be p, q, r and let the perpendicular distance from the origin to the lattice plane nearest the origin be d (see Fig. 2–17b); the plane will then have intercepts a/h, b/k, c/l where hkl are the Miller indices, and reference to Fig. 2–17b will show that

$$p = \cos \rho = \frac{d}{a/h} \qquad q = \cos \sigma = \frac{d}{b/k} \qquad r = \cos \tau = \frac{d}{c/l}$$

from which it follows that

$$h:k:l = pa:qb:rc$$

and for cubic crystals $h:k:l = p:q:r$.

9. Representation of symmetry classes The symmetry properties of a crystal may be exhibited by a stereographic projection of poles of a plane having general (not special) indices. Some of the symmetry classes of the tetragonal system appear as in Fig. 2–18, with the position of symmetry axes indicated by small geometrical figures and with poles shown by open circles when they lie in one hemisphere and filled circles when they lie in the other. Referring to the last drawing at the right, it will be seen that given one twofold axis in the plane of the drawing, the fourfold axis generates another at right angles to the first, and these in turn generate the symmetry indicated by the twofold axes inclined at 45°. This is the reason that the notation 422 can be shortened to 42, the necessary minimum to describe the class.

10. Homogeneous shearing distortions Distortions of a crystal in which all

[1] J. K. Mackenzie, *Acta Cryst.*, vol. 10, p. 61, 1957.

atom movements are in a single direction, and in which a plane of atoms remains undistorted, are handled stereographically as follows[1]: All *directions* except those in the undistorted plane move along great circles that pass through the point on the projection that represents the direction of atom movement. All *planes* except the planes parallel to the direction of atom movement, which are not rotated, rotate so that their poles move along great circles that pass through the pole of the undistorted plane. A special case of this class of homogeneous distortions, pure shear, has the shear plane as the undistorted, unrotated plane, and the shear direction as the direction of motion of the atoms.[2]

11. Crystallometry An important application of stereographic and gnomonic projections in the past and occasionally today is in connection with measurements of angles between external faces and edges of crystals exhibiting growth faces. Prior to the advent of x-ray diffraction, these measurements—"crystallometry"—provided an important method of determining crystal symmetry and axial ratios, and through these properties, determining the identity of unknown substances. In this method, the interfacial angles are measured with an optical goniometer[3] and plotted, the zone relationships are noted, and then (provided crystal perfection and operator skill are sufficient) the symmetry elements, crystal class, and axial ratios are worked out. These and other distinguishing characteristics are then compared with compilations of such data,[4] or certain specific angles are selected and used to search for an identical substance in "The Barker Index of Crystals."[5] Although this method is still employed, it has now largely been replaced by diffraction methods, and is not even mentioned in most of the recent books on crystallography.

Other perspective projections The stereographic projection is but one of a series of projections that are "perspective," i.e., that represent what the eye sees when placed at a definite position with respect to the reference sphere. When the eye (or a light source) is placed on the *surface* of the reference sphere it gives the *stereographic* projection. When it is at infinity, it gives the *orthographic* projection so widely used in mechanical drawing. And when it is at the *center* of the sphere, it gives the gnomonic projection.[6]

[1] J. S. Bowles, *Acta Cryst.*, vol. 4, p. 162, 1951.

[2] An example of the stereographic treatment of pure shear is given in A. B. Greninger and A. R. Troiano, *Trans. AIME*, vol. 185, p. 590, 1949. The problem can also be handled by vector analysis (and, of course, with much greater precision); see, for example, J. S. Bowles, C. S. Barrett, and L. Guttman, *Trans. AIME*, vol. 188, p. 1478, 1950.

[3] Modern designs of two-circle goniometers are about the size of an ordinary microscope and are particularly appropriate for measuring very small crystals and for aligning them for x-ray work; one is manufactured by the firm Nedinsko of Venlo, The Netherlands; their design and use is discussed by P. Terpstra and L. W. Codd, "Crystallometry," Longmans, New York, 1961.

[4] P. Groth, "Chemische Krystallographie" (7200 crystal descriptions). J. D. H. Donnay (ed.), "Crystal Data," 2d ed., American Crystallographic Society Monograph no. 5, Polycrystal Book Service, Pittsburgh, Pa., 15238.

[5] M. W. Porter and R. C. Spiller (eds.), Heffer, Cambridge, England, 1951. Vol. I explains the method and contains the more symmetrical crystals, vol. II contains monoclinic crystals, and vol. III contains the triclinic crystals.

[6] The radial distance of a pole from the center in the gnomonic projection is $r \tan \phi$ rather than $r \tan (\phi/2)$ as in the stereographic. Zone lines are straight lines, not circles.

3

X-rays, films, and counters

When electrons are driven at high speed into the metal target of an x-ray tube, about 2 percent of their energy is converted into x-rays; the balance is converted into heat in the target. The radiation consists of a continuous spectrum—radiation spread over a wide band of wavelengths—and a superimposed line spectrum of high-intensity single-wavelength components. The former corresponds to white light and is frequently called *white radiation*. The latter corresponds to monochromatic light, and because the wavelength of each component is characteristic of the metal emitting the rays, it is called *characteristic radiation*. The continuous spectrum can be produced without the characteristic if the tube is operated at a low voltage, but as soon as the voltage is increased beyond a critical value the characteristic spectral lines appear in addition to the white radiation.

The continuous spectrum Figure 3-1 shows the distribution of energy in the continuous or white radiation emitted from a tungsten-target tube operating at a series of voltages. The important features of the spectra in Fig. 3–1 are the abrupt ending of each spectrum at a minimum wavelength, the maximum at longer wavelengths, and the gradual decrease in intensity on the long-wavelength end of the spectrum. Increasing the operating voltage shifts both the minimum wavelength (the *short-wavelength limit*) and the point of maximum intensity to the left and increases the intensity of all wavelengths.

The radiation originates when an electron that is moving with a high velocity encounters an atom in the target. If it converts its entire kinetic energy into x-rays at a single encounter, the frequency of the rays produced will be given by the quantum relation

$$eV = h\nu$$

where e is the charge on the electron, V is the voltage applied to the tube, h is a universal constant (Planck's constant), and ν is the frequency of the radiation. While all electrons strike the target with energy eV, only rarely are they stopped in a single encounter so as to convert their entire energy into one quantum. More frequently they dissipate their energy in a series of glancing encounters with a number of atoms and generate heat or quanta of lower frequency than the maximum. From varied encounters an entire continuous spectrum is produced which extends from the limiting frequency given in the above equation down to very low frequencies. In terms of the operating potential of the tube in volts and the wavelength of the radiation

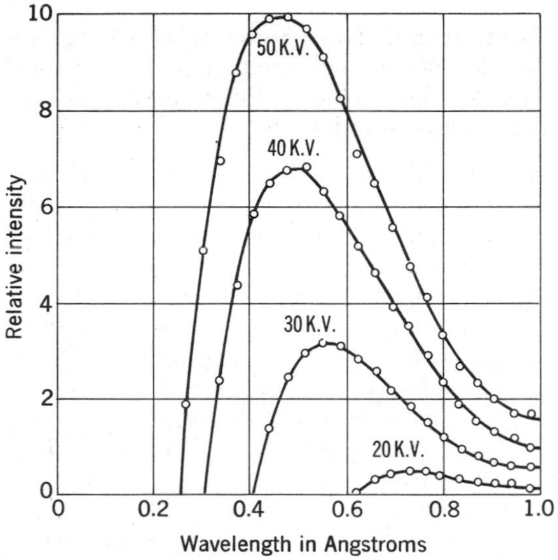

Fig. 3-1 Distribution of energy in the continuous x-ray spectrum of tungsten at different voltages. (Ulrey.)

(which is equal to the velocity of light divided by the frequency) the above relation for the short-wavelength limit of the continuous spectrum in angstroms[1] is

$$\lambda_{min} = \frac{12{,}398}{V}$$

If the voltage impressed on the tube is pulsating between limits or is alternating, it is the peak value and not the root-mean-square value that must be used in this formula to give the minimum wavelength. For x-ray diffraction the wavelengths used are 0.2 A or more and thus do not require potentials in excess of 60,000 volts (60 kv); but for industrial radiography much shorter waves are needed, and voltages are raised to 200, 400, and even thousands of kilovolts.

The efficiency of an x-ray tube as a generator of white radiation is greatly influenced by the metal used for the target and the voltage employed. The total energy emitted increases directly as the electron current and as the atomic number of the target material, and roughly as the square of the applied voltage.[2] The wavelength at which the intensity is a maximum is about 1.5 times the minimum wavelength and shifts to shorter wavelengths as the voltage on the x-ray tube is increased. For applications in which con-

[1] The wavelength in angstroms as obtained from values in kX units multiplied by a conversion factor $\sim$1.00202 (see discussion of units on p. 150).

[2] W. W. Nicholas finds that the energy emitted is proportional to $V^{1.5}$ for high voltages and a tungsten target. (*J. Res. Natl. Bur. Std.*, vol. 5, p. 853, 1930.)

tinuous radiation is used (radiography and Laue photographs) a tungsten-target tube is advantageous because of the high melting point and good thermal conductivity of tungsten and because the atomic number is high (74), which makes it an efficient generator of radiation.

The characteristic spectrum In addition to the continuous spectrum, an x-ray tube operating at a sufficiently high potential also emits a line spectrum that is characteristic of the kind of atoms in the target. The nature of the characteristic spectrum is illustrated in Fig. 3-2, which indicates some of the important peaks in intensity that are found in the radiation from a molybdenum-target tube. The peaks, or lines, are classified into the *K* series, the *L* series, the *M* series, etc., and the individuals in each series are designated by Greek letters and subscripts. Ordinarily, only the intense *K* lines are used in x-ray diffraction work. From a molybdenum target the *K* lines have wavelengths of about 0.7 A and the *L* lines about 5 A; the *M* lines have still longer wavelengths. In the order of decreasing wavelengths, the *K* series consists of the $K\alpha$ line, which is actually a close doublet composed of $K\alpha_2$ and a stronger line $K\alpha_1$; a weaker $K\beta_1$, and a very weak $K\beta_2$ (which is a close doublet, rarely resolved).

The intensity ratio $K\alpha_1/K\alpha_2 = 2{:}1$, is almost independent of the atomic number of the target. The ratio $K\alpha_1/K\beta_1$ varies from element to element,

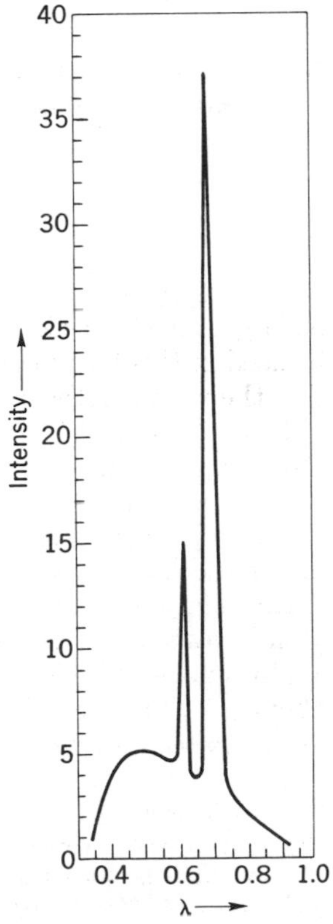

Fig. 3-2 Spectrum of molybdenum at 35,000 volts; $K\alpha$ and $K\beta$ lines and continuous spectrum.

with values from 6:1 to 3.5:1.[1] The intensities of the very weak $K\beta_2$ lines are of the order of 1 percent of the corresponding $K\alpha_1$ line. Wavelengths used in diffraction work are tabulated in the Appendix and in standard references.[2]

There are many lines in the L and M series, but these are largely absorbed in the windows of the x-ray tube and in the air, and are of importance in diffraction work only with targets of very high atomic number, such as tungsten. They are used frequently, however, in chemical analysis by x-ray fluorescence.

The targets most useful in crystallographic work are Ag, Mo, Cu, Ni, Co, Fe, and Cr, in order of decreasing atomic number. The wavelengths of the lines in the characteristic spectra vary in a regular manner from one element to another, as was first shown by Moseley. He discovered that for corresponding lines of the spectrum, the higher the atomic number of the emitting atom, the shorter the wavelength (the higher the frequency).

Origin of characteristic radiation and absorption The remarkable regularities in x-ray spectra are best understood by considering the states of energy in which an atom can exist, after the manner proposed by Bohr. In the Bohr theory an atom is capable of remaining indefinitely in a state of minimum energy unless an amount of energy is imparted to it that is capable of raising it to one of a set of higher-energy states. In a higher-energy state (excited state) there is no radiation or loss of energy until the atom reverts suddenly to a lower-energy state. At this time, the atom throws out a unit of energy, a quantum, in the form of radiation, and the frequency of the radiation, ν, is related to the loss in energy of the atom, E, by the equation

$$h\nu = E$$

where h is Planck's constant. Since the energy states, or "levels," are discrete and sharp, the transitions between them give rise to sharp lines in the spectrum.

The customary way of showing these relations on a diagram is to make a one-dimensional plot of the energy states of an atom with energy increasing vertically above the normal state (Fig. 3–3). Reference to this simplified figure shows that ejection of a K electron raises the atom to the K energy level; filling the K-shell vacancy from the L shell lowers the atom to the L energy level and produces $K\alpha$ radiation, or filling the K-shell vacancy from the M shell lowers the atom to the M level and produces $K\beta$ radiation. All lines of the K series are emitted whenever the electrons from the filament of an x-ray tube are driven into the target with sufficient energy to eject an electron from the K shell.

[1] J. H. Williams, *Phys. Rev.*, vol. 44, p. 146, 1933.

[2] "International Tables for X-ray Crystallography," vol 3, p. 60, Kynoch, Birmingham, England, 1962. Y. Cauchois and H. Hulubei, "Tables de Constantes et Données Numériques, I. Longueurs d'Onde des Emissions X et des Discontinuités d'Absorption X," Hermann, Paris, 1947. These tables, however, are now superseded by the table referred to and used in the Appendix.

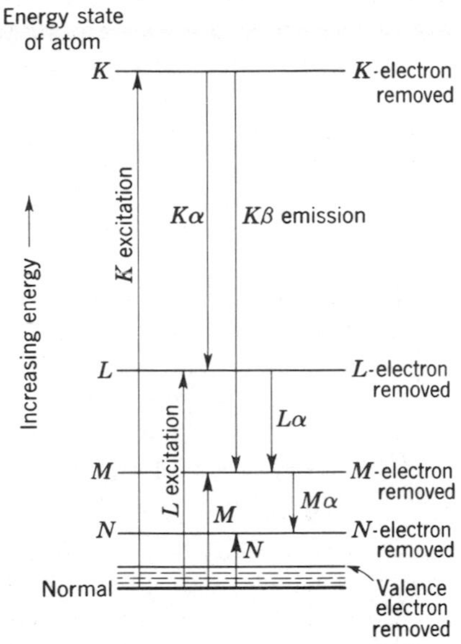

Fig. 3-3 Energy-level diagram for an atom (schematic). Excitation and emission processes indicated by arrows.

The same characteristic radiation is emitted from an atom if it is put into an excited state by the absorption of a quantum of x-rays, for if the quantum has sufficient energy (high enough frequency) it will eject an electron from the atom. If a beam of x-rays is passed through an absorbing substance and the rate of intensity diminution of the transmitted beam is measured as the wavelength of the constant-intensity initial beam is steadily decreased, it will be found that an abrupt change in absorption occurs when the frequency of the beam is such as to excite the atoms of the absorber. This will occur when the energy in a quantum of radiation, given by the equation $E = h\nu$, is sufficient to eject an electron from one of the shells of the absorbing atom. Thus, the K absorption process results in K excitation, and the discontinuity in absorption is called the *K absorption edge*. Reference to Fig. 3-3 will show that the energy change for K absorption is greater than the energy change in the emission of any K series line; i.e., the wavelength of the K absorption edge is less than any K emission wavelength. A similar relation holds for the various L and M absorption processes and the emission lines resulting from them. The frequency of each emission line is equal to the difference in frequency between two absorption edges. When radiation results from the excitation of an atom by the absorption of x-rays, it is known as *fluorescent radiation*, just as in the analogous case with visible light, and is composed almost exclusively of line radiation.[1] It would serve as an excellent source of rays for diffraction if it were not of such low intensity compared with other sources.

Thorough investigation of the absorption edges and emission lines for the elements has disclosed a complex array of energy levels; while there is

[1] A. H. Compton, *Proc. Natl. Acad. Sci.*, vol. 14, p. 549, 1928.

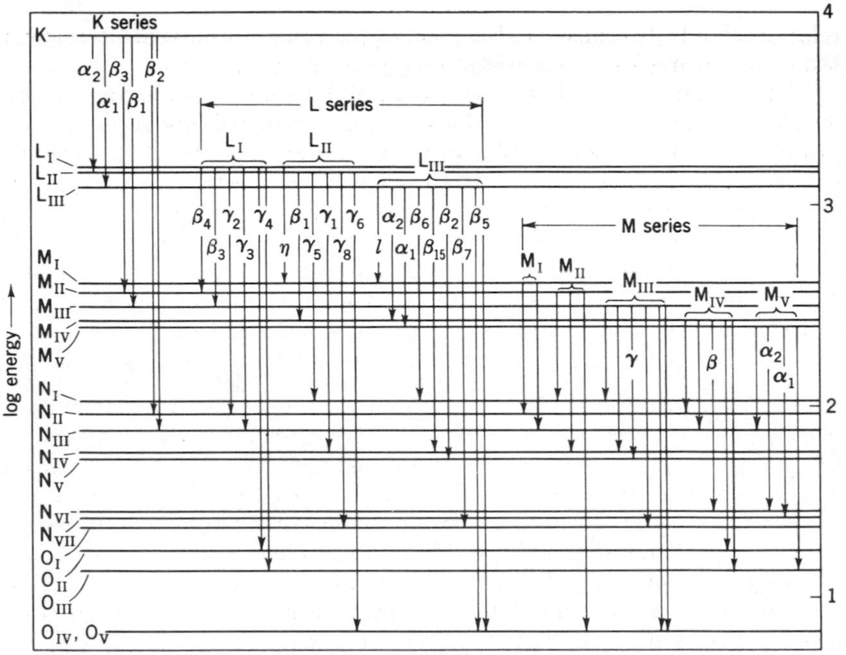

Fig. 3-4 Energy-level diagram for uranium with emission transitions indicated by arrows.

only a single K absorption edge, there are three L absorption edges, five M absorption edges, and many N and O levels (in the heavier atoms). The complete energy-level diagram for one of the heavy elements is reproduced in Fig. 3–4 with the transitions marked that give rise to the x-ray emission lines.

Dependence of line intensities on voltage Emission lines are not excited unless the voltage exceeds the critical value V_0 necessary to remove an electron entirely from an atom of the target. This occurs when, in volts,

$$V_0 = \frac{12{,}398}{\lambda_a}$$

where λ_a is the wavelength in angstroms of the absorption edge concerned (K edge in the case of the K series, etc.). The intensity I of a spectral line increases with voltage V and the current i through the x-ray tube approximately according to the relation

$$I = ci(V - V_0)^n$$

where c is a proportionality constant and V_0 is the excitation voltage for the line. The value of n is slightly less than 2 if a tube is operated with constant-potential direct current and moderate voltages; i.e., the intensity is nearly proportional to the square of the voltage in excess of the critical. The exponent decreases toward unity at higher voltages, and the squared relation should be considered a good approximation only at voltages less than two or three times the excitation potential.

The critical excitation voltage for molybdenum K radiation is about 20,000 volts, but because of the relations discussed above it is advantageous to operate considerably above this voltage to obtain a line spectrum that is intense in comparison with the continuous spectrum, which is always present. Molybdenum tubes are usually operated between 35 and 60 kv, while tubes with targets similar to copper and chromium are operated near 35 kv. The ratio of line intensities to white radiation at a given voltage becomes more favorable as the atomic number decreases, since the lighter elements are inefficient generators of white radiation. This is one of the reasons why copper target tubes are used so much in crystallographic work.

Absorption of x-rays An x-ray beam loses intensity in traversing matter both by "true" absorption, which is a transformation from x-rays into kinetic energy of ejected electrons and atoms, and by *scattering*, a transfer of radiant energy from the primary beam to scattered beams originating in the atoms of the absorbing matter. An understanding of these processes is important both in radiography and in x-ray diffraction.

Consider a monochromatic beam of x-rays penetrating a sheet of material of thickness x. If a beam of energy I traverses a thin layer of thickness dx and diminishes in energy by the fraction dI/I, then

$$\frac{dI}{I} = -\mu\, dx$$

where μ is a constant, the *linear absorption coefficient*, which depends on the wavelength of the rays and the nature of the absorber. Integration of this equation gives the absorption equation

$$I = I_0 e^{-\mu x}$$

The energy, or intensity, decreases from the initial value I_0 exponentially, and the more rapidly it decreases the greater is the linear absorption coefficient. It is convenient to put this equation in terms of the mass traversed rather than the thickness. This can be done by replacing the term x by ρx, where ρ is the density of the absorbing material. The quantity μ must then be replaced by μ/ρ, which is the *mass absorption coefficient*, and the equation becomes

$$I = I_0 e^{-\frac{\mu}{\rho}\cdot \rho x}$$

Most tables list μ/ρ rather than μ because μ/ρ is independent of the physical state (solid, liquid, or gas), whereas μ is not. The mass absorption coefficient of an alloy can readily be calculated from the weight percentages $w_1, w_2, \cdots$ and the values $(\mu/\rho)_1, (\mu/\rho)_2, \cdots$ for the individual elements in the alloy by the formula

$$\frac{\mu}{\rho} = \frac{w_1}{100}\left(\frac{\mu}{\rho}\right)_1 + \frac{w_2}{100}\left(\frac{\mu}{\rho}\right)_2 + \cdots$$

In the absorption formula this overall absorption coefficient μ/ρ is multiplied by the density of the alloy and the thickness penetrated.

These absorption equations are fundamental to the practice of *radiography*. When a beam of x-rays or gamma-rays from radioactive material passes through an object, it emerges with an intensity that is dependent upon the thickness and absorption coefficient of the material it penetrates. A beam that encounters a cavity in a metal casting, forging, or weld emerges with greater intensity than a beam that encounters only sound metal. An image of the cavity is registered on a photographic film placed behind the object, and the nature of a defect is recognized from the appearance of its image on the radiograph.

Since the loss of intensity is due to the combined effects of true absorption and scattering, μ/ρ is the sum of two separable terms, the *true absorption coefficient* τ/ρ, and the *scattering coefficient* σ/ρ. The scattering term is the less important of the two and contributes a relatively small amount to the total absorption for elements of greater atomic number than iron (26). It does not vary greatly with changes in wavelength or atomic number.

The true absorption coefficient varies markedly with wavelength and atomic number, for it depends on the efficiency of the rays in ejecting photoelectrons. Between absorption edges, τ/ρ varies with the fourth power of the atomic number and the cube of the wavelength:

$$\frac{\tau}{\rho} = cZ^4\lambda^3$$

At each absorption edge there is an abrupt change in the constant c. This results from the fact that an absorption edge marks the place on the frequency scale where the radiation is just able to eject an electron from one of the electron shells. For example, radiation of longer wavelength than the K edge cannot eject K electrons, while waves shorter than the K edge are able to do so. Owing to the fact that true absorption predominates over scattering in all except the light elements and the very short x-ray wavelengths, the total absorption varies approximately in the same way as τ/ρ. In the curve of μ/ρ against wavelength reproduced in Fig. 3–5, the absorption edges are very prominent; in most elements μ/ρ differs by a factor of about 5 on the two sides of the K absorption edge. The Appendix lists values of μ/ρ for a number of wavelengths frequently encountered.

The chemical bonds of an element have an influence on the absorption spectrum. The position of the absorption edges depends slightly on the

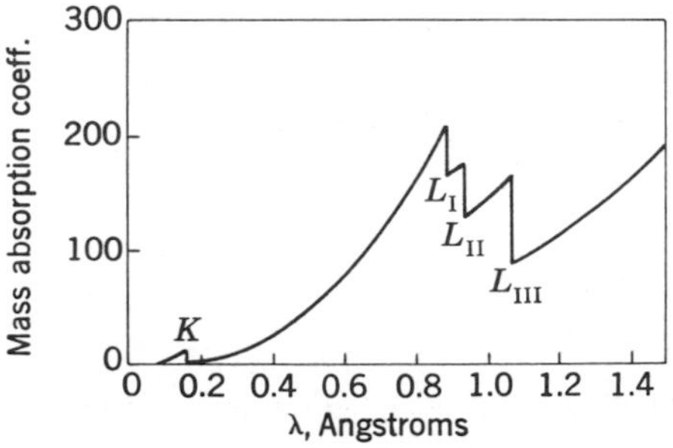

Fig. 3-5 Variation of absorption in platinum with wavelength, showing *K* and *L* absorption edges.

Table 3–1 Thickness of absorber in centimeters to reduce intensity to half value

Wavelength, angstroms	Absorber: Air at 0°C, 760 mm	Cellophane	Al	Cu	Pb
0.1	...	4.3	1.6	0.21	0.016
0.7	410	0.4	0.050	0.0016	0.00044
1.5	62	0.11	0.0056	0.0016	
2.0	26	0.049	0.0025	0.00071	

chemical binding of the atom concerned, and in addition there is a fine structure at the short-wavelength side of each main absorption edge.[1] The fine structure is due to electrons being raised to unoccupied energy levels above the lowest ones; these are influenced somewhat by the atomic arrangement in the neighborhood of the atom that is responsible for the absorption edge.

It is often convenient to know for common materials the thickness that will reduce the intensity of a beam to a certain fraction of its initial value, say one-half. The *half-value thickness* that will reduce I to $\frac{1}{2}I_0$ can be computed from the absorption equation by taking natural logarithms; this yields the equation

$$x_{\frac{1}{2}} = \frac{0.69}{\mu}$$

where $x_{\frac{1}{2}}$ is the half-value thickness in centimeters. Some values are given in Table 3–1.

Filtering Filtering is frequently used in diffraction work to remove unwanted components of the characteristic spectrum together with some of the white radiation. For this purpose, a *β filter* is used that consists of a sheet of material relatively transparent to the $K\alpha$ and opaque to the $K\beta$ radiation. Proper choice of the atomic number of the filter makes this possible, because of the abrupt change in absorption at the K absorption edge, as illustrated in Fig. 3–6 for the case of a zirconium filter to be used with the molybdenum spectrum shown just above it. A β filter for any target can be chosen by referring to wavelength tables (see the Appendix) and picking a filter with a K edge between the $K\alpha$ and $K\beta$ emission wavelengths. Metal foils frequently used for filters are listed in Table 3–2.[2] The filters may be rolled foil, or (somewhat less efficiently) electrodeposits of the filter metal on

[1] A. H. Compton and S. K. Allison, "X-rays in Theory and Experiment," p. 667, Van Nostrand, New York, 1935. L. V. Azároff, *Rev. Modern Phys.*, vol. 35, p. 1012, 1963. L. G. Parratt, *Rev. Modern Phys.*, vol. 31, p. 616, 1959.

[2] These data and additional data for other intensity ratios are given in "International Tables for X-ray Crystallography," vol. 3, p. 75, Kynoch, Birmingham, England, 1962.

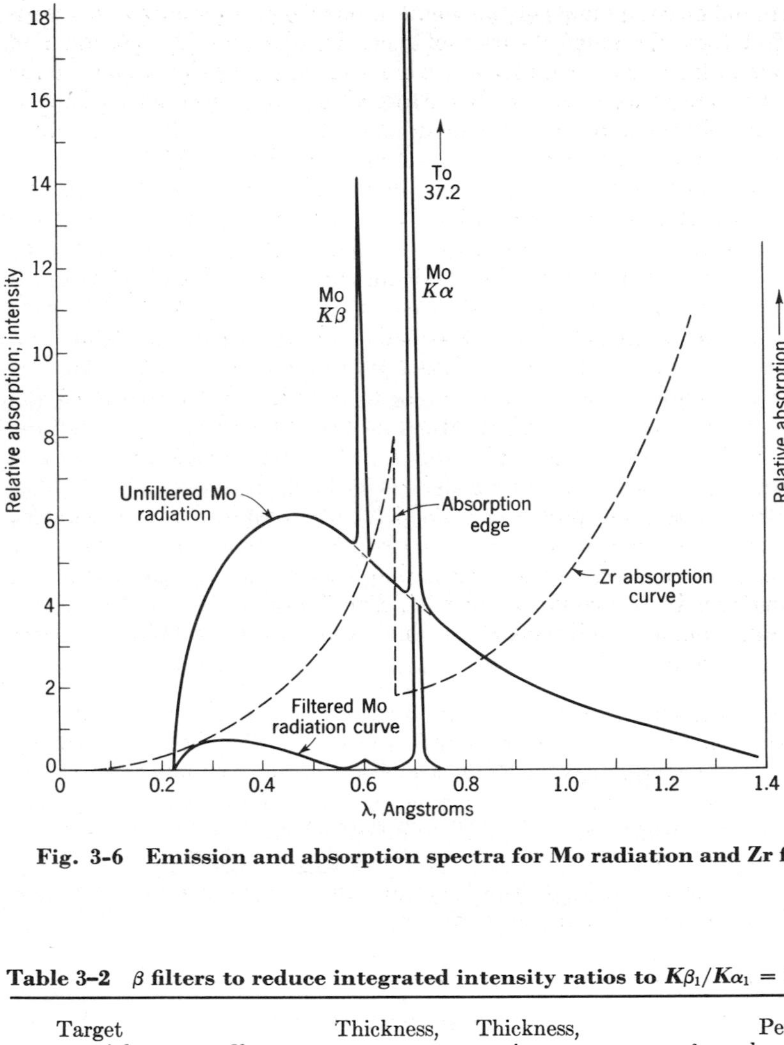

Fig. 3-6 Emission and absorption spectra for Mo radiation and Zr filter.

Table 3–2 β filters to reduce integrated intensity ratios to $K\beta_1/K\alpha_1 = 1/500$

Target material	β filter	Thickness, mm	Thickness, in.	g per cm²	Percent loss, $K\alpha_1$
Ag	Pd	0.092	0.0036	0.110	74
	Rh	0.092	0.0036	0.114	73
Mo	Zr	0.120	0.0047	0.078	71
Cu	Ni	0.023	0.0009	0.020	60
Ni	Co	0.020	0.0008	0.017	57
Co	Fe	0.019	0.0007	0.015	54
Fe	Mn	0.018	0.0007	0.013	53
	Mn_2O_3	0.042	0.0017	0.019	59
	MnO_2	0.042	0.0016	0.021	61
Cr	V	0.017	0.0007	0.010	51
	V_2O_5	0.056	0.0022	0.019	64

aluminum foil or compounds of the metal in powder form mounted on cardboard. No filter will remove all traces of general radiation, $K\beta$ radiation, and characteristic lines from impurities within the target or deposited on the target of the x-ray tube during its life. More highly monochromatized radiation can be obtained by employing double filters in a differential filter technique, or by employing a crystal monochromator.

Differential filtering is accomplished by preparing two filters of proper thickness from adjacent elements in the periodic table. Their thicknesses are adjusted (by thinning and by adjusting their angle with respect to the transmitted x-ray beam) until they transmit nearly equal intensities of all wavelengths except those between their K absorption edges. Intensity measurements of a diffracted beam are made with one filter in the beam and then with the other, and the two intensity readings are subtracted. The intensity difference is due almost entirely to radiation between the two K absorption edges.[1] If this wavelength band is made to include the $K\alpha$ emission line from the target, a strong monochromatic beam is obtained with only a small amount of white radiation content.[2] Filters balanced at the $K\beta$ wavelength are not perfectly balanced at all undesired wavelengths, however, and the use of a correcting filter superimposed on one of the other filters can improve the monochromatization,[3] as can also the use of a pulse-height analyzer in the counting circuits. For balanced filters suitable for Cu $K\alpha$ radiation, a Ni foil 0.00038 in. thick and a Co foil 0.00042 in. thick have been recommended.[4]

Monochromators Reflection from a single-crystal face is commonly used to eliminate undesired wavelengths in diffraction work. The most strongly reflecting crystals and crystal planes are used because the loss of intensity by reflection is severe. Special geometries, involving singly and doubly bent crystals, are also used, when possible, to increase reflected intensities (these are discussed in connection with the powder method, Chap. 7).

Among the strongest reflecting crystals and planes useful for monochromators should be listed the following:

Crystal	LiF	Aluminum	Fluorite		NaCl	Quartz
Reflection	200	111	111	220	200	$10\bar{1}1$
Spacing in angstroms	2.01	2.33	3.16	1.94	2.82	3.35

[1] P. A. Ross, *Phys. Rev.*, vol. 28, p. 425, 1926; *J. Opt. Soc. Am.*, vol. 16, pp. 375, 433, 1928.

[2] C. S. Barrett, *Proc. Natl. Acad. Sci. U. S.*, vol. 14, p. 20, 1928. J. A. Soules, W. L. Gordon, and C. H. Shaw, *Rev. Sci. Instr.*, vol. 27, p. 12, 1956. K. Tanaka et al., *Rev. Sci. Instr.*, vol. 30, p. 430, 1959.

[3] P. Kirkpatrick, *Rev. Sci. Instr.*, vol. 10, p. 186, 1939.

[4] Recommendations are given for other targets in "International Tables for X-ray Crystallography," vol. 3, loc. cit.

Pentaerythritol 002 reflection has also been used, as well as urea nitrate 002, calcite 200, gypsum 020, potassium bromide 200, and others that are less efficient. Most monochromators reflect not only the desired wavelength but also any components of the beam that are submultiples of this wavelength; particularly important are the components of half the wavelength of the $K\alpha$. These can be avoided by lowering the voltage on the x-ray tube sufficiently, or by using a crystal with negligible reflecting power for a second-order reflection, such as can be obtained by monochromatizing with the fluorite 111 reflection or the silicon, germanium, or diamond 111.[1]

The scattered radiation Two scattering processes contribute to the total absorption coefficient. *Coherent scattering* results from the back-and-forth acceleration of an electron by the primary radiation and is identical in wavelength with the original radiation. When this coherent radiation from the electrons of one atom is superimposed on the rays from other atoms arranged on a space lattice, reinforcement occurs and diffracted beams are formed (see Chaps. 4 and 8). A second type of scattering also occurs which is not coherent and which does not take part in diffraction; this is *modified radiation.*

A. H. Compton showed that modified radiation may be understood as being the result of an encounter of a quantum with a loosely bound or a free electron, with the result that the electron recoils under the impact and the quantum is deflected with a partial loss of energy (the Compton effect). The laws of conservation of energy and of momentum govern the encounter.[2]

It is necessary to subtract Compton modified scattering from the total measured intensities of scattering in some experimental work, particularly in diffraction from liquids, thermal vibrations in crystals, and in diffuse scattering from other causes such as local order in superlattices. Detailed calculations of the intensity distribution of the modified scattering are therefore of value in interpreting certain diffraction problems.[3] Freeman has undertaken the accurate computation of incoherent scattering vs. $(\sin \theta)/\lambda$ for atoms and ions of many of the elements.[4]

Modified radiation forms a decreasing proportion of the total scattered radiation as the wavelength of the primary beam is increased and as the angle between the primary beam and the direction of measurement of the scattered ray is decreased, other conditions remaining the same. In the heavier elements more of the electrons are tightly bound and contribute to coherent rather than to modified scattering.

X-ray tubes A great variety of x-ray tubes are used for diffraction purposes and still others for radiography, but all are of one of two general types.

[1] H. Lipson, J. B. Nelson, and D. A. Riley, *J. Sci. Instr.*, vol. 22, p. 184, 1945.

[2] A. H. Compton and S. K. Allison, "X-rays in Theory and Experiment," Van Nostrand, New York, 1926, 1935.

[3] A review from the standpoint of the crystallographer will be found in "International Tables for X-ray Crystallography," vol. 3, op. cit., p. 247.

[4] A. J. Freeman, *Phys. Rev.*, vol. 113, p. 176, 1959; *Acta Cryst.*, vol. 12, pp. 274, 929, 1959; vol. 13, p. 190, 1960; vol. 15, p. 682, 1962.

1. Gas tubes Gas tubes, in which electrons are supplied by the electrical discharge through low-pressure gas, are operated with vacuum pumps attached and are usually maintained at the proper gas pressure, near 0.01 mm of mercury, by balancing the rate of pumping against the rate of influx of air through a controlled leak, or by providing a large reservoir that is kept at operating pressure. They are the most inexpensive tubes that can be built,[1] and they produce the purest spectra, since the target is not contaminated with materials evaporated from a hot filament, but unfortunately, they are notoriously difficult to keep operating satisfactorily.

2. Electron tubes Electron tubes ("Coolidge tubes") utilize electrons emitted from a hot filament in a high vacuum. A *demountable* electron tube offers the advantages over a gas tube of better control of x-ray intensity and of focal-spot size and shape and may be operated at larger currents, usually 20, 30, or 40 ma, the limit being set by melting or excessive pitting of the target. On the other hand, the equipment requirements are greater than for the gas tube, and there is a need for periodic cleaning of the target to avoid danger of contamination of the spectrum by tungsten *L* lines (and mercury *L* lines if a mercury pump is used). Electrostatic or electromagnetic focusing of the electron beam may be used to obtain small focal spots if desired[2]; less exact focusing is ordinarily obtained by controlling the shape of the focusing cup and the position of the filament within it when cup and filament are at the same potential.[3] It is common for focal spots to have a very uneven distribution of intensity, and for much of it to be outside the area that furnishes the rays through the slits of a diffraction camera. It is well to survey the focal spot with a pinhole camera in which the rays pass from the target through a pinhole in a lead sheet placed an inch or two in front of an x-ray film. In addition to demountable tubes of conventional design,[4] occasional tubes have been built with *rotating targets* that are capable of operation at upwards of five or ten times the usual input.[5]

[1] R. W. G. Wyckoff and J. B. Lagsdin, *Radiology*, vol. 15, p. 42, 1930; *Rev. Sci. Instr.*, vol. 7, p. 35, 1936. C. J. Ksanda, *Rev. Sci. Instr.*, vol. 3, p. 531, 1932. G. Hägg, *Rev. Sci. Instr.*, vol. 5, p. 117, 1934. I. Fankuchen, *Rev. Sci. Instr.*, vol. 4, p. 593, 1933. R. E. Clay, *Proc. Phys. Soc. (London)*, vol. 40, p. 221, 1928.

[2] A. Guinier, "Théorie et Technique de la Radiocristallographie," Dunod, Paris, 1956.

[3] N. C. Breese, *Rev. Sci. Instr.*, vol. 8, p. 258, 1937. J. S. Thorp, *J. Sci. Instr.*, vol. 26, p. 201, 1949.

[4] V. E. Pullin and C. Croxson, *J. Sci. Instr.*, vol. 8, p. 282, 1931. W. M. Roberds, *Rev. Sci. Instr.*, vol. 1, p. 473, 1930. E. A. Owen and G. D. Preston, *J. Sci. Instr.*, vol. 4, p. 1, 1926. L. G. Parratt, *Phys. Rev.*, vol. 41, p. 553, 1932.

[5] A. Müller, *Proc. Roy. Soc. (London)*, vol. A117, p. 30, 1927; vol. A125, p. 507, 1929; vol. A132, p. 646, 1931. J. W. M. DuMond, B. B. Watson, and B. Hicks, *Rev. Sci. Instr.*, vol. 6, p. 183, 1935. A. Bouwers, *Physica*, vol. 10, p. 125, 1930. W. T. Astbury and R. D. Preston, *Nature*, vol. 133, p. 460, 1934. J. E. deGraaf and W. J. Oosterkamp, *J. Sci. Instr.*, vol. 15, p. 293, 1938. A. Taylor, *J. Sci. Instr.*, vol. 26, p. 225, 1949; *Rev. Sci. Instr.*, vol. 27, p. 757, 1956. A review: *Brit. J. Appl. Phys.*, vol. 1, p. 305, 1950. Taylor reports a brilliancy of the focal spot $\sim$15 times that of a conventional sealed-off tube.

Sealed-off tubes are manufactured in quantity by a number of concerns for medical purposes, industrial radiography, and diffraction; the diffraction tubes are available in a number of target materials and are made to be quickly interchanged in the x-ray unit when a different wavelength is needed. The line-focus principle is used to spread out the focal spot into a line; when viewed end-on it appears as a point focus, and when viewed from the side it appears as a line. For diffraction work, the rays that leave the target at an angle of 4 to 6° from the target face are most useful. At "take-off angles" less than 4° there is a serious loss in intensity by absorption in the target itself. The stability and great constancy of operation of sealed-off tubes make them indispensable for precise intensity measurements.

Electrical equipment for diffraction tubes Photographic diffraction work may be done with a minimum of electrical equipment; counter diffractometers require much more. Tubes can be operated with a transformer giving 30 to 50 kv, an autotransformer to regulate the voltage, a step-down transformer to heat the filament and to control the voltage, and a milliameter and voltmeter to show the operating conditions (Fig. 3-7). Both gas tubes and electron tubes can be operated in a self-rectifying manner, although the use of half-wave or full-wave rectifying circuits separate from the x-ray tube will contribute to the stability and efficiency of both types. Outmoded medical equipment has served in many laboratories. Modern diffraction installations include voltage-regulating circuits (which are necessary if experiments require precise intensity measurements), overload relays, switches to cut off power if water pressure drops too low to cool the tube properly, time switches, and various safety switches. All high-voltage circuits are completely enclosed, and shielding is provided for all x-rays other than the beam entering the diffraction equipment. The rectifier tubes are also shielded, since these may become sources of x-rays.

X-ray protection Soft radiation of the type used for diffraction work is readily absorbed in the tissues of the body and is consequently a source of grave danger to the careless operator. A brief exposure of the hands to the direct radiation can cause an x-ray burn that ultimately becomes very painful and may require years to heal, if it can be healed at all, and yet the

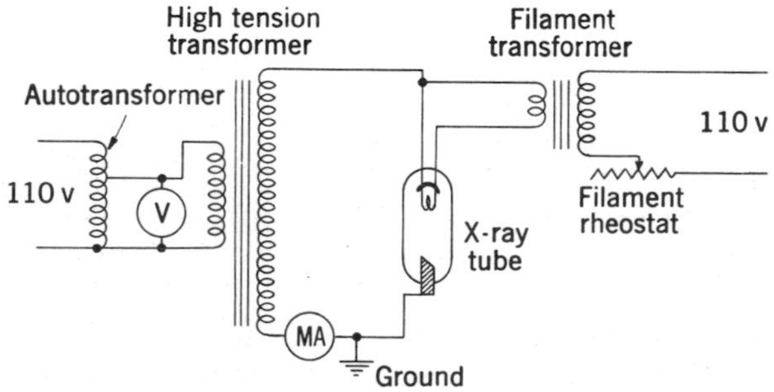

Fig. 3-7 A simple wiring diagram for a self-rectifying x-ray tube.

operator may be conscious of no sensation at the time he is exposed or for several days thereafter. Such severe burns have also been received from *scattered rays* by operators who avoided the direct beam but put their hands or faces in the way of the radiation scattered by some piece of apparatus they were demonstrating or adjusting. There is also a danger from weaker exposures over large areas of the body, and these, like the local burns, are cumulative over several weeks. A general dosage may be accumulated that will cause a serious lowering of the white blood count and other destructive effects. To guard against this danger the protective shields around the tube must be designed to absorb the unused radiation properly. A test for this is to place a piece of x-ray film in black paper near the tube during an operating time of a few days and to look for fog on the film after development. It is a rule in some laboratories that x-ray workers wear film badges while in the laboratory and have occasional blood tests. With well-designed equipment and constant caution on the part of its users, there is no need for the user to receive even the slightest exposure.

Since almost all radiation injuries to diffraction workers in the past have been on the fingers, it is now realized that the most important radiation hazards originate as follows[1]:

1. X-rays escaping from ports, in x-ray tube housings, not adequately closed off by shutters when not in use.
2. Scattered x-rays escaping from a gap between an instrument and the x-ray tube housing. Instruments should be designed so that no x-rays can escape here.
3. Scattered x-rays from the irradiated sample.
4. The direct beam passing through the instrument and emerging at the rear without being adequately stopped by absorbing lead glass or other beam-catcher device.

Photographic efficiency of x-rays The intensity of an x-ray beam can be measured in a variety of ways, including photography on x-ray films. With care, most diffraction problems can be solved with photographically measured intensities of the diffracted beams, but an understanding of the laws of film darkening by x-rays is required for this work.

The density D of a photographic emulsion after exposure to x-rays and development is measured by the absorption of light in the emulsion. If the intensity of a beam of light incident on the emulsion and transmitted through it is measured, D is given by the relation

$$D = \log_{10} \frac{\text{incident light}}{\text{transmitted light}}$$

The density is related to the *exposure*, which is defined as the product of the intensity of the rays striking the film and the time of exposure. For moderate densities there is a linear relation between exposure and the result-

[1] "International Tables for X-ray Crystallography," vol. 3, p. 336, Kynoch, Birmingham, England, 1962.

ing density of the developed film. (Severe overexposure destroys this linearity, however, and may even lead to "reversal"—the most heavily exposed spots being less dense than other places.) The characteristics of several commonly used types of x-ray films have been studied and reported by the Commission on Crystallographic Apparatus of the International Union of Crystallography.[1] The density with zero exposure is the fog in the emulsion, which varies with the type and age of the emulsion and the technique of development. The most useful range of densities lies between the fog level and densities of the order of 1.0, and most x-ray films retain their linear relationship throughout this range or even further. Linearity in fast and in screenless types of commercial x-ray films is sometimes reported up to densities of 2.0 to 4.0—ranges where the films are so dark that they must be viewed with variable high-intensity viewing screens. The contrast in the image increases with developing time[2] and varies with all factors affecting development, with different emulsions, and even with different samples of the same emulsion. Thus, a calibration curve applies strictly to a single film only.

In practice it is not usually necessary to plot the curve, for the densities to be measured can be compared directly with a graded series of densities made by known exposures on pieces of the same film and developed in an identical manner. For example, a series of spots along one edge of the film can be exposed to weak radiation of constant intensity for periods such as 2, 4, 8, 16, 32 sec, and these can be compared by eye with the spots of unknown density on the film. A rotating disk with sectors cut out so as to interrupt the rays for suitable fractions of each revolution can be placed in front of the film to provide the graded exposures without accurate timing. Calibration of the film in this way is possible because film darkening by x-rays obeys the *reciprocity law* that exposure is proportional to intensity times time, regardless of the length of exposure and whether it is continuous or interrupted. The eye is capable of judging equal densities rather well but cannot judge density ratios; when ratios are required the film must be measured on a densitometer or a microphotometer or compared with a graded series of exposures.

A series of spots in a diffraction pattern can be graded in intensity with sufficient precision for many purposes by the expedient of using two superimposed films in the same film holder. The rays will be absorbed a definite amount in the first film and will thus blacken the second film to a density that is a definite fraction of the density of the first film. The ratio of densities on the two films can be determined for a particular wavelength of radiation and can be used as a basis for judging approximate intensities of the entire series of spots. A pack of three or more films may also be used in this way to record a wider range of exposure.

When maximum detail and precision are not needed, it is possible to decrease exposure time by the use of intensifying screens placed in intimate

[1] *Acta Cryst.*, vol. 9, pp. 520, 691, 1956.

[2] Contrast is defined as the slope of the characteristic curve of D vs. $\log_{10}$ (exposure).

contact with the film. However, the advantages of intensifying screens have been much lessened by the introduction of films that even without screens compare favorably in speed with the fastest film-screen combinations. In handling films one should avoid abrasion; creasing; kinking; excessive pressure; finger marks; excessive temperature during storage or processing; incomplete developing, fixing, and washing; and water spots during drying.

Electrical methods of x-ray measurement Electrical methods permit intensities to be measured to within tenths of a percent or a few percent, an improvement by an order of magnitude over the accuracies obtained with films, which usually lie in the range 5 to 10 percent. The efficiency in registering the incident quanta of radiation is so near 100 percent for each of the better methods that other considerations govern the relative desirability of the methods for each application.

The *Geiger counter* is the simplest to apply of any of the counters. It consists of a wire electrode in a cylindrical gas chamber, with a voltage on the wire sufficient to attract an electron in the gas so strongly that the electron ionizes many atoms along its path. Each initial ionization event touches off a chain of ionizing events that culminates in a strong discharge through the gas. The current in the gas is then quenched and the counter is restored to normal conditions ready for the next count in a time interval of the order of 10^{-4} sec. The number of these pulses in a given time interval is a measure of the radiation intensity.

The counting rate depends on the voltage across the counter in the manner indicated in Fig. 3-8 when exposed to radiation of constant intensity; the counter sensitivity increases with voltage as the voltage increases beyond the threshold value *a*, but when the plateau *b-c* is reached the counter sensitivity is approximately independent of voltage. A counter should be operated on this plateau of the curve and should be reconditioned or replaced if this plateau becomes too narrow.

A counter filled with an inert gas, such as argon, to which a small percentage of organic vapor or of a halogen such as chlorine is added, may have the property of quenching itself; other counters are quenched by connecting them to the grid of a vacuum tube in a circuit that lowers the voltage on the wire abruptly when the pulse reaches the grid. When the

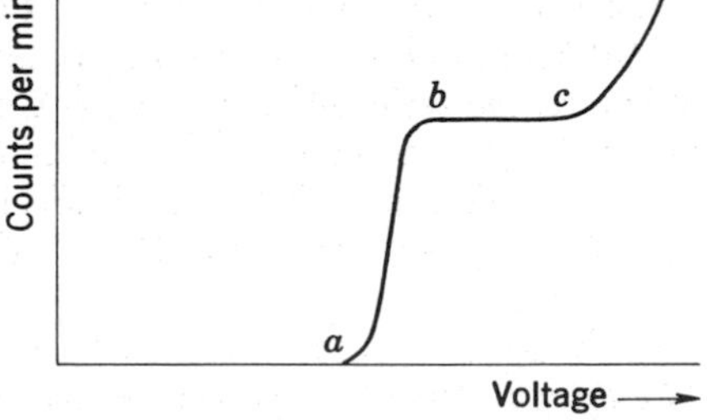

Fig. 3-8 Variation of Geiger-counter sensitivity with voltage; proper operating voltage is on the plateau from *b* to *c*.

voltage drops below the threshold value at which collision ionization is possible, the discharge ceases. It is desirable to arrange the electrical circuit and the characteristics of the counter itself so that a very short time is required for the counter to recover full sensitivity after each pulse, since the maximum number of pulses that can be counted per second is limited by this.

Detailed characteristics and principles of operation can be obtained from manufacturers and from reviews,[1] and need not be repeated here. It is sufficient here to emphasize the following characteristics that are important in diffraction applications: (1) The chief disadvantage of Geiger counters is often the loss in counts that occurs because of the "dead time" during which the counter is insensitive following a pulse. The dead time depends on voltages applied, amplification which follows, gas content, etc., but is in the range 50 to 300 μsec, and losses in counting randomly spaced quanta may exceed 8 percent even with counting rates as low as 500 counts per sec. Although the loss encountered at each *constant* intensity of an x-ray beam can be determined and corrected for, it is impossible to correct for losses during the periods in which the diffracted-beam intensity is continuously changing, as it does in the measurement of integrated intensities. (2) Geiger counters usually have counting efficiencies that vary from point to point over the window of the counter. (3) The efficiency of a counter varies with the wavelength of the x-rays in a manner that depends on the gas with which the counter is filled. A common design of xenon-filled counter has highest efficiencies between about 1.2 and 2.5 A (commonly 40 to 65 percent efficiency in this range), which is the range for the $K\alpha$ radiations from Cu, Ni, Co, Fe, and Cr. On the other hand, it counts 10 percent or less of the quanta of Mo $K\alpha$ and wavelengths shorter than 0.6 A.

The *scintillation counter* has become a favorite in many diffraction installations for detecting x-rays of 0.2- to 2.0-A wavelength. It consists of a fluorescent crystal cemented to the end of a photomultiplier tube. The crystal is usually a plate of NaI that contains around 1 percent Tl in solid solution and is sealed to exclude moisture and light but has a window to admit x-rays without appreciable absorption.

An x-ray quantum absorbed in the crystal ejects photoelectrons which energize a large number of fluorescent centers located at the thallium ions. The fluorescent light enters the photomultiplier tube and ejects electrons which are multiplied in number by 10^5 or more in the tube before passing on to other amplifiers, scalers, and rate meters.

The important characteristics of these counters are the following: (1) The counting efficiency is approximately 100 percent throughout the range of wavelengths used in diffraction. (2) The time for the decay of the fluorescence from a quantum absorption event is only about 10^{-7} sec, so that by

[1] L. F. Curtiss, *Natl. Bur. Std.* (*U. S.*), *Circ.* 490, U.S. Department of Commerce, Washington, D.C., 1950. H. Friedman, *Electronics*, vol. 18, p. 132, 1945. H. Friedman, *Proc. IRE*, *Australia*, vol. 37, p. 791, 1949. P. H. Dowling et al., *Philips Tech. Rev.*, vol. 18, p. 262, 1956/1957. "International Tables for X-ray Crystallography," vol. 3, p. 144, Kynoch, Birmingham, England, 1962.

being connected to electrical circuits with resolving times sufficiently short, it is possible to count at very high rates without losses. Counting x-ray quanta at rates of 10,000 per sec with losses of 1 percent or less is possible in modern equipment. (3) The sensitivity of many counters is uniform over the surface of the counter window. (4) The voltage of the output pulse is proportional to the energy of the incoming quantum (i.e., it varies with the x-ray wavelength). It is therefore possible to feed the pulses into a single-channel pulse-height analyzer and adjust it so as to eliminate most unwanted wavelengths and circuit "noise." The analyzer passes only pulses of a selected range of amplitude and rejects the rest. The "window" of the analyzer can thus be set to pass the diffracted $K\alpha$ radiation and to reject much of the radiation of other wavelengths, including general radiation from the target, fluorescent radiation from the sample, gamma radiation from a radioactive sample, cosmic radiation, etc. This is an advantage not obtained in Geiger counter operation unless the counter is used as a *proportional counter* (see below) rather than as a Geiger counter. (5) The abruptness with which a scintillation counter falls to zero sensitivity as the limits of the "window" are approached (i.e., the energy resolution) is inferior to the proportional counter but nevertheless is adequate for most research work. (6) The electronic noise generated in most photomultiplier tubes restricts their use to x-ray wavelengths shorter than about 2.0 A, but the range of usefulness includes the range appropriate for most diffraction work.[1]

A preamplifier unit is attached to the photomultiplier tube housing, and the output pulses are fed from the preamplifier through a cable to a linear amplifier with adjustable gain. At a given amplifier gain and a constant-intensity monochromatic x-ray beam, the counting rate increases with applied voltage on the photomultiplier between a threshold value (in the vicinity of 700 or 800 volts) and a plateau, of nearly constant rate, which begins 100 volts or so higher, depending on the wavelength. The counter is operated on the plateau, where a slight variation in the applied voltage does not change the counting rate appreciably, but below a voltage where the noise (with x-rays turned off) begins to increase markedly. The window of the pulse-height analyzer is set wide enough to accept a high percentage of the counts of the $K\alpha$ radiation being used (perhaps 90 or 95 percent). The width of the energy distribution of pulses is so great that only a slight reduction in counts due to the $K\beta$ radiation will occur, but the counts from radiation of half the $K\alpha$ wavelength will be eliminated, which is an important feature if a crystal monochromator is used. The window may also serve to eliminate much of the fluorescent radiation from the irradiated sample, unless this is very nearly of the same wavelength as the primary x-ray beam.

The *proportional counter* is also widely used, for in some respects it is superior to the scintillation counter. In construction it resembles a Geiger counter, although it is commonly built to ensure proportionality between the energy of the absorbed quantum and the resulting pulse of electrons and is operated at lower voltages, with smaller avalanches of electrons. The amplification within the counter is limited to about 10^4 rather than the 10^6

[1] For longer-wavelength x-rays, extremely thin counter windows are required, such as mica sheets 1 to 2 μ thick; or the window of a proportional counter is eliminated entirely and a *continuous flow* of gas is used, such as a mixture of 90 percent argon and 10 percent methane flowing at about 6 liters per hour.

to 10^8 that is obtained in a Geiger counter. The gain in the preamplifier and the linear amplifier must therefore be higher than that of a scintillation counter or a Geiger counter.

The important characteristics of the proportional counter are as follows: (1) The pulses are proportional in amplitude to the energy of the absorbed quantum. (2) The dead time is very short, about 0.2 μsec, so that the counter response is proportional to incident intensities up to very high counting rates, which are often limited by the electrical circuits rather than by the counter itself. (3) The requirements regarding constancy of applied voltage are more rigorous than with other counters. (4) With a monochromatic beam of a given wavelength, the distribution in amplitude of the resulting pulses is narrower with proportional counters than with scintillation counters, and "noise" is low. When used with a properly adjusted pulse-height analyzer, the proportional counter therefore permits better separation of a weak diffracted beam from "background" radiation. (5) The sensitivity is similar to that of the Geiger counter, varies similarly with wavelength, and depends similarly on the gas in the counter. The fraction of incoming quanta that is counted is therefore less at all wavelengths than with scintillation counters, but appreciable sensitivity is retained at 10 A and even at 200 A.

Amplitude-distribution curves for the output pulses are helpful in comparing different types of counters. In Fig. 3–9 typical curves are shown for the pulses generated by Cu $K\alpha$ radiation, reduced to equal maximum counting rate. The width W at half height is seen to be narrowest for the proportional counter. A convenient measure of the energy resolution is the ratio W/A where A is the average pulse amplitude (roughly the

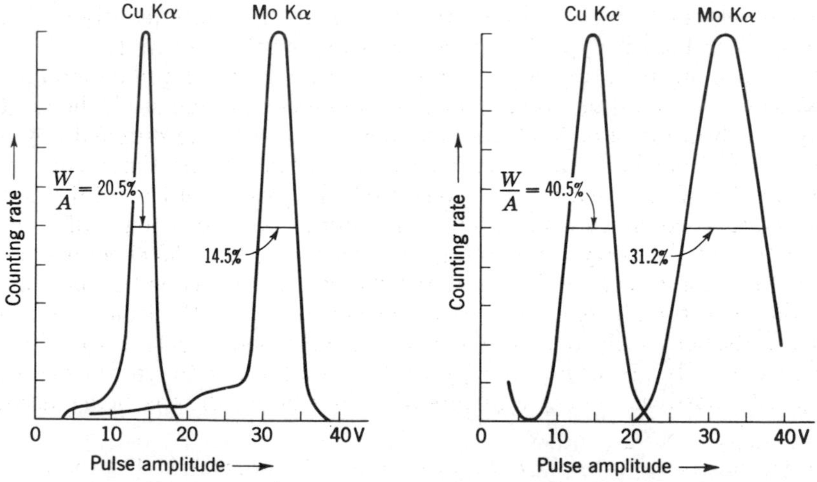

(*a*) Proportional counter (*b*) Scintillation counter

Fig. 3-9 Pulse-amplitude distribution curves of xenon-filled proportional counter and scintillation counter for Cu $K\alpha$ and Mo $K\alpha$ radiation, reduced to equal maximum rates. Background counts at the left of the main peaks are caused by fluorescence in the counter ("escape peaks").

voltage at the center of the peak). The energy resolution with the proportional counter is roughly a factor of 2 better than the scintillation counter because a greater number of events is counted for each x-ray quantum absorbed.[1] The pulse-height analyzer cannot be adjusted to exclude more than about 5 percent of the $K\alpha$ pulses without disadvantage: a counter that excludes more becomes sensitive to small variations in voltage, and further narrowing of the window also reduces the intensity of the diffracted beam at a faster rate than it reduces the background.

The small peak shown at the left of the main peak in Fig. 3–9*a* is an "escape peak" caused by fluorescent radiation escaping from atoms ionized by the incoming quantum and in turn acting as an incoming quantum of different wavelength (lower energy).[2] The xenon escape peak of Fig. 3–9*a* is small, but some peaks of this kind are large—for example, the krypton escape peak in a krypton-filled proportional counter used with Mo $K\alpha$ radiation. This peak is at 12.6 volts, and since the Mo $K\alpha$ is at 17.4 volts, it is possible to cut out the counts of the escape peak with a pulse-height analyzer.

Pulses from a counter are fed through pulse-equalizing circuits and amplifier circuits, and into a *rate meter* and/or a scaler. A rate meter consists essentially of a condenser shunted by a resistor and a voltmeter. The equalized charges of the pulses charge up the condenser until the rate of charging is balanced by the current draining from the condenser through the resistor. When this balance is attained, the condenser voltage is proportional to the number of pulses received per second and can be recorded on a strip-chart recorder. Values of the resistance in the circuit can be easily changed so as to provide a "time constant" of the circuit suitable for averaging out most of the statistical fluctuations in the count rate without introducing inconvenient sluggishness—it is often necessary to measure continuously changing x-ray intensities. The counts may be registered directly by using a *scaler*. The total count in a predetermined interval is displayed on the front of the counter and in some installations is printed on tape or otherwise recorded. If a recording rate meter is used simultaneously with a scaler it will display any spurious bursts that would falsify a count.

Losses in counting x-rays depend not only on the counter and its circuits, but also on the excitation voltage for the x-ray emission line used, the peak voltage on the x-ray tube, and the wave form of the voltage supplied to the tube. For example, in one instance[3] with a xenon-filled proportional counter measuring Cu $K\alpha$ intensities, the losses reached 1 percent of the total count at an average count rate of about 2500 counts per sec when half-wave rectification of the x-ray tube voltage was used, at about 5000 counts per sec with full-wave rectification, and at 8300 counts per sec with constant-potential operation. The differences are accounted for by the time in which the $K\alpha$ radiation is emitted weakly or not at all in each cycle of the alternating current. In precision-intensity work, it is advisable to test the degree of linearity between counts and intensity for the conditions being used.

[1] P. H. Dowling et al., *Philips Tech. Rev.*, vol. 18, p. 262, 1956/1957. C. F. Hendee and S. Fine, *Phys. Rev.*, vol. 95, p. 281, 1954. "International Tables for X-ray Crystallography," vol. 3, p. 144, Kynoch, Birmingham, England. H. Neff, "Grundlagen und Anwendung der Röntgen-Feinstruktur-Analyse," p. 196, Oldenbourg, Munich, Germany, 1962.

[2] Dowling et al., loc. cit.

[3] S. Bernstein and M. Canon, *Rev. Sci. Instr.*, vol. 33, p. 112, 1962.

A series of absorbing foils of equal thickness can be used for this purpose. First the intensity through a pack of these is measured, and then the intensities are obtained when successive foils are removed. With a truly monochromatic beam the same fractional loss in intensity will occur in each foil; therefore, a plot of the observed counting rate (on a logarithmic scale) will be linear with the number of foils removed (plotted on a linear scale) up to the point where losses become appreciable.[1]

Overall constancy; monitors With many modern commercial designs of stabilized x-ray generators and detector circuits, the constancy of the counting rate can be expected to lie within the range ±1 to 3 percent over long periods of time. For short periods, perhaps ±0.2 to 0.6 percent can be achieved. However, to benefit from such constancy requires counting long enough to reduce statistical errors to within these limits. It is usually advisable, in striving for accuracies of this order, to "monitor" the x-ray intensity by standardizing the diffracted beam count against the count taken simultaneously on a stationary monitor counter that is mounted in such a way as to register the intensity of the beam incident on the specimen.

[1] An alternate procedure is given by R. D. Burbank, *Rev. Sci. Instr.*, vol. 32, p. 368, 1961.

4

Diffraction of x-rays by crystals

This chapter presents a very brief survey of the fundamental principles and the more important methods by which x-rays are employed to investigate the inner structure of crystals. Subsequent chapters then take up the individual methods and instruments in more detail and show how they are applied to particular problems. A beam of x-rays is diffracted from a crystal when certain geometrical conditions are fulfilled. To make these clear they are stated first in a very simple way; then after the reciprocal lattice is introduced, they are treated again on the basis of this important concept. It is emphasized that the size and shape of the unit cell of a crystal determine the *positions* of the diffracted beams, whereas the distribution of atoms within the unit cell determines the *intensities* of the diffracted beams, and whether or not there will be missing reflections. The fundamentals of the relationship between atom distributions in the unit cell and these intensities are presented briefly in this chapter and in more detailed form in later chapters.

Scattering of x-rays by atoms When a beam of x-rays passes over an atom, the electric field of the beam acts upon each electron of the atom, accelerating each with a vibratory motion. Any electric charge undergoing an oscillation of this sort becomes the source of a new set of electromagnetic waves, just as the alternating electric current in the antenna of a radio transmitter sends out electromagnetic waves of radio frequency. The waves radiating out from the vibrating electrons have the same frequency as the incident beam that is responsible for the vibration, and they are coherent, unlike the incoherent radiation that results from the Compton scattering process.

In effect, each electron subtracts a small amount of energy from the impinging beam and broadcasts it in all directions, scatters it. The various scattered waves from the individual electrons of an atom combine and may be treated as a single set of radiating waves which, for most purposes, can be considered as originating from a point. But since the different electrons in an atom do not scatter exactly in phase with each other except in the direction of the incoming beam, the scattering power of an atom is a function of the scattering angle.

The superposition of waves scattered by individual atoms results in diffraction. The waves that radiate from the atoms of a crystal combine in an additive way in certain directions from the crystal but annul one another in other directions, the intensity in any direction depending on whether or not the crests of the waves from each of the atoms superpose, i.e., whether or not the individual scattered waves are in phase. The relation of the regularly repeating pattern of atoms in a crystal to the directions of the reinforced diffracted beams is given in the sections immediately following.

Bragg's law Consider a set of parallel planes of atoms in a crystal, two of which are represented by the lines AA and BB in Fig. 4–1, and suppose that a beam of monochromatic x-rays is directed at the planes in the direction LM, which makes an angle θ with the planes. The line LL_1 is drawn to represent one of the crests in the approaching waves and is perpendicular to the direction of propagation of the waves. As this crest reaches each of the atoms in the crystal, it generates a scattered wave crest. We shall see that the various scattered waves reinforce in the direction MN, their crests coinciding along the line N_1N_2. There will be a certain number of complete wavelengths in the path LMN along which a ray proceeds that is scattered by an atom at M. If the ray scattered by an atom at M_1 travels the same distance (i.e., if the distance $L_1M_1N_1$ is equal to LMN), then the scattered rays from the two atoms will be in phase and will reinforce each other. It will also be true that any other atom lying anywhere in the plane AA will also reinforce the beam in this direction. This reinforcement will take place when the incident and the scattered rays make equal angles with the atomic plane. It is then possible to regard the plane of atoms as a mirror that is *reflecting* a portion of the x-rays at an angle of reflection equal to the angle of incidence. Let us now consider the condition for reinforcement of the waves from successive planes in the crystal that lie parallel to AA. The requirement to be met is that the difference in the length of the path for rays reflected from successive planes be equal to an integral number of wavelengths. In Fig. 4–1 this corresponds to the condition that the distance PM_2Q is one wavelength or a multiple of it, since PM is drawn perpendicular to LM and MQ is drawn perpendicular to MN, making the paths of the rays from L and L_1 the same except for the distance

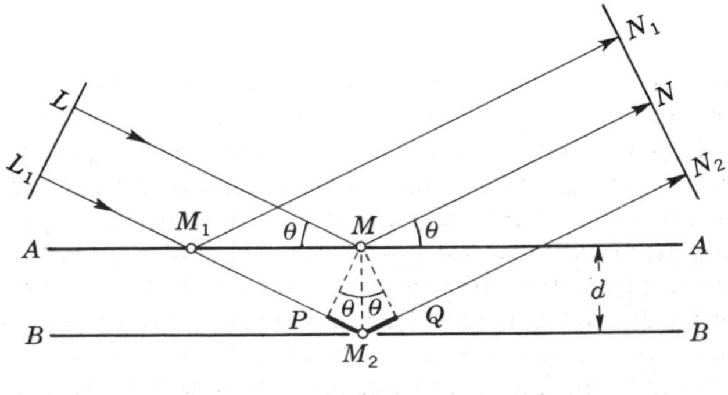

Fig. 4–1 Illustrating Bragg's law of reflection from planes of atoms in a crystal.

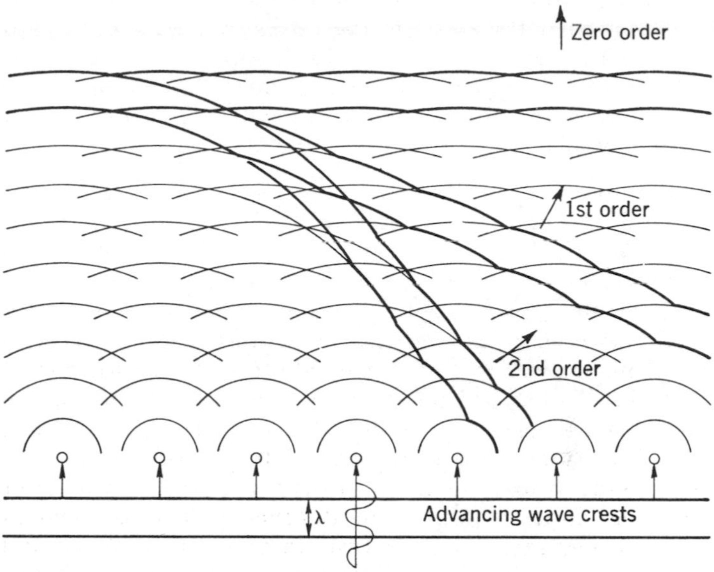

Fig. 4-2 Reinforcement of scattered waves producing diffracted beams in the different orders.

PM_2Q. It will be seen from Fig. 4–1 that

$$PM_2 = M_2Q = d \sin \theta$$

Thus, the condition for reinforcement of all the reflected rays is

$$n\lambda = 2d \sin \theta \tag{4–1}$$

where $n = 0, 1, 2, 3$, etc.; λ is the wavelength; and d is the spacing of the planes. This is Bragg's law. The integer n, which gives the number of wavelengths in the difference in path for waves from successive planes, is the *order of reflection.*

The Laue equations Diffraction from a crystal, as indicated above, is analogous to reflection from a series of semitransparent mirrors, but it is also to be understood as diffraction from a three-dimensional grating, analogous to the diffraction of light from a one-dimensional optical grating.

If an x-ray beam is directed at a row of equally spaced atoms, as represented in Fig. 4–2, each atom will be a source of scattered waves spreading spherically, which reinforce in certain directions to produce the zero-, first-, second-, and higher-order diffracted beams. Successive waves are indicated on the drawing by concentric arcs, which are linked together to show how the various orders are built up. The condition for reinforcement can be derived from Fig. 4–3, which shows the path difference for rays scattered by two adjacent atoms in the row. If the incident beam makes an angle

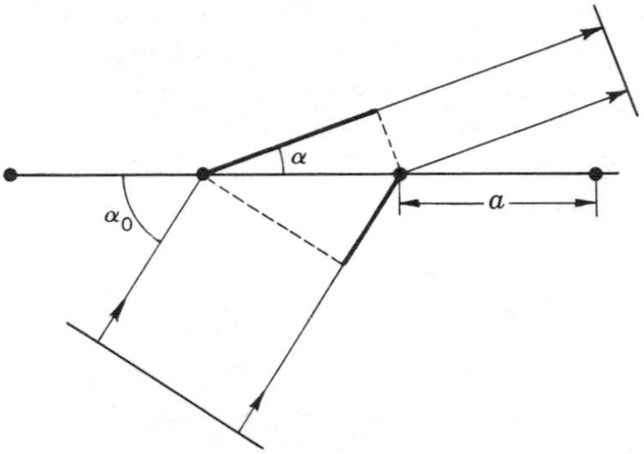

Fig. 4-3 Conditions for reinforcement leading to one of the Laue equations.

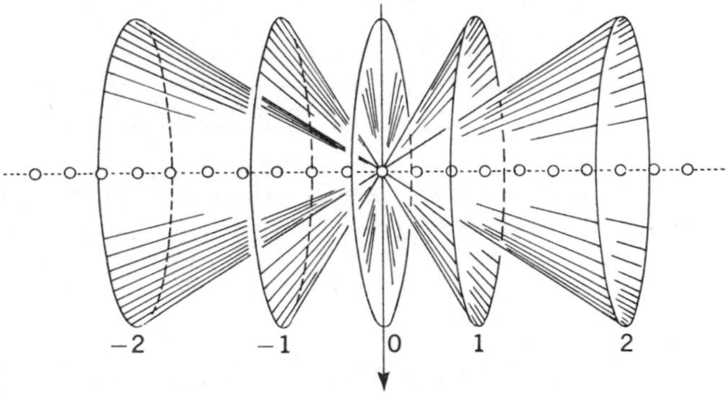

Fig. 4-4 Cones of diffracted beams around a row of atoms, with direction of primary beam indicated by the arrow. Orders of diffraction indicated by numbers.

α_0 with the row, and the diffracted beam leaves at the angle α, then the path difference is $a(\cos\alpha - \cos\alpha_0)$. This path difference must be an integral number of wavelengths if the scattered waves are to be in phase, and the following relation must therefore hold:

$$a(\cos\alpha - \cos\alpha_0) = h\lambda$$

where h is an integer and λ is the wavelength. This equation will be satisfied by all the generators of a cone that is concentric with the line of atoms and that has the semiapex angle α. Thus, for any given angle of incidence there will be a series of concentric cones surrounding the row of atoms, each cone being made up of one order of diffracted rays. A set of cones of this type is indicated in Fig. 4–4.

If there is a two-dimensional network of atoms with spacings a in one direction and b in another, there will be two simultaneous equations to be fulfilled for intense diffracted beams:

$$a(\cos\alpha - \cos\alpha_0) = h\lambda$$
$$b(\cos\beta - \cos\beta_0) = k\lambda$$

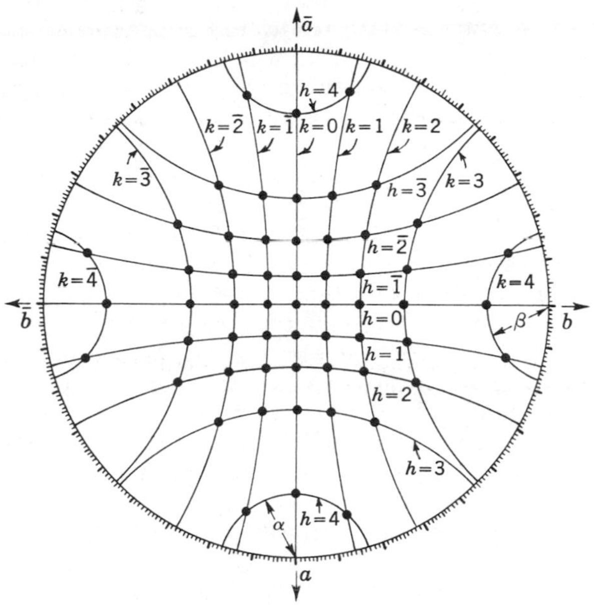

Fig. 4-5 Stereographic projection of diffracted beams from a two-dimensional square network. Cones are concentric with *a* and *b* axes; intersections are strong diffracted beams.

In these relations, α_0 and α are the angles the incident and diffracted beams make with the a rows, while β_0 and β are the corresponding angles for the b rows; h is the integer giving the order of reflection with respect to the a rows; and k is the order for the b rows. These equations correspond to two sets of cones, a set around the a axis and another concentric with the b axis. The most intense diffracted beams will travel out along the intersections of these two sets. The geometry of these beams can be illustrated neatly on a stereographic projection. The projection of Fig. 4–5 illustrates diffraction from a square network of atoms in the plane of the projection, with a and b axes as indicated; the incident beam is coming toward the observer and striking the atom plane perpendicularly. The diffracted rays coming out of the atom plane toward the observer are indicated by dots at the intersections of the cones. Projections of this type aid in visualizing what changes would result from various alterations in the quantities of the two equations; for instance, increasing the wavelength spreads out the pattern of cones and intersections and reduces the number of diffracted beams.

A crystal is a three-dimensional network of atoms, and there are, accordingly, three conditions to be met simultaneously for diffraction:

$$
\begin{aligned}
a(\cos\alpha - \cos\alpha_0) &= h\lambda \\
b(\cos\beta - \cos\beta_0) &= k\lambda \qquad (4\text{–}2) \\
c(\cos\gamma - \cos\gamma_0) &= l\lambda
\end{aligned}
$$

These are the Laue equations; the first two have the same significance as before, and the third relates to the periodicity in the third dimension, the c axis of the crystal, with which the incident beam makes the angle γ_0 and the diffracted beam the angle γ. The third equation corresponds to a set of cones concentric with the c axis. The integer l is the order of diffraction with respect to the third axis, and hkl can be called the indices of the diffracted beam. These indices are not enclosed in parentheses ( ) as Miller indices are, for these may differ from the Miller indices of the reflecting plane by a common factor (to be discussed later). The sets of cones around a, b, and c all have a common generator—a common line of intersection—only if there are special values for the variables. The requirement that all three equations be satisfied simultaneously acts as a severe limitation to the number of diffracted beams, for a strong beam is only obtained when the equations are simultaneously satisfied.

Within a few months after Laue made his discovery in 1912, W. L. Bragg showed that fulfilling the Laue equations is equivalent to reflecting from a lattice plane. To see this equivalence, consider the lattice in Fig. 4–6 to be diffracting x-rays and producing a beam with indices hkl. This means that a strong reinforcement of diffracted rays is occurring, and from the principles explained above it will be seen that in producing the hkl beam the hth cone of the nest of cones around OA, the kth cone of the nest around OB, and the lth cone around OC have a common line of intersection. Therefore, the ray scattered by point A is h wavelengths ahead of the ray from the point at the origin, i.e., the phase difference between the diffracted ray from point A and from point O is $2\pi h$. Consider now point A' at a distance from the origin at O given by a/h, a point with coordinates $1/h$, 0, 0. This point contributes a ray that is one wavelength ahead of the ray from O, i.e., its phase is 2π. Analogously, the phase of the ray from B is $2\pi k$ and from B' is 2π, and the phase of the ray from C is $2\pi l$ and from C' is 2π. Thus, points A', B', and C' each contribute rays that are in phase. It is clear from the figure that these points lie on the crystal plane that has Miller indices (hkl), and in fact every point on this plane also scatters in phase with the three just mentioned. This is the condition for reflection from a mirror, and the x-rays thus act as if they are "reflecting" from the (hkl) plane.

It is a great convenience to distinguish between *indices of a reflected*

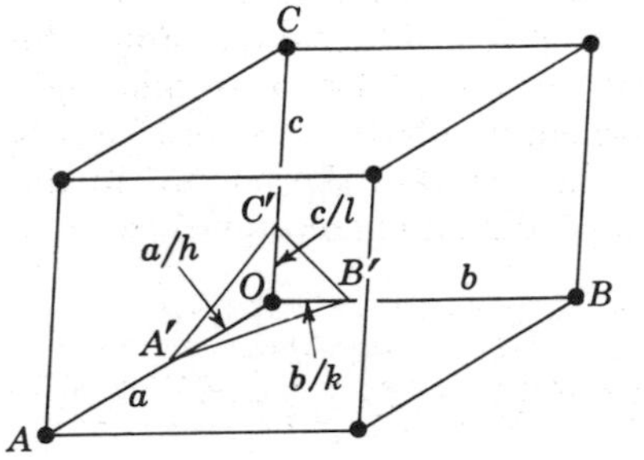

Fig. 4-6 A reflecting plane (*hkl*) in a unit cell.

beam and *indices of a reflecting plane* in a crystal. It will be remembered that indices of crystal planes never have a common factor. We are therefore free to use indices with a common factor to express the order of reflection, n. The first-order reflection is given the same indices as the plane. For example, the first-order reflection from (110) is written 110. The second order is written with the indices of the reflecting plane multiplied by 2, which gives 220. The third order is 330, etc. To avoid confusion, *the indices of the reflections are written without parentheses.*

This scheme of notation for reflections is more than a mere convention, for it simplifies the interpretation of diffraction patterns. Its usefulness depends on the fact that nth-order reflection from a plane (hkl) is equivalent to first-order reflection from an imaginary set of planes spaced $1/n$th the spacing of the (hkl) planes. If the intercepts of the (hkl) plane nearest the origin are $1/h$, $1/k$, and $1/l$, then the intercepts of the first plane of the imaginary set will be $1/nh$, $1/nk$, and $1/nl$. All higher-order reflections can similarly be regarded as first-order reflections from such imaginary sets of planes, and if the spacing of these planes is $d' = d/n$, where d is the true spacing of the lattice planes, then the Bragg equation can be simplified to

$$\lambda = 2d' \sin \theta \tag{4-3}$$

The reflecting angles are computed by the equations in the following section, in which this convention is used.

Interplanar spacings From the foregoing it is apparent that the directions of the reflected beams are governed entirely by the *geometry of the lattice* (by the orientation and spacing of the planes of atoms, from the point of view of Bragg's law, and by the periodicity of the lattice in three dimensions, from the point of view of the Laue equations). In other words, the size and shape of the unit cell determine where the diffracted beams will go. On the other hand, the *distribution of atoms within the unit cell* governs the intensities of the diffracted beams, as will be discussed later.

When the unit cell is known, it is easy to calculate the angles θ for all the diffracted beams that could occur with a given wavelength of radiation, for it is merely necessary to compute all the interplanar spacings d in the crystal and insert these in the Bragg law. Referring to Fig. 4–6, the interplanar spacing for the planes of index (hkl) is the same as the distance from the origin, which is on one plane, to the nearest plane of this set. As indicated in the figure, the plane (hkl) intersects the a axis at a/h from the origin. The distance d of this plane from the origin is measured along the normal to the plane and amounts to $(a/h) \cos \delta_1$, where δ_1 is the angle between the a axis and the (hkl) plane normal. Similarly $d = (b/k) \cos \delta_2$ and $d = (c/l) \cos \delta_3$, where δ_2 and δ_3 are the angles of the normal to the b and c axes, respectively. By squaring and adding these equations, one obtains

$$\cos^2 \delta_1 + \cos^2 \delta_2 + \cos^2 \delta_3 = d^2 \left(\frac{h^2}{a^2} + \frac{k^2}{b^2} + \frac{l^2}{c^2}\right)$$

and for orthogonal crystals

$$\frac{1}{d^2} = \frac{h^2}{a^2} + \frac{k^2}{b^2} + \frac{l^2}{c^2}$$

since $\cos^2 \delta_1 + \cos^2 \delta_2 + \cos^2 \delta_3 = 1$ for any orthogonal axis system. We use the convention mentioned earlier: hkl are indices that can have a common factor n which is the order of reflection. Substituting in the Bragg law then gives for orthogonal crystals

$$\sin^2 \theta = \frac{\lambda^2}{4}\left(\frac{h^2}{a^2} + \frac{k^2}{b^2} + \frac{l^2}{c^2}\right) \tag{4-4}$$

This formula applies to orthorhombic crystals and to hexagonal crystals referred to orthohexagonal axes. It also applies to tetragonal crystals if a is set equal to b. Still further simplification results for cubic crystals, since for these $a = b = c$ and

$$\sin^2 \theta = \frac{\lambda^2}{4a^2}(h^2 + k^2 + l^2) \tag{4-5}$$

The corresponding formula for crystals with hexagonal axes is

$$\sin^2 \theta = \frac{\lambda^2}{4}\left[\frac{4}{3}\frac{(h^2 + k^2 + hk)}{a^2} + \frac{l^2}{c^2}\right] \tag{4-6}$$

All these formulas are special cases of a more general formula which applies to lattices with any interaxial angles and axial lengths, which is presented later in connection with the reciprocal lattice concept. These equations are of great value in assigning indices to observed reflections, but because of overlapping reflections and inaccuracies in measuring the θ values, it is not always possible to determine a unit cell from these relations alone.

Reinforcement of scattered waves The intensities of the diffracted beams are governed by the distribution of scattering matter within the unit cell; for x-ray diffraction the scatterers are the individual electrons, and the intensities are governed by the combination of the waves scattered by all the electrons in the cell. (The nuclei of the atoms do not contribute to the scattering of x-rays although they do scatter neutrons.) The scattering power of an individual electron is the natural unit on which to base all quantitative treatments of intensities. It is sufficiently accurate for most crystallographic work to regard the atoms as spherical clusters of electrons with a definite scattering power that is a function of the diffraction angle, and to regard the scattering power of a unit cell as the combination of the waves from the individual *atoms* of the cell rather than as a combination of waves from individual electrons. This point of view is used in the present chapter; the methods of determining the actual *electron* distribution are discussed in Chap. 8.

The efficiency of an atom in scattering x-rays is expressed by the *atomic scattering factor f*, sometimes called the *atom form factor*. The atomic scattering factor is simply the ratio of the amplitude of the wave scattered

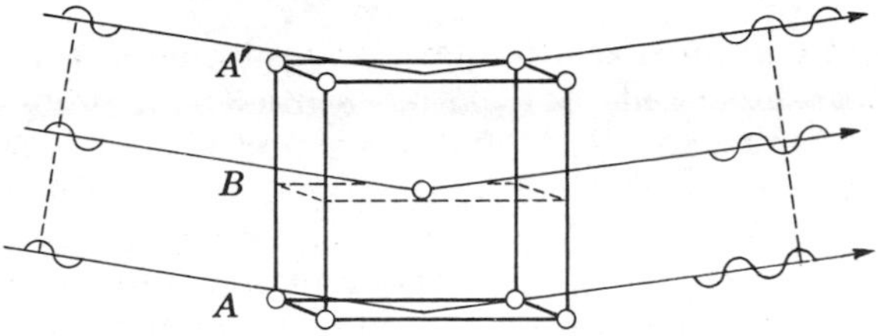

Fig. 4-7 Interference in 100 reflection from a b.c.c. lattice.

by an atom to that scattered by an electron under the same conditions. Since the *intensity* of a wave is the square of its amplitude, f^2 gives the ratio of the intensity scattered by an atom to the corresponding intensity from an electron. The atomic scattering factor varies with angle—an atom scatters less efficiently at large angles from the incident beam than at small. This is because the individual scattered waves from the various electrons in the atom are nearly in phase and reinforce each other for directions near the incident beam, while they are out of phase and reinforce each other less effectively at the larger angles. The curve of f vs. θ—or f vs. $(\sin\theta)/\lambda$—falls from a value equal to the atomic number, at $\sin\theta = 0$, in a manner determined by the radial distribution of the electrons in the atom, and can be computed with reasonable accuracy with modern theoretical methods. Tables of f values have been compiled for all the atoms of the periodic table.[1]

Before introducing the quantitative expressions for diffracted intensities, let us consider a simple case illustrated by the atomic arrangement in Fig. 4-7. If a beam reflects in the first order from the planes of atoms marked A and A', the wave from A' will be exactly one wavelength ahead of the wave from A and will completely reinforce it. On the other hand, the scattering from the plane marked B which is midway between A and A' will be exactly out of phase with those from A and from A'. Thus, the reflections from alternate planes annul each other, B canceling A, etc., throughout the crystal—provided that the scattering power of plane B is equal to that of A. This cancellation will not occur for the second-order reflection, for then the scattering from plane B will be in phase with that from A and A'. The same is true for all even-order reflections. Thus, for example, in a body-centered lattice—regardless of whether the lattice is cubic, tetragonal, or orthorhombic—the 001 reflection will be extinguished, 002 will occur, 003 will be extinguished, etc.; in general the condition for the occurrence of $00l$ will be that l must be an even integer. Note also that if the scattering power of layer B is not equal to that of A (because of a different number or kind of atoms on B than on A), there will not be complete annihilation in the first order; the same will be true if layer B is not midway between A and A'.

[1] A very complete collection of these for both atoms and ions will be found in "International Tables for X-ray Crystallography," vol. 3, Kynoch, Birmingham, England, 1962.

The structure factor The problem of computing the intensity of a particular reflection *hkl* is essentially the problem of adding sine waves of different amplitude and phase and identical wavelength. Each atom contributes a scattered wave having an amplitude proportional to the value of f for that atom at the corresponding angle θ_{hkl} and having a phase that is determined by the position of the atom in the unit cell. The phase of a scattered wave from the given atom is related to the coordinates of the atom with respect to an origin of coordinates chosen arbitrarily—perhaps at the corner of the unit cell, or more conveniently at a center of symmetry or on some other symmetry element. The summation of the waves scattered by each of the atoms in the unit cell gives the amplitude of the resulting wave, which is called the *structure factor* or *structure amplitude* **F**. The intensity of the observed reflection is then proportional to the square of the absolute value of this resultant amplitude $|\mathbf{F}|^2$, modified by certain other factors, such as absorption in the crystal, that are discussed later.

The relation between the coordinates of an atom and the phase of a wave scattered by the atom is simple. It is obvious (see text accompanying Fig. 4–6) that if the atom is at the coordinate position 100 its scattering will be one wavelength out of phase with a wave scattered at the origin when the scattered beam is the first-order reflection with respect to the periodicity along the [100] direction; i.e. the phase from the given atom will be 2π. Also, the phase of the wave from the atom at 100 will be $2\pi h$ for the h-order diffracted beam. And if the atom is at $u00$, where u is any fraction of the unit cell edge along the [100] direction, the phase of the wave scattered in the h order by this atom will be $2\pi hu$. It can be easily shown that any atom with the general coordinates uvw will contribute to the diffracted beam *hkl* a wave having a phase ϕ given by the relation

$$\phi = 2\pi(hu + kv + lw)$$

This relation, which holds for any crystal in any crystal system, is most easily derived by using the reciprocal lattice, a concept introduced in the following section.

The structure factor **F** is the sum of waves of amplitude $f_1, f_2, f_3, \cdots$ and phase $\phi_1, \phi_2, \phi_3, \cdots$ from atoms 1, 2, 3, $\cdots$ in the unit cell. If the amplitude of each wave is represented by the length of a vector, and the phase by

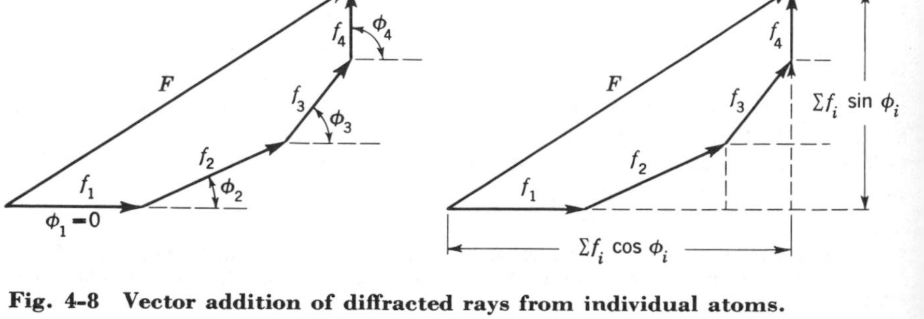

Fig. 4-8 Vector addition of diffracted rays from individual atoms.

the direction of the vector, then the resultant diffracted beam is represented by the vector sum of the waves (Fig. 4–8). As indicated in this figure, the vector for the wave from the jth atom may be resolved into a horizontal component of length $f_j \cos \phi_j$ and a vertical component $f_j \sin \phi_j$; these components, when added, give two sides of a right triangle whose hypotenuse is **F**. Consequently, we may write

$$|\mathbf{F}|^2 = \Big[\sum_j f_j \cos 2\pi(hu_j + kv_j + lw_j)\Big]^2 + \Big[\sum_j f_j \sin 2\pi(hu_j + kv_j + lw_j)\Big]^2 \quad (4\text{–}7)$$

where the summations are taken over all the atoms in the unit cell. This formula applies to all crystals, provided one assumes that the phase shift between incident and scattered waves for any atom in the crystal is the same as that for every other atom; this assumption is good to a very close approximation under the usual conditions, but is not necessarily true if the incoming radiation has a wavelength very near that of an absorption edge of one of the varieties of atoms in a crystal. Simplification of the formula is possible when a crystal has a center of symmetry, for if this center is chosen as the origin of coordinates, then for every atom at xyz there is an equivalent atom at $\bar{x}\bar{y}\bar{z}$ and the sine terms cancel out, reducing the formula to

$$|\mathbf{F}|^2 = \Big[\sum_j f_j \cos 2\pi(hu_j + kv_j + lw_j)\Big]^2 \quad (4\text{–}8)$$

In general, simplifications are always possible if atoms in the unit cell are divided into sets that are related to each other by the symmetry of the space group, for then a summation equation can be written which takes account of the waves from all atoms of the set when the coordinates of any one of the set are inserted.

Using complex numbers (involving the specification of a vector by a quantity such as $a + ib$, where $i = \sqrt{-1}$), the vector addition leading to Eq. (4–7) can be stated simply as

$$\mathbf{F} = \sum f_j \exp\,[2\pi i(hu_j + kv_j + lw_j)]$$

which is a complex number representing the resultant vector in amplitude and phase. The value of $|\mathbf{F}|^2$ is then obtained by multiplying this quantity by its complex conjugate, which is obtained by replacing i by $-i$.[1]

The additional factors that govern the intensities of the diffracted beams are presented in Chap. 8, and the methods of measuring the intensities are discussed in the chapters covering the individual diffraction methods. The

[1] It is helpful in computing intensities with complex numbers to remember the following relations: $e^{n\pi i} = +1$ if n is an even integer and $e^{n\pi i} = -1$ if n is an odd integer; $e^{n\pi i} = e^{-n\pi i}$ where n is any integer; also that $e^{ix} = \cos x + i \sin x$; $e^{ix} + e^{-ix} = 2 \cos x$; $e^{ix}e^{-ix} = 1$.

interpretation of the intensities in terms of atomic positions is discussed in the chapter on crystal-structure determination (Chap. 8).

We can illustrate the use of the structure-factor equation by applying it to a body-centered lattice having identical atoms at the coordinates 000 and $\frac{1}{2}\frac{1}{2}\frac{1}{2}$ in the unit cell, as in Fig. 4–7. Without simplifying, the formula for this case becomes

$$F^2 = f^2\left[\cos 2\pi \cdot 0 + \cos 2\pi\left(\frac{h}{2} + \frac{k}{2} + \frac{l}{2}\right)\right]^2 + f^2\left[\sin 2\pi \cdot 0 + \sin 2\pi\left(\frac{h}{2} + \frac{k}{2} + \frac{l}{2}\right)\right]^2$$

$$= f^2[1 + \cos \pi(h + k + l)]^2 + f^2 \sin^2 \pi(h + k + l)$$

and it will be seen that the intensity will be zero for every reflection for which $(h + k + l)$ is an odd number. This is true regardless of the crystal system to which the body-centered crystal belongs. Thus, if indices have been assigned to a list of reflections from a crystal and it is noticed that no reflections for which $h + k + l = 0$ occur in the list, the implication is that the crystal lattice is body-centered. Other extinction rules apply to lattices that are centered on one or more faces, and to crystals in which certain symmetry elements cause certain sets of reflections to be absent.

If we apply the formula to a crystal of cesium chloride, in which there is a cesium ion at 000 and a chlorine ion at $\frac{1}{2}\frac{1}{2}\frac{1}{2}$, quite a different result is obtained, for although the structure somewhat resembles a b.c.c. structure, the f values of the two ions are not the same and the lattice is not body-centered. The formula for the structure factor is

$$F^2 = [f_{Cs} + f_{Cl} \cos \pi(h + k + l)]^2 + [f_{Cl} \sin \pi(h + k + l)]^2$$

which reduces to $(f_{Cs} + f_{Cl})^2$ when $(h + k + l)$ is even and $(f_{Cs} - f_{Cl})^2$ when $(h + k + l)$ is odd. Since $f_{Cs} \neq f_{Cl}$, the latter reflections are not of zero intensity as they are for a body-centered lattice.

THE RECIPROCAL LATTICE

Vector algebra is almost universally used in discussions of space lattices and diffraction, and the concept of the reciprocal lattice with its vector notation is indispensable not only for crystallography but for nearly all work in the field of the solid state.

Consider a space lattice (or a crystal) that has fundamental translations specified by vectors **a**, **b**, and **c**. These vectors will be the edges of a unit cell, and will have lengths a, b, and c, respectively. Since the unit cells stack tightly together in identical orientation throughout the space lattice, the vector to any unit cell corner can be written

$$\mathbf{r} = u\mathbf{a} + v\mathbf{b} + w\mathbf{c} \tag{4–9}$$

where u, v, w, are integers and are the coordinates of the point at the end of the vector **r**.

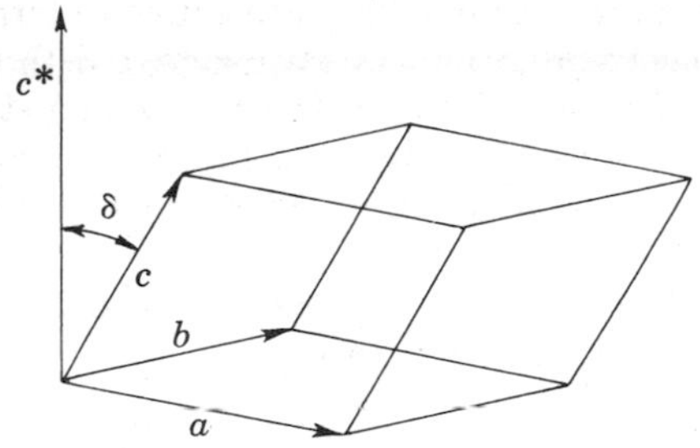

Fig. 4-9 Axis c* of the reciprocal lattice is normal to a and b of the crystal.

We now define a reciprocal lattice that is related to this space lattice in a particular way. Let the reciprocal lattice be one with cell edges $\mathbf{a}^*$, $\mathbf{b}^*$, and $\mathbf{c}^*$ so that a vector to any cell corner in this lattice is

$$\mathbf{r}^* = h\mathbf{a}^* + k\mathbf{b}^* + l\mathbf{c}^* \tag{4–10}$$

where h, k, and l are integers. Erect the reciprocal axis $\mathbf{a}^*$ normal to $\mathbf{b}$ and $\mathbf{c}$ of the space lattice, $\mathbf{b}^*$ normal to $\mathbf{a}$ and $\mathbf{c}$, and $\mathbf{c}^*$ normal to $\mathbf{a}$ and $\mathbf{b}$. Since the scalar product ("dot product") of two vectors is the product of their magnitudes times the cosine of the angle between them, it follows that

$$\mathbf{a}^*\cdot\mathbf{b} = \mathbf{a}^*\cdot\mathbf{c} = \mathbf{b}^*\cdot\mathbf{a} = \mathbf{b}^*\cdot\mathbf{c} = \mathbf{c}^*\cdot\mathbf{a} = \mathbf{c}^*\cdot\mathbf{b} = 0 \tag{4–11}$$

Finally, choose the length of $\mathbf{c}^*$ such that $\mathbf{c}^*\cdot\mathbf{c} = 1$, which means that $c^*c \cos \delta = 1$, the angle δ being indicated in Fig. 4–9. From this figure it will be seen that $c \cos \delta$ is the spacing of the (001) planes of the space lattice, so that the length of $\mathbf{c}^*$ is the reciprocal of this spacing. Let the other reciprocal lattice axes be similarly defined:

$$\mathbf{a}^*\cdot\mathbf{a} = \mathbf{b}^*\cdot\mathbf{b} = \mathbf{c}^*\cdot\mathbf{c} = 1 \tag{4–12}$$

(In occasional applications a constant other than unity has been used on the right-hand side of this equation, but here we use the equation as stated.)

From these equations it is clear that if the crystal axes are at right angles to each other the reciprocal lattice axes will also be at right angles and will be parallel to the crystal axes; in this special case the axes of the reciprocal lattice will have lengths that are the reciprocals of the lengths of the crystal axes, but this will not be true in general for all crystal systems. Equations (4–9) to (4–12) and those derived below apply to *all* crystal systems.

The reciprocal axes will now be related to the volume of the unit cell of the crystal. The vector product $\mathbf{a} \times \mathbf{b}$ is a vector, normal to $\mathbf{a}$ and $\mathbf{b}$, of magnitude $ab \sin \gamma$, where γ is the angle between $\mathbf{a}$ and $\mathbf{b}$. In Fig. 4–9 the area of the base of the unit cell is $ab \sin \gamma$ and is therefore equal to the magnitude of $\mathbf{a} \times \mathbf{b}$; the altitude of the cell is $c \cos \delta$, and the volume V may therefore be written as

$$V = (\mathbf{a} \times \mathbf{b})\cdot\mathbf{c} = (\mathbf{b} \times \mathbf{c})\cdot\mathbf{a} = (\mathbf{c} \times \mathbf{a})\cdot\mathbf{b} \tag{4–13}$$

The vectors $\mathbf{a}^*$, $\mathbf{b}^*$, $\mathbf{c}^*$ are in the directions of $\mathbf{b} \times \mathbf{c}$, $\mathbf{c} \times \mathbf{a}$, $\mathbf{a} \times \mathbf{b}$, respectively, and to be consistent with Eqs. (4–12) they must be of lengths such that

$$\mathbf{a}^* = \frac{(\mathbf{b} \times \mathbf{c})}{V} \qquad \mathbf{b}^* = \frac{(\mathbf{a} \times \mathbf{c})}{V} \qquad \mathbf{c}^* = \frac{(\mathbf{a} \times \mathbf{b})}{V} \qquad (4\text{–}14)$$

as can be shown by taking suitable scalar products of both sides of these equations; for example,

$$\mathbf{a}^* \cdot \mathbf{a} = \mathbf{a} \cdot (\mathbf{b} \times \mathbf{c})/V = 1$$

Interplanar spacings The equations giving the interplanar spacings for orthogonal crystals were presented earlier in this chapter, but we now give the general formula for these in terms of the reciprocal lattice. In any crystal system, the spacing d_{hkl} of plane (hkl) is the reciprocal of the length of the reciprocal lattice vector $\mathbf{r}^*_{hkl}$. To prove this, let the origin of crystal space be on one of the (hkl) lattice planes and consider the distance from the origin to the next plane of the (hkl) set. Since $\mathbf{r}^*_{hkl}$ is erected normal to these planes, the interplanar distance will be equal to the distance along the direction of $\mathbf{r}^*_{hkl}$. Now the interplanar distance, when measured in the direction $\mathbf{a}$, is $|\mathbf{a}|/h$. The planes we are considering intersect the $\mathbf{a}$ axis at 000 and $\mathbf{a}/h$, respectively, and their spacing is obtained by taking the component of the vector $\mathbf{a}/h$ in the direction normal to the planes. This is done by taking the dot product of the vector $\mathbf{a}/h$ and the unit vector $\mathbf{r}^*_{hkl}/|\mathbf{r}^*_{hkl}|$:

$$d_{hkl} = (\mathbf{a}/h) \cdot \mathbf{r}^*_{hkl}/|\mathbf{r}^*_{hkl}| = 1/|\mathbf{r}^*_{hkl}|$$

The convenient quantity $1/d^2_{hkl}$ is given by the relation

$$\frac{1}{d^2_{hkl}} = |\mathbf{r}^*_{hkl}|^2 = (h\mathbf{a}^* + k\mathbf{b}^* + l\mathbf{c}^*) \cdot (h\mathbf{a}^* + k\mathbf{b}^* + l\mathbf{c}^*)$$

$$= h^2a^{*2} + k^2b^{*2} + l^2c^{*2} + 2hka^*b^* \cos \gamma^*$$

$$+ 2klb^*c^* \cos \alpha^* + 2lhc^*a^* \cos \beta^* \qquad (4\text{–}15)$$

where α^*, β^*, γ^* are the angles between the axes of the reciprocal lattice, which can be expressed in terms of the interaxial angles of the crystal lattice by the formulas given in the Appendix. For crystals in the orthorhombic, tetragonal, and cubic systems, $\alpha^* = \beta^* = \gamma^* = 90°$ and the cosine terms in Eq. (4–15) are equal to zero. The special forms for the various systems are given in the Appendix.

Conditions for diffraction Now consider two atoms separated by the vector $\mathbf{r}$ (Fig. 4–10). There will be a difference in path for rays scattered in a given direction by the two atoms. The scattered waves will be in phase and maximum reinforcement will occur if the phase difference is 2π, which will occur if the path difference is equal to the wavelength λ, but in general there will be a phase difference ϕ given by $2\pi/\lambda$ times the path difference. If the direction of the incident beam is represented by the unit

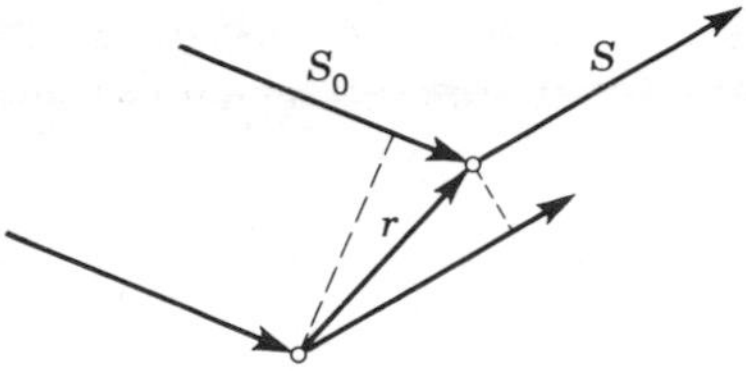

Fig. 4-10 Scattering by a pair of atoms.

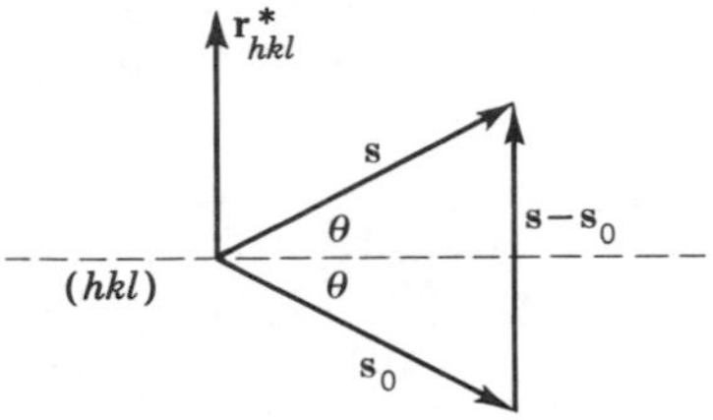

Fig. 4-11 Illustrating Bragg reflection.

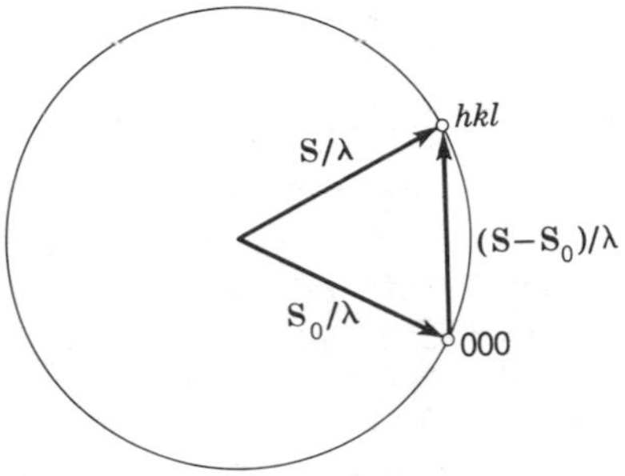

Fig. 4-12 The reflecting sphere.

vector $\mathbf{S}_0$ and the direction of the diffracted beam by the unit vector $\mathbf{S}$, then the path difference for the rays scattered by the two atoms will be the difference between the projection of $\mathbf{r}$ on $\mathbf{S}$ and the projection of $\mathbf{r}$ on $\mathbf{S}_0$; thus, the path difference is $\mathbf{r}\cdot\mathbf{S} - \mathbf{r}\cdot\mathbf{S}_0 = \mathbf{r}\cdot(\mathbf{S} - \mathbf{S}_0)$. Multiplying this path difference by $2\pi/\lambda$ gives the phase difference

$$\phi = \frac{\mathbf{r}\cdot(\mathbf{S} - \mathbf{S}_0)2\pi}{\lambda} \qquad (4\text{–}16)$$

When this phase difference is a multiple of 2π the scattered waves will reinforce each other and produce a diffracted beam of maximum intensity, and if the atoms are in a crystalline array the diffraction is equivalent to a reflection from a crystal plane, as we have seen.

Now the unit vectors $\mathbf{S}$ and $\mathbf{S}_0$ are at equal angles to the reflecting plane, as indicated in Fig. 4–11; therefore $\mathbf{S} - \mathbf{S}_0$ is a vector parallel to the vector $\mathbf{r}^*_{hkl}$ that represents the reflecting plane in reciprocal space. It is possible to express the condition for reflection, Bragg's law, in terms of these vectors. Consider two vectors of length $1/\lambda$ in the directions $\mathbf{S}$ and $\mathbf{S}_0$, i.e., the vectors $\mathbf{S}/\lambda$ and $\mathbf{S}_0/\lambda$ of Fig. 4–12. We will show that the condition for reflection from the (hkl) plane is met if the vector $(\mathbf{S} - \mathbf{S}_0)/\lambda$ extends from the origin to the reciprocal lattice point hkl, i.e., if $(\mathbf{S} - \mathbf{S}_0)/\lambda = \mathbf{r}^*_{hkl}$. Since the vector $\mathbf{r}^*_{hkl}$ has components $h\mathbf{a}^*$, $k\mathbf{b}^*$, $l\mathbf{c}^*$, we can specify it by the equation

$$\mathbf{r}^*_{hkl} = \frac{\mathbf{S} - \mathbf{S}_0}{\lambda} = h\mathbf{a}^* + k\mathbf{b}^* + l\mathbf{c}^* \qquad (4\text{–}17)$$

Substituting in Eq. (4–16) gives

$$\phi = 2\pi(u\mathbf{a} + v\mathbf{b} + w\mathbf{c})\cdot(h\mathbf{a}^* + k\mathbf{b}^* + l\mathbf{c}^*)$$
$$= 2\pi(hu + kv + lw) \qquad (4\text{–}18)$$

Reflection from the hkl plane occurs when $\phi = 2\pi n$, where n is the integer giving the order of reflection. Therefore, from Eq. (4–18) we see that reflection occurs when $(hu + kv + lw) = n$. It is always possible to choose axes for the space lattice of a crystal in such a way that the vector $\mathbf{r}$ to any lattice point in the crystal has components $u\mathbf{a}$, $v\mathbf{b}$, $w\mathbf{c}$ where u, v, w are whole numbers. With u, v, w as integers, the reflection condition $(hu + kv + lw) = n$ means that if h, k, l are also integers the reflection condition is met. Thus, the reflection condition is satisfied if the reciprocal lattice vector $(\mathbf{S} - \mathbf{S}_0)/\lambda$ touches the reciprocal lattice point hkl, which represents the reflecting plane (hkl).

Bragg's law can be shown to be equivalent to the condition just stated. Consider the plane in the crystal that is normal to $\mathbf{r}^*_{hkl}$ (Fig. 4–11). If this is the reflecting plane, the unit vectors $\mathbf{S}$ and $\mathbf{S}_0$ will make equal angles θ with this plane, which has Miller indices (hkl). From Fig. 4–11 it will be seen that $(\mathbf{S} - \mathbf{S}_0) = 2 \sin \theta$, or $(\mathbf{S} - \mathbf{S}_0)/\lambda = (2 \sin \theta)/\lambda$. But $(\mathbf{S} - \mathbf{S}_0)/\lambda = \mathbf{r}^*$ and $|\mathbf{r}^*| = 1/d_{hkl}$; therefore,

$$(2 \sin \theta)/\lambda = 1/d_{hkl}$$

which is Bragg's law.

The Laue equations, which express the conditions for diffraction in terms of the periodicities of a space lattice, follow immediately from the principles just presented. We have seen that maximum reinforcement of the scattered waves requires that the phase differences between the waves from the different lattice points be multiples of 2π and that this requirement is met by the relation of Eq. (4–18) when h, k, l are integers. Therefore, if there is a periodic distribution of scatterers along, say, the $\mathbf{a}$ axis, the condition for reinforcement can be obtained from Eq. (4–17) by taking the scalar product of each side of the equation with the vector $\mathbf{a}$:

$$\frac{\mathbf{a}\cdot(\mathbf{S} - \mathbf{S}_0)}{\lambda} = \mathbf{a}\cdot(h\mathbf{a}^* + k\mathbf{b}^* + l\mathbf{c}^*)$$

thus giving

$$\mathbf{a}\cdot(\mathbf{S} - \mathbf{S}_0) = h\lambda \qquad (4\text{–}19)$$

Similarly, the other Laue equations are obtained by scalar products of Eq. (4–17) with $\mathbf{b}$ and $\mathbf{c}$, respectively.

Ewald's sphere of reflection Ewald showed that a simple geometrical construction in reciprocal space gives the condition for reinforcement of the scattered waves in a way that proves to be extremely useful in diffraction work. It expresses the condition for diffraction in terms of a sphere in reciprocal space, the *reflection sphere*, or *Ewald's sphere*. The sphere is erected with the vector representing the incident beam as a radius. This vector is S_0/λ; therefore, it is drawn with a length $1/\lambda$ because S_0 is a unit vector. The origin of the reciprocal lattice (coordinates 000) is located on the sphere at the end of this vector, as indicated in Fig. 4–12. (One imagines the crystal to be located at this point.) Since the vector S/λ has the same length as S_0/λ, it also can be drawn as a radius of the sphere. Subtracting one vector from the other gives the vector $(S - S_0)/\lambda$, which is a vector that is a chord of the sphere as shown in the figure. As explained earlier, the vector $(S - S_0)/\lambda$ extends from the origin of the reciprocal lattice to the reciprocal lattice point *hkl* when conditions are right for the diffracted beam of indices *hkl* to occur. Therefore, the conditions for diffraction are that the sphere of reflection touch a reciprocal lattice point and that, when this occurs, the diffracted beam goes in the direction of the radius vector S/λ (but as we have imagined the crystal to be at 000, the diffracted ray, of course, goes out from 000).

X-ray diffraction methods The orientation and unit cell dimensions of a crystal determine the arrangement of the reciprocal lattice points, as we have seen, and if a beam of x-rays enters the crystal in an arbitrary direction there may be no intersection of the reflection sphere with any reciprocal lattice point, in which case there can be no diffracted beams. But if the crystal is rotated with respect to the incoming beam, then the reciprocal lattice points touch the sphere one after another. This principle is the basis for various *rotating-crystal methods.* Alternately, if the incoming beam contains many different wavelengths, the reflection sphere will be replaced by many spheres, all touching 000, and all with their centers on the same line, some of which are likely to meet the condition for diffraction even with a stationary crystal in any arbitrary orientation. This method is called the *Laue method.* A third way to ensure that conditions for diffraction are met is to have the incoming beam strike a powdered crystal or a fine-grained polycrystalline solid, as it does in the *powder method.* Various crystallites will then meet the conditions for reflection even if the x-ray beam has only a single wavelength. The individual methods are discussed in the following chapters. Because of the fact that the powder method is used in many more instances than any other, it is presented in greater detail.

5

Laue and divergent beam methods

The Laue method is now used chiefly for determining the orientation of crystals by methods presented in Chap. 9, and in metallurgical work for revealing crystalline imperfection, distortion, and recrystallization. In earlier times it was also important in crystal-structure determination, and it is still used occasionally for determining crystal symmetry, although other methods are better for this, in general.

The Laue method requires very simple equipment (a crystal mounted with an adjustable orientation in a beam of general radiation, and a flat film in a lightproof holder mounted normal to the beam a few centimeters from the crystal). Several instruments are available commercially, in most instances as an attachment or integral part of an x-ray goniometer that is also capable of taking rotating-crystal or oscillating-crystal photographs. The better Laue cameras and goniometers include a detachable combination telescope-microscope with which the sample can be aligned in the beam and oriented so that symmetrical patterns will be obtained. For transmission patterns, it is desirable to have pinholes of several sizes available; and to place a small metal disk or cup in front of the film holder, or to fasten it to the holder, to prevent fogging of the film by the central beam (0.5-mm-thick copper is recommended). A convenient crystal holder for back-reflection cameras is the *barrel holder* designed by Bond.[1]

High-intensity fluorescent screens permit visual observation of Laue patterns, provided that the observer remains in darkness or deep red light for 10 to 30 min for eye adaptation. It is then possible to orient the crystal quickly to give a symmetrical pattern.[2] Care to avoid exposure of the hands and face to the scattered rays is very necessary. Intensifying screens can be used with x-ray film or with Polaroid film to shorten exposure times. X-ray images can now be intensified to the extent that they require no darkened room, by using certain light-amplifying electronic tubes.[3]

[1] W. L. Bond, *J. Sci. Instr.*, vol. 38, p. 63, 1961. E. A. Wood, "Crystal Orientation Manual," Columbia, New York, 1963.

[2] The rods in the retina are responsible for vision in this range of intensities; compared with vision at normal intensities (using the cones), the rods are inferior in resolving power for detail and in ability to distinguish differences in light intensity. The fluorescent screens also have low resolving power.

[3] G. W. Goetze and A. Taylor, *Rev. Sci. Instr.*, vol. 33, p. 353, 1962. Much effort is also being expended to produce a spark-chamber method of recording diffraction patterns; a really successful instrument based on this principle could not be cited at the time of this writing.

Samples for the transmission Laue method are thin enough to transmit the beam without excessive absorption. There is no similar restriction to specimen dimensions when back-reflection Laue photographs are made. The radiation from a tube with a tungsten target is usually employed, with operating voltages of 40 to 60 kv, since the "white" radiation is then very intense, but other targets (e.g., molybdenum) will also serve. It should be remembered that K and L radiation will usually be superimposed on the necessary white radiation.

Prominent features of transmission Laue photographs are the ellipses of spots, each of which contains the spots from various planes belonging to an individual zone. All Laue spots, in fact, both in transmission and in back-reflection photographs, lie on conic sections. In either type of photograph a straight line of spots signifies a zone axis normal to the beam.

Determination of symmetry When either a transmission or a back-reflection Laue photograph is made with the incident beam along an n-fold axis of symmetry, the pattern of spots on a (flat) film will show n-fold symmetry around the center. The pattern reproduced in Fig. 5–1, for example, exhibits obvious rotational symmetry around the center because the beam was directed along an axis of symmetry in the crystal. If there is a plane of symmetry parallel to the beam, the pattern of spots will have a line of symmetry such that the pattern on one side of the line will be the mirror image of the pattern on the other side. Laue photographs cannot distinguish between the presence and the absence of a center of symmetry in a crystal; all crystals appear to have a center of symmetry when Laue photographs are inspected (Friedel's law). Only 11 classes of symmetry, the Laue symmetry groups, instead of the entire 32 classes, can

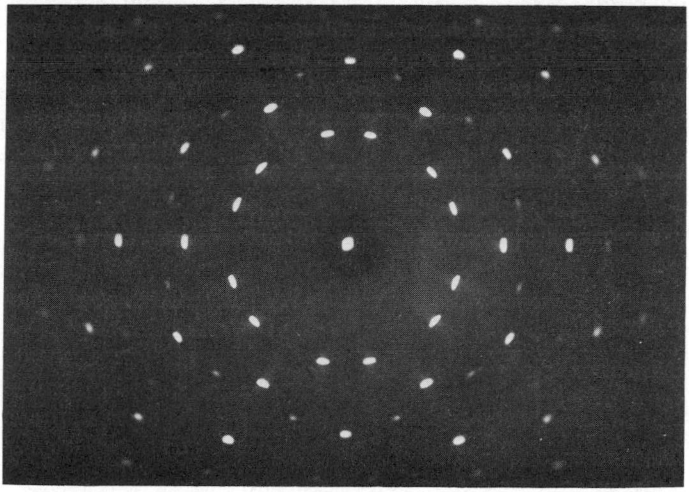

Fig. 5-1 Laue photograph showing sixfold symmetry. Transmission photograph of a magnesium crystal with x-ray beam parallel to hexagonal axis.

be distinguished by such photographs (unless special techniques involving additional information are resorted to). Symmetry that is not immediately obvious from inspection of a Laue pattern may be recognized by making a stereographic or a gnomonic projection of the spots of the pattern (see below, also Chap. 9).

The gnomonic projection of Laue patterns Indices can be readily assigned to all spots in a symmetrical Laue pattern by making a gnomonic projection of the pattern. There is a simple relation between the reciprocal lattice and the gnomonic projection of a crystal. The gnomonic projection of a crystal plane is the point where a line from the origin of the projection, drawn normal to the crystal plane, intersects the projection plane. Since the reciprocal lattice point for a crystal plane also lies on a line from the origin normal to the plane, one can construct the gnomonic projection for all points in the reciprocal lattice by merely drawing lines from the origin to the points. The intersection of each of these lines with a projection plane is the gnomonic projection of the corresponding plane. The straight rows of points in the reciprocal lattice project into straight rows on the projection.

Converging rows replace parallel rows in the projection if a crystal is oriented so that the incident beam makes a small angle with one of the crystal axes. It is often possible to estimate by inspection how to rotate the crystal to bring it into a more symmetrical orientation with respect to the beam. A stereographic projection of a Laue pattern can be constructed, and the necessary rotation can be determined with a stereographic net.[1] Laue patterns do not yield useful *intensities* of individual reflections.

Asterism The analogy between the diffraction of x-rays by atomic planes and the reflection of light by plane mirrors extends also to the diffraction by bent planes of atoms and the corresponding optical reflection from curved mirrors. It also applies to the reflection from crystal fragments arranged on a curved surface, corresponding to reflection from small mirrors arranged on a curved surface.

The optical analogy thus provides a convenient way of interpreting Laue patterns of distorted crystals. Visualizing the atomic planes as mirrors, one sees at once that the Laue spots from a perfect crystal should be sharp, their size determined by the size and divergence of the x-ray beam and the crystal-to-film distance, whereas Laue spots from bent planes or grains with substructure will be elongated. The direction and amount of the elongation will depend on the orientation of the axis of bending with respect to the x-ray beam and on the range of orientation of the planes within the area struck by the beam. The elongation is chiefly in radial or near-radial directions, in transmission patterns, giving rise to the appearance known as *asterism*, which is illustrated in Fig. 5–2.

[1] For details of graphical procedures, including the gnomonic, see N. F. M. Henry, H. Lipson, and W. A. Wooster, "The Interpretation of X-ray Diffraction Photographs," Macmillan, London, 1951, 1961; or "International Tables for X-ray Crystallography," vol. 2, Kynoch, Birmingham, England, 1959.

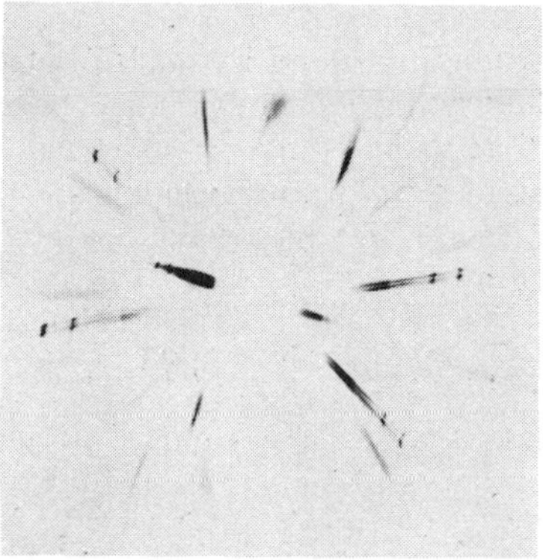

Fig. 5-2 Transmission Laue photograph of a thin crystal of alpha-iron bent cylindrically, showing asterism and sharp spots from molybdenum characteristic radiation.

Laue photographs can be used in the study of the kinetics of recrystallization of deformed sheets,[1] since the progress of recrystallization is revealed by the incidence and growth of sharp spots from strain-free grains growing at the expense of the asterism streaks (or background fog from unresolved streaks) characteristic of the deformed material. Laue photographs of the ordinary type are not suitable for detailed studies of the formation and growth of small subgrains, as in polygonization research, since they do not resolve the spots from individual subgrains. They also fail to reveal internal stresses.

Characteristics of Laue streaks If a small crystal is placed in a beam of white radiation and rotated about an axis, the Laue spots will trace out curves on the film. Charts have been published that show these curves and give the amount of movement of the spots per degree of rotation (or, equivalently, the lengths of the streaks per degree range of orientation of the irradiated portion of a crystal),[2] but in general it is more instructive to make a stereographic projection of the streaks if they are studied in detail.

The construction of a stereographic projection of a crystal plane causing a Laue spot or streak is very simple. If the projection plane is placed normal to the incident beam and parallel to the plane of the photographic film, it will be seen from simple geometry that: (1) The Laue spot and the corresponding point on the projection representing the pole of the reflecting plane have the same azimuthal position around the incident beam. (2) The radial distance of the Laue spot in a transmission pattern from the point where the direct beam strikes the film is $r = R \tan 2\theta$, where θ is the Bragg angle and R is the distance from crystal to film. The radial distance on the stereographic projection to the corresponding pole is

$$r_s = R_s \tan (90 - \theta)/2$$

where R_s is the radius of the stereographic sphere. The angle θ can be read

[1] G. Masing, K. Lücke, and P. Nölting, *Z. Metallk.*, vol. 47, p. 65, 1956.

[2] "International Tables for X-ray Crystallography," vol. 2, loc. cit.

directly from the film by using a chart made up for the crystal-film distance used.

In practice, a Laue streak may be prevented from reaching the full length corresponding to the orientation range involved. This occurs because in general each streak is a spectrum of the primary x-rays, i.e., each part is formed by x-rays of different wavelengths. A streak is therefore terminated by the short-wavelength limit of the spectrum if the orientation range includes the Bragg angle for this wavelength. On the long-wavelength end a streak may be prevented from developing its full length by the lack of intensity in the long-wavelength components.

When a transmission photograph of a polycrystalline specimen is made with radiation containing both characteristic and white radiation, the pattern is a superposition of a powder pattern of Debye rings and a Laue pattern.

The distribution of intensity along a Laue streak is the combined result of the distribution in orientation of the diffracting material, the intensity distribution in the spectrum, the efficiency of the photographic emulsion and the absorption characteristics of the specimen for the different wavelength components along the streak.

If a photograph such as Fig. 5–3 is used to judge the presence or absence of a preferred orientation of the grains in the sample, the appearance of the pattern may be misleading for two reasons. (1) There may be too few spots in the pattern to clearly reveal the clustering of spots at certain azimuths that would signify a preferred orientation, especially if the degree of orientation is weak. (2) The shapes of the spots may be such that they show more clearly in some directions from the center than in others, as in Fig. 5–3, and falsely give the pattern an appearance of preferred orientation. The spots in this photograph are elongated as a result of a converging beam of x-rays striking the small grains of the sample, the convergence being greater in the horizontal plane than in other directions. (The grains themselves were strain-free in this experiment and did not produce asterism.) Fig. 5–4 illustrates the directions of the streaks that result from such

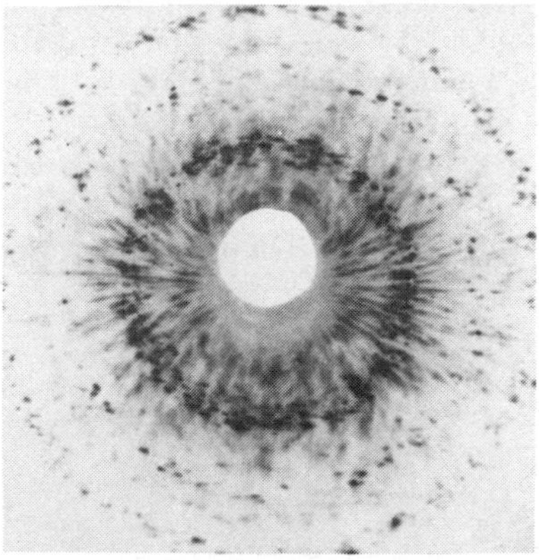

Fig. 5-3 Transmission Laue photograph of annealed steel sheet. Beam incompletely filling pinholes caused horizontal elongation of spots. Debye rings are from Mo $K\alpha$ radiation.

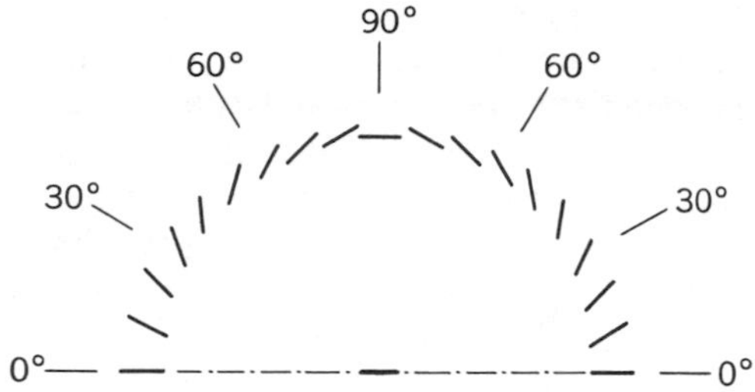

Fig. 5-4 Directions of elongation of Laue spots when primary x-ray beam converges fanwise in a horizontal plane to a small crystal.

fanwise convergence when it is in the horizontal plane. This type of pattern is to be expected whenever a pinhole collimator is incompletely filled with radiation. For example, if the beam through a pinhole system leaves the target of an x-ray tube at too small an angle, the focal spot on the target may be foreshortened too much to fill the pinholes in both horizontal and vertical directions. A fluorescent screen placed at the position of the film will then show that the beam is elliptical rather than circular in cross section. Laue spots are elongated radially when the rays converge on a small crystal equally in all directions, as with a well-filled pinhole system. On the other hand, the spots show elongation tangentially if a divergent beam strikes a large crystal.[1]

Focused Laue spots In the usual Laue patterns, asterism cannot be detected if the orientation range in the crystal is less than $\frac{1}{2}°$ or so, but the sensitivity to disorientation may be greatly increased by providing for the focusing of rays at a Laue spot.

If rays diverge from a point source, indicated in Fig. 5–5 by S, and pass through a thin crystal plate C, the rays reflected by a set of parallel atomic planes will form a focused Laue spot at P on a film placed at the proper distance (a distance which differs for different reflections). This property has been shown by Guinier and Tennevin to be useful in disclosing disorientations of the order of 10 sec of arc when the source is placed a meter or so from the crystal.[2] In their technique a line-focused x-ray tube serves as the source, and the line of the focal spot is oriented normal to the plane containing the reflecting-plane normal and the reflected ray (i.e., the source is normal to the plane of the drawing in Fig. 5–5). The high resolution for subgrain disorientation achieved in this line-source arrangement applies only to reflected rays lying in the plane of the drawing; other Laue spots are not focused.

[1] J. Leonhardt, *Z. Krist.*, vol. 63, p. 478, 1926.

[2] A. Guinier and J. Tennevin, *Rev. Met.*, vol. 45, p. 277, 1948. A. Guinier, "Théorie et Technique de la Radiocristallographie," 2d ed., Dunod, Paris, 1956.

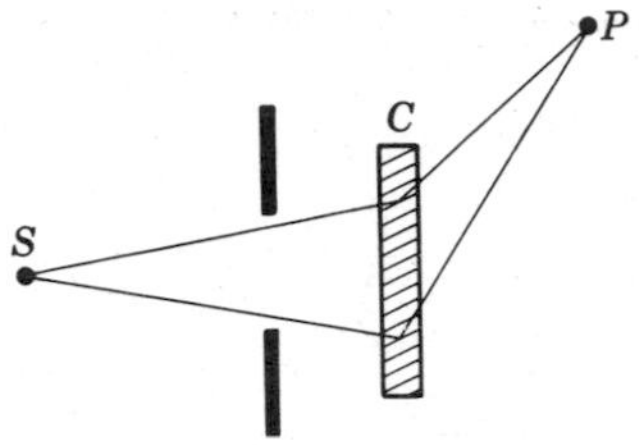

Fig. 5-5 Focusing of a Laue spot at *P* with a point source of rays at *S* (or a line source normal to the plane of the drawing) and a thin crystal at *C*.

Divergent beam x-ray photographs X-ray photographs of single crystals can be made with very divergent beams from a point source of monochromatic radiation located at or near the surface of the crystal itself. Rays are reflected and registered on a flat film placed either in the forward-reflection or back-reflection position, but because of the wide divergence of the rays the diffraction pattern does not consist of spots. The pattern consists of curved lines, called *Kossel lines*, which have some useful properties.[1] The photographs permit the determination of lattice constants with precision approaching 1 part in 100,000 without precision instruments; they have also been used in studies of crystal perfection and internal strains.[2]

Several different experimental arrangements have been used to provide the widely divergent rays needed. In some the specimen was placed within the x-ray tube and served as the anticathode,[3] while in others the crystal was placed in contact with a thin anticathode but outside the x-ray tube. Back-reflection photographs have been made with a point-focus capillary x-ray tube.[4] A method that can be used with slightly modified standard x-ray equipment consists in placing a layer of fluorescing material on the surface of the crystal and exciting it with a pinhole beam of x-rays from an ordinary x-ray tube.[5] The surface layer of the crystal itself serves as the fluorescing material for some crystals. The very sharp focus of the electron beam in an electron probe analyzer provides an effective source of x-rays.[6]

As indicated in Fig. 5–6, the rays spread out from the point source and strike a given set of parallel atomic planes (hkl) at a continuous range of angles. Those striking the planes at the Bragg angle for a particular monochromatic component in the beam will be reflected. There will be a cone of

[1] W. Kossell and H. Voges, *Ann. Phys.* (*Leipzig*), vol. 23, p. 677, 1935; vol. 25, p. 512, 1936; vol. 26, p. 533, 1936. G. Borrmann, *Ann. Phys.* (*Leipzig*), vol. 27, p. 669, 1936; *Naturwissenschaften*, vol. 23, p. 591, 1935.

[2] K. Lonsdale, *Phil. Trans. Roy. Soc. London*, vol. 240, p. 219, 1947.

[3] W. Kossel, *Ergeb. Exakt. Naturw.*, vol. 16, pp. 296, 353, 1937.

[4] T. Fujiwara and I. Takesita, *J. Sci. Hiroshima Univ. Ser. A*, vol. 11, p. 93, 1941. T. Imura, S. Weissmann, and J. J. Slade, Jr., *Acta Cryst.*, vol. 15, p. 786, 1962.

[5] G. Borrmann, *Ann. Phys.*, vol. 27, p. 669, 1936. A. H. Geisler, J. K. Hill, and J. B. Newkirk, *J. Appl. Phys.*, vol. 19, p. 1041, 1948.

[6] B. H. Heise, *J. Appl. Phys.*, vol. 33, p. 938, 1962. R. E. Hanneman, R. E. Ogilvie, and A. Modrzejewski, *J. Appl. Phys.*, vol. 33, p. 1429, 1962.

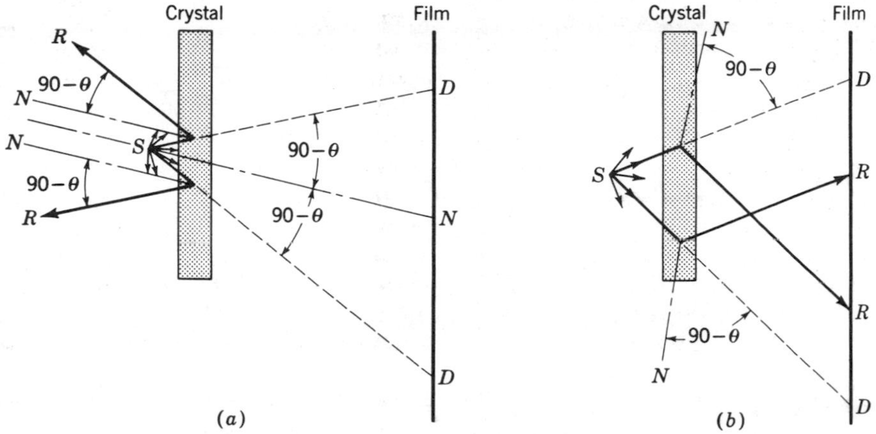

Fig. 5-6 Origin of Kossel lines. Divergent monochromatic rays from point S reflect to form the cones RR and give black lines in photographs. Deficiency cones DD give white lines. The position of the source and film determines the separation of the black and white lines in arrangement b.

rays meeting this requirement, namely, the rays forming the generators of a cone whose axis is the normal to the reflecting plane. The semiapex angle of the cone will be $90 - \theta_{hkl}$. The reflected rays will also form a cone, and will appear as a black arc on a photographic film. Depending on the angle θ, the cone of reflected rays may appear in either back-reflection or transmission photographs. Rays not reflected continue through the crystal and darken the transmission photograph fairly uniformly. If the crystal thickness is small so that absorption of the rays in the crystal is not extreme, the uniform background darkening will contain white arcs. These are formed because the energy going into the diffracted cone is subtracted from the original divergent beam, leaving a cone of rays deficient in energy. For each diffracted cone there is a deficiency cone in the general blackening of a transmission pattern. The curved-line pattern made with x-rays is called a *Kossel line pattern* and is analogous to the Kikuchi line pattern obtained from single crystals with electron diffraction.

The various arcs can be assigned indices that correspond to the indices of the Bragg reflection involved. Indexing may be carried out as follows: (1) The value θ obtained from the semiapex angle $(90 - \theta)$ of a deficiency cone may be compared with a list of values of θ computed for the crystal lattice and the wavelength of the characteristic radiation forming the Kossel line. (2) A stereographic projection of the pattern may be prepared and compared with a standard projection for the crystal. Or if a standard projection is prepared in which the calculated positions of all the expected cones are plotted, each cone having its axis coincident with the pole of a corresponding reflecting-plane normal, a direct comparison of a pattern with the projection leads to a rapid identification of all lines. Figure 5–7

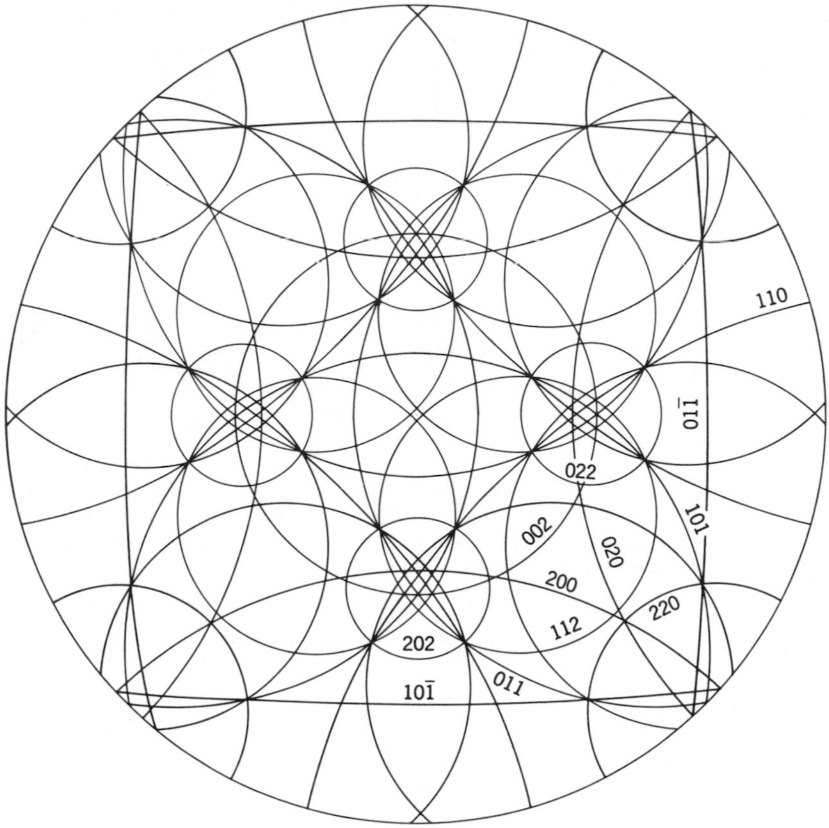

Fig. 5-7 Stereographic projection of white Kossel-line pattern for Fe $K\alpha_1$ radiation on alpha-iron crystal. (**A. H. Geisler, J. K. Hill, and J. B. Newkirk, J. Appl. Phys., vol. 19, p. 1041, 1948.**)

shows such a projection for Fe $K\alpha_1$ radiation and a single crystal of alpha-iron (b.c.c. assuming $a_0 = 2.861$ A). (The orientation of the crystal is such that [001] is in the center of the projection and [100] is at the bottom.)

Since the apex angles of the cones depend upon the lattice dimensions, the intersections and near approaches alter with the lattice constants. A few measurements of the distances between selected, closely spaced arcs, computed with spherical trigonometry, can provide a high-precision determination of the lattice constants. The method can be surpassed in precision by others, however, and because good patterns cannot be obtained if the crystals are too thick, too thin, too perfect, or too imperfect, relatively few laboratories have made use of the method.

A thickness of $1/0.2\ \mu$ is recommended, where μ is the linear absorption coefficient for the radiation used. The wavelength of the x-rays in the *crystal*, not in air, should be used in the calculations of the lattice constants.

The sharpness of the Kossel lines depends upon the perfection of the diffracting crystal. As perfection increases, the lines sharpen and eventually may become too narrow to be seen. In a limited range of perfection, therefore, the method is available for indicating the degree of perfection. Subgrains cause abrupt displacements of segments of the Kossel lines and provide another means of detecting crystal imperfections.[1]

[1] T. Fujiwara, *Mem. Defense Acad. Math. Phys. Chem. Eng.* (*Yokosuka, Japan*), vol. 2, p. 127, 1963.

6

Rotating-crystal methods

The identification of individual reflections and the measurement of their intensities are accomplished with greater convenience and certainty by using one of the rotating-crystal methods than by any other method; consequently, the rotating-crystal methods are preferred in most crystal-structure determinations. Apparatus of the photographic type can be very simple, for it may consist merely of a rotating spindle on which a single crystal is mounted, a cylindrical film around the crystal (the axis of the cylinder coinciding with the spindle axis), and a set of pinholes to collimate the x-ray beam. It is essential to place an important zone axis of the crystal parallel to the axis of rotation, and for this purpose the apparatus should have a goniometer head that provides angular and lateral adjustments of the crystal. A telescope and collimator are usually employed so that the small crystals (preferably 0.1 to 0.5 mm in average dimension) can be visually adjusted to a setting in which an important crystal edge lies along the axis of rotation, or planes of an important zone lie parallel to the axis. Optical adjustment of the crystal is often supplemented by x-ray patterns; one or two Laue photographs (preferably made on the same instrument) serve to determine the orientation of the crystal and to predict the angle through which the crystal must be turned to reach the desired setting.

An x-ray goniometer with provision for optical adjustment of the crystal and for Laue, oscillating-crystal, and rotating-crystal photographs is shown in Fig. 6–1. Any camera of the Weissenberg type, discussed on page 108, serves not only for the moving-film methods for which it is designed, but also for stationary-film rotation and oscillation photographs. A number of suitable designs of this type have been published,[1] and the most convenient is probably that of Buerger.[2] The construction and operation of many types of stationary-film and moving-film cameras are covered in detail in Buerger's books.[3] Several cameras and a large number of other crystallographic supplies are available on the market.[4]

[1] K. Weissenberg, *Z. Physik*, vol. 23, p. 229, 1924. J. Bohm, *Z. Physik*, vol. 39, p. 557, 1926.

[2] M. J. Buerger, *Z. Krist.*, vol. 94, p. 87, 1936.

[3] M. J. Buerger, "X-ray Crystallography," Wiley, New York, 1942; "The Precession Method," Wiley, New York, 1964.

[4] Comprehensive lists have been compiled by the International Union of Crystallography, A. J. Rose (ed.), "Index of Crystallographic Supplies," Société française de Minéralogie et de Cristallographie, Paris.

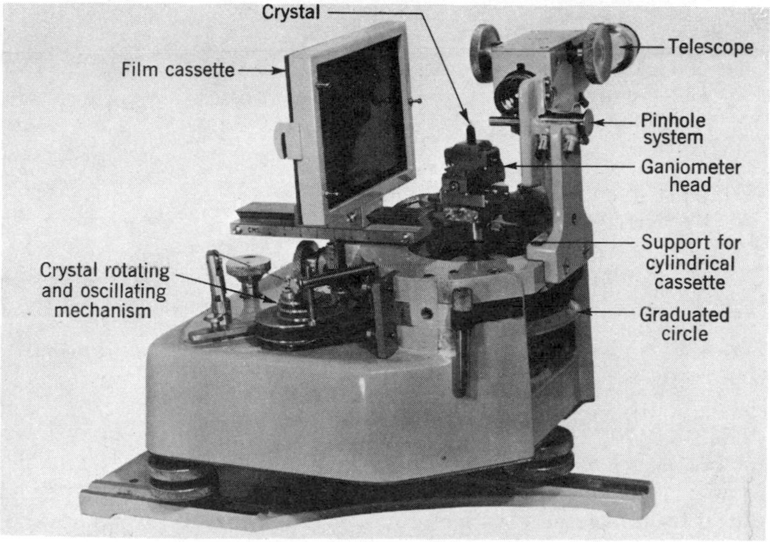

Fig. 6-1 An x-ray goniometer for Laue, oscillating-crystal, rotating-crystal, and powder methods. A flat-film cassette is shown (Unicam Instruments, Ltd., Cambridge, England.)

If the specimen is prepared in the shape of a small sphere or a cylinder, the correction for absorption of the incident and diffracted rays in the specimen is known completely[1]; the corrections have also been worked out for more complex shapes but they are extremely laborious to use for such shapes, especially when hundreds or thousands of reflections are being measured. Crystals are often ground to spherical shape by very convenient and simple methods[2] and mounted on the end of a glass fiber.[3] Care must be taken in choosing and preparing crystals to avoid any twins in the final specimen; consequently, methods involving surface abrasion cannot be used with crystals that form deformation twins easily or that are too soft.

Copper $K\alpha$ radiation is frequently used (it gives relatively high dispersion); usually use of a filter is unnecessary since the $K\beta$ spots can be recognized by the fact that they lie on the streaks made by the general radiation at θ positions where $(\sin \theta_{K\beta})/\sin \theta_{K\alpha} = \lambda_{K\beta}/\lambda_{K\alpha}$.

Interpretation of rotation photographs Figure 6–2 is a typical rotation photograph showing prominent horizontal layer lines of spots and approximately vertical rows of spots (row lines). If a crystal is mounted for

[1] Absorption for several shapes is thoroughly treated in M. J. Buerger, "Crystal Structure Analysis," Wiley, New York, 1960; and in "International Tables for X-ray Crystallography," vol. 2, Kynoch, Birmingham, England, 1959.

[2] W. L. Bond, *Rev. Sci. Instr.*, vol. 22, p. 344, 1951. K. W. Revell and R. W. H. Small, *J. Sci. Instr.*, vol. 35, p. 73, 1958.

[3] A list of recommended adhesives is given in "International Tables for X-ray Crystallography," loc. cit.

Fig. 6-2 Rotating-crystal pattern of quartz. Cylindrical film. Axis of rotation vertical. Filtered radiation. (B. E. Warren.)

rotation around the *a*, *b*, or *c* axis, the spacings of the layer lines give immediately the spacing between lattice points in the direction of the rotation axis. This can be seen from the Laue equations, for each layer line is produced by diffracted rays forming the generators of a cone coaxial with the rotation axis (Fig. 6–3). The cone is actually the cone along which reinforcement occurs according to one of the Laue equations, viz., the equation concerned with the periodicity along the axis of rotation. If the crystal is rotated around the *a* axis, the Laue equation that applies is

$$a\,(\cos\alpha - \cos\alpha_0) = h\lambda$$

and if the incident beam is perpendicular to the axis of rotation, this reduces to $a = h\lambda/\cos\alpha$, where a is the identity distance along the rotation axis, α is the semiapex angle of the diffracted cone, λ is the wavelength, and h is an integer (the order of reflection for the cone) which is 0 for the horizontal layer line through the central spot, 1 for the first layer line above this, 2 for the second layer line, etc. Similar equations hold for rotation about any other direction in the crystal. We may therefore generalize the equation to apply to the identity distance in *any* direction that is chosen as the rotation axis, and if we insert the measured distance S_n on the film (Fig. 6–3) from the zero line to the nth layer line, we have the formula

$$I = \frac{n\lambda}{\cos\alpha_n}$$

where α_n is determined by the relation $\cot\alpha_n = S_n/R$, R is the radius of the cylindrical film, and I is the identity distance in the chosen direction.

Perhaps the greatest advantage of the rotating-crystal method is the fact that the dimensions of the unit cell can be obtained unequivocally from photographs taken with each of the unit cell axes in turn serving as the rotational axis. With other diffraction methods it is not uncommon to derive a unit cell that is half the true unit cell in some dimension. In rotation photographs this error could be made only by overlooking entire layer lines, which is easily avoided, particularly if attention is directed to the higher-order layer lines. The precision of the determinations is usually inferior to

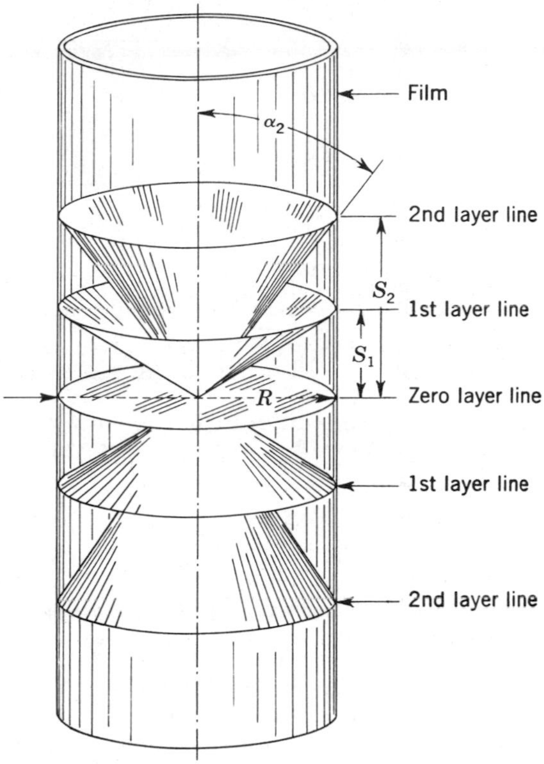

Fig. 6-3 Formation of layer lines from cones of diffracted rays.

determinations by the powder method, although it can be much improved by operating in the back-reflection region with precision cameras.[1]

Each spot on a rotation photograph is located on an invisible Debye ring for which the Bragg angle θ is given by an appropriate formula (see Appendix, or page 80). Thus, indices can be assigned to the spots by the method of calculating the θ values for all possible reflections and then comparing these calculated values with ones read from the film by the aid of an appropriate chart.

A great advantage of the rotating-crystal method is its separation of the reflections into layer lines. Not only does this prevent some overlapping of reflections that would superimpose in powder photographs, but it makes assigning indices much easier. All planes that are parallel to the rotation axis reflect to the zero layer line. On a photograph with the a axis as the axis of rotation in the crystal, these will be planes of the type $(0kl)$. Similarly, planes of the type $(1kl)$ will reflect the first layer line above the central one, $(2kl)$ to the second, $(\bar{1}kl)$ to the first one below, etc. The general rule is that reflections on the nth layer line will have indices hkl that satisfy the

[1] M. J. Buerger, "X-ray Crystallography," Wiley, New York, 1942.

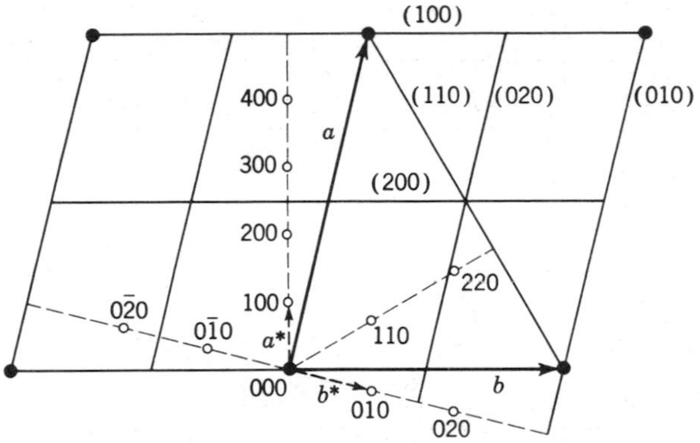

Fig. 6-4 A portion of a single layer of a reciprocal lattice (○) and a crystal lattice (●), for a monoclinic crystal.

equation

$$hu + kv + lw = n$$

where $[uvw]$ are the indices of the axis about which the crystal is rotated.

The use of the reciprocal lattice The reciprocal-lattice concept is indispensable for interpreting rotating-crystal and oscillating-crystal diffraction patterns, because of the convenient way in which the points in reciprocal space represent the orientation and spacing of the crystal planes, the conditions under which any plane will reflect, and the position on a film at which any reflection will be found.

It was explained in Chap. 4 that the reciprocal lattice points lie on lines through the origin perpendicular to crystal planes and at distances from the origin that are the reciprocals of the interplanar spacings of the corresponding crystal planes, i.e., $r^* = 1/d$. Several planes of a crystal lattice are illustrated by the solid lines and filled circles of Fig. 6–4. The dashed lines in this figure are drawn perpendicular to these planes and contain rows of open circles, the reciprocal lattice points that represent the different orders of reflection from these planes. Thus, along the a^* direction in the reciprocal lattice are the points 100, 200, 300, etc., at multiples of the reciprocal unit cell edge a^*, which represent the 100, 200, 300, ··· reflections.

A typical reciprocal lattice for an orthorhombic crystal is shown in Fig. 6–5; note that in this lattice—and indeed in the reciprocal lattice for any orthogonal crystal—the successive layers of points are arranged exactly above each other.

Reciprocal lattices are triclinic for triclinic crystals, monoclinic for monoclinic crystals (with $\mathbf{b}^*$ parallel to $\mathbf{b}$ of the crystal and with the angle β^* equal to $180° - \beta$), hexagonal for hexagonal crystals (with the sixfold axis parallel to the sixfold axis of the crystal, but with the axes $\mathbf{a}_1^*$ and $\mathbf{a}_2^*$ at 60° instead of the 120° between $\mathbf{a}_1$ and $\mathbf{a}_2$ in the crystal), rhombohedral for rhombohedral crystals, and cubic for cubic crystals.

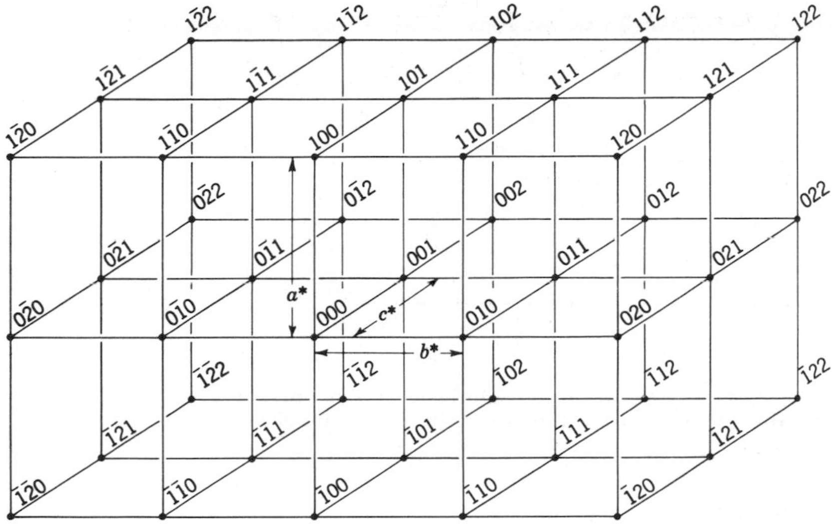

Fig. 6-5 Reciprocal lattice for an orthorhombic crystal.

If one plots only the reciprocal lattice points for which reflections can be observed, omitting the reflections that are extinguished by the periodicities of the space lattice, it will be found that body-centered crystals have points that make up a face-centered type of array in reciprocal space, and that face-centered crystals have a body-centered type of array.

It has been explained in Chap. 4 in connection with Fig. 4–12 that the condition for reflection from a plane is that the reciprocal lattice point representing some order of reflection from the plane must lie on the surface of a properly drawn reflection sphere. The sphere is drawn with a radius parallel to the incident beam, of length $1/\lambda$, where λ is the wavelength employed, and with the radius terminated at the origin 000 in the reciprocal lattice. The sphere is indicated in Fig. 6–6, which represents a set of crystal planes at O having its reciprocal lattice point for first-order reflection at P, and an incident beam along QO. As reference to the figure will show

$$\sin\theta = r^*/(2/\lambda) = \lambda/2d$$

which is Bragg's law. The reflected ray lies in the plane containing Q, O, and P and leaves the crystal at an angle of 2θ from the incident beam; from Fig. 6–6 it is obvious that the direction of the reflected beam is along the line from the center of the reflection sphere through the point P.

At any one setting of the crystal, few (if any) points will touch the sphere of reflection, but if the crystal is rotated, the reciprocal lattice will rotate with it and will bring many points into contact with the sphere. At each contact of a point with the sphere an x-ray reflection will occur. All planes that can be made to reflect have their reciprocal lattice points within the distance $2/\lambda$ from the origin, i.e., within the "limiting sphere" shown.

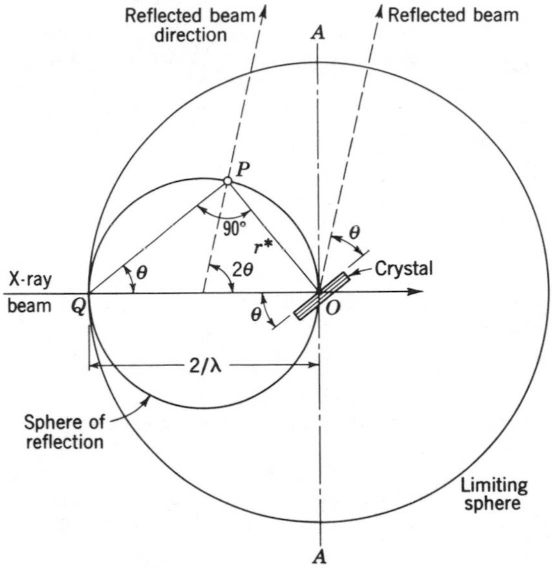

Fig. 6-6 Sphere of reflection in a reciprocal lattice.

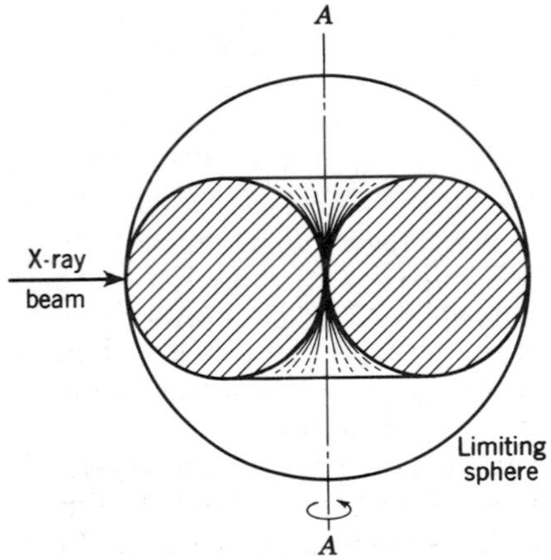

Fig. 6-7 Section through tore swept out by sphere of reflection rotating through reciprocal lattice.

Rotation of the crystal with respect to the beam can be represented by rotating the sphere of reflection and keeping the lattice fixed; if the axis of rotation is AA, the sphere of reflection will sweep out a tore whose cross section is shown in Fig. 6–7. During a complete rotation of the crystal each point within this tore will pass through the surface of the sphere twice and will produce two reflections, one on the right and one on the left side of the rotating-crystal pattern. For each point hkl that yields spots in the upper quadrants of the film, there will be a corresponding point $\bar{h}\bar{k}\bar{l}$ on the opposite

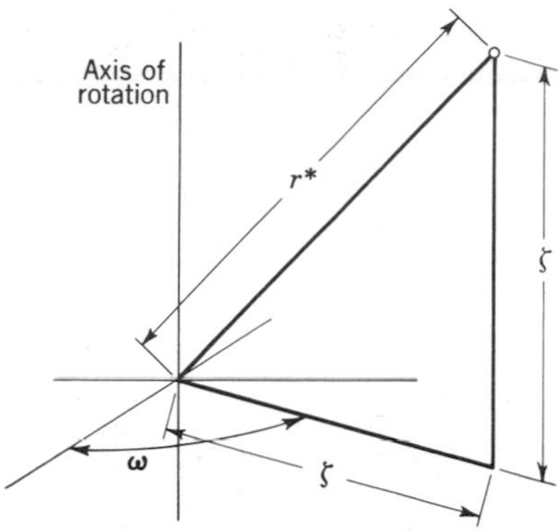

Fig. 6-8 Cylindrical coordinates for reciprocal lattice points.

side of the origin, which will also pass through the sphere twice and yield similar reflections in the lower quadrants of the rotation photograph.

Indices of spots on a diffraction pattern A rotating-crystal pattern made with very short-wavelength rays or electrons is readily visualized, for it is the pattern that would be made by reciprocal lattice points passing through a plane as they rotate and leaving spots where they hit the plane. (The reflection sphere is a plane when the wavelength is infinitesimal.) For ordinary wavelengths this pattern is distorted, since the reflection sphere has a finite diameter.

Spots on a rotating-crystal (or oscillating-crystal) photograph are conveniently referred to a rectangular coordinate system that is related in a simple way to the cylindrical coordinates of a reflecting point in the reciprocal lattice. If the cylindrical coordinates of a reciprocal lattice point are ξ/λ, ζ/λ, and ω, where ω is the angle setting of the crystal around the axis of rotation with respect to some reference orientation and λ is the wavelength of the radiation, then the coordinates of the reflection spot on the film will be ξ and ζ (Fig. 6–8). The angle ω is undetermined by the photograph, since all values of ω are passed through in each revolution. The film coordinates ξ and ζ can be read directly from the film by superimposing the film on a "Bernal chart" ruled with lines of constant ξ and ζ values.[1]

If a certain zone axis is placed in a vertical position and the crystal is rotated around this axis, it is obvious that the central layer line of the film—the zero layer line—will contain only reflections from the given zone. Each layer line of spots will be caused by one layer of reciprocal lattice points, such as the layers seen in Fig. 6–5. On a pattern made by rotation around **a***, **b***, or **c***, the mere fact that a given spot lies on a certain layer line identifies one of the indices of the spot (in the example mentioned, it is the h index that is thus determined).

[1] J. D. Bernal has published such a chart on tracing paper for a 10-cm-diameter camera, and also the equivalent chart for flat films in *Proc. Roy. Soc.* (*London*), vol. A113, p. 117, 1926. Charts are sold by the Institute of Physics, London, England; also by Polycrystal Book Service, Pittsburgh, Pa., 15238.

The remaining indices of a spot may be found by employing the coordinates defined in Fig. 6–8, which may be read from the film by using a Bernal chart. The distance ξ/λ is the radial distance from the axis of rotation in the reciprocal lattice to the given spot.[1]

If the orientation of a crystal in the camera is not exact, so that a zone axis deviates somewhat from the axis of rotation, the spots on the film will not be perfectly aligned on the layer lines. This fact may be made use of in correcting the setting of the crystal[2] and in avoiding overlapping reflections.

Oscillating-crystal photographs Overlapping reflections lead to uncertainties in indexing and in determining the intensities of reflections. Limiting the range through which the crystal is rotated reduces the number of reciprocal lattice points that come into contact with the reflection sphere and so reduces the number of reflections. The crystal is set as for a rotation photograph, and is oscillated through a 5, 10, or 15° range of angles; then a new azimuth setting is made and a new oscillation photograph is made with it, etc. until all typical reflections have been obtained. The ranges of oscillations are made small enough to prevent overlapping spots.

To determine graphically which reflections will occur in a given oscillation photograph, a plot is made of each reciprocal lattice layer normal to the axis of rotation. The reflection sphere cuts each layer in a circle, and as the crystal oscillates, the motion of the layer relative to the circle is represented by oscillating the circle through the plot of the layer. The circle will cut through each point that will give a reflection for the given range of oscillation. The circles representing the higher layer lines will be smaller than the one for the central zero layer. The centers of all circles will be at equal distances from the point representing the axis of rotation.

Weissenberg cameras Several moving-film cameras have been devised that yield diffraction patterns in which all avoidable overlapping of spots is eliminated. Full details of the common types of moving-film cameras and of the techniques involved in their use are available elsewhere and need not be repeated here.[3]

The principle of the Weissenberg camera is illustrated in Fig. 6–9. A

[1] Detailed instructions for the less symmetrical crystals are given by: J. D. Bernal, *Proc. Roy. Soc. (London)*, vol. A113, p. 117, 1926. C. W. Bunn, "Chemical Crystallography," Oxford, New York, 1945. C. W. Bunn, H. S. Peiser, and A. Turner-Jones, *J. Sci. Instr.*, vol. 21, p. 10, 1944. Analytical expressions covering the diffraction geometry are summarized in "International Tables for X-ray Crystallography," vol. 2, pp. 175–184, Kynoch, Birmingham, England, 1959.

[2] O. Kratky and B. Krebs, *Z. Krist.*, vol. 95, p. 253, 1936. O. P. Hendershot, *Rev. Sci. Instr.*, vol. 8, p. 436, 1937. A. Bairsto, *J. Sci. Instr.*, vol. 25, p. 213, 1948. J. W. Jeffery, *Acta Cryst.*, vol. 2, p. 15, 1949. (Jeffery includes comparison of his method with the earlier methods and a test of its accuracy, which may reach 0.05°.)

[3] M. J. Buerger, "X-ray Crystallography," Wiley, New York, 1942; "Crystal Structure Analysis," Wiley, New York, 1960. N. F. M. Henry, H. Lipson, and W. A. Wooster, "The Interpretation of X-ray Diffraction Photographs," Macmillan, New York, 1951, 1961.

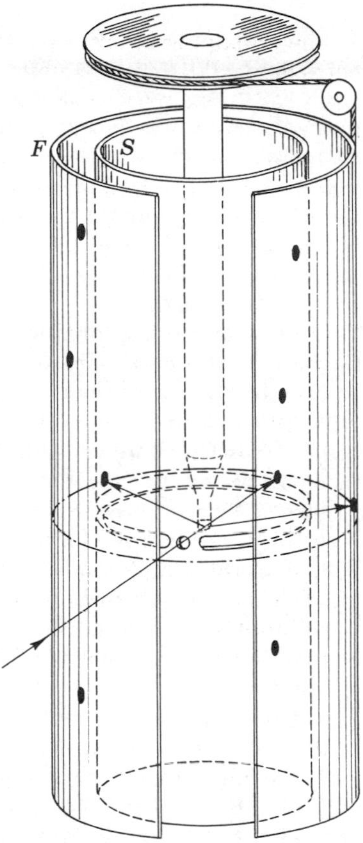

Fig. 6-9 Illustrating the principle of the Weissenberg x-ray goniometer.

cylindrical film moves back and forth along the axis of the camera in exact synchronism with the oscillation of the crystal; this is accomplished by a set of gears or pulleys. A stationary metal shield surrounds the crystal and contains a circumferential slot that permits the incoming beam to strike the crystal and the diffracted beams that form a *single* layer line to strike the film. The shifting of the film during the oscillation of the crystal causes the spots from the chosen layer line to be spread out over the whole of the film.

Before making a Weissenberg photograph, it is usually desirable to make a rotating-crystal pattern of the crystal with the same apparatus, by removing the shield and uncoupling the film-shifting mechanism from the shaft that rotates the crystal. The correctness of the crystal orientation and positioning can then be judged and corrected if necessary by adjusting the goniometer head that supports the crystal. With a stationary crystal and film it is also possible to prepare Laue photographs on the same instrument if desired.

One type of Weissenberg photograph is made with the incoming beam *normal* to the axis of rotation of the crystal. This is the *normal beam method.* It is always used for the zero layer-line photographs, and it can also be used for the other layer lines by placing the shield at the right position to pass the cone of diffracted rays for the selected layer. However, there are advantages in using the *equi-inclination method* for the nonzero layers. In this method the camera is turned with respect to the incident beam collimator to a setting such that the beam strikes the crystal at an angle μ from the normal to the rotation axis. The angle μ is chosen so that the incident beam as well as the diffracted rays that are being photographed are generators of the same cone (a cone coaxial with the rotation axis). The incident beam and the diffracted beams being photographed are then all inclined at $90 - \mu^\circ$ to the rotation axis.

In all Weissenberg photographs, a row of points in a given layer of the reciprocal lattice appears as a curved line of spots on the film, but in the equi-inclination photographs only, these curved lines have the same shape regardless of which layer is being photographed. For equi-inclination photographs it is therefore possible to lay a transparent template over the film in such a way as to show the curved rows of spots that correspond to parallel reciprocal lattice rows—and the same template can be used for all different layer-line photographs (although the meaning of the interval between ruled curves on the template changes from one layer to another).[1] Since a photograph is merely a distorted map of a layer of the reciprocal lattice, it is often possible to index the spots merely by inspection. Procedures have also been worked out to determine the polar coordinates of Fig. 6–8 from the photographs.[2]

A third method of making Weissenberg photographs is the *flat-cone method*. For this the incident beam collimator is set at an angle chosen so that the diffracted rays passing through the slit in the shield are only those that lie on a flat cone; i.e., they lie in a plane that is normal to the axis of rotation. These strike the film perpendicularly, which is an advantage in judging their intensities relative to the intensities of other layers. In some recent models mechanical devices are attached that shift the film small distances during exposure, in order to spread each spot into a patch of uniform density. The purpose of having these large uniform spots is to enable more accurate measurement of spot densities.[3]

The de Jong–Bouman camera and the precession camera The desirability of having a camera that would reproduce a layer of the reciprocal lattice without the distortion of the pattern which is characteristic of all Weissenberg photographs has led to the development of two ingenious instruments. The first of these, the de Jong–Bouman camera, rotates a flat

[1] Reproductions of the template are given in: M. J. Buerger, "X-ray Crystallography," Wiley, New York, 1942. "International Tables for X-ray Crystallography," vol. 2, Kynoch, Birmingham, England, 1959. They have also been marketed by the Institute of Physics, London, England, and by Polycrystal Book Service, Pittsburgh, Pa.

[2] See the preceding references of this section.

[3] E. H. Wiebenga and D. W. Smits, *Acta Cryst.*, vol. 3, p. 265, 1950, use two-dimensional shifting ("integration"), but one-dimensional is adequate for many problems: L. H. Jensen, *J. Am. Chem. Soc.*, vol. 78, p. 3993, 1956. D. A. Davies and A. M. Mathieson, *Rev. Sci. Instr.*, vol. 33, p. 1106, 1962.

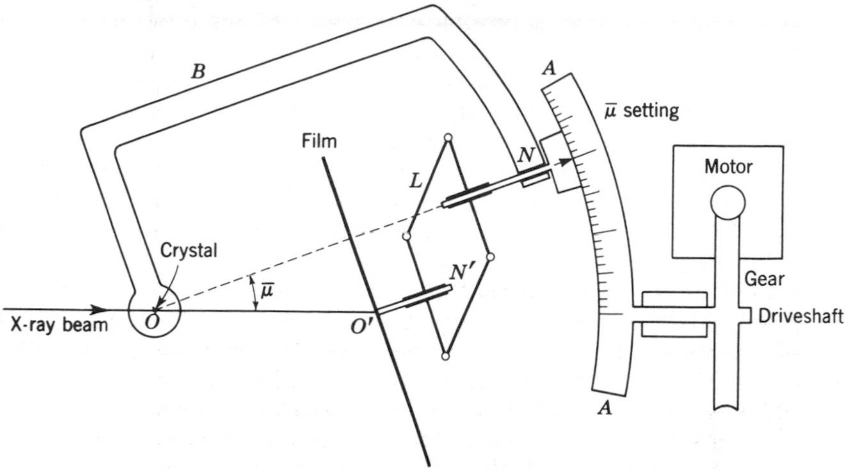

Fig. 6-10 Buerger's precession camera (schematic). Steel shafts ON and $O'N$ are maintained in parallelism by the parallelogram linkage L and arm OBN while crystal and film pivot about O and O', respectively, and precess around the x-ray beam, driven by the motor. The precession angle $\bar{\mu}$ is set at the desired angle on the arc AA. An absorbing screen (not shown), fastened to arm B, allows a single layer line of reflections to reach the film.

film in its own plane about an axis parallel to the rotation axis of the crystal and in synchronism with the latter. A shield located between the crystal and the film permits only a single layer of the reciprocal lattice to be photographed at each exposure.[1]

The second instrument is Buerger's precession camera, illustrated in Fig. 6–10. It maintains the normal to the photographic film parallel to an axis in the crystal, while the crystal axis is caused to rotate with a precessing motion.[2] The reflections from a chosen single layer of the reciprocal lattice reach the film through an annular slit in a shield which precesses with the film. Both of these cameras are most useful for crystals with large unit cells which yield patterns with a great many spots; but both have the disadvantage that all photographs made with them—except zero layer-line photographs—have an area in the center where no spots can be recorded. (Similar regions are missing also in rotating-crystal patterns, but not in equi-inclination Weissenberg patterns.) The missing spots can be recorded in a separate exposure for which the specimen is tilted differently.

[1] J. Bouman, "Selected Topics in X-ray Crystallography," Interscience, New York, 1950. References to the original papers of W. F. de Jong and J. Bouman, together with operating instructions for their camera and for the precession camera, are given in M. J. Buerger, "X-ray Crystallography," Wiley, New York, 1942. See also the following reference.

[2] M. J. Buerger, "The Photography of the Reciprocal Lattice," Murray, Cambridge, Mass., 1944; "The Precession Method," Wiley, New York, 1964. Corrections for absorption of the rays in the specimen and for geometrical factors are covered in M. J. Buerger, "Crystal Structure Analysis," Wiley, New York, 1960.

Instruments employing counters Greater precision in the measurement of intensities than is possible with photographic methods can be obtained if a counter is used. Mounting and turning the counter become quite simple if the flat-cone Weissenberg geometry is used, for then each reflection of a layer line enters the counter in turn as the counter is turned to different positions around the axis of rotation of the instrument.[1] Some instruments permit both flat-cone and normal-beam operation.[2] An attachment for a diffractometer permits single-crystal measurements to be made by bringing a desired *hkl* reflection into the counter by adjusting three angular scales after the crystal is mounted and adjusted on a standard goniometer head.[3] In another diffractometer modification, the counter is mounted on a rider that slides upward from its normal position, along an arc that is centered on the specimen. The arc lies in a plane that contains the axis of rotation of the specimen.[4]

Counter diffractometers are being designed and marketed in increasing numbers. Probably the simpler types will continue to be favored by many metallurgical and solid-state laboratories, whereas the highly automated and complex instruments will be found in the laboratories devoted to organic (particularly biochemical) crystal-structure determination, where the number of reflections to be measured in a single structure determination may run into the thousands.

Among the simpler devices are those employing the addition of automatic stepping, with controlled stepping increments and range, to standard diffractometers.[5]

Such devices are desirable for studying integrated intensities of reflections, accurate locations of diffraction peaks, and accurate determination of peak shapes, and for

[1] D. F. Clifton, A. Filler, and D. McLachlan, *Rev. Sci. Instr.*, vol. 26, p. 1024, 1951.

[2] H. T. Evans, Jr., *Rev. Sci. Instr.*, vol. 24, p. 156, 1953. Apparatus of this type has been marketed by the North American Phillips Company.

[3] One of the angles, χ, is set by sliding a block around an arc that lies in a vertical plane; another, ω, involves turning the plane of this arc around a vertical axis; and a third, ϕ, turns the crystal about an axis extending from this arc radially inward to the crystal located at the center of the arc. Detailed manipulations are given in T. C. Furnas, "Single Crystal Orienter Instruction Manual," General Electric Company, X-ray Dept., Milwaukee, Wisc., 1957, and the principles are given in T. C. Furnas and D. Harker, *Rev. Sci. Instr.*, vol. 26, p. 449, 1955. A further development of this "goniostat" incorporates 360° ϕ motion, 360° χ motion, 0.01° angular setting accuracy (or better), digitizer coupling and readout for each angle, ability to automatically program the angle setting, and minimum areas shaded by the component parts: M. H. Mueller, L. Heaton, and S. S. Sidhu, *Rev. Sci. Instr.*, vol. 34, p. 74, 1963.

[4] This design has been used for several years by W. H. Zachariasen, private communication.

[5] One example (M. H. Mueller, L. Heaton, and E. W. Johanson, *Rev. Sci. Instr.*, vol. 32, p. 456, 1961) is a stepping mechanism based on a "Slo-Syn" motor that advances 0.01 revolution per pulse of direct current applied to the windings, the pulses being supplied in preset numbers by a switching circuit. Equivalent instruments and other designs are now available commercially. Historical notes on developments in this field are included in P. P. Ewald, "Fifty Years of X-ray Diffraction," N. V. A. Oosthoek's Uitgeversmaatschappij, Utrecht, The Netherlands, 1962.

diffuse scattering studies. A steadily increasing number of laboratories are installing instruments that move crystal and counter from one diffraction peak to another automatically and record the intensity of each reflection either by reading the peak intensity or by scanning through the reflecting position so as to obtain the integrated intensity (i.e., the area under the peak in a plot of intensity vs. θ). The movements are controlled by punched cards or tapes, which are usually prepared with the aid of a computing machine.[1] The output of automatic diffractometers may be in the form of printed information, or in the form of punched cards or tape that can subsequently be used in computers with programs that apply intensity-correction factors and carry out Fourier syntheses or other computations with the data. Automatic diffractometers are particularly valuable for neutron diffraction.

Measuring intensities with counters Care must be taken if intensities measured with a counter diffractometer are to be meaningful. The counter must be uniformly sensitive over the area of the aperture receiving the diffracted ray. A small crystal in the incoming x-ray beam must be uniformly irradiated (or a large flat face of the crystal must receive and absorb *all* the beam). Every element of volume in the crystal must diffract and must send its diffracted beam into the counter during the measurement. If a stationary-crystal method is used, the x-ray source as seen from the crystal must be of uniform intensity; if a moving-crystal technique is used, the movement must be with uniform velocity or uniform step-by-step rotation (and with small enough steps). With the *moving-counter–moving-crystal technique* the counter moves through 2θ while the crystal moves through θ; the angular range is chosen large enough to accumulate the entire integrated reflection being measured. This range increases with increasing mosaic spread in the crystal, and the crystal must be rotated enough to include this spread; since the diffracted beam does not rotate during this crystal and counter rotation, the counter window must be wide enough to accommodate the *mosaic spread*. With the *moving-crystal–stationary-counter technique* similar requirements must be met, though the counter-window width must be able to accommodate the entire *wavelength spread* used (the range between the absorption edges of the balanced filters, or the spread being passed by the monochromator if one is used). With the *stationary-crystal–stationary-counter* technique, the width of the source as seen by the crystal must exceed that required by the sum of the entire mosaic spread in the crystal, the size of the crystal, and the wavelength range being used; the angular width of the counter window must exceed the width of the source as seen by the

[1] Some instruments automatically seek out the settings for peak positions so as to avoid errors in calculating these positions in advance; in general, however, this type of design runs into difficulties when weak peaks are encountered. An alternate type involves the computation of peak positions in advance by the use of measured peak positions that are spread over the entire range of angular settings involved, in which the computations and the programming of orders to the diffractometer are planned to minimize systematic and cumulative errors in the settings. Some designs also compare actual settings of the angle scale with the settings called for by the program so as to detect and warn of any malfunctioning of the circuitry. The peak intensities are measured in some designs, and integrated intensities are measured in others. A description of one instrument that employs commercially available components, and references to other designs, will be found in S. C. Abrahams, *Rev. Sci. Instr.*, vol. 33, p. 973, 1962.

crystal; and rather severe requirements must be met regarding the accuracy of the angle setting of the crystal for each reflection.[1]

It is nearly always assumed that the background intensity is the average of the backgrounds on the two sides of a peak, although in fact there is a small peak in the background under each Bragg peak. In a few instances the peak in the temperature-diffuse scattering that lies under the Bragg peak has been calculated.[2] A difficulty that has been too little recognized in the past is caused by multiple reflections (*Umweganregung*); in some types of photographs and diffractometer techniques these are common and when present they interfere with the determination of the true intensities of reflections.[3] If a reflection that has indices HKL acts as an initial beam and is again reflected, and the second reflection has indices hkl, then the doubly reflected beam appears as if it were a single reflection that has indices

$$h' = H - h \qquad k' = K - k \qquad l' = L - l$$

Strong reflections are weakened when multiple reflection occurs (because of enhanced extinction), and weak reflections are intensified (by the added multiple-reflected beams).[4]

Operation at low temperatures Most installations for low-temperature diffraction from single crystals and polycrystalline materials use conventional cameras and diffractometers modified to provide a stream of cold gas impinging on the single-crystal (or polycrystalline) specimen at a controlled velocity.[5] If the cold stream is surrounded by a stream of very dry air at room temperature, issuing from a nozzle that is concentric with the cold-gas nozzle, the danger of formation of ice on the specimen is lessened. The specimen should be mounted in such a way that the conduction of heat from the surroundings is negligible.

Cooling can be provided very simply if the specimen is merely bathed continuously in a stream of cold liquid dripping over it from an insulated reservoir. Refrigerated alcohol or liquid nitrogen serves well in this way. A dry atmosphere around the specimen can be provided by evaporating liquid nitrogen.

[1] L. E. Alexander and G. S. Smith, *Acta Cryst.*, vol. 15, p. 983, 1962, have published a discussion of the requirements.

[2] N. Nilsson, *Arkiv Fysik*, vol. 12, p. 247, 1956. D. R. Chipman and A. Paskin, *J. Appl. Phys.*, vol. 30, p. 1998, 1959.

[3] H. Cole, F. W. Chambers, and H. M. Dunn, *Acta Cryst.*, vol. 15, p. 138, 1962. W. H. Zachariasen, *Acta Cryst.*, vol. 18, p. 205, 1965.

[4] If a crystal is oriented with a symmetry axis parallel to, or a symmetry plane perpendicular to, the axis of crystal rotation, multiple reflections are likely to be common.

[5] E. F. Kaelble (ed.), "Handbook of X-rays," McGraw-Hill, 1966. H. S. Peiser, H. P. Rooksby, and A. J. C. Wilson, "X-ray Diffraction from Polycrystalline Materials," Institute of Physics, London, 1955. A. R. Ubbelohde and I. Woodward, *Proc. Roy. Soc.* (*London*), vol. A185, p. 448, 1946. J. Thewlis and A. R. Davey, *J. Sci. Instr.*, vol. 32, p. 79, 1955. H. S. Kaufman and I. Fankuchen, *Rev. Sci. Instr.*, vol. 20, p. 733, 1949. B. Post, R. S. Schwartz, and I. Fankuchen, *Rev. Sci. Instr.*, vol.22, p. 218, 1951. D. F. Clifton, *Rev. Sci. Instr.*, vol. 21, p. 339, 1950.

The widest range of temperatures and the most accurate control at any arbitrary temperature is possible with apparatus in which the specimen is cooled by conduction from a bath of cryogenic liquid. For operation at temperatures above the boiling point of the liquid, a higher-resistance link can be inserted between the specimen and the coolant, and the specimen can be warmed electrically.[1]

Operation at high temperatures The most common x-ray diffraction instruments for use at high temperatures are powder cameras, referred to in Chap. 7. However, many of the furnaces and control devices are also applicable to rotating-crystal cameras and single-crystal diffractometers. Extensive reviews of the many designs have been published.[2]

Bond's method for precision lattice-constant determination A method of determining lattice constants has been presented by Bond[3] which minimizes various errors and permits the attainment of remarkably high precision in the determination of unit cell dimensions in favorable cases. The method requires that the sample be a crystal of *high perfection*, but when this condition is met and high-angle reflections are used, it should be possible to measure lattice constants with an accuracy comparable to the accuracy with which x-ray wavelengths are measured; experience suggests that errors can be reduced to a few parts per million.

The method is based on a measurement of two angular settings of the reflecting crystal, indicated by the solid lines and dashed lines, respectively, of Fig. 6–11, which send the center of the diffraction peak into a counter located first on one side, then on the other side of the direct beam—or better, into a *pair* of counters mounted at the positions shown. The counter windows are made rather large and the counter angle settings are not used in the method since the measurements are based entirely on the crystal angle readings R_1 and R_2, indicated in the figure. A highly perfect crystal with a face parallel to the atomic plane being used is placed with the reflecting surface on the axis of rotation of a counter diffractometer and is aligned accurately parallel to this axis. The direct beam is accurately perpendicular to the axis. The diffraction angle θ is obtained from the scale

[1] Kaelble, loc. cit. Clifton, loc. cit. B. A. Calhoun and S. L. Abrahams, *Rev. Sci. Instr.*, vol. 24, p. 397, 1953. R. Keeling, B. C. Frazer, and R. Pepinsky, *Rev. Sci. Instr.*, vol. 24, pp. 1087–1095, 1953. E. O. Wollan, W. C. Koehler, and M. K. Wilkinson, *Phys. Rev.*, vol. 110, p. 638, 1958. I. A. Black et al., *J. Res. Natl. Bur. Std.*, vol. 61, p. 367, 1958. F. A. Mauer and L. H. Bolz, *J. Res. Natl. Bur. Std.*, vol. 65c, p. 225, 1960. D. N. Batchelder and R. O. Simmons, *J. Chem. Phys.*, vol. 41, p. 2324, 1964; private communication, 1965. See also Chap. 7.

[2] H. J. Goldschmidt in H. S. Peiser, H. P. Rooksby, and A. J. C. Wilson (eds.), "X-ray Diffraction from Polycrystalline Materials," chap. 9, Institute of Physics, London, 1955. W. M. Mueller (ed.), "Advances in X-ray Analysis," vol. 5, Plenum, New York, 1962. E. F. Kaelble (ed.), "Handbook of X-rays," McGraw-Hill, New York, 1966. A. Taylor, "X-ray Metallography," Wiley, New York, 1961. H. P. Klug and L. E. Alexander, "X-ray Diffraction Procedures," Wiley, New York, 1954.

[3] W. L. Bond, *Acta Cryst.*, vol. 13, p. 814, 1960.

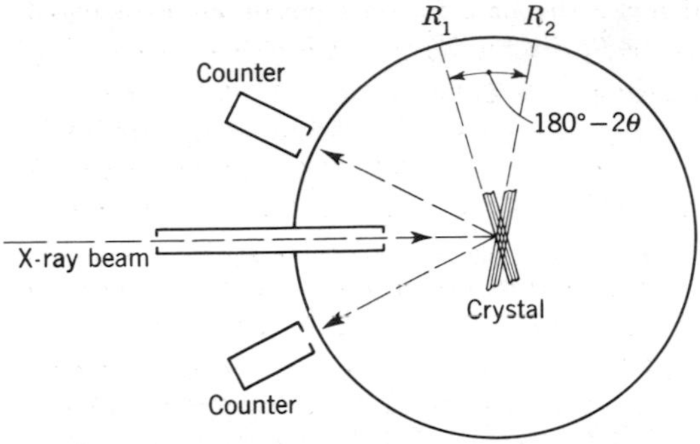

Fig. 6-11 Single-crystal diffractometer as used by Bond for precision lattice-constant measurement. The crystal is turned from angle setting R_1 to R_2 to give maximum reflection from the same plane in each position; the angle measured is R_1–R_2, not the angle between counter settings. A crystal free of subboundaries is required.

readings by one of the following relations, depending upon the position of the point $R = 0$:

$$\theta = 90° - |R_2 - R_1|/2 \text{ if } R = 0 \text{ is not between } R_1 \text{ and } R_2$$

$$\theta = |R_2 - R_1|/2 - 90° \text{ if } R = 0 \text{ is between } R_1 \text{ and } R_2$$

Bond's analysis of the errors involved shows that the following must be considered:

1. If the reflecting plane is tilted by a small angle Δ from parallelism with the rotation axis, the resulting error cannot be determined from a lack of consistency of d spacings computed from different orders of reflection. However, spherical trigonometry involving the apparent Bragg angle θ' and the tilt angle Δ gives the true angle from the relation $\sin \theta = \cos \Delta \sin \theta'$ and putting this into Bragg's law gives

$$d = n\lambda/(2 \cos \Delta \sin \theta')$$

From this it follows that for an accuracy in d of say 1 part in a million, Δ must not exceed 4.8 min. The collimator must be normal to the axis of rotation to the same angular accuracy.

2. Divergence of the incident beam causes a small error that can be evaluated by averaging the expression for the tilt error between the limits $\pm\Delta_0$, which gives $\sin \bar{\theta} = \sin \theta \, (1 + \Delta_0^2/6 + \cdots)$ and leads to the expression for the correct d spacing:

$$d = d'(1 + \Delta_0^2/6)$$

where d' is the apparent, uncorrected d spacing. If slits are 1 mm long and 215 mm apart ($\Delta_0 = 1/215$), the correction would amount to 3.5 parts per million.

3. Errors from penetration of the rays into the crystal and for displacement of the crystal from the axis of rotation do not enter, provided the crystal is perfect; however, if a range of orientation is present in the crystal these can be very serious—and even may go unrecognized. Therefore, it is important to test a crystal for substructure and distortion.

4. In high-precision work a correction must be made for the index of refraction of the rays in the crystal; the correction is made by the equation[1]

$$d = d'(1 + \delta/\sin 2\theta)$$

where δ is unity minus the refractive index;

$$\delta = ne^2\lambda^2/(2\pi mc^2)$$

where λ is the wavelength of the characteristic radiation used, e is the charge on the electron, m is the mass of the electron, and n is the number of electrons per cm^3 in the crystal. This may be used in the form

$$\delta = 4.48 \times 10^{-6} n_0 \lambda_0^2$$

where λ_0 is the wavelength in kX units and n_0 is the number of electrons per cubic kX unit.

5. The Lorentz-polarization factors distort the diffraction peaks to a degree that may become appreciable at high values of θ. A correction for this is

$$x = w^2\{\cot 2\theta/4 + \sin 4\theta/[2(1 + \cos^2 2\theta)]\}$$

where w is the width of the peak at half maximum.
The absolute value of x is to be added to θ.

6. The errors in reading θ alter the spacing computation to an extent that is seen by differentiating Bragg's law:

$$\Delta\theta = -(\Delta d/d) \tan \theta$$

In one example the angular width at half maximum was about 700 sec. The center of such a line must be found to 2.4 sec if $\theta = 85°$ and if an accuracy of 1 part in a million is to be attained; similarly, a line 230 sec wide at half maximum with $\theta = 75°$ would have to be measured to 0.77 sec.[2]

[1] A. H. Compton and S. K. Allison, "X-rays in Theory and Experiment," p. 674, Van Nostrand, New York, 1935.

[2] Bond's instrument is built around a Hilger Watts microptic clinometer reading directly in seconds (but with irregular inaccuracies of the order of 5 sec, which are minimized by averaging readings obtained on different ranges of the angle scale).

7

The powder method

The powder method, devised by Debye and Scherrer[1] and independently by Hull,[2] is the most widely used method in the field of applied x-rays, and has also contributed to fundamental research in countless ways. Since entire books are available on this method alone,[3] and rather detailed summaries are given elsewhere,[4] only the more widely used instruments[5] and techniques will be presented here.

Powder cameras A good x-ray camera for taking powder photographs has several basic requirements. The film must be held accurately cylindrical and protected from light and from unwanted x-rays. There must be provision for centering the specimen at the axis of the cylindrical film and for rotating the specimen. X-rays must be directed accurately normal to the axis of the cylinder through carefully designed pinholes or slits. Scattering from the air should be minimized by carefully designed entrance and exit tubes that surround the beam throughout most of its path within the camera. Diffraction patterns caused by the beam striking the pinholes or exit ports must be prevented from reaching the film.[6] In cameras of the Debye-Scherrer type, it is convenient to make the size such that the film is bent to a radius of 57.4 or 114.8 mm, for then the circumferential distances in millimeters transform directly into θ or 2θ in degrees.[7]

[1] P. Debye and P. Scherrer, *Z. Physik*, vol. 17, p. 277, 1916; vol. 18, p. 291, 1917.

[2] A. Hull, *Phys. Rev.*, vol. 10, p. 661, 1917.

[3] H. S. Peiser, H. P. Rooksby, and A. J. C. Wilson, "X-ray Diffraction by Polycrystalline Materials," Institute of Physics, London, 1955. L. V. Azároff and M. J. Buerger, "The Powder Method," McGraw-Hill, New York, 1958.

[4] E. F. Kaelble, "Handbook of X-rays," McGraw Hill, New York, 1966. H. P. Klug and L. E. Alexander, "X-ray Diffraction Procedures," Wiley, New York, 1954. Numerous individual papers in "Advances in X-ray Analysis," vol. 1, Plenum, New York, 1957 et seq., and in "Encyclopedia of X- and Gamma-Rays," Reinhold, New York, 1964. H. Neff, "Grundlagen und Anwendung der Röntgen-Feinstruktur-Analyse," Oldenbourg, Munich, Germany, 1962. A. Guinier, "Théorie et Technique de la Radiocristallographie," Dunod, Paris, 1959.

[5] A list of marketed instruments may be found in the International Union of Crystallography's "Index of Crystallographic Supplies," 2d ed., Société de Minéralogie et de Cristallographie, Paris, 1959.

[6] M. J. Buerger, *J. Appl. Phys.*, vol. 16, p. 501, 1945. W. Parrish and E. Cisney, *Philips Tech. Rev.*, vol. 10, p. 157, 1948.

[7] Sets of transparent scales for reading d values directly from the films, when radiations from certain common targets are used, are sold by N. P. Nies, Laguna Beach, Calif. These are printed in sets of slightly varying length to account for film shrinkage; sets are also available that read shrinkage-corrected $\sin^2 \theta$ values.

Fig. 7-1 Powder cameras. (*A*) plunger for centering specimen on axis of rotating shaft, (*B*) movable finger for expanding film against cylindrical surface and clamping it there, (*C*) rotating adjustable specimen holder, (*D*) pinhole collimator for x-ray inlet, (*E*) exit tube for beam, ending in fluorescent screen and lead glass, (*F*) film. (North American Phillips Company, Inc.)

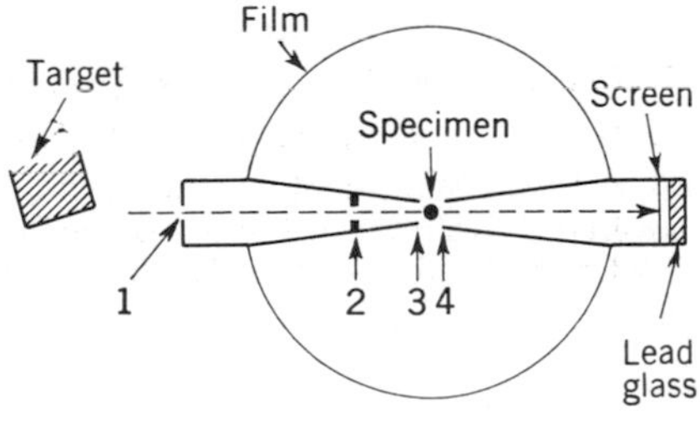

Fig. 7-2 Collimator design. Beam is collimated by openings 1 and 2 so as to avoid striking the metal at 3 and 4. Diffracted rays from the edges of 2 do not pass 3, and back scatter from screen is shielded by exit tube.

Cameras of modern design are illustrated in Fig. 7–1. The film is pressed against the inside surface of the lightproof camera; the specimen is centered on the axis of the camera by a "pusher" *A* that extends in radially to bear against the movable base of the specimen holder. The specimen is viewed through the collimator with a small magnifying lens while it is being rotated for centering. A powder camera designed by Bradley and coworkers is widely used in various sizes.[1]

Beam collimation in any camera is important. As indicated in Fig. 7–2,

[1] A. J. Bradley, W. L. Bragg, and C. Sykes, *J. Iron Steel Inst.* (*London*), vol. 141, p. 63, 1940. A. J. Bradley, H. Lipson, and N. J. Petch, *J. Sci. Inst.*, vol. 18, p. 216, 1941. H. Lipson and A. J. C. Wilson, *J. Sci. Instr.*, vol. 18, p. 144, 1941. See standard design recommendations in *J. Sci. Instr.*, vol. 22, p. 57, 1945.

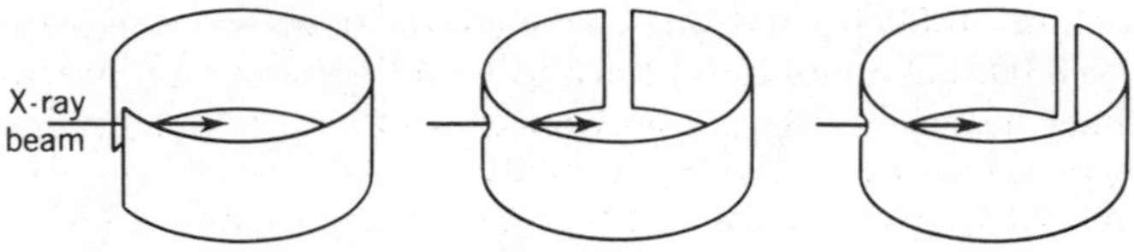

Fig. 7-3 Methods of loading films in Debye cameras. (*a*) Symmetrical, (*b*) Straumanis method, (*c*) Straumanis-Wilson method.

whether pinholes or slits are used, openings 1 and 2 should be designed to prevent the x-ray beam from striking the edges of openings 3 and 4, and inlet and exit tubes should shield a minimum of the circumference of the film. Holes are punched in the film for incoming and outgoing beams in one of the ways shown in Fig. 7–3. The methods of Fig. 7–3*b* and *c* permit the effective diameter to be determined from measurements of the Debye rings.

Specimen preparation The usual specimen for Debye cameras is a thin-walled capillary tube, 0.3 to 0.5 mm in diameter, of lithium borate glass, silica, polystyrene, or cellophane, which is filled with powder that has been passed through a sieve of about 250 or 325 mesh. Filling may be done by tapping the end into the powder or by attaching the capillary to a vacuum line and sucking the powder in until it strikes a plug of glass wool. Specimens are also prepared by mixing the powder with a small amount of collodion or other adhesive and then rolling or extruding it into small cylinders. Powders may also be coated on the outside of fibers of glass, silica, or hair. With care, patterns can be obtained when only a microgram or so of powder is available. Wires of polycrystalline metals in the recrystallized condition usually give spotty lines unless rotated. It should be remembered that a preferred orientation of grains will alter the relative intensities of the Debye lines. For rapid exposures and for materials of poorly developed crystallinity a small camera is favored, usually one of about 5 cm diameter, but for complicated patterns it is better to use larger ones.

Operation at high temperatures Diffraction patterns recorded at elevated temperatures provide information on (1) expansion coefficients, (2) transformations, (3) recovery and recrystallization, (4) thermal vibration amplitudes, and (5) equilibrium phases and equilibrium degrees of order of the atom arrangement in phases existing only at elevated temperatures.

A few camera designs will be mentioned here. Hume-Rothery and Reynolds have employed a camera in which a specimen is sealed in a capillary tube and mounted on the axis of a wire-wound furnace in a water-cooled camera body. An aluminum shield protects the film from the heat. With care, the temperature gradient in the irradiated portion of the specimen is reduced to 1° at 1000°C, and lattice constants are determined with a reproducibility of 0.0001 A.[1] Goldschmidt and Cunningham[2] employ two hemispherical furnace halves in vacuum or a gas atmosphere, which operate up to a maximum temperature in the range 1400 to 1700°C.[3] For reaching higher temperatures it is possible to use induction heating of a metal cylinder surrounding the specimen.[4] The temperature of the diffracting material is measured by thermocouples, but for calibration some workers use the known transition temperature of polymorphic compounds such as ammonium nitrate, ammonium perchlorate, potassium perchlorate, and quartz,[5] or mix silver powder with the sample and use its known expansion coefficient as a means of determining the actual temperature in the irradiated portion of the specimen. Diffraction by reactive metals at temperatures up to 2200°C is discussed by Hanak and Daane.[6] Gradual shifting of a film axially can be the basis for a camera that records temperature-induced changes—up to the equivalent of 120 separate exposures on one film in Warlimont's design.[7] Some cameras employ cylindrical or conical heaters of nickel around the specimen, which act simultaneously as filters for copper-beta radiation and as electric furnace elements.[8] In some cameras these radiant heaters are halved longitudinally or transversely to lessen absorption and scattering of the direct beam.

For use with diffractometers, a flat specimen may be placed in contact with an electric heater or hot plate[9] or surrounded by a furnace. A selected angular range can be repeatedly scanned during a heating and cooling cycle to record phase changes as they occur.[10]

Heating may also be by means of focusing heat from a remote heat

[1] W. Hume-Rothery and P. W. Reynolds, *Proc. Roy. Soc. (London)*, vol. A167, p. 25, 1938.

[2] H. J. Goldschmidt and J. Cunningham, *J. Sci. Instr.*, vol. 27, p. 177, 1950; "User Specification for High-Temperature Powder Cameras," Institute of Physics, London, 1950.

[3] Unicam Instruments Ltd., Cambridge, England, has marketed this type, and other related types are also commercially available.

[4] J. W. Edwards, R. Speiser, and H. L. Johnston, *Rev. Sci. Instr.*, vol. 20, p. 343, 1949.

[5] M. J. Buerger, N. W. Buerger, and F. G. Chesley, *Am. Mineralogist*, vol. 28, p. 285, 1943.

[6] J. J. Hanak and A. H. Daane, *Rev. Sci. Instr.*, vol. 32, p. 712, 1961.

[7] H. Warlimont, *Z. Metallk.*, vol. 50, p. 708, 1959.

[8] L. S. Dent and H. F. W. Taylor, *J. Sci. Instr.*, vol. 33, p. 89, 1956; A. Barclay and J. D. Donaldson, *J. Sci. Instr.*, vol. 38, p. 286, 1961; J. Spreadborough and J. W. Christian, *J. Sci. Instr.*, vol. 36, p. 116, 1959.

[9] L. S. Birks and H. Friedman, *Rev. Sci. Instr.*, vol. 18, p. 576, 1947. R. A. Rowland, E. J. Weiss, and D. R. Lewis, *J. Am. Ceramic Soc.*, vol. 42, p. 133, 1959. W. A. Bassett and D. M. Lapham, *Am. Mineralogist*, vol. 42, p. 548, 1957. S. Hurwitt and A. Appel, *Norelco Reporter*, vol. 8, p. 90, 1961. D. K. Smith, *Norelco Reporter*, vol. 10, p. 19, 1963.

[10] Y. Shimura, *Rev. Sci. Instr.*, vol. 32, p. 1404, 1961; commercially available from Rigaku-Denki Co., Ltd., Tokyo, Japan.

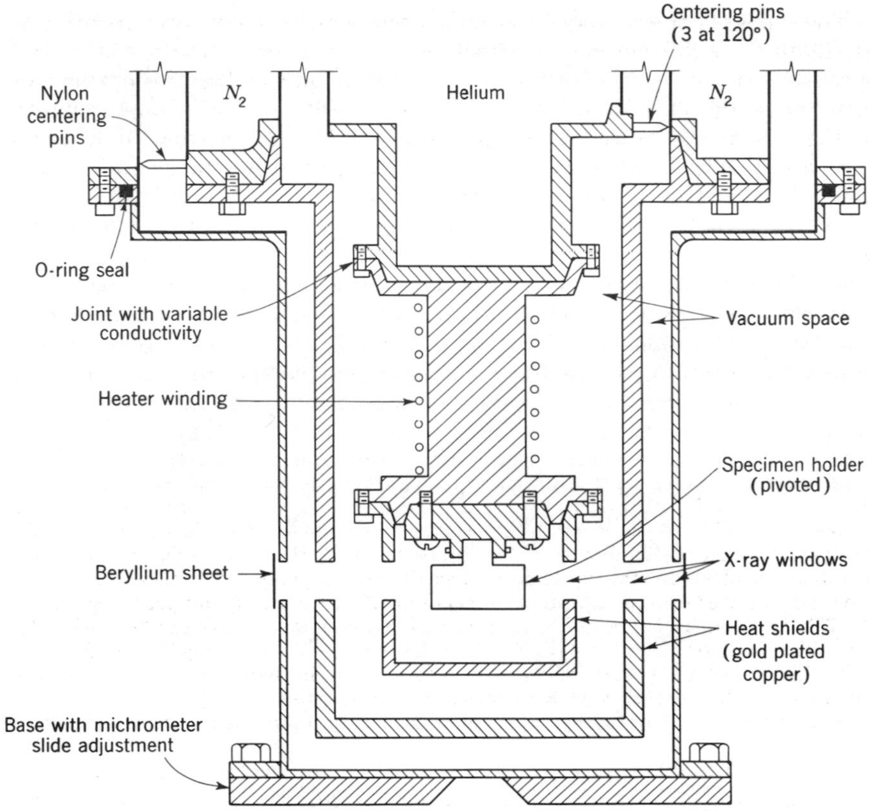

Fig. 7-4 Internal construction of the lower part of a cryostat designed for cooling a specimen on a diffractometer to liquid N_2, H_2, and He temperature ranges.

source,[1] by electron bombardment, or simply by blowing a stream of hot gas over the sample.[2]

Operation at low temperatures The principal uses of diffraction at low temperatures are as follows: (1) determining the structures of substances that are not solid at room temperature, (2) studying phase changes, (3) determining thermal contraction and precision lattice constants for use in other physical and chemical research, (4) studying freshly condensed films, (5) studying defects that would not be fully retained at higher temperatures, (6) using the shifting of lines upon cooling to aid in assigning indices to them, (7) increasing the intensity of the reflections of high θ values.

[1] Employed in an instrument made by Materials Research Corp., Yonkers, N.Y.

[2] G. K. Williamson and A. Moore, *J. Sci. Instr.*, vol. 37, p. 107, 1956.

The types of cameras most used at low temperatures have been mentioned in Chap. 6. A diffractometer attachment that permits cold-working of a specimen at low temperatures (for inducing phase transformations or introducing defects) is sketched in Fig. 7–4.[1] The vacuum chamber shown in the drawing is bolted with an O-ring seal to the bottom of a metal cryostat; a stainless-steel rod, used for scraping and hammering the specimen when cold, is brought into the chamber through a ball-and-socket joint, with flexibility being provided by a sylphon bellows and an O-ring seal.

An alternate way of cold-working and x-raying at low temperatures is to work the specimen at 78°K while immersed in liquid nitrogen and then to transfer it (while cooled with a stream of the liquid) to a low-temperature camera or diffractometer attachment in which the specimen is cooled by flowing a small stream of cryogenic liquid over it.[2]

An effective device for cooling a specimen by a cold liquid or gas has been perfected by Shull[3] and used by him in his neutron diffraction work at helium temperatures. A stream of coolant liquid flows through pipes to the specimen. A $\frac{3}{32}$-in. tube with 0.010-in. wall carries the liquid to the specimen from a storage flask; the exiting cold vapor leaves the specimen through a $\frac{1}{4}$-in.-diameter tube concentric with the liquid tube, which serves as a radiation shield for the liquid tube. A small vacuum pump acting through a needle valve controls the flow rate; a liquid helium flow of 45 cc per hour maintains the specimen at 4.2°K for about 12 hours with a 1.5-liter flask. Other temperatures can be reached by controlling the pressure above the liquid in the storage vessel, using a second needle valve. The transfer tubes are enclosed in a tube $\frac{7}{16}$ in. in diameter which serves as a radiation shield, and this in turn is enclosed in a vacuum tube $\frac{11}{16}$ in. in diameter; nylon star-shaped spacers are used to separate the concentric tubes from each other.

Operation at elevated pressures X-ray diffraction provides more meaningful and more fundamental information regarding structural changes at high pressures than any other methods currently being used, such as measurements of volume and of electrical resistance. Pressures up to 5000 kg per sq cm were attained in the 1930s and 1940s[4]; higher pressures were obtained by constructing miniature bombs out of beryllium and diamond crystals,[5] the beryllium serving up to 10,000 atmospheres and the diamonds up to about 25,000 atmospheres. The specimen was contained in a small drilled hole about 0.015 in. in diameter and compressed by a piston. Photo-

[1] C. S. Barrett, *Acta Cryst.*, vol. 9, p. 621, 1956 (designed by E. A. Long, L. Meyer, C. B. Walker, and C. S. Barrett).

[2] K. Lonsdale and H. Smith, *J. Sci. Instr.*, vol. 18, p. 133, 1941. D. F. Clifton, *Rev. Sci. Instr.*, vol. 21, p. 339, 1950. Liquid H_2 is not advisable for this work without special precautions, as it is explosive when mixed with air throughout the concentration range 4 to 96 percent.

[3] C. G. Shull, private communication, 1963.

[4] R. B. Jacobs, *Phys. Rev.*, vol. 54, p. 325, 1938; vol. 56, p. 211, 1939. For a review of the early methods, see J. C. Jamieson and A. W. Lawson in R. H. Wentorf, Jr. (ed.), "Modern Very High Pressure Techniques," p. 70, Butterworth, London, 1962.

[5] A. W. Lawson and N. A. Riley, *Rev. Sci. Instr.*, vol. 20, p. 763, 1949. A. W. Lawson and T. Y. Tang, *Rev. Sci. Instr.*, vol. 21, p. 815, 1950.

graphs were obtained up to 60,000 atmospheres by compressing a specimen between two diamond pistons and passing the x-ray beam through the diamonds.[1]

Still higher pressures have been reached in cameras in which a small specimen is surrounded by an amorphous or nearly amorphous material acting as a pressure gasket between compression plates. The plates are squeezed until the tendency of the material to extrude outward between the plates is halted by friction. Operating pressures up to about 250,000 atmospheres have been reached between anvils ("Bridgman anvils") of cemented tungsten carbide with gaskets of amorphous boron and have been extensively used by Jamieson in x-ray studies of phase changes.[2] A gradient of pressure throughout the specimen is to be expected in this type of camera, leading to difficulties in interpreting the patterns of some materials. Preferred orientations may be induced as the sample flattens out, which make intensity measurements unreliable. An added difficulty encountered with these devices arises from sample compressibilities; a given applied load generates different pressures in different samples, making it necessary to recalibrate the apparatus for each sample material.

These difficulties with Bridgman-anvil devices are minimized in designs in which the sample is contained in a three-dimensional rather than a two-dimensional space. Of the various modifications possible, the most desirable ones for x-ray work are judged to be multianvil types; Barnett and Hall[3] have designed a tetrahedral-anvil press that permits x-ray powder diffractometry at pressures up to at least 75,000 atmospheres with temperatures up to 1000°C.

The fundamental principles of the press are similar to those in Hall's original tetrahedral press:[4] four rams of 600-ton capacity press four tungsten-carbide anvils against the sample chamber, which is a tetrahedron 1 in. on an edge molded from material relatively transparent to x-rays, such as LiH or 50–50 mixtures of LiH and boron. X-rays enter the chamber through a hole along the axis of a ram and a hole through the anvil. Diffracted rays (in the θ range 0 to 110°) pass out through a compressible gasket and a series of slits to a scintillation counter. The heating element and thermocouple are molded into the tetrahedral cell together with the x-ray sample. The patterns obtained are of a quality adequate to permit structure analysis, at least for relatively simple structures. For example, tin II is found at 34 kb to have a b.c.t. lattice with $a = 3.81$ A and $c = 3.48$ A, with single atoms at each lattice point.

Focusing cameras Cameras in which the slits, specimen, and film all lie on the circumference of the same cylinder, so that focusing of the diffracted rays occurs, have been much used since their development by

[1] G. J. Piermarini and C. E. Wier, *J. Res. Natl. Bur. Std.*, vol. A66, p. 325, 1962.

[2] J. C. Jamieson and A. W. Lawson, *J. Appl. Phys.*, vol.33, p. 776, 1962.

[3] J. D. Barnett and H. T. Hall, *Rev. Sci. Instr.*, vol. 35, p. 175, 1964.

H. T. Hall, *Rev. Sci. Instr.*, vol. 29, p. 267, 1958.

Seeman and Bohlin.[1] They offer shorter exposure times and higher dispersion than do Debye cameras of comparable size. Westgren and Phragmén in their extensive studies of alloy constitution used three separate focusing cameras to cover the range from 16 to 82° in order to obtain optimum conditions in each range,[2] but a fairly wide range can be covered adequately with a single camera. It is well to have the specimen mounted on an oscillating arm, pivoted at the axis of the cylinder, in order to give smoother diffraction lines. For precision lattice-constant determination, it is desirable to use a camera in which the film is placed symmetrically around the incoming beam, i.e., in the symmetrical back-reflection position.[3] In symmetrical cameras, θ can be determined from measurements of a Debye ring on two sides of the direct beam, but in other cameras θ is measured from the shadow of a knife edge and corrected for film shrinkage. Single-emulsion films should be used (or one-side development of double-emulsion films) because of the oblique incidence of the diffracted rays. It is customary to calibrate unsymmetrical cameras with the diffraction pattern of standard samples. The geometry for focusing has been discussed by Brentano.[4] An approximation to true focusing (*parafocusing*) is employed in most diffractometers: a flat plate of powder is located on the axis of rotation and is turned at half the angular rate of the counter. Approximate focusing is also obtained with a back-reflection camera by using flat films or plates; by setting the pinhole collimator at the proper position in such a camera it is possible to focus one line at a time (the pinhole, specimen, and line on film must lie on the circumference of a circle for this). When a beam is taken from a fine-focus tube[5] at a low angle so that the apparent width of the focus is only about 5 μ, it can be used with a miniature Debye camera or a focusing camera to give extremely sharp lines; the filtered Cu $K\alpha$ doublet diffracted by a ZnO sample has been resolved down to $\theta = 18°$ in a 2-min exposure.[6]

Monochromators The use of a crystal monochromator with a powder camera or diffractometer greatly increases the contrast between the diffraction lines and the background, but when a flat-crystal monochromator

[1] H. Seeman, *Ann. Physik*, vol. 59, p. 455, 1919. H. Bohlin, *Ann. Physik*, vol. 61, p. 421, 1920.

[2] A. F. Westgren, *Trans. AIME*, vol. 93, p. 13, 1931.

[3] E. R. Jette and F. Foote made good use of this type in their alloy studies, *J. Chem. Phys.*, vol. 3, p. 605, 1935.

[4] J. C. M. Brentano, *J. Appl. Phys.*, vol. 17, p. 420, 1946; *Proc. Phys. Soc. (London)*, vol. 37, p. 184, 1925; vol. 49, p. 61, 1937. Brentano developed a cylindrical camera that focuses rays from a curved surface that oscillates at the center of the camera.

[5] W. Ehrenberg and W. E. Spear, *Proc. Phys. Soc. (London)*, vol. B64, p. 67, 1951; available from Hilger and Watts, Ltd., London (Jarrel-Ash Co., Newtonville, Mass., agents).

[6] H. S. Peiser, H. P. Rooksby, and A. J. C. Wilson, "X-ray Diffraction from Polycrystalline Materials," Institute of Physics, London, 1955.

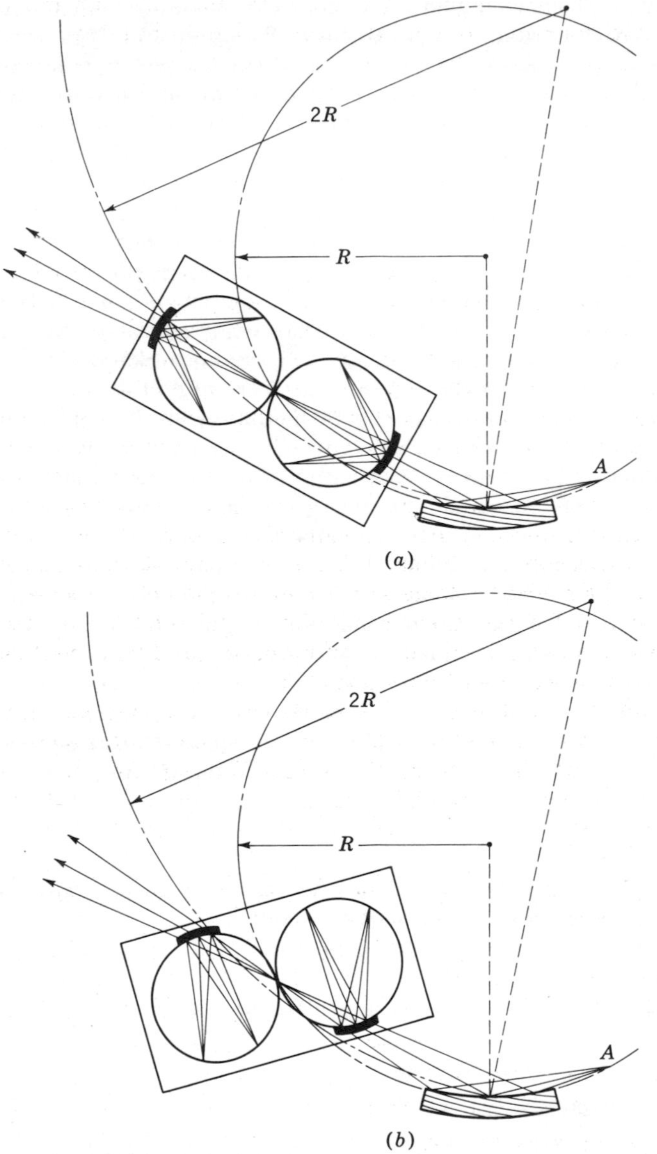

Fig. 7-5 Focusing geometry for a Guinier-Jagodzinski double-cylinder camera operating with a bent crystal monochromator of quartz. (*a*) two cameras in symmetrical position; (*b*) in asymmetrical position. R = radius of curved surface of monochromator. $2R$ = radius of curvature of lattice planes. A = line focus (normal to paper). (A. Taylor, "X-ray Metallography," Wiley, New York, 1961.)

is used with a Debye-Scherrer camera the exposure times become much longer than usual. With focusing cameras, on the other hand, the advantages of monochromatized radiation may be had without excessive exposure times. The converging rays from the monochromator can be brought to a focus at the entrance slit of the focusing camera, or can be transmitted through the specimen and come to a focus on the film; the different possible arrangements will be seen in the individual cameras of Fig. 7–5. To cover the entire range of θ values effectively, it is desirable to use a set of cameras, and by building a pair of cameras together, as indicated in the figure, the various ranges are covered by the two cameras operating simultaneously.[1] When used with an x-ray tube with a narrow line focus this camera pair permits isolation of the $K\alpha_1$ line from the $K\alpha_2$ (cut off by the slit system), and with exposures of an hour or two, gives high-resolution photographs over the full θ range. The intermediate range can be covered in two ways, as indicated; one gives maximum, the other minimum spread between $K\alpha_1$ and $K\alpha_2$ (i.e., additive, or minimal, wavelength dispersion). Another advanced form of focusing camera is based on the fundamental designs of Guinier[2] and De Wolff.[3] It combines an elastically bent monochromator crystal with four thin transmission focusing cameras stacked side by side.

The idea of using elastically curved crystals for such purposes was introduced by Johann[4] and Cauchois[5]; Johansson[6] showed that better focusing and intensity are obtained if the monochromator crystal—say a flat slab of quartz—is ground to a cylindrical surface of radius $2R$ and then bent so that the ground face is concave with a radius of R.[7]

Still greater intensity is obtainable if a monochromator is plastically bent about two axes so as to focus in vertical as well as horizontal planes. For studies of diffuse scattering, which require maximum-intensity monochromatic beams but do not require high angular resolution in the diffraction pattern, effective monochromators may be made by plastically bending

[1] E. G. Hofmann and H. Jagodzinski, *Z. Metallk.*, vol. 9, p. 601, 1955; manufactured by R. Seifert and Co., Hamburg, Germany, and perhaps others.

[2] A. Guinier, *J. Sci. Instr.*, vol. 22, p. 139, 1945; *Ann. Phys.*, vol. 12, p. 161, 1939; "Radiocrystallographie," Dunod, Paris, 1945, 1956.

[3] P. M. De Wolff, *Acta Cryst.*, vol. 1, p. 207, 1948; available from N. V. Nederlandsche Röntgen Apparatenfabriek ("Nonius") Evershed-Enraf, Delft, The Netherlands.

[4] H. H. Johann, *Z. Physik*, vol. 69, p. 185, 1931.

[5] Y. Cauchois, *J. Phys. Radium*, vol. 3, p. 320, 1932.

[6] T. Johansson, *Z. Physik*, vol. 82, p. 507, 1933.

[7] For example, a quartz plate cut to the face $(10\bar{1}1)$ may be ground to 600 mm radius and bent around a 300-mm cylindrical support; this reflects Cu $K\alpha$ with an angle of incidence of 13° 21′ to a focus 142 mm from the crystal. This may also be used for Fe $K\alpha$. For Mo $K\alpha$ the corresponding radii are 1200 and 600 mm. Crystal monochromators are sometimes equipped with means of varying the curvature so as to focus different wavelengths. W. A. Wooster, G. N. Ramachandran, and A. R. Lang, *J. Sci. Instr.*, vol. 26, p. 156, 1949. Crystals cut and bent asymmetrically are also used: A. Guinier, *Compt. Rend.*, vol. 223, p. 31, 1946.

to a certain doubly curved contour.[1] Optimum conditions then may require that the focus in the horizontal plane be at a different place than the focus in the vertical plane.

Undesired wavelengths from monochromators Most monochromators supply, in addition to the desired radiation of wavelength λ, one or more components with wavelengths that are submultiples of λ: particularly $\lambda/2$ and sometimes $\lambda/3$. This occurs because the reflecting angle for each of these components of the beam is the same when $\lambda/2$ and $\lambda/3$ components reflect in the second and third orders, respectively; submultiples of higher-order reflections can also occur. This feature is of importance when intensities are being measured, for it leads to false apparent intensities, and may even produce apparent reflections from planes having a true reflecting power of zero. The voltage on the x-ray tube can be used to discriminate against these components, or balanced filters or a monochromator crystal for which the second-order reflection of some plane has near-zero intensity can be used (see Chap. 3).

Choice of radiation Several factors should be considered in choosing the characteristic radiation to be used. Long-wavelength radiation, such as Cr $K\alpha$ and Fe $K\alpha$, spreads the diffraction pattern over a wide range of diffraction angles but limits the number of reflections recorded, while short-wavelength radiation, such as Mo $K\alpha$, gives more reflections but compresses the pattern. A compromise frequently chosen is Cu $K\alpha$. For precision lattice-constant determinations, it is best to choose $K\alpha$ or $K\beta$ radiation that will produce reflections at very high θ values.

It is important to avoid as much as possible the effects of fluorescent radiation coming from the irradiated specimen. If the $K\alpha$ radiation from the x-ray tube is of shorter wavelength than the K absorption edge of the specimen, a fluorescing occurs that may submerge even the strongest lines of the pattern; therefore, the general rule is to choose a $K\alpha$ radiation of wavelength *longer* than the K absorption edge of the principal constituents of the specimen. An alternate technique sometimes used is to choose Mo $K\alpha$ radiation, which excites fluorescence in many of the elements, but then to place a filter between the sample and the counter or film. In this location a filter of zirconium or niobium removes not only the Mo $K\beta$ radiation but also much of the fluorescent radiation from the specimen.

Radiations of Cr, Mn, Fe, Co, Ni, Cu, Mo are used the most, and some use has been made occasionally of Ag and Rh, and also of Au $L\alpha_1$ and $L\alpha_2$ with Au $L\beta$ and $L\gamma$ lines removed with a Ga filter.

Diffractometers The photographic methods of recording diffraction patterns are now supplemented in almost every x-ray laboratory by the x-ray diffractometer. Several manufacturers have placed on the market well-stabilized x-ray sources, counter tubes, counting circuits, and recording

[1] D. R. Chipman, *Rev. Sci. Instr.*, vol. 27, p. 164, 1956. L. H. Schwartz, L. A. Morrison, and J. B. Cohen in "Advances in X-Ray Analysis," vol. 7, p. 281, Plenum, New York, 1963. H. J. Garrett and H. A. Lipsitt, *Rev. Sci. Instr.*, vol. 32, p. 942, 1961.

devices combined in convenient units with diffractometers of reasonably high accuracy. Operating manuals and short courses of instruction by the manufacturers have speeded the adoption of the instruments. Some electrical circuits permit discrimination against unwanted counts, such as the count from radiation having submultiples of the desired wavelength and the counts from cosmic rays and fallout. Automatic sweep scanning, step-by-step scanning, and oscillation through a chosen angular range are possible with some of the instruments, and some provide for both powder work and single-crystal work.

The value of the diffractometer in many applications consists in the possibility of scanning a diffraction pattern or the most significant portion of a pattern in a short time and providing an easily read chart of the results. In other problems it is used to determine θ values of the powder pattern with precision—but not necessarily with greater precision than can be obtained with powder photographs. However, for equivalent effort and time on commercially available equipment, a higher precision is obtainable with the diffractometer.[1] For the accurate measurement of intensities of the reflected beams, the diffractometer excels all photographic methods.

Diffractometers are much used for qualitative and quantitative chemical analysis,[2] as in problems of inspection and control of industrial materials. They are almost indispensable in the analysis of line profiles such as that required in studies of defects in crystals, microstresses, and particle size. They are useful in preferred orientation determination (Chap. 9) and in crystal-structure analysis, particularly when precise intensity data are required. The ability of the diffractometer to record the changing intensity of a diffraction line rapidly while changes are occurring in the sample is a valuable asset in following chemical reactions and phase changes as a function of time.

Details of the construction and operation of the commercially available instruments are given in the manufacturers' literature and vary for the different instruments; we are concerned here with the fundamental principles common to all.

Characteristic radiation diverging from a slit at the x-ray tube (or from the line focus in the tube) falls on a flat plate of powder and is received by a slit at the counter; the two slits and the specimen are located on the circumference of a circle in order that the conditions of parafocusing will be met, as indicated in Fig. 7–6. The parafocusing is retained by having the plane of the powder rotate at half the angular speed of the counter, i.e., retaining a $\theta/2\theta$ relationship with each other. The angular divergence of the rays leaving the slit at the x-ray tube should be under the control of the operator, since he chooses wide divergence when he wants high intensities and rapid scanning rates, or he chooses narrow slits, more limited divergence, and slow scanning when he wants the sharper diffraction

[1] L. F. Vassamillet and H. W. King, *Advan. X-Ray Analy.*, vol. 6, p. 142, 1962.

[2] C. L. Christ, R. Bowling Barnes, and E. F. Williams, *Anal. Chem.*, vol. 20, p. 789, 1948 (organic materials). J. L. Abbott, *Iron Age*, Feb. 13, 20, 27, 1947 (metallurgical applications). See later section of this chapter.

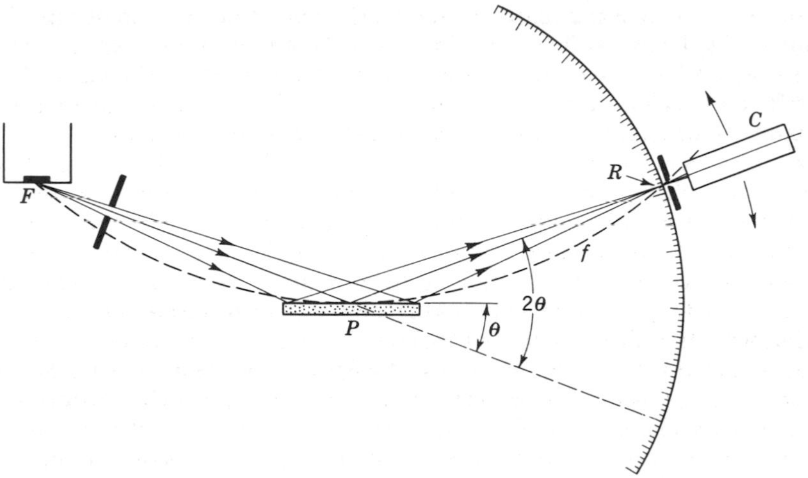

Fig. 7-6 Focusing geometry of a diffractometer. F **= focal spot.** P **= powder specimen.** f **= focusing circle.** R **= receiving slit.** C **= counter.**

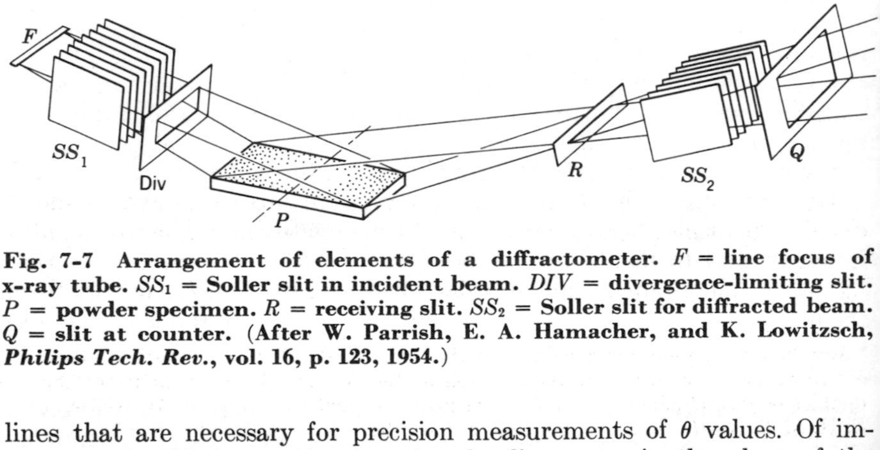

Fig. 7-7 Arrangement of elements of a diffractometer. F **= line focus of x-ray tube.** SS_1 **= Soller slit in incident beam.** DIV **= divergence-limiting slit.** P **= powder specimen.** R **= receiving slit.** SS_2 **= Soller slit for diffracted beam.** Q **= slit at counter. (After W. Parrish, E. A. Hamacher, and K. Lowitzsch, *Philips Tech. Rev.*, vol. 16, p. 123, 1954.)**

lines that are necessary for precision measurements of θ values. Of importance in this connection is not only divergence in the plane of the drawing (Fig. 7–6) but also divergence out of the plane of the drawing, for this also tends to broaden the diffraction lines. To minimize the effect of the latter divergence and still retain high intensities, it is common to use *Soller slits*, as shown in Fig. 7–7. These consist of thin metallic sheets stacked parallel to each other and separated by spacers. The rays passing through the stack are only those lying within $\frac{3}{4}$ to 1° of perpendicularity to the axis of the diffractometer.

To be properly aligned, a diffractometer must have the beam-defining slits at tube and counter accurately parallel to the axis of the diffractometer and must have the Soller slits normal to the axis; the tube slit should be uniformly filled with radiation, and the portion of the counter that receives the diffracted beam should have a uniform sensitivity to the rays. The center of the incident radiation should pass over the axis of the diffractometer, and the face of the powder specimen should lie on this axis.

For experiments in which it is necessary to reduce the background as much as possible, balanced filters or a monochromator can be inserted in the path of the rays. The beam from a curved-crystal monochromator (see page 127) can be focused at the first slit of Fig. 7–6, or the rays passing through the second slit and diverging toward the counter can be reflected from a curved-crystal monochromator placed between the slit and the counter, with the slit at its focus.[1] Increased accuracy in the zero θ reading can be obtained if the counter arm can swing through a range of angles on *both* sides of the position for which $\theta = 0$. A rate meter which registers the counting rate is connected to a strip-chart recorder, with synchronous motor drives such that marked intervals on the chart correspond to degrees on the 2θ scale on the diffractometer circle. A scaler is also needed to allow the actual count in accurately timed intervals to be read off or printed out.

Many specific operating suggestions will also be found in the recent volumes of *Advances in X-ray Analysis*, *Advances in X-ray Diffractometry and X-ray Spectrography*,[2] and the *Norelco Reporter*, as well as in the books cited at the beginning of this chapter.

Sample preparation for the diffractometer In preparing samples for the diffractometer, attention should be given to particle size, sample thickness, preferred orientation, strain and cold-working, and surface flatness. A common technique is to employ a rectangular metal plate with a hole cut through it, and to tamp the powder into the hole.[3]

Preferred orientation in the particles will alter intensities and should be avoided as much as possible. Particle size in the sample must be small if intensity measurements are to be reproducible and accurate.[4] Surface flatness is important, for an irregular surface on the sample contributes to

[1] Any fluorescent radiation generated at the sample and radiation emanating from radioactive samples are then removed, as well as the $K\beta$ line. Either member of the $K\alpha$ doublet may be eliminated by the use of precisely positioned slits. D. M. Koffman and S. H. Moll, *Norelco Reporter*, vol. XI, p. 95, 1964.

[2] A book of reprints of selected journal articles, edited by W. Parrish, published by Centrex Publishing Company, Eindhoven, the Netherlands, 1962. Other volumes edited by W. Parrish and M. Mack, *Data for X-ray Analysis*, are in preparation.

[3] C. L. McCreery, *J. Am. Chem. Soc.*, vol. 32, p. 141, 1949. H. P. Klug and L. E. Alexander, "X-ray Diffraction Procedures," Wiley, New York, 1954. With a glass cover slide temporarily covering one side of the cavity, powder is tamped in from the other side. Small amounts of powder can be held in place by using dilute collodion, or sifted onto a glass slide wet with grease or adhesive.

[4] L. Alexander, H. P. Klug, and E. Krummer, *J. Appl. Phys.*, vol. 19, p. 742, 1948. P. M. De Wolff, J. Taylor, and W. Parrish, *J. Appl. Phys.*, vol. 30, p. 63, 1959.

broadening and displacing of peaks, and the increased absorption of the rays in the projecting portions of the sample causes incorrect intensities.[1]

Sample thickness should be such as to give reflections of full intensity, if possible. Usually the specimen thickness need not exceed 1 mm, but it depends upon the linear absorption coefficient μ, the true density of a particle, ρ, and the actual density of the powder compact, ρ'. For maximum intensity at Bragg angle θ from a flat sample, the thickness t must satisfy the relation[2]

$$t \geqq \frac{3.2}{\mu} \frac{\rho}{\rho'} \sin \theta$$

Cold work introduced into metallic materials when a sample is prepared by filing can cause severe broadening of peaks. The residual strains left in brittle materials by crushing and grinding are small, although not always negligible. For most studies the strains should be removed by an annealing treatment, and to minimize the final effective size of the crystallites, the annealing is usually done at a temperature that will permit recovery of the peak sharpness without causing recrystallization and grain growth (if the sample permits such separation of recovery from recrystallization). The possibility that the annealing may cause other changes in the specimen should always be considered—there may be contamination, selective volatilization, solution or precipitation of phases, etc.

Intensity measurements with the diffractometer Fig. 7–8 is a strip-chart record of a diffraction pattern showing typical peaks, background, and $K\alpha$ doublet resolution representative of good conditions and slow scanning rates. The height above background is often assumed to be proportional to the integrated intensity in charts of this type; however, a peak-height measure of intensity, while it is useful in much of the ordinary work with powder patterns, should be regarded as approximate only.

The most accurate intensity measurements must be made with the *integrated intensity* technique. For this, the area under a peak and above background intensity may be measured with a planimeter on a linear-scale strip-chart record. Or better, the total count on a scalar may be obtained when the diffractometer is turned at a uniform rate from one side of a peak through the peak to the other side, either continuously or stepwise. The scanning is started and ended far enough from the peak to avoid missing any appreciable part of the total area under the peak, and the total integrated count is corrected by subtracting the background from it. For lack of a convenient and precise way of making the background correction, the assumption is nearly always made that the background intensity at all points underneath a peak may be taken as a straight line drawn between

[1] B. W. Batterman, D. R. Chipman, and J. J. DeMarco, *Phys. Rev.*, vol. 122, p. 68, 1961, have shown that tests for surface roughness can be made with fluorescence radiation, comparing a powdered sample with a highly polished solid sample of the same material.

[2] H. Klug and L. Alexander, "X-ray Diffraction Procedures," Wiley, New York, 1954.

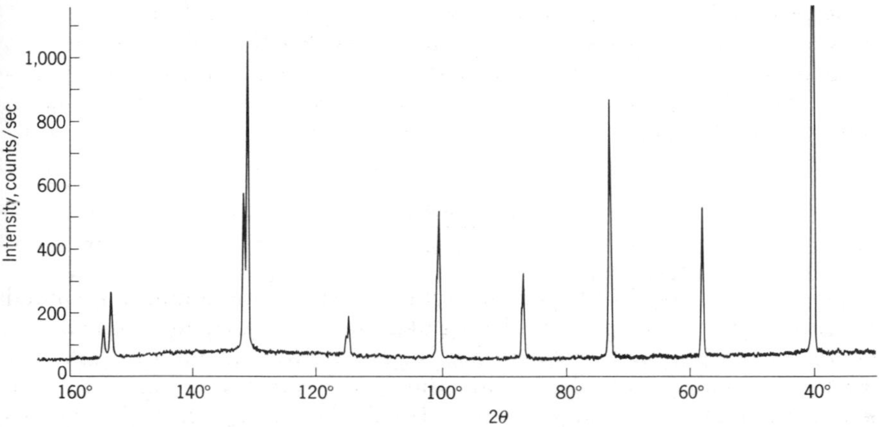

Fig. 7-8 Diffractometer strip-chart record of a powder-diffraction pattern of stress-free tungsten powder. Filtered Cu $K\alpha$ radiation; 3° beam slit at x-ray tube; 0.05-in. detector slit; Soller slits; 2° 2θ per min scanning speed. (Courtesy H. Pickett, General Electric X-ray Company.)

points on each side of the peak that appear to be uninfluenced by the peak in question or by any neighboring peaks, or by any absorption edges that abruptly alter background intensities. (This convenient assumption is not, however, theoretically correct; it is only approximate.) Therefore, if background readings of B_1 and B_2 counts per min were found on the two sides of a peak, and if m minutes were required to rotate the counter from one side of the peak to the other, the total count accumulated in this m minute scanning would be corrected by subtracting a background count of $m(B_1 + B_2)/2$ from the total.

In well-designed commercial instruments, the constancy of the tube current and voltage is so high that fluctuations in x-ray beam intensity over periods of hours are negligible (0.1 to 1 percent at most, after a warm-up period) even though the line voltage fluctuates much more widely. However, for extreme accuracy in the measurement of intensities, or for less well-regulated x-ray sources, a monitor counter can be placed in a beam from the same x-ray source and used as a standard measure of the incident-beam intensity. The monitor can be made to turn off the main counter when a given count on the monitor has been reached, permitting the integrated count in the main counter to be independent of intensity fluctuations in the incident beam.

Counting statistics X-ray quanta arrive at a counter randomly spaced in time; consequently, as mentioned in the section on counters in Chap. 3, counters begin to lose counts at rates much lower than they would if the quanta were evenly spaced. Nevertheless, it is possible to retain a linear relation between average counting rate and intensity throughout the range

Table 7–1 Percentage error as a function of total number of counts, computed for the 90 percent confidence level

Count	Error, percent	Count	Error, percent
100	16	1,000	5.2
200	12	10,000	1.6
300	9.6	27,000	1.0
500	7.4	100,000	0.52

of intensities used in most crystallographic work if either proportional or scintillation counters are used. Individual determinations of the number of counts N in a given time t are subject to statistical variation and will be found to have a Gaussian distribution about the "true" value $\bar{N}$ obtained by averaging many determinations. The *standard deviation* of the distribution is given by $\sigma = \bar{N}^{1/2}$. The *probable error* of a single determination (the error which is exceeded in 50 percent of the determinations) is $\epsilon_{50} = 0.67/N^{1/2}$. Thus, with a count of $N = 4500$ there is a 50–50 chance that the true count lies within the range $\bar{N} \pm \sqrt{4500}$, i.e., $N \pm 67$, and the probable error is thus $\epsilon_{50} = 0.67/67$, or 1 percent. If the count is only 1000, the probable error is 2 percent. In practice it is often preferable to consider the count that is necessary to achieve a given proportionate error a much higher percent of the time, say 90 percent (ϵ_{90}), instead of half the time as in computing the probable error. The formula then is $\epsilon_{90} = 1.64/N^{1/2}$, and the relation of count to percentage error for this 90 *percent confidence level* is then shown by the examples in Table 7–1. If the total count including background is N and the background count is N_B in the same time interval, then the relative error in the *corrected intensity* $N - N_B$ is given by

$$\epsilon_{90} = \frac{1.64\sqrt{N + N_B}}{N - N_B}$$

for the 90 percent confidence level, provided $N - N_B$ is not too small a number. In terms of standard deviations, if σ_N and σ_B are the standard deviations in total actual counts and background counts, respectively, then the standard deviation of the difference is

$$\sigma = \sqrt{\sigma_N^2 + \sigma_B^2}$$

To retain the relative precision of intensity determinations when weak peaks are measured, or when peak-to-background ratios are small, longer counting times must be used, and if a large number of measurements are to be made, it may save time to work out a "counting strategy" based on statistical laws in order to gain maximum precision in a given counting interval.[1]

[1] W. Parrish, *Philips Tech. Rev.*, vol. 17, p. 206, 1955/1956. M. Mack and N. Spielberg, *Spectrochim. Acta*, vol. 12, p. 169, 1958. L. J. Rainwater and C. S. Wu, *Nucleonics*, vol. 1, no. 2, p. 60, 1947; vol. 2, no. 1, p. 42, 1948. Formulas that apply when monitors are used are given in H. S. Peiser, H. P. Rooksby, and A. J. C. Wilson, "X-ray Diffraction by Polycrystalline Materials," Institute of Physics, London, 1955.

Statistics with a rate meter A rate meter consists of a resistance R connected across a condenser of capacitance C which receives pulses from the counter and amplifying circuit. The pulses are "shaped"—reduced to constant voltage and duration—before being fed to the RC circuit of the rate meter, so that the total current received by the rate meter per second is proportional to the number of pulses per second. The current of puless charges the rate meter condenser, which discharges through the resistance. The voltage on the condenser resulting from a single pulse, V_O, decays exponentially with time, falling to $V_0/2.72$, i.e., 37 percent of V_0, in a time given by the product RC and it drops to 1 percent of the initial value after 4.6 RC sec, if R is expressed in ohms and C in farads. Similarly, if any steady counting rate is suddenly changed to another, the rate meter will have moved 99 percent of the way to the new position in 4.6 RC sec after the change has occurred. The product RC is the *time constant* of the circuit; an understanding of the effect of this variable on the rate meter behavior is most important. When RC is small, fluctuations in the counting rate are prominently displayed by the meter; when RC is large, these fluctuations are smoothed out.

The precision in the rate meter reading is equal to that which would be obtained by a scaler that accumulates counts for a time interval given by $2RC$.[1] Thus, after the meter has reached equilibrium at an average counting rate of n per sec, the standard deviation for the instantaneous readings of the meter is given by $\sigma = 1/\sqrt{2nRC}$, the probable error is given by $\epsilon_{50} = 0.67\sigma$, and the rélative error for the 90 percent confidence level is given by $\epsilon_{90} = 1.64\sigma$. If a strip-chart record is made of the rate meter readings and a straight line is drawn on the chart averaging the readings of a constant-intensity beam for m sec, this average will have a probable error given by $\epsilon_{50} = 0.67/\sqrt{2mnRC}$.

It is important to realize that if a peak is being scanned at a constant rate to produce a strip chart such as that shown in Fig. 7–8, the recorder lags behind the input by a time interval of roughly RC, resulting in an asymmetric distortion of the peak, displacement of the peak, and consequent errors in the measurement of diffraction angles. The displacement can be corrected by scanning in both directions and averaging the two peak positions.

To reduce the displacement to negligible magnitude, it should be remembered that the displacement depends on the product of the scanning speed v and the time constant RC.[2] A widely used rule is to make the time constant less than half the *time width* of the slit at the counter, the time width being defined as the time for the slit to travel its own width.

[1] L. I. Schiff and R. D. Evans, *Rev. Sci. Instr.*, vol. 7, p. 458, 1936.

[2] One published recommendation to make the displacement small is that the product vRC should be about 1° per min sec, v being expressed in degrees per minute and RC in seconds. However, a shorter time constant is necessary for distortion-free scanning of sharp peaks than for broad peaks, since at a given scanning speed the counting rates change faster with the sharp peaks.

Interpretation of powder patterns A powder or polycrystalline specimen has crystalline particles at all orientations. Consequently, the diffracted rays from it travel outward in all directions that make angles of 2θ with the direct beam, where the θ's are the Bragg angles. Thus, each order of reflection from each set of planes forms a cone of semiapex angle 2θ concentric with the primary beam.

Indexing the lines is a simple matter with cubic crystals but is difficult with crystals of lower symmetry and generally impossible with monoclinic or triclinic crystals. Fortunately, most of the metals and alloys are cubic, tetragonal, or hexagonal. It should be emphasized that great caution is necessary when determining crystal structures solely from powder data, for several incorrect unit cells have been deduced in this way.

Indexing the lines in a powder pattern is always based on comparing the measured list of θ, $\sin\theta$, or $\sin^2\theta$ values with a list predicted for a unit cell of known or assumed dimensions. Given the parameters $a, b, c, \alpha, \beta, \gamma$ of a cell and the wavelength to be used, one can quickly compute all possible reflections by using the appropriate spacing equation (Chap. 4, pages 80 and 86) and inserting all possible values of h, k, and l. With *cubic* crystals the equation is

$$\sin^2\theta = K(h^2 + k^2 + l^2) \tag{7–1}$$

where $K = \lambda^2/4a^2$. It will be noted that reflections 100, 010, and 001 all fall on the same diffraction ring; likewise, reflections from all planes of the general form $\{hkl\}$ superimpose. However, this is not true of noncubic crystals.

The quantity $(h^2 + k^2 + l^2)$ in Eq. (7–1) has small integral values—in fact, with the proper indices this quantity can have any integral value except 7, 15, 23, 28, 31, 39, 47, 55, $\cdots$. To identify the lines of a cubic pattern, it is necessary merely to choose a set of these integers so that $(\sin^2\theta)/(h^2 + k^2 + l^2)$ will have the same value K for every line in the pattern. The value of the constant K then permits a computation of a since $K = \lambda^2/4a^2$. Noncubic formulas do not permit the lines to be indexed so easily.

To facilitate computation of the lattice constant or, conversely, to speed the calculation of cubic patterns when a is known, a table of reflecting planes is included in the Appendix.[1]

Only with *simple* space lattices can reflections be expected for any values whatever of $(h^2 + k^2 + l^2)$; other space lattices will have *characteristic absences* that serve to identify the type of space lattice of the specimen. These characteristic absences occur similarly in *all* crystal systems, a consequence of the zero value of the structure factor, as explained in connection with Eqs. (4–7) and (4–8) (page 83) and as discussed in Chap. 8. They occur as a consequence of the body-centered or face-centered positions of *lattice points* in addition to the points located at unit cell corners, and they apply whether a single atom or a group of atoms is located at each

[1] For more extensive tables, see "International Tables for X-ray Crystallography," vol. 2, p. 124, Kynoch, Birmingham, England, 1959.

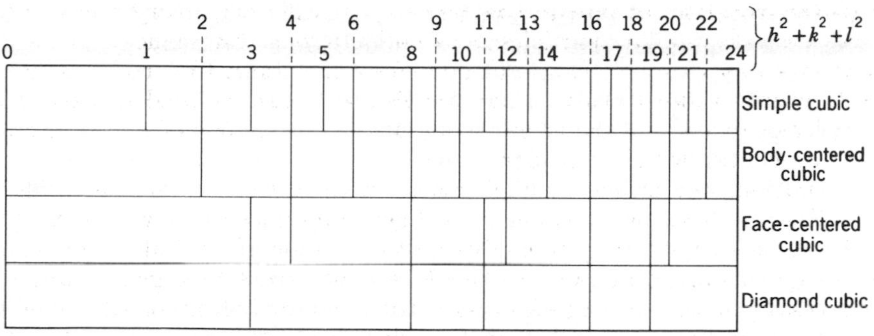

Fig. 7-9 Powder-diffraction patterns for different cubic crystals, illustrating systematic absences.

point. (Other absences may occur as a result of the values of the position parameters of atoms when there is more than one atom per lattice point.) Any crystal with a body-centered cell of any crystal system will have $|F|^2 = 0$ for all reflections with $(h + k + l)$ an odd number; for example 100, 111, 210, etc. (Fig. 7–9). Any face-centered crystal will have all reflections missing for which the indices h, k, l are mixed odd and even; for example, 100, 110, 210, etc. The atom arrangement in diamond shows absences of the face-centered type, Fig. 7–9, and additional ones caused by the presence of two atoms (rather than one atom) per lattice point of the cubic unit cell.

If the dimensions of the unit cell are known, tetragonal crystal patterns may be indexed by using the formula

$$\sin^2\theta = A(h^2 + k^2) + Bl^2 \tag{7–2}$$

where $A = \lambda^2/4a^2$ and $B = \lambda^2/4c^2$. A table is made up of the constant A multiplied by all values of $(h^2 + k^2)$ and another table of all values of Bl^2; the predicted $\sin^2\theta$ values are then merely the sums of all combinations of the numbers in the two tables.

Similar computations will serve to index orthorhombic crystals, using the formula

$$\sin^2\theta = Ah^2 + Bk^2 + Cl^2 \tag{7–3}$$

where $A = \lambda^2/4a^2$, $B = \lambda^2/4b^2$, and $C = \lambda^2/4c^2$. Hexagonal crystals and rhombohedral crystals referred to hexagonal axes require the formula

$$\sin^2\theta = A(h^2 + hk + k^2) + Bl^2 \tag{7–4}$$

where $A = \lambda^2/3a^2$ and $B = \lambda^2/4c^2$. All these formulas may be derived by considering the values of r^*_{hkl} from the origin to the point hkl in the reciprocal lattice.

Computations for monoclinic and triclinic crystals can best be made by computing directly the distances in the reciprocal lattice; since the coordinates ξ and ζ, illustrated in Fig. 6–8 (page 107), are normal to each other, $r^* = \sqrt{\xi + \zeta}$. Computations with these formulas are not difficult if one makes use of existing tables (see Appendix, Table A-3) and hand computers; they can very rapidly be run on high-speed computers.[1]

Interpretation when the unit cell is unknown It is rather surprising that thus far no direct method or computer program has been developed that is generally acknowledged to be successful for solving powder diffraction patterns of crystals of unknown symmetry. Difficulties are encountered with the inaccuracies of the $\sin^2 \theta$ data, the prevalence of partially and fully overlapped lines, and the many extinctions resulting from symmetry and from accidental values of atom-position parameters. Powder patterns continue to be solved by trial-and-error procedures, starting with the assumption that the symmetry is the simplest (cubic), then trying the more complex systems in turn until predicted and observed data agree or until it is concluded that it is hopeless to obtain a solution without using single-crystal data. These trials are vastly aided by careful inspection of the relationships between observed $\sin^2 \theta$ values[2] and by the graphical and analytical methods discussed below. Many crystals in all symmetries except the monoclinic and triclinic have been solved in this way—and on occasion even a monoclinic crystal has been solved.[3] The strategy is to use the lines with low θ values to suggest possible indexing schemes, then to use the lines at higher angles to test the proposed schemes and to determine the lattice constants with precision.

Numerical methods for solving patterns A systematized numerical method of indexing a pattern of an unknown crystal has been presented by Hesse.[4] It is best that the original paper be consulted for most of the details; we give here only the simple suggested test for whether a crystal is tetragonal or hexagonal. Both have $\sin^2 \theta$ formulas of the type $\sin^2 \theta = AM + Bl^2$, as mentioned in the preceding section, where M depends upon h and k and can only have certain values (see Table A-3 in the Appendix). These values of M are 0, 1, 2, 4, 5, 8, 9, etc., for tetragonal crystals, and 0, 1, 3, 4, 7, 9, etc., for hexagonal crystals. Provided the diffraction data are sufficiently precise ($\sin^2 \theta$ errors of 0.0001 to 0.0005) and assuming that A/B is not a ratio of small integers, it is possible to find pairs of planes of the $hk0$ type by applying the requirement that $c_1 \sin^2 \theta_1 = c_2 \sin^2 \theta_2$ for these, where c_1 and c_2 are small integers with no common factor and are not squares of integers. For these pairs of planes, only certain ratios of c_2/c_1

[1] See the computing programs listed in "International Union of Crystallography World List of Crystallographic Computer Programs."

[2] N. F. M. Henry, H. Lipson, and W. A. Wooster, "The Interpretation of X-ray Diffraction Photographs," p. 178, Macmillan, London, 1951.

[3] W. H. Zachariasen has solved a monoclinic structure (one phase of plutonium) for which only powder patterns could be obtained and in which the first 20 lines, of lowest θ values, were absent. An extremely difficult task like this is not undertaken if single-crystal data can be obtained. W. H. Zachariasen and F. H. Ellinger, *Acta. Cryst.*, vol. 16, p. 369, 1963. W. H. Zachariasen, ibid., vol. 16, p. 784, 1963.

[4] R. Hesse, *Acta Cryst.*, vol. 1, p. 200, 1948.

can occur in the tetragonal (T) or hexagonal (H) systems. The simpler ratios, with c_1 and c_2 less than 10, will then be found to be as follows, provided that both planes of the corresponding pair reflect.

c_2/c_1 for T	2, 4, 5, 8, 9, 2/5, 2/9, 4/5, 4/9, 5/8, 5/9, 8/9
c_2/c_1 for H	3, 4, 7, 9, 3/4, 3/7, 4/7, 4/9, 7/9

The Hesse-Lipson method[1] can be used with the orthorhombic system. Use is made of the fact that in the formulas for $\sin^2 \theta$, given in the preceding section, the value of $\sin^2 \theta_{hk0}$ is the sum of two other $\sin^2 \theta$ values, namely, $\sin^2 \theta_{h00}$ and $\sin^2 \theta_{0k0}$, and analogous relationships hold for $\sin^2 \theta_{h0l}$ and $\sin^2 \theta_{0kl}$, etc. It is essentially a search for often-repeated differences that suggest values of $\sin^2 \theta_{100}$, $\sin^2 \theta_{010}$, and $\sin^2 \theta_{001}$ to be tried as a possible fundamental triple to account for all reflections.

Vand[2] has modified this procedure to apply to *long-spacing compounds*, such as soaps, fats, and fatty acids, that have unit cells in which one dimension is much longer than the other two. He makes use of the fact that the long dimension, say c, causes a close grouping of lines having indices in which h and k are the same and only l is different. The method has been developed to apply to crystals of *any* symmetry.

It is well to remember that anisotropic thermal expansion can be useful as an aid to indexing complex diffraction patterns, since line shifts are quickly converted to linear expansion data normal to the individual reflecting planes; and a line that changes in width or becomes split with changing temperature can be presumed to be a superposition of two or more lines of different indices.[3]

Patterns of many alloys and of relatively simple compounds are often solved by direct comparison with patterns of common structure types. An experienced crystallographer can recognize several types (in addition to the cubic ones) just by inspection of a film or strip chart. The American Society for Testing Materials (ASTM) file, mentioned later, is an aid to this.

Ito has developed a different analytical procedure for indexing crystals of any system,[4] based on the principle that any three noncoplanar reciprocal lattice vectors,

[1] R. Hesse, *Acta Cryst.*, vol. 1, p. 200, 1948. H. Lipson, *Acta Cryst.*, vol. 2, p. 43, 1949. Also summarized in L. V. Azároff and M. J. Buerger, "The Powder Method in X-ray Crystallography," p. 80, McGraw-Hill, New York, 1958, and N. F. M. Henry, H. Lipson, and W. A. Wooster, "The Interpretation of X-ray Diffraction Photographs," p. 181, Macmillan, London, 1951.

[2] V. Vand, *Acta Cryst.*, vol. 1, pp. 109, 290, 1948.

[3] Additional information of value in indexing powder patterns becomes available if the specimen has a *preferred orientation or a fiber texture*, for then the Debye rings are not equally intense at all azimuths around the incident beam but instead consist of arcs. The more limited is the spread in orientation, the more nearly the diffraction pattern becomes equivalent to a rotating crystal or an oscillating crystal and consequently the easier may the methods of Chap. 6 be applied. Further discussion is given in H. S. Peiser, H. P. Rooksby, and A. J. C. Wilson, "X-ray Diffraction by Polycrystalline Materials," p. 358, Institute of Physics, London, 1955.

[4] T. Ito, "X-ray Studies on Polymorphism," Maruzen Co., Tokyo, 1950.

read from three observed reflections chosen arbitrarily from the list of reflections (perhaps the first three), serve to define a possible unit cell, both as to size and interaxial angles. Then, bv a procedure of Delaunay,[1] the *reduced unit cell* is determined—the unit cell that has the shortest sides and all three angles obtuse, and that is primitive, *P*. Criteria are then applied to test whether the correct unit cell should be one that contains more than one lattice point (*I*, *A*, *B*, *C*, or *F*), and an appropriate transformation of axes is made if necessary. The method requires highly precise measurements (errors of less than 0.0003 in $1/d^2$). Attempts to use this or other schemes on a low-symmetry crystal consume time that is usually much better invested in growing a single crystal—no matter how small—and using single-crystal methods. Although the Ito method is very attractive from a theoretical standpoint, experienced crystallographers have varied success with it; some call it "OK," others, "No good in practice." Presumably the range of opinion depends on chance encounters with the various difficulties we have mentioned on page 138. Zsoldos[2] has presented a modification of Ito's method based on a principle employed by Vand, which should be helpful with crystals of low symmetry.

Graphical methods for solving patterns Convenient graphical methods exist for solving cubic, tetragonal, hexagonal, and rhombohedral crystal patterns. The charts devised by Hull and Davey[3] for lattices having two parameters are based upon a logarithmic scale of *d* spacings. These are plotted along the axis of abscissas, as indicated in Fig. 7–10, and different axial ratios c/a appear at different vertical levels. At a given axial ratio the distances from right to left correspond to decreasing log *d* values for the various *hkl* planes (and the higher-order reflections of these, indicated by parentheses). A strip of paper is marked with the log *d* values observed on a pattern, and is moved over the chart to various horizontal positions until the marks coincide with curves on the chart (see Fig. 7–10). Doing so amounts to trying various c/a and a values. Indices are then read from the labels on the curves.

The disadvantage of Hull-Davey charts is the crowding of lines in some regions of the charts. The charts developed by Bunn[4] are better in this respect. Although they are based on a different form of the spacing equation than the Hull-Davey charts, the procedure in using them is the same: a strip with the observed data marked on it is moved parallel to itself until the observed lines coincide with the plotted curves.

Bunn charts have been published in conveniently large size by the Institute of Physics.[5] In using charts, more than one matching position can be found; the position giving the simplest indices corresponds to the smallest unit cell. Several curves will correspond to missing reflections in

[1] B. Delaunay, *Z. Krist.*, vol. 84, p. 132, 1933. Finding reduced cells is also discussed in L. V. Azároff and M. J. Buerger, "The Powder Method in X-ray Crystallography," chaps. 11, 12, McGraw-Hill, New York, 1958.

[2] L. Zsoldos, *Acta Cryst.*, vol. 11, p. 835, 1958.

[3] A. W. Hull and W. P. Davey, *Phys. Rev.*, vol. 17, p. 549, 1921. W. P. Davey, *Gen. Elec. Rev.*, vol. 25, p. 564, 1922.

[4] C. W. Bunn, "Chemical Crystallography," Oxford, New York, 1945. "International Tables for X-ray Crystallography," vol. 2, p. 184, Kynoch, Birmingham, England, 1959.

[5] Charts in sets of five are sold by the Institute of Physics, London, England, and also by Polycrystal Book Service, Pittsburgh, Pa. Tables by which one can plot them directly are given in "International Tables for X-ray Crystallography," vol. 2, op. cit., p. 207.

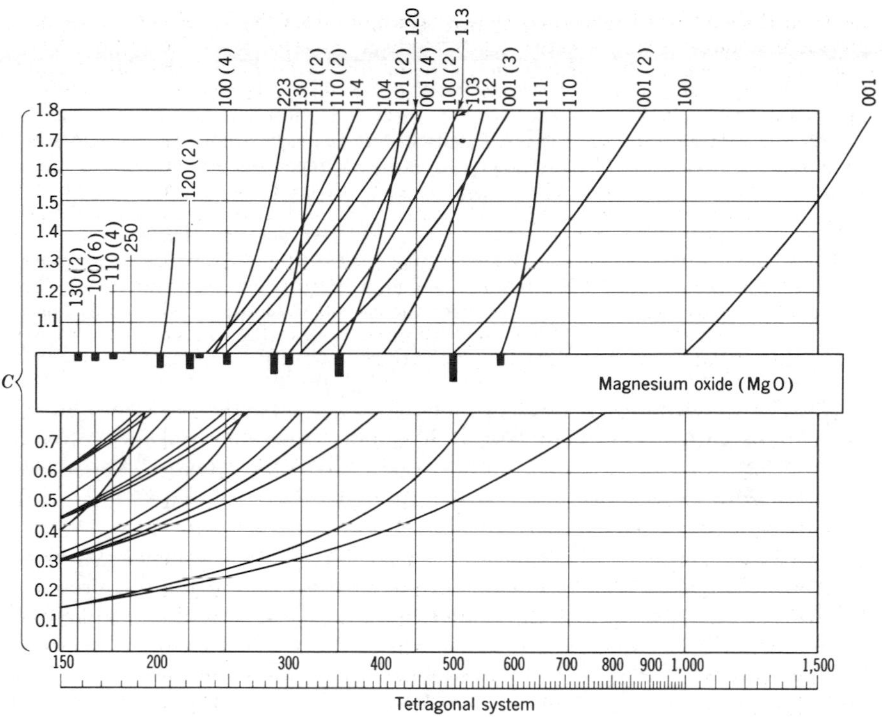

Fig. 7-10 Hull-Davey chart for predicting powder patterns of tetragonal crystals. Its use in indexing an MgO pattern is illustrated. (R. W. G. Wyckoff.)

any example. Small errors in the observations must be expected. After finding a probable match, numerical computations should be made, since graphical methods should not be trusted for the final assignment of indices (especially at high *hkl* values) or for accurate determinations of lattice constants.

Precise measurement of lattice constants Precise determinations of unit cell dimensions are used in the identification of unknown materials, in developing concepts of interatomic bonds and effective valences of elements in compounds, in determining solubility limits, in determining residual stresses in solid materials, in computing theoretical densities and expansion coefficients (especially when these quantities are relatively inaccessible to direct measurement), in determining the type and composition of solid solutions, in studying effects of clustering, ordering, and precipitation in solid solutions, and in studying effects induced by special treatments, such as radiation damage.

In the early years attempts were made to do precise work with forward-reflection Debye cameras (Fig. 7–3*a*) but all recent developments have utilized the higher dispersion of the back-reflection region, including 2θ values as high as conveniently possible. The importance of the high-angle region is illustrated by the fact that measurement of 2θ with a precision of

0.01° corresponds to a precision in d spacings of 0.005 percent (1 part in 20,000) when 2θ is near 120°, and 0.0005 percent (1 part in 200,000) when 2θ is near 175° (the latter angle is about the upper practical limit for Debye cameras).

Precision (i.e., reproducibility) approaching a few parts per million can be attained with both photographic and diffractometer methods using powders, and certainly with highly perfect crystals (see Bond's method, Chap. 6); such precision is adequate for many uses. However, when absolute accuracy is to be combined with precision, the task is much more difficult, for all systematic errors must then be somehow eliminated. The prevalence of unsuspected systematic errors was dramatically exposed by an interlaboratory comparison organized by the International Union of Crystallography,[1] in which portions of the same homogeneous samples of silicon, tungsten, and diamond were studied by various precision methods in 16 laboratories around the world. Although some laboratories estimated their errors to be as low as 1 part in 550,000 the agreement between the different laboratories was only about 1 part in 10,000 (0.01 percent).

Minimizing errors in Debye cameras Photographic film undergoes expansion and shrinkage with changes in its moisture content. This *film shrinkage* may occur even during the time it is being read, if a lamp is near (though changes are lessened if the film is sealed between glass plates). It is usually assumed that unless the film has been mishandled or has dried unevenly the shrinkage is uniform over the length of the film strip. Correction can then be determined by measuring the distance between marks that are put on the film at the time of the x-ray exposure.

One method of correcting for film shrinkage is to wrap the film around the camera as indicated in Fig. 7–3b (or c). The position of $\theta = 0$ is then determined to be midway between right- and left-hand arcs of small-angle Debye rings, and the position $\theta = 90°$ is similarly determined from the large-angle rings. The distance between these two positions then determines the effective diameter of the camera. Alternatively, the film may be cut long enough to overlap a little; the inner end is then in position to cast a sharp shadow on the overlapping end at a distance along the film that equals the effective circumference of the camera.

Individual films may be calibrated by printing a complete scale on the film before developing.[2] It can also be done by using a homogeneous mixture of a standard substance with the substance being studied.[3] Recommended standard substances include high-purity silicon ($a_0 = 5.41959kX$ at 25°C) and tungsten ($a_0 = 3.15884kX$ at 25°C).[4]

[1] W. Parrish, *Acta Cryst.*, vol. 13, p. 838, 1960.

[2] G. Hägg, *Rev. Sci. Instr.*, vol. 18, p. 371, 1947.

[3] G. E. Bacon, *Acta Cryst.*, vol. 1, p. 337, 1948. K. W. Andrews, *Acta Cryst.*, vol. 4, p. 562, 1951.

[4] Thermal expansion values, $\Delta a_0/°C$, are 22.5×10^{-6} for silicon and 14.5×10^{-6} for tungsten. Other standards are listed by G. D. Rieck and K. Lonsdale in "International Tables for X-ray Crystallography," vol. 3, p. 122, Kynoch, Birmingham, England, 1962.

The finite *radius of the specimen* introduces an error if the rays undergo appreciable absorption in the specimen, or if the specimen is mounted on a glass fiber that is too large in diameter. Formulas have been derived for correcting this error.[1] This error, which is very common in practice, may be reduced by keeping the specimen diameter as small as conveniently possible, and sometimes also by diluting the specimen with a weakly absorbing noncrystalline substance.

The divergence of the rays in the direction of the specimen axis (*axial divergence*) shifts the back-reflection lines to higher angles and the forward-reflection lines to lower angles. This error is normally small and is minimized by using a well-constructed incident beam collimator that permits only slight axial divergence.

If the axis of rotation of the specimen is not *concentric* with the axis of the film, the lines are shifted in a manner dependent upon the direction and magnitude of the specimen displacement. Errors from this cause should be negligible in any good camera, as are errors caused by the incident beam not being perpendicular to the specimen axis. The specimen should be on the axis of rotation so it will not *wobble* during rotation.

Accurate control of specimen temperature is essential; to keep errors from this smaller than other errors, the coefficient of expansion should be considered, but in general, temperatures should be known and held constant to about 0.1°C in high-precision work.

Extrapolation methods for back-reflection cameras The percentage error in d-spacing measurement caused by a given error in diffraction angle measurement approaches zero as $\cot\theta$ approaches zero (as θ approaches 90°).[2] Many errors may therefore be eliminated by plotting the apparent lattice constant determined from each line against some function of θ and extrapolating to $\theta = 90°$. The difficulty is to decide what function of θ should give a straight line over a considerable range of θ values all the way to $\theta = 90°$. In general, each source of error requires a different function.[3]

It is best in precise work to load a Debye camera as a back-reflection camera, i.e., with the film continuous across the *inlet* hole (see Fig. 7–3*b* or *c*), since the back-reflection region contains the most valuable data. On a back-reflection film the measurement of the distance between two sides of a Debye ring gives an angle 2ϕ rather than the angle 4θ of the corresponding measurement in forward-reflection films, where $2\phi = 360° - 4\theta$.

If the largest source of error is known, the extrapolation function should be the one corresponding to this type of error. General practice, however,

[1] F. Wever and O. Lohrmann, *Mitt. Kaiser-Wilhelm-Inst. Eisenforsch, Düsseldorf*, vol. 14, p. 137, 1932.

[2] Differentiating the Bragg law $d \sin\theta = n\lambda/2$, one obtains $\Delta d/d = -\cot\theta\,\Delta\theta$.

[3] These are treated in detail in M. J. Buerger, "X-ray Crystallography," Wiley, New York, 1942, and summarized by W. Parrish and A. J. C. Wilson in "International Tables for X-ray Crystallography," vol. 2, p. 216, Kynoch, Birmingham, England, 1959. A recent discussion is given by K. E. Beu in G. L. Clark (ed.), "Encyclopedia of X- and Gamma Rays," Reinhold, New York, 1963.

is to use a function that is found by experience to give straight-line extrapolation over large ranges of θ; such a function, which has some theoretical justification, has been worked out by Nelson and Riley[1] and independently by Taylor and Sinclair[2]; the function is

$$\frac{1}{2}\left(\frac{\cos^2\theta}{\sin\theta} + \frac{\cos^2\theta}{\theta}\right)$$

Convenient tables of this function are printed frequently,[3] and values can be obtained easily from a program on a computer—but computer accuracy is by no means necessary for plotting purposes. It is important, however, to have high-angle lines, preferably at least one above $2\theta = 160°$; this is usually possible if wavelengths are suitably chosen. An example of a graphical extrapolation using this function on a cubic material is given in Fig. 7–11.

An extrapolation function introduced by Bradley and Jay[4] for *forward-reflection loading* of Debye cameras was $\cos^2\theta$; it was designed to extrapolate to a value that contained no errors from incorrect centering of the specimen or from absorption of rays in the specimen; film-shrinkage and camera-radius errors were to be corrected *before* plotting the measurements.

With *back-reflection loading,* film shrinkage errors approach zero as θ approaches 90°, i.e., as ϕ approaches 0, where $\phi = 90 - \theta$. Both shrinkage and radius errors are proportional to ϕ, and the error due to displacement of the sample from the center of the camera is proportional to

$$\sin(\phi/2)\cos(\phi/2) = (\sin\phi)/2$$

which for small values of ϕ is approximately proportional to ϕ. With highly absorbing samples, the errors introduced by absorption are also proportional to ϕ at small values of ϕ.[5] When the position of maximum intensity is measured, these systematic errors are additive, giving an approximate combined error which can be written $\Delta\phi = -2\Delta\theta = D\phi$, where D is a constant of proportionality. Substitution in the differentiated form of Bragg's law, $\Delta d/d = -\cot\theta\,\Delta\theta$, then gives $\Delta d/d = D(\phi/2)\tan(\phi/2)$. For small values of ϕ, this is closely approximated by

$$\frac{\Delta d}{d} = D\sin^2\frac{\phi}{2} = D\cos^2\theta \qquad (7\text{–}5)$$

[1] J. B. Nelson and D. P. Riley, *Proc. Phys. Soc. (London)*, vol. 57, p. 160, 1945.

[2] A. Taylor and H. Sinclair, *Proc. Phys. Soc. (London)*, vol. 57, p. 126, 1945.

[3] Tables are given in many of the books listed in the references of this chapter: "International Tables for X-ray Crystallography," vol. 2, p. 228, Kynoch, Birmingham, England, 1959; Henry, Lipson, and Wooster; Peiser, Rooksby, and Wilson; and Taylor. Most books give tables in which misprints of the original Nelson-Riley table are corrected.

[4] A. J. Bradley and A. H. Jay, *Proc. Phys. Soc. (London)*, vol. 44, p. 563, 1932.

[5] M. J. Buerger, "X-ray Crystallography," Wiley, New York, 1942.

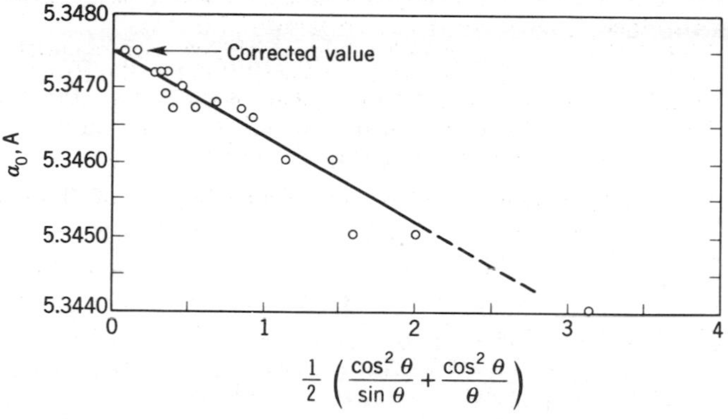

Fig. 7-11 Linear extrapolation of lattice-constant data for a cubic crystal when plotted vs. the Nelson-Riley-Taylor-Sinclair function. (After J. B. Nelson and D. P. Riley.)

Therefore, $\cos^2\theta$ serves as an appropriate linear extrapolation function for these back-reflection Debye measurements with θ greater than 30°. The same function also serves for *back-reflection symmetrical focusing cameras*. Since $\sin^2\theta + \cos^2\theta = 1$, extrapolation using $\sin^2\theta$ is equivalent to extrapolation using $\cos^2\theta$ and may be more convenient if the investigator has already tabulated his observed $\sin^2\theta$ values.

Extrapolation is regarded as impractical for *flat-film back-reflection cameras*. It is best to correct individual Debye rings in these cameras by using calibration lines from a standard substance, by minimizing errors in the specimen-to-film distance, and by using only a single-emulsion film.[1]

Many other theoretical treatments of functions for linear extrapolation of errors could be cited, but each one is based on certain assumptions regarding the effective size of the source, the distribution of intensity in the source, the amount of divergence, and perhaps other parameters as well. Current practice generally involves using a function that empirically is found satisfactory; in the high-angle region either $\cos^2\theta$, $(\cos^2\theta)/\sin\theta$, or $\phi \cot\theta$ serves well, but for wider ranges the Nelson-Riley-Taylor-Sinclair function is better. The latter may also be applied very effectively to *diffractometer measurements* of the center-of-gravity positions of peaks from rotating cylindrical samples.[2]

Precise determinations of unit cell constants for hexagonal and tetragonal crystals can be carried out by a successive-approximation procedure, starting from approximate values of the constants and noting the changes required to move points of certain indices closer to the average extrapolation

[1] "International Tables for X-ray Crystallography," vol. 2, p. 220, Kynoch, Birmingham, England, 1959. D. E. Thomas, *J. Sci. Instr.*, vol. 25, p. 440, 1948.

[2] H. M. Otte, *J. Appl. Phys.*, vol. 32, p. 1536, 1961.

line on a Nelson-Riley-Taylor-Sinclair plot.[1] The procedure does not require a computer.

The above discussion relates to graphical extrapolation but much use is recently being made of least-squares solutions of the same problems, especially since computing machines have become widely available. These methods have the advantage that less subjective error enters. The methods will be introduced by describing Cohen's treatment, which is based on one type of extrapolation function, but the same method can be directly extended to any of the other functions.

Cohen's least-squares method[2] Taking the logarithm of Bragg's law and differentiating leads to the expression $(\Delta \sin^2 \theta)/\sin^2 \theta = -2\Delta d/d$, and substituting the $\cos^2 \theta$ relation of Eq. (7–5) in this expression leads to the statement that values of $\sin^2 \theta$ are incorrect by an amount

$$\Delta \sin^2 \theta \cong -2D \sin^2 \theta \cos^2 \theta = -D \sin^2 2\theta \tag{7–6}$$

where D is constant for any given film but differs from film to film. Bragg's law may then be written with a correction term in the form

$$\frac{\lambda^2}{4d^2} = \sin^2 \theta - D \sin^2 2\theta \tag{7-7}$$

An equation of this type is written for each individual Debye ring that is found in the high-angle region. Each equation for a given film contains the same D and contains the hkl indices of a particular reflection. The least-squares solution is obtained for the series of equations. For *cubic* crystals the individual equations may be written in the form

$$K(h^2 + k^2 + l^2) + D \sin^2 2\theta = \sin^2 \theta \tag{7–8}$$

where $K = \lambda^2/4a_0^2$. The normal equations then are

$$\begin{aligned} K \Sigma \alpha_i^2 + D' \Sigma \alpha_i\delta_i &= \Sigma \alpha_i \sin^2 \theta_i \\ K \Sigma \alpha_i\delta_i + D' \Sigma \delta_i^2 &= \Sigma \delta_i \sin^2 \theta_i \end{aligned} \tag{7–9}$$

where $\alpha_i = h^2 + k^2 + l^2$, $D' = D/10$, and $\delta_i = 10 \sin^2 2\theta_i$. The factor 10 is used merely so that the coefficients of the normal equations will be of the same magnitude. The summations extend over all the equations for the

[1] A procedure is presented by T. B. Massalski and H. W. King, *J. Inst. Metals*, vol. 89, p. 169, 1960–1961, which is a modification of that of H. Lipson and A. J. C. Wilson, *J. Sci. Instr.*, vol. 18, p. 144, 1941, and which has proven to be convenient in use, particularly in connection with a table of computed functions prepared by H. W. King and T. B. Massalski (mentioned in *J. Inst. Metals*, vol. 90, p. 486, 1961–1962). A method leading directly to a solution is given in E. J. Meyers and F. C. Davies, *Acta Cryst.*, vol. 14, p. 194, 1961.

[2] M. U. Cohen, *Rev. Sci. Instr.*, vol. 6, p. 68, 1935; vol. 7, p. 155, 1936; *Z. Krist.*, vol. 94, pp. 288, 306, 1936. See also the following discussions: E. R. Jette and F. Foote, *J. Chem. Phys.*, vol. 3, p. 605, 1935. M. J. Buerger, "X-ray Crystallography," p. 426, Wiley, New York, 1942. B. E. Warren, *J. Appl. Phys.*, vol. 16, p. 614, 1945.

individual lines that lie in the appropriate range of θ. Solution of the two simultaneous normal equations gives the corrected value of K and thus the lattice constant a. This method of calculation is readily extended to other systems with additional parameters; for hexagonal crystals, Eq. (7–8) is replaced by

$$\alpha K_1 + \gamma K_2 + \delta D = \sin^2 \theta \tag{7–10}$$

where $\alpha = h^2 + hk + k^2$, $\gamma = l^2$, $\delta = 10 \sin^2 2\theta$, $K_1 = \lambda^2/3a_0^2$, and $K_2 = \lambda^2/4c_0^2$. *Three* normal equations must be solved simultaneously for the three unknowns. Orthorhombic crystals are handled in a similar way. Jette and Foote[1] have discussed the evaluation of standard errors and fiduciary limits in such results.

Hess's modification Hess has shown[2] that Cohen's method assigns equal weight to the *lattice constants* obtained from each of the diffraction lines. But equal weights should, instead, be given the individual measurements of line positions on the *film*, provided that the lines are sharp enough to permit precision work.[3] If this is done, the lattice constants computed for the lines near $\phi = 0$ will be given much greater weight than the others, which is proper. This principle amounts to weighting the individual observation equations (such as Eq. (7–8) for cubic crystals) by the weighting factor $\operatorname{cosec}^2 \phi$.

Least-squares computer programs Lattice-constant computation programs can have correction terms included. The current edition of "World List of Crystallographic Computer Programs," issued by the International Union of Crystallography, should be consulted for these. To give an idea of some of the potentialities of such programs, we mention two comprehensive programs.[4] Input data are the reflections and their indices; output data are the cell parameters, determined by least-squares procedures, and the uncertainties in these parameters.

In the Argonne National Laboratory program there is also provision for as many as three correction terms, separately or together, which can be any of the commonly used functions. If desired, also, a weighting factor may also be used for each reflection, or a trigonometric weighting factor such as Hess's, or both together. The computer prints out the observed and calculated $\sin^2 \theta$ values, the least-squares values of the lattice constants,

[1] E. R. Jette and F. Foote, *J. Chem. Phys.*, vol. 3, p. 605, 1935.

[2] J. B. Hess, *Acta Cryst.*, vol. 4, p. 209, 1951.

[3] H. Ekstein and S. Siegel, *Acta Cryst.*, vol. 2, p. 99, 1949.

[4] M. H. Mueller, L. Heaton, and K. T. Miller, *Acta Cryst.*, vol. 13, p. 828, 1960. A later improved version of this, "B106," originally in Fortran 63, by L. Heaton, J. Gvildys, and M. H. Mueller, which applies to every crystal system, has been put into use at the Argonne National Laboratory. H. G. Norment, "A Collection of Fortran Programs for Crystal Structure Analysis," NRL Report 5885, AD404 653, U.S. Department of Commerce, Office of Technical Services, Washington, D.C., 1963 (no corrections are provided).

the computed standard error, and an error function that takes the weighting into account. With the results of several drift constants and weighting schemes available for comparison, it is then possible to judge the combinations that give the best fit with the data. When the exact correction function is not known because of a lack of knowledge of the exact distribution of intensity in the incident x-ray beam and other factors, perhaps this "best-fit" criterion is as likely as any to indicate the most probable of the available answers.

Another strategy that has been given much thought in spite of its relative complexity is to attempt to identify every source of systematic error, to correct for these by the appropriate functions, and then to apply a statistical test to ascertain that no systematic errors remain uncorrected.[1]

Diffractometer correction and extrapolation procedures In contrast to the usual Debye-Scherrer film procedures, the identification of the "position" of a diffraction profile must be made relative to the spectral distribution of the incident radiation.[2] Whereas it has been argued by Wilson and his collaborators[3] that for the highest accuracy the use of *centroids* has some advantages, it is somewhat easier to measure *peak positions*, and at higher diffraction angles these are less affected by the Lorentz factor and dispersion. Peak positions have been used in most wavelength determinations.

For most of the reasonably high-precision work with diffractometers, either peak or centroid measurements are satisfactory. If certain defects such as stacking faults are present in the specimen, they may affect peak shape as well as position, thus shifting the peaks differently from the centroids. It is usually felt that extrapolation with $\cos^2 \theta$ or with the Nelson-Riley-Taylor-Sinclair function removes most of the errors, and if peaks are scanned on both sides of the direct beam and at high angles, extremely good reproducibility can be attained.[4]

Precise alignment of a filtered beam or, alternately, a monochromated beam can be conveniently done if the diffractometer equipment is mounted on a milling machine bed which provides rotation plus x and y translation.[5]

[1] Among contributions to this approach, the work of A. J. C. Wilson and of E. R. Pike should be mentioned; recent work by K. E. Beu has appeared in United States Atomic Energy Reports from Goodyear Atomic Corporation, Portsmouth, Ohio, and in F. Kaelble, "Handbook of X-rays," chap. 10, McGraw-Hill, New York, 1966. A. J. C. Wilson, "Mathematical Theory of X-ray Powder Diffractometry," Centrex, Eindhoven, The Netherlands (not available when this chapter was written).

[2] E. R. Pike and A. J. C. Wilson, *Brit. J. Appl. Phys.*, vol. 10, p. 57, 1959.

[3] A. J. C. Wilson, *J. Sci. Instr.*, vol. 27, p. 321, 1950. E. R. Pike, *J. Sci. Instr.*, vol. 34, p. 355, 1957.

[4] H. W. King and L. F. Vassamillet, *Advan. X-Ray Anal.*, vol. 5, p. 78, 1961. L. F. Vassamillet and H. W. King, *Advan. X-Ray Anal.*, vol. 6, p. 142, 1962. H. W. King and C. M. Russell, *Advan. X-Ray Anal.*, vol. 8, p. 1, 1965. The extension to high negative angles is accomplished by using a microfocus tube of small dimensions, and reproducibility to 1 part in 100,000 is reported.

[5] L. H. Schwartz, L. A. Morrison, and J. B. Cohen, private communication, 1964.

For attempts at the very highest accuracy, the detailed studies of individual errors may be consulted; some of the more recent work is listed below.

Flat specimen and specimen transparency A. J. C. Wilson, *J. Sci. Instr.*, vol. 27, p. 321, 1950; *Proc. Phys. Soc. (London)*, vol. 78, p. 249, 1961. M. E. Milberg, *J. Appl. Phys.*, vol. 29, p. 64, 1958. C. G. Vonk, *Norelco Reporter*, vol. 8, p. 92, 1961. J. I. Langford and A. J. C. Wilson, *J. Sci. Instr.*, vol. 39, p. 581, 1962. B. W. Delf, *Brit. J. Appl. Phys.*, vol. 12, p. 421, 1961.

Axial divergence J. N. Eastabrook, *Brit. J. Appl. Phys.*, vol. 3, p. 349, 1952. E. R. Pike, *J. Sci. Instr.*, vol. 34, p. 355, 1959; *Acta Cryst.*, vol. 12, p. 87, 1959. J. I. Langford, *J. Sci. Instr.*, vol. 39, p. 515, 1962.

Refraction A. J. C. Wilson, *Proc. Cambridge Phil. Soc.*, vol. 36, p. 485, 1940. M. E. Straumanis, *Acta Cryst.*, vol. 8, p. 654, 1955.

Dispersion and Lorentz-polarization factors A. R. Lang, *J. Appl. Phys.*, vol. 27, p. 485, 1956. J. Ladell et al., *Acta Cryst.*, vol. 12, p. 567, 1959. E. R. Pike, *J. Sci. Instr.*, vol. 34, p. 355, 1959; *Acta Cryst.*, vol. 12, p. 87, 1959. E. R. Pike and J. Ladell, *Acta Cryst.*, vol. 14, p. 53, 1961. E. R. Pike and A. J. C. Wilson, *Proc. Phys. Soc. (London)*, vol. 72, p. 908, 1958. A. J. C. Wilson, "Mathematical Theory of X-ray Powder Diffractometry," Philips Technical Library, Eindhoven, the Netherlands.

Quantum-counting efficiency W. Parrish, "International Tables for X-ray Crystallography," vol. 3, p. 144, Kynoch, Birmingham, England, 1962. A. J. C. Wilson and B. W. Delf, *Proc. Phys. Soc. (London)*, vol. 78, p. 1256, 1961.

Line broadening from refraction A. J. C. Wilson, *Proc. Phys. Soc. (London)*, vol. 80, p. 303, 1962.

Cylindrical specimen corrected by Nelson-Riley extrapolation H. Otte, *J. Appl. Phys.*, vol. 32, p. 1536, 1961.

Alignment and zero determination A. Smakula and J. Kalnajs, *Phys. Rev.*, vol. 99, p. 1737, 1955. H. W. King and L. F. Vassamillet, *Advan. X-ray Anal.*, vol. 5, p. 78, 1961. W. Parrish and K. Lowitzsch, *Am. Mineralogist*, vol. 44, p. 765, 1959. H. W. King and C. M. Russell, *Advan. X-ray Anal.*, vol. 8, p. 1, 1965.

Extrapolation procedures L. F. Vassamillet and H. W. King, *Advan. X-ray Anal.*, vol. 6, p. 142, 1962.

General summaries W. Parrish, "Advances in X-ray Diffractometry and X-ray Spectroscopy," Centrex, Eindhoven, The Netherlands, 1962. "International Tables for X-ray Crystallography," vol. 2, p. 216, Kynoch, Birmingham, England, 1959. A. J. C. Wilson, *Proc. Phys. Soc. (London)*, vol. 78, p. 249, 1961 (peak displacement). A. J. C. Wilson, "Mathematical Theory of X-ray Powder Diffractometry," Centrex, Eindhoven, The Netherlands.

Wavelength and refraction *Refraction* corrections are necessary for highest accuracy, not only with single-crystal work (see Chap. 6) but also with powder work (see references above). Two effects occur. X-rays are altered in direction on entering and leaving a crystal, but if a powder sample is composed of roughly spherical particles that absorb the beam negligibly, this effect averages out. A second refraction effect, which remains regardless of the specimen shape and absorption coefficient, arises from the fact that the wavelength of the rays in a crystal differs from that in air. For a substance of index of refraction N, and density ρ in grams per cubic centimeter, the apparent spacing d' should be increased by

$$(1 - N)d' = 2.70 \times 10^{-6}\, \lambda^2 \rho d\, \mathbf{\Sigma}\, Z / \mathbf{\Sigma}\, A \qquad (7\text{–}11)$$

which for cubic crystals reduces to

$$(1 - N)a_0' = 4.48 \times 10^{-6} (\lambda / a_0)^2 \mathbf{\Sigma}\, Z \qquad (7\text{–}12)$$

where λ is the wavelength used in angstroms, $\mathbf{\Sigma} Z$ is the sum of the atomic numbers, and ΣA is the sum of the atomic weights of the atoms in the unit cell. The correction is small; for example, with Cu $K\alpha$ on tungsten the correction $(1 - N)a'$ is 1.6×10^{-4} and on aluminum it is 3.4×10^{-5}. Publications should state whether or not a refraction correction has been made.

Uncertainty in the effective *x-ray wavelengths* remains a serious limitation to the *accuracy* of lattice-constant determinations, regardless of the *precision* that is reached in θ values. The wavelengths most used have not always been measured with the highest accuracy, and even when they have, the asymmetric spectral distribution across the individual narrow $K\alpha$ and $K\beta$ lines is usually unknown and may appear to be different when different diffraction techniques are employed.

A further uncertainty in wavelengths arises from the fact that most of them were determined by crystal diffraction. Siegbahn, who pioneered in these measurements,[1] pointed out that the *absolute* values of his wavelengths were less accurate than were the *relative* values of the different emission lines, since the absolute values involved inaccuracies in the knowledge of Avogadro's number, atomic weights, densities, and systematic errors of the experiment. He used a unit of length, the X unit, and 1000 times X, the kX unit, which must be converted to the angstrom by a conversion factor. The problem of determining the best value of this conversion factor is a perennial one. Much work was directed toward accurate determinations of wavelengths with ruled gratings and the determination of the conversion factor by comparing these with the crystal-spectrometer values. By international agreement in 1947 the conversion factor 1 kX = 1.00202 A was adopted, which was believed to be accurate to about 0.00003 A. Since that time various modifications of this factor have seemed more probable as more systematic errors have been discovered.[2] The most probable value as of 1965 appears to be 1 kX = 1.002076 ± 0.000007 A. Obviously, further changes can be expected; therefore, in future precision measurements it is important that the *exact value of the wavelengths assumed in making the computations be explicitly stated.* A common practice has been to use the wavelengths published by Cauchois and Hulubei[3] and to report them in kX units (except for a few long wavelengths that were measured by gratings and are properly in angstroms). The editors of the "International Tables for X-ray Crystallography" decided to convert the kX values to angstroms by using the factor 1.00202. Newer values of this factor have since appeared, and Sandström has published new tables of wavelengths.[4] Still more recently, however, Bearden has reevaluated the entire

[1] M. Siegbahn, "Spectroscopie der Röntgenstrahlen," Springer, Berlin, 1931.

[2] "International Tables for X-ray Crystallography," vol. 3, p. 41, Kynoch, Birmingham, England, 1962.

[3] Y. Cauchois and H. Hulubei, "Tables de Constantes et Données Numériques. Longueurs d'Onde des Émissions X et des Discontinuités d'Absorption X," Hermann et Cie, Paris, 1947.

[4] A. E. Sandström, "Flügge's Encyclopedia of Physics," vol. 30, p. 162, Springer, Berlin, 1957.

list and has based the set on a better wavelength standard, as explained in the Appendix where they are reproduced.

Chemical analysis by powder diffraction *Qualitative analysis* is carried out with powder-diffraction cameras and diffractometers by making use of the fact that a crystalline substance can be identified by the d spacings and relative intensities of the reflections in its diffraction pattern. Identification of several substances in a mixture is often possible, since the diffraction pattern for the mixture is a superposition of the patterns for the individual constituents of the mixture. *Quantitative analysis* is also possible, since the intensity of each constituent of a mixture is (normally) proportional to the amount of that constituent of a mixture in the specimen.

Since the pattern is characteristic of the crystal structure or structures present in a specimen, the pattern discloses information about the state of combination of the chemical elements present that cannot be obtained by chemical analysis. For example, a substance that has a given ratio of the elements such as A_xB_y may exist in two or more polymorphic forms which may be identified by their patterns whether they are present singly or as a mixture of different polymorphs, and whether or not the crystal structures have ever been worked out.

The powder pattern of a substance effectively serves as a fingerprint by which it can be identified. As a result of a cooperative effort of a great many investigators around the world, a file has been assembled, analogous to a fingerprint file, containing the diffraction patterns of over 10,000 substances. An unknown substance may be identified by comparing its diffraction pattern with those in the file to find one that matches. A mixture of two, three, or sometimes even more substances can be analyzed similarly, as the superposition of patterns in the file. The fundamental principles involved in analysis are thus very simple, though detailed techniques require thought,[1] and methods of arranging the file have been given extensive study.

The powder-data file, which was initiated by Hanawalt and coworkers,[2] is now being continuously expanded; additions are being published yearly by the ASTM. About 1000 to 1500 new substances are added to the list each year, and many corrections to earlier data are being inserted, including data of precision quality produced at the National Bureau of Standards that have also appeared in a series of NBS circulars.[3]

[1] Detailed discussions of procedures are given in: L. V. Azároff and M. J. Buerger, "The Powder Method in X-ray Crystallography," chap. 13, McGraw-Hill, New York, 1958. E. Kaelble, "Handbook of X-rays," McGraw-Hill, New York, 1966. A particularly thorough treatment of quantitative analysis by powder methods is given in H. P. Klug and L. E. Alexander, "X-ray Diffraction Procedures," Wiley, New York, 1954.

[2] J. D. Hanawalt, H. Rinn, and L. K. Frevel, *Ind. Eng. Chem. Anal. Edition*, vol. 10, p. 457, 1938.

[3] These circulars are obtainable from the U.S. Government Printing Office, Washington, D.C.

The ASTM Powder Diffraction Data File[1] consists of file cards available as 3- by 5-in. separate cards as shown in Fig. 7–12 or as copies of these mounted in various ways mentioned below. Each card contains a "section number" (a section of cards has been issued yearly), a serial number for the card within the section to which it belongs, and data for a single substance. At the top of the card the spacings of the three strongest lines of the substance d_1, d_2, and d_3 are given, together with the intensities of these three reflections relative to the intensity of the strongest line, i.e., I/I_1. All cards with d_1 (the strongest line) lying within a certain range are grouped together, and the groups are arranged in the order of decreasing d_1 values. Within a given group, the cards are arranged in the order of decreasing d_2 values, and when d_1 and d_2 are similar, the cards are in the order of decreasing d_3 values. Since the relative intensities of these three may be judged differently in different experiments, additional cards are inserted in which the order of d_1, d_2, and d_3 is permuted. Thus, the same substance will be filed in the group for the d_2 spacing as the strongest line, and in the group for d_3, as well as in the group for d_1. (In one form of the file, eight rather than three d values are permuted in this way.) A data card also contains a list of *all* observed d spacings and their intensities relative to the strongest line, and statements regarding the substance and the technique used to obtain the data. For example, the wavelength of the x-rays used in obtaining the diffraction pattern is included, an important factor since different wavelengths yield different intensity ratios for a given substance. An introductory card explains the conversion of intensity data from one wavelength to intensities for another wavelength.

Samples to be analyzed should be prepared in such a way as to be free from preferred orientations, and should produce smooth Debye lines. If a sample is highly absorbing, it is wise to dilute it with flour, cornstarch, or similar material so as to avoid doubled lines or displaced lines in a powder photograph. Intensities estimated from a film, with or without calibration exposures, are usually adequate; more precise intensities may be obtained conveniently with a diffractometer. In comparing an experimental list of d spacings with a tabulated list, the possibility should be kept in mind that an error of perhaps $\pm$ 0.01 A may exist in either the experimental or the tabulated d values.

With a mixture of substances in a sample, any pair of d's may belong to one substance and the third d of the strongest three may belong to another substance; or each of the three may come from a different substance; or still other substances may be present that are not represented in the three strongest lines. Complications of this kind do not necessarily defeat the search although they add measurably to the difficulty of a complete analysis. The file with the eight strongest lines listed is an aid in such cases. When one substance of a mixture is identified, the lines attributable to that substance are deleted from the experimental list and the search is continued for cards to explain the remaining data.

[1] Available from the American Society for Testing Materials, X-ray Dept., Philadelphia, Pa.

d	2.94	4.74	3.75	5.71	$FeWO_4$ ★
I/I_1	100	40	35	15	Iron Tungstate (Ferberite)

Rad. CuKα λ 1.5405 Filter Ni Dia.
Cut off I/I_1 Geiger Counter Diffractometer
Ref. Sasaki, Mineral. J. (Japan) **2** 375 (1959)

Sys. Monoclinic S.G. $P2/c$ (13)
a_0 4.734 b_0 5.708 c_0 4.965 A 0.8294 C 0.8698
α β 90°00′ γ Z 2 Dx 7.518
Ref. Ibid.

$\xi\alpha$ $n\omega\beta$ $\xi\gamma$ Sign
$2V$ D mp Color BLACK
Ref. Ibid.

Material prepared by fusion of sodium tungstate dihydrate and ferrous chloride tetrahydrate at 800°C

d Å	I/I_1	hkl	d Å	I/I_1	hkl
5.71	15	010			
4.736	40	100			
3.745	35	011			
3.644	35	110			
2.940	100s	111, $11\bar{1}$			
2.856	25	020			
2.481	25	002			
2.474	30	021			
2.443	10	120			
2.367	15	200			
2.194	20	$12\bar{1}$			
2.051	5	112			
2.000	10	211, $21\bar{1}$			
1.902	5	030			
1.873	10	022			
1.823	10	220			
1.765	20	130			
1.712	30	221, $22\bar{1}$			

Fig. 7-12 Card 12–729 from ASTM Powder Diffraction Data File. (**Courtesy American Society for Testing Materials.**)

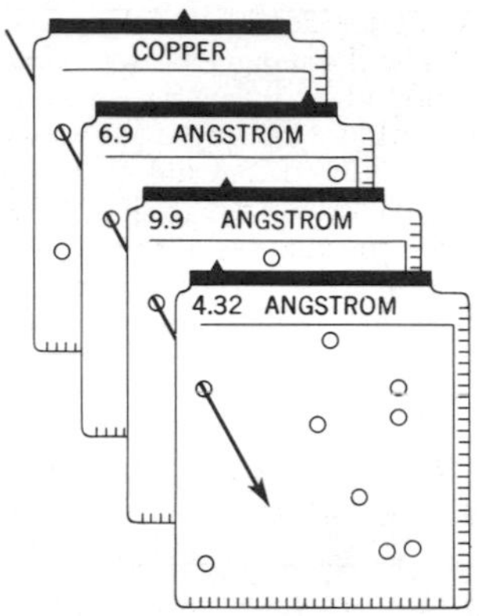
Each of these cards
has holes at position
representing substance
No. 8273 in the file
COPPER
6.9 ANGSTROM
9.9 ANGSTROM
4.32 ANGSTROM

Fig. 7-13 Illustrating the principle of coincidence of holes used in Termatrex cards for the ASTM Powder Diffraction Data File. Substance no. 8273 containing Cu and with *d* values 6.9, 9.9, and 4.32 causes coincidence of punched holes and is thereby identified as a possible source of the diffraction pattern.

It is now possible to obtain the ASTM Powder Diffraction Data File in several forms, listed below. Some of these forms, though simple and compact, are adequate for the needs of a small laboratory that refers to the data only occasionally, and others are valuable for frequent use in larger laboratories. The forms are:

1. Cards based on the *d*'s of the three strongest lines, with sets (sections) issued yearly.
2. A book containing the revised and corrected cards of the first five sections (additional books are to follow).
3. Books in which the eight strongest lines are used in the filing system (not needed for many simple analyses).
4. An index listing chemical data for the file cards. The kinds of books mentioned above, which apply to each section, are included with the purchase of each new section of the cards.
5. IBM cards punched with chemical and diffraction data according to the syst~m of the Wyandotte Chemical Corp., and suitable for searching methods appropriate to IBM machines.
6. "Keysort" cards containing the same information that is given on the standard set of cards, but with wide margins that can be slotted in such a way that searches can be conducted by the use of needles inserted through the slots and holes in the margins.
7. The "Matthews Coordinate Index," developed by F. W. Matthews to use an optical coincidence scheme with plastic cards known as "Termatrex" cards.[1] In this index there is a "standard card" for each individual chemical element. The cards are large enough so that a given coordinate position on a card is alloted to a given substance in the file; i.e., each substance in the file is represented individually by some position on every card. The card for a given chemical element contains holes representing every substance in the index that contains that element. Another card shows every substance

[1] The cards are sold by the American Society for Testing Materials, Philadelphia, Pa. and several forms of reading devices are sold by Jonkers Business Machines, Gaithersburg, Md.

of the file that has one of its strongest reflections within a certain range of d spacing. (A supplementary set of cards reduces the danger that experimental error in d spacings might move the spacing over the boundary into a wrong range and defeat the search.) Superposition of a card for copper with a card for 6.9 A, for example, would then bring holes into coincidence for all compounds of the file that contain copper and have a strong line at $d = 6.9$ A plus or minus a small d range (Fig. 7–13). If the cards are aligned on a reader with a light behind it, the coordinates of these compounds appear as spots of light. The coordinate numbers at these spots lead the operator via an index book to the ASTM cards for the compounds that combine these two properties. By superposing additional Termatrex cards representing additional diffraction lines or elements, the number of coincidences is much reduced.

Crystallite-size determination from line broadening Many factors contribute to the widths of the diffraction lines from powders: (1) inhomogeneous strains; (2) composition variations; (3) the range of wavelengths in the incident radiation; (4) instrumental factors involving (*a*) the thickness of the diffracting layer in a diffractometer specimen or the diameter of the specimen in a Debye camera and (*b*) the widths and heights of the slits defining the incident and diffracted beams; and (5) the mean dimension of the diffracting crystallites in the direction normal to the reflecting plane. Factor 5, *particle-size broadening*, is an analog of the broadening of diffraction lines from an optical grating that has a limited number of ruled lines, a problem which is treated in elementary physics texts. If all other causes of broadening are negligible or are corrected for, the mean crystallite dimension D can be determined—if it lies between about 20 and 3000 A—by using Scherrer's equation for particle-size broadening,

$$D = \frac{k\lambda}{\beta \cos \theta} \qquad (7\text{–}13)$$

where β is a measure of the amount to which a line is broadened by particle-size widening, if the line is for a reflection at Bragg angle θ with wavelength λ, and k is a constant of the order of unity. It is convenient to take β as the width in radians (in terms of 2θ, by convention) of a diffraction peak at half peak height above background intensity—after the peak has been corrected for instrumental and other contributions to width. The value of k depends somewhat upon the shape of a particle. It is 0.9 for spherical particles of uniform size, but varies with shape up to values of over 1.0 for some common shapes; for approximate determinations of size and for relative determinations of size in a series of samples, it may be assumed that $k = 1.0$. The central problem in using Eq. (7–13) for particle-size determination is the problem of correcting the measured peak to remove the influence of factors 1 to 4, mentioned above. Various methods have been proposed which will be found in more extensive treatments of the subject[1]; most are based on rather inaccurate assumptions as to the shape of the diffraction peaks.

Scherrer assumed that $\beta = B_o - b$, where B_o is the observed width and b is the breadth due to factors 1 to 4. This is adequate only for high-angle lines and geometrical condi-

[1] H. P. Klug and L. E. Alexander, "X-ray Diffraction Procedures," Wiley, New York, 1954. E. F. Kaelble, "Handbook of X-rays," chap. 17, McGraw-Hill, 1966.

tions in which b is small compared with B_o. If the peaks are Gaussian (i.e., if the intensity varies according to the function $e^{-k^2x^2}$, where x is the distance from the maximum of the peak), then it has been shown[1] that $\beta^2 = B_o^2 - b^2$, but some investigators have concluded that the Gaussian function is generally not a very good description of the peaks, and procedures for somewhat improved functions have been worked out.[2]

Uncertainties in the meaning of the x-ray results may arise if a considerable range of sizes is present in the sample. Inaccuracies become severe if the mean size is less than about 25 A or more than about 2000 A. Metallic powders and samples in fine-grained polycrystalline form are particularly difficult to study because the broadening from strain gradients and composition gradients is difficult to avoid. Several studies of the domain size of superlattices have been made with line-width measurements of superlattice lines, and here additional possible causes of broadening must be dealt with: stacking faults, anisotropic shapes of ordered domains, internal stresses, and coexisting phases of slightly altered unit cell dimensions.

More elaborate analysis of particle-size broadening is possible if an "unfolding" procedure is used in correcting the observed peak. The Stokes method[3] uses a Fourier treatment of the broadened and standard peaks. Let $h(2\theta)$ be the diffracted-intensity profile of the broadened peak and $g(2\theta)$ the profile of the standard peak, both having been corrected for the $K\alpha_2$ contribution to the $K\alpha_1$ peak by the method of Rachinger.[4] Then the convolution relation[5]

$$h(2\theta) \propto \int_{-\infty}^{+\infty} f(2\theta') \cdot g(2\theta - 2\theta')d(2\theta')$$

yields the true diffraction profile, $f(2\theta')$. The profiles $f(2\theta')$, $g(2\theta)$, and $h(2\theta)$ are each expressed as Fourier series, and the unfolding is accomplished by the use of the Fourier transforms $F(\mathrm{x})$, $G(\mathrm{x})$, and $H(\mathrm{x})$ of these intensity functions, respectively, and the relationship $H(\mathrm{x}) = G(\mathrm{x})\cdot F(\mathrm{x})$. This Stokes correction process, which has been programmed for a computer,[6] provides coefficients that are used in an analysis of peak broadening by the method of Warren and Averbach.[7]

Grain-size determination from spots on Debye rings Grains larger than about 10 μ produce discontinuous Debye rings. The *size* of the individual spots on the rings may be used to determine the grain size in the sample, for with a given camera the average spot size is proportional to the average grain size, provided strain broadening and composition gradients are negligible.[8] Similarly, the *number* of spots recorded under a given set of conditions can be calibrated to yield the average grain size. The lower

[1] B. E. Warren, *J. Appl. Phys.*, vol. 70, p. 679, 1946.

[2] L. E. Alexander, *J. Appl. Phys.*, vol. 25, p. 155, 1954.

[3] A. R. Stokes, *Proc. Phys. Soc. (London)*, vol. 61, p. 382, 1948.

[4] W. A. Rachinger, *J. Sci. Instr.*, vol. 25, p. 254, 1954.

[5] A. Guinier, "Théorie et Technique de la Radiocrystallographie," chap. 13, Dunod, Paris, 1956. A. Guinier, "X-ray Diffraction in Crystals, Imperfect Crystals and Amorphous Bodies," Freeman, San Francisco, 1963. H. Neff, "Grundlagen und Anwendung der Röntgen Feinstruktur-Analyse," Oldenbourg, Munich, Germany, 1962. J. B. Cohen, "Diffraction Methods in Material Science," Macmillan, New York, 1965.

[6] E. N. Aqua, "Computer Programs for the Analysis of Positions and Profiles of X-ray Powder Patterns" (for the IBM 709), in press.

[7] B. E. Warren, *Progr. Metal Phys.*, vol. 8, p. 147, 1959.

[8] R. Glocker, "Materialprüfungen mit Röntgenstrahlen," Springer, Berlin, 1949. G. L. Clark, "Applied X-rays," McGraw-Hill, New York, 1955.

limit for grain-size determinations by spot counting can be extended to 0.5 μ, below the limit obtainable with ordinary Debye cameras or back-reflection cameras, by using microbeams (beams collimated by capillary tubes) and small cameras.[1]

Another method uses a pair of photographs made with identical beam divergence but with different small angles of oscillation of the specimen about an axis normal to the beam.[2]

For a transmission pattern made with a beam of cross section A penetrating a specimen of uniform thickness t, the mean grain volume $\bar{v}$ is given by

$$\bar{v} = \frac{\psi_1 - \psi_2}{N_1 - N_2} \frac{ptA \cos \theta}{\pi}$$

where ψ_1 and ψ_2 are the oscillation ranges that give the numbers of spots N_1 and N_2, respectively, on a Debye ring at Bragg angle θ caused by a reflecting plane with multiplicity factor p.

Exposure times for the two photographs should be in the ratio $\psi_1 : \psi_2$ (e.g., if $\psi_1 = 10°$ and $\psi_2 = 5°$, the exposure for ψ_1 should be twice that for ψ_2). Furthermore, the sample should be thin enough and exposures long enough so that doubling either exposure does not appreciably increase the number of visible spots. The area A is determined by exposing a film at the position of the specimen. The relation of $\bar{v}$ to the linear dimensions of the reflecting grains is discussed in detail in Andrews and Johnson's paper.[3]

Grain sizes can also be estimated from the (calibrated) amplitudes of peaks on a diffractometer rate meter record obtained when the counter is stationary at a position to receive a strong reflection and the specimen is slowly oscillated a few degrees through the parafocusing position.[4]

Small-angle scattering of x-rays Any sample with microscopic regions that differ in density from their surroundings will scatter x-rays in a way that is characteristic of the *size, shape,* and *number* of the regions. If the intensity distribution of the scattered rays is measured at small angles, it is possible to analyze the data to give values for the mean dimensions and the number of the particles—although ambiguity in the interpretation of the data is encountered if a range of sizes or shapes is present.

Small-angle scattering has been found useful for dilute colloidal suspensions with particles $\sim$ 20 to 500 A in size, particularly when the colloids have a high electron density, such as Au, Ag, and many metallic oxides and hydrides. Powdered catalysts and colloidal preparations are studied in this way to investigate the effects of variables in manufacturing processes even when they have a range of sizes that precludes a full and rigorous analysis. Gels, finely divided solids, and precipitates from solid solution give small-angle scattering, but with these the interpretation of the intensity distribution is complicated by interparticle scattering if the concentration

[1] P. B. Hirsch, in "X-ray Diffraction by Polycrystalline Materials," Institute of Physics, London, 1955. A summary has also been published by Hirsch in *Progr. Metal Phys.*, vol. 6, p. 236, 1956.

[2] K. W. Andrews and W. Johnson, *British J. Appl. Phys.*, vol. 10, p. 321, 1959.

[3] Ibid.

[4] C. S. Barrett in J. B. Newkirk and J. H. Wernick (eds.), "Direct Observation of Imperfections in Crystals," Interscience, New York, 1962.

becomes high. A simple interpretation is possible only when interparticle effects are negligible.

An additional complication may arise, which has been troublesome, for example, in metallic samples: a beam that has undergone Bragg reflection may pass from one region into another where the lattice is slightly misoriented with respect to the first and may also undergo a Bragg reflection in the second region. The double-reflected beam then appears to be typical small-angle scattering although its true interpretation is very different, since the intensity distribution is actually related to the preferred orientation of the subgrains rather than to density differences within the sample.[1]

Extensive discussions of small-angle scattering have been published.[2] The intensity distribution of the small-angle scattering does not depend upon the internal structure of the particles, provided it is of uniform density; the particles can be either crystalline or amorphous and their electron density may be either higher or lower than that of their surroundings. The intensity distribution is determined by the external dimensions of the particles; the total intensity is determined by the number and volume of the particles and the difference in electron density between the particles and their surroundings.

With randomly oriented identical particles scattering independently, Guinier has shown that the scattered intensity $I(h)$, where $h = (4\pi \sin \theta)/\lambda$, is given approximately by the following expression, provided $h < 1/R$:

$$I(h) \propto V^2 N \,|\, \rho - \rho_0 \,|^2 \exp\,(-\,h^2R^2/3) \tag{7-14}$$

where R is the radius of gyration of the particle (defined as the root mean square (rms) of the distances from all the electrons in a particle to its center of gravity). The slope of a "Guinier plot" of log I vs. h^2 in the very small 2θ range thus gives R and is widely used for this purpose in small-angle studies. Beyond this linear portion of the plot, the curve depends upon the shape of the particles. For a particle of known shape, R^2 is easily calculated. For spheres, $R^2 = 3a^2/5$ where a is the radius of the spheres. For ellipsoids of revolution, with axes a, a, and va, $R^2 = a^2(2 + v^2)/5$. The effective radius of gyration for samples in which there is a range of particle sizes is given by

$$\bar{R}^2 = \Sigma_m p_m n_m{}^2 R_m{}^2 / \Sigma_m p_m n_m{}^2 \tag{7-15}$$

where p_m is the number of particles with a given size and n_m is the number of electrons in each particle of that size. The larger particles are obviously weighted heavier than the smaller ones in Eq. (7–15).

In practice it is necessary in general to subtract the scattering from slits, windows, and

[1] R. H. Nanahbor, W. G. Brammer, and W. W. Beeman, *J. Appl. Phys.*, vol. 30, p. 656, 1959.

[2] A. Guinier and G. Fournet, "Small-Angle Scattering of X-rays," Wiley, New York, 1955; "X-ray Diffraction in Crystals, Imperfect Crystals and Amorphous Bodies," Freeman, San Francisco, 1963. M. M. Beeman et al., "Size of Particles and Lattice Defects, Part A, Handbook der Physik," vol. 32, p. 348, Springer, Berlin, 1957. H. P. Klug and L. E. Alexander, "X-ray Diffraction Procedures," Wiley, New York, 1954. E. F. Kaelble, "Handbook of X-rays," McGraw-Hill, New York, 1966. C. G. Shull and L. C. Roess, *J. Appl. Phys.*, vol. 18, pp. 295, 308, 1947. K. L. Yudowitch, *Rev. Sci. Instr.*, vol. 23, p. 83, 1952.

solvent from the total measured intensities before analyzing the intensity distribution; and if slits are used it is also necessary to correct for the angular divergence of the rays—a procedure discussed in detail in the original papers.

Since many different shapes have the same R and since a sample may contain a range of sizes and shapes of particles, the exact interpretation of the x-ray intensity distribution may be ambiguous unless information from other sources, such as electron micro-

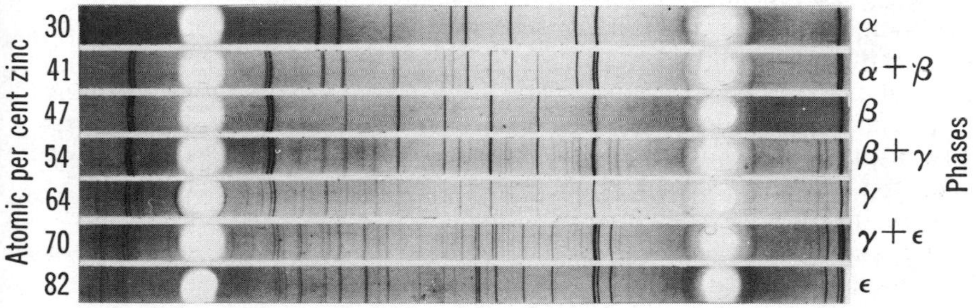

Fig. 7-14 Powder-diffraction patterns of alternating one-phase and two-phase regions in the Cu-Zn system. (Courtesy H. Pops.)

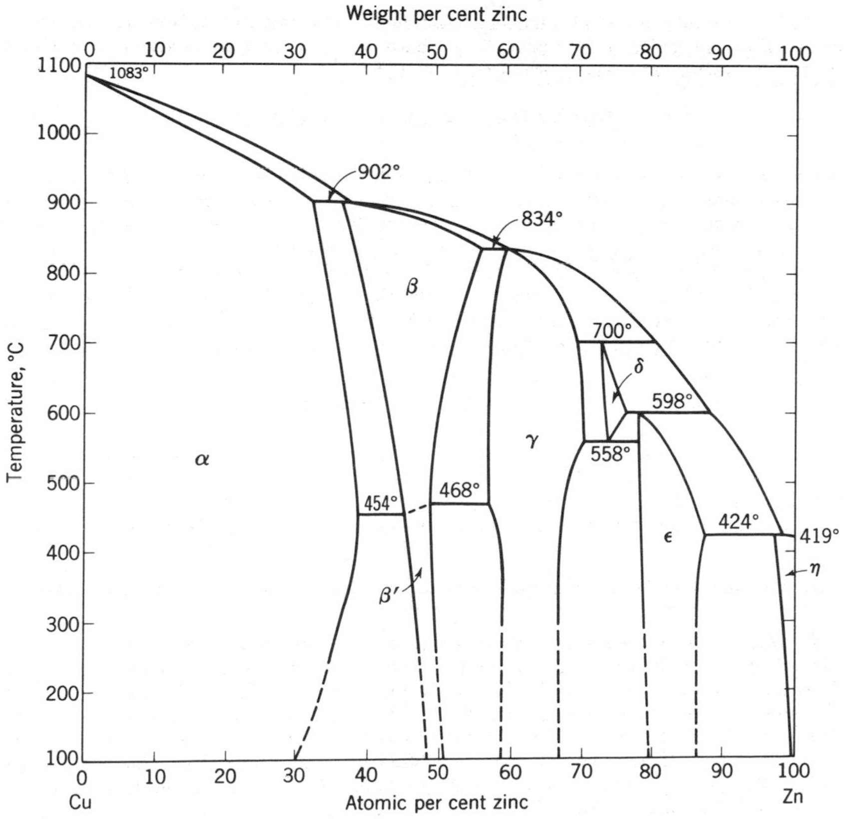

Fig. 7-15 Constitution diagram for copper-zinc alloys.

scopy, is available. A Guinier plot that is not linear at small angles is a clear indication that a range of particle sizes may be present.

Both filtered radiation and crystal-monochromatized radiation have been used for work in this field, and equipment for both counter diffractometer and photographic recording has become commercially available. The equipment is also useful for studies of crystalline samples and fibers with unusually large periodicities—say greater than 25 A—as is common, for example, with polymers.

Phase-diagram determination The powder method is particularly useful for determining phase diagrams, as will be immediately obvious from Figs. 7–14 and 7–15. In two-phase regions of a binary equilibrium diagram or in two-phase or three-phase regions of a ternary equilibrium diagram, the separate diffraction patterns of the individual phases superimpose (see Fig. 7–14). As compositions approach a boundary of one of these regions at a given temperature, the intensity of one phase increases at the expense of the other(s) in accord with the "lever rule" for the relative amounts of the phases: along a tie line such as $X - Y$ in Fig. 7–16 the α phase of composition X and the β phase of composition Y coexist in the proportions given by the relative lengths of the lines PY and PX; i.e., PY/XY of the total alloy is α, and PX/XY is β. Extrapolation to the point of disappearance of one phase on a plot of intensity vs. composition thus locates a boundary. This is the "disappearing-phase method," which is often less precise than the "parametric method" given below.

Within single-phase regions the lattice parameters change gradually with changing composition. For all compositions along a tie line in a two-phase region at a given temperature, the parameters of each phase are the same as the values corresponding to the ends of the tie line. Thus, each parameter of the α phase in alloy P (Fig. 7–16) will match some value on a curve that shows the composition dependence of that parameter at the same temperature in the single phase α region; the position of the $\alpha/\alpha + \beta$ boundary is located by the composition of this matching point. An analogous principle holds for the tie lines in two-phase regions of three-component and n-component systems; also for the three-phase regions of a ternary system (since all three phases are then invariant) and for the n-phase regions of an n-component system.

Long experience has shown the precautions that should be taken in

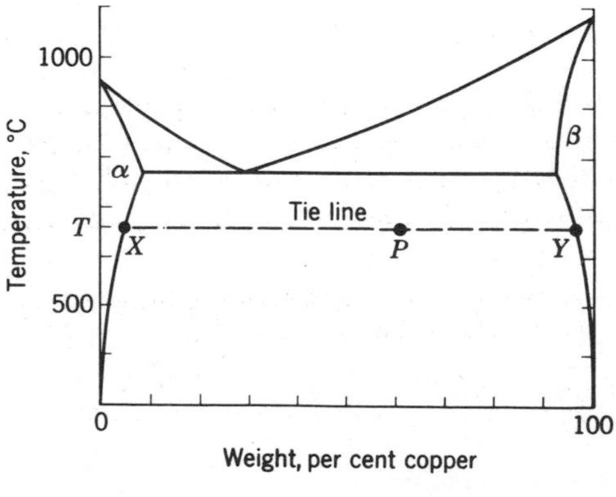

Fig. 7-16 The lever principle illustrated in the two-phase region of the copper-silver diagram.

using these methods.[1] In brief, errors should be avoided that arise from (1) incomplete removal of segregation in the original ingot, (2) oxidation, (3) sublimation, (4) contamination by diffusion from adjoining solids or by reaction with vapors, (5) alteration of surface composition by the selective action of an etchant, and (6) internal strains resulting from quenching, filing, or grinding of the samples and conceivably also due to retained defects.

Usually the parametric method is employed at room temperature using quenched samples. A technique that is generally satisfactory is as follows. The alloy is first homogenized in a single phase field (if possible), and after quenching, small portions of it are annealed in a two-phase region at a number of successively lower temperatures. After quenching, the samples are powdered by filing or crushing and are reannealed in small capsules of glass or silica at the same temperatures at which the respective bulk specimens were annealed and for times sufficient to produce stress relief. (A small flame is used to avoid heating the powder, and minimum unoccupied space is left in order to minimize the loss of a volatile constituent.) The capsules are plunged into water and smashed immediately, thus cooling the particles very fast.

Determining compositions by chemical analysis also requires care.[2] Computing the percentage of one element by difference may lead to errors of as much as 10 percent due to dust, moisture, etc., when filing is done under ordinary conditions. Filing under argon is recommended.

When quenched samples are likely to undergo, during cooling or thereafter, any decomposition, phase transformation, or change in atomic distribution within a phase, the x-ray work should be done at elevated temperatures.[3] Precipitation of a supersaturated concentration of vacancies is believed to have caused some variation of lattice parameters at room temperature; vacancy migration accelerates any tendency for solute atoms to order or to cluster and thereby to alter lattice parameters.[4]

Rapid cooling; "splat" and "crusher" cooling Enhanced solubilities, nonequilibrium phases and unusual amorphous and crystalline struc-

[1] See, for example, a discussion by W. Hume-Rothery, J. W. Christian, and W. B. Pearson, "Metallurgical Equilibrium Diagrams," Institute of Physics, London, 1952. A. Taylor, *Phil Mag.*, vol. 35, p. 215, 1944. G. W. Brindley, *Phil. Mag.*, vol. 36, p. 347, 1945. The pitfalls and safeguards connected with the x-ray methods and the question of the relative advantages of the x-ray and chemical metallographic methods are fully discussed and illustrated in a number of discussions published in *J. Inst. Metals*, vol. 69, p. 1, 1943; vol. 76, p. 679, 1950. The nature of metallurgical phase diagrams, both binary and ternary, is discussed in detail in F. N. Rhines, "Phase Diagrams in Metallurgy," McGraw-Hill, New York, 1956. A comprehensive review of phase diagrams and their determination is given by G. V. Raynor in R. W. Cahn (ed.), "Physical Metallurgy," North Holland Publishing, Amsterdam, 1965.

[2] W. Hume-Rothery and G. V. Raynor, *J. Sci. Instr.*, vol. 18, p. 74, 1941.

[3] For example, in Al-Ag alloys, M. Simerska, *Czechoslovak J. Phys.*, vol. B12, p. 54, 1962, found positive deviations from Vegard's law at a high temperature, whereas a negative deviation has been reported at room temperature with quenched alloys.

[4] R. Feder, A. S. Nowick, and D. R. Rosenblatt, *J. Appl. Phys.*, vol. 29, p. 784, 1958. H. M. Otte, *J. Appl. Phys.*, vol. 32, p. 1536, 1961; vol. 33, p. 1436, 1962. T. B. Massalski and J. E. Kittl, *J. Australian Inst. Metals*, vol. 8, p. 91, 1963.

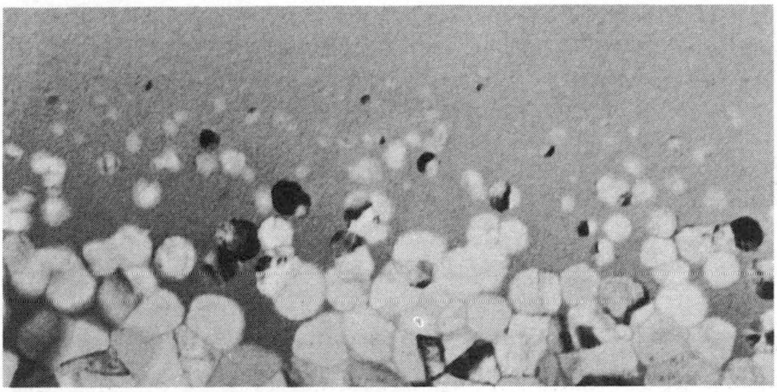

Fig. 7-17 Thin foil of Pd–20 atomic percent Si in amorphous and fine-grained states resulting from splat cooling from the liquid state. Electron transmission micrograph. ×120,000. (P. Duwez.)

tures have been obtained when small amounts of liquid alloys were cooled at extreme rates. Rapid cooling is achieved in a variety of ways: injecting a small droplet of a molten alloy into a liquid quenching bath,[1] propelling a small liquid droplet (about 25 mg) by means of a shock wave against a solid copper target that can be moving or stationary and can be cooled to cryogenic temperatures,[2] and firing a small droplet from a pistol.[3] Colloquially, the term *splat technique* is sometimes used for these methods.

Transmission electron microscopy shows the small grains (often of 0.3-μ size) produced by splat cooling; in Fig. 7–17 the upper part of the micrograph is of amorphous material, while the lower part developed small crystalline grains because it happened to be less drastically quenched.

Fast cooling is also accomplished by a *crusher apparatus*.[4] A drop of liquid falling from a crucible passes through a beam of light and triggers a mechanical device. Two flat metal compression plates are then slammed together to crush the drop between them. The resulting specimens are more uniform in thickness than those obtained with splat devices. Extreme quenching rates are also obtained when vapor is deposited on cold substrates.[5]

[1] W. T. Olsen, Jr., and R. Hultgren, *Trans. AIME*, vol. 188, p. 1223, 1950.

[2] P. Duwez, R. H. Willens, and W. Klement, Jr., *J. Appl. Phys.*, vol. 31, p. 1136, 1960. P. Duwez and R. H. Willens, *Trans. AIME*, vol. 227, p. 362, 1963. P. Duwez in T. B. Massalski (ed.), "Alloying Behavior in Concentrated Solid Solutions," Gordon and Breach, New York, 1965.

[3] P. Predecki, B. C. Giessen, and N. J. Grant, *Trans. AIME*, vol. 233, p. 1438, 1965; P. Predecki, A. W. Mullendore, and N. J. Grant, *Trans. AIME*, vol. 233, p. 1581, 1965.

[4] P. Duwez and R. H. Willens, private communication, 1964; the principle was suggested by P. Pietrokowsky, *Rev. Sci. Instr.*, vol. 34, p. 445, 1963.

[5] Reviews of this field giving details of the techniques used, the amorphous and crystalline structures that are obtained, and various properties of such phases may be found in publications by P. Duwez in footnote 2, above, and in articles that are in press at the time of writing of this chapter: *Prog. Chem. Solids*, 1966; and in P. S. Rudman and J. Stringer (eds.) "Phase Stability in Metals and Alloys," McGraw-Hill, 1966.

8

The determination of crystal structure

Metals and alloys tend to assume the simpler crystal structures, which can be solved by relatively simple x-ray investigations. It is sometimes possible to guess the structure and to confirm the guess by a few photographs. On the other hand, many organic and inorganic compounds have crystal structures so complex that they resisted years of research without being deciphered. Intermediate between these two extremes are the many structures that can be solved by the methods discussed in the preceding chapters. The steps that are generally followed in such a crystal-structure determination are briefly outlined in this chapter.

There is no completely standardized procedure for determining a structure, for each crystal presents new difficulties that may have to be overcome by devising new methods of attack. To do effective work in this field of research requires a critical attitude toward experimental data and inferences drawn from them and a generous expenditure of patience, understanding, and ingenuity. In solving crystal structures of great complexity it is possible to obtain great benefits from the mathematical device of using Fourier series to analyze the diffraction data.

The usual steps in determining the structure of a crystal are as follows:

1. Determination of symmetry class (macroscopic symmetry).
2. Determination of unit cell (space lattice and unit cell dimensions).
3. Determination of space group.
4. Calculation of number of atoms or molecules in the unit cell.
5. Determination of the sets of equivalent points in the space group that could be occupied by each kind of atom.
6. Determination of approximate parameters for the positions of the atoms.
7. Refinement of parameters to give the best agreement between calculated and observed intensities. Normally "best agreement" is taken to mean the minimum value of the R factor defined by the relation

$$R = \frac{\Sigma \mid F_{obs} - F_{calc} \mid}{\Sigma \mid F_{obs} \mid}$$

i.e., the sum of the relative discrepancies between calculated and observed values of $\mid F_{hkl} \mid$—usually with each term in the sum weighted according to an arbitrary weighting scheme.

8. Computation of important interatomic distances in the structure, and

identification of configurations or groups of atoms that may relate the structure to the structure of other crystals.

9. Identification of any important content of imperfections, such as stacking faults.

It is not always possible to complete all these steps, but at least the first three to five can usually be completed if suitable specimens are available. On the other hand, it is often possible to arrive at the final structure much more directly if a similarity is detected between the diffraction pattern of a sample and the pattern for a crystal of similar or closely related structure, when this structure has previously been fully determined.

Determination of the symmetry class The symmetry of a crystal is determined from an optical and/or a physical examination of the crystal or from the diffraction patterns. When well-formed growth faces or cleavage faces are present, the angles between them may be determined by the use of an optical goniometer; the macroscopic symmetry of the crystal may then be recognized if the poles of the faces are plotted on a stereographic projection.[1] Measurements of interfacial angles must be interpreted with care, however, for certain faces of low indices may not appear, or the sample may contain twins such that it appears to have higher symmetry than it actually possesses. A study of etch pits on the surfaces may aid in recognizing symmetry elements. Transparent crystals may be examined by transmitted polarized light in a petrographic microscope to detect the presence of twins and perhaps to determine some of the symmetry properties. Other physical properties also yield some information on symmetry (e.g., pyroelectricity, piezoelectricity, etc.).

The most effective determination of symmetry, however, is based upon the symmetry that is evident in diffraction patterns. For this purpose Laue photographs have sometimes been used, but recently Weissenberg or precession photographs are most commonly used, or data obtained on a diffractometer. An important principle of diffraction symmetry is that in general a diffraction pattern appears always to come from a crystal containing a center of symmetry; special techniques are necessary to disclose the absence of a center of symmetry. A crystallographer can usually determine not only the macroscopic elements but also the microscopic elements (e.g., screw axes and glide planes) and the one or more space groups to which a crystal belongs.

Determination of the unit cell The rotating-crystal method is preferred when choosing the correct unit cell from the various cells that can be imagined, for there is then less likelihood of adopting a cell in which some edge is a multiple of the true unit cell. Rotation photographs around each cell edge in turn give the edge lengths directly from layer-line spacings. Laue photographs may also be used and serve as a check on the determi-

[1] Further remarks on crystallometry will be found on p. 50.

nation. For precise measurement of axial lengths the powder method is usually preferred, although it is possible to achieve precision in a rotating-crystal photograph with back-reflection technique.

To prevent confusion it has been necessary to adopt conventional rules regarding the choice of axes for the unit cell and the labeling of these axes. It is the custom among crystallographers to choose the shortest three axes that will give a unit cell having the symmetry of the crystal. If orthogonal axes are desired in spite of the symmetry (e.g., orthohexagonal axes), the shortest three orthogonal axes are chosen. A test for the proper unit cell is that each cell edge must be shorter than the face diagonals of all faces touching it. Base-centered and f.c.t. cells are discarded in favor of smaller cells that are simple or b.c.t.; hexagonal and rhombohedral cells are always chosen so as to be simple, and monoclinic cells to be either simple or base-centered. Triclinic crystals are referred to axes giving the smallest simple cells that permit α, β, and γ to be equal to or greater than 90° and the direction cosines of [111] to be all positive or zero.[1]

Determination of space-lattice and space group If a unit cell is not simple (i.e., primitive) but is centered on one or more faces or is body-centered, then certain reflections will be absent. From the structure-factor equation, it can be seen that the following criteria hold:

I—body-centered lattices: reflections absent if $h + k + l$ is odd.
F—face-centered lattices: reflections absent if h, k, l are mixed odd and even.
Base-centered lattices:
A—face-centered: reflections absent if $k + l$ is odd.
B—face-centered: reflections absent if $h + l$ is odd.
C—face-centered: reflections absent if $h + k$ is odd.
P—simple primitive space lattices: no systematic absences.

These characteristics result from the fact that in the directions corresponding to missing reflections the waves scattered by body- or face-centered atoms are exactly out of phase with those scattered by the atoms at the cell corners. That is, the spacings of certain planes are halved, and odd-order reflections from such planes are consequently destroyed.

Microscopic symmetry elements likewise reduce certain spacings and destroy corresponding reflections. Thus, a glide plane halves the spacings in the glide direction. A twofold screw axis halves the spacings along the screw axis, while a threefold screw axis reduces spacings along itself to thirds, etc.

A systematic application of these principles to each of the 230 space

[1] "International Tables for X-ray Crystallography," vol. 2, Kynoch, Birmingham, England, 1959, discusses this and other conventions in choosing axes and settings. A convention often used is to choose a setting for triclinic or monoclinic crystals such that $c < a < b$ when determining a crystal structure, but when reporting the structure determination this choice of axes may be changed to a different one because of symmetry considerations; for example, in a monoclinic crystal the twofold axis should be taken as the b axis.

groups has been completed, and characteristic extinctions or nonextinctions have been tabulated for each.[1] To determine the space group or the several possible space groups to which a crystal belongs, the crystallographer first assigns indices to his observed reflections, lists them, notes the characteristic absences, and then searches the tables for space groups having similar absences. In so doing he must bear in mind that his choice of a, b, and c edges may be interchanged with respect to those listed in the tables and must take precautions to avoid being misled by this. (For example, he may list the absences in terms of every possible permutation of a, b, and c axes.) A large number of indexed reflections should be available in order to show the characteristic absences well.

Properly done, this procedure yields a list of every possible space group to which the crystal could belong. Prism-face reflections (one index zero) are studied carefully to distinguish between systematic absences and "accidental" extinctions arising from the particular position of certain atoms in the cell, and an attempt is made to single out the space group that explains all systematic absences. If it is known to which of the 32 classes a crystal belongs, it is sometimes possible to establish the space group unequivocally, but when the symmetry class is in doubt this may not be possible. Reflection planes and rotation axes do not cause extinctions, and therefore the presence or absence of them cannot be determined from lists of reflections. Another source of ambiguity should be mentioned: the shielding of some characteristic extinctions by a more general class of extinctions. When there is a characteristic set of extinctions in some general class of reflections, this set includes corresponding extinctions in a less general class. For example, a body-centered lattice produces extinctions in hkl reflections whenever $h + k + l$ is odd, but it also extinguishes $h0l$ reflections when $h + l$ is odd and $h00$ reflections when h is odd. If the body-centered crystal in this example contains a screw axis 2_1, this axis cannot be detected, for it also extinguishes $h00$ reflections when h is odd.

Certain sets of special atomic positions occur in many different space groups. Such a set is called a *lattice complex*. For example, the b.c.c. set of points (coordinates 000, $\frac{1}{2}\frac{1}{2}\frac{1}{2}$) is a special position of the space group O_h^9—$Im3/m$, but it is found also in space group O_h^1—$Pm3/m$. Heavy atoms may be located at such a set of points and give strong reflections, while light atoms, scattering weakly, may be at places corresponding to only one of these space groups. Lists of lattice complexes are used by some crystallographers to aid in recognizing the various possibilities.

Another convenience in space-group tables is a list of *point symmetries*, which gives the symmetry of the group of atoms that exists around a given point position in the crystal. A study of such a list shows possible positions for a group of atoms, such as a silicate group, and eliminates incompatible positions. Other tables, together with all well-established formulas and

[1] "International Tables for X-ray Crystallography," vol. 1, op. cit., 1952. W. T. Astbury and K. Yardley, *Phil. Trans. Roy. Soc.* (*London*), vol. A224, p. 221, 1924. C. Hermann, *Z. Krist.*, vol. 68, p. 257, 1928. J. D. H. Donnay and D. Harker, *Nat. Can.*, vol. 67, p. 33, 1940.

charts that have proved useful in crystallographic work, will be found in "International Tables for X-ray Crystallography."

Number of atoms or molecules per unit cell The number of molecules in the unit cell is determined from the measured density of the substance, ρ, the known mass of each molecule, M, and the volume of the cell, V. The obvious relation is $\rho = nM/V$, where n is the number of molecules in the cell and M is the molecular weight multiplied by the mass of an atom of unit molecular weight (1.66×10^{-24} g). In a chemical compound with definite molecular formula, n is integral and is determined as the integer that most nearly satisfies the relation $n = \rho V/M$. Not all substances, however, have the ideal structure implied by this statement. If a substance has a "defect structure," there may be a nonintegral number of atoms of one kind or of several kinds per unit cell. In the case of alloys the concept of "molecules" is of no value, and the density formula should be put in terms of the number of atoms of each kind per unit cell.

Determination of atomic positions For many crystals for which full structure determinations have not been completed, the unit cells and space groups have been determined. It is first advisable, therefore, to study the listings available,[1] but it is wise to confirm any prior space-group identification since an error in this could waste much effort later.

If the space group or possible space groups have been identified and the number of atoms of each kind in the unit cell is known, in simple cases it may only be necessary to choose between several alternate sets of positions for the atoms, each of which is consistent with the symmetry and is listed in the space-group tables. If, for example, there are four atoms of a certain kind in a unit cell, they may be equivalent in the structure and therefore may lie on a set of equivalent positions of multiplicity four; the coordinates may then be specified exactly in the tables, or there may be one or more parameters (listed as variables x, y, z, etc.) which must be determined before the atom coordinates are completely known.

As mentioned in Chap. 4, the atom-position parameters govern the magnitude of the structure factor, F_{hkl}, and the observed intensities of the reflections, which are proportional to F^2_{hkl}. A correct set of parameters will produce agreement between observed and calculated intensities. To reach a satisfactory agreement it is necessary to make successive adjustments of the parameters, computing the intensities after each adjustment and concluding from the intensity changes what additional adjustment might improve the agreement. This process of "refinement" of the structure is continued until some "reliability index" reaches a value regarded as satisfactory. It is customary to use the index $R = \Sigma \mid F_{obs} - F_{calc} \mid / \Sigma \mid F_{obs} \mid$

[1] "Structure Reports," Oosthoek's Uitgevers Mij, Utrecht, The Netherlands or Polycrystal Book Service, Pittsburgh, Pa. For minerals: "The System of Mineralogy of J. D. Dana and E. S. Dana," 7th ed., by J. D. Dana, E. S. Dana, Charles Palache, Harvey Berman, and Clifford Frondel, Wiley, New York, 1944 and subsequent volumes. J. D. H. Donnay (ed.), "Crystal Data," 2d ed., American Crystallographic Association, Polycrystal Book Service, Pittsburgh, Pa., 15238, 1963.

for this purpose and to continue refining the parameters until R is reduced to a value of the order of 0.1 (0.05 or lower if precision intensities are available and the problem is not too difficult) or until the time and cost of further refinement become prohibitive. The sum is taken over all the observed intensities or over all the calculated intensities. Some investigators use a similar formula for R based on F^2 instead of F. Although nearly everyone uses and reports an R-value test, it is not highly informative. More detailed information on how to proceed with refinement is obtained by using a Fourier synthesis with coefficients that are the values of $F_{obs} - F_{calc}$, for this "difference synthesis" indicates how individual atoms should be moved to improve the agreement.[1]

Further remarks on factors that must be considered in deriving F values from the measured intensities are deferred to later sections of this chapter in order to preserve a better continuity of presentation of the steps in determining a structure.

In compounds of stoichiometric composition a set of equivalent points in a unit cell will be occupied, normally, by atoms of a single chemical species. In solid solutions, however, the space group remains the same over a considerable range of compositions and a set of equivalent points may be occupied by different kinds of atoms. The distribution of each kind of atoms on the various occupied sets of points must then be considered. Some degree of order of a *superlattice* may then exist. In a few alloys it has been found, also, that some of the positions of a set remain vacant, forming a *defect structure*. In compounds, the number of molecules may be greater than the number of asymmetric units that is required to build up the symmetry of the space group. For example, if a triclinic unit cell has only one position for an asymmetric group of atoms per unit cell and yet there are two or more molecules in the cell to be placed somewhere, it will be necessary in solving the structure to determine the coordinates of each of the atoms in each of the molecules making up the asymmetric unit. On the other hand, if the number of molecules and the number of asymmetric units per unit cell are equal, finding the coordinates of the atoms of one molecule is sufficient, since the space-group tables will specify the coordinates of all the others.

Crystals with one or two parameters can be solved rather directly (graphical aids are useful in this work),[2] and more parameters can be handled if they happen to be separable into independent pairs. In complicated structures it is impossible to guess or to deduce by straightforward methods what values of the many parameters are likely because there is a multiple infinity of atom positions to be considered. For these crystals the methods of Fourier series are of great value in fixing parameters and indicating atomic arrangements.

[1] H. Lipson and W. Cochran, "The Determination of Crystal Structures," pp. 298, 307, G. Bell, London, 1957. M. J. Buerger, "Crystal Structure Analysis," p. 595, Wiley, New York, 1960.

[2] W. L. Bragg, *Nature*, vol. 138, p. 362, 1936. W. L. Bragg and H. Lipson, *Z. Krist.*, vol. 95, p. 323, 1936. J. M. Robertson, *Nature*, vol. 138, p. 683, 1936.

A great many highly developed techniques and principles are now available to aid in structure analysis; these are ably presented in an expanding list of advanced books.[1] Crystals having many atoms per unit cell have often been solved, when they are one of an isomorphous series of compounds, by substituting atoms of a different kind for a set of atoms in the crystal, thus changing the intensities of certain reflections and thereby revealing the positions of the substituted atoms. In some cases this can be done with negligible distortion of the structure, but in other cases errors may be introduced if it is assumed that the structure is essentially unchanged.

In recent years computer programs are used not only for the computation of unit cell dimensions from diffraction data (see Chap. 7) but also for conducting least-squares refinement of parameters, for automatically plotting electron-density maps, for calculating interatomic distances in crystals with known atomic coordinates and for other special purposes. The International Union of Crystallography has undertaken to publish a series of such under the title "I. U. Cr. World List of Crystallographic Computer Programs."[2]

Sizes of atoms and ions A knowledge of atomic and ionic sizes and of habitual groupings and the way they can link together is of great value in determining structures and judging the validity of proposed structures. Much empirical information has been accumulated on the subject.[3]

The distances between atom centers in crystals can be interpreted as the sum of the radii of the two neighboring atoms, as if they were tightly packed spheres. After lengthy study of crystallographic data it was found that a consistent set of radii can be assumed, so that a given element always has approximately the same radius when in the same valence state and in similar surroundings. Sizes depend on whether an atom is closely bonded to 1, 2, 3, 4, 6, 8, or 12 neighbors; as the coordination (number of nearest neighbors) increases, the radius of an atom or ion increases. A further discussion of sizes and the factors that influence them is given in Chap. 13.

A major factor governing the effective sizes of atoms and ions in crystals is the nature of the binding forces that are operating—the ionic, covalent, metallic, van der Waals, or intermediate type. When electrons are added to a neutral atom, its size increases; when they are subtracted, its size decreases. These effects are not small, as will be seen from the following

[1] "International Tables for X-ray Crystallography," vols. 1, 2, 3, Kynoch, Birmingham, England, 1952, 1959, 1962. M. J. Buerger, "Crystal Structure Analysis," Wiley, New York, 1960. H. Lipson and W. Cochran, "The Determination of Crystal Structures," G. Bell, London, 1957.

[2] The first edition was published in September, 1962, and was supplied, in the United States, to members of the American Crystallographic Association; D. P. Shoemaker was editor. Correspondents in many countries are collecting additions for subsequent editions.

[3] See C. W. Bunn, "Chemical Crystallography," Oxford, New York, 1945. A. F. Wells, "Structural Inorganic Chemistry," 3d ed., Clarendon, Oxford, 1962. Interatomic distances in inorganic compounds are tabulated in "International Tables for X-ray Crystallography," vol. 3, p. 257, Kynoch, Birmingham, England, 1962.

radii for atoms having various positive and negative charges:

O^{--}	1.32	S^{2-}	1.74	Fe	1.24	Cu	1.28
O	0.60	S	1.04	Fe^{2+}	0.83	Cu^{+}	0.96
		S^{6+}	0.34	Fe^{3+}	0.67		

Variations are also introduced by the valence state of neighboring ions and by their sizes (radius ratio effect).

Example of structure determination Jacob and Warren's determination[1] of the structure of uranium will serve to illustrate many of the steps in a determination of crystal structure. Powder-diffraction data alone were sufficient to solve the structure, even though it is orthorhombic.

Thirty-nine lines were measured on a powder pattern[2] and were graded in intensity as strong, medium, weak, or very weak. Simple tests showed that the structure was not cubic, hexagonal, or tetragonal; it could be solved on the basis of an orthorhombic cell. The orthorhombic reciprocal lattice points are related to the spacings by the equation

$$\frac{1}{d^2} = h^2a^{*2} + k^2b^{*2} + l^2c^{*2}$$

Any network of points in this lattice that contains the a^*, b^*, or c^* axes will be an orthogonal net, and the lattice can be considered as constructed of a series of these nets. A Hull-type chart was made for the two-dimensional nets by using the logarithm of this expression with $k = 0$. The experimental d values were plotted on a strip, and by moving the strip over the chart in the usual way several positions were found at which the d values fitted the lines of the chart. Each of these indicated two possible axes of the reciprocal lattice, at least one of which was a^*, b^*, or c^*, and consideration of these tentative solutions led to assuming a unit cell with $a = 2.852$, $b = 5.865$, and $c = 4.945$, which was found to predict all observed spacings satisfactorily.

After assigning indices to each reflection, it was noticed that the general reflections hkl occur only for $h + k$ even (111, 131, etc.); and among the prism reflections, $h0l$ occurs only for h and l even, $0kl$ only for k even, and $hk0$ only for $h + k$ even. Reference to tables showed that these characteristics are found in the orthorhombic space groups only in D_{2h}^{17}—$Cmcm$. The number of atoms in the cell was computed and found to be 4. If these were placed at the special positions $4a$ or $4b$ they would halve the c spacings and cause hkl to occur only if l were even; it was therefore concluded that the atoms were in the third set of special positions, $4c$, with coordinates $0y\frac{1}{4}$; $0\bar{y}\frac{3}{4}$; $\frac{1}{2},y+\frac{1}{2},\frac{1}{4}$; $\frac{1}{2},\frac{1}{2}-y,\frac{3}{4}$ and the value of y was then determined. The structure factor for the space group with atoms in $4c$ reduces to $F = 4f \sin 2\pi ky$ when l is odd and to $F = 4f \cos 2\pi ky$ when l is even. From the fact that 020 and 022 are very weak, it follows that $\cos 2\pi 2y$ is nearly zero and y is approximately $\frac{1}{8}$.

By trial it was found that assuming $y = 0.105 \pm 0.005$ gave satisfactory agreement with observed intensities when the relative intensities of pairs of neighboring lines were considered (in order to minimize errors from the strong absorption in the sample). A plot of the structure is shown in plan and elevation in Fig. 8–1, with the height of atoms above the plane of the drawing indicated by the fractions of the cell edges marked at the atom positions. Plots and models aid in locating important interatomic distances and in understanding the structure; in this case a study of drawings disclosed that the structure can be considered as distorted c.p.h. in which 4 of the 12 nearest neighbors are moved in to appreciably closer distances.

[1] C. W. Jacob and B. E. Warren, *J. Chem. Soc.*, vol. 59, p. 2588, 1937.

[2] Copper radiation was filtered through nickel to remove $K\beta$, and 0.002-in. aluminum covered the film to remove fluorescent M radiation from the sample. Copper filings were mixed with the uranium filings in some exposures for calibration.

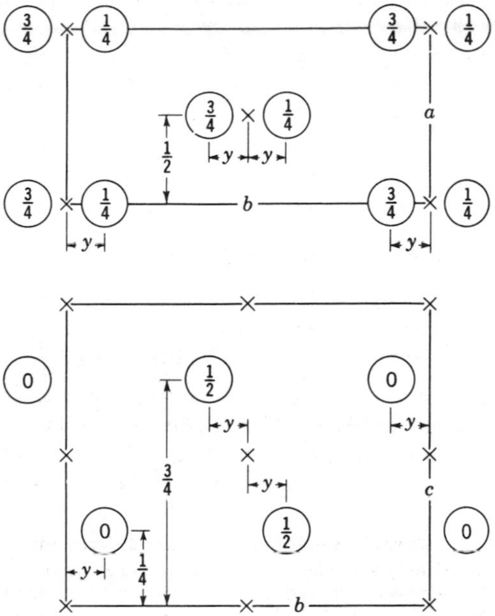

Fig. 8-1 Alpha-uranium. (C. W. Jacob and B. E. Warren.)

Factors governing intensities of x-ray reflections The measured intensities of reflections depend upon many factors, each of which must be given consideration if a structure determination is to be reliable. Effects arise from the state of polarization of the x-rays, absorption of the rays, and geometrical factors, as well as the mean position and thermal motion of each of the atoms. Some fundamentals of these subjects are presented below, starting with the scattering from a single electron.

Scattering by an electron J. J. Thomson[1] derived the *classical-theory formula* for the intensity of x-rays that would be scattered by a free electron or by an electron that is held under negligible constraining forces. Consider a polarized beam of x-rays falling on an electron at *e*, Fig. 8–2. The electric field of the incident beam will accelerate the electron with a vibratory motion and cause it to radiate electromagnetic waves of the same wavelength as the original beam. The intensity I_e of this secondary radiation at the point P will be related to the intensity of the primary beam, I_0, the distance r, and the angle α between the direction of observation and the direction of the electric vector of the incident radiation. The relation derived by Thomson is

$$I_e = \frac{I_0 e^4}{r^2 m^2 c^4} \sin^2 \alpha \tag{8–1}$$

[1] J. J. Thomson, "Conduction of Electricity through Gases," 2d ed., p. 325, University Press, Cambridge, 1928, reviewed in A. H. Compton and S. K. Allison, "X-rays in Theory and Experiment," p. 117, Van Nostrand, New York, 1935.

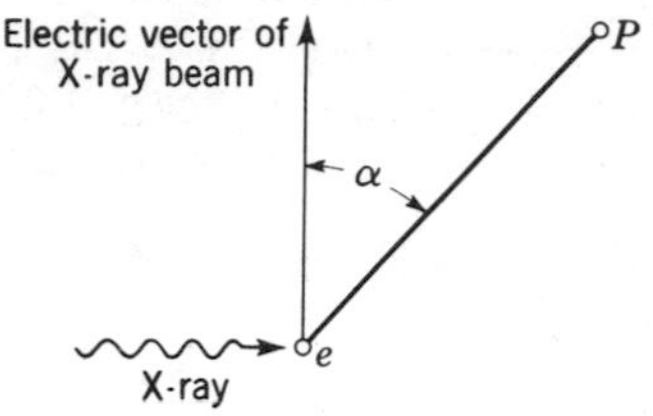

Fig. 8-2 Scattering from an electron.

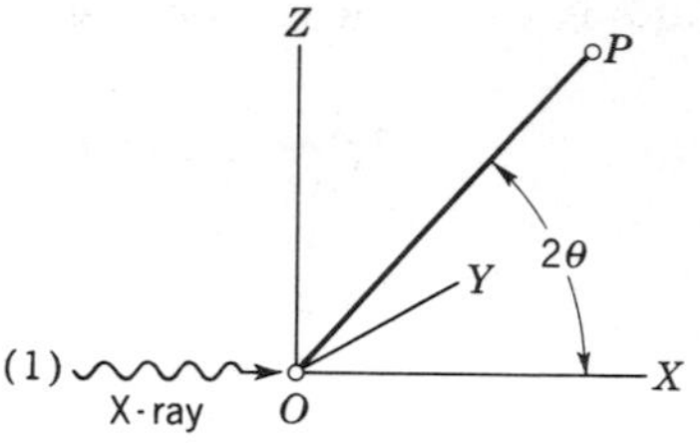

Fig. 8-3 Coordinates for scattered ray.

where e and m are the charge and mass of the electron, c is the velocity of light, and r is the distance from the electron to the point P.

We shall apply Thomson's formula to an electron at the origin of the coordinate axes of Fig. 8–3, with the incident beam proceeding along the X axis and the scattered rays observed at point P in the XZ plane. Let the angle between OP and the X axis be 2θ (since θ will then correspond to the angle in Bragg's law). If the electric vector of a polarized beam is parallel to Z and if the beam has intensity I_z, the intensity at P according to Eq. (8–1) will be

$$I_e = \frac{I_z e^4}{r^2 m^2 c^4} \cos^2 2\theta \tag{8–1a}$$

while if the electric vector of the polarized beam is normal to X and Z, and of intensity I_y, the intensity at P will be

$$I_e = \frac{I_y e^4}{r^2 m^2 c^4} \tag{8–1b}$$

Under ordinary conditions the incident beam is unpolarized, and the electric vector E_0 of the primary ray may be resolved into components E_y along Y and E_z along Z such that

$$E_y^2 + E_z^2 = E_0^2$$

With unpolarized rays the electric vector occurs with equal probability at all angles; therefore, on the average, its component along Y is equal to its component along Z. On the average, therefore,

$$E_y^2 = E_z^2 = \tfrac{1}{2}E_0^2$$

and since the intensity is equal to the square of the amplitude, it follows that

$$I_y = I_z = \tfrac{1}{2}I_0$$

The intensity for an unpolarized beam is obtained by adding the contributions of the two components specified in (8–1*a*) and (8–1*b*), giving

$$I_e = \frac{e^4}{r^2m^2c^4}(I_y + I_z \cos^2 2\theta) = \frac{e^4}{r^2m^2c^4}(\tfrac{1}{2}I_0 + \tfrac{1}{2}I_0 \cos^2 2\theta) \quad (8\text{–}2)$$

$$= \frac{I_0 e^4}{r^2m^2c^4}\frac{(1 + \cos^2 2\theta)}{2}$$

The factor $\frac{1}{2}(1 + \cos^2 2\theta)$ in this equation appears in subsequent formulas for intensities of reflection from crystals and is known as the *polarization factor*.

Scattering by an atom In an atom that is much smaller than the wavelength of the incident x-rays, the electrons oscillate back and forth together so that the atom acts as a unit of mass Zm and charge Ze, where Z is the number of electrons in the atom. Equation (8–2) then becomes

$$I_a = \frac{I_0(Ze)^4}{r^2(Zm)^2c^4}\frac{(1 + \cos^2 2\theta)}{2} = Z^2 I_e \quad (8\text{–}3)$$

In x-ray diffraction work, however, the wavelengths used are of the same order of magnitude as the atomic diameters. Consequently, the electrons within an atom do not scatter in phase and the intensity is less than the value predicted by Eq. (8–3).

It is customary to use a quantity f, the *atomic scattering factor*, by which the efficiency of the cooperation among the electrons in the atom may be expressed. The definition of f is given by the relation

$$f^2 = \frac{I_a}{I_e} \quad (8\text{–}4)$$

or by the equivalent statement—since the intensity of a wave is the square of its amplitude—that *f is the ratio of the amplitude scattered by the atom to that scattered by an electron*. When θ is very small, f approaches the atomic number Z, because the electrons scatter nearly in phase, as assumed in Eq. (8–3). But f falls as θ increases because the waves from the individual electrons must traverse increasingly unequal paths. Some of the radiation is scattered incoherently with a modified wavelength (Compton scattering); this forms an increasing fraction of the total intensity at the larger angles.

The atomic scattering factor is directly related to the distribution of electrons in the atom. Every part of the electron cloud surrounding the nucleus of the atom scatters radiation in proportion to its density. For the atom at rest (not "blurred" by thermal motion in the crystal) the formula for the atomic scattering factor is

$$f_0 = \int_0^\infty U(r)\frac{\sin kr}{kr}\,dr \quad (8\text{–}5)$$

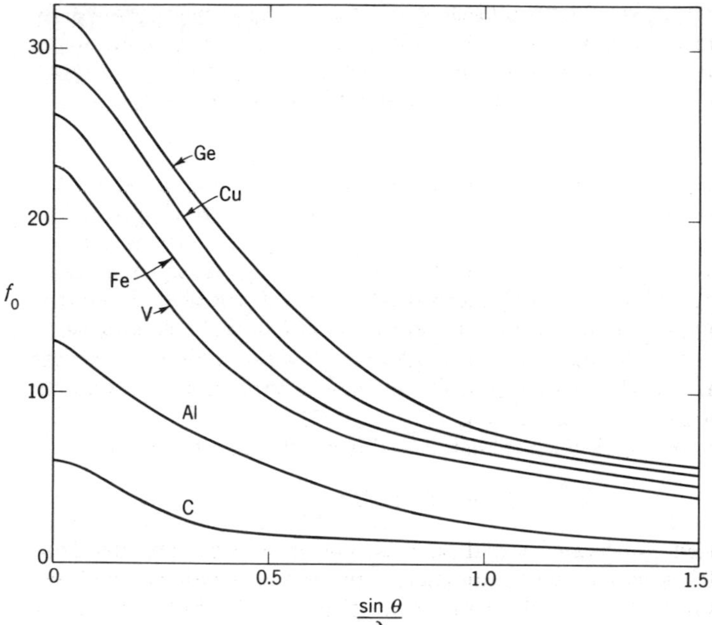

Fig. 8-4 Typical scattering-factor curves.

where $k = 4\pi(\sin\theta)/\lambda$, λ is the wavelength of the radiation, and $U(r)$ represents the radial distribution of electric-charge density. $U(r)dr$ is the number of electrons between r and $r + dr$ from the center of the atom, which is assumed to be spherical,[1] and

$$\int_0^\infty U(r)\,dr = Z$$

Calculations of atomic scattering factors Equation (8–5) indicates that f_0 is a function of $(\sin\theta)/\lambda$, so that a single table of values of f_0 for different values of $(\sin\theta)/\lambda$ will serve for any wavelength. This is true to an approximation that is close enough for many purposes, and very valuable tables of f_0 have now been worked out theoretically for each of the chemical elements, for some different stages of ionization, and for the entire range of $(\sin\theta)/\lambda$ values employed in crystallographic work. Examples are shown in Fig. 8–4.

[1] It is immaterial whether $U(r)$ is considered as the probability of finding an electron at a radius between r and $r + dr$ or whether it is assumed that a continuous charge is distributed in the atom with a density that varies as $U(r)$. In the language of wave mechanics, the function ψ of Schrödinger's wave equation is such that $|\psi|^2\,dv$ is the probability of finding an electron in the element of volume dv at the point considered, and it has been shown that Eq. (8–5) gives the coherent scattering from an atom if we let $U(r) = 4\pi r^2\,|\psi|^2$. Thus, wave-mechanical models of an atom can predict f_0, and, conversely, experimental f_0 curves afford a test of atomic models.

There has been a steady improvement in tables of f_0 in recent years as the approximate electron-density distributions for the atoms have improved. Extensive tables of atomic scattering factors have been published, not only for x-rays, but also for electrons and neutrons [1] In computing intensities it is necessary to interpolate between tabulated values (computer programs can do this automatically when the tabulated values have been inserted).

Dispersion corrections for x-ray scattering The atomic scattering factor f_0 is independent of the frequency of the diffracted radiation only if the frequency is very large compared with the frequency of every absorption edge of the scattering atom. Although this is normally a good approximation in most crystallographic work, there is actually a deviation from tabulated values caused by dispersion, which is large enough to be measurable in precision work with all wavelengths customarily used, a deviation that becomes quite large if the incoming radiation has a wavelength near one of the absorption edges of the scattering atom.

The atomic scattering factor corrected for dispersion has the form

$$f_{corr} = f_0 + \Delta f' + i\Delta f''$$

since the correction consists of a real part $\Delta f'$ and an imaginary part $i\Delta f''$ with a phase 90° ahead of the real part. Approximate values for these correction terms have been worked out theoretically and tabulated for the individual elements at various $(\sin \theta)/\lambda$ values for the radiations most commonly used. Dauben and Templeton[2] published tables for $K\alpha$ radiation from Cr, Cu, and Mo, for $(\sin \theta)/\lambda = 0$, which were extended to other $(\sin \theta)/\lambda$ values.[3] Cooper[4] further extended these to Ag $K\alpha$ and Co $K\alpha$. Since relatively few of the published values have been verified experimentally, caution is advised when high accuracy is needed.

Dispersion has proved useful in at least three applications: (1) It has made possible the determination by x-rays of atom distributions in crystals composed of atoms that normally have almost indistinguishable f_0's. (2) It has been used to determine the absolute configuration of acentric crystals, in that if a radiation is chosen just to the short-wavelength side of an absorption edge of a suitable atom in an acentric crystal, Friedel's law no longer holds, and $F_{hkl} \neq F_{\bar{h}\bar{k}\bar{l}}$; therefore, the crystal can be distinguished from its enantiomorph. (3) The dispersion terms can be made the basis of an unusual type of Fourier synthesis that has recently been used in crystal-structure determination.

Scattering from a unit cell Each atom in a unit cell scatters with an amplitude proportional to f and with a certain phase dependent upon its

[1] "International Tables for X-ray Crystallography," vol. 3, pp. 201–246, Kynoch, Birmingham, England, 1962.

[2] C. H. Dauben and D. H. Templeton, *Acta Cryst.*, vol. 8, p. 841, 1955.

[3] "International Tables for X-ray Crystallography," vol. 3, op. cit., p. 213.

[4] M. J. Cooper, *Acta Cryst.*, vol. 16, p. 1067, 1963.

position in the cell. The intensity of the coherently scattered radiation from the unit cell as a whole is the result of adding the sine waves from the individual atoms, each with its proper amplitude and phase. This has been treated vectorially in Chap. 4. The resultant vector gives the amplitude and phase of the resultant wave, the "structure amplitude" (or structure factor) F_{hkl} for the *hkl* reflection, and the intensity is proportional to $| F_{hkl} |^2$.

The integrated reflection Because real crystals lack perfection and reflect over varying orientation ranges, and because the incident beam is never fully parallel and monochromatic, it is not always meaningful to measure the intensity of an *hkl* reflection from a crystal by holding it in a stationary position. Special precautions are necessary if this technique is to be used. In general it is preferable to measure a quantity called the *integrated reflection.* A small crystal completely bathed in the beam is rotated through its reflecting range from background intensity through peak intensity and on until the background is reached again in a manner such as to record the entire Bragg reflection from all irradiated portions of the crystal. If the angular velocity of rotation is ω, if the incident-beam intensity is I_0, and if the total energy of the reflected beam during this rotation is E, then the integrated reflection is given by $E\omega/I_0$. It has been shown that an infinitely small block of crystal in which x-ray absorption is negligible has an integrated reflection proportional to its volume δV and given by

$$\frac{E\omega}{I_0} = \frac{1}{\sin 2\theta} \frac{N^2 e^4 \mid F \mid^2}{m^2 c^4} \lambda^3 \frac{1 + \cos^2 2\theta}{2} \delta V \tag{8-6}$$

when the incident radiation is unpolarized. The term $(1 + \cos^2 2\theta)/2$ is the *polarization factor* appropriate to the unpolarized case; it must be modified when a crystal monochromator is used since this provides a beam that is partially polarized. The factor $1/\sin 2\theta$, known as the *Lorentz factor*, depends upon geometrical relationships such as the velocity with which the *hkl* reciprocal lattice point moves through the reflecting sphere, and therefore this factor differs for different techniques. These two trigonometric terms are conveniently treated together as a single *Lorentz-polarization factor*; values of this are given below and have also been published elsewhere.[1] In practice, additional factors alter the integrated reflection: the *absorption factor*, the *multiplicity factor*, and a factor to account for *extinction*.

The Lorentz-polarization factors Each x-ray diffraction method requires its particular form of the Lorentz-polarization factor; a lengthy discussion of these is unnecessary here, for those most commonly used can be stated very simply.

Powder method, Debye-Scherrer lines on a cylindrical film For unpolarized

[1] "International Tables for X-ray Crystallography," vol. 2, pp. 265–290, Kynoch, Birmingham, England, 1959.

incident radiation,

$$\text{Lorentz-polarization factor} = \frac{1 + \cos^2 2\theta}{\sin^2 \theta \cos \theta}$$

and for radiation monochromatized by reflection at Bragg angle θ_M from a crystal monochromator,

$$\text{Lorentz-polarization factor} = \frac{1 + \cos^2 2\theta_M \cos^2 2\theta}{\sin^2 \theta \cos \theta (1 + \cos^2 2\theta_M)}$$

Additional formulas for single crystals are given in the international tables cited; tables and charts have also been published.[1]

The absorption factor The Lorentz-polarization factor must be multiplied by an absorption factor to take account of the loss of intensity of the incident and reflected beams within the crystal. This has now been evaluated for various crystal and powder-sample shapes.[2] Some of the absorption factors A are given by a simple function of the absorption coefficient, μ, but for many shapes A can be obtained only with computer programs if at all. Given a specific and simple shape, it is convenient to apply the appropriate correction by a computer program, for the correction may be different for each of the many different *hkl* reflections.

A narrow beam incident on an extended face of a crystal. If the crystal is thick enough to absorb essentially all the initial beam, and if the reflecting planes are parallel to the surface, the absorption does not change from one *hkl* to another and the correction is the simplest possible:

$$A = \frac{1}{2\mu}$$

but if the transmission is not negligible, for a crystal of thickness t the factor is

$$A = \frac{1 - \exp(-2\,\mu t \operatorname{cosec} \theta)}{2}$$

A thick extended block of powder (t large enough to give negligible transmission) Reflections from planes parallel to the extended surface, as in the usual diffractometer, have the absorption factor $A = 1/2\mu$, and as in the symmetrical case of the thick single crystal, no correction is needed for intensities of reflection *relative* to each other. Here μ is the linear absorption coefficient for the powder compact, not the crystal.

Cylindrical powder samples, cylindrical single-crystal samples, and small spherical samples These samples have absorption factors that have been

[1] M. J. Buerger and G. E. Klein, *J. Appl. Phys.*, vol. 17, p. 285, 1946. W. Cochran, *J. Sci. Instr.*, vol. 25, p. 253, 1948. G. Kaan and W. F. Cole, *Acta Cryst.*, vol. 2, p. 38, 1949. C. S. Lu, *Rev. Sci. Instr.*, vol. 14, p. 331, 1943.

[2] Tabulated values are included in "International Tables for X-ray Crystallography," vol. 2, op. cit., pp. 291–312.

thoroughly worked out and tabulated.[1] Spherical crystals 0.1 to 0.2 mm in diameter are often prepared by Bond's abrasive method: they are driven around in a cylindrical cavity by a tangential stream of air.[2]

Weissenberg and equi-inclination Weissenberg photographs from many-faceted crystals have absorption factors that vary from region to region on the film, for which approximate corrections can be worked out, for instance by using the variations in the background darkening.[3]

With specimens of relatively low absorption coefficient, it is frequently possible to reduce absorption corrections to a negligible magnitude by using small samples and penetrating radiation such as Mo $K\alpha$.

The multiplicity factor p The number of different equivalent sets of planes that reflect to the same spot on a photographic film is a factor the value of which varies with the method used in photographing the diffraction pattern and the symmetry of the reflecting crystal. In the powder method all planes of the same spacing superimpose to the same diffraction ring, and the multiplicity factor is then the number of permutations of *hkl* that give identical $\sin^2 \theta$ values in the quadratic form for the crystal. In powder patterns of cubic crystals of the common classes O_h, O, and T_d, which have high symmetry, the value of p is 6 for $\{100\}$ planes, 8 for $\{111\}$, 12 for $\{110\}$, 24 for $\{hk0\}$ and for $\{hhl\}$, and 48 for $\{hkl\}$. In rotating-crystal and oscillating-crystal patterns the value of p depends not only on the permutations of *hkl* indices but also on the orientation of the crystal in the camera, and in the oscillating-crystal method also on the range of oscillation (which can be made so small that all overlapping is avoided). The Weissenberg and other moving-film methods make $p = 1$ for all spots.

The temperature factor If an atom were at rest in a crystal it would scatter with an amplitude given by f_0, but at all temperatures it oscillates about a mean position. The excursions of an atom from its mean position have the effect of smearing out the electron cloud and reducing f. The higher the temperature, the more blurred the atom will appear and the more f will decrease as the diffraction angle θ is increased.

If each atom is assumed to vibrate independently of the others with a mean-square displacement $\bar{U}^2$ at right angles to the reflecting plane (hkl), the effect will be to reduce the scattering factor for each from f_0 to f_0e^{-M} where

$$M = 8\pi^2\bar{U}^2(\sin\theta)^2/\lambda^2$$

[1] A. J. Bradley, *Proc. Phys. Soc. (London)*, vol. 47, p. 879, 1935 (cylinders). F. C. Blake, *Rev. Mod. Phys.*, vol. 5, p. 169, 1933 (cylinders). H. T. Evans and M. G. Eckstein, *Acta Cryst.*, vol. 5, p. 540, 1952 (spheres). "International Tables for X-ray Crystallography," vol. 2, pp. 292–305, Kynoch, Birmingham, England, 1959 (tables that also include transmission pole-figure absorption factors). M. J. Buerger, "Crystal Structure Analysis," pp. 210–215, Wiley, New York, 1960 (tables for cylinders and spheres).

[2] W. L. Bond, *Rev. Sci. Instr.*, vol. 22, p. 344, 1951. K. S. Revell and R. W. H. Small, *J. Sci. Instr.*, vol. 35, p. 73, 1958. For cylindrical crystal preparation, see F. Barbieri and J. Durand, *Rev. Sci. Instr.*, vol. 27, p. 871, 1956.

[3] M. J. Buerger, *Z. Krist.*, vol. A99, p. 189, 1938; "Crystal Structure Analysis," Wiley, New York, 1960. C. H. MacGillavry and H. J. Vos, *Z. Krist.*, vol. A1050, p. 257, 1943. R. G. Howells, *Acta Cryst.*, vol. 3, p. 366, 1950.

The value of M may be different for each atom, and will be anisotropic if the atom vibrates anisotropically rather than isotropically. In most crystals there are thus a great number of thermal parameters to be determined if the structure-factor formula is to be evaluated rigorously. Frequently, however, it is a reasonably good approximation to assume isotropic displacements, and to assume that all atoms of one variety in a crystal vibrate equally. If the crystal structure appears to be one in which the further simplifying assumption can be made that *all* atoms vibrate with about equal amplitudes, the factor e^{-M} may be taken as equal for each of the atoms with the result that $|F|^2$ becomes simply $|F|^2e^{-2M}$.

It is useful to consider a displacement parameter B such that $B = 8\pi^2\bar{U}^2$ and $M = B(\sin^2\theta)/\lambda^2$. The mean-square displacement of an atom, $\bar{U}^2$, is not zero even at $T = 0°K$ because of "zero-point energy," and at any other temperature the total B may be treated as the sum of the zero-point vibration contribution B_0 and the thermal contribution B_T:

$$B = B_0 + B_T$$

For a cubic crystal containing one kind of atom only, a relationship of B to the Debye temperature, Θ, of the crystal has been derived theoretically.[1]

Normally in crystal-structure analysis B is taken as a parameter to be refined together with atom-position parameters, using a least-squares computer program for isotropic B. A more accurate structure determination may be obtained, in general (particularly when the anisotropy is considerable), if the temperature factor is considered as anisotropic—provided enough precise values of F_{obs} are available. There are then two or more factors to be determined, designated as β_{ij}, for each atom or at least for each set of atoms that are presumed to oscillate similarly, because M in the temperature factor for a particular atom that is to be used in computing the intensity of the hkl reflection is given by

$$M_{hkl} = \beta_{11}h^2 + \beta_{12}hk + \beta_{13}kl + \beta_{21}kh + \beta_{22}k^2 + \beta_{23}kl + \beta_{31}lh + \beta_{32}lk + \beta_{33}l^2$$

However, not all β_{ij} values are independent; it is always true that $\beta_{ij} = \beta_{ji}$, and other relationships exist that depend upon symmetry considerations.[2]

Extinction A highly perfect crystal that is so oriented that a strong reflection is occurring acts as if it has an abnormally high absorption coefficient and the reflected intensities are abnormally low. The phenomenon involves two different processes known as *primary extinction* and *secondary extinction*. Primary extinction results from the fact that within a perfect crystal or a perfect mosaic block inside a crystal the reflected ray is incident upon the reverse side of the reflecting planes at the exact angle to be reflected again. A twice-reflected ray is therefore formed, which travels along with the primary and cancels a portion of the primary-ray intensity.

[1] "International Tables for X-ray Crystallography," vol. 2, p. 264, Kynoch, Birmingham, England, 1959; vol. 3, 234–244, 1962. The appropriate Θ for x-rays is only approximately equal to Θ obtained from specific heats and otherwise.

[2] H. A. Levy, *Acta Cryst.* vol. 9, p. 679, 1956. R. W. James, "The Optical Principles of the Diffraction of X-rays," G. Bell, London, 1948.

(This is because there is a shift of phase of 90° upon reflection, a shift of 180° with two reflections.) Further reflections also occur. Secondary extinction is due to the upper blocks of a mosaic crystal, when they reflect, shielding to some extent the lower blocks.

In a perfect crystal the integrated reflection is proportional to $|F|$; in an ideally imperfect crystal—with negligible extinction—it is proportional to $|F|^2$. In ordinary crystals neither condition holds, but it is often possible to approach the ideally imperfect state closely enough to make extinction corrections small for all but the strongest reflections. It is helpful in decreasing extinction to powder a sample, or to subject a crystal to thermal shock by immersing it in liquid nitrogen. If the size of the mosaic block can be made as small as a few thousand angstroms, primary extinction should be negligible, but this is not always feasible. Theories have been developed for the corrections,[1] but corrections are not easy to make, and the usual procedure is to avoid using the measured values of the integrated reflection for the strongest reflections. In Fourier syntheses the values suspected of having large extinction effects may be replaced by calculated values.

Scale factor; refinement of scale and temperature factors The measurement of integrated reflections on an *absolute* basis, that is, relative to the intensity of the primary beam, is difficult. Many errors have to be guarded against. Extinction effects can be troublesome. With powders there may be absorption errors from particle roughness, and there is often some degree of preferred orientation. The ratio of intensities of primary and reflected beams is very large, which makes the problem of avoiding counting losses or accurately correcting for them a serious one. It is common practice in structure analysis to measure relative rather than absolute intensities, and to determine the *scale factor* that relates the group of relative measurements to the primary-beam intensity by a refinement procedure that also refines the temperature factor. Although modern techniques have succeeded in obtaining absolute intensities and f_0 values with an error of the order of 1 percent,[2] it is much easier to obtain the scale factor by refinement.

After correcting for Lorentz-polarization, absorption, and multiplicity factors, the measured intensities or integrated intensities yield a list of KF_{obs} values (K is the scale factor) which are to be compared with calcu-

[1] C. G. Davrin, *Phil. Mag.*, vol. 43, p. 88, 1922. V. Vand, *J. Appl. Phys.*, vol. 26, p. 1191, 1955. Reviewed in M. J. Buerger, "Crystal Structure Analysis," Wiley, New York, 1960. A more recent treatment by W. H. Zachariasen corrects some deficiency in the original theory: W. H. Zachariasen, *Acta Cryst.*, vol. 16, p. 1139, 1963. W. H. Zachariasen and H. A. Plettinger, *Acta Cryst.*, vol. 18, p. 710, 1965.

[2] An example is B. W. Batterman, D. R. Chipman, and J. J. DeMarco, *Phys. Rev.*, vol. 122, p. 68, 1961. The ratios of the measured f_0's for powders of Fe, Cu, and Al agreed with the ratios calculated from Hartree-Fock theory to within 1 percent, but the Hartree-Fock theory gives absolute values about 4 percent higher than the observed values in the region of low $(\sin\theta)/\lambda$, possibly because of differences in the atoms of a gas and a solid.

lated values. But the calculated values are obtained from a structure-factor formula based on $f_0 \exp -(B \sin^2 \theta)/\lambda^2$ where B is initially unknown, and f_0 is the atomic scattering factor at absolute zero, the value given in the tables. If all atoms in a crystal are assumed to have the *same isotropic vibrations*, then the observed and calculated structure factors are related by the equation

$$KF_{obs} = F_{calc}\, e^{-(B \sin^2 \theta)/\lambda^2}$$

and therefore

$$\ln(F_{obs}/F_{calc}) = \ln(1/K) - (B/\lambda^2) \sin^2 \theta$$

which is of the form $y = c + ax$. A plot of $\ln(F_{obs}/F_{calc})$ vs. $\sin^2 \theta$ should therefore be a straight line. The slope of the line gives B, and the intercept at $\sin^2 \theta = 0$ provides the scale factor K. Approximate values of B and K are first used in a calculation of F's.[1] The plot of $\ln(F_{obs}/F_{calc})$ (or a least-squares solution) then yields new values of B and K. Successive refinement cycles are carried through until B and K and all other parameters cannot be further refined. The power of modern computer programs is such that all parameters, even the anisotropic factors β_{ij}, in very complex crystals can be refined simultaneously by least-squares procedures.

Electron density expressed by Fourier series The density of the diffracting material in a crystal (the electron density) varies periodically along any direction through the lattice, with the same spacial periodicity as the distribution of atoms. It is therefore possible to describe the electron distribution by a Fourier series, as can be done for any periodic distribution. Various special Fourier series have proved to be extremely useful in crystal-structure analysis over the years, and have become increasingly so since digital computers have become common and very rapid. Fourier-series methods are particularly valuable for the more complex crystals. The many details of the subject that have now been developed and the limitations and possible ambiguities of each type of series when applied to crystallographic problems cannot be adequately treated in a brief space; fortunately this is unnecessary as excellent books on the subject are now available.[2]

The general case, rather than any of the many special cases applicable to specific crystal symmetries and structure-analysis applications, should be presented first. Let $\rho(xyz)$ represent the electron density at a point having the coordinates xyz in the unit cell. A Fourier series can be set up,

[1] A method of obtaining these is available. A. J. C. Wilson, *Nature*, vol. 150, p. 152, 1942. D. Harker, *Am. Mineralogist*, vol. 33, p. 764, 1948. Review in M. J. Buerger, "Crystal Structure Analysis," Wiley, New York, 1960.

[2] H. Lipson and W. Cochran, "The Determination of Crystal Structures," G. Bell, London, 1953, 1957. M. J. Buerger, "Crystal Structure Analysis," Wiley, New York, 1960. R. W. James, "The Optical Principles of the Diffraction of X-rays," G. Bell, London, 1948, 1954. A. D. Booth, "Fourier Technique in X-ray Organic Structure Analysis," Cambridge, New York, 1948. H. Lipson and C. A. Taylor, "Fourier Transforms and X-ray Diffraction," G. Bell, London, 1958 (introductory monograph).

each term of which represents a stationary system of electron-density waves. Each set of such waves is capable of diffracting x-rays with a certain intensity, $|F_{hkl}|^2$, since the successive parallel sheets of density act as reflecting planes. A set of standing waves parallel to the (hkl) plane of the crystal is represented by a Fourier term that has F_{hkl} as a coefficient. The density $\rho(xyz)$ is obtained by summing, over all hkl values, the series

$$\rho(xyz) = \frac{1}{V} \sum_{h=-\infty}^{\infty} \sum_{k=-\infty}^{\infty} \sum_{l=-\infty}^{\infty} F_{hkl}\, e^{-2\pi i(hx+ky+lz)} \qquad (8\text{–}7)$$

where V is the volume of the unit cell. Carrying out the summation is equivalent to superimposing the differently oriented sheets of electron density. Consider a single term in this series, $F_{hkl} \exp - 2\pi i(hx + ky + lz)$; this can be written as

$$[A_{hkl} + iB_{hkl}]\,[\cos 2\pi(hx + ky + lz) - i \sin 2\pi(hx + ky + lz)]$$

where A and B are the components of the complex quantity F_{hkl}.

To put the series in a form more suitable for computation, each term may be combined with its conjugate in which hkl is replaced by $\bar{h}\bar{k}\bar{l}$ to give

$$\begin{aligned} &[A_{hkl} + iB_{hkl}][\cos 2\pi(hx + ky + lz) - i \sin 2\pi(hx + ky + lz)] \\ &\quad + [A_{\bar{h}\bar{k}\bar{l}} + iB_{\bar{h}\bar{k}\bar{l}}][\cos 2\pi(hx + ky + lz) + i \sin 2\pi(hx + ky + lz)] \\ &\quad\quad = 2A_{hkl} \cos 2\pi(hx + ky + lz) + 2B_{hkl} \sin 2\pi(hx + ky + lz) \qquad (8\text{–}8) \end{aligned}$$

provided we assume Friedel's law and set $A_{hkl} = A_{\bar{h}\bar{k}\bar{l}}$ and $B_{hkl} = -B_{\bar{h}\bar{k}\bar{l}}$.

The factor 2 in Eq. (8–8) does not apply to the term with $hkl = 000$ since this term is its own conjugate; therefore, the F_{000} term is not multiplied by 2. It also is necessary to sum the series, in general, only over half the complete reciprocal lattice. Eq. (8–8) therefore can be put in the form

$$\rho(xyz) = \frac{1}{V}\left[F_{000} + 2\sum_{h=0}^{\infty} \sum_{k=-\infty}^{\infty} \sum_{l=-\infty}^{\infty} A_{hkl} \cos 2\pi(hx + ky + lz) + B_{hkl} \sin 2\pi(hx + ky + lz)\right] \qquad (8\text{–}9)$$

where the summations exclude the term for which $h = k = l = 0$.

Various simplifications are possible when a crystal possesses symmetry elements. If a crystal has a center of symmetry, $F_{hkl} = F_{\bar{h}\bar{k}\bar{l}}$ and the structure factors are all real numbers; Eq. (8–9) then reduces to

$$\rho(xyz) = \frac{1}{V}\left[F_{000} + 2\sum_{h=0}^{\infty} \sum_{k=-\infty}^{\infty} \sum_{l=-\infty}^{\infty} F_{hkl} \cos 2\pi(hx + ky + lz)\right] \qquad (8\text{–}10)$$

Convenient forms of the series have been worked out for each of the space groups.[1] Since any set of sinusoidally varying density waves can be dis-

[1] "International Tables for X-ray Crystallography," vol. 1, Kynoch, Birmingham, England, 1952.

placed perpendicular to its wave front without altering either the angular position or the intensity of the diffracted wave, the relative positions of the waves are not directly determined by the x-ray measurements, i.e., the *phases* are unknown.

There is an analog with optical diffraction gratings: the positions and intensities of the spectra produced by a diffraction grating are not in any way changed by displacing the grating in its own plane perpendicular to its rulings. In the formula for centrosymmetrical crystals (Eq. 8–10), the uncertainty in phase is an uncertainty in the *sign* of each term, which can be indicated by writing the coefficient as $\pm F_{hkl}$. This frustrating ambiguity of phase has stimulated an immense amount of research directed toward removing it, and a series of important advances have recently been made, but the ambiguity is not a serious obstacle in most of the simpler structure determinations and even in some very complex ones.

If the unit cell contains a heavy atom that stands out from the rest in scattering power, Patterson methods (mentioned below) should give the coordinates of this atom. In compounds that do not have a heavy atom it may be possible to obtain crystalline derivatives that do. Certain phases of the Fourier terms are fixed by the fact that the heavy-atom position determines them. Other phases are then determined in successive refinement cycles. If isomorphous compounds are available, the change in F_{hkl} on passing from one compound to another is a great help in determining the signs of terms. And in any structure determination the crystallographer's knowledge of empirical atom and ion sizes and coordination tendencies can be put to use in suggesting trial structures and in eliminating unlikely ones.

Several special types of Fourier series are widely used that are simpler to deal with than those involving the three-dimensional summation mentioned above. For example, the electron density can be projected onto one of the principal planes of the crystal (or even onto an inclined plane, if desired); (see, for example, Figs. 8–5 and 8–6). Alternately, a series can

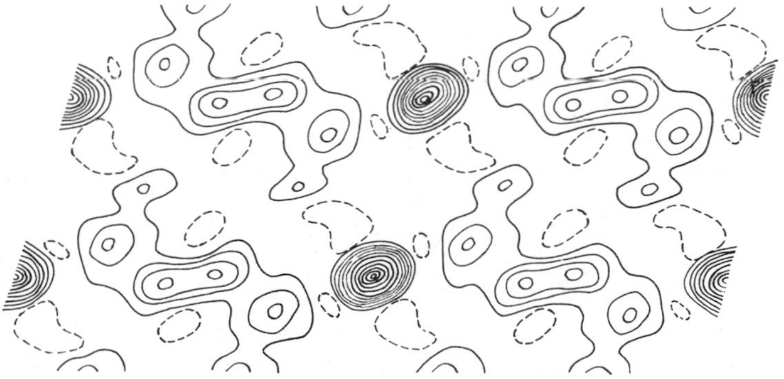

Fig. 8-5 Electron density of diopside (calcium-magnesium silicate) projected upon (010) plane. (W. L. Bragg.)

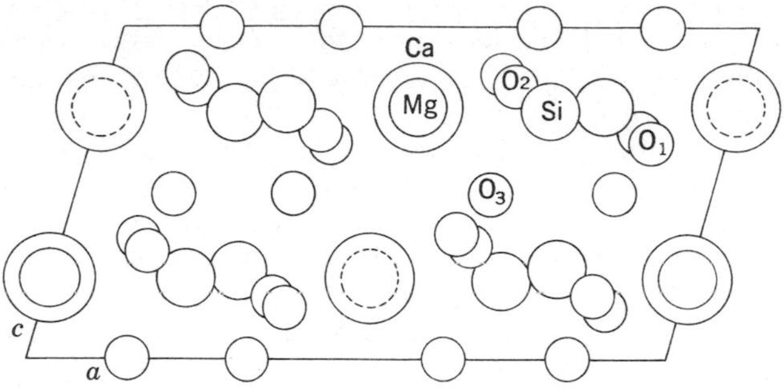

Fig. 8-6 Atom positions in diopside, projected as in Fig. 8-5. (W. L. Bragg.)

be set up to give the density distribution over any arbitrary plane cutting through a unit cell, or the density of a slice of a given thickness projected onto a plane.

As a result of carrying out summations for many points on a plane, a map is constructed on which magnitudes can be recorded by numbers written at each point. Contours can then be drawn on the map along lines of constant density by interpolating between the calculated values. Digital and analog computers have been devised to do plotting of this type. A succession of contour plots on plastic sheets stacked above each other can serve as a three-dimensional model of a crystal.

The F^2 series of Patterson A later development by Patterson[1] and its modification by Harker[2] stem from the same fundamental principles and find wide usage in solving complex structures. It is possible to set up a Fourier series in which $|F_{hkl}|^2$ occurs, and information about the atomic positions can be derived from it *without making assumptions as to the signs of the coefficients* as is necessary in $F(hkl)$ series. Patterson's series for three dimensions is

$$P(uvw) = \frac{1}{V} \sum_{h=-\infty}^{+\infty} \sum_{k=-\infty}^{+\infty} \sum_{l=-\infty}^{+\infty} |F_{hkl}|^2 e^{-2\pi i(hu+kv+lw)} \tag{8–11}$$

This reduces to the cosine form (either with or without a center of symmetry in the crystal), with $F_{hkl} = F_{\bar{h}\bar{k}\bar{l}}$:

$$P(uvw) = \frac{1}{V} \sum_{h=-\infty}^{+\infty} \sum_{k=-\infty}^{+\infty} \sum_{l=-\infty}^{+\infty} |F_{hkl}|^2 \cos 2\pi(hu + kv + lw) \tag{8–12}$$

The function $P(uvw)$ represents the product of the electron density at any point in the unit cell whose coordinates are xyz and the electron density at another point whose coordinates are $x + u$, $y + v$, $z + w$. Thus, if

[1] A. L. Patterson, *Z. Krist.*, vol. 90, pp. 517, 543, 1935; *Phys. Rev.*, vol. 46, p. 372, 1934.
[2] D. Harker, *J. Chem. Phys.*, vol. 4, p. 381, 1936.

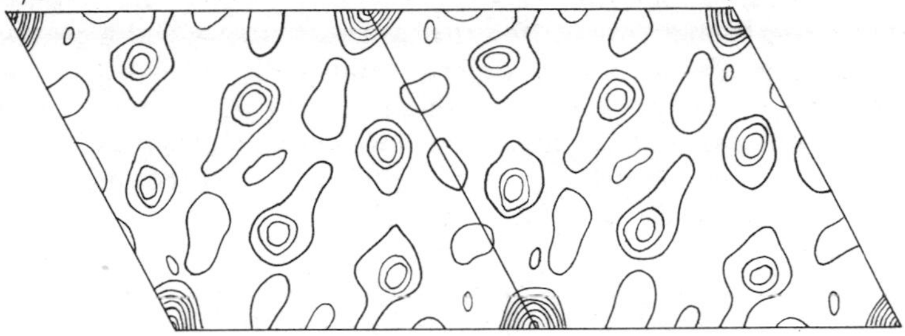

Fig. 8-7 Contour-line plot of F^2 function for C_6Cl_6. (A. L. Patterson.)

there is an atom in the crystal at xyz and another at $x + u$, $y + v$, $z + w$, there will be two peaks in the electron density $\rho(xyz)$, their distance apart will be given by the vector whose components are uvw, and there will be a maximum in $P(uvw)$ at the point uvw corresponding to this pair of atoms. In other words, Eq. (8–12) defines an electron-density product $P(uvw)$ that has maxima at distances and directions from the origin corresponding to the distances and directions between pairs of atoms in the crystal. The amplitudes of the peaks of $P(uvw)$ correspond to the products of the electron densities at the two points considered (roughly products of the two atomic numbers).

This triple Fourier series becomes more manageable if $P(uvw)$ is projected on one of the faces of the unit cell. With projection along the c axis onto the (001) plane, for instance, the projected function is

$$p(uv) = \int_0^1 P(uvw)\,dw = \frac{1}{V} \sum_{h=-\infty}^{+\infty} \sum_{k=-\infty}^{+\infty} |F_{hk0}|^2 \cos 2\pi(hu + kv) \qquad (8\text{–}13)$$

A two-dimensional "Patterson plot" can then be made of the function $p(uv)$ with contours drawn to show the distribution of this function throughout the projection, as shown in Fig. 8–7.[1] A peak at uv on this plot corresponds to an interatomic distance in the crystal whose components along the x and y axes are u and v.

The Patterson-Harker F^2 series Harker[2] has applied Eq. (8–12) in such a way as to make use of knowledge of the symmetry elements in crystals, thus simplifying the method. When preliminary analysis has determined the space group to which a crystal belongs, a knowledge of the symmetry elements makes it possible to write down all the possible equivalent positions that atoms can occupy. Suppose there is a *twofold axis* parallel to the b axis of the crystal; then an atom at xyz has an equivalent

[1] A. L. Patterson, *Z. Krist.*, vol. 90, pp. 517, 543, 1935. It is frequently unnecessary to plot more than a half or a quarter of a unit cell, if symmetry elements supply the rest.

[2] D. Harker, *J. Chem. Phys.*, vol. 4, p. 381, 1936.

atom at $\bar{x}y\bar{z}$, and the vector between these two has components $2x$, 0, $2z$. These values substituted in Eq. (8–12) will yield a maximum value of $P(uvw)$ at the point $u = 2x$, $v = 0$, $w = 2z$, which will be a point in the plane $v = 0$. Every other atom will also be paired with an equivalent atom in the same way and will lead to maxima in the same plane. Therefore, the u and w coordinates of every atom in the crystal can be found by evaluating $P(uvw)$ for the special case of $v = 0$.

The table below gives the form of series required for each type of symmetry axis parallel to b and for each type of symmetry plane perpendicular to b. Cyclic interchange will yield the corresponding cases for elements parallel and perpendicular to a and c axes.

Symmetry element	Form of $P(uvw)$
(1) Axes parallel to b:	
$2, 4, 4_2, \bar{4}, 6, 6_2$	$P(u0w)$
$2_1, 4_1, 4_3, 6_1, 6_5$	$P(u\frac{1}{2}w)$
$3_1, 6_2, 6_4$	$P(u\frac{1}{3}w)$
(2) Planes perpendicular to b:	
Reflection plane	$P(0v0)$
Glide plane with glide of $\frac{1}{2}a_0$	$P(\frac{1}{2}v0)$
Glide plane with glide of $\frac{1}{2}c_0$	$P(0v\frac{1}{2})$
Glide plane with glide of $\frac{1}{2}(a_0 + c_0)$	$P(\frac{1}{2}v\frac{1}{2})$
Glide plane with glide of $\frac{1}{4}(a_0 + c_0)$	$P(\frac{1}{4}v\frac{1}{4})$
Glide plane with glide of $\frac{1}{4}(3a_0 + c_0)$	$P(\frac{3}{4}v\frac{1}{4})$

A Patterson-Harker plot effectively focuses all the diffraction data from a crystal on a certain feature in the structure, for instance, on the interatomic distances for atoms on a certain plane in the unit cell. The resolving power that results is correspondingly higher than is found in a two-dimensional Patterson plot which uses only reflections from planes of a single zone. Furthermore, this resolving power can be brought to focus on any plane the investigator believes will be most illuminating; or on a whole series of planes in turn that are spaced at intervals through the cell, giving a series of cross sections; or on a line that is run through the cell at any important position where a hint as to the structure might be found.

A serious disadvantage of F^2 projections is the overlapping of peaks when there are many atoms in the unit cell. (If the crystal cell contains N atoms, the Patterson cell, which is of the same size, contains N^2 peaks each of which is roughly twice the breadth of the atomic peaks.) Some improvement in resolution of the Patterson peaks can be obtained by "sharpening" methods—multiplying the coefficients of the series terms by suitable factors.[1]

Vector sets; superposition methods Suppose the Patterson projections could be sharpened so effectively that the peaks became points (which cannot be done in practice), and suppose the crystal structure consisted of

[1] A. L. Patterson, *Z. Krist.*, vol. 90, p. 517, 1935. For a review, see H. Lipson and W. Cochran, "The Determination of Crystal Structure," chap. 6, G. Bell, London, 1953, 1957.

point atoms. Then the crystal could be described as atom positions at the ends of vectors, the jth atom being at $\mathbf{r}_j$ and of weight Z_j. The Patterson function would then have finite value only at points $\mathbf{r}_i - \mathbf{r}_j$ and at these points it would have weight Z_iZ_j. To determine the crystal structure from the Patterson synthesis, then, means to deduce the "fundamental set" of points $\mathbf{r}_j$ from the "vector set" of points $\mathbf{r}_i - \mathbf{r}_j$. Various methods for doing this have been devised, of which the "superposition methods" should be mentioned here.[1] If a vector set is copied on a second sheet and if this copy is displaced parallel to itself until its origin lies on an $\mathbf{r}_i - \mathbf{r}_j$ peak, then other peaks will also superimpose, and these will be the fundamental set plus a set related to it by inversion at a symmetry center. Except for certain accidental coincidences and this ambiguity of the congruent inverted set, the problem is thereby solved (in principle), and ambiguities can be attacked by repeating the superposition operation by other displacements, $\mathbf{r}_{i'} - \mathbf{r}_{j'}$. In general,[2] a synthesis involving m superimposed identical Pattersons can be used, with origins for this "mixed series" at $x_1y_1z_1$, $x_2y_2z_2$, $\cdots x_my_mz_m$.

Error synthesis An effective way of finding out what atoms are misplaced in the early stages of working out a crystal structure, and of learning what changes in their positions could be expected to improve the agreement between F_{obs} and F_{calc}, is to compare the Fourier syntheses given by the observed and the calculated structure factors. This may conveniently be done by using as Fourier coefficients $(F_o - F_c)$ in a series to give $\rho_o - \rho_c$ at points on a two-dimensional map. Using projection along the a axis onto the bc plane, the series would be

$$\rho_o - \rho_c = \frac{1}{A} \sum_k \sum_l (F_o - F_c) e^{-2\pi i(ky+lz)} \qquad (8\text{–}14)$$

If all atoms were correctly placed and all phases were correct, there would be a uniform value of $\rho_o - \rho_c$ everywhere, but with atomic coordinates slightly in error, the gradients that develop in $\rho_o - \rho_c$ will indicate in what direction and what distance to move atoms to improve the model of the structure—provided enough signs of the F's are correct. This method proved to be useful in the famous work on penicillin for which Crowfoot won a Nobel Prize.[3] Applications of this and related methods in refinement are amply illustrated by recent papers in *Acta Crystallographica*.

Least-squares refinement by digital computers has become almost standard as the most acceptable way of finding the best fit of the list of F_o's to the list of F_c's. Least-squares refinement has the advantage that it is free

[1] M. J. Buerger, *Acta Cryst.*, vol. 3, p. 87, 1950; vol. 4, p. 531, 1951. J. Clastre and R. Gay, *Compt. Rend. Acad. Sci.* (*Paris*), vol. 230, 1976, 1950; *J. Phys. Radium*, vol. 11, p. 75, 1950. C. A. Beevers and J. H. Robertson, *Acta Cryst.*, vol. 3, p. 164, 1950. Reviewed in Lipson and Cochran, chap. 6, loc. cit., and in M. J. Buerger, "Vector Space," Wiley, New York, 1959.

[2] D. McLachlan, *Proc. Natl. Acad. Sci.*, vol. 37, p. 115, 1951.

[3] D. Crowfoot et al., "The X-ray Crystallographic Investigation of the Structure of Penicillin," Oxford, Fair Lawn, N. J., 1949.

from errors induced by termination of the Fourier series. Another advantage is that F_o's that are doubtful (say because of extinction) can be omitted. Temperature and scale factors can also be simultaneously refined, and a weight can be assigned to each term used in determining the residual quantity R' which is being minimized:

$$R' = \Sigma\, w_{hkl}(|F_o| - |F_c|)^2_{hkl} / |F_o|^2$$

where w_{hkl} is assigned by some currently favored scheme.

Computer programs The available programs are rapidly changing and the "World List" should be consulted for specific needs; however, it is felt that the following selections may be of such general usefulness as to warrant mention here.

1. IBM 7090 program ERFR2, for two- and three-dimensional straight or difference Fourier summations for any space group. Beevers-Lipson principle. This is a reedited and modified version of their IBM 704 program MIFR1, W. G. Sly and D. P. Shoemaker, Massachusetts Institute of Technology, Cambridge, Massachusetts, June 1, 1962.
2. IBM 7090 program ERBR1, to compute structure factors and least-squares refinement for any space group and to perform diagonal-terms-only regression on the postulated structure parameters using the observed data. Intended for IB Fortran Monitor. Output suitable for use with Sly-Shoemaker program. Jan H. van den Hende, Esso Research and Engineering Co., Central Basic Research Laboratory, Linden, New Jersey.
3. IBM 7090 (IBM 704 with minor changes) program ERBR2, for absorption and Lorentz-polarization corrections for normal beam and equi-inclination data from cylindrical or spherical specimens. Generation of indices optional. J. H. van der Hende.
4. IBM 704 program ORXLS, in Fortran for isotropic or anisotropic temperature factor, least-squares full matrix refinement and structure factor calculation for all space groups. Central Files No. ORNL 59-4-37. W. R. Busing and H. R. Levy, Oak Ridge National Laboratory, Oak Ridge, Tennessee, April 14, 1960. Rewritten for IBM 7090 in Fortran.
5. IBM 7090 program ORFFE, Fortran II program for computing distance between two atoms, angle defined by three atoms, magnitude and direction of principal axes of the anisotropic temperature factor, components of thermal displacement in direction defined by two atoms, dihedral angle between planes each defined by three atoms, when given unit cell parameters, atomic coordinates, and/or anisotropic temperature factors. Central Files No. ORNL-TM-306. Subroutines must be recompiled if they are to be used on IBM 704. [This is a revision of Busing-Levy ORNL-CF 59-12-3 (1959).] W. R. Busing, K. O. Martin, and H. A. Levy, Oak Ridge National Laboratory, Oak Ridge, Tennessee, March, 1964.

See also Chap. 7 for powder-method programs.

Other techniques Some crystals have now been determined directly; i.e., the phases of the terms in Fourier syntheses have been determined without indirect procedures, as can be seen from the recent books cited and some of the recent journals. The effective methods have been based on relations among certain F's and F^2's—various inequalities—or upon certain statistical relations among the signs of terms. Recently, an entirely new direct method has come into use which involves the real and imaginary components of the dispersion correction; the method makes use of the fact that when the dispersion correction is large enough to be important in precise measurements, the intensity from (hkl) planes is no longer equal to the intensity from $(\bar{h}\bar{k}\bar{l})$ in noncentrosymmetric crystals.[1]

[1] Y. Okaya, Y. Saito, and R. Pepinski, *Phys. Rev.*, vol. 98, p. 1857, 1955; discussed by W. H. Zachariasen, *Acta Cryst.*, vol. 18, p. 714, 1965. A review of anomalous scattering for determining the absolute configuration of a molecule is given in G. N. Ramachandran (ed.), "Advanced Methods of Crystallography," Academic, New York, 1964.

Methods of summing Fourier series have changed in popularity so rapidly that only the latest books approach a reasonable view of current practice. Although optical methods were in favor for a time, very few laboratories use them now. Analog computers have largely been superseded by high-speed digital computers. Beevers and Lipson strips[1] and various modifications of these are used with a desk calculator—but only when faster computers are not available.

Diffraction from liquids and vitreous solids The structure of a liquid is somewhat analogous to the arrangement of ball bearings in a box when it is continuously vibrated. At any instant a number of balls will be in contact or near contact with any given ball, as there is some tendency toward close packing, but the coordination number in the liquid is not the same as in the solid. No interatomic distances can be expected that are less than the sum of the effective radii of the nearest-neighbor atoms. The preference for atoms to have certain interatomic spacings and the tendency to avoid others causes diffraction effects. Since there is no periodicity in the atomic arrangement in a liquid (or glass or amorphous bodies), the diffraction effects are not sharp spots or rings as they are in diffraction patterns of crystals, but there are diffuse maxima which nevertheless contain information on the statistical distribution of interatomic distances in the specimen.[2]

This information is best derived from the distribution of intensity of the scattering from a liquid by Fourier methods that are mentioned below. Diffraction studies have been made not only of liquids with spherically symmetrical atoms or molecules but also of liquids with molecules of elongated or flattened shapes in which there are varying degrees of alignment of the individual molecules. Investigations also have explored the influence of temperature and pressure on the structure of liquids.

The radial distribution method The Fourier method for determining the average radial distribution of atoms around an atom was suggested by Zernike and Prins[3] and was first successfully used in a study of liquid mercury by Debye and Menke.[4] In this method a monochromatic beam of x-rays is diffracted from, say, a small capillary tube or a stream of the liquid, and the intensity diffracted at different angles is measured (in recent years by a counter diffractometer). The measured intensities are corrected for absorption in the sample, for polarization in the beam, and for the presence of incoherent radiation; they are then placed on an absolute-intensity scale so that the intensity is known relative to the intensity of scattering of a single electron. The curve is then analyzed by a Fourier

[1] C. A. Beevers and H. Lipson, *Proc. Phys. Soc. (London)*, vol. 48, p. 772, 1936. C. A. Beevers, *Acta Cryst.*, vol. 5, p. 670, 1952 (Fourier strips at a 3° interval).

[2] N. S. Gingrich, *Rev. Modern Phys.*, vol. 15, p. 90, 1943. R. W. James, "The Crystalline State, Vol. 2, The Optical Principles of Diffraction of X-rays," G. Bell, London, 1948. A. Guinier, "Théorie et Technique de la Radiocristallographie," Dunod, Paris, 1956; "X-ray Diffraction in Crystals, Imperfect Crystals, and Amorphous Bodies," Freeman, San Francisco, 1963. J. M. Ziman, *Phil. Mag.*, vol. 6, p. 1013, 1961.

[3] F. Zernike and J. A. Prins, *Z. Physik*, vol. 41, p. 184, 1927.

[4] P. Debye and H. Menke, *Ergeb. techn. Röntgenkunde*, vol. 2, p. 1, 1931.

method, yielding a curve that gives the average number of atoms between r and $r + dr$ from any atom in the liquid.

If $\rho(r)dv$ is the probability of finding an atom in the volume element dv at the distance r from the center of any given atom, the probability of finding an atom in a spherical shell of radius r and thickness dr from the center of the given atom is $4\pi r^2\rho(r)$. If the mean number of atoms per unit volume is ρ_0, then $4\pi r^2\rho(r)$ approaches $4\pi r^2\rho_0(r)$ for large values of r.

Monochromatic x-rays are required. If these are obtained by a crystal monochromator with incidence angle θ_M, the Lorentz-polarization correction is made by dividing the measured intensities at scattering angle 2θ by the quantity

$$P = (1 + \cos^2 2\theta_M \cdot \cos^2 2\theta)/(1 + \cos^2 2\theta_M) \tag{8-15}$$

or if they are obtained by a balanced filter,

$$P = (1 + \cos^2 2\theta)/2 \tag{8-16}$$

The absorption correction is made by dividing the measured intensities by a factor dependent upon sample geometry; if incident and diffracted rays make equal angles with a flat sample so that parafocusing is obtained, no absorption correction is necessary.

The Compton modified scattering for the specimen is next computed for the entire angular range or taken from published tables, and the normalized intensities for the coherent scattering, I_{coh}, are obtained: $I_{coh} = \beta I_{corrected} - I_{incoherent}$, where β is the normalization constant and both I_{coh} and I_{inc} are expressed in terms of the classical scattering of an electron.

To determine β, two methods are available: (1) The incoherent contribution can be computed at high angles; also the coherent scattering of N atoms when scattering independently, Nf^2, is computed from tables of the atomic scattering factor. By assuming that at high angles the atoms are scattering independently, the total intensity at high angles may be taken as the sum of the coherent and incoherent contributions and the value of β is then determined. (2) The Krogh-Moe-Norman method may be used,[1] which is based on the value of the radial distribution function near the origin. The value of β is obtained from experimentally determined or available quantities by a relation[2] which gives smallest weight to the measurements at high angles where the measurements are least accurate,

$$\beta = \left\{\int_0^{S_{max}} [(I_{inc}/f^2) + 1]S^2e^{-\gamma S^2}\,dS - 2\pi^2\rho_0\right\} \Big/ \int_0^{S_{max}} (I_{corr}/f^2)S^2e^{-\gamma S^2}\,dS \tag{8-17}$$

where $S = 4\pi(\sin\theta)/\lambda$.

The radiation distribution curve may then be computed from the equation[3]

$$4\pi r^2\rho(r) = 4\pi r^2\rho_0 + \frac{2r}{\pi}\int_0^\infty [S\cdot i(S)]\sin rS\,dS \tag{8-18}$$

where $S\cdot i(S) = S[(I_{coh}/nf^2) - 1]$. Diminishing the weighting of the high-angle, low-accuracy measurements by the factor $\exp(-\gamma S^2)$, as in Eq. 8-17, causes only a slight widening of the peaks in the radial distribution curve[2] and eliminates series-termination

[1] J. Krogh-Moe, *Acta Cryst.*, vol. 9, p. 951, 1956. N. Norman, *Acta Cryst.*, vol. 10, p. 370, 1957.

[2] B. E. Warren, private communication.

[3] R. W. James, "The Optical Principles of the Diffraction of X-rays," G. Bell, London, 1948, 1954.

errors that would otherwise be prominent. A linear interpolation is assumed between $S = 0$ and $S = S_{min}$.

The radial distribution function is then obtained by evaluating the Fourier sine coefficients of the $S \cdot i(S)$ curve and carrying through the computations at values of r given by $r = nr_0$, where r_0 is a constant usually chosen to be 0.1 A and n takes integral values. A Fortran program compiled for use with an IBM 709 computer has been used to carry out the computations.[1]

Warren and coworkers and many others have made numerous and effective applications of this general type of analysis, not only to monatomic liquids[2] and solids[3] but also to polyatomic vitreous substances.[4] The application of Fourier analysis to polyatomic liquids and glasses, however, encounters serious difficulties, and a rigorous and full solution of a diffraction pattern has not been obtained; a simple Fourier inversion such as is used to derive Eq. (8–18) is not possible.

Warren's techniques have been improved more recently in ways which sometimes can double the resolution in the distribution curves over that of earlier techniques. This is done by (1) using shorter-wavelength radiation, such as Rh $K\alpha$ which allows a wider range of $(\sin\theta)/\lambda$ to be reached and (2) suppressing most of the Compton modified radiation which, for light elements, is several times as intense as the unmodified, coherent radiation.[5] A beam of curved-crystal monochromated Rh $K\alpha$ is diffracted by the specimen through a slit ($\sim 1°$ wide) onto a Mo plate. The fluorescence that is excited by the unmodified radiation and not by the modified (which is of longer wavelength than the Mo K absorption limit) is passed through a thin Zr filter to suppress diffracted modified radiation.

There is an effect of interatomic distances in a molecule on the diffuse intensity modulations from molecular liquids, and in the case of amorphous powders the size and shape of the powder particles also have their effects. For an introduction to this field the reader is referred to James's book.[6]

Advances in the theory of the electrical resistivity of liquid metals[7] have caused an increase in interest in the interference function I_{coh}, and methods for increasing the accuracy of the measurements of the function have steadily increased. Careful attention is currently given to the accuracy of alignment in diffractometry with liquids, and to the accuracy of the

[1] H. Ocken, Tech. Rept. no. 3, U.S.A.E.C. Contract AT(30–1) 2560, September, 1963; Thesis, Yale University, New Haven, Conn., 1963.

[2] L. P. Tarasov and B. E. Warren, *J. Chem. Phys.*, vol. 4, p. 236, 1936 (sodium).

[3] B. E. Warren and N. S. Gingrich, *Phys. Rev.*, vol. 46, p. 368, 1934 (sulfur).

[4] B. E. Warren, H. Krutter, and O. Morningstar, *J. Am. Ceramic Soc.*, vol. 19, p. 202, 1936 (SiO_2). The absence of small-angle scattering in vitreous silica indicates that the structure is continuous without numerous voids in the 10- to 100-A range, but silica gel does have small-angle scattering.

[5] B. E. Warren, private communication, 1965.

[6] James, *loc. cit.*

[7] J. M. Ziman, *Phil. Mag.*, vol. 6, p. 1013, 1961. C. C. Bradley et al., *Phil. Mag.*, vol. 7, p. 865, 1962. W. A. Harrison, *Phys. Rev.*, vol. 129, p. 2912, 1963.

correction factors and the normalization procedures.[1] Spurious ripples in the experimental radial distribution curves that are due to series-termination errors can sometimes be identified by terminating the data at different values of S and are lessened by the use of a damping factor.

For elements having close-packed structures in the solid state, there is practically no change in interatomic spacing during the solid-to-liquid transition; but for mercury and tin the results indicate that the open-packed structures of the solid state alter and become closer to an ideal monatomic structure, characteristic of a metal, in the liquid state.[2] If one arbitrarily defines the coordination number of the liquid as the area under the first peak of the radial distribution function, the number of nearest neighbors has recently been found to be 11.5 for both Cu and Ag, 8.5 for Sn, and 10 for Hg.

[1] C. N. J. Wagner, H. Ocken, and M. L. Joshi, *Z. Naturforsch.*, vol. 20*a*, p. 325, 1965. R. Kaplow and B. L. Averbach, *Rev. Sci. Instr.*, vol. 34, p. 579, 1963 (includes description of a diffractometer). A review of the field is: F. Furukawa, *Rept. Progr. Phys.*, vol. 25, p. 395, 1962.

[2] Wagner, Ocken, and Joshi, *loc. cit.* R. Kaplow, S. L. Strong, and B. L. Averbach, *Phys. Rev.*, vol. 138A, p. 1336, 1965.

9

Pole figures and orientation determination

A knowledge of the orientation of a single crystal or the preferred orientation (texture) of a polycrystalline sample is frequently of fundamental importance in science and in industry. It is appropriate, therefore, to devote a chapter to the methods of determining orientations and textures. The textures that are found in wires, rods, and compressed specimens, referred to as *fiber textures*, are of simple types; the methods of analyzing these are discussed first. More complex textures, such as are found in rolled metals, are commonly specified by means of *pole figures*, which are stereographic projections showing the density of crystallographic poles of selected planes as a function of orientation; methods appropriate to these textures are presented next. The texture portion of the chapter concludes with a discussion of *inverse pole figures*, which are stereographic projections of a principal direction in a sample—for example, the axis of a wire—on a standard projection of the crystal lattice. The rest of the chapter is devoted to the methods for determining the orientation of *single crystals*.

Detection of preferred orientations When the grains of a polycrystalline metal are oriented at random, a photograph made by passing a pinhole-collimated beam through the metal will show Debye rings of uniform intensity all around their circumference. On the other hand, if the grains cluster around certain orientations so that the metal has a *texture* or *preferred orientation*, then more grains will be in position to reflect to certain segments of the diffraction rings than to others, and the rings will have maxima and minima.

Figure 9–1 illustrates the pattern that is obtained with monochromatic

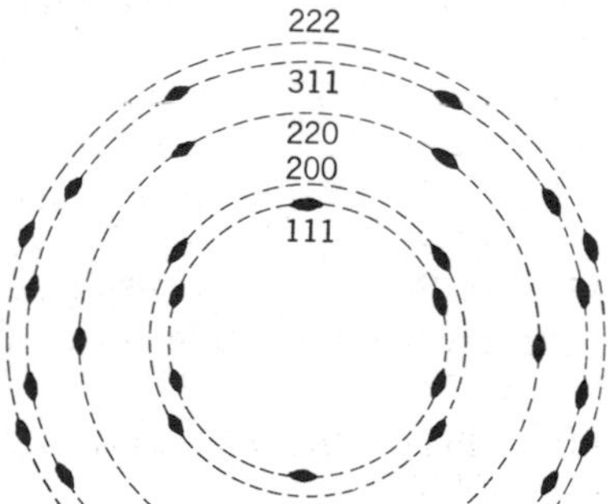

Fig. 9-1 Pattern of cold-drawn aluminum wire. Radiation of a single wavelength incident perpendicular to the wire axis, which is vertical.

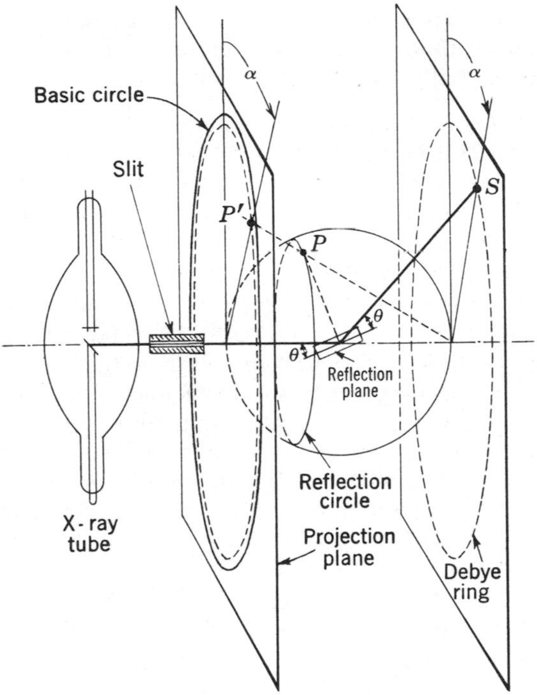

Fig. 9-2 Relation between crystal plane, diffracted beam, and stereographic projection. Pole *P*, diffracted spot *S*, and projection *P′* all lie in a plane containing the incident beam.

x-rays from a cold-drawn wire of aluminum, which is a f.c.c. metal that becomes oriented with the [111] direction in each grain parallel to the axis of the wire. The dashed circles represent Debye rings that would be present if the metal had a random orientation, with all orientations present. The limited number of orientations that are present in the wire permit only a few spots in the Debye rings to appear strongly, as indicated in the figure. A certain crystallographic axis in many grains or crystalline fragments is parallel to the wire, but all other crystallographic axes have a random distribution around the wire; i.e., all azimuthal orientations around the axis are equally probable. An x-ray photograph of the wire is in effect a rotating-crystal photograph in which the axis of rotation of the crystal corresponds to the axis of the wire.[1] The textures of wires are appropriately called *fiber textures*, for they resemble the structure of fibrous materials. The axis of symmetry of the texture (the longitudinal axis of a wire) is the *fiber axis*. In a rolled sheet, on the other hand, there is not an equal

[1] The intense spots on the Debye rings are actually on layer lines. If the axis of the wire has indices [*uvw*] in each oriented grain, then the *n*th layer line will contain the spots *hkl* for which $hu + kv + lw = n$.

probability for all orientations around the direction of elongation (the direction of rolling), and a more detailed analysis is necessary. The stereographic projection is almost universally used for this purpose.

Stereographic projection of data A simple and direct relation exists between spots on a preferred orientation pattern and their stereographic projection. The relationship is illustrated in Fig. 9–2. A crystal plane is here shown in position to reflect a beam of x-rays to form a spot S on the film. The plane normal intersects the reference sphere, which is inscribed about it, at the point P, which projects stereographically to the point P' in the projection plane. The incident beam, the reflected beam, and the pole of the reflecting plane all lie in the same plane tipped at an angle α from the vertical. Thus, when the film and projection plane are placed normal to the beam as shown, it will be seen that the angle α on the projection will be exactly equal to α on the film.

Since the angle of incidence, θ, of the beam on an (hkl) plane is determined by the Bragg law $n\lambda = 2d \sin \theta$ whenever reflection occurs, it follows that the poles of all such (hkl) planes capable of reflecting must lie at a constant angle $90 - \theta°$ from the incident beam and must intersect the reference sphere only along the circle known as the "reflection circle," a circle on the projection $90 - \theta°$ from the centrally located beam. Both the azimuthal and radial positions on the pole figures are thus determined.

Fiber textures Let us consider an ideal fiber texture in which all grains have a certain crystallographic direction, say [111], parallel to the fiber axis. The stereographic projection of the (100) planes would then form the pole figure of Fig. 9–3, in which (100) poles of various grains lie at various spots along the latitude lines shown as heavy lines at an angle $\rho = 54°44'$ from the fiber axis, since this is the angle between [111] and [100] poles. Now if a beam entered the reference sphere at B, perpendicular to the fiber axis, and reflected from the planes at the Bragg angle θ, re-

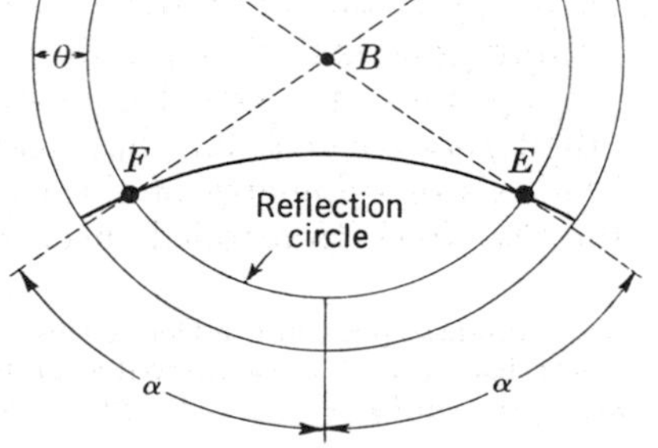

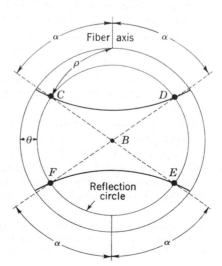

Fig. 9-3 Ideal pole figure for (100) planes in a wire of a cubic metal having a [111] fiber texture. Poles at *C*, *D*, *E*, and *F* can reflect when x-ray beam direction is *B*.

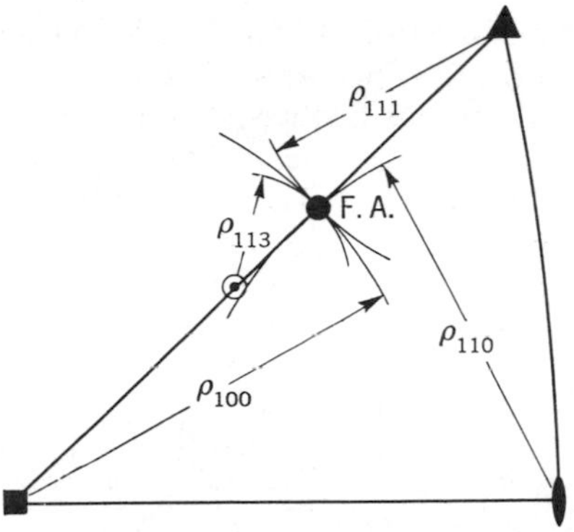

Fig. 9-4 Method of determining fiber axis (F.A.) on a standard stereographic projection of a cubic crystal. Angle ρ for each reflection is laid off from a corresponding pole of the projection.

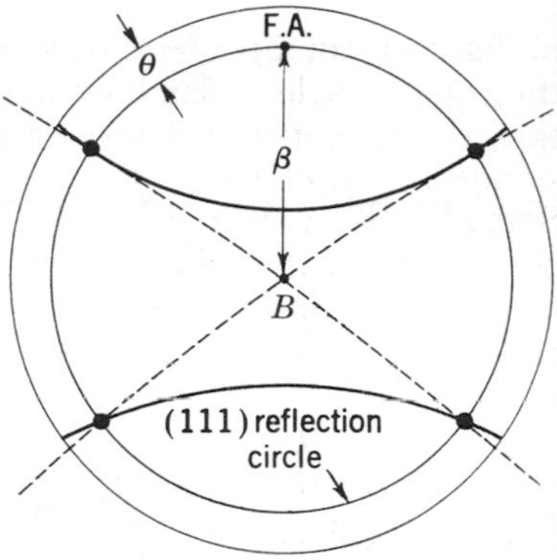

Fig. 9-5 Stereographic projection showing fiber axis tilted to cause strong reflection from planes normal to F.A.; beam direction is B.

flections would occur at the intersections C, D, E, and F. All reflections would fall upon the film at equal angles α from the projection of the fiber axis.

The angle α was stated to be identical on projection and film, but Fig. 9-3 shows that this is not exactly equal to the angle ρ between the pole of the reflecting plane and the fiber axis. From the spherical trigonometry of the figure when the beam is normal to the fiber axis, it follows that

$$\cos \rho = \cos \alpha \cos \theta$$

which approaches the relation $\alpha = \rho$ as θ approaches 0. If the fiber axis is inclined so that it makes an angle β with the incident beam, the corresponding equation is

$$\cos \rho = \cos \beta \sin \theta + \sin \beta \cos \theta \cos \alpha$$

Thus, a series of ρ values can be computed for the observed spots on a film, and from these the indices of the fiber axis may be deduced with the aid of tables of angles between crystallographic directions.[1] Some textures are composed of two fiber textures superimposed; for example, iron after compression has some grains with [111] parallel to the fiber axis and others with [100], forming a *duplex fiber texture* [111] + [100].

A graphical determination of fiber axis from a series of ρ values is shown

[1] Tables of angles are given in Chap. 2. Curves have been published that make possible a graphical computation of the data from rotating-crystal patterns and fiber patterns. When many orientations of crystals of a given substance are to be determined, the work can be accelerated by using these. J. Thewlis, *Z. Krist.*, vol. 85, p. 74, 1933. R. M. Bozorth, *Phys. Rev.*, vol. 23, p. 764, 1924.

in Fig. 9–4. On a stereographic projection the point (F.A.) is found that is the proper angle ρ from the pole of each reflecting plane; such a point is then the projection of the fiber axis and can be identified by reference to a standard projection. In Fig. 9–4 the fiber axis coincides with [112]. Note that the arcs in the figures are loci of points at equal *angles* from reflecting poles, not equal *distances*. If there is a considerable range of orientation in the specimen, both maximum and minimum ρ values may be plotted and an *area* may be determined on the standard projection rather than a *point*.

If a plane (*hkl*) is perpendicular to the fiber axis in each grain, it is advantageous to tilt the wire with respect to the x-ray beam so as to reflect from this plane. A tilt of θ_{hkl} will obviously do this, since the reflection circle will then touch the fiber axis. For the case of aluminum wire, the fiber axis is [111] and for molybdenum rays $\theta_{111} = 10.1°$. The tilted condition is represented in the (111) pole figure of Fig. 9–5. The reflection in the vertical plane is very strong, since all grains of the texture contribute to it, while only a small fraction of the grains contribute to the other reflections. Thus, for example, a weak texture in iron after compression can readily be shown by this kind of tilting of the specimen[1], although it has been overlooked in some less sensitive tests.

Plotting of pole figures Preferred orientations produced by deformation other than simple uniaxial elongation or compression are complex. While they may be specified in an approximate way in terms of *ideal orientations* with certain crystal axes parallel to the principal axes of strain, the choice of indices is often arbitrary and the description is incomplete. *Pole figures* provide a more complete description of the texture and a safer basis for studies of the underlying mechanism, for they represent the observational data in an unprejudiced manner.[2]

To plot a pole figure, the relations of Fig. 9–2 must be kept in mind. All the intensity maxima on a single diffraction ring (Debye ring) are plotted on a single reflection circle on the pole figure. When the circle cuts through heavily populated regions on the pole figure, the diffraction ring will show intense blackening at the same azimuth; when it cuts through lightly populated regions, the corresponding arc of the diffraction ring will be weak. To determine the true extent of the areas on the pole figure, one must plot a series of reflection circles that form a network covering the projection. This may be accomplished by taking a series of diffraction patterns with the specimen tilted increasing amounts in steps of 5 or 10°. The number of exposures in such a series may vary from 5 to 20, depending upon the detail required in the pole figure. An oscillating-film arrangement can be used in texture cameras so that one exposure gives the information ordinarily obtained from many individual exposures of the stationary type (see later section on cameras, page 199).

[1] C. S. Barrett, *Trans. AIME*, vol. 135, p. 296, 1939.

[2] F. Wever, *Mitt. Kaiser-Wilhelm-Inst. Eisenforsch. Düsseldorf*, vol. 5, p. 69, 1924; *Z. Physik*, vol. 28, p. 69, 1924; *Trans. AIME*, vol. 93, p. 51, 1931.

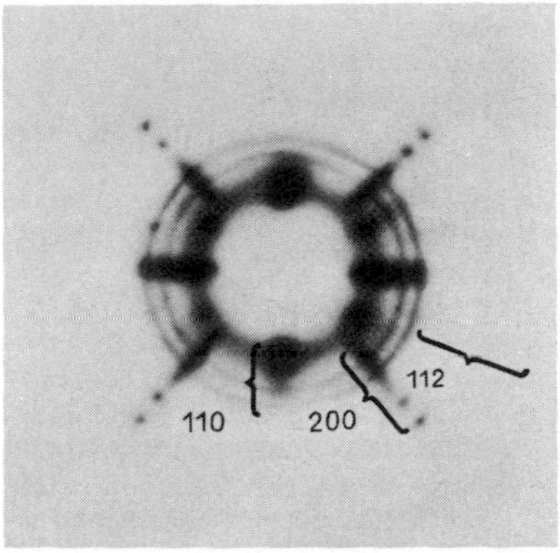

Fig. 9-6 Diffraction pattern of cold-rolled steel. Mo radiation perpendicular to rolled surface, rolling direction vertical. Shows $K\alpha$, $K\beta$, and white radiation reflections from (110), (200), and (112) planes.

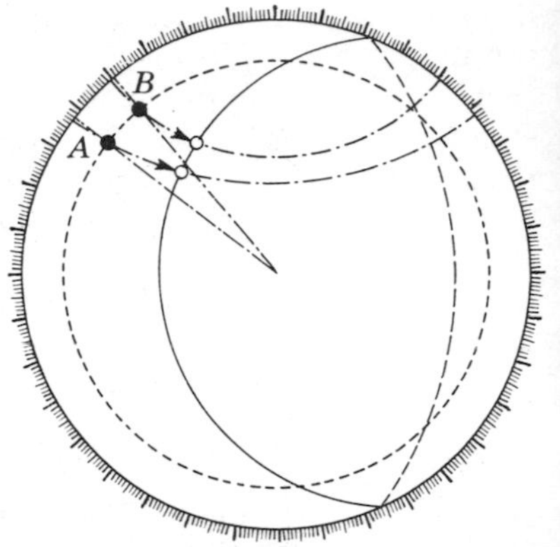

Fig. 9-7 Plotting a pole figure; (- - - -), reflection circle with beam normal to projection plane; (——) and (— —), reflection circles with plane of specimen as projection plane.

The plotting of the data is illustrated by using rolled steel as an example. If the radiation from molybdenum is used, the photographs will resemble Fig. 9–6, with concentric Debye rings (incomplete) from $K\alpha$ and $K\beta$ characteristic radiation and an inner band from white radiation. The diffraction from the ferrite {110} planes will occur at $\theta \simeq 10°$, and the reflection circle will lie $90 - \theta° = 80°$ from the center of the pole figure. If the first photogram is made with the beam normal to the surface of the specimen, the surface will appear on the projection as the basic circle and the reflection circle will lie concentric with it.

Let us suppose that the specimen is then rotated about a vertical axis and a new photogram taken with the specimen turned, say 30°, the near side rotating to the left as we look away from the x-ray tube. We first plot the data as before, with the projection plane again normal to the beam and the reflection circle again concentric. This is illustrated in Fig. 9–7, where the reflection circle is shown as a dashed circle and an intensity maximum is indicated between the limits *A* and *B*. But with this setting of the specimen its surface will lie at a tilt of 30° from the projection plane (the right side tilted up from the projection plane), and all data from this setting must be rotated 30° back to the right in order to plot a pole figure in which the plane of the specimen is the projection plane. The rotation is done with the Wulff net, each point on the reflection circle being moved to the right along its latitude line a distance of 30° of longitude. This is shown in Fig. 9–7, where the reflection circle with the intensity maximum on it is plotted after rotation as a full line. Part of the rotated reflection circle on the right-hand side passes to the negative hemisphere and is shown as a broken line. The process is repeated with different settings of the specimen and corresponding rotations of the data until the areas on the pole figure are sufficiently well defined.

Much of the labor of plotting may be eliminated if a chart is made up in which the series of reflection circles are shown in the position they would have after rotation back to the normal setting, i.e., their position with respect to the surface of the specimen as the projection plane. A different chart is required, of course, for every different value of θ that is used, thus for different wavelengths, different specimen materials, and different reflecting planes. Wever[1] has published charts for copper $K\alpha$ radiation reflecting

[1] F. Wever and W. E. Schmid, *Mitt. Kaiser-Wilhelm-Inst. Eisenforsch., Düsseldorf*, vol. 11, p. 109, 1929. F. Wever, *Trans. AIME*, vol. 93, p. 51, 1931.

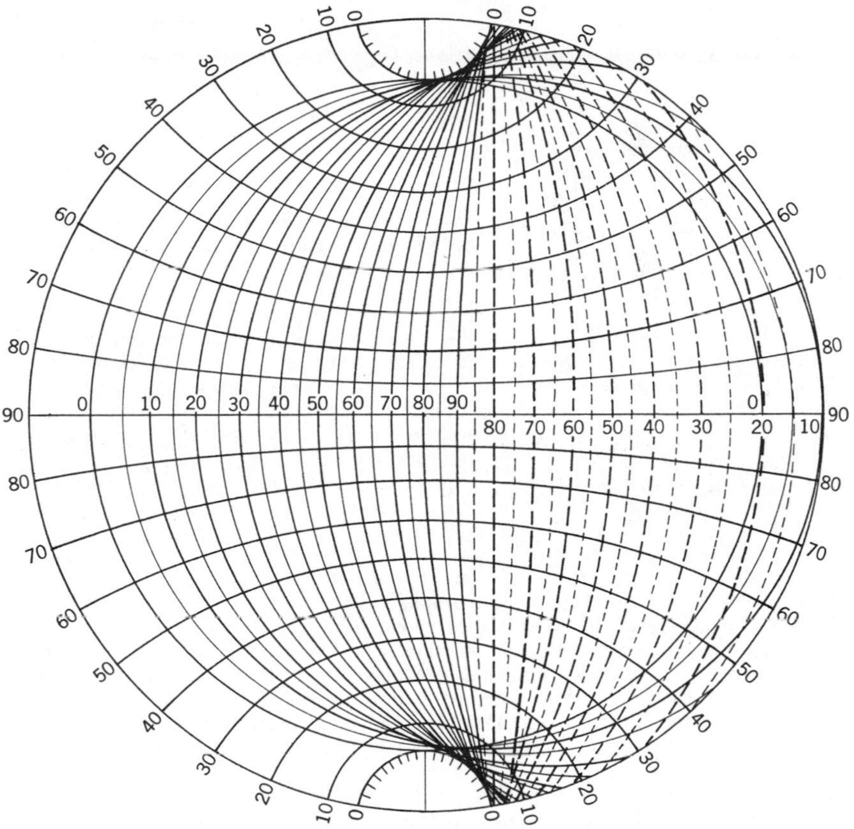

Fig. 9-8 Pole-figure chart for Mo $K\alpha$ radiation reflecting from (110) planes of iron ($\theta = 10°$).

from {001} and from {111} planes of aluminum and for iron $K\alpha$ radiation reflecting from {110} of iron. Figure 9–8 presents a chart for molybdenum $K\alpha$ radiation reflecting from {110} planes of iron ($\theta = 10°$) with reflection circles plotted for every 5° rotation interval from 0 to 90°. The azimuthal positions on all the circles are given by their intersections with the latitude lines that are drawn on the chart and labeled with values of the angle α (the numbers around the circumference). Reflection circles on the back hemisphere are shown as dashed lines, but the same lines of constant α apply as on the near hemisphere. This same chart will serve for plotting the {200} reflections from iron if one reads the intensity maxima on the broad ring caused by general (white) radiation. In this instance, the most intense portion of the {200} reflection of the general radiation from a tube operating at 30 or 40 kv is in the neighborhood of $\theta = 10°$. The maximum intensity of general radiation from {110} planes is around $\theta = 7°$ and is particularly suitable for pole-figure work, for it is free from overlapping {200} reflections and shows clearly the slight differences of intensity that are sometimes important.

Areas near the top and bottom of the pole figure are not crossed by any reflection circles on the chart. To fill in these areas the specimen may be turned 90° on its own plane and then rotated a small amount about the vertical axis, as before.

Specimens and cameras for texture studies A sheet specimen used for pole figure studies requires an absorption correction for the observed intensities that depends

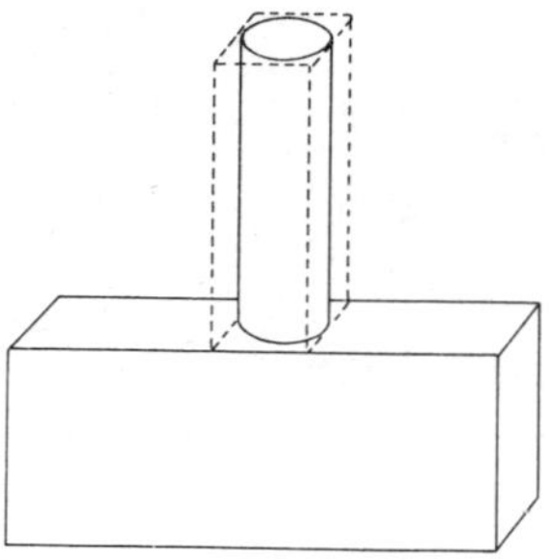

Fig. 9-9 Specimen ground to a shape to have unchanging absorption when rotated around its axis. (P. W. Bakarian.)

markedly on the angle of setting in the camera.[1] Bakarian[2] has improved the technique by cutting a stem out of the sheet metal with a jeweler's saw, then grinding the stem to a cylindrical shape, as in Fig. 9–9. Rotation of this specimen around its axis does not change the absorption. With adequate calibration of the films for exposure, it is then possible to make a fairly reliable estimate of the relative number of poles in each orientation region of the pole figure, but counter diffractometry is used for precise measurements.

Large-grained specimens must be exposed in cameras that bring more than the usual number of grains into reflecting position in the beam. This increase can be provided by oscillating the specimen through a slight range of angles on an oscillating-crystal camera or by shifting the specimen in its own plane during the exposure. The importance of having many grains reflect is not often appreciated by beginners, and several incorrect conclusions have resulted from photographs of specimens in which the grains were too large to provide adequate sampling and averaging. Devices to move large areas of a specimen into the beam, "integrating cameras," have been of many kinds, both simple and complex.[3] A camera of simple design may merely have motor-driven rolls which serve to translate a strip-shaped sample across the x-ray beam.

Texture goniometers have been designed in which a single exposure with an oscillating film gives the information ordinarily obtained from many individual exposures.[4]

Use of x-ray diffractometers for texture studies The intensities of the various areas of a pole figure can best be measured with an x-ray diffractometer. The visual estimation of film darkening and of absorption that is characteristic of early pole-figure work can be replaced by quantitative measurement, and pole figures can now be drawn with precision contours when desired.

A method developed by Norton[5] employs a series of cylindrical specimens

[1] This has been computed by J. F. H. Custers, *Physica*, vol. 14, pp. 453, 461, 1948, and is most easily applied when a sample is first tilted $90 - \theta°$ from its position normal to the beam, then rotated in a series of steps about its surface normal. Equations for variable tilt angle are given by R. Smoluchowski and R. W. Turner, *Rev. Sci. Instr.*, vol. 20, p. 173, 1949.

[2] P. W. Bakarian, *Trans. AIME*, vol. 147, p. 266, 1942.

[3] F. Pawlek, *Z. Metallk.*, vol. 27, p. 160, 1935. B. F. Decker, *J. Appl. Phys.*, vol. 16, p. 309, 1945. F. H. Wilson and R. M. Brick, *Trans. AIME*, vol. 161, p. 173, 1945. J. Thewlis and A. R. Pollock, *J. Sci. Instr.*, vol. 27, p. 72, 1950.

[4] O. Kratky, *Z. Krist.*, vol. 72, p. 529, 1930. F. E. Haworth, *J. Sci. Instr.*, vol. 11, p. 88, 1940. W. E. Dawson, *Physica*, vol. 7, p. 302, 1927; *Phil. Mag.*, vol. 5, ser. 7, p. 756, 1928. C. S. Barrett, *Trans. AIME*, vol. 93, p. 75, 1931. A. Guinier and J. Tennevin, *Rev. Met.*, vol. 45, p. 277, 1948. R. Smoluchowski and R. W. Turner, *Physica*, vol. 16, p. 397, 1950. W. A. Wooster, *J. Sci. Instr.*, vol. 25, p. 129, 1948.

[5] J. T. Norton, *J. Appl. Phys.*, vol. 19, p. 1176, 1948.

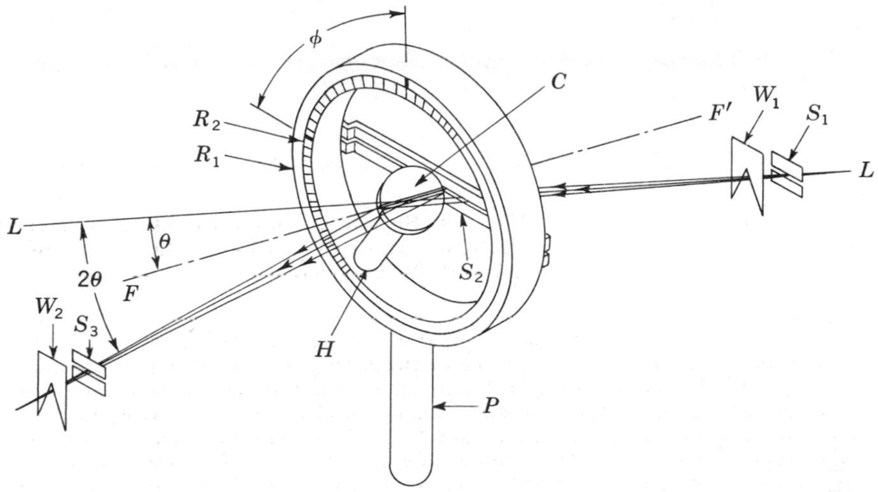

Fig. 9-10 Slits and goniometer head for Schulz's reflection method of texture determination in an x-ray spectrometer. Beam through S_1, W_1, and S_2 reflects from specimen at C and into counter through S_3 and W_2.

(Fig. 9–9) for the purpose, each having been cut at a different angle in the rolled sheet or other object being studied and mounted, in turn, on the axis of a diffractometer. The specimen mounted on the diffractometer is turned slowly about the specimen axis while the counter is set to receive a reflection of low indices; the intensity of the reflected beam is recorded automatically on a strip-chart recorder. On a pole figure of a rolled sheet in which the plane of the sheet is the plane of projection, each specimen provides data for points along a diameter of the pole figure. A disadvantage of the method is that a new specimen is required for each new diameter that is to be plotted. No correction for absorption is required.

The most convenient method proposed to date appears to be that of Schulz,[1] since a flat specimen is used and no corrections for absorption or for changes in geometry during rotation of the specimen are necessary from the center of the pole figure (normal to the surface of the sample) out to about 70°.

In Schulz's method, a reflection specimen is mounted in a goniometer head of the type sketched in Fig. 9–10 on plate C. The post P of the goniometer head is mounted on the axis of the x-ray spectrometer. This post supports a ring R_1 in which rests a smaller ring R_2 that can be turned through angles ϕ indicated by graduations on the ring. A second post, H, supports the specimen, which is cut to circular shape. The post is adjustable so that the specimen surface coincides with the ring axis FF'. The collimating system is composed of slits S_1 and S_2 which permit negligible divergence in the vertical direction.[2] The horizontal lengths of the slits S_1 and S_3 are limited by wedges W_1 and W_2. Slit S_2 permits radiation to strike only a narrow strip along the diameter

[1] L. G. Schulz, *J. Appl. Phys.*, vol. 20, p. 1030, 1949.

[2] For use with the original model Norelco spectrometer, S_1 and S_2 are about 0.020 in. and S_3 is between 0.020 and 0.050 in.

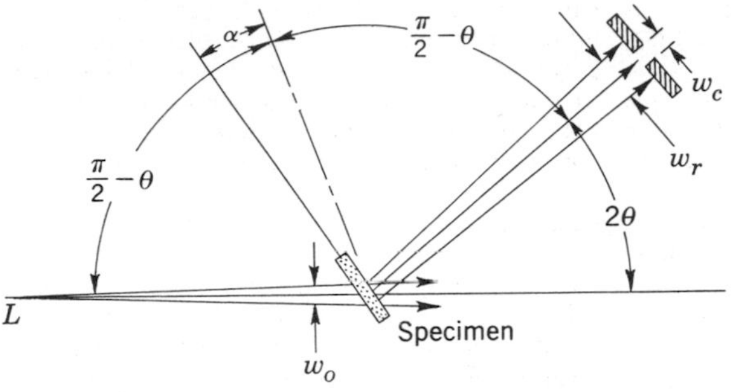

Fig. 9-11 Geometry of Schulz's transmission method for texture studies. Counter slit width w_c is small compared with w_r; diverging rays from line source L pass through slit of width w_o at the specimen.

of a circular specimen at the position of the ring axis FF'. Wedge W_2 permits the entire width of a Debye ring to enter the counter, and S_3 sharply limits the length of the arc that enters. Readings are taken with varying angles ϕ to give data along one diameter of the pole figure. The setting of the specimen on the circle C is then changed by rotating the specimen in its own plane, to obtain data for another diameter.

Certain limitations of the Schulz method should be kept in mind. The arrangement is intended for specimens thick enough to provide effectively complete absorption of the beam. Since the area irradiated is small—of the order of 0.020 by 0.3 in.—there is danger that orientations in the irradiated area are not a representative sampling of the average orientations over the surface (unless an integrating mechanism is provided); in addition, any reflection method such as this is limited to the surface layers reached by the beam.

Since the above technique applies without correction only out to 70 or 80°, Schulz employs a transmission method for the outer rim of the pole figure,[1] in which corrections are avoided throughout a wide band near the rim by using a thin specimen and appropriate slits. Rays from a point source or from a slit not more than 0.0005 in. wide and 0.050 in. high at L (Fig. 9–11) diverge to a width w_o at the specimen and to a width w_r at the counter. No correction is required in a band about 25° wide around the outside of the pole figure if the counter slit width w_c is made small compared with the reflected beam width, provided the specimen is thin compared with w_o (say 0.005 in. or less compared with w_o of the order of 0.050 in.) and if μt, the product of the linear absorption coefficient and the specimen thickness, lies in the neighborhood of 0.3 to 0.4. Suitably thin specimens can be used in both transmission and reflection positions without appreciable intensity errors.[2] However, errors arising from defocusing and faulty positioning of the specimen can become important even when the specimen is tilted only 40 or 50°. These errors are lessened by narrowing slit S_2 and by accurate specimen alignment.[3] The spectrometer is adjusted to the maximum-intensity position in the center of the reflected beam. The specimen is then turned to different angles α from the position in which it bisects the angle between incident and reflected beams; α is the angle between the pole of the reflecting plane and the basic circle of the pole figure.

One method suggested for correcting diffracted intensities in pole-figure determina-

[1] L. G. Schulz, *J. Appl. Phys.*, vol. 20, p. 1033, 1949.

[2] J. Grewen, A. Segmüller, and G. Wassermann, *Arch. Eisenhuttenw.*, vol. 29, p. 115, 1958.

[3] W. P. Chernock and P. A. Beck, *J. Appl. Phys.*, vol. 23, p. 341, 1952.

tion is based on the intensities of Compton modified scattering. If background intensities can be measured between diffraction peaks without contributions from the tails of the peaks (which is not always possible), and if precautions are taken to remove all fluorescence and all other "parasitic" scattering, the remainder of the scattering will be modified scattering, which varies with the effective scattering volume of the crystal as a function of angle setting in the diffractometer in the same way as the coherent radiation; it will also be absorbed similarly. It can therefore provide a correction factor for the coherent radiation.[1]

It is desirable to use integrated intensities and to express these in terms of the intensities that correspond to a random sample. This may be done by preparing a random sample of the same material (which is not always easy), or by computing the intensities that would be obtained from such a sample. Counter diffractometers[2] have now replaced photographic methods for nearly all quantitative work. Improvements in specimen-holder design permit a more complete coverage of the area of the pole figure,[3] and slow continuous rotation of the specimen about two axes permits the reflecting condition to sweep over the pole figure in a spiral path which makes possible fully automatic plotting of the data.[4] Automatic operation and plotting have proved useful in laboratories where large numbers of pole figures are studied each year. Randomly oriented samples to serve as standards for pole densities can be prepared from powder compacts held together by a binder or by sintering; a standard random sample should have the same μt as the textured sample to which it is compared.[5] Quantitative pole figures can also be standardized by comparing measured intensities with intensities averaged over a sphere surrounding the specimen.[6] In quantitative measurements it is necessary to guard against false readings due to the superposition of general radiation reflected by one set of planes on the $K\alpha$ reflection from another set. Pulse-height analyzers used with proportional or scintillation counters can reduce this danger.[7] Care should be taken, also, to see that in setting slits to receive *hkl* reflections, they exclude reflections from neighboring Debye rings.

Inverse pole figures The projections discussed thus far in this chapter are direct pole figures, in that x-ray data are plotted directly on the projections. An inverse pole figure represents the density distribution of an important direction in the polycrystalline sample (for example, the axis of a wire sample) on a stereographic projection of the crystal lattice in some standard orientation. Thus, in a direct pole figure the basic circle of the projection might represent the normal to the wire axis, whereas in the inverse pole figure it would represent a plane through perhaps the poles of (100) and (010) planes.

[1] R. H. Bragg and C. M. Packer, *Rev. Sci. Instr.*, vol. 34, p. 1202, 1963; *J. Appl. Phys.*, vol. 35, p. 1322, 1964.

[2] B. F. Decker, E. T. Asp, and D. Harker, *J. Appl. Phys.*, vol. 19, p. 388, 1948. J. T. Norton, *J. Appl. Phys.*, vol. 19, p. 1176, 1948. An instrument for obtaining pole-figure data automatically (and also for shifting the specimen with a reciprocal motion to provide integration) was developed by A. H. Geisler, "Modern Research Techniques in Physical Metallurgy," p. 131, American Society for Metals, Cleveland, Ohio, 1953.

[3] M. Schwartz, *J. Appl. Phys.*, vol. 26, p. 1507, 1955. J. B. Newkirk and L. Bruce, *J. Appl. Phys.*, vol. 29, p. 151, 1958.

[4] A. N. Holden, *Rev. Sci. Instr.*, vol. 24, p. 10, 1953. H. Neff, *Z. Metallk.*, vol. 47, p. 646, 1956. Apparatus and techniques are reviewed in detail in G. Wassermann and J. Grewen, "Texturen metallischer Werkstoffe," Springer, Berlin, 1962.

[5] Newkirk and Bruce, loc. cit.

[6] J. Grewen, A. Segmüller, and G. Wassermann, *Arch. Eisenhüttenw.*, vol. 29, p. 115, 1958. H. Möller and H. Stäblein, *Arch. Eisenhüttenw.*, vol. 29, p. 377, 1958.

[7] A. H. Geisler, *J. Appl. Phys.*, vol. 25, p. 1245, 1954.

Inverse pole figures have an advantage over the ordinary direct pole figures chiefly for fiber textures, such as wire textures or compression textures. They are not necessarily free from ambiguity when used for rolling textures. Furthermore, their precise determination by measurements of reflected intensities depends upon the availability of intensity measurements for a considerable number of reflections; therefore they are appropriate only when relatively short-wavelength x-rays are used, such as Mo $K\alpha$ or Ag $K\alpha$.

Plots based on orientations of individual grains Inverse pole figures can be plotted by the laborious method of determining a large number of orientations of individual grains by methods appropriate to single crystals. This has been done for rolled and annealed silicon-steel sheet, for example, using transmission Laue photographs of individual grains.[1] A plot is made on a standard stereographic triangle showing the density distributions of three axes, individually, in a rolled sheet; i.e., plots of densities showing the distribution of the rolling direction, the transverse direction, and the normal direction. With such "axis density plots" it is possible to derive the presence of a number of individual components which would be unresolved in ordinary pole figures. Electron diffraction has also been used to determine the individual orientations, as has optical goniometer data: (see page 209).

The method of Harris for inverse pole figures of fiber textures An inverse pole figure can also be plotted from a polycrystalline specimen measured on a diffractometer. One way of doing this has been published by Harris.[3] A sample ground to a spherical shape can be used, or a flat sample with the reflecting surface of the sample normal to the fiber axis. The sample is placed with the fiber axis bisecting the incident and reflected beams. The intensities of a number of different *hkl* reflections are measured by changing the angle 2θ for the counter position and simultaneously changing the angle of incidence of the primary beam onto the reflecting plane (i.e., changing θ). The intensities from the various planes of a similar specimen that has been prepared with a *random* distribution of grain orientations are also measured; these intensities are used as the units with which to measure the intensities from corresponding planes of the sample having a texture. In the corrected form, as used by Mueller, Chernock, and Beck,[4] the integrated intensity of the *hkl* reflection may be written

$$I_{hkl} = CI_oALN_{hkl}\,|\,F_{hkl}\,|^2\,p_{hkl} \qquad (9\text{–}1)$$

where C is a constant for a given sample, I_o is the intensity of the incident beam, A is the absorption factor multiplied by a factor to take account of the geometry of the beam as it strikes the specimen in the setting required for the reflection, L is the Lorentz-polarization factor for the reflection, $|\,F_{hkl}\,|$ is the structure amplitude for the reflection, N is the multiplicity of the planes $\{hkl\}$, and p_{hkl} is the fraction of the crystals that have $\{hkl\}$ plane normals lying parallel to the fiber axis. The values of p_{hkl} are in normalized units so that the mean value averaged over all orientations is equal to unity:

$$\bar{p} = \frac{1}{4\pi}\int p_{hkl}\,d\Omega = 1 \qquad (9\text{–}2)$$

[1] C. G. Dunn, *J. Appl. Phys.*, vol. 30, p. 850, 1959.

[2] C. S. Barrett and L. H. Levenson, *Trans. AIME*, vol. 127, p. 112, 1940.

[3] G. B. Harris, *Phil. Mag.*, vol. 43, p. 113, 1952.

[4] M. H. Mueller, W. P. Chernock, and P. A. Beck, *Trans. AIME*, vol. 212, p. 39, 1958.

A sample with randomly oriented grains would then have p everywhere equal to $\bar{p}$, and in a textured sample the pole densities are expressed in terms of the density in a random sample of the same material.

For a random sample Eq. (9–1) becomes

$$I_{r,hkl} = C_r I_o A L N_{hkl} \mid F_{hkl} \mid^2 \tag{9–3}$$

where C of Eq. (9–1) is replaced by C_r for the random sample and p is set equal to unity. The ratio C/C_r can be determined as follows. Dividing Eq. (9–1) by Eq. (9–3), one has

$$\frac{I_{hkl}}{I_{r,hkl}} = \frac{C}{C_r} p_{hkl} \tag{9–4}$$

If Eq. (9–4) is applied to a sufficiently large number n of reflections and summed, one can evaluate the quantity

$$\frac{1}{n} \sum \frac{I_{hkl}}{I_{r,hkl}} = \frac{C}{C_r} \frac{\Sigma p_{hkl}}{n} \tag{9–5}$$

and since the value of $\Sigma (p_{hkl}/n)$ can be taken as approximately equal to unity when n is large, Eq. (9–5) becomes

$$\frac{C}{C_r} = \frac{1}{n} \sum \frac{I_{hkl}}{I_{r,hkl}} \tag{9–6}$$

and Eq. (9–4) then becomes

$$p_{hkl} = \frac{I_{hkl}}{I_{r,hkl}} \Big/ \frac{1}{n} \sum \frac{I_{hkl}}{I_{r,hkl}} \tag{9–7}$$

If only a few reflections are available, Eq. (9–6) becomes inaccurate, and also the points plotted on the inverse pole figure become too few to be useful. It is then important to maximize the number of reflections by using a short wavelength such as Mo $K\alpha$ or Ag $K\alpha$.

The method of Jetter, McHargue, and Williams The inverse pole figure for a fiber texture may be derived from a curve of measured intensities of reflections from a given set of planes, $\{hkl\}$, as a function of ϕ, the angle between the fiber axis and the diffraction-plane pole, and the data may be normalized, without the use of a random sample, by a method that is essentially a trial-and-error procedure.[1] The values for the intensities of the $\{hkl\}$ reflections corresponding to randomness are obtained by graphical integration of $I_{hkl} \sin \phi d\phi$ from $\phi = 0$ to 90°. Here I is the reflected intensity corrected for any geometrical factors that may have entered as a specimen of a certain shape is tilted in a diffractometer to different ϕ positions. The exact values of the fiber-axis densities at the points on the inverse pole figure corresponding to the reflecting-plane normals are obtained directly from the normalized intensities. But for other points on the inverse pole figure, a trial-and-error method is used, leading to successively improved approximations, and a final fiber-axis distribution is accepted when an assumed distribution reproduces the observations for at least three different I_{hkl} vs. ϕ curves. With this method Jetter, McHargue, and Williams were able to account for the observed pole distributions in an extruded aluminum sample by an inverse pole-figure chart which indicated that 47.5 percent of grains constituted the $\langle 111 \rangle$ component and 52.5 percent the $\langle 100 \rangle$ component of the two-component texture.

This method was not reduced to a *direct* computation at the time it was published, but it was pointed out that if the pole-distribution function is written as a series of linear equations, a matrix inversion on a computer should yield another series of linear equations, the solutions of which would give the fiber-axis distribution. Since that time, direct computation methods involving desk computers or high-speed digital computers

[1] L. K. Jetter, C. J. McHargue, and R. O. Williams, *J. Appl. Phys.*, vol. 27, p. 368, 1956.

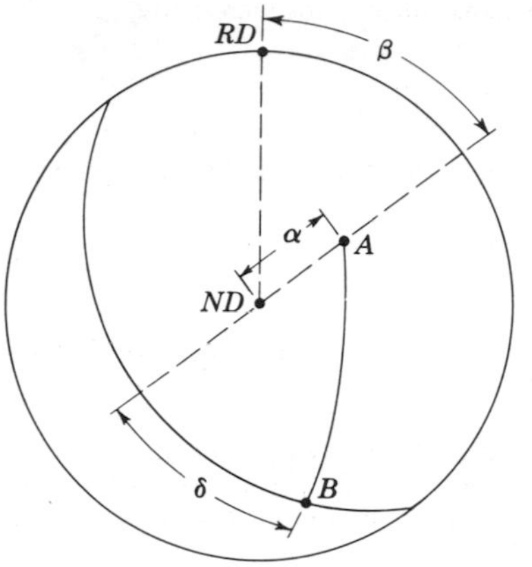

Fig. 9-12 Relations between the angles, the poles *A* and *B* of a crystal, and the principal directions in a rolled sheet: RD = rolling direction, ND = normal direction.

have been preferred over trial-and-error methods as computing facilities have become increasingly available.

Extending their fiber-texture method to rolling textures, Jetter, McHargue, and Williams prepared samples by bonding pieces of a sheet together with an adhesive and cutting out a sphere from the composite. They then obtained pole-distribution data for nine different sets of planes, with the specimen rotating rapidly, in successive experiments, around the rolling direction, the transverse direction, and the normal direction as an axis, while the specimen was scanned slowly through a range of ϕ from 0 to 90° in each of the three experiments. The inverse pole figures derived from each of these three were used to specify the rolling texture in terms of the individual components, the volume fractions of each, and the nature of the scatter around each of the ideal orientations.

It has been pointed out that a rigorous statement of the orientations in a rolled sheet or in any sample lacking full axial symmetry cannot in general be fully represented by a group of plots of this sort.[1] When the texture of a rolled sheet is described in terms of three independent inverse pole figures for rolling, transverse, and normal directions, the orientation distribution of the grains is not fully specified, since the description ignores *correlation* between the plots.

Williams points out in defending the method, however, that what must be guarded against is merely the ambiguity arising when peaks from two or more different components of a texture superimpose on a direct pole figure, and that even when superposition occurs it need not lead to ambiguity if the number of components is not too large and if their spread around a mean orientation (an "ideal orientation") is not too large.[2] The crystals are usually highly symmetrical, and if only one of three or more symmetrically equivalent poles in one component is involved in such a superposition, the other peaks from this component may be clearly separated, disclosing the relative amount of the component.

The method of Mitchell and Rowland[3] Let the angles α and β define the position of a pole A of a crystal plane, with the angles related to the principal directions in a rolled sheet as indicated in Fig. 9–12. Let a second pole B be located on a circle around A at the proper angle from A. (The method was originally applied to orthorhombic uranium with A normal to (001) and B normal to (100) planes, the angle between A and B therefore being 90°.) The position of pole B is specified by the angles α, β, and δ.

The pole density for pole A is a function of α and β and is represented by $p_A(\alpha,\beta)$,

[1] M. H. Mueller, W. P. Chernock, and P. A. Beck, *Trans. AIME*, vol. 212, p. 39, 1958.

[2] R. O. Williams, *Trans. AIME*, vol. 215, p. 646, 1959.

[3] C. M. Mitchell and J. F. Rowland, *Acta Met.*, vol. 2, p. 559, 1954.

the function being normalized so that its average over all orientations is unity:

$$\bar{p} = \frac{1}{2\pi} \int p(\alpha,\beta) \sin \alpha \, d\alpha d\beta = 1 \tag{9–8}$$

The number of A poles within an element of solid angle $d\Omega$ surrounding the point at α, β is then

$$p_A(\alpha,\beta) d\Omega = p_A(\alpha,\beta) \sin \alpha \, d\alpha d\beta \tag{9–9}$$

and would be $\bar{p} d\Omega$ for a randomly oriented sample.

Given a grain with pole A in element $d\Omega$, pole B will lie on the appropriate circle around A at a position specified by δ; the pole density here can be represented by $p_B(\alpha,\beta,\delta)$.

To determine the inverse pole figure for one of the directions in a rolled sheet such as the normal direction ND, the density distribution for this direction must be determined as a function of orientation with respect to specified crystal directions such as poles A and B.

The relative abundance of crystals with pole A in the area element $d\Omega$ and with pole B on its circle around A and within the angular range $d\delta$ at position δ is σ_{AB}. It is assumed that σ_{AB} is proportional to the product of p_A and p_B. In terms of normalized values such that the mean value of σ_{AB} is equal to unity,

$$\sigma_{AB}(\alpha,\beta,\delta) d\Omega d\delta = \frac{p_A p_B \sin \alpha \, d\alpha d\beta d\delta}{\int_0^\pi \int_0^\pi p_A \left(\int_0^\pi p_B d\delta \right) \sin \alpha \, d\alpha d\delta} \tag{9–10}$$

The fraction of the crystals in which the direction ND lies within the element of solid angle $d\Omega'$ at α; β is given by the expression

$$p_{ND}(\alpha,\beta) d\Omega' = \frac{1}{2\pi} \left[\int_0^\pi \sigma_{AB}(\alpha,\beta,\delta) d\beta \right] \sin \alpha \, d\alpha d\delta \tag{9–11}$$

where $d\Omega'$ is an element of area on the inverse pole figure at position ND. This is equivalent to the following operation: In Fig. 9–12 the triangle A, B, ND is rotated around the corner ND and at each successive position of the triangle the values p_A and p_B are multiplied together. These products are summed for the complete rotation of the triangle; the result gives the density in the sheet-normal distribution for the point α, β on the standard projection of the crystal, and when repeated for different points serves to build up the inverse pole figure for ND. A similar procedure gives the inverse pole figure for the rolling direction or for the transverse direction in the rolled sheet.

Again there is the question of whether three independent inverse pole figures adequately specify the crystallite orientations (see page 206 above). The method involves the product of two densities, p_A and p_B, and assumes that the density p_A at point A is caused by the *same* crystals that caused density p_B at point B. In simple, well-defined textures this may well be true, in which case the product yields the required orientation density. But in complex textures another texture component may interfere; in this case the product of the two *independent* probabilities is not equal to the *correlated* probability that defines the true orientation distribution. The contribution from an interfering component must therefore be removed if a false result is to be avoided.

The method of Roe and Krigbaum The most complete analysis for fiber textures is that of Roe and Krigbaum,[1] who express the crystallite orientation function in terms of a series of Legendre polynomials, the coefficients of which are determined from the experimental diffraction data by means of a set of simultaneous linear equations. (The expansion of the distribution function in a series of spherical harmonics was also done

[1] R. J. Roe and W. R. Krigbaum, *J. Chem. Phys.*, vol. 40, p. 2608, 1964.

for cubic crystals by Bunge.)[1] Because the resultant distribution function specifies fully the orientation distribution of crystallites in the fiber and not simply an incompletely correlated set of distribution functions for individual poles, it is possible then to predict the distribution function for poles other than those used in the analysis. It is also possible to use more data in the analysis than the minimum needed in the determination; a weighting and least-squares analysis of the data is then possible. Intensities from a random sample are not needed. The resulting crystallite orientation function is equivalent to an inverse pole figure and can be plotted to give such a pole figure. In carrying out the various steps in the analysis, attention is given to the two kinds of symmetry properties that enter the problem: the symmetry of the point group to which the crystal belongs, and the symmetry of the distribution of the crystallites in the sample. Roe and Krigbaum discuss these in some detail, and point out that in fiber textures of polymers some special symmetry problems are of importance—a long needle-shaped crystallite may be aligned preferentially along the axis of the sample, but it may or may not have a random orientation around the needle axis. Randomness around the needle axis results in certain coefficients vanishing; a lack of randomness is an important parameter in specifying the orientation characteristics of polymer samples.

This method provides an analysis that is somewhat analogous to the crystal-structure analysis of x-ray diffraction data from a single crystal in terms of electron-density maps. Electron densities are represented by Fourier series derived from observed intensities; the probability density of crystallite orientations is represented by a series of spherical harmonics. Symmetry properties in crystals cause certain systematic absences of reflections, while symmetry elements in a crystal or in the statistical orientation-distribution function cause the systematic absence of certain coefficients in the series of spherical harmonics. Series termination errors occur in both maps. A striking difference between the two exists, however: in crystal-structure analysis the phase angles of the terms in the Fourier series are not immediately evident from the observed data, whereas in the texture analysis all the coefficients can, in principle, be derived directly from the data.

DETERMINING THE ORIENTATION OF INDIVIDUAL CRYSTALS

Scientists and engineers are increasingly making use of single crystals with known orientations or known crystal faces; for some applications an uncertainty of 1 to 2° in orientation is tolerated, but for others the uncertainty should not be more than a few hundredths of a degree. A number of methods of varying speed and accuracy are presented in the following paragraphs; others use electron diffraction and neutron diffraction, which are discussed in other chapters.

The determination of orientations by etch pits Various orientation methods have been employed that use the reflection of light from crystallographic etch pits. Bridgman's method[2] consists in attaching the specimen to a sphere (a transparent sphere is convenient[3]), holding it at arm's length, and marking a spot on it when it has been turned so that the etch pits reflect light into the eye of the observer from a lamp standing behind him. It has an accuracy of about 2° and is extremely rapid. A greater accuracy can be obtained by causing the reflected light to fall on a screen,

[1] H. J. Bunge, *Monatsber. Deut. Akad. Wiss. Berlin*, vol. 1, pp. 27, 400, 1959; vol. 3, p. 97, 1961.

[2] P. W. Bridgman, *Proc. Am. Acad. Arts Sci.*, vol. 60, p. 305, 1925.

[3] J. B. Baker, B. B. Betty, and H. F. Moore, *Trans. AIME*, vol. 128, p. 118, 1938.

as has been done by Czochralski,[1] Chalmers,[2] and Schubnikov.[3] Tammann and his coworkers[4] developed a method for estimating the number of grains in a polycrystalline aggregate having orientations in a given region of the stereographic triangle, and thus obtained statistical information about deformation and recrystallization textures. While the method has had some use,[5] it is not applicable to the majority of problems in physical metallurgy. Smith and Mehl[6] determined orientations by plotting the directions of the sides of individual etch pits on a stereographic projection. The accuracy is limited to about 3°, and the method requires careful polishing and high magnification.

The most useful methods involve the measurement of the angles between the etch-pit faces using a goniometer. Weerts[7] has given a lengthy discussion of a technique using a three-circle goniometer and polarized light. A less elaborate method, which gives an accuracy that is sufficient for the great majority of investigations, consists simply in measuring the orientations of the etch-pit faces on a two-circle optical goniometer.[8] It has an accuracy of $\frac{1}{2}$ or 1°. The method operates satisfactorily with grains as small as 0.1 mm diameter and has been used with cold-worked and annealed grains and single crystals. An inexpensive goniometer for this work may be constructed from an old surveyor's transit[9] or, more conveniently, from an old astro-compass.[10]

Etching technique An ideal technique is one that develops etch pits or facets with plane faces accurately parallel to crystallographic planes of low index. The etched metal specimen then appears intensely bright when the normals to these planes bisect the angle between telescope and collimator and perfectly dark in all other positions. Actually, however, the etch-pit faces are always more or less rounded, causing the intensity of the reflected light gradually to build up to and drop off from a maximum as the reflecting plane approaches and recedes from the reflecting position. For the same metal and etchant the position and sharpness of the maxima vary with the etching

[1] Czochralski, *Z. Anorg. Allgem. Chem.*, vol. 144, p. 131, 1925.

[2] B. Chalmers, *Proc. Phys. Soc. (London)*, vol. 47, p. 733, 1935.

[3] A. Schubnikov, *Z. Krist.*, vol. 78, p. 111, 1931.

[4] G. Tammann, *J. Inst. Metals*, vol. 44, p. 29, 1930. G. Tammann and H. H. Meyer, *Z. Metallk.*, vol. 18, p. 339, 1926.

[5] K. J. Sixtus, *Physics*, vol. 6, p. 105, 1935.

[6] D. W. Smith and R. F. Mehl, *Metals Alloys*, vol. 4, pp. 31, 32, 36, 1933.

[7] J. Weerts, *Z. Tech. Physik*, vol. 9, p. 126, 1928.

[8] L. W. McKeehan, *Nature*, vol. 119, p. 705, 1927. L. W. McKeehan and H. J. Hodge, *Z. Krist.*, vol. 92, p. 476, 1935. H. H. Potter and W. Sucksmith, *Nature*, vol. 119, p. 924, 1927. C. S. Barrett and L. H. Levenson, *Trans. AIME*, vol. 137, p. 112, 1940. See also "Crystallometry" in Chap. 2.

[9] C. S. Barrett, *Trans. AIME*, vol. 135, p. 296, 1939. C. S. Barrett and L. H. Levenson, *Trans. AIME*, vol. 137, p. 76, 1940.

[10] The vertical circle is detached from the horizontal circle and mounted far enough back from the center of the instrument to provide room for a specimen holder and specimen; the surface of the specimen is brought to the position at the intersection of the axes of the two circles. (G. T. Gow, private communication.)

Table 9-1 Etching for orientation determination*

Metal	Purity, percent	Etchant	Etching time in minutes	Planes developed	Remarks
Ag		$HNO_3:H_2O = 55:45$		{111}, {100}, {110},	Composition is critical.
Al	99.95	$HCl:HNO_3:HF:H_2O = 9:3:2:5$	2	{100}	Avoid heating etchant.
Al		Dry HCl gas.	30	{111}	Wash in acetone.
Cu	OFHC	$HCl:H_2O = 1:1$; mixture saturated with $FeCl_3 \cdot 6H_2O$	10	{100} and {110}	
Brass	70–30	Above etch for Cu plus H_2O, 1:1	20	{100} and {111}	Agitate at 2-min intervals. {110} also reported.
Fe		$HNO_3:H_2O = 1:4$	4	{100}	Wipe surface during etching.
Ge		(a) $HNO_3:HF:5\%\ AgNO_3$ solution = 2:4:4	1	{111}	
		(b) $H_2O_2:HF:H_2O = 1:1:4$	2	{100}	
In		Conc. HCl + $KClO_4$.	5	{100}, {110}	
Mg		27% NH_4Cl solution.	5	{0001}	
Ni		Saturated $FeCl_3$ solution.		{111}, {100}	
Pb	99.9	H_2O_2:glacial acetic acid: $H_2O = 3:2:2$	10	{100}	
Si		(*a*) 1% NaOH at 90°C	5	{111}	For surfaces near {111}.
		(*b*) 10% NaOH at 90°C	5	{111}	Gives also {100} when surface is near {100}.
Sn	99.99	Etch for Cu + H_2O = 1:1.	10	{100}, {110}	Wipe surface at 3-min intervals.
W		Saturated $K_3Fe(CN)_6$: saturated $KOH:H_2O$ = 100:5:95	15	{110}	Agitate at 2-min intervals. Etchant composition is critical.
Zn		Saturated $CuCl_2 \cdot 2H_2O$: $HCl:H_2O = 7:3:90$	3	$\{10\bar{1}l\}$	l has many values. Wipe surface at 3-min intervals.

* This table originated in C. S. Barrett and L. H. Levinson, *Trans. AIME*, vol. 137, p. 76, 1940, and J. S. Bowles and W. Boas, *J. Inst. Metals*, vol. 74, p. 501, 1948. Since it was well received, it was repeated and expanded in previous editions of this book and in K. Lark-Horowitz and V. A. Johnson (Eds.), "Methods of Experimental Physics, vol. 6, Solid State Physics," Academic, New York, 1959. References and alternate

time and the temperature of the etchant. When the maxima are sharp enough to be reproducible within $\frac{1}{2}°$, their positions usually coincide with the crystallographic poles, but sometimes they tend to fall a degree or two away from true crystallographic poles. For example, maximum reflections from cubic etch pits are often 88 or 89°, rather than 90°, apart. Since only two poles are necessary to determine the orientation and three can usually be measured, the observed poles can be corrected by moving them a minimum amount to make them mutually perpendicular. Occasionally, a few large noncrystallographic pits are developed; these may be readily recognized as such, since they have no definite maximum. In Table 9–1 directions are given for etching various metals for orientation work. The etching times are for etchants at room temperature and for strain-free metals of the purity indicated; for cold-worked metals the etching time usually must be reduced. Nonmetallic crystals can also be etched.[1]

Method of plotting The readings on the two circles of the goniometer corresponding to a reflection maximum (a cube pole in this discussion) can be plotted directly as a point on a stereographic projection. The data are plotted on tracing paper fastened over a stereographic net by a pin at the center of the net, the paper being left free to turn about the pin.

Transferring data from the goniometer to the projection is direct and simple, for readings on the horizontal circle of the goniometer correspond to radial distances from the center of the projection and readings on the vertical circle correspond to azimuthal positions around the center if one makes the plane of projection parallel to the surface of the specimen. Taylor has worked out an instrument that plots the data from an optical goniometer automatically.[2]

Application to deformed metals An optical goniometer has especial value in the study of lattice orientations in deformed grains and single crystals, fields in which x-ray work is difficult. In general, with either single crystals or grains of a polycrystalline aggregate, the orientation after deformation varies considerably and in an irregular manner over each grain. In determining the range of orientation, a grain is divided into a number of areas, and the orientation of each area is measured separately and transferred to a standard projection.[3] The grouping of points on the final projection gives a pole figure of the preferred orientations present. This technique has been used to prepare pole figures of polycrystalline specimens, in both cold-worked and recrystallized states,[4] the individual grains of which were a few millimeters in diameter; but it does not, of course, detect the fragments of microscopic or submicroscopic size that can be registered by x-rays or the electron microscope.

Back-reflection Laue method for determining crystal orientation

One of the most convenient and accurate methods of determining the orientation of a single crystal or an individual grain in an aggregate is the back-reflection Laue method.[5] It is useful not only for laboratory research but also in the production of commercial crystals, such as quartz crystals cut for piezoelectric control of radio circuits. An exceedingly simple camera will yield accuracies of the order of $\frac{1}{2}°$, and cameras for more precise work

[1] "Methods of Experimental Physics," vol. 6, pt. A, p. 160, Academic, New York, 1959.

[2] A. Taylor, *J. Sci. Instr.*, vol. 25, p. 301, 1948.

[3] C. S. Barrett and L. H. Levenson, *Trans. AIME,* vol. 137, p. 112, 1940.

[4] C. S. Barrett, *Trans. AIME,* vol. 137, p. 128, 1940.

[5] L. Chrobak, *Z. Krist.*, vol. 82, p. 342, 1932. W. Boas and E. Schmid, *Metallwirtschaft*, vol. 10, p. 917, 1931. A. B. Greninger, *Trans. AIME*, vol. 117, p. 61, 1935; *Z. Krist.*, vol. 91, p. 424, 1935. Elementary instructions are given in B. D. Cullity, "Elements of X-ray Diffraction," Addison-Wesley, Reading, N.Y., 1956, and in Elizabeth A. Wood, "Crystal Orientation Manual," Columbia University Press, New York, 1963.

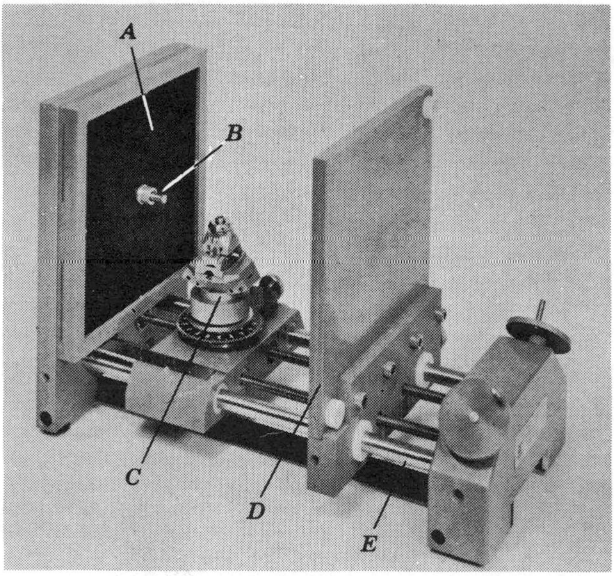

Fig. 9-13 Laue camera. *A*, Cassette for back-reflection photographs; *B*, pinhole collimator for beam; *C*, goniometer head with crystal; *D*, cassette for transmission photographs; *E*, instrument track. (Courtesy E and A Company.)

are not elaborate, as will be seen from the one illustrated in Fig. 9–13. The interpretation of a photograph may be carried out in a few minutes by making use of the chart developed by Greninger,[1] together with a stereographic net and a standard projection of the crystal. When the incident beam in a back-reflection Laue camera is made to straddle a boundary between two grains, two components of a twin, or two imperfect regions of a crystal, so that the two are jointly recorded on the film, it is possible to determine their relative orientations to within a few minutes of arc.[2]

Tungsten-target x-ray tubes are best suited for the work, but any tube that gives appreciable general radiation with or without characteristic radiation will serve.[3] A back-reflection pattern of a b.c.c. crystal (alpha-iron) is shown in Fig. 9–14. The circle at the center indicates the hole that is punched to enable the film to slip over the pinhole system. It will be seen that the Laue spots lie on hyperbolas. The spots on each of these rows are reflections from various planes of a given zone, i.e., planes parallel

[1] A. B. Greninger, *Trans. AIME*, vol. 117, p. 61, 1935; *Z. Krist.*, vol. 91, p. 424, 1935.

[2] A. B. Greninger, *Trans. AIME*, vol. 117, p. 75, 1935; vol. 122, p. 74, 1936.

[3] In typical experiments a tungsten tube operating at 15 to 20 ma, 40 to 60 kv, requires 5- to 30-min exposures with a pinhole 0.8 mm diameter and 6 cm long, and a crystal-to-film distance of 3 cm. A 5- by 7-in. film is convenient at this distance. For metals such as Cu, Zn, Fe, etc., there is considerable fogging from fluorescent radiation, but this may be reduced by a filter, for example, by an 0.01-in. aluminum sheet placed in front of the film. For maximum speed, Polaroid films have been used.

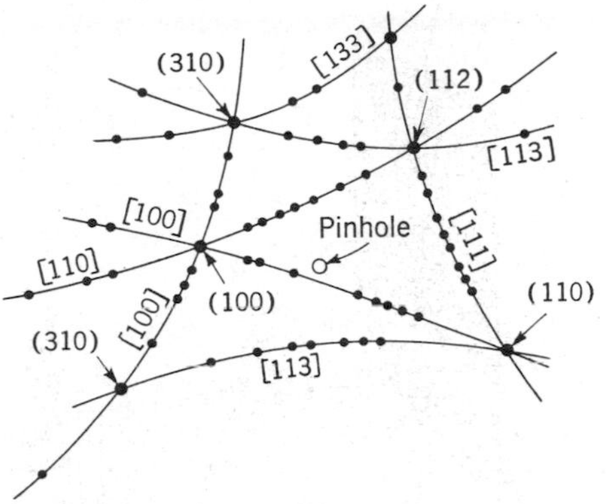

Fig. 9-14 Back-reflection Laue pattern of alpha-iron with principal zones and spots identified.

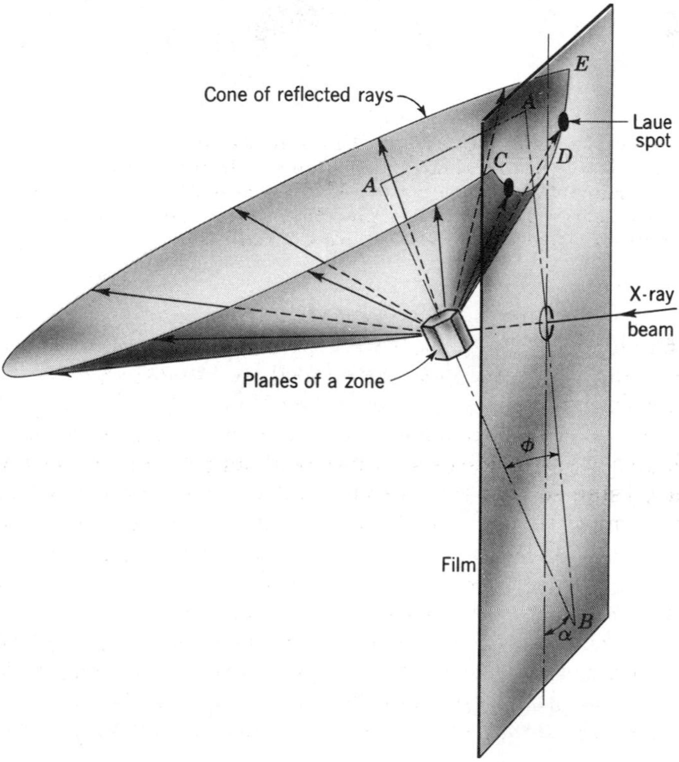

Fig. 9-15 Back-reflection Laue pattern of planes of a zone in a crystal.

to a line that is the zone axis. The geometrical conditions are sketched in Fig. 9–15. A cone of reflected rays is formed for each zone of planes in the crystal, such as that shown in the drawing, and each cone intersects the flat film to form a row of spots (C, D, E) along a hyperbola. If the zone axis AB is inclined to the film, the hyperbola will lie at a distance from the

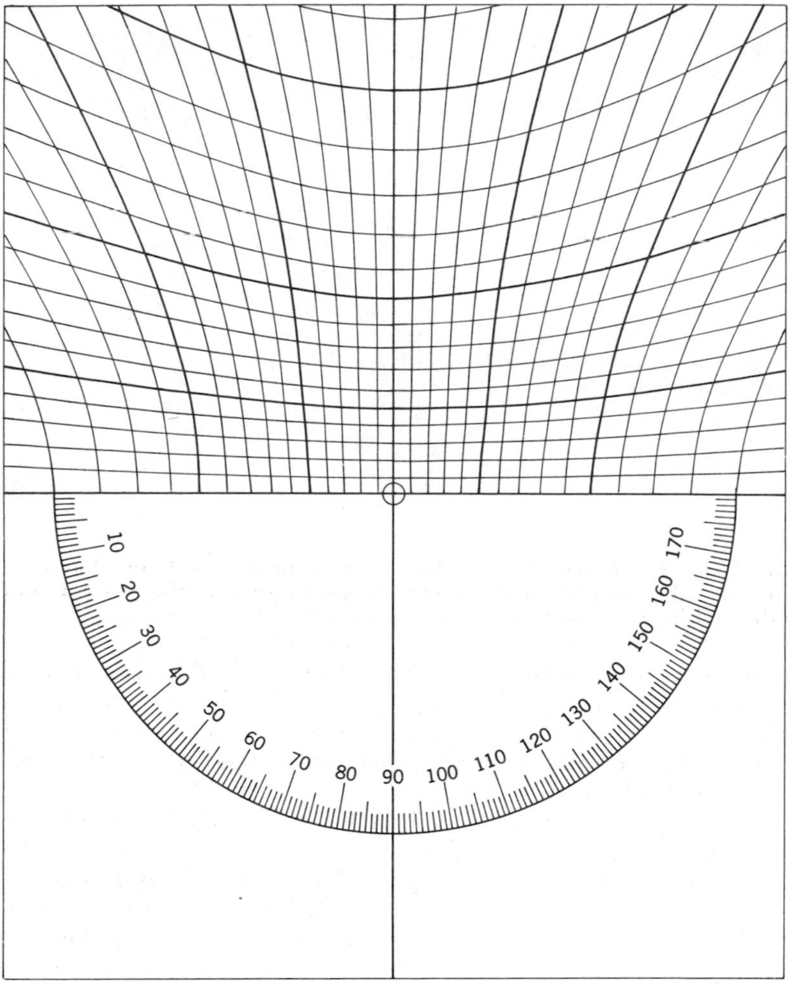

Fig. 9-16 Chart for back-reflection photographs. Printed in the size for 3-cm distance from specimen to film; graduated in 2° intervals.

center of the film, a distance that is related to the angle of inclination, ϕ[1]; if AB is parallel to the film ($\phi = 0$), the hyperbola will degenerate into a line passing through the center of the film. If the axis of the cone is projected perpendicularly upon the film, forming the line $A'B$, it is obvious that the hyperbola will be symmetrical with respect to this line.

In determining the orientation of a crystal from a pattern of this type, it is necessary to assign the proper indices to some of the zones causing

[1] The closest approach of the hyperbola to the center of the film is given by the relation $S = R \tan 2\phi$, where R is the distance from the specimen to the film and S is the minimum distance of the hyperbola from the center.

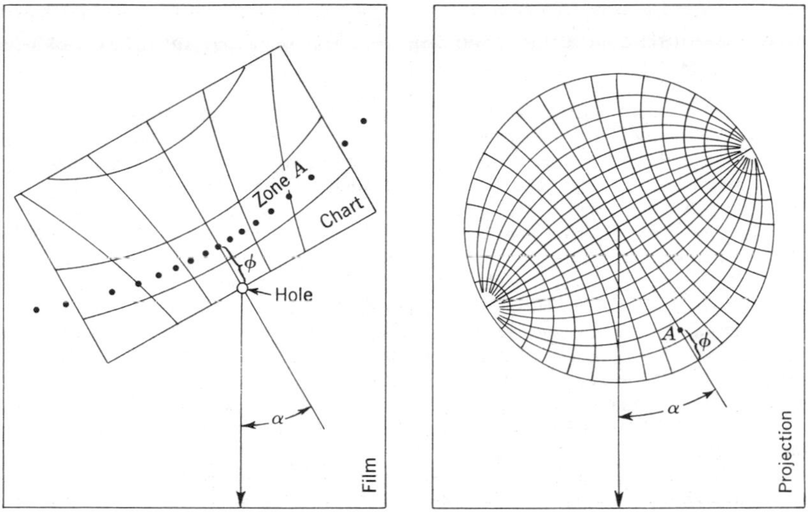

(a) (b)

Fig. 9-17 (*a*) **Back-reflection Laue film superimposed on chart.** (*b*) **Stereographic projection paper superimposed on net; the zone of spots indicated in** (*a*) **is plotted at** *A*.

these hyperbolas or to some of the individual spots. The technique for doing this is much simplified if attention is directed merely to the hyperbolas which are the most densely packed with spots and to those spots which lie at the intersections of three or more prominent hyperbolas, for then only low indices are involved.

On every film of a cubic crystal will be found prominent rows of spots from the zones [100], [110], and [111] and less prominent ones from the zone [113]. Many other zones also appear but are not easily recognized. The spots at the intersections of these principal hyperbolas are reflections from the planes common to the principal zones in the crystal; in the b.c.c. patterns they are (100), (110), (112), and others of higher indices, which are less prominent. In the f.c.c. patterns, (111) appears in addition to these. These important spots with low indices are always of high intensity and are somewhat isolated from their neighbors on the hyperbola (see Fig. 9–14).

Figure 9–16 is Greninger's chart for reading angular relations on back-reflection films and is reproduced in the proper size for a 3-cm distance from specimen to film.[1] Any row of spots on the film can be made to coincide with a hyperbola extending across the chart horizontally by suitably turning the film about its center when it is placed centrally on the chart. The angle of inclination, ϕ, of the zone axis to the film is given directly from the chart; hyperbolas have been drawn for each 2° of ϕ, with heavy lines each 10°. The amount of rotation of the film that is required to get a

[1] This net has been marketed by N. P. Nies, Laguna Beach, Calif.

row to coincide with a horizontal hyperbola on the chart is a direct measure of the azimuthal angle α (see Fig. 9–15), for it is the angle between the line $A'B$ and the vertical reference line.

It is most convenient to solve the films by plotting a stereographic projection; this can be done rapidly by a method which will be described with reference to Fig. 9–17. The plane of the projection is made perpendicular to the primary x-ray beam (parallel to the plane of the film), and the film is read from the side opposite to that on which the reflected rays were incident, so as to give a projection of the crystal that corresponds to viewing it from the position of the x-ray tube. An arrow is marked on the film parallel to some fiducial mark on the specimen, and the film is superimposed on the chart, as in Fig. 9–17*a*, with its central hole at the center of the chart. A piece of tracing paper is then placed on a stereographic net and pivoted at the center of it with a pin. An arrow is drawn on the paper radially outward from the central pin. The zones recorded on the film are then plotted on the paper, as indicated in Fig. 9–17*b*, by turning the paper and film together so that the two arrows are parallel and point to the same angle α, and by plotting the angle ϕ upward from the basic circle of the projection. The point A thus plotted is the projection of the zone responsible for the entire row of spots on the film.

Tentative indices are assigned by determining the angles between prominent zones and comparing the angles with a table (see the table on page 40). This tentative assignment of one or more possible sets of indices is merely for the purpose of saving time in the final operation, which is that of finding a position of a standard projection of zone axes and an amount of rotation of the standard such that its points will coincide with all the plotted ones and thus will disclose their indices. When a match has been found between each zone and a corresponding zone from a rotated standard projection, it is possible, of course, to rotate any other zone or any pole of the standard by the same amount and thus to show its position in the crystal. The final plot expresses the crystal orientation graphically; in addition, it can also be described in terms of the angles between the crystallographic axes and convenient marks or surfaces on the crystal. It is convenient in this work to have accurate standard projections at hand of a size matching the stereographic nets being used, standard projections for several standard orientations of the same crystal (see Chap. 2).

The above technique is but one of several by which the indices of the zones and spots may be deduced. No use has been made in this procedure of the vertical curves of the chart (Fig. 9–16). These have been drawn to measure angular relations between spots on a given horizontal hyperbola and are spaced at 2° intervals just as are the others; i.e., two planes of the same zone in a crystal that lie at an angle of 2° to each other will reflect to form two spots separated by one of the intervals marked out by two adjacent vertical curves. The angle between the two principal spots of Fig. 9–14, for example, can be measured directly from the chart and will be found to be 45°; this suggests at once that the indices of one of the spots are (100) and of the other are (110), although it does not tell which spot is which.

If the spots thus partly identified are plotted on the projection, they will greatly reduce the number of trial rotations of the standard projection that are necessary to find the complete solution. The pole of a plane causing a spot is plotted by a procedure analogous to that of Fig. 9–17, except that the spot is made to lie on the *central vertical line* of the chart, and the corresponding pole is plotted $\phi°$ *up from the center* of the projection, rather than $\phi°$ up from the circumference (since the pole is normal to its zone axis). Poles thus plotted are useful for a rapid solution of the film, but zones can be read from the film more accurately. Much time can be saved if, after prominent zones are plotted, the important great circles shown in the standard projection, such as those shown for cubic crystals in Fig. 2–9, are looked for.

Vector relations in the reciprocal lattice are useful in determining the orientations of crystal axes from the coordinates of spots on a photograph.[1]

[1] B. F. Decker, *J. Appl. Phys.*, vol. 15, p. 610, 1944.

Relationships are particularly simple with cubic crystals; computations may be made directly from coordinates of film spots measured with respect to orthogonal X and Y axes drawn with arbitrary orientation on a flat film, with the direct beam spot as origin of coordinates.

For orientations to be determined with greater accuracy than about $\frac{1}{2}°$ it is necessary to supplement the graphical methods of this chapter or to abandon them altogether and resort to direct computation of the trigonometric relationships, using precise camera or diffractometer equipment and techniques.[1] For some purposes it is convenient to describe orientations by using a set of direction cosines of the angles between crystal axes and reference axes.

The transmission Laue method It was mentioned in Chap. 5 that a transmission Laue pattern has a symmetry around the center similar to the symmetry of the axis or plane in the crystal along which the x-ray beam is directed. If a cubic crystal is mounted on a goniometer head and turned until the Laue pattern has fourfold symmetry, it becomes at once obvious that the incident beam is parallel to a cube axis [100]; similarly, threefold symmetry indicates that the beam is parallel to a cube diagonal [111]. With crystals of high reflecting power and with intense radiation, it is possible to see these patterns on a fluorescent screen and to adjust the crystal quickly to a symmetrical position, though it is necessary to work in a darkened room with the eyes fully accommodated to the dark, and with the hands protected from direct and scattered radiation.

To find the orientation of an unsymmetrically oriented crystal, one can transfer each of the spots on a transmission photograph to a stereographic projection. The prominent zones then appear as points along great circles of the projection and can be identified either by the symmetry of their arrangement or by the angles between them. Figure 9–18 is a chart that aids this plotting.[2] When enlarged to the size that corresponds to the correct crystal-film distance D, the ellipses and hyperbolas in the upper half of the chart represent zones of Laue spots, as they would be registered on a flat film. If the chart is centered on the direct beam spot of a Laue pattern and pivoted about the center, any zone of spots can be made to superimpose on one of the curves (or on a curve interpolated between these); the angle between the zone axis and the beam is then read directly from the chart by means of the numbered curves and the 2° interpolated curves. The lower half of the chart consists of circles numbered with values of the Bragg angle θ, which may be used for plotting individual spots. On a stereographic projection with the beam at the center, a zone axis is plotted by going radially outward from the center the measured angular distance, and a plane is plotted by going $\theta°$ inward from the basic circle, in each case along a radius parallel to the radius on the film.

When many orientations are to be determined, it may lessen the work if a set of standard photographs is prepared with which to compare the

[1] J. Barraud, *Compt. Rend.*, vol. 217, p. 683, 1943.

[2] C. G. Dunn and W. W. Martin, *Trans. AIME*, vol. 185, p. 417, 1949.

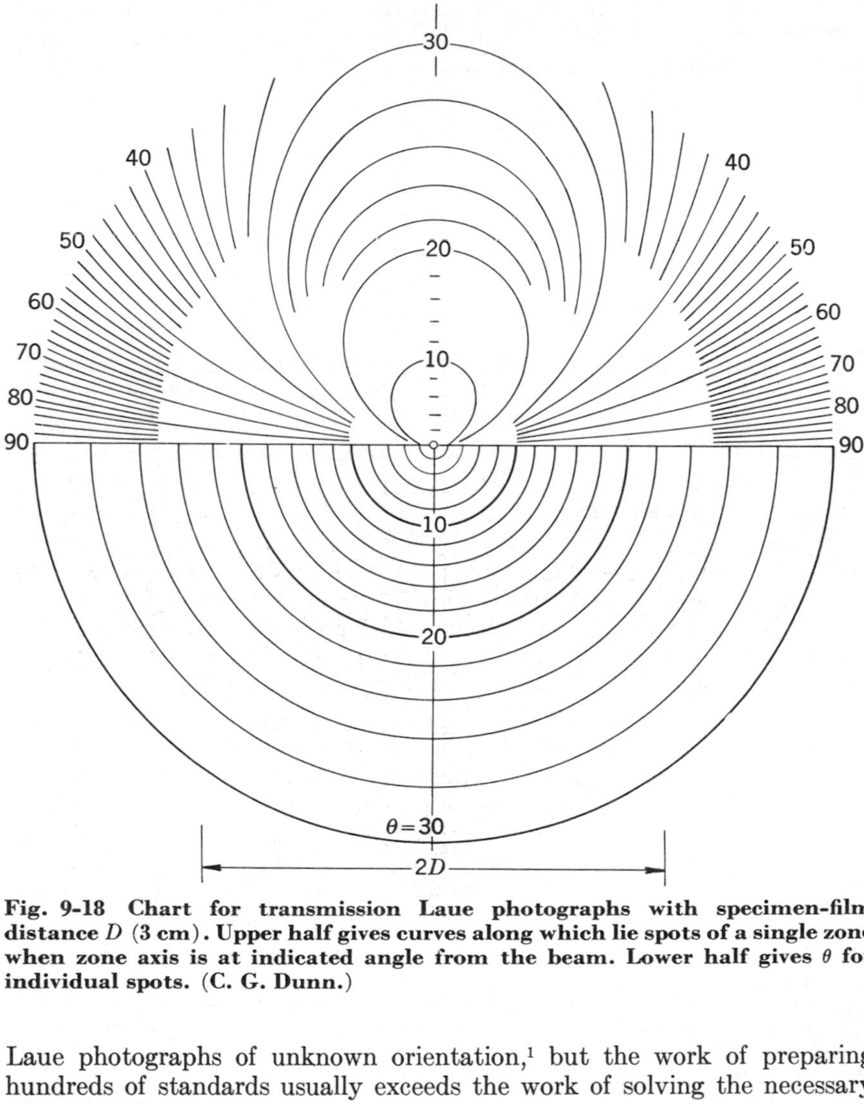

Fig. 9-18 Chart for transmission Laue photographs with specimen-film distance D (3 cm). Upper half gives curves along which lie spots of a single zone when zone axis is at indicated angle from the beam. Lower half gives θ for individual spots. (C. G. Dunn.)

Laue photographs of unknown orientation,[1] but the work of preparing hundreds of standards usually exceeds the work of solving the necessary films directly.

Crystal orienting methods using $K\alpha$ radiation A number of methods have been devised in which monochromatic or semimonochromatic radiation is used.[2] They are particularly useful for samples in which several

[1] M. Majima and S. Togino, *Sci. Papers Inst. Phys. Chem. Res. (Tokyo)*, vol. 7, pp. 75, 259, 1927. C. G. Dunn and W. W. Martin, *Trans. AIME*, vol. 185, p. 417, 1949.

[2] References are given in F. M. Henry, H. Lipson, and W. A. Wooster, "The Interpretation of X-ray Diffraction Photographs," Macmillan, London, 1951, 1960. A familiar problem with crystallographers is that of aligning a crystal accurately in a rotating-

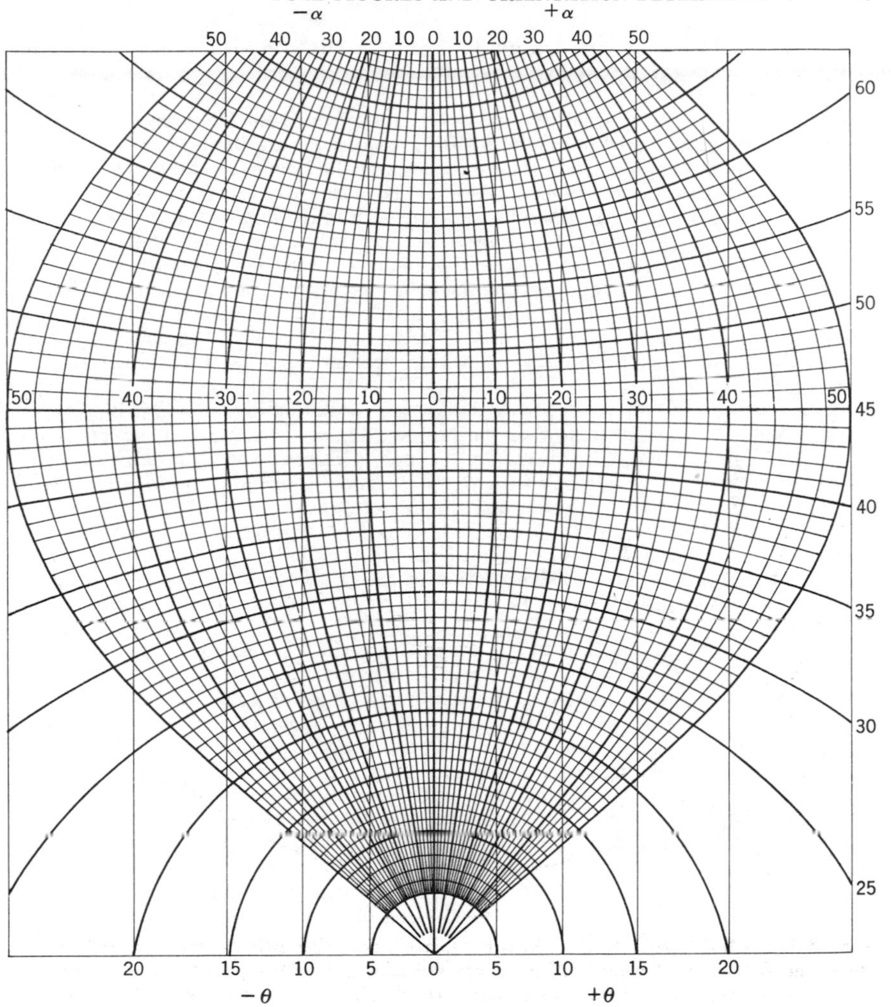

Fig. 9-19 Chart for reading α and θ from stationary cylindrical film. Axis of cylinder is horizontal. Chart must be enlarged until distance from $\theta = 0$ to $\theta = 45°$ is $\frac{1}{4}$ of film circumference.

crystal orientations or crystal phases are simultaneously present, or in which reflections are seriously distorted by imperfections from growth or from plastic deformation. Laue photographs are often inadequate to solve

crystal camera or a Weissenberg or Buerger-precession camera. Discussions of ways of doing this will be found in: J. W. Jeffrey, *Acta Cryst.*, vol. 2, p. 15, 1949. "International Tables for X-ray Crystallography," vol. 3, p. 35, Kynoch, Birmingham, England, 1962. Directions for a commercial orienter ("goniostat") are given in T. C. Furnace, "Single Crystal Orienter Instruction Manual," General Electric Company, Milwaukee, Wisc., 1956.

problems of this sort. The major features of these methods are presented in the following paragraphs.

An oscillating-crystal method that we may call the *Davey-Wilson method*[1] is based on principles that are also used in modified form in others. By oscillating a crystal, several $K\alpha$ spots are photographed; from their θ values and the known crystal structure of the sample their indices are determined. Then by recording the same spots on a moving film, the amount of crystal rotation at the time each individual reflection occurred can be determined. These various rotations, together with the angular positions of the poles of the reflecting planes at the time they reflected, are then combined on a stereographic projection and yield the crystal orientation as it was at one end of the oscillation. In the Davey-Wilson camera, the moving film is carried on a cassette behind the stationary film and is coupled to the crystal so as to oscillate around the same axis as the crystal; the beam (usually Mo $K\alpha$) penetrates the inner film and exposes both films simultaneously.

Similar results can be obtained with the crystal oscillating in a standard *Weissenberg* camera. A film is exposed first with the film stationary, positioned at one end of its travel. The spots produced in this exposure supply θ values which, by comparison with a list of θ's for the sample, identify the reflecting planes. A second exposure is then made with the film shifting in synchronism with the rotation of the crystal in the manner of Weissenberg photographs. The same reflections will be photographed, but each spot of the second exposure will be displaced from its position in the first exposure by a distance that is proportional to the rotation of the crystal between the two. A stereographic projection can then be made with a technique related to that used for plotting pole figures.

The poles of reflecting planes are first plotted with the θ and azimuthal positions each had individually at the time its reflection occurred, and the poles are labeled with their indices. Each pole is then rotated with the stereographic net a number of degrees indicated by the distance between the corresponding spots of the first and second exposures and in a direction to carry each pole back to the point on the projection where it was at the end of the oscillation range, thereby forming a projection of known poles at this particular orientation of the crystal. The chart reproduced in Fig. 9–19 or an enlargement of it to proper size is an aid in plotting the data of the first exposure (for either Weissenberg or Davey-Wilson cameras). If the film is cylindrical with its axis horizontal and the beam is normal to it, θ values for each spot from the stationary film exposure can be read with the curves labeled $\theta = 0$ to $60°$; the azimuthal angle α around the beam from the vertical $\alpha = 0°$ for a pole of a reflecting plane can be measured by the curves labeled $\alpha = -50°$ to $+50°$.[2] Each spot is plotted as indicated by the example in Fig. 9–20 for $hkl = 123$ on a projection with the beam entering the projection sphere at point B. Crystal rotation of $\phi°$ moves a pole along one of the latitude lines indicated by the dashed arcs.

A method *combining a Laue and an oscillating-crystal photograph*[3] is convenient. An oscillating-crystal exposure is made with *unfiltered* radiation and stationary film; each Laue spot then produces a streak on the film, and on a few of the streaks the $K\alpha$ (and $K\beta$) components produce strong maxima at θ values which serve to identify the streaks. After the oscillating-crystal exposure is finished, the crystal is left stationary at one end of its oscillation range and exposed until Laue spots are recorded at one end of each of the Laue streaks. Laue spots that have been identified by the streaks containing $K\alpha$ reflections just mentioned are plotted according to the θ and α values of the stationary

[1] W. P. Davey, *Phys. Rev.*, vol. 23, p. 764, 1924. T. A. Wilson, *Gen. Elec. Rev.*, vol. 31, p. 612, 1928.

[2] A chart with wider ranges of angles is given in "International Tables for X-ray Crystallography," vol. 2, p. 174, Kynoch, Birmingham, England, 1959, and various crystallographic texts.

[3] Henry, Lipson, and Wooster, loc. cit.

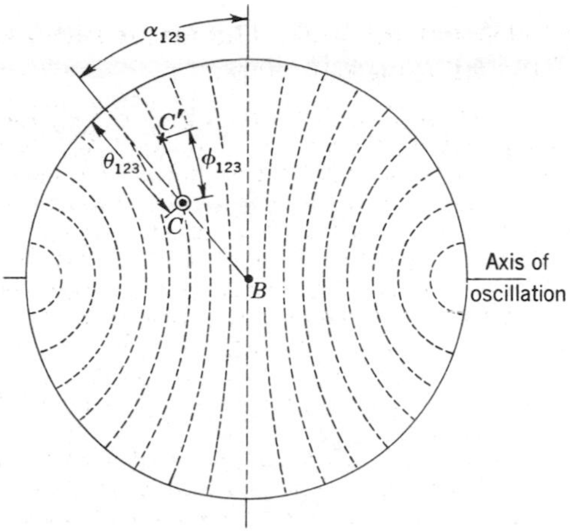

Fig. 9-20 Plotting a (123) reflection from known θ and α values. Position after rotating $\phi°$ is shown by a cross.

exposure (by using the chart of Fig. 9–19). No further rotation is necessary, for the plot refers to the crystal orientation during the stationary exposure.

If only the orientation of the *axis* of a crystal in the form of a rod or wire is needed, it can be determined by taking a photograph with the crystal rotating about its axis with the film stationary. Since the axis of rotation may be a crystallographic direction of high or irrational indices, the usual methods for solving rotating crystal patterns may be unsatisfactory, but the graphical analysis mentioned in connection with fiber textures (Fig. 9–4) can be used.

Crystal orienting with a diffractometer Precision orientation determination is often accomplished with a counter diffractometer. The counter is set at a 2θ angle for one of the low-index reflections. The crystal is mounted on a goniometer and adjusted so that an irradiated spot is on the axis of rotation of the diffractometer. The angular setting of the crystal is then altered slowly with the counter stationary until a $K\alpha$ reflection is received in the counter. When reflection occurs, it is then known that the normal of an identified plane bisects the angle between the incident and reflected beams. The procedure is then repeated for other reflections of the same or different indices. The accuracy is limited only by the collimation of the beam, the accuracy of alignment, and the accuracy of the angle scales used.

The detection of subgrains differing from each other so little (minutes or seconds of arc) that ordinary photographs do not disclose them is sometimes of importance in research. These can best be studied by using a double-crystal spectrometer,[1] or by focused Laue spots (see Chap. 5).

Cutting crystals to known orientations It is frequently necessary to cut or grind a face of certain indices on a crystal. For such purposes the crystal should be mounted on a goniometer head by means of which it can

[1] A. H. Compton and S. K. Allison, "X-rays in Theory and Experiment," Van Nostrand, New York, 1935. W. L. Bond, *Proc. IRE,* vol. 38, p. 886, 1950. W. L. Bond and J. Andries, *Am. Mineralogist,* vol. 37, p. 622, 1952.

be rotated a known number of degrees about two (or more) axes, as required by graphical or trigonometric solutions of the problem, so as to bring a desired plane into position normal to a reference axis. The crystal in its goniometer head is then transferred to a cutting wheel or grinding machine on which the faces are cut. Various designs make such operations easy either with photographic or diffractometer orienting.[1]

[1] W. L. Bond, *J. Sci. Instr.*, vol. 38, p. 63, 1961. J. G. Walker, H. J. Williams, and R. M. Bozorth, *Rev. Sci. Instr.*, vol. 20, p. 947, 1949. H. Sato, J. Getsko, and W. E. Hickman, *Rev. Sci. Instr.*, vol. 28, p. 58, 1957.

10

The structure of metals and alloys

Interest in the crystal structures of metals, alloys, and other substances has been steadily increasing as phase diagrams have become better known, as the relationship of properties to crystal structure has been studied in greater detail, and as attempts have continued to understand the basic principles governing the stability of structures.

This chapter reviews the more important structures of metals and alloys, with emphasis on important interrelated families of structures and with due regard to the position of the participating elements in the periodic table (when significant), but without attempting to present complete lists of structures. Superlattices are discussed in Chap. 11 and theoretical matters in Chaps. 12 and 13; structures of the elements are listed in the Appendix.

Numerous summaries of crystal structures have been published, the most extensive being the various volumes of "Strukturbericht" and "Structure Reports,"[1] which contain abstracts of all x-ray crystal-structure determinations. More condensed listings of structures are available in recent books.[2] Phase diagrams for metallic systems are covered together with some structure data in the extensive book by Hansen and Anderko[3] and elsewhere.[4]

[1] Published serially in *Z. Krist.* and also as separate volumes by Akademische Verlagsgesellschaft m.b.H., Leipzig, from 1913 to 1939 and continued as "Structure Reports" by the International Union of Crystallography. Many of the data on metallic crystals up to 1936 are summarized by M. C. Neuburger, *Z. Krist.*, vol. 93, p. 1, 1936.

[2] R. W. G. Wyckoff, "Crystal Structures," 2d ed., 1963, Wiley, Interscience Publishers, New York. W. B. Pearson, "A Handbook of Lattice Spacings and Structures of Metals and Alloys," Pergamon, New York, 1958. K. Schubert, "Kristallstrukturen, Zweikomponentiger Phasen," Springer-Verlag, Berlin, 1964. G. B. Bokii, "Introduction to Crystal Chemistry," Moscow University Publishing House, Moscow, 1954, English trans. by United States Joint Publications Research Service, Oak Ridge, Tenn. E. S. Makarov (trans. by E. B. Uvarov), "Crystal Chemistry of Simple Compounds of Uranium, Thorium, Plutonium, Neptunium," Consultants Bureau, New York, 1963. J. D. H. Donnay, "Crystal Data," American Crystallographic Association, 1963. K. A. Gschneidner, Jr., "Rare Earth Alloys," Van Nostrand, New York, 1961. E. Parthé, "Crystal Chemistry of Tetrahedral Structures," Gordon and Breach, New York, 1964.

[3] M. Hansen and K. Anderko, "Constitution of Binary Alloys," McGraw-Hill, New York, 1958. R. P. Elliot, First Supplement, McGraw-Hill, 1965.

[4] "Metals Handbook," American Society for Metals, Cleveland, Ohio, 1948. "Strukturbericht," loc. cit., and "Structure Reports," loc. cit. H. J. Goldschmidt, "The Structure of Carbides in Alloy Steels," *J. Iron Steel Inst. (London)*, vol. 160, p. 345, 1948. C. J. Smithells, "Metals Handbook," 3d ed., Butterworth, London, and Interscience Publishers, New York, 1963.

For a treatise on the structures of inorganic compounds and the structural chemistry of the elements, the reader may refer to the latest edition of Wells' book.[1]

Crystals may be classified according to types of binding forces as follows:

1. *Metallic crystals* consist of positive ions immersed in a "gas" of negative electrons. The attraction of the positive ions for the negative electrons holds the structure together and balances the repulsive forces of the ions for one another and of the electrons for other electrons. The electrons move freely through the lattice and provide good electrical and thermal conductivity.
2. *Ionic crystals* are bound together by the electrostatic attraction between positive and negative ions. They are combinations of strongly electronegative and electropositive elements. In NaCl, for example, the electron affinity of chlorine atoms causes a transfer of electrons from the electropositive sodium atoms to yield Na^+ and Cl^- ions.
3. *Covalent crystals* ("homopolar," or "valence," crystals) are held together by the sharing of electrons between neighboring atoms. Diamond is a typical example, in which each carbon atom shares its four valence electrons with the four nearest neighbors and thus completes an outer shell of eight electrons in each atom. The crystals are characterized by poor conductivity and great hardness.
4. *Molecular crystals* are composed of inactive atoms or neutral molecules bound by weak van der Waals forces. They have low melting points. Typical examples are the rare gases.

Many crystals are intermediate between these "ideal" types. For example, some alloy phases have metallic conductivity and other properties associated with metallic binding and yet at the same time resemble covalent crystals. It is possible for some interatomic bonds in a crystal to be of one type, say ionic, and others to be of another type, say covalent; or each bond may be somewhat intermediate in type between some of those listed above.[2]

STRUCTURES OF THE ELEMENTS

The crystal structures of the elements are listed in the Appendix. We are concerned in this section with the more common types and with their possible relation to the Periodic Table.

The typically metallic structures are f.c.c. (Strukturbericht *A*1), b.c.c. (Strukturbericht *A*2), and c.p.h. (Strukturbericht *A*3), introduced in Chap. 1; they result from simple ways of packing spheres. In order to show atomic positions in crystals, it is convenient to illustrate the unit cells in perspective and to indicate atomic positions by means of spheres drawn purposely not in contact (Fig. 10–1). This way of drawing does not show the effective sizes of the atoms in relation to their distance apart, but it permits a clear identification of atomic planes in the structure, and of

[1] A. F. Wells, "Structural Inorganic Chemistry," Clarendon, Oxford, 3d ed., 1962.

[2] For a more detailed survey of the bond types and their characteristics, see R. C. Evans, "An Introduction to Crystal Chemistry," 2d ed., Cambridge, London, 1964.

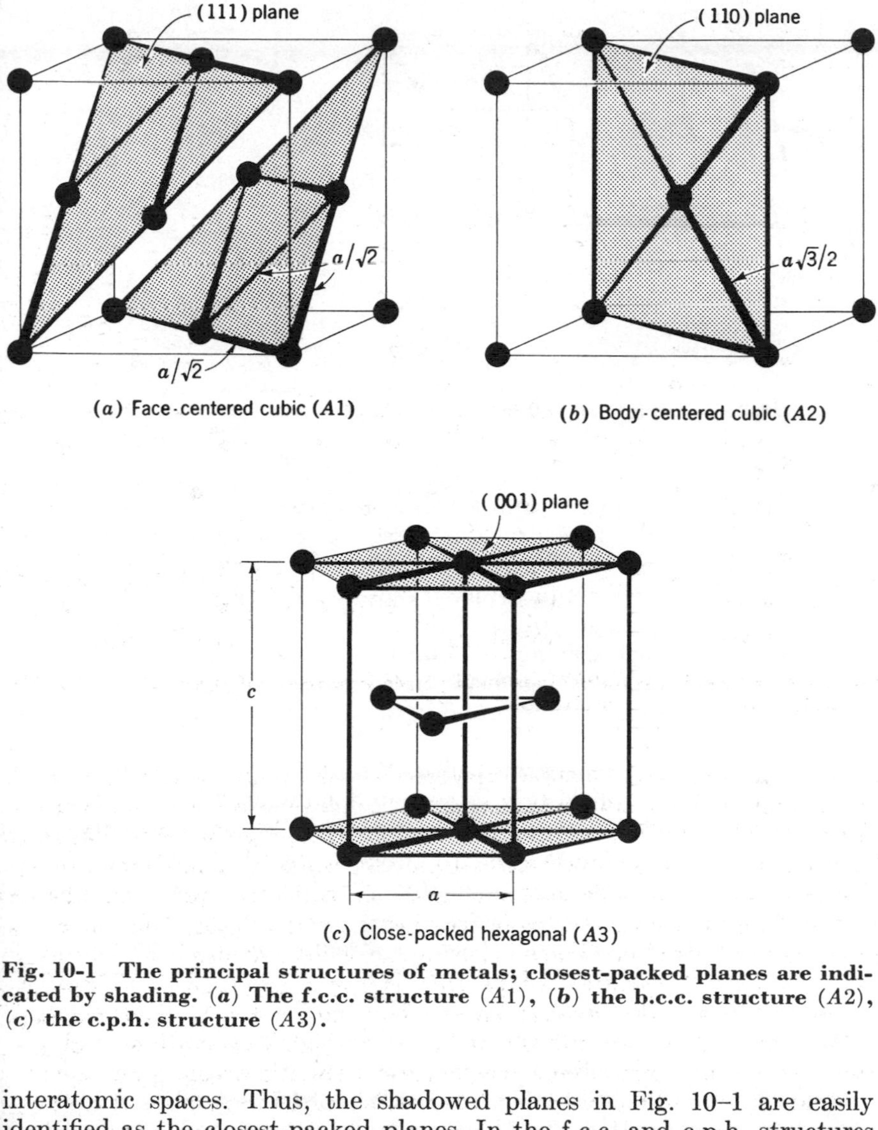

(*a*) Face-centered cubic (*A*1)

(*b*) Body-centered cubic (*A*2)

(*c*) Close-packed hexagonal (*A*3)

Fig. 10-1 The principal structures of metals; closest-packed planes are indicated by shading. (*a*) The f.c.c. structure (*A*1), (*b*) the b.c.c. structure (*A*2), (*c*) the c.p.h. structure (*A*3).

interatomic spaces. Thus, the shadowed planes in Fig. 10–1 are easily identified as the closest-packed planes. In the f.c.c. and c.p.h. structures the *coordination number* is 12, each atom being surrounded by 12 nearest neighbors. In the b.c.c. structure the coordination number is 8, with 8 nearest neighbors at a distance $a\sqrt{3}/2 = 0.866a$ (where a is the length of the cube edge), and with an additional 6 neighbors at a distance a, only slightly farther away. For this reason the b.c.c. structure is sometimes discussed in terms of coordination number 14 (see also Chap. 13, page 376).

As noted earlier, the same structure may be represented in a number of ways. For example, the primitive cell in the hexagonal system is a parallelepiped with lattice points at corners only (Fig. 1–4, Chap. 1), but another

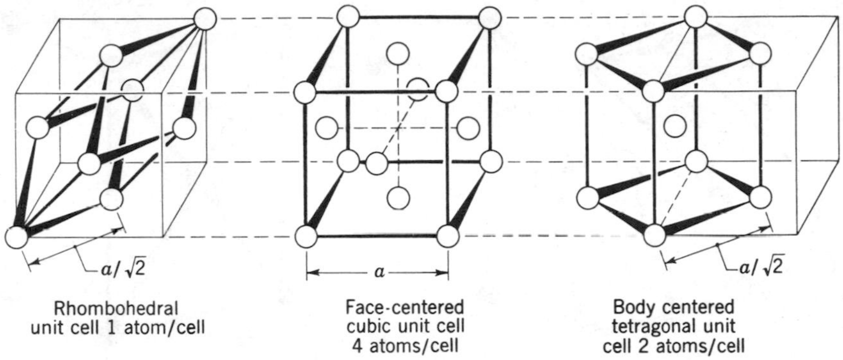

Fig. 10-2 Possible unit cells of the f.c.c. structure.

cell can be drawn in this system which is orthohexagonal (Fig. 1–9, Chap. 1). Similarly, in the f.c.c. structure, various possible unit cells can be drawn, as shown in Fig. 10–2, but only the f.c.c. cell with four atoms per cell exhibits clearly the full symmetry of the arrangement. If the x-ray reflections are indexed according to this cell, the lattice spacing a is associated with the average spacings of atoms located at the corners of the cube and is larger than the spacings between the neighboring atoms of the other possible unit cells, and therefore exceeds the *closest distance of approach* ($a\sqrt{2}$).

As will be frequently shown below, the atomic positions in the more complex structures can also be drawn in a number of alternative ways, each revealing different structural relationships. Such a flexibility in the diagrammatic representation of structures frequently brings out striking similarities among a number of related structures that would otherwise not be apparent.

There is a clear relationship between structure and the position of an element in the Periodic Table. The "classical table" as originated by Mendeleieff is arranged in the form of horizontal *rows* (or *periods*) and vertical *groups* as shown in Table 10–1. Elements in the same group tend to have the same structure at ordinary (room) temperature; for example, the *alkali metals* Li, Na, K, Rb, and Cs (group IA) are all b.c.c.; Be, Mg, Zn, and Cd (groups IIA and IIB) are c.p.h.; Cu, Ag, and Au (group IB) are f.c.c., as are most of the elements in the adjoining group VIII in at least one of their modifications. Elements to the left in the Periodic Table show electropositive behavior and tend to have the simple metallic structures. Elements to the right are sometimes called *metalloids* and tend to have more complex structures. In the intermediate groups occupied by the *transition elements* and the *rare earths*, the process of filling inner shells while the number of electrons in outer shells remains constant tends to encourage again the typically metallic structures. On the whole, although these structural relationships clearly exist in the periodic table, their inter-

	IA	IIA	IIIA	IVA	VA	VIA	VIIA	VIII	VIII	VIII	IB	IIB	IIIB	IVB	VB	VIB	VIIB	O
1st								1 H A3 A1	2 He (A3) (A2)									
2nd	3 Li A2 A1 A3	4 Be (A2) A3											5 B H T R	6 C R H A4	7 N H C	8 O C (R)	9 F	10 Ne A1
3rd	11 Na A2 A3	12 Mg A3											13 Al A1	14 Si A4	15 P C O C	16 S O M R	17 Cl A1 A3 A2	18 A A1
4th	19 K A2	20 Ca A2 A1	21 Sc (A2) A3	22 Ti A2 A3	23 V A2	24 Cr (A1) A2	25 Mn A2 A1 C C	26 Fe A2 A1 A2	27 Co A1 A3	28 Ni A1	29 Cu A1	30 Zn A3	31 Ga O	32 Ge A4	33 As A7	34 Se A8 M	35 Br O	36 Kr A1
5th	37 Rb A2	38 Sr A2 A3 A1	39 Y A2 A3	40 Zr A2 A3	41 Nb A2	42 Mo A2	43 Tc A3	44 Ru A3	45 Rh A1	46 Pd A1	47 Ag A1	48 Cd A3	49 In A6	50 Sn A5 A4	51 Sb A7	52 Te A8	53 I O	54 Xe A1
6th	55 Cs A2	56 Ba A2 (T) (H)	57 La A2 A1 H	72 Hf A2 A3	73 Ta A2	74 W A2	65 Re A3	76 Os A3	77 Ir A1	78 Pt A1	79 Au A1	80 Hg R T	81 Tl A2 A3	82 Pb A1	83 Bi A7	84 Po R C	85 At	86 Rn
7th	87 Fr	88 Ra	89 Ac A1															

6th (continued)	58 Ce A2 A1 H	59 Pr A2 H	60 Nd A2 H	61 Pm	62 Sm (A2) R	63 Eu A2	64 Gd (A2) A3	65 Tb (A2) A3	66 Dy (?) A3	67 Ho (?) A3	68 Er A3	69 Tm A3	70 Yb A2 A1	72 Lu (?) A3
7th (continued)	90 Th A2 A1	91 Pa T	92 U A2 T O	93 Np A2 T O	94 Pu A2 T A1 O M M	95 Am H	96 Cm	97 Bk	98 Cf	99 Es	100 Fm	101 Md	102 No	103 Lw

These Strukturbericht and other designations are used to denote the structures of the allotropic forms which are given in the order of their appearance. *A*1 = f.c.c.; A2 = b.c.c.; *A*3 = c.p.h.; *A*4 = diamond cubic; *A*5 = b.c.t.; *A*6 = f.c.t.; *A*7 = rhombohedral; *A*8 = trigonal; H = hexagonal (usually *ABAC*···close-packed); R = rhombohedral; O = orthorhombic; C = complex cubic; T = tetragonal; M = monoclinic; ( ) = uncertain.

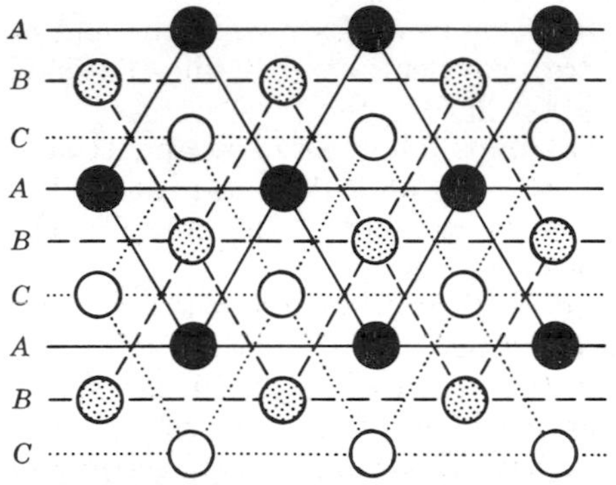

Fig. 10-3 Stacking of closest-packed planes *ABCABC* . . . in the f.c.c. structure. Atom positions are projected onto the (111) plane.

pretation in terms of details of bonding and band structure is still chiefly empirical.

Close-packed hexagonal and f.c.c. structures are much more closely related than would appear from Fig. 10–1. Both are structures that represent spheres of uniform size in closest packing. In the c.p.h. crystal structure the close-packed (0001) layers are stacked above each other in the sequence *ABAB*···; i.e., atoms of the third layer are directly above those of the first. In the f.c.c. structure the (111) planes have the same c.p.h. array of atoms and are stacked in the sequence *ABCABC*··· so that the fourth layer is directly above the first. This is illustrated in Fig. 10–3. The atoms in each layer fit in the hollows of the layer beneath.

The beginning elements of the rare-earth series crystallize in the f.c.c. structure, the c.p.h. structure, or in a close-packed structure with a more complex stacking sequence, i.e., of the form *ABACABAC*··· as observed in samarium. The rare-earth elements, after europium which has the b.c.c. structure, have c.p.h. structures at room temperature with the exception of ytterbium which is f.c.c.[1] Many rare earths have allotropic transformations at high or low temperatures.[2]

In the *actinide series*, thorium crystallizes in two modifications, f.c.c. and b.c.c., and protoactinium is b.c.t. The following three elements, uranium, neptunium, and plutonium, all possess the b.c.c. structure at high temperatures but, as indicated in the Appendix, they form other complex structures. For example, the tetragonal structure of beta-uranium is similar to that of the σ phase in the iron-chromium system (see page 266).

The 8 − *N* rule At the right of the periodic table the *B*-subgroup elements, in which the outer shells are again in the process of being filled, are in general less closely packed. The binding forces become more covalent between certain neighboring atoms, instead of a binding among positive

[1] α-La, α-Pr, and α-Nd have a c.p.h. structure with c spacing double the usual value and numerous stacking faults. This distinguishes them from the elements in the latter part of the series which have the usual c.p.h. structure.

[2] F. H. Spedding and A. H. Daane summarize data on the structures and properties of the rare earths and their alloys in *Met. Rev.*, vol. 5, p. 297, 1960.

ions and electrons. The homopolar character increases toward the end of the horizontal periods until it is a true covalent bond in the diatomic molecules of group VII*B*, the *halogens* F_2, Cl_2, Br_2, and I_2.

Most of the structures in the *B* subgroups follow Bradley and Hume-Rothery's interesting "8 − *N* rule," which states that each atom has 8 minus *N* close neighbors, where *N* is the number of the group in the Periodic Table to which the element belongs.

Consider, for example, carbon (group IV*B*) with the diamond structure (Strukturbericht *A*4). The atoms are arranged on two interpenetrating f.c.c. lattices. A perspective view of the structure (Fig. 10-4*a*) shows clearly that each atom has four near neighbors. Gray tin, silicon, and germanium, also of group IV*B*, likewise have this diamond cubic structure

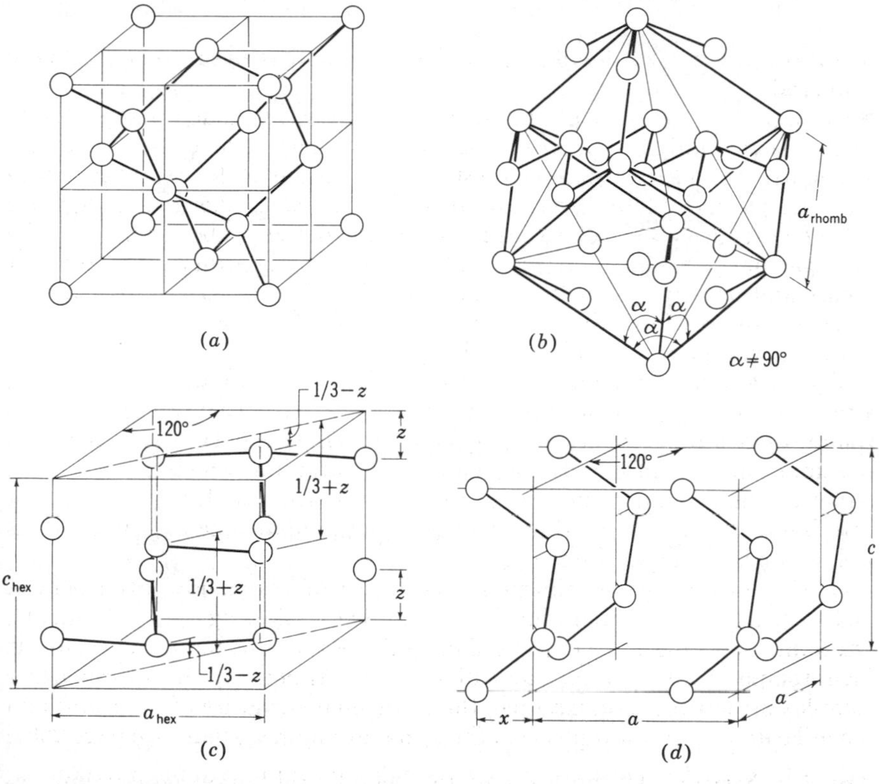

Fig. 10-4 The structures of the *B*-subgroup elements. (*a*) The cubic structure of diamond, germanium and silicon, (*A*4). Each atom has *four* nearest neighbors. (*b*) The rhombohedral structure of As, Sb, and Bi (*A*7). Each atom has *three* nearest neighbors; the shortest bonds are indicated. The structure consists of two interpenetrating face-centered sublattices, nearly cubic, one sublattice being slightly displaced from the mid-edge position of the other. (*c*) The As, Sb, Bi (*A*7) structure referred to hexagonal axes, with the parameter *z* indicated. In units of *c*, *z* = 0.226, 0.2349, 0.23389 for As, Sb, and Bi, respectively. (*d*) The structure of Te and Se (*A*8). Atoms are arranged in spiral chains along the *c* axis; each atom has *two* nearest neighbors. Parameter *x* is 0.217 in Se, 0.269 in Te, in units of the *a* axis of the hexagonal cell.

with coordination number 4. The heavier elements of this group tend to have more metallic structures; white tin is b.c.t. (Strukturbericht *A*5) and lead is f.c.c.

In group V*B*, As, Sb, and Bi have rhombohedral structures (Strukturbericht *A*7) (a slightly distorted simple cubic structure) with each atom surrounded by three near neighbors (Fig. 10–4*b*). These more closely bonded atoms form puckered layers (horizontal in Fig. 10–4*b*); within the layers the bonding is covalent, and between the layers it is partially metallic. Group VI*B* has Se and Te with the trigonal structure (Strukturbericht *A*8) of Fig. 10–4*c*, in which the atoms are linked in chains with each atom having two neighbors. There is only one near neighbor to each atom in solid iodine, of group VII*B*, for which $8 - N = 1$. The pairs of iodine atoms correspond to the diatomic molecules characteristic of the other elements of this group. The structure is orthorhombic.

The $8 - N$ rule does not apply to elements of group III*B*. Gallium has an orthorhombic structure with eight atoms per unit cell and four different interatomic distances between adjacent atoms. Indium has a structure in which all atoms have four near neighbors: it is a tetragonal structure that may be derived from the f.c.c. by elongating a cube edge until the axial ratio c/a is 1.08. Thallium is c.p.h. ($c/a = 1.59$) at room temperature and b.c.c. above ~230°C; aluminum is f.c.c., and the structure of boron is still uncertain.[1] The failure of these elements to have coordination number 5 is attributed by Hume-Rothery to incomplete ionization of the atoms.[2]

Polymorphism Many of the elements have different structures or allotropic forms when the temperature or pressure is varied, or when they are subjected to unusual thermal or mechanical treatments. Details are discussed in a later chapter on transformations, and a listing of the allotropic forms will be found in the Appendix; the more important ones will be mentioned here. The room-temperature b.c.c. form of iron (α-Fe) changes to f.c.c. on heating above 910°C (γ-Fe) and reverts to b.c.c. again at about 1400°C (δ-Fe). Manganese exists in four forms, α, β, γ, and δ, and plutonium in six, α, β, γ, δ, δ', and ϵ. Tin has a metallic form at ordinary temperatures (white, or β) but a nonmetallic form (gray, or α) with the structure of diamond, below about 13°C. This low-temperature form is the cause of "tin pest" to which some tinned articles are susceptible when exposed to cold weather (see also Chap. 18, page 490).

Several metals that are c.p.h. at lower temperatures transform to b.c.c. at higher temperatures; these include zirconium, titanium, thallium, lithium, and beryllium.[3] Metals that are c.p.h. at low temperatures sometimes transform to f.c.c. at higher temperatures, e.g., calcium, lanthanum,

[1] Several forms exist; a common one is hexagonal.

[2] W. Hume-Rothery, "The Structure of Metals and Alloys," Institute of Metals, London, 2nd ed., 1944, and subsequent editions.

[3] The structure of Li changes from b.c.c. to c.p.h. at cryogenic temperatures (see Chap. 18, p. 491) and the structure of Be becomes b.c.c. near the melting point.

and scandium.[1] Similarly, cobalt, which is stable in the c.p.h. form (commonly with frequent errors in the stacking of atom layers) at low temperatures, is f.c.c. at high temperatures.

Among the nonmetallic elements, striking polymorphism is found in phosphorus (white, black, red, and yellow forms) and carbon. The diamond structure of carbon is very different from the layer structure of graphite. The familiar form of graphite has a hexagonal structure with atoms at (or very close to) the positions 000, $00\frac{1}{2}$, $\frac{1}{3}\frac{2}{3}0$, $\frac{2}{3}\frac{1}{3}\frac{1}{2}$, and the basal planes consist of a hexagonal network in which each atom is linked with three neighbors (not with four as in diamond). Interatomic bonds within the basal planes are strong, but those between the layers are very weak so that they glide easily over each other and give graphite the quality of being a good lubricant. The second form of graphite has a unit cell three layers high instead of the two-layer cell of the normal form, an $ABCABC\cdots$ sequence instead of $ABAB\cdots$, and is therefore rhombohedral. Atoms in this form are at 000, $\frac{1}{3}\frac{2}{3}0$, $00\frac{1}{3}$, $\frac{2}{3}\frac{1}{3}\frac{1}{3}$, $\frac{1}{3}\frac{2}{3}\frac{2}{3}$, $\frac{2}{3}\frac{1}{3}\frac{2}{3}$. Most samples are predominantly of the hexagonal form.

Polymorphic transformations are usually brought about by temperature changes, but several have been induced by high hydrostatic pressures, or by high pressures combined with drastic cold-work, as Bridgman's extensive research has shown. The transformations with increasing pressure usually yield the more dense modifications, which is an expression of Le Châtelier's principle. An unusual transformation in which both high- and low-pressure forms are f.c.c. has been discovered in cerium.[2] The high-pressure form has 16.55 percent smaller volume than the form at atmospheric pressure. In general, however, when polymorphic changes occur at normal pressures, even though near-neighbor distances often vary quite markedly from one allotropic form to another, the atomic volumes, and hence also the total energies, are surprisingly similar.[3] This is evident, for example, in the case of iron (Fig. 10–5).

As a result of improvements both in high-pressure techniques and in x-ray diffraction techniques,[4] the number of newly reported allotropic forms obtained under high pressure in different elements is steadily in-

[1] Ca is b.c.c. $> 450°C$ (c.p.h. 450 to 250°C in an impure sample), f.c.c. $< 250°C$; La is cph $< 340°C$, f.c.c. $> 340°C$; the existence of the b.c.c. form in Sc is in some doubt; (the "International Tables," vol. 3, suggest a b.c.c. structure above 1335°C).

[2] A. W. Lawson and Ting-Yuan Tang, *Phys. Rev.*, vol. 76, p. 301, 1949. The transformation is thought to be due to the 4*f* electron being literally squeezed into a 5*d* state. Cerium is the first element in the Periodic Table to have a 4*f* electron, and the energy difference between 4*f* and 5*d* levels is small. The transformation also occurs on cooling at ordinary pressures: A. F. Shuck and J. H. Sturdivant, *J. Chem. Phys.*, vol. 18, p. 145, 1950.

[3] N. F. Mott, *Rept. Progr. Phys.*, vol. 25, p. 218, 1962. H. W. King in R. W. Cahn (ed.), "Physical Metallurgy," North Holland Publishing, Amsterdam, 1965. P. Rudman, *Trans. AIME*, vol. 233, p. 864, 1965.

[4] J. C. Jamieson and A. W. Lawson in R. W. Wentorf, Jr. (ed.), "Modern Very High Pressure Techniques," Butterworth, Washington, D.C., 1962.

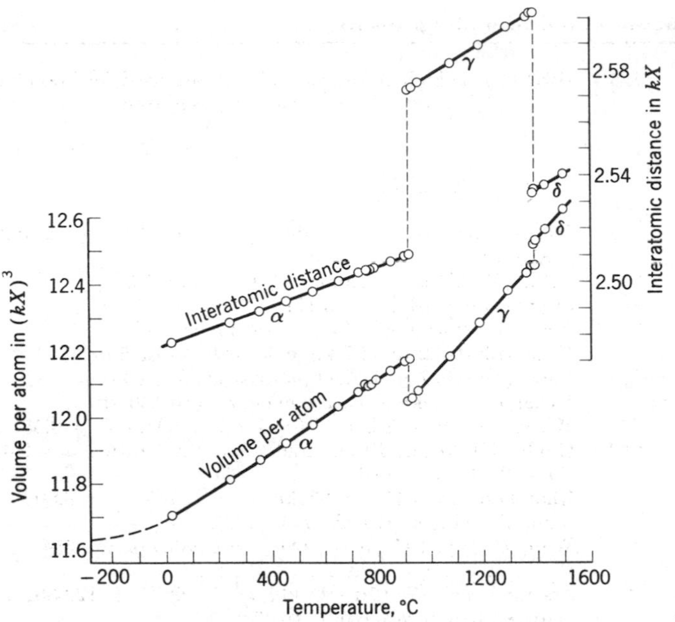

Fig. 10-5 Temperature dependence of lattice parameters (a) and volume per atom (Ω) in pure iron. [After W. Hume-Rothery, Z. S. Basinski, and A. L. Sutton, Proc. Roy. Soc. (London), vol. A229, p. 459, 1955.]

creasing. A list that is believed to have been complete at the time of writing is presented in Table 10–2. (For the apparatus used in obtaining the data, see Chap. 7.) The table does not list structures that have been determined at room pressure after treatment of samples at high pressures. Patterns are often poor and structure determinations are correspondingly uncertain. Generalizations would be difficult to state with confidence at this time; however, in general, the high-pressure forms are more densely packed structures than the ordinary structures. Several transitions occur among the III-V and II-VI compounds, and the elements Si and Ge transform to the metallic white-tin structure.

SOLID SOLUTIONS

Types of solid solutions When atoms of two or more elements are able to share together, and with changing proportions, various sites of a given crystal structure, a *solid solution* is formed. The replacement of nickel atoms by copper atoms on the lattice of nickel is an example of a *substitutional* solid solution. Nickel can dissolve copper at all proportions, providing an example of *complete solid solubility*. This is usually only possible if the sizes of the atoms differ by no more than about 15 percent. There are many examples of restricted mutual solid solubilities even between elements with similar crystal structures and atom sizes.

Table 10–2 Structures studied under pressure

Substance	Ref.	Structure at high pressure (and pressure used, in kilobars); unit cell dimensions in angstroms
		Elements
Antimony	*v,aa*	Diffuse patterns[v]; simple cubic 50 to ≈ 85 kb[aa]; c.p.h. above 85 kb[aa]
Barium	*m*	C.P.h., 62 kb, $a = 3.90$, $c = 6.15$, $c/a = 1.58$
Bismuth	*gg*	Simple cubic BiII, $a = 3.177 \pm 0.009$ at 25 kb
Cerium	*y*	F.c.c. transforming to f.c.c. with smaller cell
Cesium III	*l*	F.c.c. from 42.2 to 42.7 kb, $a = 5.800$ at 52.5 kb
Cesium II	*l*	F.c.c. from 23.7 to 42.2 kb, $a = 5.984$ at 41 kb
Dysprosium	*j*	Hexagonal stacking rearrangement 50 to 160 kb
Germanium	*a*	White tin type ($A5$), $a = 4.884$, $c = 2.692$, $c/a = 0.551$
Iron	*e,o,bb*	C.p.h., 130 kb; at 192 kb and room temperature $a = 2.45$, $c = 3.93$, $c/a = 1.61$[bb]
Phosphorus (black)	*a*	Rhombohedral ($A7$) at 83 kb, $a = 3.377$, $c = 8.806$; to simple cubic at 124 kb, $a = 2.377$
Silicon	*a*	White tin type ($A5$), $a = 4.686$, $c = 2.585$, $c/a = 0.554$
Strontium	*s*	B.c.c., 42 kb, $a = 4.43$
Tellurium	*cc*	Arsenic type, $A7$, 15 to 42 kb, $a = 4208$, $c = 12.036$; unsolved structure above 42 to 45 kb
Tellurium	*dd*	Selenium type, $A8$, to ∼ 40 kb; $A7$ type not observed; unsolved structure 40 to ∼ 70 kb; simple rhombohedral ($\beta-$Po type) $a_r = 3.002$, $\alpha = 103.3°$ above 70 kb
Thallium	*f*	F.c.c. above 60 kb, $a = 4.778$
Tin	*z*	B.c.t. phase at 39 kb and about 310°C, $a = 3.881$, $c = 3.483$, $c/a = 0.914$
Tin	*a,v*	B.c.c. ($A2$) at 115 kb, room temperature; b.c.t. above 115 kb at room temperature[v]
Titanium	*d*	Hexagonal (ω-phase[w] of Ti-V) above 90 to 130 kb, $a = 4.625$, $c = 2.813$, $c/a = 0.608$
Ytterbium	*k*	B.c.c., also second-order transitions, 40 kb
Zirconium	*d*	Distorted b.c.c. (ω-phase[w] of Ti-V), 90 to 130 kb, $a = 5.036$, $c = 3.109$, $c/a = 0.617$
		Compounds
InSb	*h*	White tin type, $a = 5.537$, $c = 2.970$, $c/a = 0.536$ (con firmed by several investigators)
InSb	*b*	Simple orthorhombic, $a = 2.92$, $b = 5.56$, $c = 3.06$ at >30 kb, room temperature
GaSb	*h*	White tin type, $a = 5.348$, $c = 2.937$, $c/a = 0.549$
AlSb	*h*	White tin type, $a = 5.375$, $c = 2.892$, $c/a = 0.538$
InAs	*h*	NaCl type, $a = 5.514$
InP	*h*	NaCl type, $a = 5.310$
RbF	*n*	CsCl type, 9 to 15 kb
KNO_4	*i*	Orthorhombic 5,000 kg per cm^2, $a = 16.12$, $b = 10.12$, $c = 7.75$
NaCl	*x*	CsCl type, $a = 3.35$, 17.7 kb per cm^2 (not confirmed by several laboratories)
RbCl	*p*	CsCl type, 7,500 to 11,000 kb per cm^2
RbBr	*y*	CsCl type

Table 10–2 Structures studied under pressure (Cont.)

Substance	Ref.	Structure at high pressure (and pressure used in kilobars); unit cell dimensions in angstroms
RbI	p	CsCl type, 7,500 to 11,000 kb per cm²
KI	y,f	CsCl type, $a = 4.093$ above 20 kb[f]
SnTe	q	Orthorhombic (*Pnma*) 18 kb
AgI	r	Orthorhombic 3 kb
AgI	y,f	NaCl type, $a = 6.067$ above 3.3 kb[f]
$CaCo_3$	t	Orthorhombic 5 kb and elevated temperature
SiO_2	y	Quartz transforms to coesite
KNO_3 III	y	Aragonite transforms to rhombohedral
KNO_3 IV	y	Aragonite transforms to orthorhombic
$CaCo_3$	y	Calcite structure, anion disorder
H_2O II	y	Cubic
H_2O III	y	Tetragonal
PbS, PBSe, PbTe	dd	NaCl type transforms to orthorhombic SnS type (*Pmcn*)
CdS	ff	Wurtzite form transforms to NaCl type >25 kb, $a = 5.42$
CdTe	ff	Zinc blende type transforms to NaCl type >40 kb, $a = 5.92$

High-pressure structures not yet determined because of diffuseness or complexity of patterns[v]: As, Pb, S, ZnO, HgO, ZnTe, CdTe, ZnS, WO_3, $CsClO_4$, BiSn.

No transitions noted by x-rays until time of writing: Be, Cd, C, Dy, Hf, Mg, Se, Zn, Fe_2O_3, Fe_3O_4, NiO_2, MgO, KCl.[u,v]

[a] J. C. Jamieson, *Science*, vol. 139, pp. 762, 1291, 1963. [b] J. S. Kasper and H. Brandhorst, *J. Chem. Phys.*, in press. [c] J. C. Jamieson in *Abstracts for 1962, Geol. Soc. Am. Special Paper*, vol. 73, p. 178, 1963. [d] J. C. Jamieson, *Science*, vol. 140, pp. 72–73, 1963. [e] J. C. Jamieson and A. W. Lawson, *J. Appl. Phys.*, vol. 33, p. 776, 1962. [f] G. J. Piermarini and C. E. Wier, *J. Res. Natl. Bur. Std.*, vol. 66, p. 325, 1962. [g] J. C. Jamieson in A. A. Giardini and E. C. Lloyd (eds.), "*High Pressure Measurements 1962*," Butterworth, Washington, 1963. [h] J. C. Jamieson, *Science*, vol. 139, p. 845, 1963. [i] J. C. Jamieson, *Z. Krist.*, vol. 107, p. 65, 1956. [j] J. C. Jamieson, *Science*, vol. 145, pp. 572–574, 1964. [k] H. T. Hall, J. D. Barnett, and L. Merrill, *Science*, vol. 139, pp. 111–112, 1963. [l] H. T. Hall, L. Merrill, and J. D. Barnett, *Science*, vol. 146, pp. 1297–1299, 1964. [m] J. D. Barnett, R. B. Bennion, and H. T. Hall, *Science*, vol. 141, pp. 534–535, 1963. [n] G. J. Piermarini and C. E. Weir, *J. Chem. Phys.*, vol. 37, pp. 1887–1888, 1962. [o] R. L. Clendenen and H. G. Drickamer, private communication. [p] L. T. Vereshchagin and S. S. Kabalkina, *Dokl. Akad. Nauk, SSSR*, vol. 113, p. 797, 1957. [q] J. A. Kafalas, A. N. Marino, *Science*, vol. 143, p. 952, 1964. [r] B. L. Davis and L. H. Adams, *Science*, vol. 146, pp. 519–521, 1964. [s] D. B. McWhan and A. Jayaraman, *Appl. Phys. Letters*, vol. 3, p. 129, 1963. [t] B. L. Davis, private communication. [u] J. G. Kereiakes, *Phys. Rev.*, vol. 98, pp. 553–554, 1955. [v] J. C. Jamieson, private communication, 1965. [w] J. M. Silcock, M. H. Davies, and H. K. Hardy in "The mechanism of Phase Transformations in Metals," p. 93, Institute of Metals, London, 1956. [x] V. V. Evdokimova and L. F. Vereshchagin, *Soviet Phys. JETP*, vol. 16, p. 855, 1963. [y] Quoted in R. H. Wentorf, "Modern Very High Pressure Techniques," Butterworth, Washington, D.C., 1962. [z] J. D. Barnett, R. B. Bennion, and H. T. Hall, *Science*, vol. 141, p. 1041, 1963. [aa] S. S. Kabalkina and V. P. Mylov, *Soviet Phys. Dokl.*, vol. 8, pp. 917–918, 1964. [bb] T. Takahashi and W. A. Bassett, *Science*, vol. 145, p. 483, 1964. [cc] S. S. Kabalkina, L. F. Vereschchagin, and B. M. Shulenin, *JETP(USSR)*, vol. 45, p. 2073, 1963. [dd] D. B. McWhan and J. C. Jamieson, private communication. [ee] T. Takahashi, W. A. Bassett, and J. S. Weaver, "Abstracts of 1963 Annual Meeting of the Geological Society of America. [ff] N. B. Owen et al., *J. Phys. Chem. Solids*, vol. 24, p. 1519, 1963. [gg] R. Jaggi, private communication, 1965.

If the mutual solid solubility is restricted, as it is in the Cu-Ag system, to only those portions of the diagram that are linked to the pure elements, the resulting phases are known as *primary* (*or terminal*) *solid solutions*. Such solutions have, of course, the same structure as the elements on which they are based. If other phases are present in the system, they are usually known as *intermediate phases* or *intermetallic phases* and they frequently possess structures that are quite different from the structure of either of the component elements.

Interstitial solid solutions are formed when atoms with small radii are accommodated in the interstices of the lattice of a solvent. The solid solution of carbon in γ-Fe (austenite) is an example of this type; the iron atoms are on f.c.c. lattice points, and the carbon atoms occupy interstitial positions.

In general, because of the restricted size of the interstices, only rather small atoms dissolve interstitially. The radii of atoms known (or likely) to form interstitial solutions are all less than 1.0 A:

H	B	C	N	O
0.46	0.97	0.77	0.71	0.60

When a solvent accepts one of these interstitially, there is always an expansion of the unit cell.

In multicomponent systems some atoms may be dissolved interstitially and others substitutionally, as in manganese steel, where manganese atoms replace iron atoms on lattice points and carbon atoms enter the interstices. Both interstitial and substitutional solid solutions can be random, with a statistical distribution of atoms on the atom sites, or they may be partially or completely ordered, forming a superlattice, in which case the unlike atoms show preference for being neighbors of one another. Alternatively, the like atoms may tend to associate together to form clusters within the solid solution. Again, the clusters may be dispersed randomly or they may be ordered or oriented in various ways, producing a variety of complex arrangements within the solid solution (superlattices are discussed in Chap. 11, clusters in Chap. 18).

While it is possible to consider a completely random solid solution as an idealized example, the mounting experimental evidence, based mainly upon diffuse x-ray scattering, suggests that complete randomness (like perfect crystallinity) is probably never found in nature. Hence, solid solutions that are in a thermodynamical equilibrium may be considered to be truly *homogeneous* on a macroscopic scale, but they need not be homogeneous on the submicroscopic scale where atoms are considered individually.

Determination of the type of solid solution The types of solid solutions may be distinguished from each other by density comparisons. The density of an alloy computed from an x-ray measurement of the unit cell size will agree with the observed density if the correct type of solution is assumed. For a substitutional solid solution the calculated density ρ_c is given by the relation

$$\rho_c = \frac{n\bar{A}}{VN}$$

where n is the number of atoms contained in the unit cell of volume V, N is Avogadro's number, and $\bar{A}$ is the mean atomic weight of the atoms.[1]

Arrangement of atoms of different elements in simple metallic structures In a truly interstitial solid solution the smaller atoms are merely deposited in the interstitial voids (holes) between the bigger atoms, which may be assumed to be in contact. There is then a clear geometrical restriction associated with the available size of the interstitial holes. However, this restriction becomes less distinct if the bigger atoms relax and move apart a little, without altering the structure, thus allowing for a bigger hole than if they were in contact. Atoms with widely differing sizes and electrochemical affinities can then be accommodated within the holes, resulting in a variety of new structures. This possibility is further enhanced by the fact that typical metallic lattices have at least two different kinds of interstices that can be occupied by additional atoms. The correlations between the basic structures, the types of voids being filled, the degree of occupancy of such voids, and the various types and sizes of participating atoms constitute one of the more fascinating aspects of the crystal chemistry of alloy phases.

Of special interest are, for example, some typical alloy structures that can be derived from the three basic metallic structures, f.c.c., b.c.c., and c.p.h.; these and other common types illustrated in the preceding figures and in Figs. 10–6 to 10–13 are relatively simple.

In the f.c.c. structure the two types of interstitial voids are identified in Fig. 10–6*a* and *b*. The larger voids, known as *octahedral*, are surrounded by six atoms situated at the corners of a regular octahedron (Fig. 10–6*a*). The centers of the voids are at the midpoints of the unit cell edge, $00\frac{1}{2}$, etc., and at the center, $\frac{1}{2}\frac{1}{2}\frac{1}{2}$. Smaller voids, known as *tetrahedral*, are surrounded by a tetrahedron of four atoms (Fig. 10–6*b*), with centers at positions such as $\frac{1}{4}\frac{1}{4}\frac{1}{4}$, etc. In γ-Fe, if one considers the iron atoms as spheres in contact with each other, the larger interstices have room for a

[1] The mean atomic weight is computed from the atomic percentages p_1, p_2, etc., of the elements comprising the alloy and the atomic weights A_1, A_2, etc., with the formula

$$\bar{A} = \frac{p_1A_1}{100} + \frac{p_2A_2}{100} + \cdots$$

The atomic percentages are obtained from weight percentages w_1, w_2, etc., by the relation

$$p_1 = \frac{w_1/A_1}{w_1/A_1 + w_2/A_2 + \cdots} \cdot 100$$

Cyril S. Smith has published logarithm tables for interconversion of atomic and weight percentages of binary alloys: "Metals Handbook," p. 196, American Society for Metals, Cleveland, Ohio, 1948.

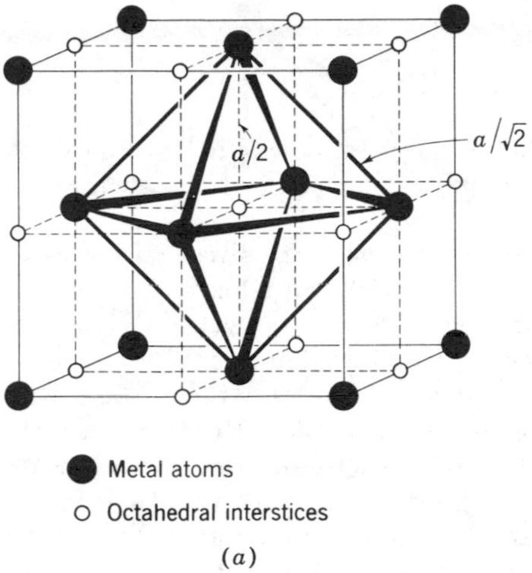

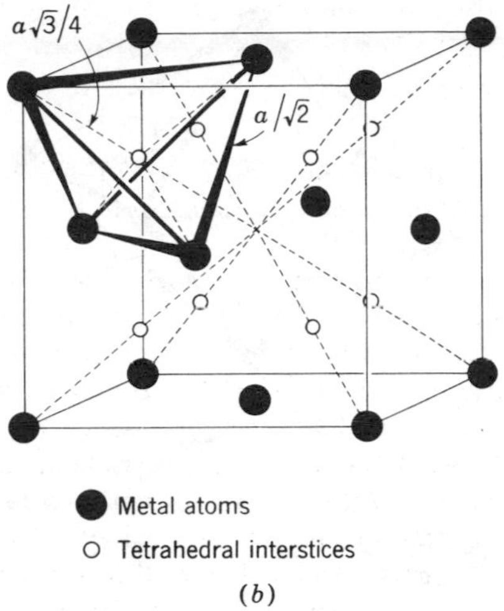

Fig. 10-6 The interstitial voids in the f.c.c. structure. (*a*) Octahedral voids (as in austenite); (*b*) tetrahedral voids.

spherical atom of 0.52 A radius, and the smaller for one of only 0.28 A.[1] The larger interstices therefore accommodate a carbon atom (0.8-A radius) or a nitrogen atom (0.7 A radius) only with an expansion of the lattice, and the smaller holes are probably inaccessible to the dissolved atoms.

In the b.c.c. metals there are also two types of voids to be considered. The larger of the two is now the tetrahedral one situated at $\frac{1}{2}\frac{1}{4}0$ and at equivalent positions (Fig. 10–7*a*). A spherical atom here would touch four spherical atoms of the solvent. In α-Fe with spherical iron atoms in contact with each other, there would be room for an interstitial atom of 0.36 A radius in these tetrahedral voids. The smaller types of voids are found at the midpoints of the edges ($00\frac{1}{2}$, etc.) and at the equivalent positions $\frac{1}{2}\frac{1}{2}0$, etc. (Fig. 10–7*b*), where they are surrounded by six atoms at the corners of a slightly compressed octahedron, two atoms being nearer than the other four. In α-Fe there is room for a sphere of only 0.19 A radius if the iron atoms are considered as spheres in contact with each other. It therefore appears that interstitial solution in α-Fe should be more difficult than in γ-Fe (as is the case) and that severe distortion may result from interstitial solution.

There is much evidence to indicate that the carbon atoms in α-Fe are located in the ($00\frac{1}{2}$) octahedral positions rather than in the largest interstices. The insertion of an atom into a tetrahedral interstice will cause all four surrounding atoms to be displaced, whereas if an oversize atom is inserted into an octahedral interstice, only the two atoms located nearest

[1] Further discussion will be found in C. H. Johansson, *Arch. Eisenhüttenw.*, vol. 11, p. 241, 1937. L. W. Strock, *Z. Krist.*, vol. 93, p. 285, 1936. L. V. Azároff, "Introduction to Solids," McGraw-Hill, New York, 1960.

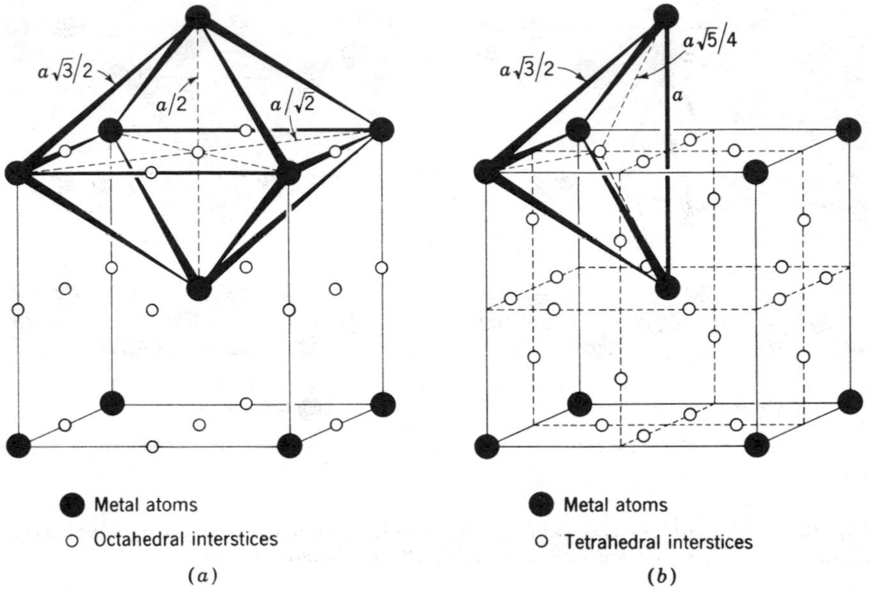

Fig. 10-7 The interstitial voids in the b.c.c. structure. (*a*) Octahedral voids; (*b*) tetrahedral voids.

to it will be displaced, causing a strain which can be relieved by an expansion of the lattice in the direction of the nearest atoms. The distortion of the α-Fe structure produced by an individual carbon atom is thus anisotropic; the effect is to distort the lattice from the cubic to tetragonal symmetry.[1] This accounts for the tetragonality of martensite, the b.c.t. structure that is formed when austenite is quenched and transforms rapidly without opportunity for the carbon atoms to form a carbide. Another consequence of the anisotropic distortion from interstitially dissolved elements is that an applied force can alter the distribution of the solute atoms among the holes, causing them to prefer holes that are enlarged by the applied stress rather than those that are contracted. The stress-induced shifting of atoms from one set of octahedral voids to another causes marked internal friction effects.[2]

The c.p.h. structure also has two types of interstitial voids, as shown in Fig. 10–8*a* and *b*. As in the f.c.c. structure the octahedral holes are larger than the tetrahedral holes. The positions of the octahedral holes are emphasized in the NiAs structure (Strukturbericht *B*8, Fig. 10–13) in which

[1] It has been suggested that this uniaxial distortion can occur more easily than isotropic expansion, so that α-Fe is able more easily to accommodate the interstitial atom in octahedral than in tetrahedral voids with their equidistant neighbors.

[2] J. L. Snoek, *Physica*, vol. 8, p. 711, 1941. L. J. Dijkstra, *Philips Res. Rept.*, vol. 2, p. 357, 1947. T. S. Kê, *Phys. Rev.*, vol. 74, pp. 7, 16, 1948. C. Zener, "Elasticity and Anelasticity of Metals," University of Chicago, Chicago, 1948. T. S. Kê, *Trans. AIME*, vol. 176, p. 448, 1948.

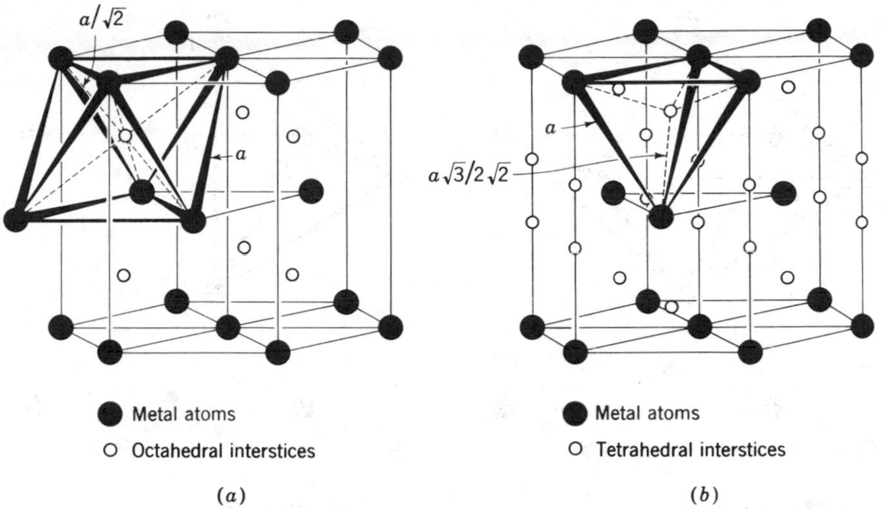

Fig. 10-8 The interstitial voids in the c.p.h. structure possessing ideal axial ratio $c/a = \sqrt{8/3}$. (a) Octahedral voids; (b) tetrahedral voids.

they are occupied by nickel atoms while the metalloid arsenic atoms lie on the close-packed planes of the c.p.h. lattice.[1]

The tetrahedral holes of the c.p.h. structure are partially occupied in the structure of wurtzite (Strukturbericht *B*4, Fig. 10–9), as are the tetrahedral holes in the f.c.c. structure partially occupied in the case of zinc blende (ZnS, Strukturbericht *B*3, Fig. 10–9). Hence, many metalliclike solid solutions simulate the well-known structures of compounds, with the frequently observed difference that the more metallic atoms are in the positions that are occupied by the nonmetallic atoms in the compounds.

The NaCl structure (Fig. 10–9) can be described as a f.c.c. packing of spherical ions of the larger variety (chlorine) with all octahedral voids being occupied by the smaller (sodium) ions. Each sodium ion is surrounded by six equidistant chlorine ions, and each chlorine ion is surrounded by six equidistant sodium ions. The CaF_2 structure (Fig. 10–10), on the other hand, is a f.c.c. arrangement of one kind of ion with the other kind occupying tetrahedral voids.

INTERMEDIATE PHASES

It is rare to find alloy systems in which intermediate phases obey the normal valences of the elements. This is a consequence of the fact that most of the phases have *metallic* bonding, which means that valence electrons are at least partially free to move about in the lattice, as contrasted with inorganic or organic compounds in which the electrons are tightly bound into stable groups of atoms. At first, attempts were made to define an

[1] The NiAs-type structures are designated by Pearson as $B8_1$, while the "filled-up" structures corresponding to the Ni_2In-type are designated as $B8_2$.

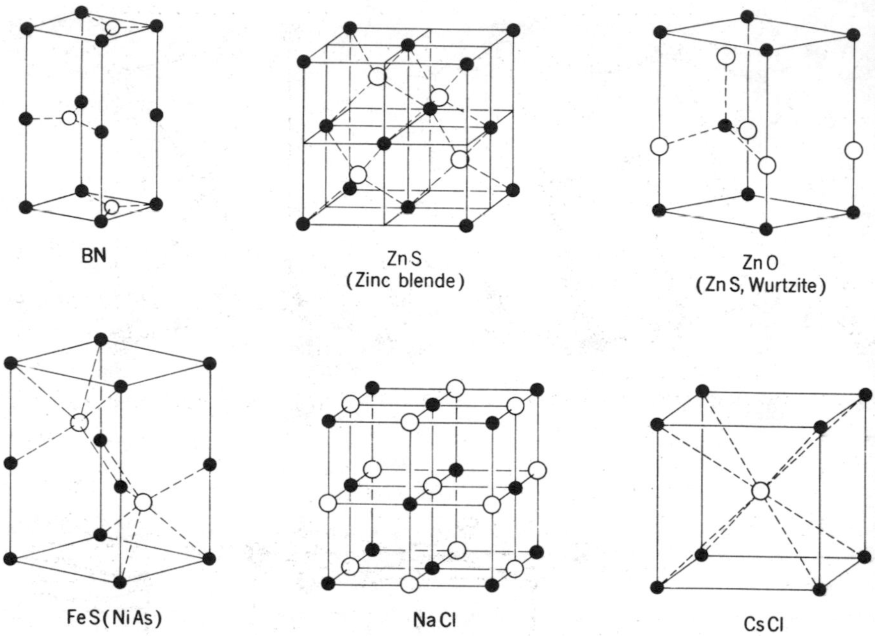

Fig. 10-9 Typical crystal structures of compounds of type *AX*. (**V. M. Goldschmidt.**)

intermetallic compound and to distinguish it from an intermediate solid solution, but the attempts were a source of confusion rather than of clarification. It was later realized that some phases had chemical formulas that followed valence rules and yet had the metallic conductivity and reflectivity characteristic of metallic bonding, and consequently that such phases possessed an intermediate type of bonding between covalent and metallic, resulting in a partial breakdown of covalent bonds and also in an increased solid solubility. Intermediate types of bonding can exist, in fact, between all the ideal types of bonding: ionic, homopolar, molecular, and metallic.

An alloy may be a true solid solution at high temperatures and yet resemble a compound at low temperatures when it becomes an ordered superlattice, as is discussed later. The tendency to form an ordered superlattice may be so weak that ordering cannot be detected, or so strong that the alloy melts before thermal agitation is able to produce detectable disorder.

Nomenclature of alloy phases and compounds The symbols used to describe alloy phases and intermetallic compounds have developed historically from a number of sources such as the studies of phase diagrams, the compound-formation rules of inorganic chemistry, and the investigations of crystal structure. The system of Strukturbericht designations (*A*1, *A*2, *B*1, *B*2, etc.) used in the preceding pages has evolved from the latter. It is, however, quite arbitrary, and after symbols for simpler structures were proposed, the naming of more complex structures became confused; in many cases Strukturbericht designations are lacking. In metallurgical literature, the phases occurring in binary and more complex alloy systems are usually designated by Greek letters but sometimes also by Latin capital letters or by chemical formulas.

The above types of phase designations frequently overlap, with the result that the same phase is denoted by different symbols (for example, gamma-brass, Cu_5Zn_8, or $D8_2$), or the same symbol is used for structurally different phases (for example, μ-Ag_3Al, Strukturbericht *A*13, and μ-Fe_7W_6, Strukturbericht $D8_5$). Such a state of affairs sometimes produces confusion; hence, repeated attempts have been made to develop an improved nomenclature for metallic phases. Recently, a subcommittee of the ASTM Committee on Metallography proposed a possible new scheme[1] that employs the chemical symbols of the elements concerned and describes crystal structures by various Latin letters. The system is less arbitrary than Strukturbericht but so far has found little following in the literature; in addition, the phases with unknown crystal structure cannot be rationally described and remain a problem.

In this and following chapters Strukturbericht designations for various structures will be given whenever possible but otherwise no particular scheme will be followed, and the nomenclature will be that most commonly used in current metallurgical literature.

Structures of compounds with normal valence From the early work of Goldschmidt[2] and others it was found that the simple structures illustrated in Figs. 10–9 and 10–10 are common to a great number of compounds, including many intermetallic phases that obey normal valence laws.[3]

The structure of rock salt (NaCl, *B*1) may be considered as two interpenetrating f.c.c. lattices of the two types of atoms, with the corner of one located at the point $\frac{1}{2}00$ of the other. The fluorite structure (CaF_2, *C*1) is also cubic, with Ca atoms at cube corners and face centers and with F atoms at all quarter-way positions along the cube diagonals ($\frac{1}{4}\frac{1}{4}\frac{1}{4}$, $\frac{1}{4}\frac{3}{4}\frac{3}{4}$, etc.). These typically ionic structures are found frequently among alloys of metals with elements of groups IV*B*, V*B*, and VI*B*:

NaCl Structure (*B*1)			CaF_2 Structure (*C*1)			
MgSe	MnSe	BaTe	Mg_2Si	Li_2S	Sn_2Pt	AgAsMg
CaSe	PbSe	MnTe	Mg_2Ge	Na_2S	Al_2Ca	LiMgN
SrSe	SrTe	SnTe	Mg_2Sn	Cu_2S	CuCdSb	LiZnN
BaSe	SnTe	PbTe	Mg_2Pb	Cu_2Se	CuMgSb	
		CaTe	Cu_2Se	Be_2C	CuBiMg	

The NaCl structure is also found in many oxides, fluorides, chlorides, hydrides, and carbides, as well as in many compounds of the rare earths with As, Sb, and Bi; the CaF_2 structure is found with many oxides, fluorides, and in some intermediate phases of the noble metals with Al, Ga, and In (page 242). Although the fluorite structure is easily understood as a normal valence compound, it is of interest that the examples listed above corre-

[1] A report of Nomenclature Subcommittee, *ASTM Bull.*, no. 226, December, 1957.

[2] V. M. Goldschmidt, *Trans. Faraday Soc.*, vol. 25, p. 253, 1929.

[3] The structure cells shown in Figs. 10–9 and 10–10 show certain crystallographic relationships, but like the simple metallic structures discussed earlier (p. 224), they can be redrawn to indicate other features. For example, the same zinc blende structure is illustrated in three different ways in Figs. 10–9, 10–11, and 10–12.

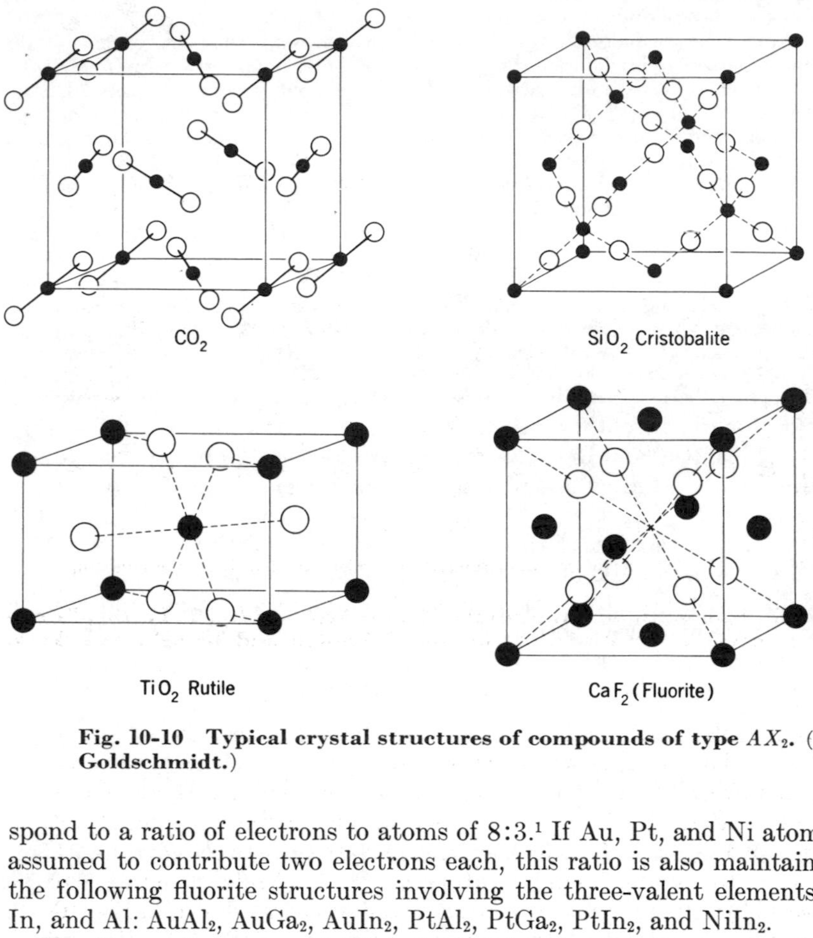

Fig. 10-10 Typical crystal structures of compounds of type AX_2. (V. M. Goldschmidt.)

spond to a ratio of electrons to atoms of 8:3.[1] If Au, Pt, and Ni atoms are assumed to contribute two electrons each, this ratio is also maintained in the following fluorite structures involving the three-valent elements, Ga, In, and Al: $AuAl_2$, $AuGa_2$, $AuIn_2$, $PtAl_2$, $PtGa_2$, $PtIn_2$, and $NiIn_2$.

The diamond structure ($A4$) is formed by elements having a ratio of 4 electrons to 1 atom (C, Si, Ge, Sn) in conformity with the $8 - N$ rule mentioned earlier. It is possible also to obtain the same kind of bonding among alloys by choosing equal amounts of two elements having valence other than four and being equally removed in the Periodic Table to the left and to the right of column IV. This is sometimes known as the *Grimm-Sommerfeld rule*,[2] and it suggests the existence of the 3–5, 2–6, and 1–7 compounds that obey normal valence rules. Some of these compounds are well known in *semiconductor* physics.[3]

[1] Alternatively, one can speak of an *electron concentration*, e/a, of 2.67. This parameter is discussed more fully in Chap. 13, p. 340.

[2] H. G. Grimm and A. Sommerfeld, *Z. Physik*, vol. 36, p. 36, 1926. See also E. Parthé, "Crystal Chemistry of Tetrahedral Structures," Gordon and Breach, New York, 1964.

[3] E. Mooser and W. B. Pearson, *J. Chem. Phys.*, vol. 26, p. 893, 1957. For a more detailed review see E. Mooser and W. B. Pearson, *Progress in Semiconductors*, vol. 5, p. 103, 1960.

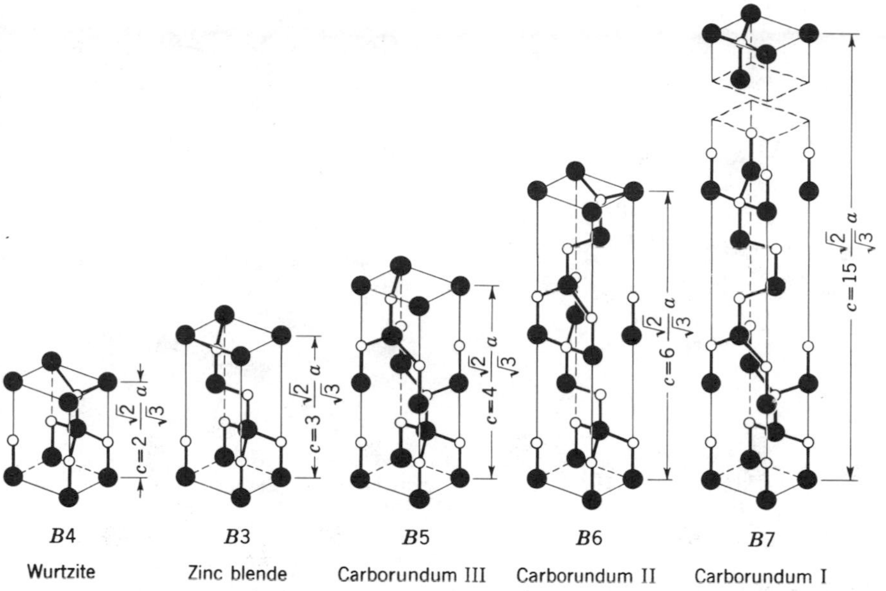

Fig. 10-11 Five common tetrahedral structure types. (**After E. Parthé, "Crystal Chemistry of Tetrahedral Structures," Gordon and Breach, New York, 1964.**)

The five relatively simple types of *tetrahedral structures* are illustrated in Fig. 10–11. The zinc blende structure (ZnS, *B*3) may be regarded as two interpenetrating face-centered lattices of the elements, with the corner of one located at the position $\frac{1}{4}\frac{1}{4}\frac{1}{4}$ of the other, as in the diamond structure. Figures 10–9 and 10–10 show the close similarity between ZnS and CaF_2 structures. In another representation the zinc blende structure can be illustrated by either a cubic or a hexagonal unit cell as shown in Fig. 10–12. It is then easily related to the wurtzite structure (ZnS, ZnO, *B*4), which has one kind of atom on c.p.h. positions and the other at intermediate points, corresponding to the tetrahedral voids, where each atom is surrounded symmetrically by four atoms of the other kind.[1] The hexagonal layers are stacked in one sequence in wurtzite, and in another in zinc blende (see Fig. 10–11), and stacking faults are frequent.[2]

The current interest in the semiconducting properties of materials has resulted in an intensive search for new compounds with wurtzite, zinc blende, and related structures. Many such compounds contain nonmetallic atoms as partners and show no obvious metallic properties. The boundary between quasi-metallic and nonmetallic behavior then becomes difficult to define. An example is the series Ge-GaAs-ZnSe-CuBr; their only similarity is that of related structure. All wurtzite and zinc blende structures, some-

[1] Coordinates are zinc at 000 and $\frac{1}{3}\frac{2}{3}\frac{1}{2}$, and oxygen at $00u$ and $\frac{1}{3}$, $\frac{2}{3}$, $u + \frac{1}{2}$, where u is approximately 3/8 and c/a is roughly 1.63.

[2] H. Jagodzinski, *Acta Cryst.*, vol. 2, p. 298, 1949.

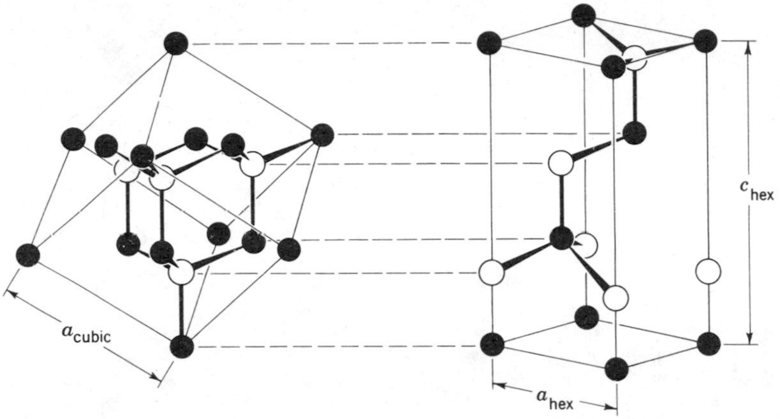

Fig. 10-12 Zinc blende structure with cubic and hexagonal unit cell. (After E. Parthé, "Crystal Chemistry of Tetrahedral Structures," Gordon and Breach, New York, 1964.)

times called the *adamantine* (related to diamond) *structures*, constitute a subgroup of the so-called tetrahedral structures. A few typical adamantine structures are listed below[1]:

Zinc blende structure (*B*3)				Wurtzite structure (*B*4)	
BeS	γ-AgI	ZnSe	GaSb	MgTe	β-CdS
AlP	BeSe	CuI	CdTe	CdSe	**β-ZnS**
GaP	AlAs	InSb	ZnTe	β-AgI	AlN
α-ZnS	GaAs	BeTe	CuBr		
α-CdS	CdSe	AlSb	GaSb		

Some generalizations are possible for the whole group of *normal valence compounds*. *Zintl's rule*[2] states that in saltlike ionic compounds only those elements that precede the noble gases by one to four places are able to become negative ions. Exceptions to the rule were found by Hellner and Laves[3] in alloys of In or Ga with Ni, Pd, Pt, Cu, Ag, and Au: many phases containing indium or gallium (five places before the noble gases) show definite tendencies for the indium and gallium to become negative ions in typically saltlike compounds (CsCl or the closely related Ni_2Al_3

[1] For a complete listing of tetrahedral structures, see E. Parthé, "Crystal Chemistry of Tetrahedral Structures," Gordon and Breach, New York, 1964.

[2] E. Zintl, J. Goubeau, and W. Dullenkopf, *Z. Physik. Chem.*, vol. A154, p. 1, 1931. E. Zintl and H. Kaiser, *Z. Anorg. Allgem. Chem.*, vol. 211, p. 113, 1933.

[3] E. Hellner and F. Laves, *Z. Naturforsch.*, vol. 2a, p. 177, 1947.

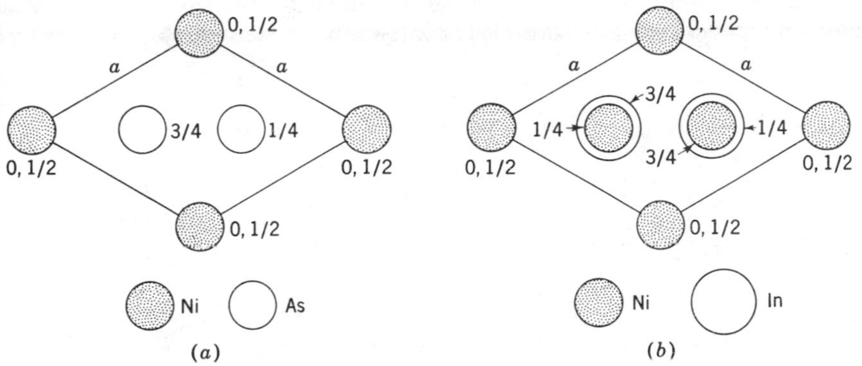

Fig. 10-13 Hexagonal structures of the $B8$ type projected on (0001) with the distance above the projection plane indicated in units of c. (a) NiAs, type $B8_1$; (b) Ni_2In, type $B8_2$, with all octahedral interstices between the Ni atoms at 000 and $00\frac{1}{2}$ filled.

type of compound), judging by abnormally short distances between unlike atoms in these phases. Thus, the rule should apparently be modified to include elements one to five places before a noble gas rather than one to four.

Hume-Rothery[1] has shown that two generalizations are possible regarding normal valence compounds: (1) There is a general tendency for all metals to form normal valence compounds with elements of groups IV*B*, V*B*, and VI*B*. (2) This tendency and the stability of the compound are greater the more electropositive the metal and the more electronegative the *B*-subgroup element.

The compounds with the tetrahedral structure have received much attention recently, since it is possible to derive correlations for these compounds among the chemical composition, the crystal structure, and the electronic structure.[2] A very large number of derivatives exist in this group of related phases since it is again possible in this case to form *defect tetrahedral structures* by omitting atoms from certain tetrahedral vertices, or to form *filled tetrahedral structures* by stuffing extra atoms into available voids.

The nickel arsenide structure ($B8_1$ and $B8_2$) The *NiAs-type structures* are frequently formed by the transition metals Cr, Mn, Fe, Co, Ni, (Cu), (Pd), and (Pt) alloyed with the metalloids S, Se, Te, As, Sb, Bi, and sometimes also Ge and Sn. The structure with full occupancy of octahedral interstitial holes is composed of alternating layers of metal and metalloid atoms, each layer being a c.p.h. (0001) plane as illustrated in Fig. 10-13. Many alloys have metallic conductivity, and there can be more than 50 atomic

[1] W. Hume-Rothery, "The Structure of Metals and Alloys," Institute of Metals, London, 1936.

[2] See, for example, Parthé, loc. cit.

percent of a metal present; for these reasons they are more metallic than ionic. The following are examples of known NiAs structures:

Nickel arsenide structure (*B*8)					
CrS	CoSe	FeTe	PtTe	NiSb	NiBi
CoS	FeSe	NiTe	AuSn	CrSb	MnBi
FeS	NiSe	CrTe	CuSn	MnSb	RhBi
NiS	CrSe	MnTe	PtSn	PdSb	PtBi
VS	TiSe	TiTe	NiSn	PtSb	InPb
TiS	VSe	VTe	CoSb	MnAs	PtPb
NbS	CoTe	PdTe	FeSb	**NiAs**	

Phases with this structure often have wide homogeneity ranges, and the range then may not include the *AB* composition. The Ni_2In-type ($B8_2$) structure is found, for example, in the following phases:

Co_2Ge	Cu_2In	Rh_3Sn	FeGeMn	CoNiSn
Ni_2Ge	**Ni_2In**	Pd_5Sb_3	CoGeMn	FeMnNi
Mn_2Sn	Fe_2Ge	Mn_3Sb_2	CoNiSb	GeMnNi

These characteristics indicate that there is much metallic character associated with many of the phases; in fact, they may be regarded in some instances as typical metallic phases obeying electron-concentration rules that will be discussed later. The metallic character increases as the metalloid element becomes less electronegative or when the percentage of the metallic elements increases.

Alloy phases with the NiAs-type structure can be considered as having an electron-atom ratio of 2:5. This corresponds to 10 electrons per unit cell (since the unit cell contains two ideal formula weights), a number that applies to normal and ideal structures, as well as to those in which atom sites are vacant. In computing this ratio, however, Raynor[1] assumes that the valence contribution per atom of the transition metals is not always zero, but depends upon the nature of the alloying partner and its relative amount: Ni contributes no electrons in NiSb, perhaps contributes one in NiSn, and appears to absorb one in NiS.

As mentioned above, phases with composition differing from *AB* may be considered to be of the defect interstitial type (excess metal atoms entering interstices). Thus, starting with NiTe, individual Ni atoms can be removed until the composition is $Ni_{0.5}Te$ without destroying the stability of the phase, and excess Co atoms can be added to CoSb with the excess going into the interstices. The NiAs structure present in the NiSb system extends between the 46.4 and 54.4 atomic percentages of nickel.

[1] G. V. Raynor, "Progress in Metal Physics," vol. 1, Interscience, New York, 1949.

Laves and Wallbaum[1] have found a continuous series of NiAs-type structures with varying c/a ratio as follows:

Structure	NiS	NiSe	NiAs	Ni_3Sb_2	Ni_3Sn_2	Ni_2Ge	Ni_2In
c/a	1.633	1.46	1.39	1.31	1.27	1.28	1.228

and with intermediate axial ratios obtainable from solid solutions between adjacent compounds in the list. At the left of this list is the NiS structure with $c/a = 1.633$ as required for the hexagonal close packing of spheres. Here the sulfur cations pack tightly together, and the nickel fills octahedral interstices between them. (All octahedral interstices are filled in NiAs, but extra Ni atoms can be accommodated in tetrahedral interstices.) The binding is chiefly ionic in character. The axial ratio decreases steadily along the series to a value $\sim \sqrt{3}/\sqrt{2} = 1.225$ that would correspond to a pseudocubic structure. Both octahedral and tetrahedral interstices are filled in Ni_2In. The metallic character increases from left to right; Ni_2In, at the extreme right, has a diffraction pattern and a structure remarkably near that of gamma-brass (see page 253); the formula Ni_9In_4 has the Hume-Rothery electron-atom ratio for γ phases if Ni is counted as univalent. There is thus a progressive alteration of structure through this series, which correlates with position of the metalloid in the Periodic Table and with the general metallic character of the metal-atom partners.

The close relationship between NiAs-type structures and the monocarbides with a simple hexagonal structure such as WC is well known. Both have a high space-filling parameter,[2] particularly if the partners have nearly the same atomic radii and the axial ratio is close to the ideal (1.633). Most of the observed related types of structure in this group involve a combination of a transition metal with a nonmetal or a semimetal. This has been shown diagrammatically by Jellinek[3] (Fig. 10–14).

Electron phases Electron phases are typically metallic and possess wide ranges of homogeneity; because of their dependence upon electron concentration, they are of special interest in the theory of alloy phases. It has long been noted that equilibrium diagrams for alloys of the "noble metals," Cu, Ag, and Au, with metals of the *B* subgroups show remarkable similarities. When the structures of the intermediate phases were determined by x-rays, largely by Westgren and Phragmén in Sweden, the similarities became even more impressive, for phases of the same structure occurred in many systems.

The system Cu-Zn is typical of these structurally analogous systems.

[1] F. Laves and H. J. Wallbaum, *Z. Angew. Mineral.*, vol. 4, p. 17, 1942.

[2] The space-filling parameter discussed by E. Parthé, *Z. Krist.*, vol. 115, p. 51, 1961, is defined as the ratio of the volume of the atoms in the unit cell to the volume of the unit cell. It may be taken as an approximate measure of the efficiency of the sphere-packing characteristic of a given structure (see also Chap. 13, p. 367).

[3] F. Jellinek, *Österreichische Chem. Ztg.*, vol. 60, p. 311, 1959.

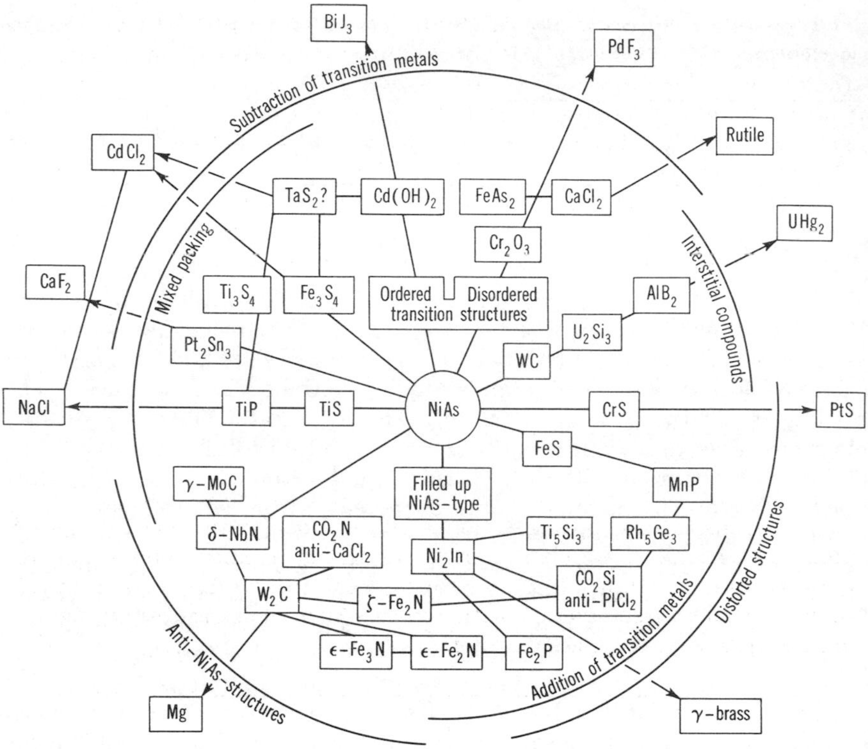

Fig. 10-14 The NiAs-type structure and its varieties. (**After F. Jellinek, Österreich. Chem. Ztg., vol. 60, p. 311, 1959.**)

As indicated on the diagram (page 159), there is a sequence of phases, all with wide solid solubility as follows:

α-terminal solid solution, Cu-rich (f.c.c.), maximum range of solid solubility 0 to 38 atomic percent Zn

β phase (disordered b.c.c.), maximum range of solid solubility 36 to 55 atomic percent Zn, or ordered (CsCl type, β′) below 454°C.

γ phase (complex-cubic, 52 atoms per cell), maximum range of solid solubility 57 to 70 atomic percent Zn

δ phase (complex-cubic, high-temperature phase with numerous voids), maximum range of solid solubility 73 to 76 atomic percent Zn

ε phase (c.p.h., axial ratio c/a ~1.58), maximum range of solid solubility 78 to 86 atomic percent Zn

η terminal solid solution (Zn-rich c.p.h., axial ratio c/a ~1.7 to 1.85), maximum range of solid solubility 97 to 100 atomic percent Zn

Similar sequences of structurally analogous phases are found in many systems. Apart from the γ phase, they are structurally simple and possess the structures of the typical metals. A given phase does not always occur at the same composition or in all systems. The ε phases, for example, are found only when Zn or Cd is alloyed with the noble metals, and they always contain less than 50 atomic percent of the noble metal "solvent."[1] In other

[1] T. B. Massalski and H. W. King, *Progr. Mater. Sci.*, vol. 10, p. 1, 1961.

systems the electron phases with the c.p.h. structure are the ζ phases, which follow the primary f.c.c. solid solutions and possess an axial ratio in the range of the ideal value ($c/a = \sqrt{8/3} = 1.633$).

Hume-Rothery[1] pointed out that some fundamental principles become apparent in a number of such related systems if the number of valence electrons per atom is computed for each phase. Systematic study of numerous other systems has indicated the existence of closely related structures at, or near, certain values of the electron concentration. This has led Bernal[2] to suggest that these phases may be conveniently called *electron compounds* since their occurrence appears to depend upon the ratio of electrons to atoms present in the structure; some authors refer to them also as Hume-Rothery compounds. Clearly, however, these structures are not compounds in a chemical sense and, since they possess wide ranges of solid solubility, the term *electron phases* seems more appropriate. Detailed accounts of the formation and properties of electron phases will be found in the literature.[3]

It is of historical interest that many theories of alloy phases have developed from the original suggestion of Hume-Rothery that certain electron-atom ratios tend to be characteristic of the β, γ, and ϵ phases in many related systems. Hume-Rothery proposed the following values of electron concentrations to correspond with the assumed chemical formulas for the phases present in the Cu-Zn systems:

Phase	β	γ	ϵ
Suggested chemical formula	CuZn	Cu_5Zn_8	$CuZn_3$
Electron/atom ratio (e/a)	3:2	21:13	7:4

In this and subsequent general schemes the valences were assumed to be the number of electrons in excess of the last completed shell, as follows:

Valence	Element
1	Cu, Ag, Au (group I)
2	Be, Mg, Zn, Cd, Hg (group II)
3	Ga, Al, In (group III)
4	Si, Ge, Sn, Pb (group IV)
5	P, As, Sb, Bi (group V)
0	Fe, Co, Ni, Ru, Rh, Pd, Pt, Ir, Os (group VIII)[1]

[1] There is basis for assuming different, or variable, valences for the transition elements Cr, Mn, Fe, Co, and Ni, as is discussed in later chapters.

[1] W. Hume-Rothery, *J. Inst. Metals*, vol. 35, pp. 295, 307, 1926.

[2] J. D. Bernal, *Ann. Rept. Progr. Chem.*, vol. 30, p. 387, 1933.

[3] W. Hume-Rothery and G. V. Raynor, "The Structure of Metals and Alloys," 4th ed., Monograph and Rept. Ser. no. 1, Institute of Metals, London, 1962. T. B. Massalski and H. W. King, *Progr. Mater. Sci.*, vol. 10, p. 1, 1961.

Table 10-3 Classification of alloy phases of the noble metals* (e/a range 1.0–2.0)

Structure	Phase type	Approximate e/a range	Copper system	Copper phases	Silver system	Silver phases	Gold system	Gold phases	Transition metals system	Transition metals phases
F.c.c.	α	1.00–1.42	noble metal primary solid solutions							
			Cu–Zn	α_1			Au–Zn	α_1		
							Au–Cd	α_1 α_2		
							Au–In	α_2		
B.c.c.	β	1.36–1.59	(Cu–Be)	β β'	(Ag–Li)	β'	(Au–Mg)	β'	Mn–Zn	β
			Cu–Zn	β β'	(Ag–Mg)	β'	Au–Zn	β'	Mn–Hg	β'
			(Cu–Al)	β	~Ag–Zn	β β'	Au–Cd	β'	Mn–Al	β'
			Cu–Ga	β	Ag–Cd	β β'	(Au–Al)	β	Fe–Al	β'
			Cu–In	β	(Ag–Al)	β	(Au–Mn)	β'	Co–Al	β'
			(Cu–Si)	β	Ag–In	β†			Ni–Zn	β'
			Cu–Sn	β					Ni–Al	β'
			(Cu–Pd)	β'					Ni–Ga	β'
									Ni–In	β'
Cubic	μ	1.40–1.54	(Cu–Si)	μ	(Ag–Al)	μ	(Au–Al)	μ	Co–Zn	μ
B.c.c.	γ	1.54–1.70	Cu–Zn	γ	(Ag–Li)	γ_{1-3}	Au–Zn	γ_{1-3}	Mn–Zn	γ_{1-3}
			Cu–Cd	γ	Ag–Zn	γ	Au–Cd	γ_{1-3} γ_0	Mn–In	γ_{1-3}
			Cu–Hg	γ_{1-3}	Ag–Cd	γ γ_0	Au–Ga	γ_{1-3}	Fe–Zn	γ_1
			(Cu–Al)	γ_3 γ_{1-3} γ_0	Ag–Hg	γ_{1-3}	Au–In	γ_{1-3} γ_0	Co–Zn	γ_{1-3}
			Cu–Ga	γ_3 γ_{1-3} γ_0	Ag–In	γ_{1-3} γ_0			Ni–Zn	γ_{1-3}
			Cu–In	γ_{1-3}					Ni–Cd	γ_{1-3}
			(Cu–Si)	γ_{1-3}					Ni–Ga	γ_{1-3}
			Cu–Sn	γ_0					Ni–In	γ_{1-3}
									Pd–Zn	γ_{1-3}
									Pt–Zn	γ_{1-3}
									Pt–Cd	γ_{1-3}

Cubic	δ	1.55–2.00	Cu–Zn	δ			Au–Zn	δ		
			Cu–In	δ_0						
			(Cu–Si)	δ_0						
			Cu–Ge	δ_0						
			Cu–Sn	δ_0						
			Cu–Sb	δ_0						
C.p.h.	ζ	1.32–1.83	Cu–Zn	ζ(cw)	Ag–Zn	ζ(cw) ζ°	Au–Cd	ζ ζ'	Mn–Zn	ζ
			Cu–Ga	ζ	Ag–Cd	ζ	Au–Hg	ζ		
			(Cu–Si)	ζ	Ag–Hg	ζ	Au–In	ζ ζ'		
			Cu–Ge	ζ	(Ag–Al)	ζ	Au–Sn	ζ		
			Cu–As	ζ	Ag–Ga	ζ ζ°				
			Cu–Sb	ζ	Ag–In	ζ' ζ' ζ°				
					Ag–Sn	ζ				
					Ag–As	ζ				
					Ag–Sb	ζ				
C.p.h.	ϵ	1.65–1.89	Cu–Zn	ϵ	Ag–Zn	ϵ	Au–Zn	ϵ	Li–Zn	ϵ
			Cu–Ge	ϵ'	Ag–Cd	ϵ	Au–Cd	ϵ ϵ'	Li–Cd	ϵ
			Cu–Sn	ϵ'	Ag–In	ϵ'				
			Cu–Sb	ϵ'	Ag–Sn	ϵ'				
					Ag–Sb	ϵ'				
C.p.h.	η	1.93–2.00	primary solid solutions based on Zn and Cd							

(cw) Phase induced by cold-working the β' phase.
* After T. B. Massalski and M. W. King, *Progr. Mater. Sci.*, vol. 10, p. 1, 1961. $\alpha = A1$; $\alpha' = L1_2$; α_1 = tetragonal (distorted) α'; $\alpha_2 = DO_{24}$; $\beta = A2$; $\beta' = B2$; $\mu = A13$. $\gamma = D8_2$; $\gamma_1 = D8_1$; $\gamma_3 = D8_3$; $\gamma_{1-3} = D8_{1-3}$; γ° = distorted $D8_{1-3}$; δ = related to $A2$ and $B8$; $\zeta, \epsilon, \eta = A3$; $\zeta' = DO_{19}$; ζ° = distorted ζ (ordered); ϵ' = ordered ϵ, usually DO_{19}.
† Phase field unusual.

The detailed list of the structures of phases formed by the noble metals with the B-subgroup elements (and occasionally with Be, Al, Si, and Mg) in the range of electron concentration between 1 and 2 is given in Table 10–3. From the last column of the table it may be seen that a few late transition elements, and Li, also occasionally form electron-phase structures when alloyed with the B-subgroup elements. Most of the phases listed can be regarded as electron phases. In addition to the more familiar α, β, γ, ζ, and ϵ phases, there are several variants, most of which possess some degree of order. Furthermore, not only the presence or absence of order, but also its precise form, is often related to the valence of the solute element. The

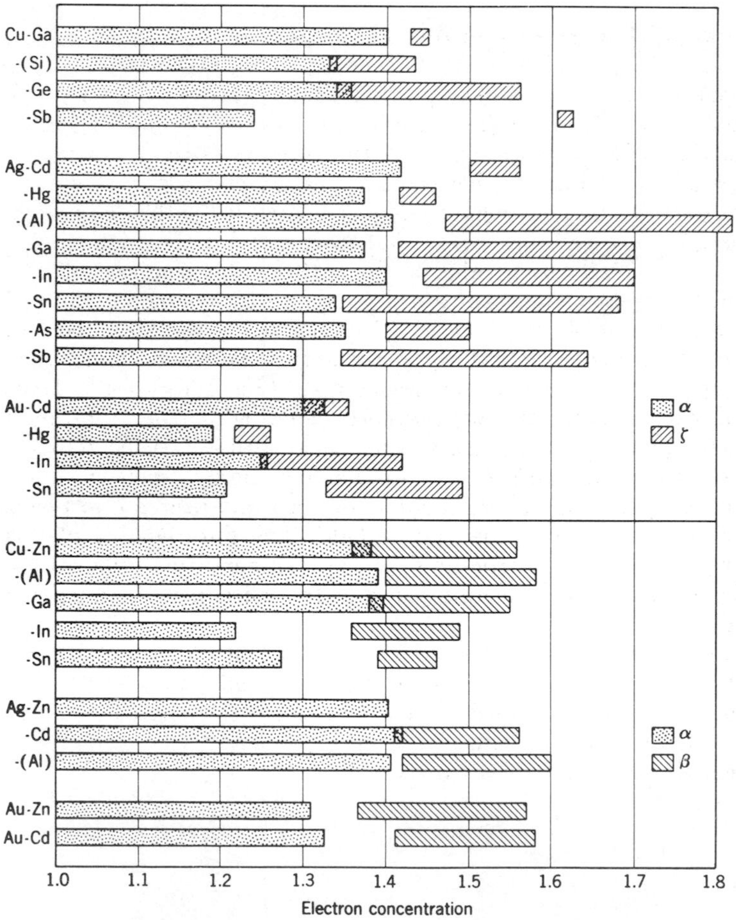

Fig. 10-15 Extent of maximum solid solubility of the primary phase, α, and of the intermediate phases ζ and β in alloys based on the noble metals Cu, Ag, and Au with the B-subgroup elements of the Periodic Table.

table also indicates that the most typical electron phases extend over wide ranges of e/a. As they are now known, the maximum ranges in terms of e/a, for the cases when the primary solid solutions (α) are followed by the c.p.h. ζ phases or the b.c.c. β phases, are plotted in Fig. 10–15. The assignment of any particular value of e/a as typical for a chemical formula for any one phase does not appear very significant and would lead to confusion. However, the general ranges of stability and their relations to e/a are clearly significant.

The systematic occurrence of the electron phases has become the basis for an electronic theory of their stability. This theory will be reviewed in Chap. 13 after the concepts of the Brillouin zones and Fermi surfaces are first discussed in Chap. 12. Here we shall deal only with certain characteristic properties of the electron phases.

Electron phases with cubic symmetry To this group belong the following structures listed in Table 10–3: β, β', μ, γ, γ_1, γ_3, γ_{1-3}, δ, and δ_0. The disordered β phases are stable only at high temperatures. They usually decompose eutectoidally unless the phase becomes ordered as in the Cu-Zn system. In all cases the range of homogeneity decreases as the temperature is lowered, causing the phase fields to have a characteristic V shape, as illustrated in Fig. 10–16. This behavior has been coupled with the unusual elastic properties of these phases, as demonstrated by the high elastic anisotropy,[1] the unusually high value of the $(110)[\bar{1}10]$ shear compliance modulus, and the positive temperature dependence of Young's modulus.[2] Zener[3] has noted that a shear in the $[\bar{1}10]$ direction in the (110) plane in a b.c.c. lattice of hard spheres that are in contact should leave the distance of the nearest neighbors practically unchanged. This allows for a large amplitude of lattice vibrations with a correspondingly high vibrational entropy which causes the free energy to decrease with increasing temperature because of the large ΔTS term in the free-energy equation. Conversely, a decrease in temperature should cause a rapid decrease in phase stability resulting in the V-shaped phase field. The disordered β phases tend to transform to other structures at low temperatures unless they become ordered,[4] and even the ordered structures frequently transform martensitically.[4,5]

The structures of the γ phases are not all identical but they are strikingly similar, and all have large unit cells. They are usually ordered, certain related atomic sites being occupied by the solute and others by the solvent. The precise form of order varies with the ratio of solvent to solute atoms, and three general types may be discussed.[6]

[1] D. Lazarus, *Phys. Rev.*, vol. 74, p. 1726, 1948; vol. 76, p. 547, 1949. S. Zirinsky, *Acta Met.*, vol. 4, p. 164, 1956.

[2] Lazurus, loc. cit. C. Zener, *Phys. Rev.*, vol. 71, p. 846, 1947.

[3] Zener, loc. cit.

[4] For details see T. B. Massalski and H. W. King, *Progr. Mater. Sci.*, vol. 10, p. 1, 1961.

[5] H. Warlimont in Iron and Steel Institute Special Report No. 93, London, 1965.

[6] Massalski and King, loc. cit.

In many of the γ phases the precise form of order has not been determined and may be a mixture of the three general types ($D8_3$, $D8_2$, and $D8_1$). These phases are referred to as γ_{1-3} in Table 10–3. The unit cell of gamma-brass in Cu-Zn contains 52 atoms and may be considered as made up of 27 unit cells of beta-brass (which would amount to 54 atoms) with 2 atoms removed and the rest shifted somewhat in position. This is shown in Fig. 10–17. The Cu-Zn γ phase has a b.c.c. space lattice; in the Cu-Al system it has a simple b.c.c. space lattice with 49 to 52 atoms in the unit cell. In

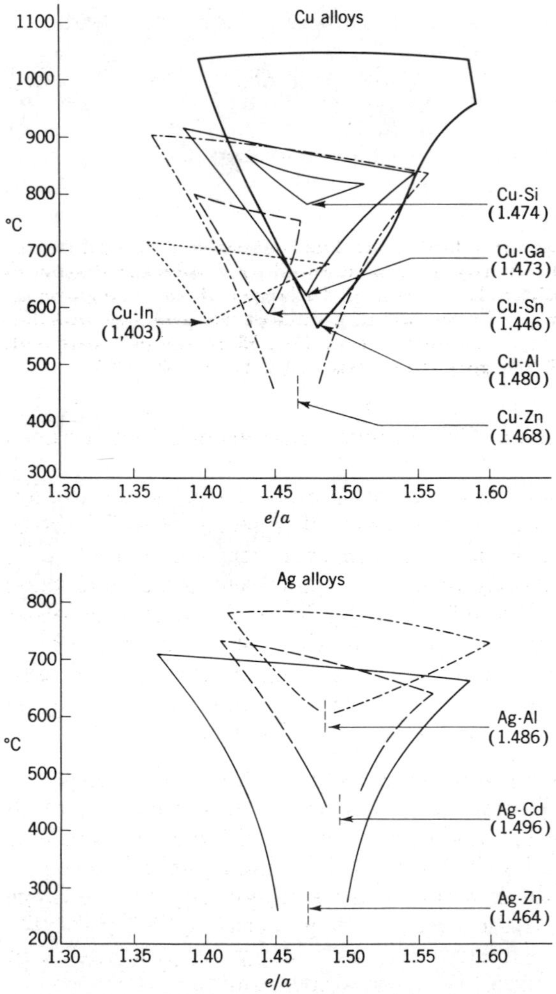

Fig. 10-16 The typical V-shaped phase fields of the disordered β phases: (*a*) in copper alloys, (*b*) in silver alloys. (T. B. Massalski and H. W. King, Progr. Mater. Sci., vol. 10, p. 1, 1961.)

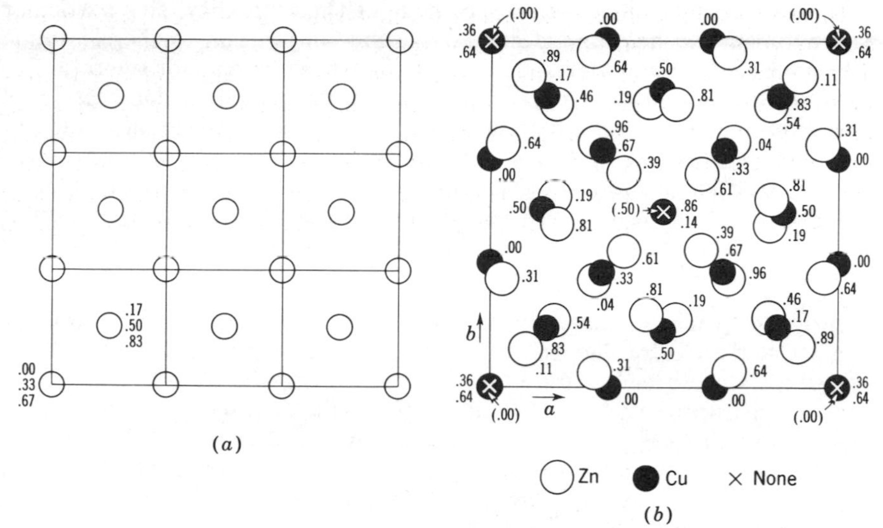

Fig. 10-17 The structure of gamma-brass: (*a*) The comparison structure consisting of 27 b.c.c. cells without any atoms removed or displaced; distances above projection plane are indicated, in terms of the large cell. (*b*) The gamma-brass structure derived from (*a*) by removing atoms at 000 and $\frac{1}{2}\frac{1}{2}\frac{1}{2}$ and displacing others. Space group $I\bar{4}3m$, structure type $D8_2$, 52 atoms per unit cell. (After A. J. Bradley and C. H. Gregory, Phil. Mag., vol. 12, p. 143, 1931.)

some systems, such as Cu-Al and Cu-Ga, numerous vacant sites have been observed in γ structures.

The μ phases, which crystallize in the cubic β-*Mn* (*A*13) structure, are found near room temperature in a number of systems (Cu-Si, Ag-Al, Au-Al, and Co-Zn) in the region of electron concentration characteristic of the β phases (which are usually present at higher temperatures in these systems). They have been studied relatively little. A number of other phases with general cubic symmetry resemble the γ structure. Some have structure cells in which the number of atoms exceeds 100 or even 200; others are characterized by the presence of ordered or randomly distributed lattice vacancies.[1] In Table 10–3 such phases are denoted by δ or $\delta°$.

Electron phases with hexagonal symmetry The c.p.h. phases, denoted usually by the Greek letters ζ, ϵ, and η, are the most numerous of all alloy phases of the noble metals. Each group occurs within a characteristic electron-concentration range (see Table 10–3) but taken together the c.p.h. structures may occur anywhere within the electron-concentration range between 1.32 and 2.00, except for a narrow range between 1.87 and 1.93.[2] Within the total range the ζ phases are particularly versatile; although many ζ phases include the value 1.48 (considered significant for the b.c.c.

[1] K. Schubert and E. Wall, *Z. Metallk.*, vol. 40, p. 383, 1949.

[2] T. B. Massalski and H. W. King, *Acta Met.*, vol. 10, p. 1171, 1962.

β phases; see page 348) within the range of their stability, this particular value appears to have no special significance in relation to the electronic structure of the c.p.h. alloys. The same appears to be true for the ratio 7:4 (e/a) in the case of the ϵ phases. In fact, for the Cu-Zn epsilon-brass, this particular value of the electron concentration falls outside the range of homogeneity.[1]

The behavior of the lattice spacings and of the axial ratio with changes of composition in the c.p.h. electron phases constitutes a very particular chapter in the theory of these alloy phases and most certainly holds a clue to their stability (see Chap. 13, page 360).

The Laves phases Many intermediate phases possess one of the three related structures of the general formula AB_2, exemplified by the three prototype structures based on magnesium: $MgCu_2$ (Strukturbericht *C*15), $MgZn_2$ (Strukturbericht *C*14), and $MgNi_2$ (Strukturbericht *C*36). Much of the original work on these structures is due to Laves, Witte, and their associates.[2] For this reason the whole group of these related phases are frequently called the *Laves phases*. A few representative binary structures are shown in the following list:

$MgCu_2$ type (cubic) (*C*15)	$MgZn_2$ type (hexagonal) (*C*14)	$MgNi_2$ type (hexagonal) (*C*36)
$AgBe_2$	$CaMg_2$	$NbZn_2$
$TiBe_2$	$ZrRe_2$	$ScFe_2$
$NaAu_2$	KNa_2	$ThMg_2$
LaMg	$TaFe_2$	$HfCr_2$
$BiAu_2$	$NbMn_2$	β-Co_2Ti
KBi_2	UNi_2	UPt_2

The partners that form Laves phases may come from distant parts of the Periodic Table or may be unusually close (viz., KNa_2). The same metal may be an *A* partner in one compound and a *B* partner in another (viz., $MgCu_2$ and $LaMg_2$) and both transition and nontransition elements may be involved. Of the 223 binary Laves phases reviewed recently by Nevitt,[3]

[1] T. B. Massalski and H. W. King, *Progr. Mater. Sci.*, vol. 10, p. 1, 1961.

[2] F. Laves and H. Witte, *Metallwirtschaft*, vol. 14, p. 645, 1935; vol. 15, p. 840, 1936. F. Laves in American Society for Metals Symposium, "Theory of Alloy Phases," p. 124, Cleveland, Ohio, 1956.

[3] M. V. Nevitt in P. A. Beck (ed.), "Electronic Structure and Alloy Chemistry of the Transition Elements," Wiley, New York, 1963. Laves phases formed between alkaline earths and Rh, Pd, Pt, and Ir are reviewed by E. A. Wood and V. B. Compton, *Acta Cryst.*, vol. 11, p. 429, 1958. Laves phases formed between rare earth, Hf and Rh, Pd, and Pt and Ir are reviewed by V. B. Compton and B. T. Matthias, *Acta Cryst.*, vol. 12, p. 651, 1959. Atomic volume in Laves phases is discussed by P. S. Rudman, *Trans. AIME*, vol. 233, p. 872, 1965.

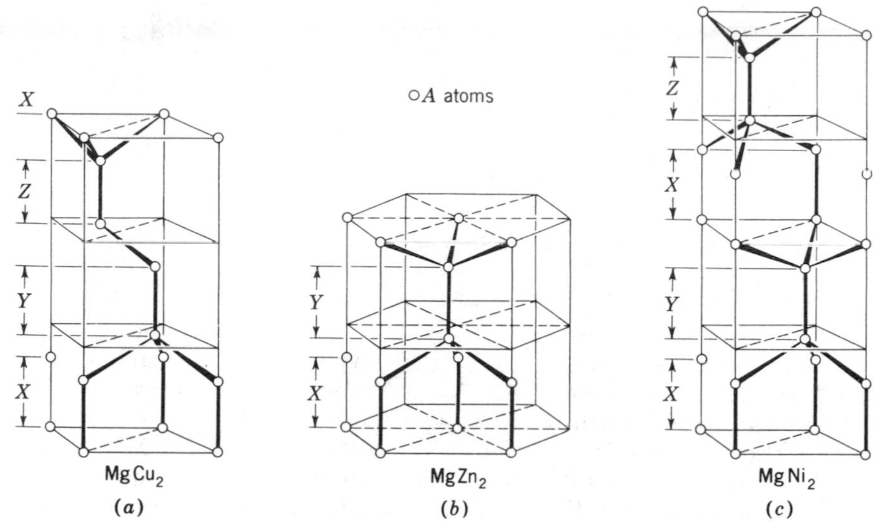

Fig. 10-18 Distribution of *A* atoms and stacking of double layers in Laves phases.

210 have a transition metal for at least one of the components, 152 have the $MgCu_2$-type structure, 67 the hexagonal $MgZn_2$-type structure, and only a few the hexagonal $MgNi_2$-type structure.

The above structures involve a special partnership between larger (*A*) and smaller (*B*) atoms and they have certain geometrical features in common that can be best illustrated diagrammatically.[1] The lattices of the larger *A* atoms are constructed from double layers with hexagonal network, with each *A* atom of the upper layer directly above one in the lower layer. Consider first the Mg atoms which represent the *A* atoms in the three prototype structures. For the $MgCu_2$ structure a portion of each hexagonal doubler layer is illustrated in Fig. 10–18*a*, and the similarity to the tetrahedral structures illustrated in Fig. 10–11 is immediately apparent. For convenience the double layers are labeled *X*, *Y*, and *Z*. The first double layer is represented by the eight atoms in *X* position and the second double layer is represented by the two atoms in *Y* position; it is identical with the first but is shifted laterally (just as is the second layer in the c.p.h. structure). The third layer is again displaced laterally. The resulting sequence may be described as *XYZXYZ*. . . . Figure 10–18*b* and *c* shows that in the same description the $MgZn_2$ and $MgNi_2$ structures will be represented by the *XYXYXY* . . . and *XYXZXYXZ* . . . stacking of the larger Mg atoms, respectively.

The smaller atoms of the other metals (the *B* atoms) are grouped around the Mg atoms in tetrahedral schemes that differ somewhat in the three types of structure. This is illustrated in Fig. 10–19. In the $MgCu_2$ structure the Cu atoms lie at the corners of the tetrahedra which are in contact and arranged into a network as shown in Fig. 10–19*a*. The larger Mg atoms lie in the holes between the tetrahedra formed by the Cu atoms. The pattern of the tetrahedra in the $MgCu_2$ structure is such that the *A* atoms in this description may be shown to lie on a cubic lattice shown in Fig. 10–20. The hexagonal arrangement of the double layers would be shown by placing the cube diagonal vertically

[1] R. L. Berry and G. V. Raynor, *Acta Cryst.*, vol. 6, p. 178, 1953. See also W. Hume-Rothery and G. V. Raynor, "The Structure of Metals and Alloys," 4th ed., Monograph and Rept. Sec. no. 1, Institute of Metals, London, 1962.

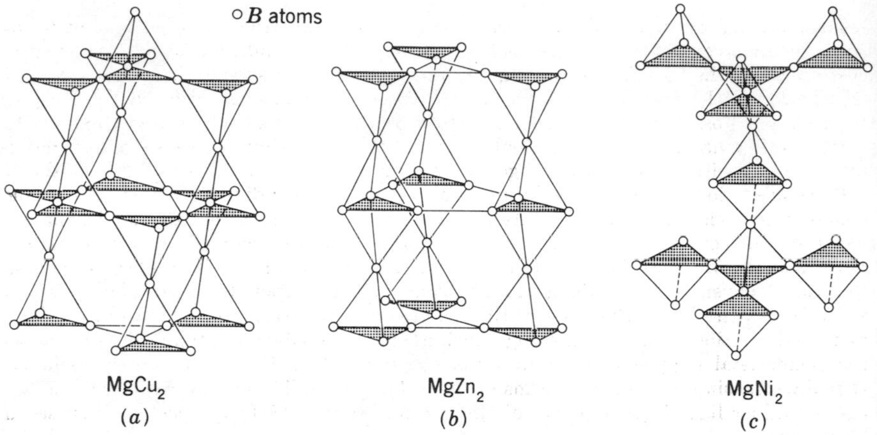

Fig. 10-19 Distribution of *B* atoms and stacking of tetrahedra in Laves phases.

in analogy to the zinc blende structure illustrated earlier in Fig. 10–12. In the $MgZn_2$-type of structure the smaller *B* atoms (or Zn) are again arranged into tetrahedra, but these are now joined alternately base-to-base and corner-to-corner, as shown in Fig. 10–19*b*. In this arrangement the *A* atoms, fitted between the tetrahedra, now occupy a wurtzite-type lattice. The arrangement of tetrahedra in the $MgNi_2$ structure represents a mixture of the other two arrangements (Fig. 10–19*c*), and the *A* atoms can be described as being located on a lattice corresponding to a mixture of diamond-cubic and wurtzite-type structures.

In the cubic arrangement shown in Fig. 10–20, the distance between the nearest *A* atoms is $a\sqrt{3}/4$, while the distance between the *B* atoms is $a\sqrt{2}/4$, where *a* is the side of the cubic cell. If, for maximum filling of space, the *A* atoms are made to contact one another and the *B* atoms are also made to contact one another, the ratio of the atomic radii that permits this is $r_A/r_B = \sqrt{3/2} = 1.225$. It is believed that one of the main

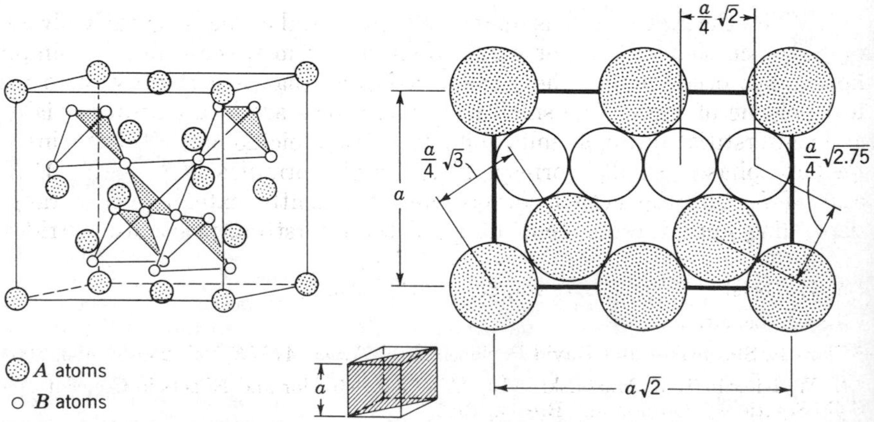

Fig. 10-20 Distribution of *A* and *B* atoms in the cubic cell of $MgCu_2$ structure: left, perspective view; right, schematic arrangement in the (110) plane; large circles *A* atoms, small circles *B* atoms. (After E. R. Schultze, Z. Elektrochem., vol. 45, p. 849, 1939.)

reasons for the existence of the Laves phases is of geometrical origin—that of filling space in a convenient way. Nevertheless, the ratios of the radii do not always seem to correspond to the most ideal packing and in many of the observed phases they deviate from 1.225 within a range of 1.05 to 1.68. Of the 164 phases surveyed by Dwight[1] the majority possess ratios between 1.1 and 1.4, but 26 phases fall outside these limits. On the other hand, many AB_2 combinations between various elements whose radius ratios lie well within the typical range do not form a Laves phase. For example, of the 45 AB_2 combinations between scandium-group elements (including the lanthanides) and copper-group elements, none crystallize in a Laves phase, although the lanthanide elements are very prolific formers of these phases.[2] Hence, geometrical considerations alone appear to be insufficient criteria for existence of the phases. It is clear, of course, that any analysis of space filling in alloy phases depends upon the definition of atomic radii. It appears likely that, where the atomic diameters seem to differ from the ideal ratio, polarization occurs on alloying which modifies the effective atomic sizes to satisfy the geometrical requirements. Some interpretations along these lines involve schemes that describe sizes of atoms in terms of two arbitrary radii,[3] one for major ligands and one for minor ligands; or in terms of their atomic volume.[4,5] This aspect is discussed in Chap. 13.

Interstitial phases and compounds The alloying of transition metals with H, B, C, N, and Si frequently produces compounds that are metallic, have high melting points, and are extremely hard. Dispersed in steel, they harden it; cemented together, they form high-speed, long-life cutting tools.

Gunnar Hägg has been largely responsible for systematizing many of such compounds (referred to as Hägg compounds by some authors), appropriately called *interstitial compounds*.[6] He has shown that they may be classified according to the relative sizes of the transition metal and metalloid atoms, i.e., the radius ratio R_X/R_M, where R_X is the radius of the small nonmetallic atom and R_M is the atomic radius of the transition metal. When R_X/R_M is less than 0.59, the structures are simple; when the ratio is more than 0.59, they are complex, except for some borides in which no B–B bonds occur.[7]

1. When the radius ratio is under 0.59, the metal atoms are nearly always on a lattice that is f.c.c. or c.p.h., or in a few instances b.c.c. or simple hexagonal; occasionally, they are on a lattice that is a slightly distorted form of one of these. The small metalloid atoms are located interstitially, as in interstitial terminal solid solutions. The stoichiometries of the intermediate phases usually correspond to simple formulas MX, M_2X, M_4X, and MX_2, but the actual compositions frequently extend over a range depending upon the degree of filling of the interstitial positions. Nitrides,

[1] A. E. Dwight, *Trans. ASM,* vol. 53, p. 477, 1960.

[2] Nevitt, loc. cit.

[3] Clara B. Shoemaker and David P. Shoemaker, *Trans. AIME,* vol. 230, p. 486, 1964.

[4] H. W. King in T. B. Massalski (ed.), "Alloying Behavior and Effects in Concentrated Solid Solutions," Gordon and Breach, 1965.

[5] P. S. Rudman, *Trans. AIME,* vol. 233, p. 872, 1965.

[6] G. Hägg, *Z. Physik. Chem.,* vol. B12, p. 33, 1931; vol. B11, p. 433, 1930. A. Westgren, *J. Franklin Inst.,* vol. 212, p. 577, 1931.

[7] R. Kiessling, *Acta Chem. Scand.,* vol. 4, p. 209, 1950.

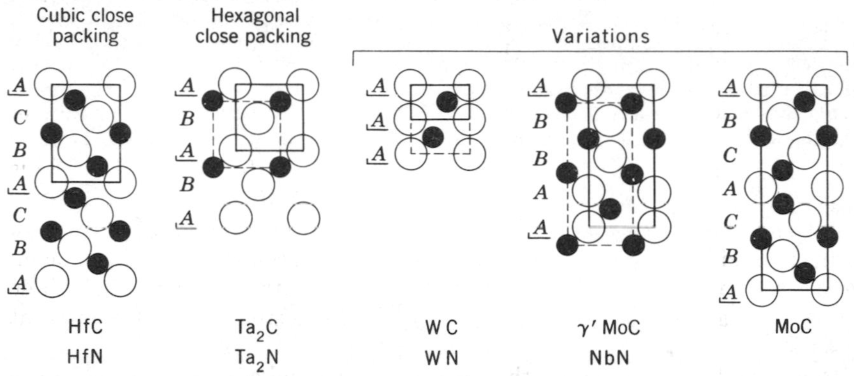

Fig. 10-21 Stacking of close-packed layers in interstitial Hägg phases. (After E. Parthé, Österreich. Chem. Ztg., vol. 56, p. 153, 1955.)

carbides, and hydrides with the formula *MX* are generally cubic with the metal atoms at f.c.c. positions and with the metalloids at interstitial positions in a structure of the NaCl type or the zinc blende type (positions indicated by open circles in Fig. 10–9). Examples are the nitrides ZrN, ScN, TiN, VN, and CrN; the carbides ZrC, TiC, TaC, and VC; and the hydrides ZrH and TiH. An exception is TaH, which is based on a b.c.c. lattice.

The compounds M_2X generally have the metal atoms arranged in hexagonal close packing[1] (Fe_2N, Cr_2N, Mn_2N, Nb_2N, Ta_2N, V_2N, Ni_3N, W_2C, Mo_2C, Ta_2C, V_2C, and Nb_2C); but sometimes they are cubic (W_2N and Mo_2N). The stacking of the metallic layers in many carbides and nitrides is illustrated in Fig. 10–21. In addition to the cubic and hexagonal close packing, a number of variants are possible that may be schematically represented as *AAA . . .*, *AABB . . .*, or *ABCACB . . .*[2]

In the system Ti-H the compound TiH apparently has the zinc blende structure, and TiH_2 has the fluorite (CaF_2) structure; the difference between these is merely that half the interstices are filled in the former, and all are filled in the latter (see Fig. 10–10). Neutron diffraction can disclose the positions of the hydrogen atoms in hydrides, so that a complete structure determination may be accomplished. This has been done, for example, with ThH_2, ThD_2, and ZrD_2, each of which has been found to be b.c.t. with atoms arranged in a distorted fluorite (CaF_2) structure.[3]

[1] Carbides and nitrides are reviewed by H. J. Goldschmidt, *J. Iron Steel Inst.*, vol. 160, p. 345, 1948, and by K. W. Andrews and H. Hughes, *J. Iron Steel Inst.*, vol. 193, p. 304, 1959. For listing of many other compounds, see: P. Schwarzkopf and R. Kieffer, "Refractory Hard Metals," Macmillan, New York, 1953. R. Kieffer and F. Benesovsky, "Hartstoffe," Springer-Verlag, Wien, 1963.

[2] H. Nowotny et al., *Monatsh. Chem.*, vol. 85, p. 255, 1954. E. Parthé, *Österreichische Chem. Ztg.*, vol. 56, p. 153, 1955.

[3] R. E. Rundle, C. G. Shull, and E. O. Wollan, *Acta Cryst.*, vol. 5, p. 22, 1952.

2. Large-radius ratios ($R_X/R_M > 0.59$) nearly always go with complex crystal structures. In this class are the carbides of Cr, Mn, Fe, Co, and Ni ($R_X/R_M = 0.60 - 0.61$) and the borides of Fe and N.

In dicarbides the C_2 groups may enter as parallel units into the largest interstices of the metal lattice instead of entering as individual atoms.[1] This results in tetragonal rather than cubic cells for the isomorphous phases CaC_2, SrC_2, BaC_2, LaC_2, CeC_2, PrC_2, and NdC_2, and the structures are a slightly deformed type of NaCl structure. The compounds of carbon, nitrogen, and hydrogen with the transition metals are more metallic in character than the similar compounds with nontransition metals. The carbides of calcium, strontium, and barium, for example, are transparent crystals.

Rundle, in his discussion of the metallic interstitial compounds of composition MX, where X is carbon, nitrogen, or oxygen, pointed out that nearly all of these have the NaCl structure, no matter what the structure or radius of the metal M.[2] Monocarbides, mononitrides, and monoxides are formed only with the A-subgroup metals of groups III, IV, V, and VI.[3] The very high melting points, the brittleness, the conductivity, and the interatomic distances in these compounds led Rundle to propose that the bonds between metal and nonmetal atoms are strong, directional (octahedral), and covalent, and are formed by electrons resonating between the different positions in the manner proposed by Pauling,[4] stealing valence electrons from the weaker metal–metal bonds.

Nevertheless, the metallic character of some interstitial phases, such as TiC, is well established,[5] and some, such as NbN, ZrN, ZrB, MoN, Mo_2N, and W_2C, show superconducting properties.[6]

Many of the interstitial phases show extended mutual solid solubilities and frequently form a continuous series of solid solutions with one another. The carbides of V, Ti, Nb, Ta, and Zr show complete solid solubility in almost all binary combinations,[7] as do nitrides of V, Ti, Nb, and Zr except in ZrN-VN.[8] A number of carbide–nitride quarternary complete solubilities have also been reported.[9]

[1] M. von Stackelberg, *Z. Physik. Chem.*, vol. B9, p. 437, 1930.

[2] R. E. Rundle, *Acta Cryst.*, vol. 1, p. 180, 1948. Exceptions occur, or possibly occur, with group VI metals, Cr and Mo.

[3] The monocarbide of technetium has been reported to be f.c.c. with $a = 3.928$ A, density = 11.5 g per cm³. See W. Trzebiatowski and J. Rudzinski, *Z. Chem.*, vol. 2, p. 158, 1962.

[4] L. Pauling, *Proc. Roy. Soc. (London)*, vol. A196, p. 343, 1949.

[5] L. E. Hollander, Jr., *J. Appl. Phys.*, vol. 32, p. 996, 1961.

[6] J. H. Wernick in R. W. Cahn (ed.), "Advanced Physical Metallurgy," North Holland Publishing, Amsterdam, 1965.

[7] J. T. Norton and A. L. Mowry, *Trans. AIME*, vol. 185, p. 133, 1949.

[8] P. Duwez and F. Odell, *J. Electrochem. Soc.*, vol. 97, p. 299, 1950.

[9] Ibid. See also R. Kieffer and F. Benesovsky, "Hartstoffe," Springer-Verlag, Wien, 1963.

Structure of phases formed by the transition elements, rare earths, etc. Alloy phases that involve the transition elements, the rare earths, the lanthanides,[1] and the actinides are extremely numerous and are often considered as separate groups.[2]

Alloys and compounds of the transition elements are of great importance because of their beneficial or detrimental influence upon the materials that involve these elements. Compounds of rare earths are often used as neutron absorbers, hydrogen-moderator carriers, diluents of nuclear fuels, and structural materials. Among the alloys of the transition elements, two subdivisions are possible:

1. Phases with more or less fixed stoichiometries at certain simple atomic ratios, such as A_3B, AB, etc.
2. Phases known as σ, μ, χ, δ, P, R, and E that resemble the electron phases of the noble metals in that they often have wide ranges of homogeneity, with compositions shifting in fairly regular manner from system to system. An example of this subdivision is illustrated in Fig. 10–22 for the binary phases of Sc (including lanthanides), Ti, V, and Cr groups (the A elements) with elements of the Mn, Fe, Co, Ni and Cu groups (the B elements).[3] From left to right the following structures with restricted solubility at simple stoichiometric ratios are observed: at A_3B the Cr_3Si type ($A15$); at A_2B the Ti_2Ni type or "η-carbide" ($E9_3$), the $MoSi_2$ type ($C11_b$), and the $CuAl_2$ type ($C16$); at AB the CsCl-type structure ($B2$); at AB_2 the Laves phases ($C14$, $C15$ and $C36$) and another group of $MoSi_2$-type phases; at AB_3 the various close-packed ordered structures with general coordination number 12; and finally, at AB_5 the UNi_5-type structures ($C15_b$) and $CaCu_5$-type structures ($D2_d$).

Phases with fixed stoichiometry Only a few of the structures mentioned above have been studied in detail. Of some interest is the cubic Cr_3Si-type structure ($A15$) that occurs in systems where the A elements

[1] Lanthanides (the elements from lanthanum to lutetium) and also scandium and yttrium are sometimes included in rare earths.

[2] More recent reviews on the structures formed by the transition elements are due to W. Hume-Rothery and B. R. Coles, *Advan. Phys.*, vol. 3, p. 149, 1954. P. Duwez in American Society for Metals Symposium, "Theory of Alloy Phases," Cleveland, Ohio, 1956. M. V. Nevitt in American Institute of Mining, Metallurgical, and Petroleum Engineers Symposium, P. A. Beck (ed.), "Electronic Structure and Alloy Chemistry of the Transition Elements," Wiley, 1963. (Many contributions in this field are due to P. A. Beck and his associates, and numerous reviews and listings of studies will be found in papers referred to in the article by Nevitt.) Rare earth intermetallic compounds are reviewed by O. D. McMasters and K. A. Gschneidner, Jr., *Nucl. Met.*, vol. 10, p. 93, 1964. K. A. Gschneidner, Jr., "Rare Earth Alloys," Van Nostrand, Princeton, N. J., 1961. A large literature now exists on the structures and alloying behavior of nuclear materials such as plutonium and other actinides, but this is considered to be beyond the scope of this book. The reader may consult: W. D. Wilkinson (ed.), "Plutonium and its Alloys," Interscience, New York, 1960. Various papers in the Nuclear Metallurgy series published by the American Institute of Mining, Metallurgical, and Petroleum Engineers.

[3] The incorporation of lanthanides into the scandium group is appropriate only to the specialized consideration of the crystal chemical behavior. The magnetic behavior of Sc is certainly not that of a rare earth; M. V. Nevitt, private communication.

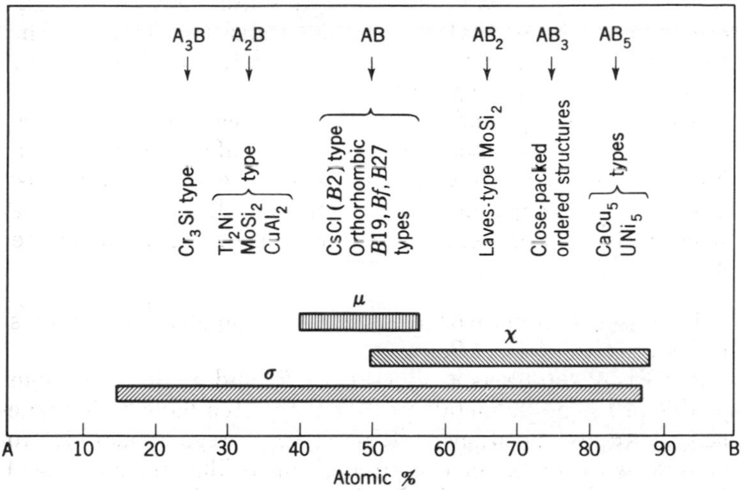

Fig. 10-22 Occurrence and distribution of intermediate phases with fixed and variable compositions in binary systems of Sc-, Ti-, V-, or Cr-group elements with Mn-, Fe-, Co-, Ni-, and Cu-group elements. (After M. V. Nevitt, in P. A. Beck (ed.), "Electronic Structure and Alloy Chemistry of the Transition Elements," Wiley, 1963.)

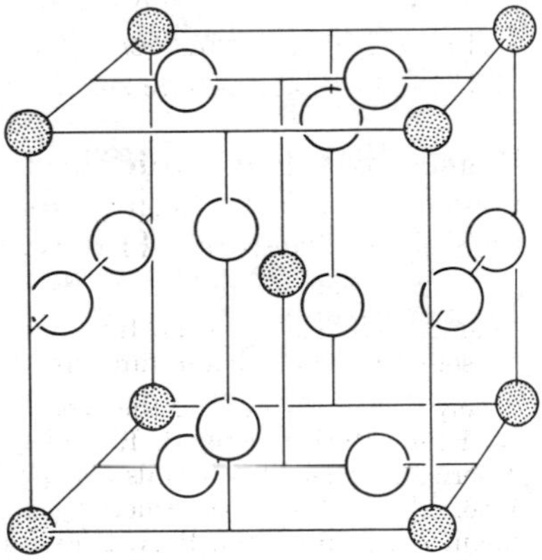

Fig. 10-23 The structure of Cr_3Si. (*A*15 type) (*a*) chains of *A* (chromium) atoms in the ⟨100⟩ directions (open circles); (*b*) cubic arrangement of the *B* (silicon) atoms (shaded circles).

come from the Ti, V, or Cr group and the *B* elements can be from the Mn, Fe, Co, Ni, Cu, Al, Si, or P group. It was commonly accepted that W forms this structure at high temperatures (the so-called β phase, or beta-wolfram structure) but it is now known that it is found only with impure W (particularly with oxygen as an impurity).[1]

The atomic arrangement in the Cr_3Si structure is shown in Fig. 10-23. *B* atoms (silicon) are in b.c.c. positions and there are no *B*–*B* nearest

[1] G. Hägg and N. Schönberg, *Acta Cryst.*, vol. 7, p. 351, 1954. T. Millner, *J. Inorg. Chem. USSR*, vol. 3, p. 946, 1958.

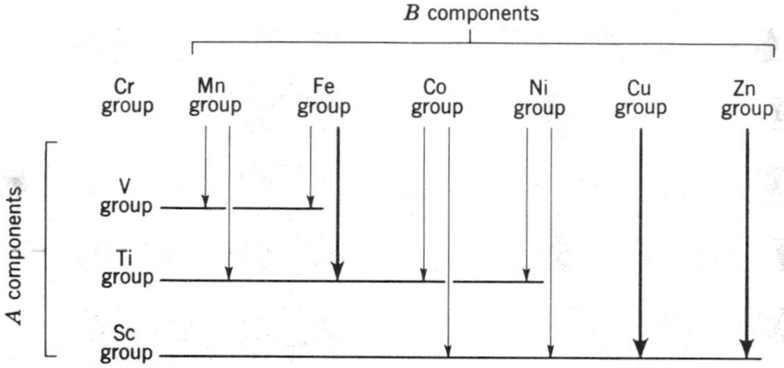

Fig. 10-24 Diagrammatic illustration of the occurrence and stability of CsCl-type phases formed between transition elements. Thick lines indicate increasing strength of bonding and the resulting large number of phases. (After M. V. Nevitt in P. A. Beck (ed.), "Electronic Structure and Alloy Chemistry of Transition Elements," Wiley, New York, 1963.)

neighbors. Each A atom (chromium) has 14 neighbors, but not all are equidistant, so that the coordination shell is not spherical. Within the structure the A atoms are arranged along criss-crossing (orthogonal) chains in $\langle 100 \rangle$ directions and strong covalent bonding is expected to exist in the chains.[1] The coordination number of each A atom may be considered to be 2, 6, or 14 depending upon the number of neighboring atoms that are considered.

Some interesting size effects exist in the Cr_3Si-type structures. If one assumes that A and B atoms are in contact and that the A–B distance is d_{AB}, the lattice parameter is given by $a = 4/d_{AB}/(\sqrt{5}) = 1.79(r_A + r_B)$, where r_A and r_B are individual atomic radii corresponding to distances derived for close packing in pure elements. With respect to the A–A distances, the r_A values are 8 to 14 percent shorter than the values for A–B distances,[2] and hence the A atoms are effectively "egg-shaped" rather than spherical. Several schemes of atomic radii have been proposed to account for this effect.

Among the Cr_3Si-type structures are some striking examples of superconductors with high T_c temperatures (see page 335), in the region of 18°K. The large number of CsCl-type ordered structures that occur between transition elements are characteristically stoichiometric, which suggests the dominance of electrochemical interaction. It is rather interesting, however, that in many cases a decreasing separation in the Periodic Table of the component elements appears to produce an increase in bond strength.[3]

[1] F. Laves, in "Theory of Alloy Phases," American Society for Metals, Cleveland, Ohio, 1956.

[2] Nevitt, loc. cit.

[3] A. E. Dwight, *Trans. ASM*, vol. 53, p. 477, 1960.

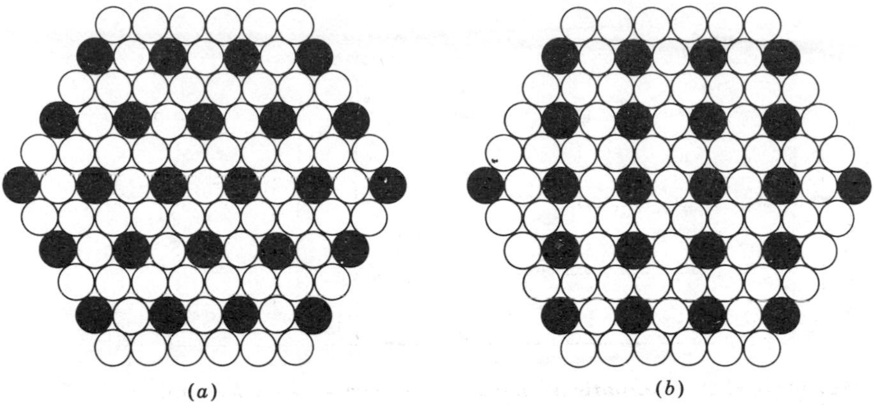

Fig. 10-25 Ordered close-packed planes of atoms that occur in hexagonal and cubic AB_3-type structures. (**After S. Saito and P. A. Beck, Trans. AIME, vol. 215, p. 938, 1959.**)

Thus, the tendency of Ti-group elements to form CsCl-type structures with other groups decreases gradually from the Fe-group elements to the Co-group elements. Occurrence of CsCl-type structures between other groups of elements is shown in Fig. 10–24. Near the other end of the row of stoichiometric ratios the close-packed ordered structures with the general formula AB_3 show an interesting example of stacking. In analogy to the structures of the close-packed elements discussed earlier, where *abcabc* and *abab* types of stacking represent the f.c.c. and c.p.h. structures, respectively, the ordered phases may be also described by a suitable stacking of close-packed layers. This was first shown by Laves and Wallbaum[1] for the structure $TiNi_3$. The close-packed layer of composition AB_3 is illustrated in Fig. 10–25*a*. Each A atom is surrounded by a complete shell of 12 nearest-neighbor B atoms. Stacking of such layers in the sequence *abab* results in the c.p.h. structures of the $MgCd_3$ type (DO_{19}) while the *abcabc* scheme produces the cubic $AuCu_3$-type structure ($L1_2$)[2]; in the same description the stacking scheme for the $TiNi_3$-type structure (DO_{24}) is *abacabac*. For more complex structures, such as VCo_3, or $PuAl_3$, the stacking scheme is *abcacb* and is composed of paired sequences *abc* and *cba*, both of which correspond to the $AuCu_3$ structure, but with orientations in twin relation to each other[3]. Structures related to the above family are the $TiCu_3$ type and $TiAl_3$ type (DO_{22}), which may be described by using ordered close-packed planes of the type shown in Fig. 10–25*b*. Neglecting distortion, the stacking of the planes in these structures is *abab* and *abcdef*, respectively.[4]

[1] F. Laves and H. J. Wallbaum, *Z. Krist.*, vol. 101, p. 78, 1939. A systematic survey of possible layer-stacking structures is given by P. A. Beck, *Zeit. Krist.*, in press.

[2] A. E. Dwight and P. A. Beck, *Trans. AIME*, vol. 215, p. 976, 1959.

[3] S. Saito, *Acta Cryst.*, vol. 12, p. 500, 1959.

[4] S. Saito and P. A. Beck, *Trans. AIME*, vol. 215, p. 938, 1959.

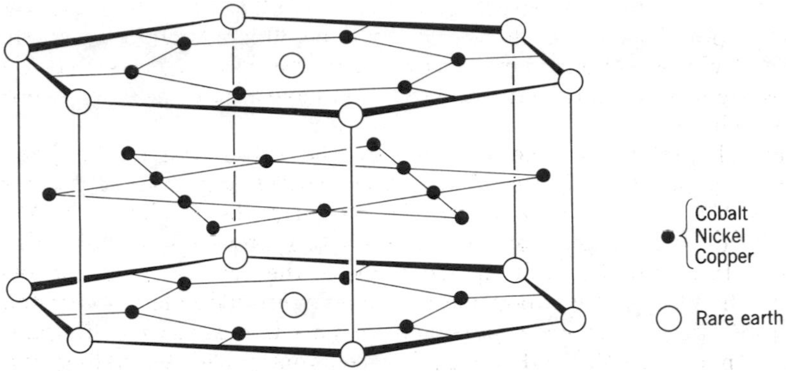

Fig. 10-26 The Cu_5Ca-type structure.

A characteristic feature of many binary systems of Co and Ni with the lanthanide elements is the simultaneous presence of a Laves phase, at composition AB_2, and a structural variant of a Laves phase at composition AB_5. The UNi_5-type structure ($C15_b$) may be derived from the cubic MgCu-type structure by a simple replacement of some A atoms by more B atoms.[1] The $CaCu_5$-type structure can be derived from the hexagonal $MgZn_2$ type, but the replacement is more complex and all A–A contacts are replaced by A–B contacts.[2] The resultant structure is like that shown in Fig. 10–26. Both size effects and magnetic effects have been discussed for a number of such structures.[3]

Phases with variable composition These phases, known by various Greek or Latin symbols as σ, μ, δ, χ, P, or R, possess complex crystal structures composed of quasi-hexagonal layers. The σ phase ($D8_b$) has received much detailed attention, chiefly because of the important effects which the formation of this phase has on the properties of industrial alloys, such as austenitic stainless steels. In the theoretical field much speculation has taken place concerning the nature of factors that are responsible for its occurrence and stability. The σ-phase unit cell is known to be tetragonal, with $c/a \cong 0.52$ and with 30 atoms per cell.[4] It is very much like the structure of beta-uranium.[5] Detailed x-ray studies have revealed that the majority of atoms are arranged in two alternating quasi-hexagonal layers, but certain atoms are displaced to positions intermediate between the

[1] N. C. Baenziger et al., *Acta Cryst.*, vol. 3, p. 34, 1950.

[2] A. E. Dwight, *Trans. ASM*, vol. 53, p. 479, 1961.

[3] M. V. Nevitt in P. A. Beck (ed.), "Electronic Structure and Alloy Chemistry of the Transition Elements," Wiley, 1963. J. H. Wernick in R. W. Cahn (ed.), "Advanced Physical Metallurgy," North-Holland Publishing, Amsterdam, 1965.

[4] J. S. Kasper, B. F. Decker, and J. R. Belanger, *J. Appl. Phys.*, vol. 22, p. 361, 1951.

[5] C. W. Tucker, Jr., *Science*, vol. 112, p. 448, 1950. G. J. Dickens, A. M. B. Douglas, and W. H. Taylor, *J. Iron Steel Inst.*, vol. 167, p. 27, 1951.

layers, thus producing increased atomic packing.[1] There are five kinds of crystallographically equivalent positions for the 30 available atoms in the unit cell. Evidence obtained with neutron diffraction[2] suggests that the σ phases are ordered and that various atoms have a preference for one of the five positions.

Structurally related to the σ phase are the (often ternary) μ phase ($D8_5$),[3] the χ phase[4] with the alpha-manganese structure ($A12$), the P phase (in Mo-Ni-Cr),[5] the σ phase (in Mo-Ni),[6] the R phase (in Mo-Co-Cr),[7] and the E phase.[8] Comparison of the σ, χ, and P phases shows that these structures can be built up from layers that closely resemble one another. The ranges of composition of σ phases are variable from system to system.[9] Their occurrence in binary systems is summarized diagrammatically in Fig. 10–27, and it may be seen that ranges extending over several percent are possible. This behavior is characteristic of all the above related phases and makes them similar in this respect to the electron phases of the noble metals. Descriptively, the sequence of phases σ, P, R, μ, and χ occurs in a relatively narrow range of electron concentration between the chromium group and the manganese group.[10] Hence, much speculation has been advanced about the electronic nature of their stability.[11] In general, for phases having the same A elements in a number of A–B systems, the composition shifts towards A as the number of electrons outside filled shells increases in the B elements. This is well evident from Fig. 10–27 for the binary σ phases. Ternary σ phases can have a nontransition element, such as Al or Si, as one of the partners,[12] which suggests complex electronic interactions. Recently, it has been also shown that

[1] B. G. Bergman and P. D. Shoemaker, *Acta Cryst.*, vol. 7, p. 857, 1954. Clara Brink and D. P. Shoemaker, *Acta Cryst.*, vol. 8, p. 734, 1955.

[2] J. S. Kasper in "Theory of Alloy Phases," American Society for Metals Symposium, Cleveland, Ohio, 1956. See also Chap. 23, p. 605.

[3] H. Arnfelt and A. Westgren, *Jernkontorets Ann.*, vol. 119, p. 184, 1935. B. N. Das and P. A. Beck, *Trans. AIME*, vol. 218, p. 733, 1960.

[4] Relationship between the σ phase and the χ phase may be more of an electronic than a structural kind; see F. C. Frank and J. S. Kasper, *Acta Cryst.*, vol. 12, p. 483, 1959. Coordination number 13, which appears in α-Mn or χ structures, is not common to other structures.

[5] S. P. Rideout et al., *Trans. AIME*, vol. 191, p. 872, 1951. C. Brink and D. P. Shoemaker, *Acta Cryst.*, vol. 8, p. 734, 1955. D. P. Shoemaker, C. B. Shoemaker, and F. C. Wilson, *Acta Cryst.*, vol. 10, p. 1, 1957.

[6] F. H. Ellinger, *Trans. ASM*, vol. 30, p. 607, 1942. C. B. Shoemaker and D. P. Shoemaker, *Acta Cryst.*, vol. 16, p. 997, 1963.

[7] Y. Komura, W. G. Sly, and D. P. Shoemaker, *Acta Cryst.*, vol. 13, p. 575, 1960.

[8] C. B. Shoemaker and D. P. Shoemaker, *Acta Cryst.*, vol. 18, p. 900, 1965.

[9] See articles by Nevitt, loc. cit., and E. Raub, *J. Less-Common Metals*, vol. 1, p. 3, 1959.

[10] B. N. Das and P. A. Beck, *Trans. AIME*, vol. 218, p. 733, 1960.

[11] See, for example, the review article by Nevitt, loc. cit.

[12] K. P. Gupta, N. S. Rajan, and P. A. Beck, *Trans. AIME*, vol. 218, p. 617, 1960.

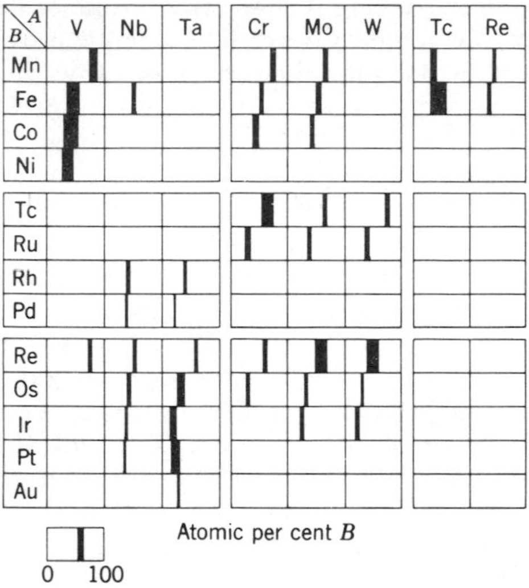

Fig. 10-27 Occurrence and ranges of stability of binary σ phases. [After M. V. Nevitt in P. A. Beck (ed.), "Electronic Structure and Alloy Chemistry of the Transition Elements," Wiley (Interscience), New York, 1963.]

several binary σ phases and some χ phases resemble the Cr_3Si-type phases and show superconductivity with the T_c temperature as high as 6.5°K.[1]

The family of the complex structures described above has also been considered in terms of packing of spheres of various sizes (see Chap. 13, page 377).

Other complex structures The structures considered in this chapter constitute only a small portion of the multitude of structures that are formed by various combinations of elements. The division between "alloy chemistry" and "crystal chemistry" is almost impossible to define since, as has been shown above, the transition is frequently a gradual one. Some 350 structures corresponding with the general formulas AB, AB_2, and AB_3 have been reviewed by Laves.[2] Various borides, carbides and nitrides, aluminides, and silicides of metals are discussed by Nowotny,[3] and carbides of the general formula M_3MeC_X (M = metal, Me = metalloid, C = carbon, $0.3 < X < 1$) are discussed by Stadelmaier.[4] Aluminum sometimes forms a cubic structure of an approximate composition MAl_{12} (M = Mn, Mo, Cr),[5] and Be forms tetragonal structures approximating the general formula

[1] R. D. Blaugher and J. K. Hulm, *J. Phys. Chem. Solids*, vol. 19, p. 134, 1961.

[2] F. Laves in American Society for Metals Symposium, "Theory of Alloy Phases," Cleveland, Ohio, 1956.

[3] H. Nowotny in P. A. Beck (ed.), "Electronic Structure of Alloy Chemistry of the Transition Elements," Interscience, New York, 1963. H. Nowotny in P. S. Rudman and J. Stringer (eds.), "Phase Stability in Metals and Alloys," McGraw-Hill, in press.

[4] H. H. Stadelmaier, *Z. Metallk.*, vol. 51, p. 758, 1961.

[5] J. Adam and J. B. Rich, *Acta Cryst.*, vol. 7, p. 813, 1954.

MBe_{12} (M = Al, Hg, Co, Cr, Fe, Mn, Mo, Nb, Pd, Pt, Ta, V, W).[1] Still richer in solvent are the compounds of Be, Zn, and Cd, represented by the prototype structure $NaZn_{13}$ or $NpBe_{13}$ with a f.c.c. structure.[2]

[1] R. F. Rauechle and F. W. von Batchelder, *Acta Cryst.*, vol. 8, p. 691, 1955; vol. 11, p. 122, 1958.

[2] N. C. Baenziger and R. E. Rundle, *Acta Cryst.*, vol. 2, p. 258, 1949. O. J. C. Runnals, *Acta Cryst.*, vol. 7, p. 222, 1954.

11

Superlattices

A large number of solid solutions become ordered at low temperatures. The process of ordering involves a change from a statistically nearly random distribution of atoms among the atom sites into a more regular arrangement, whereby designated sites are occupied predominantly by one kind of atoms. In a *disordered* alloy[1] of composition AB, for example, any given atom site is occupied indifferently by either A or B atoms, but on ordering, A and B atoms segregate more or less completely to designated atomic sites, so that the resulting arrangement can be described as a lattice of A atoms interpenetrating a lattice of B atoms. The segregation of atoms to particular atom sites may take place with little or no deformation of the lattice, creating an *ordered solid solution*, or *superlattice*, or *superstructure*, out of a random solid solution. There exist, of course, in the crystal chemistry of alloy phases, a large number of compoundlike phases which are ordered at any temperature and to which the term "superlattice" is frequently applied in spite of the absence of the corresponding disordered structure.

In a disordered solid solution, crystallographically equivalent planes of atoms are identical (statistically) with one another, but in an ordered superlattice this need not be true. For example, alternate planes of a set may become A-rich and B-rich planes, respectively, and the distance between identical planes may become twice the distance between identical planes of the disordered alloy (or some other multiple of this distance). Hence, the structures of ordered alloys usually produce diffraction patterns that have additional Bragg reflections, the *superlattice lines* associated with the new and larger spacings which are not present in patterns of the disordered alloys. An example is reproduced in Fig. 11–1. Bain[2] in 1923 and Johansson and Linde[3] in 1925 were the first to observe these lines with x-ray diffraction, though the possibility of ordering had been considered some years earlier by Tammann.[4]

The formation of superlattices, frequently described as *long-range order*, takes place at relatively low temperatures and usually at compositions expressed by a simple formula such as AB or AB_3 or at compositions near

[1] A state of perfect disorder is probably never found in practice (see Chap. 10, page 235).

[2] E. C. Bain, *Chem. Met. Eng.*, vol. 28, pp. 21, 65, 1923; *Trans. AIME*, vol. 68, p. 625, 1923.

[3] C. H. Johansson and J. O. Linde, *Ann. Physik*, vol. 78, p. 439, 1925.

[4] G. Tammann, *Z. Anorg. Chem.*, vol. 107, p. 1, 1919.

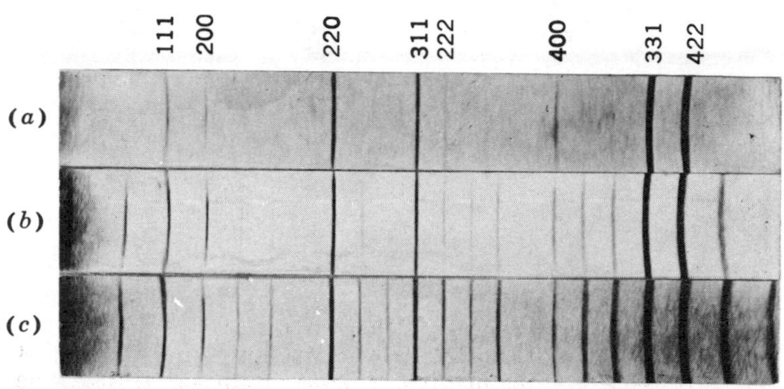

Fig. 11-1 Powder diffraction patterns of the superlattice Cu_3Au. (*a*) Disordered, (*b*) partially ordered, (*c*) highly ordered. (From C. Sykes and H. Evans, J. Inst. Metals, vol. 58, p. 255, 1936.)

these. At all temperatures above a certain critical temperature the usual randomness persists; when the temperature is lowered through the critical point, order sets in and increases as the temperature drops, approaching perfection only at low temperatures. Nevertheless, over certain ranges of composition and at temperatures above the critical temperature, the structures may at times be neither perfectly random nor perfectly ordered; even in the disordered state certain correlations may be present between nearest, second-nearest, and more distant neighbors, which may be described as *short-range order*. Such correlations frequently exist in alloys which do not show a state with long-range order at any temperature. The presence of both long-range and short-range order can be measured experimentally by their effect on the scattering of x-rays, electrons, and neutrons, and by their influence upon certain mechanical, physical, and thermal properties.

The intensive study of superlattices that followed their discovery has provided a remarkably clear view of the dynamic conditions within metallic crystals associated with the balance between the tendency of the atoms to take up regular positions and the opposing tendency of thermal agitation to maintain a chaotic arrangement. The main features of the order-disorder transformation have been worked out theoretically with considerable success and in many cases confirmed by ingenious experimentation.[1]

Common types of superlattices The majority of typical superlattices are related to the three principal metallic structures, the f.c.c. ($A1$), the b.c.c. ($A2$), and the c.p.h. ($A3$). It is becoming a generally accepted practice to refer to specific superlattice types by their Strukturbericht[2]

[1] The more recent reviews on this subject in which references to earlier publications may be found are those due to: F. C. Nix and W. Shockley, *Rev. Modern Phys.*, vol. 10, p. 1, 1938. H. Lipson, *Progr. Metal Phys.*, vol. 2, p. 1, 1950. T. Muto and Y. Takagi, *Solid State Phys.*, vol. 1, p. 194, 1955. L. Guttman, *Solid State Phys.*, vol. 3, p. 145, 1956.

[2] See page 240 and also Chap. 1, page 28.

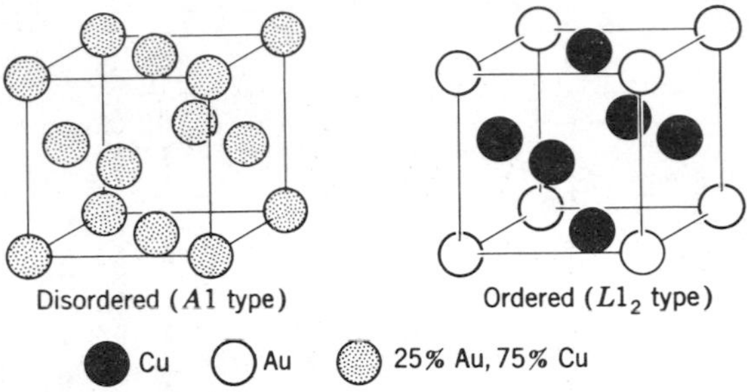

Fig. 11-2 The superlattice of Cu_3AuI.

designations although these are in general available only for the more commonly occurring types. Thus, for example, closely related superlattices based upon the disordered *A*2 structure are the *B*2, *D*0_3, and *L*2_1 types, of which typical examples are the ordered beta-brass (CuZn), the ordered Fe_3Al alloy, and the Heusler alloy (Cu_2MnAl), respectively. Superlattices based upon the random close-packed structures *A*1 and *A*3 are frequently of the *L*1_2, *D*0_{19}, and *L*1_0 types, characterized, for example, by the ordered structures $AuCu_3$, Mg_3Cd, and CuAuI,[1] respectively. Examples of these structures will be considered below.

The *L*1_2-, or Cu_3AuI-type superlattice Historically, copper-gold alloys containing about 25 atomic percent gold were among the first investigated. In the disordered state, which exists at high temperatures, Cu_3Au (now known as Cu_3AuI,[1] see page 279) has a nearly random array of Au and Cu atoms on a f.c.c. lattice (Fig. 11–2). If the alloy is annealed below a critical temperature, about 390°C (see Fig. 11–9), the atoms segregate as shown in the drawing of the ordered Cu_3AuI structure, Au atoms going to the cube corners and Cu atoms to the face centers. The unit cell shown in Fig. 11–2 can thus be thought of as having Au atoms at 000 positions and Cu atoms at $\frac{1}{2}0\frac{1}{2}$, $\frac{1}{2}\frac{1}{2}0$, and $0\frac{1}{2}\frac{1}{2}$ positions, and is a prototype of four interpenetrating simple cubic sublattices, each occupied by atoms of only one kind. This represents the condition when ordering is complete, the equilibrium condition at low temperatures.

The *L*1_2 structure has been observed in some sixty alloy systems.[2] Typical examples are: **Cu_3AuI**, α''-Au_3Cd, α'-$AlCo_3$, Pt_3Sn, Al_3U, $AlZr_3$, Co_3V, $FeNi_3$, $FePd_3$, $MnNi_3$, Si_3U, $TiZn_3$, Tl_3U, etc.

The *B*2-, or beta-brass-type superlattice The superlattice in beta-brass is illustrated in Fig. 11–3. It was first established by Jones and Sykes,[3] who used x-ray techniques. The unit cell can be represented in terms of

[1] I and II indicate Roman 1 and 2, not iodine, throughout this book, when occurring in CuAuI, Cu_3AuI, CuAuII, etc.

[2] The structures of superlattices and systems in which they occur are compiled by W. B. Pearson, "A Handbook of Lattice Spacings and Structures of Metals and Alloys," Pergamon, New York, 1958.

[3] F. W. Jones and C. Sykes, *Proc. Roy. Soc. (London)*, vol. A161, p. 440, 1937.

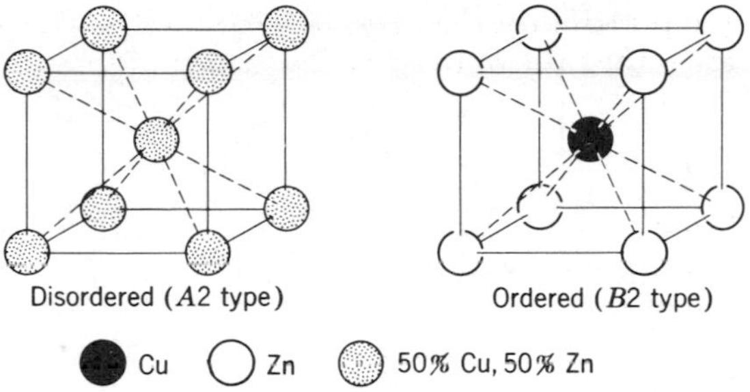

Fig. 11-3 The superlattice of beta-brass (cubic, CsCl or *B*2 type).

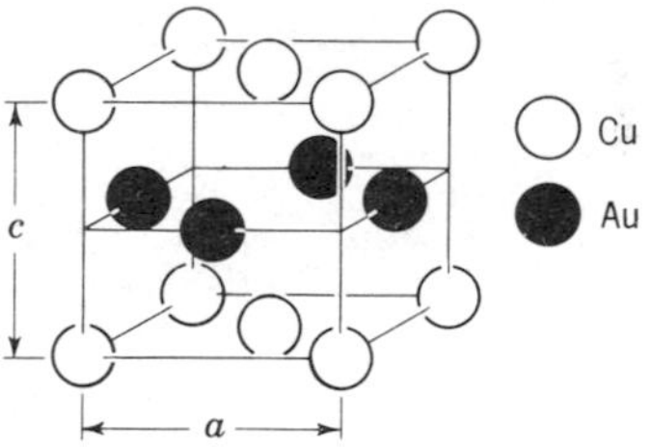

Fig. 11-4 The tetragonal superlattice CuAuI.

two interpenetrating simple cubic sublattices. The disordered crystal is b.c.c. ($A2$) with equal probabilities of having copper and zinc atoms at each lattice point; the ordered structure has copper atoms and zinc atoms segregated to cube corners and centers, respectively, in a structure of the CsCl type. This type of superlattice is very frequently encountered in alloy systems. About seventy $B2$ structures are listed by Pearson.[1] Typical examples are: β'-**CuZn**, β-AuCd, β-AlNi, β-NiZn, LiTl, etc.

The $L1_0$-, or CuAuI-type superlattice In the Cu-Au system, from 47 to 63 atomic percent, a superlattice forms in which alternate (001) planes contain only copper or only gold atoms. Hence, each atom has eight nearest neighbors of opposite kind in the adjacent (001) planes and four of the same kind in its own (001) plane. This is illustrated in Fig. 11-4. The resulting structure is tetragonal with axial ratio approximately $c/a = 0.93$,[2] which corresponds to about 7 percent deviation from unity. At the stoichiometric composition, CuAu, this ordered structure forms below approximately 385°C (see Fig. 11-9) and is commonly known as CuAuI. Its Strukturbericht designation is $L1_0$.

The origin of the tetragonal distortion has been the subject of some debate. The nominal radii of copper and gold derived from the closest spacings in their respective structures are 1.28 and 1.44 A. Thus, it may be

[1] Pearson, loc. cit.

[2] C. H. Johansson and J. O. Linde, *Ann. Physik*, vol. 82, p. 449, 1927; vol. 25, p. 1, 1936. R. Hultgren and L. Tarnopol, *Trans. AIME*, vol. 133, p. 228, 1939.

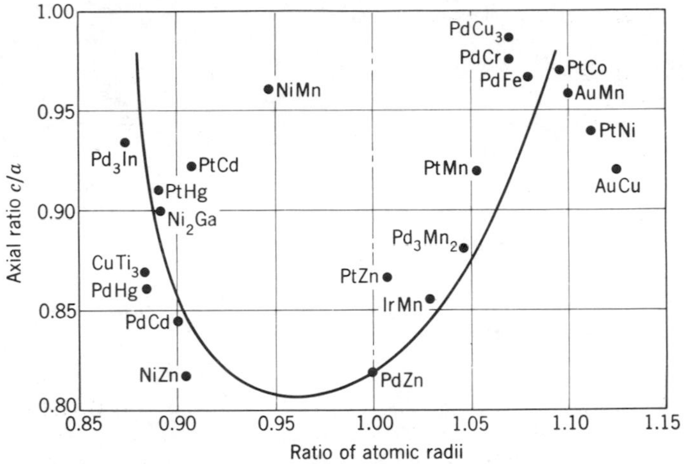

Fig. 11-5 Dependence of the axial ratio of $L1_0$ phases upon the ratio of atomic radii of the component elements. (**After K. Schubert, Z. Metallk., vol. 46, p. 43, 1955.**)

considered that the shortening of the cell edge in the tetragonal structure CuAuI, in the direction perpendicular to the (001) planes, is the result of a more convenient packing of unlike atoms on ordering. However, other superlattices of the $L1_0$ type suggest a more complex relationship. A plot of the c/a ratio vs. the radius ratio[1] of the two elements for a number of alloys with $c/a < 1$ reveals, as shown in Fig. 11–5, that the correlation between the two variables is not simple. For example, $\sim$ 18 percent deviation from unity is observed in the axial ratio of the PdZn alloy ($c/a \sim 0.81$) for which the values of the atomic radii of the elements are relatively close, as if bonds between unlike neighbors tended to be shorter than those between like neighbors. But in some examples[2] (viz., AlTi, HgZn, and HgTi) the opposite would have to be assumed for $c/a > 1$. A different approach to an understanding of the distortions resulting from ordering is to ascribe the distortions to long-range forces associated with the interaction of electrons with the lattice (more specifically, to a Fermi-surface–Brillouin-zone interaction),[3] a subject discussed later (Chap. 13, page 351).

The superlattices of the $L1_0$ type are quite numerous and have been observed in the following alloy systems:[4] AgTi (0.993), θ-CdPt (0.914),

[1] K. Schubert, *Z. Metallk.*, vol. 46, p. 43, 1955.

[2] Pearson, loc. cit.

[3] The Brillouin-zone mechanisms for superlattices have been discussed by: J. F. Nicholas, *Proc. Phys. Soc.* (*London*), vol. A66, p. 201, 1953. J. C. Slater, *Phys. Rev.*, vol. 84, p. 179, 1951. H. Sato and R. S. Toth, *Phys. Rev.*, vol. 124, p. 1833, 1961; vol. 127, p. 469, 1962, and in T. B. Massalski (ed.), "Alloying Behavior and Effects in Concentrated Solid Solutions," Gordon and Breach, New York, 1965.

[4] The number in parenthesis represents the value of the axial ratio.

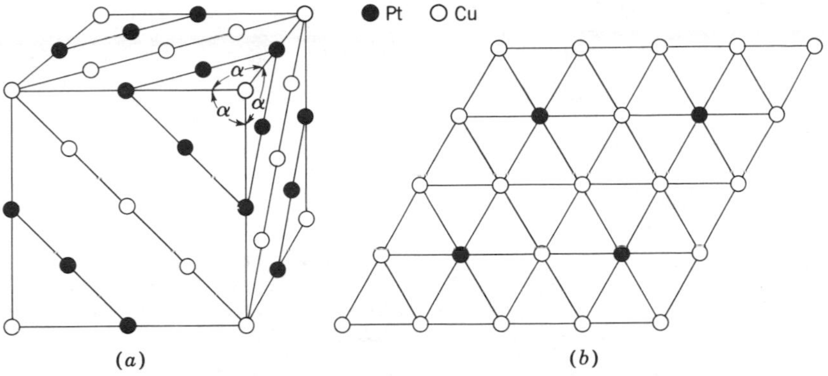

Fig. 11-6 The rhombohedral superlattice of CuPt, $L1_1$ type. (*a*) Distribution of atoms at the composition CuPt. Alternate (111) planes are occupied by Cu and Pt atoms. (*b*) Distribution of atoms on a single (111) plane, at the composition Cu_3Pt_5.

δ-HgPt (0.910), AlTi (1.020), CoPt (0.973), HgTi (1.343), **AuCuI** (0.926), δ-CuTi (0.643), HgZr (1.320), α-BiLi (0.894), FePd (0.966), β''-InMg (0.960), BiNa (0.980), FePt (0.968), NiPt (0.939), θ-CdPd (0.845), HgPd (0.862), θ-PdZn (0.816), and PtZn (0.860).

The $L1_1$ superlattice in CuPt In the Cu-Pt system near 50 atomic percent, the f.c.c. disordered lattice takes up the ordered structure shown in Fig. 11–6*a*, consisting of alternating layers of Cu and Pt atoms on (111) planes. This form of ordering produces a lattice distortion from cubic to rhombohedral[1] for which the Strukturbericht designation is $L1_1$. CuPt is the only known example of this structure.[2] Since each atom has six like and six unlike neighbors in both the ordered and the disordered state, there is no increase in the number of unlike neighbors on ordering.[3] Copper atoms in excess of the 50–50 composition displace Pt atoms at random on the Pt (111) planes, but in alloys with Pt atoms in excess of the 50:50 ratio there is an interesting tendency for them to displace certain atoms in the Cu (111) planes in the manner illustrated by the sketch of the (111) plane in Fig. 11–6*b*. Each Pt atom tends to be surrounded by Cu atoms, and a secondary type of ordering is thus produced. This arrangement would be complete in a Cu_3Pt_5 alloy.

The $D0_3$ and $L2_1$ superlattices Bradley and Jay[4] were the first to study in detail the superlattices in the system FeAl. As Al is added to the b.c.c. structure of alpha-iron, at first the Al atoms replace Fe atoms at random,

[1] J. O. Linde, *Ann. Physik*, vol. 30, p. 151, 1937.

[2] A. Schneider and U. Esch, *Z. Elektrochem.*, vol. 50, p. 290, 1944.

[3] C. B. Walker, *J. Appl. Phys.*, vol. 23, p. 118, 1952.

[4] A. J. Bradley and A. H. Jay, *Proc. Roy. Soc. (London)*, vol. A136, p. 210, 1932; *J. Iron Steel Inst. (London)*, vol. 125, p. 339, 1932.

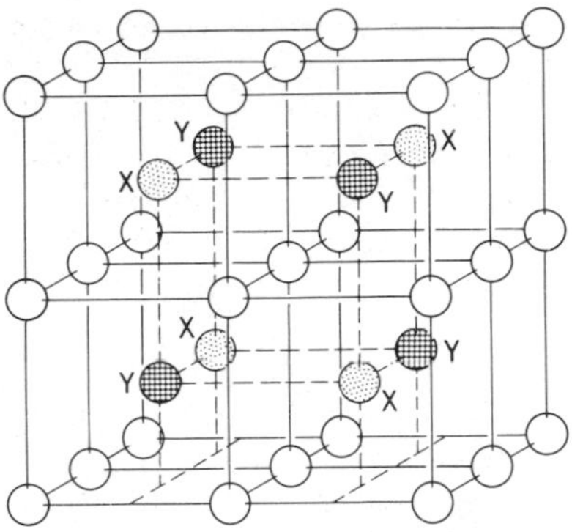

Fig. 11-7 The structure of Fe_3Al and FeAl. Al atoms are confined to the *X* positions in Fe_3Al and the *X* and *Y* positions in FeAl.

but beyond 18 atomic percent they concentrate in certain positions and desert others, giving rise to a number of ordered structures. The FeAl phase diagram is as yet not clearly established; the more recently proposed phase relationships are due to Taylor and Jones.[1] Of particular interest are the Fe_3Al and FeAl superlattices, both of which are based upon the unit cell shown in Fig. 11–7, which corresponds to the $D0_3$-type superlattice. From 18 to 25 atomic percent the Al atoms concentrate more and more in the positions labeled *X* in Fig. 11–7, this process being completed at the composition Fe_3Al. From 25 to 50 atomic percent, increasing numbers of Al atoms go to *Y* positions until both *X* and *Y* positions are filled with Al atoms at the composition FeAl. The structure is then similar to the ordered structure in β brass. In the fully ordered Fe_3Al structure, each Al atom is surrounded by eight Fe nearest neighbors. To obtain this fully ordered structure it is necessary to cool very slowly through 550°C to room temperature or to anneal for a prolonged period around 300 to 350°C. A characteristic feature of the $D0_3$ structures is that each atom is surrounded by the maximum number of unlike nearest and second-nearest neighbor atoms. Another important $D0_3$ superlattice is found in the Fe_3Si phase[2] and has been analyzed in detail.[3]

Closely related to the $D0_3$ superlattice is the ternary $L2_1$ type based on the composition A_2BC. The most representative structure is that of the *Heusler alloy* with the formula Cu_2MnAl. A related group of such structures was discovered by Heusler near the end of the nineteenth century and they possess the remarkable property of being ferromagnetic when in ordered condition despite the fact that the constituent atoms are in most cases nonferromagnetic. In terms of the cell shown in Fig. 11–7 the Al atoms are in *Y* positions, the Mn atoms in *X* positions, and the Cu atoms at the remaining points shown.[4] By analogy to the $D0_3$ structure, the atoms in

[1] A. Taylor and R. M. Jones, *J. Appl. Phys.*, vol. 29, p. 522, 1958; *J. Phys. Chem. Solids*, vol. 6, p. 16, 1958.

[2] G. Phragmén, *Stahl Eisen*, vol. 45, p. 299, 1925.

[3] M. C. M. Farquhar, H. Lipson, and A. R. Weill, *J. Iron Steel Inst.*, vol. 152, p. 457, 1945.

[4] A. J. Bradley and J. W. Rodgers, *Proc. Roy. Soc. (London)*, vol. A144, p. 340, 1944.

the X and Y positions of the $L2_1$ superlattice tend to avoid being close nearest or second-nearest neighbors.

The ferromagnetic ordered phase Cu_2MnSn is isostructural with Cu_2MnAl and the Sn atoms of the former take the positions of the Al atoms of the latter.[1] The same structure is found again at or near the compositions Cu_2MnIn and Cu_2MnGa, which are also ferromagnetic superlattices.[2] Other Heusler-type alloy structures, such as Cu_2NiAl and Zn_2CuAu, are associated with thermoeleastic martensites (see Chap. 18).

The structure of Ni_2Al_3 is based on the FeAl structure but contains vacant lattice points which increase in number and take up ordered positions as the Al content increases.[3] Thus, not only substitutional atoms, but also vacancies and interstitial atoms (as in austenite, and as in Fe_2N, Fe_3N, and Fe_4N)[4] take up superlattice positions.

Examples of $D0_3$ superlattices are found in the following alloys: β_1-AlCu, H_3La, **$AlFe_3$**, $HgLi_3$, $BiLi_3$, $LaMg_3$, β-Cu_3Sb, β-Li_3Sb, α'-Fe_3Si, and Mg_3Pr, and examples of the $L2_1$ superlattices are found in a number of ternary alloys in addition to those mentioned above: Ni_2TiAl, Cu_2CoSn, Co_2MnSn, Cu_3FeSn, Cu_2MnSn, Cu_2NiSn, Ni_2MgSb, Ni_2MgSn, and $LiMg_2Tl$.

The $D0_{19}$- or Mg_3Cd-type superlattice A superlattice closely related to the $L1_2$ type of $AuCu_3$ is the $D0_{19}$ superlattice with the unit cell shown in Fig. 11–8. Here again the cell can be described in terms of four interpenetrating simple sublattices, and the arrangement of atoms in the close-packed planes of both structures is identical. However, unlike the $L1_2$ type, each sublattice in the $D0_{19}$ type is c.p.h. with the a spacing corresponding to twice the spacing of the disordered structure and the c spacing unchanged. B atoms occupy one of the four sublattices, resulting in the general formula A_3B, for example in the case of Mg_3Cd or $MgCd_3$, which are perhaps the best-known examples of $D0_{19}$ superlattices.[5]

Over the last several years a large number of alloy phases have been discovered which possess the c.p.h. structure at high temperatures and which undergo ordering on cooling. Depending on the composition ranges at which such phases are stable (i.e., near A_3B, A_2B, AB, etc.), the number of unlike nearest neighbors and their distribution on ordering will differ with the resulting changes in the details of the ordered structure. As can be seen from Fig. 11–8, by analogy with the tetragonal distortion of the $L1_0$ structure, one might expect an orthorhombic distortion of the basic unit cell of the $D0_{19}$ superlattice, and indeed the ordered structure of MgCd is orthorhombic ($B19$),[6] and so are a number of superlattices of the noble

[1] L. A. Carapella and R. Hultgren, *Trans. AIME*, vol. 147, p. 232, 1942.

[2] S. Valentiner and I. Pusicha, *Metallforschung*, vol. II-4, p. 127, 1947. F. A. Hames and D. S. Eppelsheimer, *Trans. AIME*, vol. 185, p. 495, 1949. B. R. Coles, W. Hume-Rothery, and H. P. Myers, *Proc. Roy. Soc.* (*London*), vol. A196, p. 125, 1949.

[3] A. J. Bradley and A. Taylor, *Proc. Roy. Soc.* (*London*), vol. A159, p. 56, 1937.

[4] K. H. Jack, *Proc. Roy. Soc.* (*London*), vol. A195, p. 34, 1948.

[5] Atom positions in Mg_3Cd and similar structures are as follows: Mg: ($\frac{1}{2}$,0,0, 0,$\frac{1}{2}$,0, $\frac{1}{2}$,$\frac{1}{2}$,0), ($\frac{1}{6}$,$\frac{1}{3}$,$\frac{1}{2}$, $\frac{1}{6}$,$\frac{5}{6}$,$\frac{1}{2}$, $\frac{2}{3}$,$\frac{5}{6}$,$\frac{1}{2}$); Cd: (0,0,0, $\frac{2}{3}$,$\frac{1}{3}$,$\frac{1}{2}$).

[6] K. Riederer, *Z. Metallk.*, vol. 29, p. 423, 1937.

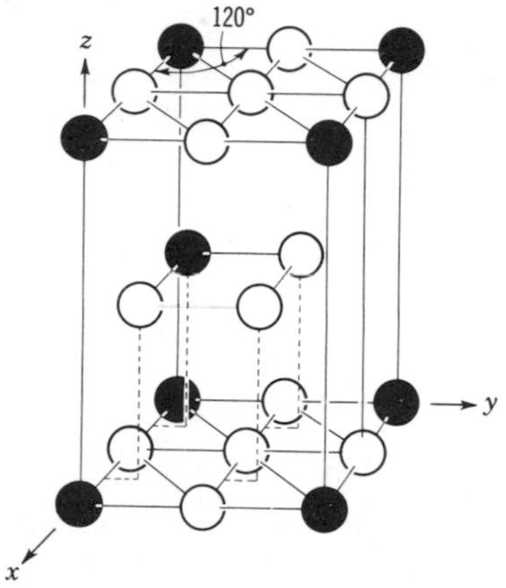

Fig. 11-8 The unit cell associated with the DO_{19}-type superlattice.

metals Cu, Ag, and Au with polyvalent solvents. In the latter cases ordering is often associated with pronounced changes of the c lattice spacings and the axial ratio,[1] and with the formation of stacking faults,[2] as well as with more complex lattice distortions and formation of "shift structures" discussed later. The crystallographic aspects of such structures and possible interpretations have been considered in detail in a number of publications by Schubert and his associates.[3]

Detailed theories of these distortions and shifts can scarcely be simple, for one would expect several tendencies to exist in varying degree: (1) minimum strain energy, (2) minimum internal energy, estimated by relative numbers of bonds between like and unlike neighbors, and (3) possible collective interaction between conduction electrons and the lattice. The known $D0_{19}$ superlattices and their axial ratios are as follows: Al_3Th (0.712), β''-Fe_3Sn (0.799), $PbTi_4$ (0.8096), Cd_3Mg (0.8093), GeMn (0.8165), Pt_3U (0.851), **$CdMg_3$** (0.8038), γ-InNi (0.798), $Sb_{0.8}Ti_{3.2}$ (0.807), Co_3Mo (0.803), $Mn_{11}Sn_3$ (0.798), $Si_{0.7}Ta_{3.2}$ (0.806), Co_3W (0.8047), β-Ni_3Sn (0.8018), and SnTi (0.805).

Less common superlattices The existence of superlattices has now been clearly established in many other complex structures of binary, ternary, and many-component systems. Not infrequently, ordering occurs by stages, so that certain lattice positions become occupied in an ordered fashion at one temperature and additional sites become ordered at some lower temperature. This apparently is the case in the Cu-Al-Sn system.[4]

[1] J. Wegst and K. Schubert, *Acta Met.*, vol. 6, p. 720, 1958. T. B. Massalski and H. W. King, *Acta Met.*, vol. 8, p. 677, 1960.

[2] A. Bystrom and K. E. Almin, *Acta Chem. Scand.*, vol. 1, p. 76, 1947. T. B. Massalski, *Acta Met.*, vol. 5, p. 541, 1957.

[3] See, for example, W. Burkhardt and K. Schubert, *Z. Metallk.*, vol. 50, p. 442, 1959. K. Schubert et al., *Z. Metallk.*, vol. 46, p. 692, 1955. E. Gunzel and K. Schubert, *Z. Metallk.*, vol. 49, p. 124, 1958. J. Wegst and K. Schubert, *Z. Metallk.*, vol. 49, p. 533, 1958. K. Schubert in "Symposium on Physical Chemistry of Metallic Solutions and Intermetallic Compounds," vol. 1, p. 58, Chemical Publishing, New York, 1960.

[4] J. S. L. Leach and G. V. Raynor, *Proc. Roy. Soc. (London)*, vol. 224A, p. 251, 1954.

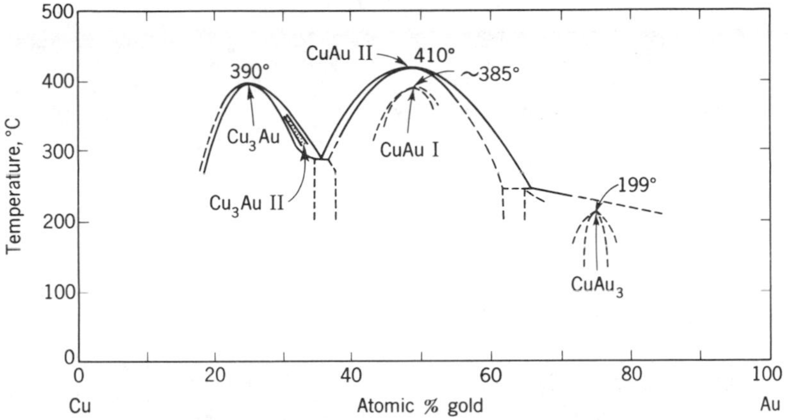

Fig. 11-9 The most probable diagram of the Cu-Au system showing various superlattices reported up to 1964. Arrows indicate maximum temperatures of long-range order.

A ternary phase observed in this system and other ternary phases such as Ag_2HgI and Cu_2HgI show complex order-disorder changes resembling somewhat those in Cu_3Au[1] but involving ordered and disordered arrangements of both metal atoms and vacant sites. Recent researches on phases formed by the transition elements with one another have disclosed various new types of superlattices, some with fairly complex atomic arrangements. The best-known example is the σ-phase structure (related to the beta-uranium structure) and its derivatives.[2]

The reasons for ordering in the above phases appear to be associated both with the tendency of certain elements to be located at positions of maximum coordination (largest possible number of nearest neighbors) and with the possibility of localized electron sharing between certain pairs of atoms.

Long-period superlattices Among the large number of superlattices there is a special group which may be classified as *long-period superlattices* or *antiphase superlattices.* The copper-gold system provides a classical example of such structures. Copper and gold form a continuous series of solid solutions at high temperatures, but several superlattices exist at low temperatures, near compositions Cu_3Au, CuAu, and $CuAu_3$, all of which have been extensively investigated.[3]

A current version of the phase diagram is shown in Fig. 11-9. The alloy CuAu, in addition to the f.c.t. superlattice CuAuI mentioned earlier, exhibits another superlattice known as CuAuII which exists between

[1] J. A. Ketelaar, *Z. Physik. Chem.*, vol. B26, p. 327, 1934; vol. B30, p. 53, 1935; *Z. Krist.*, vol. 87, p. 436, 1934.

[2] An account of neutron diffraction studies of σ phases has been given by J. S. Kasper in "Theory of Alloy Phases," American Society for Metals, Cleveland, Ohio, 1956. Earlier x-ray studies were also made: B. F. Decker, R. M. Waterstrat, and J. S. Kasper, *J. Metals*, vol. 6, p. 1406, 1954. G. Bergman and D. P. Shoemaker, *Acta Cryst.*, vol. 7, p. 857, 1955.

[3] See, for example, M. Hansen and K. Anderko, *"Constitution of Binary Alloys,"* 2d ed., p. 198, McGraw-Hill, New York, 1958.

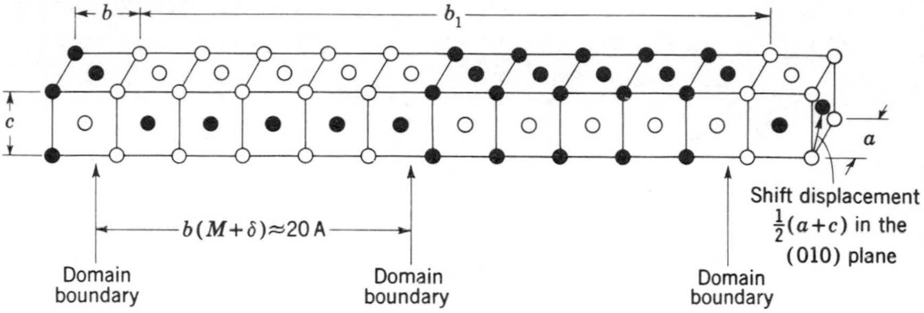

Fig. 11-10 The unit cell of the CuAuII superlattice with superperiod $M = 5$.

approximately 385 and 410°C and is orthorhombic. The crystallographic unit cell of CuAuII, as indicated in Fig. 11–10, may be depicted by stacking 10 CuAuI tetragonal cells aligned in the b direction and switching the content of (001) planes from all gold atoms to all copper atoms halfway along the new long cell. This gives rise to an *antiphase* or *out-of-step boundary* halfway along the long cell (i.e., at intervals of five unit lengths in the b direction) and at subsequent similar intervals along the b axis. The displacement of atom type that occurs at the antiphase boundaries is equivalent to a lattice shift of $\frac{1}{2}(\mathbf{a} + \mathbf{c})$ in the plane normal to the ab plane, and the distance between two antiphase boundaries is given by a distance $b(M + \delta)$, where M is the half period of the superlattice and δ is a slight

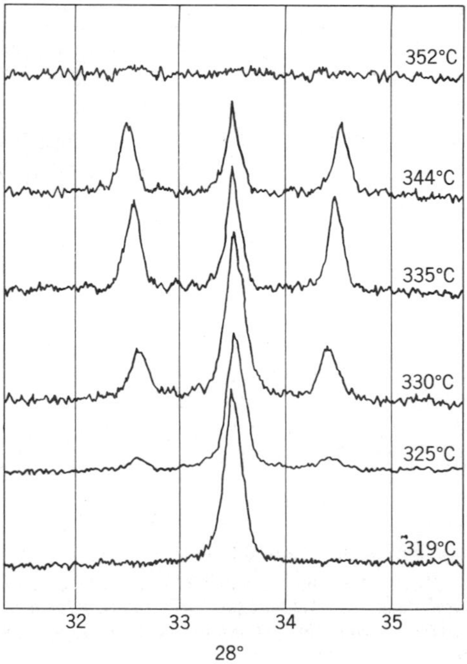

Fig. 11-11 Powder pattern recordings of the central peak and satellites for the (110) reflection from a Cu_3AuII sample containing 31.6 atomic percent gold, air-quenched from the indicated temperatures. Filtered Cu $K\alpha$ radiation. To be comparable, the peak heights should be doubled for the 319 and 325°C samples. (After R. E. Scott, J. Appl. Phys.. vol. 31, p. 2112, 1960.)

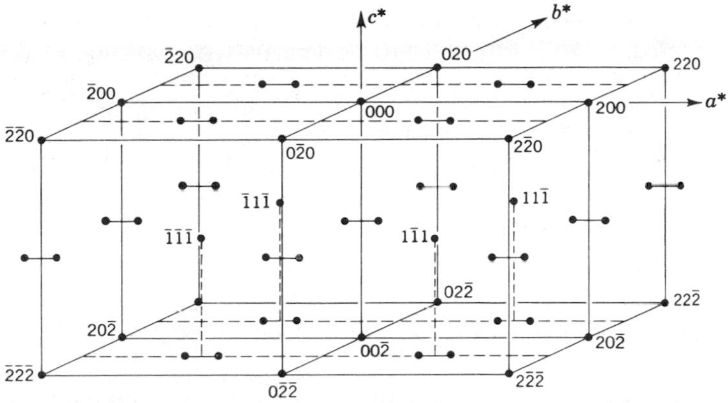

Fig. 11-12 Portion of reciprocal space for the CuAuII superlattice. Some reciprocal lattice points are doubled.

expansion of the lattice in the direction of the *superperiod*.[1] For a CuAuII alloy of stoichiometric composition, $M = 5$. Therefore, this structure may be called a *one-dimensional long-period superlattice*, with the superperiod $M - 5$ existing only in one direction. As a result of the slight expansion at the antiphase boundary and the accompanying change in the atomic arrangement, each individual original CuAuI cell may also be considered to have deformed slightly in the b direction, so that $b/a > 1$ in addition to the original deformation $c/a < 1$.

The superlattice CuAuII has been studied by numerous investigators since the classical x-ray study by Johansson and Linde,[2] who used the powder method and showed that an orthorhombic structure is formed. A characteristic feature of the x-ray pattern is the appearance of strong superlattice reflections, for example those shown in Fig. 11–11 for the Cu_3AuII superlattice, as a consequence of the periodic arrangement of the antiphase boundaries.

If a is the side of the original f.c.c. cell, then $c = 0.92$ and $b = a$ in the tetragonal cell of CuAuI and $b_1 = 2b(M + \delta)$ in the orthorhombic cell of CuAuII. The reciprocal lattice corresponding to the CuAuII superlattice is shown in Fig. 11–12, and it may be seen that instead of superlattice reflections of the form $\{110\}$, sets of double spots exist along the [001] direction. Some twenty years after the original x-ray work, electron diffraction and electron transmission microscopy were applied to evaporated thin films of the CuAu and Cu_3Au alloys suitably annealed,[3] and many fine details of the ordered structures and their growth were shown. CuAuII

[1] The length $2M$ is sometimes called the superperiod.

[2] C. H. Johansson and J. O. Linde, *Ann. Phys.*, vol. 25, p. 1, 1936.

[3] H. Raether, *Z. Angew. Phys.*, vol. 4, p. 53, 1952. S. Ogawa and D. Watanabe, *J. Phys. Soc.* (*Japan*), vol. 9, p. 475, 1954. S. Ogawa et al., *Acta Cryst.*, vol. 11, p. 872, 1958. D. W. Pashley and A. E. B. Presland, *Proc. Roy. Soc.* (*London*), vol. A250, p. 132, 1959; *J. Inst. Met.*, vol. 87, p. 419, 1958.

may form either from the disordered high-temperature form or from the ordered CuAuI form. The physical properties, the kinetics of the formation,[1] and the existence of latent heat at the transition temperatures[2] show that the long-period superlattice is an equilibrium state. However, unlike the usual order-disorder transformations where the size of the ordered regions is a combined function of the temperature and time of annealing, the domains[3] in the long-period structure remain of the same size M, irrespective of the thermal history. Their size is nevertheless sensitive to changes of composition.

Systematic studies of long-period superlattices have shown that these structures are found mostly in f.c.c. alloys having compositions AB and A_3B.[4] In the Cu-Au system the existence of a stable long-period superlattice, Cu_3AuII, has been established by x-ray methods,[5] electron diffraction, and transmission electron microscopy.[6] Diffractometer tracings for the (110) powder pattern peak at various temperatures corresponding to the disordered f.c.c., the Cu_3AuII, and the Cu_3AuI structures are shown in Fig. 11–11. Although it is still not possible to draw the precise limits of the Cu_3AuII region in Fig. 11–9, a stable region has been located by Scott around 340°C at the composition range near 31.6 atomic percent Au. For this superlattice, M is about 9.0.[7]

In addition to the one-dimensional long-period superlattices of the CuAuII or Cu_3AuII type, there exist *two-dimensional long-period superlattices*, in A_3B type alloys, with domain sizes M_1 and M_2 along two crystal axes, with M_1 not necessarily equal to M_2. Examples are found in Cu-Pd, Au-Zn, Au-Mn, etc. An illustration of the one- and two-dimensional types for an alloy of A_3B composition is shown in Fig. 11–13. No three-dimensional long-period superlattice has been reported thus far.

[1] See, for example: G. Borelius, *J. Inst. Met.*, vol. 74, p. 17, 1947. G. J. Dienes, *J. Appl. Phys.*, vol. 22, 1020, 1951.

[2] M. Hirabayashi, S. Nagasaki, and H. Maniwa, *Nippon Kinzoku Gakkaishi*, vol. B14, p. 1, 1950 (quoted by H. Sato and R. S. Toth, *Phys. Rev.*, vol. 124, p. 1833, 1961; vol. 127, p. 469, 1962).

[3] See p. 297.

[4] K. Schubert et al., *Z. Metallk.*, vol. 46, p. 692, 1955. S. Ogawa et al., *J. Phys. Soc. (Japan)*, vol. 14, p. 936, 1959. H. Sato and R. S. Toth, *Phys. Rev.*, vol. 127, p. 469, 1962. Also see a review article by H. Sato and R. S. Toth in T. B. Massalski (ed.), "Alloying Behavior and Effects in Concentrated Solid Solutions," Gordon and Breach, New York, 1965.

[5] R. E. Scott, *J. Appl. Phys.*, vol. 31, p. 2112, 1960. See also earlier work on the metastable phases by A. Guinier and R. Griffoul, *Rev. Met.*, vol. 55, p. 387, 1948.

[6] R. S. Toth and H. Sato, *J. Appl. Phys.*, vol. 33, p. 3250, 1962. S. Marcinkowski and L. Zwell, *Acta Met.*, vol. 11, p. 373, 1963. S. Yamaguchi, D. Watanabe, and S. Ogawa, *J. Phys. Soc. (Japan)*, vol. 17, p. 1030, 1962.

[7] The value of M calculated from experimental data for many long-period superlattices is *nonintegral*. This is usually interpreted as indicating a mixture of different periods. K. Fujiwara, *J. Phys. Soc. (Japan)*, vol. 12, p. 7, 1957, has shown that such a mixture can still lead to sharply defined reciprocal lattice points. Nevertheless, even in the prototype CuAuII superlattice there is still some controversy in the case of slightly nonstoichiometric alloys for which M turns out to be greater than 5. S. Ogawa et al., *Acta Cryst.*, vol. 11, p. 872, 1958, following a direct observation of domain length in an electron microscope, have suggested two maxima at $M = 5$ and $M = 6$, whereas D. W. Pashley and A. E. B. Presland, "Structure and Properties of Thin Films," Wiley, New York, 1959, found only a single maximum at 5.5.

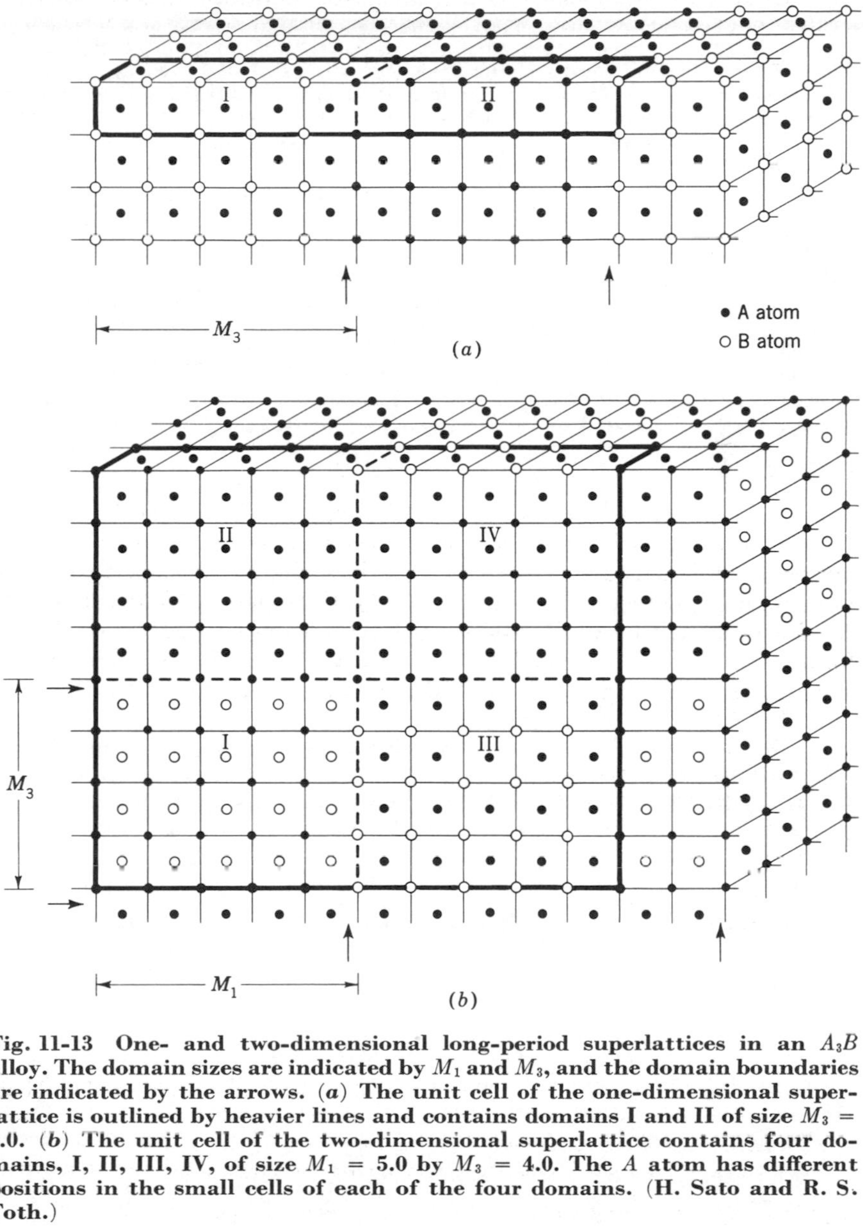

Fig. 11-13 One- and two-dimensional long-period superlattices in an A_3B alloy. The domain sizes are indicated by M_1 and M_3, and the domain boundaries are indicated by the arrows. (*a*) The unit cell of the one-dimensional superlattice is outlined by heavier lines and contains domains I and II of size $M_3 = 5.0$. (*b*) The unit cell of the two-dimensional superlattice contains four domains, I, II, III, IV, of size $M_1 = 5.0$ by $M_3 = 4.0$. The *A* atom has different positions in the small cells of each of the four domains. (**H. Sato and R. S. Toth.**)

In each type of superlattice the additional slight distortion of the original unit cell dimension and the appearance of the long period take place in the direction perpendicular to the antiphase boundary. The existence of these distortions can be confirmed from the modification of the diffraction patterns. The corresponding modification of the Brillouin zone appears to play an important role in the stability of the long-period superlattices (see Chap. 13).

In addition to the distortions of the unit cell dimensions that accompany long-period ordering, other details of these superlattices have received attention. Elongation of certain diffraction spots are observed in some experiments, and are attributable to *stacking faults*; certain intensity relationships have been attributed to small shifts in the positions of some of the atoms, constituting a *lattice modulation* that is different in type in different superlattices.[1]

Elements of superlattice theories Theoretical treatments for fully ordered superlattices have been given by Borelius,[2] Johansson and Linde,[3] Gorsky,[4] Dehlinger,[5] and Dehlinger and Graf,[6] chiefly on the basis of formal thermodynamic relations. The problem was considered anew by Bragg and Williams,[7] Williams,[8] Bethe,[9] and Peierls,[10] who as a group started with simple assumptions about atomic forces and calculated quantitative results that compared very favorably with experiment.

At high temperatures, or when the compositions deviate from the ideal values, very imperfect states of order must exist in which the crystal is only partially ordered. Recognition of this fact has created a further need for a suitable general description of imperfectly ordered crystals. In such crystals both the unit cell and the translation group lose their strict significance since the lattice translation vectors may sometimes join atoms of different kinds rather than identical atoms only, as in the perfectly ordered structures. The state of order may then be more conveniently described by a set of parameters related to *pair-density functions*, which describe the state of occupation not of single sites but in terms of pairs of sites. The point of view of the earlier theories of Bragg-Williams and Bethe (quasi-chemical theories) will now be considered in some detail and will be followed by an outline of the pair-density theories.

In a fully ordered alloy there are great distances within a crystal through which there is a perfect arrangement of A atoms on one set of lattice points and B atoms on another set. The ordering is consistent, in step, through long distances. The degree of this long-range order may be defined by a fraction S, which varies from zero at complete disorder up to unity at com-

[1] S. Ogawa and D. Watanabe, *J. Phys. Soc. (Japan)*, vol. 9, p. 475, 1954 (CuAuII, one-dimensional antiphase structure). D. Watanabe, *J. Phys. Soc. (Japan)*, vol. 15, p. 151, 1940 (Au_3Mn, two-dimensional). H. Iwasaki, *J. Phys. Soc. (Japan)*, vol. 17, p. 1620, 1962 (Au-Zn system, one-dimensional). M. Wilkens and K. Schubert, *Z. Metallk.*, vol. 48, p. 550, 1957 (Au_{3+} Zn, two-dimensional).

[2] G. Borelius, *Ann. Physik*, vol. 20, p. 57, 650, 1934.

[3] C. H. Johansson and J. O. Linde, *Ann. Physik*, vol. 78, p. 439, 1925.

[4] W. Gorsky, *Z. Physik*, vol. 50, p. 64, 1928.

[5] U. Dehlinger, *Z. Physik. Chem.*, vol. B26, p. 343, 1934.

[6] U. Dehlinger and L. Graf, *Z. Physik*, vol. 64, p. 359, 1930.

[7] W. L. Bragg and E. J. Williams, *Proc. Roy. Soc. (London)*, vol. A145, p. 699, 1934; vol. A151, p. 540, 1935.

[8] E. J. Williams, *Proc. Roy. Soc. (London)*, vol. A152, p. 231, 1935.

[9] H. A. Bethe, *Proc. Roy. Soc. (London)*, vol. A150, p. 552, 1935; *J. Appl. Phys.*, vol. 9, p. 244, 1938.

[10] R. Peierls, *Proc. Roy. Soc. (London)*, vol. A154, p. 207, 1936.

plete order; S, for an AB superlattice, is the fraction of the atoms that are in their right positions on a given sublattice minus the fraction that are in wrong positions. As a simple illustration, consider an alloy AB in which 100 A atoms and 100 B atoms are randomly arranged; just half of the A atoms are in the places they would occupy in the ordered structure, and the other half are in wrong positions; and the same would be true of the B atoms, giving $S = 0$. If 75 A atoms were right and 25 wrong, the degree of order would be $0.75 - 0.25 = 0.50$.

This definition may be generalized to cover the condition in which n atomic positions of a total number N can be occupied by either kind of atom.[1] In the structure Fe_3Al the fraction $n/N = \frac{1}{2}$, since even in the completely disordered state, half the atom positions are still occupied exclusively by one kind of atom. Suppose a fraction r of the n sites is occupied by A atoms in the state of perfect order; these rn sites are right positions for A atoms. In a partially ordered alloy, some of these positions are filled by A and some by B atoms. If p is the probability that a right position for an A atom is filled by an A atom, then the long-distance order may be defined by the relation

$$S = (p - r)/(1 - r) \tag{11-1}$$

which varies from 0 to 1 as order increases.

In an alloy of A and B atoms the energy of the crystal will be a minimum when order is complete and will be increased by an amount V if a pair of atoms is interchanged so that an A atom takes a place that should be occupied by a B atom, and a B atom moves to a position that should be occupied by an A atom. In other words, V is the net amount of work required to effect this interchange. Under thermal agitation an equilibrium will be reached such that the ratio of the number of atoms in the right positions to the number in wrong positions is proportional to Boltzmann's factor $e^{-V/kT}$, where k is Boltzmann's constant and T is the absolute temperature.

If V were a constant, say V_0, independent of the degree of order in the alloy, there would be a gradual increase in disorder with rising temperature along a curve plotted in Fig. 11–14 as a dashed line. This cannot be the case, however, for a decrease in order results in a decrease in the forces that tend to maintain order. In other words, it becomes energetically easier for an atom to disorder if some of its neighbors are already disordered. The distinction between right and wrong positions for an atom, in fact, vanishes when disorder is complete, and the energy V to effect an interchange then drops to zero. This dependence of V on order, coupled with a gain in entropy with disorder, is responsible for the decrease of order at an accelerating rate as the temperature is raised. As Bragg puts it, demoralization sets in, and there is a complete collapse of the ordered state. For this reason the order-disorder change is known as a *cooperative phenomenon.*

[1] W. L. Bragg and E. J. Williams, *Proc. Roy. Soc. (London)*, vol. A145, p. 699, 1934.

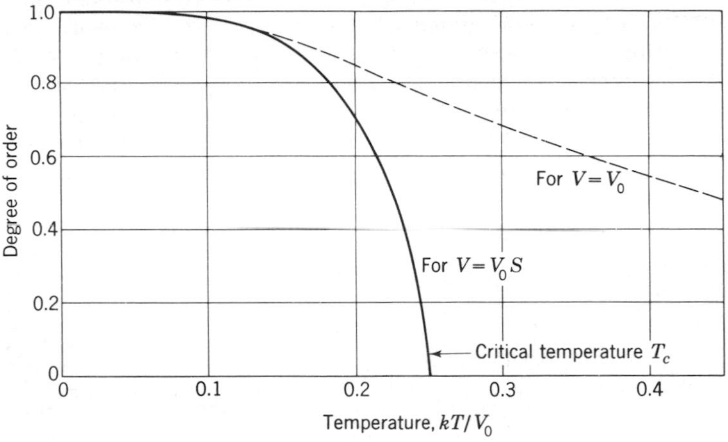

Fig. 11-14 Dependence of order on temperature according to the Bragg and Williams theory for long-distance order at equilibrium.

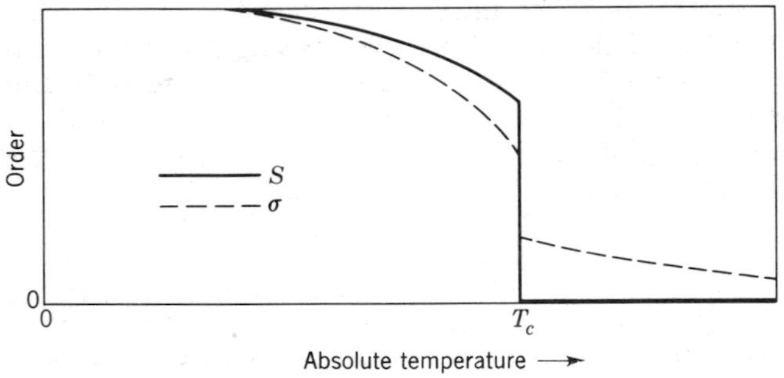

Fig. 11-15 Dependence of long-range order S and short-range order σ on temperature in an AB_3 superlattice.

To make a theoretical calculation of the degree of order in equilibrium at each temperature, therefore, it is necessary to assume a particular relation between V and the degree of order. For simplicity, Bragg and Williams[1] assumed that V is proportional to the degree of long-distance order, S, according to the relation $V = V_0S$, where V_0 is a constant representing the interchange energy when the order is complete. The curve for equilibrium order vs. temperature is shown in Fig. 11–15 by the solid line. The curve is computed for a superlattice of composition AB, such as beta-brass and AuCu; it also holds for Fe_3Al. Long-distance order decreases to zero at a critical temperature T_c that is directly analogous to the Curie temper-

[1] W. L. Bragg and E. J. Williams, *Proc. Roy. Soc.* (*London*), vol. A145, p. 699, 1934.

ature at which a ferromagnetic material loses its ferromagnetism[1] (which is another example of a cooperative phenomenon).

The critical temperature T_c is directly related to the ordering energy V_0; in a 50 atomic percent alloy the relation is approximately $V_0 = 4kT_c$, where k is Boltzmann's constant.

A similar calculation for a composition like $AuCu_3$ gives a similar curve up to a certain temperature and then an abrupt drop to zero order, as indicated by the solid curve in Fig. 11–15.

Short-range order In Bragg and Williams's theory, discussed above, it is assumed that the ordering energy is proportional to the long-distance order in a crystal, yet it seems certain that the principal interactions in crystals are between very close neighbors. A logical development of superlattice theory is therefore to consider a concept of order that concerns only nearest neighbors, and to compute ordering energies based on nearest-neighbor interactions (the quasi-chemical approach). Theories based on this point of view have been worked out by Bethe[2] and extended by Williams,[3] Peierls,[4] Easthope,[5] Kirkwood,[6] and Yang and Li.[7]

Short-range order is defined in terms of the number of "right pairs" of atoms, just as long-distance order is defined in terms of the number of right atoms on a given sublattice. A right pair is a pair of unlike atoms, an AB pair. At increasing temperatures the number (or density) of AB pairs diminishes and the number of AA and BB pairs (wrong pairs) increases until a disordered state is reached in which half the pairs are right and half are wrong. The local, or short-range, order σ may be defined as the probability of finding an unlike atom beside a given atom minus the probability of finding a like atom there. Considering a certain A atom, the probability that a nearest neighbor is a B atom is $\frac{1}{2}(1 + \sigma)$, while

[1] Each point on the curve of equilibrium order satisfies two conditions that can be stated as follows: (1) The dependence of V upon S is assumed to be $V = V_0S$ and it is assumed to be practically independent of temperature. (2) The dependence of equilibrium order upon V is dependent upon the Boltzmann factor $e^{-V/kT}$ in a way that leads to the relation $S = \tanh (V/4kT)$ when applied to a superlattice where $r = \frac{1}{2}$ (e.g., CuZn, Fe_3Al). Plotting S against V, the equilibrium degree of order at any temperature is found at the intersection of the straight line of relation 1 for the given temperature with the curve of relation 2. At this intersection the interchange energy V caused by the amount of order present just balances the shuffling tendency of thermal agitation. As the temperature is raised, the intersection of relations 1 and 2 occurs at lower and lower S values until the intersection occurs at $S = 0$ at the critical temperature. Here $S = \tanh (V/4kT)$ is approximately equal to $S = V/4kT$, and by substituting relation 1 for this temperature T_c, one obtains $V_0 = 4kT_c$.

[2] H. A. Bethe, *Proc. Roy. Soc. (London)*, vol. A150, p. 552, 1935.

[3] E. J. Williams, *Proc. Roy. Soc. (London)*, vol. A152, p. 231, 1935.

[4] R. Peierls, *Proc. Roy. Soc. (London)*, vol. A154, p. 207, 1936.

[5] C. E. Easthope, *Proc. Cambridge Phil. Soc.*, vol. 33, p. 502, 1937.

[6] J. G. Kirkwood, *J. Chem. Phys.*, vol. 6, p. 70, 1938.

[7] C. N. Yang, *J. Chem. Phys.*, vol. 13, p. 66, 1949. C. N. Yang and Y. Y. Li, *Chinese J. Phys.*, vol. 17, p. 59, 1947. Y. Y. Li, *J. Chem. Phys.*, vol. 17, p. 447, 1949.

the probability that it is an A atom is $\frac{1}{2}(1 - \sigma)$. The Boltzmann factor gives the ratio of these two when equilibrium is reached at a temperature T:

$$\frac{\frac{1}{2}(1 - \sigma)}{\frac{1}{2}(1 + \sigma)} = e^{-V/kT} \tag{11–2}$$

where V is the change in energy of the crystal when one pair is changed from an AB to an AA pair. This energy V must be positive if a superlattice is to form; if it is negative, there will be a tendency for like atoms to cluster together and precipitate from solid solution.[1] If V_{AA}, V_{BB}, and V_{AB} are the energies associated with the pairs AA, BB, and AB, respectively, then

$$V = \frac{1}{2}(V_{AA} + V_{BB}) - V_{AB} > 0 \tag{11–3}$$

If V were a constant independent of the degree of order, we should find σ decreasing slowly toward zero at high temperatures, and there would be no critical point. In Bethe's theory, V is assumed to depend on order in a manner he computes from the long-distance order that exists in the crystal, and this accounts for the accelerated decline toward zero as the temperature is raised toward the critical point.

The curve resembles the curve for the Bragg-Williams theory at low temperature, as will be seen from Fig. 11–15, but at the critical temperature Bethe's theory predicts that σ does not fall entirely to zero but to a residual value greater than zero, a value which in turn gradually decreases as the temperature is still further increased. Thus, even at high temperatures there are more than the random number of AB atom pairs; and while they are unable to link up together into a constant long-distance order in the crystal, they are able to form small domains within which there is order. At T_c the domains begin to hook together into long-distance order, and as the temperature is lowered the long-distance order increases toward perfection. Even at low temperatures, however, a crystal may be divided into domains that are out-of-step with each other.[2] The configurational energy for each degree of order may be computed by statistical mechanics from the number of different arrangements of atoms that will have that degree of order.

The existence of local order above the critical temperature has been confirmed by observations on specific heats of superlattices (see below), by the presence of certain diffuse rings and spots in electron diffraction patterns,[3] and by the diffuse scattering of x-rays.[4]

[1] The theory of precipitation from a simple binary eutectic system has been discussed by R. Becker, *Z. Metallk.*, vol. 29, p. 245, 1937. See R. F. Mehl and L. K. Jetter in "Symposium on Age Hardening," American Society for Metals, Cleveland, Ohio, 1940.

[2] The recent calculations of J. M. Cowley, *Phys. Rev.*, vol. 120, p. 1648, 1960, imply that at all temperatures below T_c the equilibrium state of order may be one in which there is a fluctuation in the average composition as the result of out-of-phase domains.

[3] L. H. Germer, F. E. Haworth, and J. J. Lander, *Phys. Rev.*, vol. 61, p. 93, 1942. H. Raether, *Z. Angew. Phys.*, vol. 4, p. 53, 1952. In a recent investigation of thin foils electrolytically etched from the bulk of Cu-Au alloys containing 22.2, 24.4, and 31.6 atomic percent Au, M. J. Marcinkowski and L. Zwell, *Acta Met.*, vol. 11, p. 373, 1963, obtained electron diffraction patterns which show diffuse superlattice reflections cor-

More recently the presence of short-range order above T_c has been deduced in beta-brass[5] by means of neutron diffraction. In alpha-brass (in which the presence of long-range order is still uncertain)[6] the presence of dislocations associated in pairs points to regions of local order,[7] and such dislocations have been observed in transmission electron micrographs.[8] Changes in other physical properties also suggest a possible presence of local order.[9]

Bethe's first treatment of the local-order problem was limited to simple structures of composition AB; this was extended to composition AB_3 by Peierls[10] and generalized to compositions other than stoichiometric ratios by Kirkwood,[11] Easthope,[12] and others.

Definition of short-range order in terms of multiple parameters The term σ, first introduced by Bethe, refers only to binary alloys and to nearest neighbors. In principle, however, one can describe any possible kind of atom pair at any interatomic distance in terms of individual parameters. Cowley[13] has devised a theory in which the interactions of an atom with all atoms in concentric shells around it are treated. Instead of referring to sublattice occupation, one considers a series of concentric shells having specified either an A atom or a B atom as center. Considering a certain B atom in an alloy of A and B atoms, there will be c_i atoms on the ith shell around it, of which n_i will be, on the average, A atoms. Then the short-range order may be considered in terms of the parameters $\alpha_i = 1 - n_i/m_A c_i$, where m_A is the fraction of atoms in the alloy that are A atoms. These parameters are zero for the completely disordered state, but

---

responding to the Cu_3AuII superlattice in samples both quenched from above T_c and also studied at temperature. Similar findings were made also by S. Yamaguchi, D. Watanabe, and S. Ogawa, *J. Phys. Soc. (Japan)*, vol. 17, p. 1030, 1962, on the Cu_3Au alloy. H. Sato, D. Watanabe, and S. Ogawa found evidence of nuclei of the CuAuII superlattice in thin evaporated films of CuAu studied about 50°C above T_c. Such experiments suggest that short chains of antiphase domains of the long-period-type superlattices exist sporadically in the disordered lattices well above T_c.

[4] Z. W. Wilchinsky, *J. Appl. Phys.*, vol. 15, p. 806, 1944. J. M. Cowley, *J. Appl. Phys.*, vol. 21, p. 24, 1950. B. W. Roberts and G. H. Vinegard, *J. Appl. Phys.*, vol. 27, p. 203, 1956. D. R. Chipman, *J. Appl. Phys.*, vol. 27, p. 739, 1956.

[5] C. B. Walker and D. T. Keating, *Phys. Rev.*, vol. 130, p. 1726, 1963.

[6] L. M. Clarebrough, M. E. Hargreaves, and M. H. Loretto, *J. Australian Inst. Metals*, vol. 6, p. 104, 1961.

[7] J. B. Cohen and M. E. Fine, *Acta Met.*, vol. 11, p. 1106, 1963.

[8] G. Thomas, *Australian J. Met.*, vol. 8, p. 80, 1963. W. Bell, W. R. Roser, and G. Thomas, *Acta Met.*, vol. 12, p. 1247, 1964.

[9] A review of these may be found in: L. M. Clarebrough, M. E. Hargreaves, and M. H. Loretto, *Proc. Roy. Soc. (London)*, vol. A257, p. 338, 1961; vol. A261, p. 500, 1961.

[10] R. Peierls, *Proc. Roy. Soc. (London)*, vol. A154, p. 207, 1936.

[11] J. G. Kirkwood, *J. Chem. Phys.*, vol. 6, p. 70, 1938.

[12] C. E. Easthope, *Proc. Cambridge Phil. Soc.*, vol. 33, p. 502, 1937.

[13] J. M. Cowley, *Phys. Rev.*, vol. 77, p. 669, 1950; *J. Appl. Phys.*, vol. 21, p. 24, 1950; *Phys. Rev.*, vol. 120, p. 1648, 1960.

for the completely or partially ordered state, they may have various positive and negative values, depending upon the alloy structure. Negative values of α_i suggest an excess of unlike neighbors in a given shell i, and positive values an excess of like neighbors. Hence, both ordering and clustering tendencies can be considered in any given solution. Cowley uses these parameters α_i because they can be estimated experimentally by an analysis of the diffuse x-ray scattering.[1] The probability that an atom at lmn will be an A atom is given by $p_{lmn} = m_A(1 - \alpha_{lmn})$, if a B atom is at the origin, and by $p_{lmn} = m_A + m_B\alpha_{lmn}$ if an A atom is at the origin, where m_B is the atom fraction of B atoms, and α_{lmn} has the same significance as α_i. These probabilities govern the x-ray scattering power, and in fact, the α_{lmn} are the coefficients of the three-dimensional Fourier series that describes the intensity of the diffuse scattering throughout the reciprocal lattice, as discussed in later paragraphs. By expressing the configurational energy and the entropy of the structure in terms of α_i, m_Am_B, kT, and energies of interaction of an atom with like and with unlike atoms in the ith shell, and by making certain approximations, the free energy is computed.[2] The principle that the free energy is a minimum at equilibrium then enables a computation of the interaction energies from the values of α_i determined by x-rays. Then equations for the long-range order S are obtained by considering the limiting case of i becoming very large, and the curve of S vs. temperature is found to fit experimental data for Cu_3Au[3] and CuZn[4] better than previous theories and, in fact, within experimental error. The curves of α_i for the short-range order vs. T are also computed, and when they are compared with experiment for the case of Cu_3Au the agreement is again surprisingly good. The theory predicts properly that there will be a "liquidlike" distribution of atoms about a given atom, in that certain shells will have an excess and others will have a deficiency of like atoms, this tendency being more dependent on the radial distance from the given atom than upon lattice coordinates. The observed variation of α_i with composition is accounted for by the theory, and the variation of T_c with composition in the Cu-Au system is fitted better by this theory than by the earlier theories. Asymmetry in the phase diagram about the 50–50 composition is predicted; this stems from the assumption that interaction between neighbors depends on interatomic distance, which increases from 2.55 to 2.88 A in going from Cu to Au. (An inverse sixth-power law for the dependence of energy on distance is assumed.) Superlattices are predicted

[1] It has been suggested that these parameters should be called "Warren s.r.o. parameters" in honor of their originator; almost all workers in this field are students or collaborators of B. E. Warren.

[2] One of the important approximations in Cowley's original theory, *Phys. Rev.*, 1950, loc. cit., was that the order parameters α_i are independent of one another. Consideration of their interdependence has produced certain modifications in the original theory (Cowley, 1960, loc. cit.).

[3] J. M. Cowley, *J. Appl. Phys.*, loc. cit., p. 25. B. W. Roberts and G. H. Vinegard, *J. Appl. Phys.*, vol. 27, p. 203, 1956. D. R. Chipman, *J. Appl. Phys.*, vol. 27, p. 739, 1956.

[4] B. E. Warren and D. R. Chipman, *Phys. Rev.*, vol. 75, p. 1629, 1949. D. R. Chipman and B. E. Warren, *J. Appl. Phys.*, vol. 21, p. 696, 1950. D. T. Keating and B. E. Warren, *J. Appl. Phys.*, vol. 22, p. 286, 1951. L. Muldawer, *J. Appl. Phys.*, vol. 22, p. 663, 1951.

at $AuCu_3$, AuCu, and Au_3Cu, with T_c being a maximum at each stoichiometric composition.

The application of neutron diffraction and electron diffraction, particularly to alloys with imperfect states of order characteristic of high temperatures, has permitted further advance in this field. Walker and Keating[1] recently made measurements of the diffuse scattering of monochromatic thermal neutrons from single crystals of beta-brass (Cu-Zn system) at temperatures above the critical temperature. Neutron scattering factors differ appreciably for these particular elements, and this difference was further enhanced by using copper enriched to about 98 percent with an isotope, Cu^{65}. The measurements, obtained at 550°C and higher temperatures, clearly showed the existence of pronounced short-range order extending over surprisingly large distances, through at least the tenth neighbors. The observed intense diffuse scattering has been called *critical scattering*, by analogy with the critical opalescence in light scattering or the critical magnetic scattering of neutrons from magnetic substances near their Curie points.

Alternate views on the theories of ordering Despite much progress and good agreement between theories of ordering and experimental observations, the origins of the forces which produce ordering (both short-range and long-range) are still far from clear. Mounting evidence suggests that in all probability there are a number of different factors involved in the driving energy which is responsible for ordering and that these predominate to a differing degree, according to each specific case. The three major approaches usually discussed in connection with order are: (1) the pairwise interaction leading to a quasi-chemical theory, in which energy is lowered due to reduced interaction energy between unlike neighbors, (2) the strain-relaxation theory, in which superlattice formation reduces strain energy in a solid solution composed of atoms of different size, and (3) the electron–Brillouin-zone interaction theory (see Chap. 13), in which energy of the conduction electrons is lowered by a specific interaction between the Fermi surface of the electrons and the Brillouin zone of the superlattice. Most of the theoretical work discussed in the previous section has been based upon the quasi-chemical nearest-neighbor theory; the alternate views are rather infrequently discussed in the literature, or avoided because of inherent difficulties.

The simplest quasi-chemical theories assume that only nearest-neighbor pairs of atoms in the crystal interact and that the interaction terms are spherically symmetrical. These theories can only deal with the formation of nearest-neighbor pairs of unlike atoms, and cannot, for example, explain the existence of certain ordered phases with very large unit cells (see below) or the ordering in CuPt in terms of alternate layers of like atoms on the (111) planes (see page 275) in which neither long-range nor short-range order produces any increase in the number of unlike nearest neighbors. Walker,[2] Fournet,[3] and Slater[4]

[1] C. B. Walker and D. T. Keating, *Phys. Rev.*, vol. 130, p. 1726, 1963.

[2] C. B. Walker, *J. Appl. Phys.*, vol. 23, p. 118, 1952.

[3] G. Fournet, *Bull. Soc. Franc. Mineral. et Crist.*, vol. 77, p. 711, 1954; *Compt. Rend.*, vol. 232, p. 155, 1951; *Compt. Rend.*, vol. 235, p. 1377, 1952.

[4] J. C. Slater, *Phys. Rev.*, vol. 84, p. 179, 1951.

have discussed the Cu-Pt system theoretically. One possibility is to consider ordering forces between layers of atoms; another is to use the quasi-chemical approach by introducing energies of interaction between more remote pairs of atoms. Nevertheless, success in accounting for observed degrees of order in terms of certain chosen energy-interaction parameters and a suitable model does not necessarily pinpoint the real driving force for ordering, which for example, might be the reduction of the strain energy.

By using the quasi-chemical approximation the equilibrium properties of a number of alloys have been calculated with varying degrees of success by the methods of statistical thermodynamics and have been compared with the experimentally measured degree of short-range order using x-rays or certain thermodynamical properties such as thermodynamic activity. The values for Cu-Zn (alpha-brass) and certain compositions in the Ag-Au, Ag-Zn, Ag-Cd, Ag-Al, and Al-Zn systems have been deduced from measurements of the vapor pressures, or by means of a galvanic cell. Negative deviations from ideal solution behavior correlate with the formation of more *A-B* bonds at the expense of *A-A* and *B-B* bonds than would be found in truly random solutions, and in each of the above cases where size disparity between the component atoms is not too severe this correlation is found.[1]

Further improvements in the interaction-energy models have been produced by allowing for more distant interactions between atoms that are not nearest neighbors[2] and by making the interaction energy concentration-dependent.[3] Other considerations involve the additional influence of the entropy of mixing, the vibrational entropy, and strain energy.[4]

The quasi-chemical approximation is quite inadequate for systems in which a considerable disparity between atomic radii may exist, such as, for example, Co-Pt or Au-Ni. This has led to the discussion of the contribution of the strain energy in solid solutions to the total energy[5] and its tendency to be reduced by either ordering or clustering in alloys.[6] The principle that superlattice formation should reduce strain energy has long been recognized.[7] For those structures in which a consistent antiferromagnetic arrangement of unpaired *d* electron spins is possible, the energy associated with the antiferromagnetism may also be a factor influencing superlattice formation.

The detection of order with x-rays: temperature dependence The degree of long-distance order in a superlattice may be determined from the intensity of the superlattice lines. For example, if the equation for the intensity of reflection (Chap. 4) is evaluated for $AuCu_3$, where atoms with

[1] A number of papers are of interest in this connection: Y. Takagi, *Proc. Phys. Math. Soc. (Japan)*, vol. 23, p. 44, 1941. C. E. Birchenall, *Trans. AIME*, vol. 171, p. 166, 1947. L. Guttman, *Trans. AIME*, vol. 175, p. 178, 1948. C. E. Birchenall and C. H. Cheng, *Trans. AIME*, vol. 185, p. 428, 1949. R. A. Oriani, *Acta Met.*, vol. 2, p. 608, 1954. J. E. Hilliard, B. L. Averbach, and M. Cohen, *Acta Met.*, vol. 2, p. 621, 1954. P. S. Rudman and B. L. Averbach, *Acta Met.*, vol. 2, p. 576, 1954. B. L. Averbach, P. A. Flinn, and M. Cohen, *Acta Met.*, vol. 2, p. 92, 1954.

[2] Reference to a number of papers by G. Fournet and H. Sato will be found in the review by Guttman, loc. cit.

[3] See, for example, M. Hillert, *J. Phys. Radium*, vol. 23, p. 835, 1962.

[4] See, for example, A. W. Lawson, *J. Chem. Phys.*, vol. 15, p. 831, 1947; *Trans. ASM*, vol. 42A, p. 85, 1950.

[5] Ibid. E. S. Machlin, *J. Metals*, vol. 6, p. 592, 1954.

[6] R. A. Oriani, *Acta Met.*, vol. 1, p. 144, 1953. B. L. Averbach, P. A. Flinn, and M. Cohen, *Acta Met.*, vol. 2, p. 92, 1954. P. A. Flinn, B. L. Averbach, and M. Cohen, *Acta Met.*, vol. 1, p. 664, 1953. L. L. Seigle, M. Cohen, and B. L. Averbach, *J. Metals*, vol. 4, p. 1320, 1952. P. S. Rudman and B. L. Averbach, *Acta Met.*, vol. 5, p. 65, 1957.

[7] See, for example, W. Hume-Rothery and H. M. Powell, *Z. Krist.*, vol. 91, p. 23, 1935.

scattering power f_{Au} are at positions with coordinates 000 and atoms with scattering power f_{Cu} are at $\frac{1}{2}\frac{1}{2}0$, $\frac{1}{2}0\frac{1}{2}$, and $0\frac{1}{2}\frac{1}{2}$, it will be seen that there are two classes of reflections. The main reflections occur when indices are all odd or all even, and have intensities proportional to $|F|^2 = (f_{Au} + 3f_{Cu})^2$. The superlattice lines occur when indices are mixed odd and even, and have intensities proportional to $|F|^2 = (f_{Au} - f_{Cu})^2$ in a completely ordered alloy and proportional to $S^2(f_{Au} - f_{Cu})^2$ in a partially ordered alloy having S as the degree of long-distance order. Thus, in a fully disordered alloy $S = 0$ and the superlattice lines vanish, since each lattice point then has, on the average, the same scattering power. Measurements of long-range order by reflected intensities require consideration of the geometry used, absorption, extinction, the multiplicity of the reflection, and the temperature factor, and have been made using both powder and single-crystal methods. If the atomic scattering factors possess nearly equal values as, for example, in the Cu-Zn system, the intensity of the superlattice reflections may be insufficient to be recorded unless special techniques are used. With powder techniques one can sometimes employ x-rays of special wavelength which enhance the difference as a result of the divergence in the anomalous scattering factors.[1] Even more helpful are single-crystal methods where the diffractometer counter can be positioned directly at the expected superlattice reflection.

Single-crystal methods have the advantage of giving greater intensities. The use of diffractometers permits a detailed study of the changes of intensity with time and the determination of the shape of the superlattice reflections. The long-range order in single crystals of Cu_3Au has been studied with the use of diffractometers[2] and also the order in single crystals of beta-brass.[3] Batterman[4] has recently shown that long-range order exists in $CuAu_3$ alloys below the critical temperature $T_c = 199°C$. He found no order at compositions richer in Au than $CuAu_3$.

If the structure of a superlattice is tetragonal, it has been shown that the axial ratio can be used as an index of order.[5] The tetragonality of the superlattices CuAu,[6] Cu_3Pd,[7] and CoPt[8] have been studied in this manner.

The nature of the short-range order in alloys has been studied by investigating the intensity of the diffuse background between the main lattice reflections in the regions where superlattice peaks would appear with long-range order. The x-ray techniques for measuring short-range order have

[1] See, for example, F. W. Jones and C. Sykes, *Proc. Roy. Soc. (London)*, vol. A161, p. 440, 1937. J. E. Kittl and T. B. Massalski, *J. Appl. Phys.*, vol. 33, p. 242, 1962.

[2] J. M. Cowley, *J. Appl. Phys.*, vol. 21, p. 24, 1950.

[3] D. T. Keating and B. E. Warren, *J. Appl. Phys.*, vol. 22, p. 286, 1951.

[4] B. W. Batterman, *J. Appl. Phys.*, vol. 28, p. 556, 1957.

[5] G. Borelius, *J. Inst. Met.*, vol. 74, p. 17, 1947. A. H. Wilson, *Proc. Cambridge Phil. Soc.*, vol. 34, p. 81, 1938.

[6] B. W. Roberts, *Acta Met.*, vol. 2, p. 597, 1954. N. N. Buinov, *Zh. Eksperim. i Teor. Fiz.*, vol. 17, p. 41, 1947.

[7] D. Madoc Jones and E. A. Owen, *Proc. Phys. Soc. (London)*, vol. B67, p. 297, 1954.

[8] P. S. Rudman and B. L. Averbach, *Acta Met.*, vol. 5, p. 65, 1957.

Table 11–1 Short-range order parameters for Cu_3Au* †

Shell number i	lmn	α_i Perfect order	Cowley	Moss
1	110	−0.333	−0.152	−0.218
2	200	+1.000	+0.186	+0.286
3	211	−0.333	+0.009	−0.012
4	220	+1.000	+0.095	+0.122
5	310	−0.333	−0.053	−0.073
6	222	+1.000	+0.025	+0.069
7	321	−0.333	−0.016	−0.023
8	400	+1.000	+0.048	+0.067
9	330	−0.333	−0.026	−0.028
	411	−0.333	+0.011	+0.004
10	420	+1.000	+0.026	+0.047

* After S. C. Moss, *J. Appl. Phys.*, vol. 35, p. 3547, 1964.

† The coordinates of an atom in the ith shell are $l_i m_i n_i$; the α_i coefficients are for equilibrium at 405°C (T_c = 390°C), and would all be zero for a disordered alloy.

been developed to a large degree by Warren and his colleagues. Wilchinsky[1] showed that short-range-order parameters could be determined from powder photographs, but because of the difficulty in accurately correcting for the diffuse scattering arising from thermal motion of the atoms, powder methods are inferior to single-crystal methods. The classical single-crystal, x-ray diffuse scattering study of short-range order in Cu_3Au was made by Cowley,[2] using single-crystal diffracted intensities with monochromatized radiation. The values of α_i obtained by Cowley and the more recent ones obtained by Moss[3] are listed in Table 11–1. A number of corrections have been introduced into the method in later studies, which will be mentioned below, but the general nature of the short-range order seems to be about the same as was indicated by the original measurements: there is an excess of Cu atoms in the first shell around an atom of Au compared with a random distribution, a deficiency in the second shell, and a very small deviation from randomness ($\alpha \cong 0$) in the third shell; nonrandomness extends out much farther than had previously been thought.

The determination of order with x-rays: other complicating factors Following the original work of Cowley, the diffraction theory describing the diffuse x-ray scattering has been modified and extended to include a

[1] Z. W. Wilchinsky, *J. Appl. Phys.*, vol. 15, p. 806, 1944.

[2] J. M. Cowley, *J. Appl. Phys.*, vol. 21, p. 24, 1950.

[3] S. C. Moss, *J. Appl. Phys.*, vol. 35, p. 3547, 1964.

number of other effects which can produce diffuse scattering contributions to the diffraction pattern of an alloy. The measured intensity is the sum of at least five major contributions:

$$I = I_{CM} + I_T + I_{LO} + I_{SE} + I_A$$

where I_{CM} is the *Compton modified scattering*,[1] I_T is the temperature-diffuse scattering, I_{LO} is the diffuse *scattering due to local order*, I_{SE} is a diffuse modulation in *scattering due to atom-size effects* in solid solutions, and I_A is *air scattering*. In addition, if measurements are made in the very low-angle region of 2θ, the *double Bragg scattering* can interfere.[2] The I_{CM} and I_A contributions are readily determined, I_A by direct measurement and I_{CM} by use of tabulated values. However, the remaining contributions present several problems. In the last decade or so the particular effects of scattering due to (1) *static displacements* resulting from disparity of atomic sizes[3] and (2) *dynamic displacements* produced by thermal vibration of the lattice[4] have been given much attention.

The existence of short-range order in an alloy produces diffuse peaks at the positions of the sharp peaks that occur with an ordered alloy, which are superimposed on a diffuse background that is modulated by the existence of differing interatomic distances between nearest neighbors—the so-called *size-effect modulation* I_{SE}. By measuring diffracted intensities from powders, it is possible to make corrections and compute some approximate short-range-order coefficients[5]; but a more reliable procedure is to use intensities from a single crystal, with measurements throughout the volume of an unsymmetrical portion of a unit cell in the reciprocal lattice, determining the short-range-order coefficients (α_i) and similar coefficients (β_i) in a series representing the size effect, combining the data from peaks and from diffuse scattering.

In the 1950s many investigations of local order and size effect were based only on measurements of the diffuse scattering between fundamental

[1] R. W. James, "The Optical Principles of the Diffraction of X-Rays," G. Bell, London, 1950.

[2] B. E. Warren, *Acta Cryst.*, vol. 12, p. 837, 1959. G. Nagorsen and B. L. Averbach, *J. Appl. Phys.*, vol. 32, p. 688, 1961. Double scattering has caused not only forbidden spots but also stacking-fault streaks at forbidden positions: S. Fujime, D. Watanabe, and S. Ogawa, *J. Phys. Soc. (Japan)*, vol. 19, p. 711, 1964 (electron diffraction from Co).

[3] K. Huang, *Proc. Roy. Soc. (London)*, vol. A190, p. 102, 1947. B. E. Warren, B. L. Averbach, and B. W. Roberts, *J. Appl. Phys.*, vol. 22, p. 1493, 1951. B. Borie, *Acta Cryst.*, vol. 10, p. 89, 1957; vol. 12, p. 280, 1959. S. C. Moss, *J. Appl. Phys.*, vol. 35, p. 3547, 1964. References on experimental work done by a number of Russian workers will be found in the papers by B. Borie.

[4] A. Münster and K. Sagel, *Z. Phys. Chem.*, vol. 12, p. 147, 1957. C. B. Walker and D. T. Keating, *Acta Cryst.*, vol. 14, p. 1170, 1961. Moss, loc. cit.

[5] B. E. Warren and B. L. Averbach in "Modern Research Techniques in Physical Metallurgy," p. 95, American Society for Metals, Cleveland, Ohio, 1953. B. E. Warren, B. L. Averbach, and B. W. Roberts, *J. Appl. Phys.*, vol. 22, p. 1493, 1951.

peaks,[1] but in some investigations the diffuse scattering data were supplemented by integrated intensities of the peaks, treating the displacements due to atom-size disparity as a quasi temperature reduction in the intensities of the fundamental reflections.[2] A further improvement is possible if the different kinds of atoms are treated individually, as has recently been done, and if the corrections for I_T and I_{SE} are minimized by measuring the intensities in a unit cell of reciprocal space that is near the origin, where these corrections are smaller than they are at larger distances from the origin.[3]

An experimental determination of short-range-order coefficients involves the measurement of weak intensities at many orientations, preferably at temperatures above the critical temperature for long-range ordering; it also involves standardizing the intensities with respect to the intensity of the incident beam, and it requires corrections that frequently amount to a considerable fraction of the entire measured intensity. An experimental determination of size-effect modulation and the determination of size-effect coefficients[4] encounter similar difficulties. A method of suppressing the modified radiation is mentioned on page 191, Chap. 7.

It is apparent that significant, reliable work in this field can only be expected from investigators well equipped with apparatus, with a fundamental and thorough understanding of the methods, approximations, and assumptions involved, and with the time and patience for the many precision measurements and corrections necessary.

Disorder in superlattices is not the only form of crystalline disorder that can be effectively studied by diffraction; a bibliography on diffraction from disordered crystals in general is given by Jagodzinski,[5] and a treatment of disorder scattering from a different standpoint is given by Beeman et al.[6]

[1] See, for example, P. A. Flinn, B. L. Averbach, and M. Cohen, *Acta Met.*, vol. 1, p. 664, 1953 (Au-Ni system). P. S. Rudman and B. L. Averbach, *Acta Met.*, vol. 2, p. 576, 1954 (Al-Zn and Al-Ag systems). P. S. Rudman and B. L. Averbach, *Acta Met.*, vol. 5, p. 65, 1957 (Co-Pt system). B. W. Roberts, *Acta Met.*, vol. 2, p. 597, 1954 (Cu-Au system). F. H. Herbstein and B. L. Averbach, *Acta Met.*, vol. 4, p. 414, 1956 (Li-Mg system). J. M. Dupuoy and B. L. Averbach, *Acta Met.*, vol. 9, p. 755, 1961 (Mo-Ti system). E. Suoninen and B. E. Warren, *Acta Met.*, vol. 6, p. 172, 1958 (Ag-Zn system). C. R. Houska and B. L. Averbach, *J. Appl. Phys.*, vol. 30, p. 1525, 1959 (Cu-Al system).

[2] For example, see F. H. Herbstein, B. S. Borie, and B. L. Averbach, *Acta Cryst.*, vol. 9, p. 466, 1956. B. S. Borie, *Acta Cryst.*, vol. 10, p. 89, 1957; vol. 12, p. 280, 1959. C. R. Houska and B. L. Averbach, *J. Chem. Phys. Solids*, vol. 23, p. 1763, 1961.

[3] E. Suoninen and B. E. Warren, *Acta Met.*, vol. 6, p. 172, 1958 (βAgZn). S. C. Moss, loc. cit. (Cu_3Au). The intensities were reduced to absolute values by a method believed to be more accurate than formerly used; namely, by reference to the scattering from polystyrene, C_8H_6, at $2\theta = 100°$, an intensity computed with scattering factors corrected for dispersion.

[4] F. H. Herbstein, B. Borie, and B. L. Averbach, *Acta. Cryst.*, vol. 9, p. 466, 1956. B. Borie, *Acta Cryst.*, vol. 10, p. 89, 1957; vol. 12, p. 280, 1959. B. L. Averbach in "The Theory of Alloy Phases," p. 301, American Society for Metals, Cleveland, Ohio, 1956.

[5] H. Jagodzinski in G. N. Ramachandran (ed.), "Advanced Methods of Crystallography," p. 181, Academic, New York, 1964, and further work, to be published.

[6] W. W. Beeman et al. in S. Flügge (ed.), "Handbuch der Physik," vol. 32, p. 321, Springer, Berlin, 1957.

Superlattice domains In a superlattice in which the long-range order is not perfect, many different atomic configurations are possible. One possibility is a single long-range scheme that has occasional atoms out of place on the atom sites of the crystal. Another possibility is that of many *domains* each of which has a long-range scheme of perfect or partial order but an arrangement of atoms that is out of step with the arrangement in each adjacent domain. Also, there may or may not be a two-phase arrangement, in which a partially ordered phase coexists with a disordered phase (or with a partially ordered phase of a different superlattice structure).

If the domains come in contact, the region of contact between them is a surface or *domain wall*. A certain amount of disorder exists there, since atoms in the vicinity of the wall will have some neighbors out of step with their other neighbors. Such atoms will find it easier to disorder than those inside the domains, causing the disordered band to widen as the temperature increases. A number of calculations of the surface energy of domain walls, and their dependence upon orientation and composition, have been made for a variety of structures.[1]

X-ray investigations have not always been adequate to distinguish between various possible models of the scheme of ordering. Some x-ray results have been interpreted solely in terms of one scheme arbitrarily chosen. An important feature of incompletely ordered samples is the tendency for domain boundaries to lie along certain planes so as to minimize domain-boundary energies. The preference for one type of boundary over another influences the short-range-order parameters; likewise, the effective mean size of the domains and the effective thickness of the domain walls influence the parameters. Thus, although much can be deduced from x-ray data alone,[2] a full specification of the state of order in a partially ordered superlattice is a complex matter that cannot always be readily deduced from x-ray studies alone. A marked advance in the knowledge of partially ordered structures occurred when it was found that the boundaries could be seen in electron micrographs, and the crystallographic nature of each boundary could be interpreted by studying contrast effects in the pictures.[3]

Antiphase domain boundaries can result not only from the growing together of different nuclei of order, but also from the passage of dislocations through an ordered crystal. A perfect dislocation having a unit Burgers vector in the disordered alloy is only a partial dislocation in the ordered alloy, since the unit cell for the ordered state is larger than for the disordered. A movement of such a dislocation across a slip plane in a superlattice therefore leaves a plane behind it that is an antiphase boundary. In order to avoid the extra energy involved in producing this boundary, dislocations in a superlattice tend to travel in pairs, with only a narrow

[1] N. Brown, *Phil Mag.*, vol. 4, p. 693, 1959. P. A. Flinn, *Trans. AIME*, vol. 218, p. 145, 1960. J. W. Cahn and R. Kikuchi, *J. Chem. Phys. Solids*, vol. 20, p. 94, 1961; vol. 23, p. 137, 1962.

[2] A. J. C. Wilson, "X-Ray Optics; the Diffraction of X-Rays from Finite and Imperfect Crystals," Methuen, London, 1960.

[3] This is discussed further on p. 301 and in Chap. 14.

ribbon of antiphase boundary extending from one to the other.[1] The passage of the second dislocation of the pair restores the ordered structure.

Electron diffraction from superlattices The distribution of points in the reciprocal space of the CuAuII superlattice shown in Fig. 11–12 is strikingly illustrated by a transmission electron diffraction pattern obtained from a thin film of the alloy as shown in Fig. 11–16.[2] By using a technique in which evaporated films are grown epitaxially on cleavage faces of rock salt and examined after removal, it is found that there is a very strong preference for the *c* axis of the long cell to be perpendicular to the plane of the film, so that the resulting diffraction pattern essentially corresponds to a pattern from a single crystal. With the incident beam along the [001] axis, the central spot corresponds to the undeviated beam, the four sets of intense fourfold satellite spots correspond to the (110) point of the fundamental reciprocal lattice, being split into four by the new periodicities; the (020) point is also surrounded by new spots. The fourfold satellite spots are sometimes referred to as crosses. When the specimen is annealed at a slightly lower temperature, a spot appears in the center of each cross indicating that a mixture of CuAuII and CuAuI exists. The crosslike spots are superlattice reflections, and the splitting is a direct consequence of the periodicity of the domain boundaries ($\frac{1}{2}b_1 \approx$ 20 A in Fig. 11–10). The domain spacing M is directly related to the separation of these spots and can be calculated quite accurately. As M increases the separation diminishes, until with "infinite" M the crosses would merge to become the {100} superlattice reflections of CuAuI. The occurrence of crosses rather than pairs of spots is due to the division of the specimen into zones within each of which the antiphase boundaries are all parallel but perpendicular to those in adjoining "unidirectional zones" (see Fig. 11–19). The resulting pattern is the superposition of patterns from a number of such zones. The satellites around the main beam are less clearly understood.[3]

Early electron diffraction work on superlattices was done by Raether[4] on Cu_3AuII and by Japanese and English investigators[5] on CuAuII. The list of other superlattices having related structures is steadily growing.[6]

[1] This principle was first suggested by J. S. Koehler and F. Seitz, *J. Appl. Mech.*, vol. A217, p. 14, 1947. Electron micrographs give direct evidence of it whenever the separation between the dislocations of a pair exceeds a few atom diameters.

[2] S. Ogawa and D. Watanabe, *J. Phys. Soc. (Japan)*, vol. 9, p. 475, 1954.

[3] A. B. Glossop and D. P. W. Pashley, *Proc. Roy. Soc. (London)*, col. vol. A250, p. 132, 1959. S. Ogawa, *J. Phys. Soc. (Japan)*, vol. 17, Suppl. B-II, p. 253, 1962.

[4] H. Raether, *Acta Cryst.*, vol. 4, p. 70, 1951; *Z. Angew, Phys.*, vol. 4, p. 53, 1952.

[5] S. Ogawa and D. Watanabe, *J. Phys. Soc. (Japan)*, vol. 9, p. 475, 1954. D. P. W. Pashley and A. E. B. Presland, *J. Inst. Metals*, vol. 87, p. 419, 1958–1959. A. B. Glossop and D. P. W. Pashley, *Proc. Roy. Soc. (London)*, vol. A250, p. 132, 1959.

[6] For references, see: S. Ogawa, *J. Phys. Soc. (Japan)*, vol. 17, Suppl. B-II, p. 253, 1962. D. P. W. Pashley and A. E. B. Presland, *Proc. Roy. Soc. (London)*, vol. A250, p. 132, 1959; *J. Inst. Met.*, vol. 87, p. 419, 1958. S. M. Marcinkowski and L. Zwell, *Acta Met.*, vol. 11, p. 373, 1963. H. Sato and R. S. Toth in T. B. Massalski (ed.), "Alloying Behavior and Effects in Concentrated Solid Solutions," Gordon and Breach, New York, 1965.

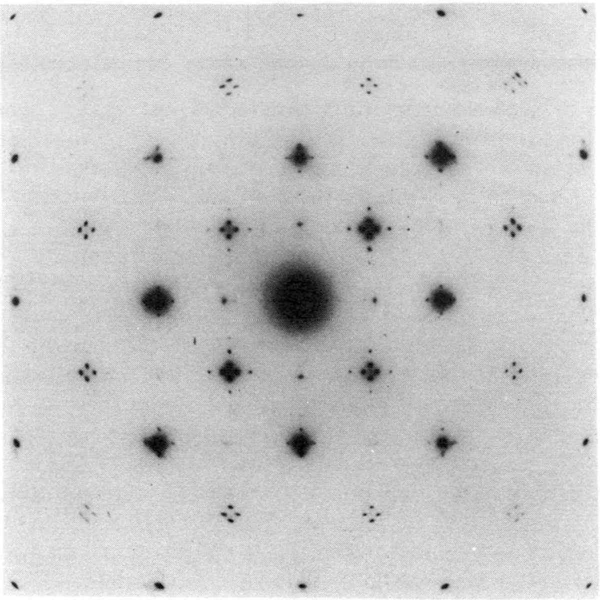

(*a*)

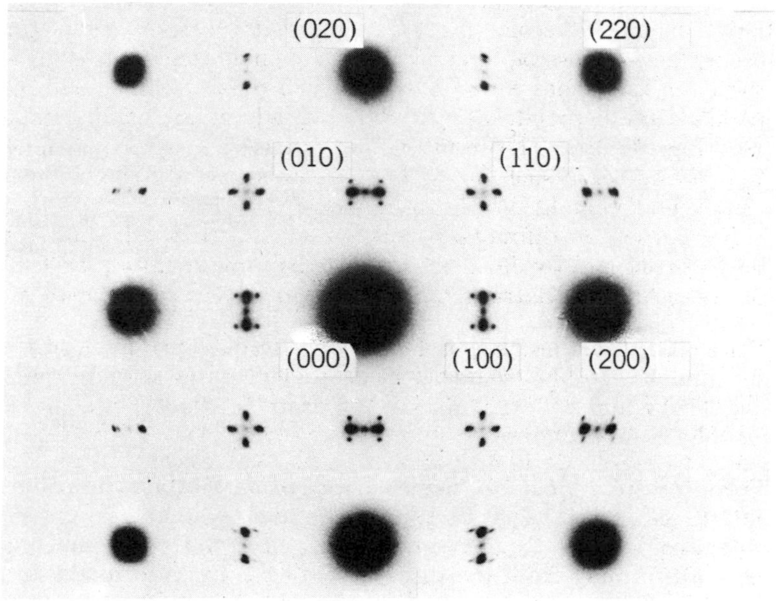

(*b*)

Fig. 11-16 (*a*) **Electron diffraction pattern of CuAuII superlattice one-dimensional antiphase domains.** (*b*) **Electron diffraction pattern of $Cu_3Pd(\alpha'')$ superlattice with two-dimensional antiphase domains. (Courtesy S. Ogawa).**

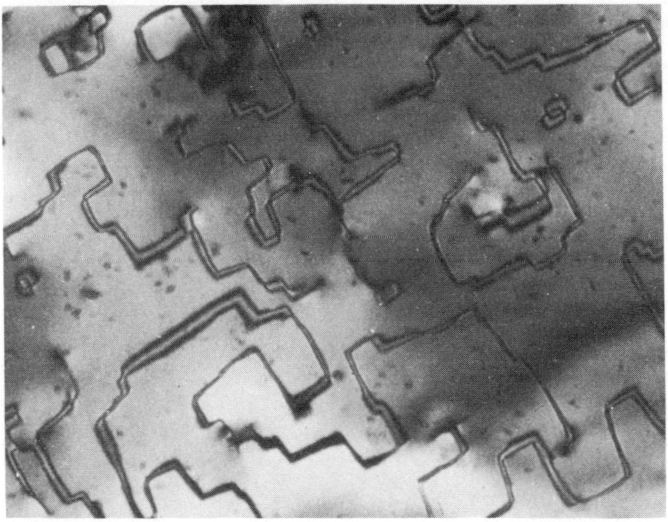

Fig. 11-17 Electron micrograph of antiphase boundaries in CuAuI superlattice prepared by evaporation techniques. (D. P. W. Pashley and A. E. B. Presland, J. Inst. Metals, vol. 87, p. 419, 1958–1959.)

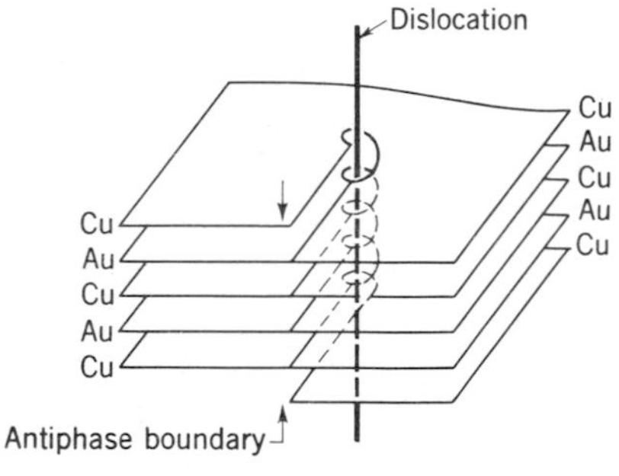

Fig. 11-18 Suggested mechanism whereby a dislocation introduces an antiphase domain boundary in the CuAuI superlattice. (After A. B. Glossop and D. P. W. Pashley, Proc. Roy. Soc. (London), vol. A250, p. 132, 1959.)

Electron diffraction patterns obtained from structures with antiphase domains often show symmetrical arrangements of satellite spots of great complexity and beauty, such as, for example, that shown in Fig. 11–16*b* for the Cu_3Pd structure.

Transmission electron microscopy of antiphase boundaries The first direct observations of antiphase boundaries in the CuAuI superlattice were made on thin evaporated films using electron transmission microscopy.[1] On transforming from the disordered f.c.c. structure to the tetragonally distorted CuAuI, the superlattice aligns itself in the evaporated film with the [001] direction (and hence the *c* axis) perpendicular to the substrate surface. A pattern of domain configurations results, in which domains of only two kinds are formed, one kind entirely surrounded by those of the

[1] Glossop and Pashley, loc. cit. Pashley and Presland, loc. cit.

Table 11–2 Displacement vectors at domain boundaries in common types of superlattices

Superlattice type	Examples	Displacement vector **p** at domain boundaries*
$B2$	CuZn	$(\frac{1}{2})a_0 \langle 111 \rangle$
DO_3	Fe_3Al, Fe_3Si	$(\frac{1}{4})a_0' \langle 111 \rangle$, $(\frac{1}{2})a_0' \langle 100 \rangle$†
$L2_1$	Cu_2MnAl	
$L1_2$	Cu_3Au	$(\frac{1}{2})a_0 \langle 101 \rangle$, $(\frac{1}{6})a_0 \langle 112 \rangle$
DO_{19}	Mg_3Cd	$(\frac{1}{2})a_0 \langle 11\bar{2}0 \rangle$, $(\frac{1}{6})a_0' \langle 10\bar{1}0 \rangle$†
$L1_0$	CuAuI	$(\frac{1}{2}) \{a_0 [\bar{1}00] + c_0 [001]\}$ etc. $\cong (\frac{1}{2})a_0 \langle 101 \rangle$ $(\frac{1}{6})a_0 \langle 112 \rangle$ (assuming $c_0 \cong a_0$)

* Indices are in terms of the unit cell of the ordered superlattice, not the smaller unit cell of the disordered alloy.

† $a_0' = 2a_0$ = length of the cell edge in the superlattice.

other kind, as shown in Fig. 11–17. Since the thin foil was slightly tilted from an orientation with [001] parallel to the electron beam, the domain boundaries appear as pairs of lines; some boundaries end on dislocations, as indicated schematically in Fig. 11–18. In both the CuAuI and Cu_3AuI superlattices there is a marked tendency for the domain boundaries to lie on cube planes of the original disordered lattice.

An important characteristic of an antiphase boundary is the displacement by which the atomic arrangement on one side of the boundary can be brought into alignment with the arrangement on the other side, i.e., the shift by which the antiphase relationship can be removed. This shift can be represented by a vector **p** that has a certain *direction* in the crystal, [*uvw*], and any magnitude that is not a complete period in the superlattice in the [*uvw*] direction. In the CuAuI and Cu_3AuI superlattices, for example, which are based on a disordered f.c.c. lattice, a typical value of **p** is $(\frac{1}{2})a_0[110]$. A table of values of **p** for some common superlattices is given in Table 11–2. If the **p** vectors are recognized, the contrast characteristics of different domain boundaries can be determined,[1] and given a crystallographic plane for the boundary, the possible changes it makes in the number of unlike first- and second-nearest neighbors can be computed.

A striking accomplishment of electron microscopy is the direct observation of the regularly spaced antiphase boundaries of long-period superlattices[2] in CuAuII. The pattern of antiphase domains in a thin film deposited from the vapor is shown in Fig. 11–19, from Pashley and Presland[3]; the *c* axis is perpendicular to the foil and the *a* axes have two different orientations normal to each other in different domains. Within each of the differently oriented domains, regularly spaced straight lines are seen which are the images of the long-period antiphase boundaries spaced, in this case,

[1] M. J. Whelan and P. B. Hirsch, *Phil. Mag.*, vol. 2, p. 1121, 1303, 1957. R. M. Fisher and M. J. Marcinkowski, *Phil. Mag.*, vol. 6, p. 1385, 1961.

[2] S. Ogawa et al., *Acta Cryst.*, vol. 11, p. 872, 1958.

[3] D. P. W. Pashley and A. E. B. Presland, *J. Inst. Met.*, vol. 87, p. 419, 1958–1959.

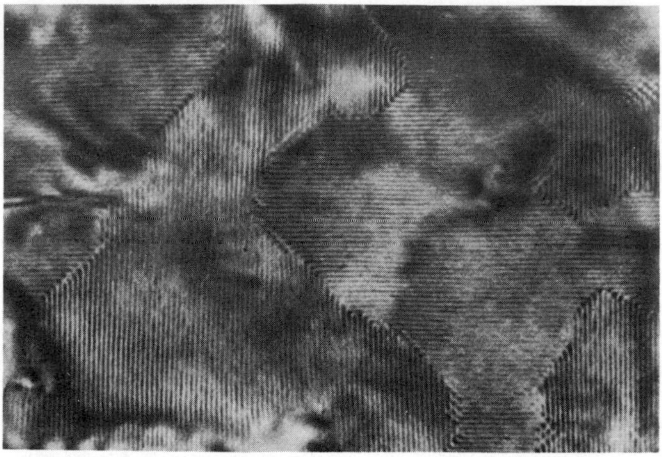

Fig. 11-19 Electron micrograph of one-dimensional periodic antiphase domain structure in CuAuII. (D. P. W. Pashley and A. E. B. Presland, J. Inst. Metals, vol. 87, p. 419, 1958–1959.)

about 20 A apart—a spacing that agrees with results obtained by x-ray diffraction.

Studies of thin foil specimens have revealed additional crystallographic features, such as the existence of twinning on an extremely fine scale in CuAuI and CuAuII,[1] the details of growth of ordered nuclei, substantial differences in the structure of samples etched from bulk samples, and deposited films.[2] Current developments in this rapidly advancing field will be found in the transactions of various conferences.[3]

Monte Carlo calculations The availability of high-speed and large memory-capacity electronic computing machines has attracted interest towards their use as a tool to attack certain problems in statistical mechanics. The *Monte Carlo techniques* involve generating a mathematical "sample" ensemble of several hundred (or several thousand) interacting atoms and allowing the energy state of this ensemble to change by computing tentative random transitions to nearby energy states. In each transition the corresponding macroscopic energy change ΔE is calculated, and the next state of the sequence of possible transitions becomes the new

[1] D. P. W. Pashley and A. E. B. Presland, "Proceedings of the European Regional Conference in Electron Microscopy," Nederlandse Vereniging voor Electronenmicroscopie, Delft, The Netherlands, 1960, vol. I.

[2] See, for example, A. M. Hunt and D. P. W. Pashley, *J. Phys. Radium*, vol. 23, p. 846, 1962.

[3] See, for example, G. Thomas and J. Washburn (eds.), "Electron Microscopy and the Strength of Crystals," Interscience, New York, 1963. J. B. Newkirk and J. H. Wernick (eds.), "Direct Observation of Imperfections in Crystals," Interscience, New York, 1962.

state or the previous (old) state according to whether or not the energy change obtained satisfies a predetermined pattern. The process amounts, therefore, to a mathematical sampling technique. The sample ensemble can be used to compute the mean values of certain desired thermodynamical functions. Monte Carlo sampling methods have been used to treat a two-dimensional hard-sphere gas[1] and, more recently, to investigate order-disorder phenomena using essentially the Ising model of interatomic interactions.[2]

Such calculations point satisfactorily to the existence of order-disorder transitions in a number of binary systems based on f.c.c. and b.c.c. structures; they also give a reasonable agreement with experimental data on long- and short-range order, heat capacities, and transition temperatures. With the present limitation of the capacity of the available computers the generated samples are just about large enough to show the cooperative behavior of atoms in the order-disorder transitions but not yet large enough to show more complex details, such as the change in the transition temperatures with the departure of the sample from the ideal composition. Their greatest use has been in the provision of a physical model for the behavior of systems with relatively strong interatomic interactions. Recent calculations by Gehlan and Cohen have developed three-dimensional models of the regions with short-range order that are consistent with measured short-range-order coefficients (without considering interaction energies).[3]

The thermodynamic order of superlattice transitions Phase transitions may be classified in the thermodynamic sense according to the degree of the lowest derivative of the free energy that shows a discontinuity. At the equilibrium temperature the (Gibbs) free energy G itself is always continuous, since the two phases must have the same free energy. The *first* and *second* derivatives of the free energy with respect to temperature T and pressure P are: $(\partial G/\partial T)_P = -S$; $(\partial G/\partial P)_T = V$; $(\partial^2 G/\partial T^2)_P = -(\partial S/\partial T)_P = -C_P/T$; and $(\partial^2 G/\partial P^2)_T = (\partial V/\partial P)_T = -\beta V$, where S and V are entropy and volume, C_P is specific heat at constant pressure, and β is *isothermal compressibility*.

Thus, a transformation is said to be of the first order (or first degree) if the free energies are equal at the transformation temperature but the entropies and volumes of the two phases differ, and of the second order if the free energies, the entropies, and the volumes are equal but the heat capacities or compressibilities differ. For a third-order transition, C_P and β should be continuous, but a discontinuity should occur in $(\partial^3 G/\partial T^3)_P$ or $(\partial^3 G/\partial T^3)_T$. Similarly, an nth-order transition is one in which a discontinuity occurs in the nth derivative of the free energy but not in a lower derivative. An example of a second-order transition in a one-component

[1] N. Metropolis et al., *J. Chem. Phys.*, vol. 21, p. 1087, 1953.

[2] Z. W. Salsburg et al., *J. Chem. Phys.*, vol. 30, p. 65, 1959. L. D. Fosdick, *Phys. Rev.*, vol. 116, p. 565, 1959. P. A. Flinn and G. M. McManus, *Phys. Rev.*, vol. 124, pp. 54–59, 1961. L. Guttman, *J. Chem. Phys.*, vol. 34, p. 1024, 1961.

[3] P. C. Gehlan and J. B. Cohen, *Phys. Rev.*, vol. 139A, p. 844, 1965.

system is the λ transformation in liquid helium. In a two-component system a second-order transformation requires not only equality of entropy and volume of the two phases but also identical composition of the two phases.[1] Thus, a first-order transformation appears on a phase diagram as two lines bounding the region where two phases of different composition coexist in equilibrium, but a second-order transformation appears as a single line. In distinguishing between first- and second-order transformations in alloys, it is usual to choose the presence or absence of a *latent heat of transformation* ΔH as the criterion for the degree of the transition. However, one encounters the difficulty that this heat is spread over a temperature range (i.e., the two-phase range). If sufficient care is taken, the presence or absence of the two-phase region may be established by x-ray investigations of equilibrated alloys. A third method is to determine by precision x-ray measurements whether or not there is a discontinuity in volume during transformation; and in general both volume and entropy will show discontinuities together. All such methods encounter the difficulty that equilibrium is reached very slowly in solid–solid reactions, particularly near transformation points. Another difficulty is connected with the procedures in which one tries to extrapolate the properties of the phases of a system into regions where they do not normally exist. In the case of a first-degree transition, extrapolation presents no problems because the existence of the transition arises only from the equality of the free energies of the two phases, and is not connected with any peculiar properties of either phase separately. For example, when a liquid is undercooled, all its properties are found to be continuous with, and predictable from, those of the liquid just above the melting point; the fact that the liquid is metastable with respect to the solid has no effect on its properties. In a second-order transition, on the other hand, such as may be associated with certain kinds of ordering processes in alloys, a well-defined ordered phase is not being replaced by a well-defined disordered phase sharply at a critical temperature T_c; rather, order decreases with increasing temperature in a single phase at a rate which is so rapid that at T_c a singularity appears, for example, in $(\partial^2G/\partial T^2)_P$, i.e., in the heat capacity.

As heat is supplied in a first-order transition, the relative amounts of the two phases involved will change but the temperature will remain constant until the transition is completed. In the second-order transition the addition of any finite quantity of heat raises the temperature of the system by a finite amount at all temperatures, including T_c; hence, the two phases never coexist. The transition from the ordered phase to the disordered phase is a continuous process involving larger and larger fluctuations, and near the T_c temperature it resembles critical *opalescence*, as mentioned earlier (page 291). The common property of the systems which may exhibit second-order transitions is that they are based on a b.c.c. lattice in which no nearest neighbors of a given atom are also nearest neighbors of one another. Systems showing first-order transitions usually involve a close-packed structure in one or both phases, in which the atoms closest to a

[1] J. W. Stout, *Phys. Rev.*, vol. 74, p. 605, 1948.

central atom are sometimes closest to each other as well.[1] Both kinds of transitions have been predicted and discussed theoretically (for additional comments, see Chap. 18).

There has been considerable discussion of whether or not an order-disorder transformation in alloys is of the first or the second order. Experimental evidence indicates at present that both kinds of transition occur, but that the second-order transition is relatively rare. The superlattice studies on the ordered phases CuAu, Cu_3Au, CoPt, $MgCd_3$, and Mg_3Cd have established these as first-order transitions.[2] The neutron scattering experiment of Walker and Keating, mentioned earlier,[3] now conclusively establishes the order-disorder transformation in beta-brass as a second-order (or higher-order) transition. Similar conditions may exist also in the case of FeCo, Fe_3Al, and Fe_3Si ordering reactions[4] (see also Chap. 18).

[1] This point is discussed in more detail by L. Guttman, *J. Chem. Phys.*, vol. 34, p. 1024, 1961.

[2] F. Rhines and J. B. Newkirk, *Trans. ASM*, vol. 45, p. 1029, 1945. R. A. Oriani, *Acta Met.*, vol. 2, p. 608, 1954. J. B. Newkirk et al., *J. Appl. Phys.*, vol. 22, p. 290, 1951. D. A. Edwards, W. E. Wallace, and R. S. Craig, *J. Am. Chem. Soc.*, vol. 74, p. 5256, 1952. K. F. Sterrett et al., *J. Phys. Chem.*, vol. 64, p. 705, 1960. G. S. Kamath, R. S. Craig, and W. E. Wallace, *Trans. AIME*, vol. 227, p. 26, 1963.

[3] C. B. Walker and D. T. Keating, *Phys. Rev.*, vol. 130, p. 1726, 1963.

[4] Guttman, loc. cit.

12

Electrons in metallic crystals

The structure of solids as well as many of the fundamental properties of solids, such as free energies, elastic constants, vibrational spectra, electrical and thermal conductivities, specific heats, etc., are related to their *electronic structure,* i.e., to the motions and distributions of electrons in crystal lattices. Very complex problems are involved in determining these parameters, and much effort is being directed toward formulating theories that might accurately predict the details of the electronic energies in crystals. Although the successes of the theories to date are rather limited, the field is developing rapidly.

As indicated in the preceding chapters, factual knowledge concerning crystal structures has grown enormously, and from this knowledge have sprung many empirical correlations and semiquantitative rules, particularly for alloy phases. To understand in a qualitative way the current interpretations of these rules requires an acquaintance with fundamentals concerning the behavior of electrons in solids such as are outlined in this chapter and the following. The theoretical prediction of what type of crystal structure will be most stable for any element or alloy is in most cases a task too difficult for theory in its present state. In many cases, however, one can discuss interactions that must be assumed in order to account for the structures that are already known to exist.

The metallic bond It is now well established that the atoms in metallic crystals are ionized and that a metal should be thought of as an assemblage of positive ions immersed in a cloud of electrons. The electrons of this cloud are relatively "free": the majority are not bound to any particular ion but move rapidly through the metal in such a way that there is always an approximately uniform density of them throughout the interior between the ions. Metallic crystals are held together by the electrostatic attraction between this "gas" of negative electrons and the positively charged ions. The binding forces in metals are thus of a different character from those in nonmetallic substances, in which the predominating forces are between neighboring atoms or between positive and negative ions, and are strongly dependent upon the interatomic distances.

Cohesive and repulsive forces in metals Wigner and Seitz[1] have shown a convenient way to calculate the attractive forces between the

[1] E. Wigner and F. Seitz, *Phys. Rev.*, vol. 43, p. 804, 1933; vol. 46, p. 509, 1934. E. Wigner, *Phys. Rev.*, vol. 46, p. 1002, 1934.

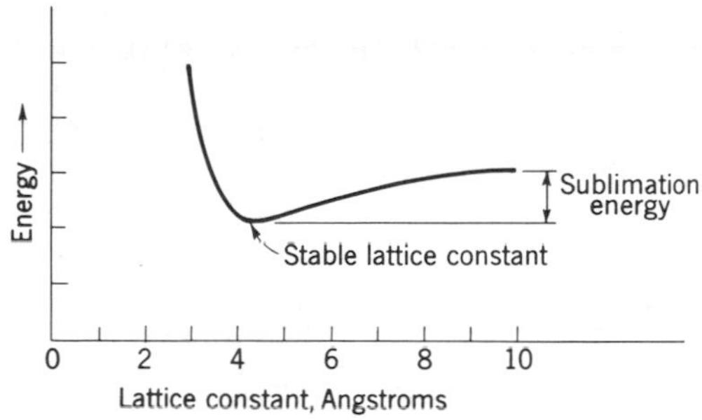

Fig. 12-1 Variation of energy with lattice constant.

positive ions and the electrons of a *metal.* Planes are drawn to bisect perpendicularly the lines joining an ion to each of its neighbors; these planes then form polyhedra surrounding each ion of the crystal. The polyhedra are convenient units for the calculation of the energy of the lattice, for each polyhedral cell contains one ion and, on the average, one free electron (in monovalent metals). The method, sometimes called *cellular*, suggests from the outset that the cohesive energy is likely to depend on the atomic volume and not on interatomic distance or the nature and number of atoms that surround a given atom.[1]

Balanced against the attraction of each positive ion for the electrons that happen to be within its electrostatic field are two principal repulsive forces: (1) the mutual repulsion of the electrons in the electron gas and (2) the repulsion between ions that are in contact.

It has been shown[2] that for monovalent crystals such as Li or Na the electronic energy E is approximately equal to the total energy of the crystal when other terms are small. If computations are carried out for various assumed values of the lattice constant of a metal, E is found to vary in the manner illustrated in Fig. 12–1 (a plot for sodium). The minimum on a curve of this type gives the theoretical prediction of the lattice constant. The difference in energy between this minimum and the value of E for an infinitely large lattice constant is the energy that would be required to remove the atoms from the crystal and scatter them to infinity; this is the heat of sublimation. The shape of the curve near the minimum is directly related to the compressibility β of the crystal: $\beta^{-1} \propto \partial^2 E/\partial V^2$, where V is the volume of the crystal.

In a detailed quantum mechanical calculation of the cohesive energy, a large number of small corrections are introduced. For lithium the results

[1] This point has been emphasized further by N. F. Mott, *Rept. Progr. Phys.*, vol. 25, p. 218, 1962.

[2] For details, see: N. F. Mott and H. Jones, "The Theory of the Properties of Metals and Alloys," chap. IV, Oxford, New York, 1936. F. Seitz, "The Modern Theory of Solids," chap. X, McGraw-Hill, New York, 1940.

of calculations[1] are in especially good agreement with experimental data: the theoretical cohesive energy at 0°K is 36.1 kcal per mole and the experimental value is 36.5. Nearly as good agreement has been obtained for the other alkali metals, the divalent metals Mg and Ca, the trivalent metal Al, and the monovalent noble metals Cu, Ag, and Au.[2] The agreement is least satisfactory in the case of the three noble metals. In these the atom cores with their *d* electrons as well as the valence electrons must be taken into account; otherwise the predicted values are too low—the inner shells of the atoms come into contact and begin to overlap before the lattice constant is reduced to the minimum point on the curve corresponding to Fig. 12–1. A strong repulsion then sets in, and the energy curve turns upward more steeply than in Fig. 12–1; therefore, the compressibility is lower. The fact that the difference between the theoretical and experimental values of the cohesive energy is smallest for Ag, in which chemical evidence (single valence) and the color of the metal indicate that electrons in the *d* shell are most firmly bound, suggests that the energy discrepancy in the calculation cited is associated with the *d* shells.[3]

The repulsive force between ions varies more rapidly with distance than does the attractive force. Consequently, when the atoms of a crystal vibrate about their mean position with an amplitude of vibration that increases with increasing temperature, the ions move away from their low-temperature positions more on the far side than on the near side of their low-temperature positions (the curve of Fig. 12–1 is not symmetrical about the minimum which represents the stable lattice constant at the lowest temperatures). Consequently, the increased amplitudes of thermal oscillations at higher temperatures result in *thermal expansion.*

Metallic valence The concept of *valence* dates from the early studies of compound formation. The established rules require that molecules form by the union of elements at integral ratios, governed by the valence principles of electroneutrality for closed-shell configurations. For example, in CaF_2, calcium plays the role of a positive divalent element, and fluorine that of a negative monovalent element. The compound is an insulator and an ionic salt. The valence electrons are exchanged between the ions, and bonding results. On the other hand, in the case of metallic compounds, such as Mg_2Sn, Mg_2Si, and Mg_2Pb, even though their structures are similar to CaF_2, the presence of a certain amount of metallic conductivity suggests the presence of partially free electrons. Here the valence electrons responsible for bonding may differ in number from those responsible for electrical conductivity.

Concepts of valence are further complicated by the extensive solid solubilities and the common occurrence of nonstoichiometric compositions of alloy phases. Chemical-type formulas given to typical alloy phases with

[1] C. Herring, *Phys. Rev.*, vol. 84, p. 283, 1951. For Na and Cs, see H. Brooks, *Trans. AIME*, vol. 227, p. 546, 1963.

[2] H. Brooks, in "Theory of Alloy Phases," p. 199, American Society for Metals, Cleveland, Ohio, 1956.

[3] N. F. Mott, *Rept. Progr. Phys.*, vol. 25, p. 218, 1962.

metallic luster and reasonable conductivity, such as KHg_{11}, Cu_5Zn_8, and Fe_5Zn_{21}, certainly do not resemble compounds that follow simple rules of chemistry. In KHg_{11} the valence of potassium may be reasonably taken as 1 but this would make the valence of mercury 1/11. Both Cu_5Zn_8 (gamma-brass) and Fe_5Zn_{21} show wide solid solubility which would permit a number of other formulas for these phases. Hence, the notion of valence needs a careful definition for the special type of bonding between metallic elements. A basically chemical approach to the problem is due to Pauling,[1] who proposed that each atom in a metal shares its valence electrons with the surrounding neighbors, forming covalent links which resonate among the available interatomic positions. The term "valence" in this sense describes the number of *bonding electrons* per atom, a number which may differ from the usually accepted valence familiar in the alloy chemistry of the element in question. For example, in a discussion based on a review of cohesion and magnetic and mechanical properties, Pauling ascribes a fractional valence of 5.56 to copper and the *diminishing* valences of 4.56, 3.56, 2.56, and 1.56 to Zn, Ga, Ge, and As, respectively, which follow copper in the Periodic Table.

A rather different view of the metallic linkage, that of an array of positive ions held together by a gas of freely moving *conduction electrons*, has been developed throughout the various stages of the electron theories described in the following sections. The electrons are treated collectively and are described by Bloch functions associated with waves extending throughout the entire crystal. Mainly those electrons which contribute to *electrical conduction* and to the Fermi surface are considered. Development of this point of view is mostly due to Mott and Jones.[2] By contrast with the Pauling scheme, the valences for Cu, Zn, Ga, Ge, and As are *increasing* from 1 to 5. Therefore, on alloying, electron concentration will be expected to increase according to one scheme and decrease according to the other. Clearly, the exact meaning of *valence* in any discussion should be specified in detail to avoid confusion.

In alloy systems that involve transition elements, rare earths, actinides, lanthanides, and transuranic elements, the assessment of valence presents further problems and may depend upon the nature of the particular properties being considered. Thus, striking regularities in a given property may be revealed in a group of related elements or alloy systems provided that some valence scheme is adopted, but consideration of another property may suggest a different scheme of valences. It is becoming increasingly clear also that further complications arise from the fact that various elements can show *variable valence*, changing throughout any given alloy system, owing to a transfer of electrons between neighboring atoms.

Despite these considerable difficulties, however, progress has been

[1] L. Pauling, "The Nature of the Chemical Bond," Cornell University, Ithaca, N. Y., 1940, 1960. L. Pauling, *Proc. Roy. Soc.* (*London*), vol. A196, p. 343, 1949. L. Pauling, in "Theory of Alloy Phases," p. 220, American Society for Metals Symposium, Cleveland, Ohio, 1956.

[2] N. F. Mott and H. Jones, "The Theory of the Properties of Metals and Alloys," Clarendon, Oxford, 1936; Dover, New York (reprint), 1958.

achieved in a systematic approach to metallic valence by considering the relative *change* of valence from one element to another within a related group of elements in the Periodic Table rather than by considering the *absolute values* of individual valences.

Because of valence uncertainties there is some advantage at times in using the concept of *average group number* (AGN), as has been done by Hume-Rothery in a discussion of intermetallic chemistry of some transition metal alloys.[1] Thus, an equiatomic alloy of Mo (group VI) and Ru (group VIII) has an AGN of 7.0, regardless of the valence scheme preferred by the reader. The AGN, integral or nonintegral, can then be interpreted by one or another valence scheme if desired. (For example, the 1960 Pauling valences[2] would assign a valence of 6 to an AGN of 7.0, although it would be 7.0 if all the outer electrons were considered.) The boundaries of single-phase regions in various alloy systems are found to occur at or near certain characteristic values of the AGN—in the past usually cited as values of the valence electron concentration.

Electronic theories of metals The application of quantum mechanics to the problems of the solid state has brought about immense advances in understanding. A brief review of the subject will be included here to form a basis for discussing modern theories and concepts of metallic structures. Detailed treatments will be found in all recent advanced books on the solid state, and intermediate treatments appear in most introductory texts on solid-state physics, physical chemistry, physical metallurgy, and science of materials.[3]

As is natural, the development of electronic theories has proceeded by

[1] W. Hume-Rothery, *J. Less-Common Metals*, vol. 7, p. 152, 1964.

[2] L. Pauling, "The Nature of the Chemical Bond," Cornell University, Ithaca, N. Y., 1960.

[3] Some elementary texts: Charles Kittel, "Elementary Solid State Physics," Wiley, New York, 1962; "Introduction to Solid State Physics," 2d ed., Wiley, New York, 1956. T. S. Hutchison and D. C. Baird, "The Physics of Engineering Solids," Wiley, New York, 1963. A. J. Dekker, "Solid State Physics," Prentice-Hall, Englewood Cliffs, N.J., 1957. L. V. Azároff, "Introduction to Solids," McGraw-Hill, New York, 1960. L. V. Azároff and J. J. Brophy, "Electronic Processes in Materials," McGraw-Hill, New York, 1963. R. B. Leighton, "Principles of Modern Physics," McGraw-Hill, New York, 1959. W. Hume-Rothery, "Atomic Theory for Students of Metallurgy," enlarged ed., Institute of Metals, London, 1960. W. Hume-Rothery, "Electrons, Atoms, Metals and Alloys," Cornwall, London, 1958. L. Brillouin, "Wave Propagation in Periodic Structures, Electric Filters and Crystal Lattices," 2d ed., Dover, N.Y., 1953. J. M. Ziman, "Electrons in Metals: A Short Guide to the Fermi Surface," Taylor and Francis, Ltd., London, 1963. D. F. Gibbons in R. W. Cahn (ed.), "Advanced Physical Metallurgy," North Holland Publishing, Amsterdam, 1965.

More advanced texts: J. M. Ziman, "Electrons and Phonons," Clarendon, Oxford, 1960. G. H. Wannier, "Elements of Solid State Theory," Cambridge, New York, 1959. R. Peierls, "Quantum Theory of Solids," Clarendon, Oxford, 1955. J. C. Slater, "Quantum Theory of Matter," McGraw-Hill, New York, 1951. F. Seitz, "The Modern Theory of Solids," McGraw-Hill, New York, 1940. N. F. Mott and H. Jones, "Theory of the Properties of Metals and Alloys," Clarendon, Oxford, 1936. A. H. Wilson, "Theory of Metals," 2d ed., Cambridge, New York, 1953. R. A. Smith, "Wave Mechanics of Crystalline Solids," Chapman and Hall, London, 1961.

stages, from a treatment of electrons as if they constituted a gas of minute particles to a quantum-mechanical description of electron motion in terms of wave functions in a periodic potential. The first important theory of metals was proposed by Drude and developed further by H. A. Lorentz soon after the discovery of the electron. The electric potential in the interior of a metal was assumed to be uniform. The conduction electrons in this uniform field moved about obeying the laws of the kinetic theory of gases (Maxwell-Boltzmann statistics), but when a voltage was applied to the metal they drifted down the electric potential gradient, carrying electric current through the metal. This is the basis of the *classical free-electron theory*. The introduction of wave mechanics, which took account of the fact that a moving electron behaves as if it were a system of waves, permitted a major advance in the theory of metals, leading to the *quantum free-electron theory* proposed by Sommerfeld in 1928,[1] after the wave nature of the electron had been established by the electron diffraction experiments of Davisson and Germer and of G. P. Thompson. The theory described the velocities that a wavelike electron could have as it moved around in a metal, provided the electric potential everywhere in the metal was uniform and not perturbed by the atomic arrangement. The wavelength of the electron (or more properly, the wavelength that describes the motion of the electron) had been shown by de Broglie to depend on the velocity of the electron according to the relation $\lambda = 2\pi\hbar/mv$, where $2\pi\hbar = h$, h is the Planck's constant, m is the mass of the electron, and v is its velocity.[2] A much-used quantity in electron theories is the *wave number* k, which in a one-dimensional case is the number of waves in a length 2π. Thus, $k = 2\pi/\lambda = mv/\hbar$, which shows that the wave number of a free electron is proportional to its momentum mv.

Since the kinetic energy of a free electron with wave number k is

$$E_k = \frac{mv^2}{2} \tag{12–1}$$

it can be expressed as

$$E_k = \frac{\hbar^2}{2m} k^2 \tag{12–2}$$

in which form it is clear that the kinetic energy is proportional to k^2. Since the potential energy of the electron within the metal is assumed to be constant, it may be taken arbitrarily to be zero, and the total energy is then equal to the kinetic energy. The relation between energy and wave number is then parabolic, as shown in Fig. 12–2. The curve is symmetrical about $k = 0$ because electrons moving in opposite directions with the same velocity would have the same energy. In many problems it is the wave vector $\mathbf{k}$ rather than the wave number k which must be considered; this vector gives the direction of motion of the electron and has the magnitude

[1] A. Sommerfeld, *Z. Physik*, vol. 47, p. 1, 1928.

[2] $\hbar = h/2\pi = 6.624 \times 10^{-27}/2\pi = 1.054 \times 10^{-27}$ erg sec; $m = 9.1 \times 10^{-28}$ gram

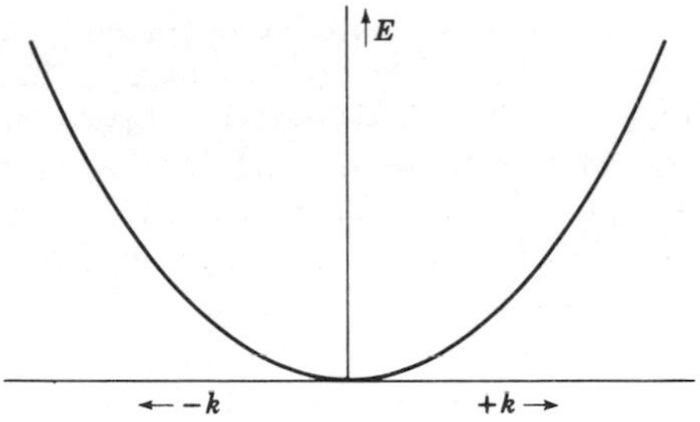

Fig. 12-2 Parabolic relationship between energy and wave number.

$|\mathbf{k}| = k$. The free electrons are taken to be the valence electrons of the atoms composing the solid. Outside the metal the electrons have a potential energy several electron volts higher than inside. The above model is therefore often described as "an electron in a potential box."

Problems dealing with wave motion are handled by studying the solutions to an appropriate "wave equation." Here the equation needed is the Schrödinger equation,

$$\nabla^2\psi + \frac{\hbar^2}{2m}[E - V(\mathbf{r})]\psi = 0 \qquad (12\text{–}3)$$

where $V(\mathbf{r})$ is the potential energy of the electron as a function of its position $\mathbf{r}$, and E is the total energy of the electron.[1] The quantity ψ is the "wave function" or "eigen function"; it is a complex quantity best understood in terms of the square of its absolute value, since $|\psi|^2 dv$ is defined as the probability that an electron is present in the element of volume dv.

The solutions of the one-dimensional Schrödinger equation for electrons in a box (of "length" L) are a set of values of wave functions, each of which satisfies certain boundary conditions. These conditions are frequently taken as the requirement that $\psi = 0$ beyond the boundaries of the box. The solutions of Eq. (12–3) then are a set of wave functions, ψ, representing standing waves with nodes at the boundaries of the box. Since we are concerned here not only with standing waves but with traveling waves, we take an alternate set of boundary conditions suitable for the latter. We do not require that $\psi = 0$ at the boundaries, but we require that, for a box of length L in the x direction, the curve of $\psi(x)$ where it leaves the box on one side join smoothly with the $\psi(x)$ curve entering the box on the other side (as if the ends of the box were bent around and joined as atoms). Thus, we require that the wave function be periodic in $\psi(x)$ with period L, and we require analogous periodicity in the y and z directions. These "periodic boundary conditions" can be written

$$\psi(x + L, y, z) = \psi(x, y, z)$$

The solutions then represent traveling waves. The allowed energy levels E_n are

$$E_n = \frac{\hbar^2}{2m}\left(\frac{2\pi}{L}\right)^2 n^2 \qquad (12\text{–}4)$$

where n has the integral values $n = \pm1, \pm2, \pm3, \ldots$, which are quantum numbers. By Eq. (12–2) these allowed energy values correspond to allowed values of k given by

$$k = \frac{2\pi n}{L} \qquad (12\text{–}5)$$

[1] The first term, $\nabla^2\psi$, written out in Cartesian coordinates, is $d^2\psi/dx^2 + d^2\psi/dy^2 + d^2\psi/dz^2$.

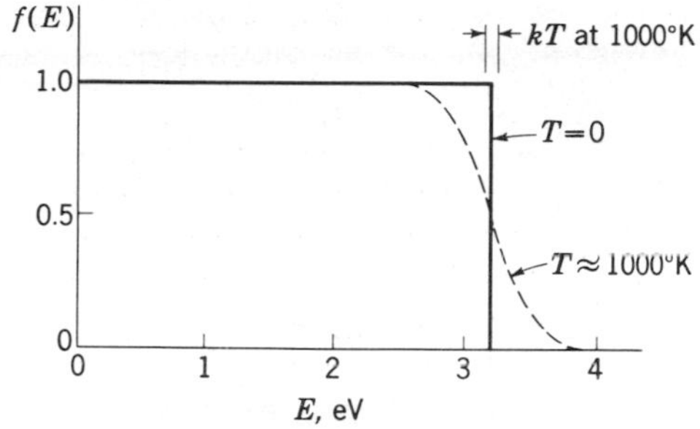

Fig. 12-3 The Fermi-Dirac distribution function at absolute zero and at temperature T.

The corresponding three-dimensional case would be for an electron confined to a cubic box of edge length L. The allowed energies are then given by

$$E_n = \frac{\hbar^2}{2m}\left(\frac{2\pi}{L}\right)^2(n_x^2 + n_y^2 + n_z^2) = \frac{\hbar^2(2\pi)^2n^2}{2mV^{2/3}} \tag{12-6}$$

where the volume of the box is $V = L^3$. The quantum numbers n_x, n_y, and n_z have positive or negative integral values that define the various possible electron *states*. In the absence of a magnetic field, the spin of the electrons need not be represented by an additional quantum number $m_s = \pm\frac{1}{2}$ because electrons with plus and minus spins have the same energy. Since the energy of any state is proportional to $n^2 = (n_x^2 + n_y^2 + n_z^2)$, several choices of the individual quantum numbers n_x, n_y, and n_z result in the same electron energy, i.e., the energies are degenerate.

At 0°K all electrons reside in the lowest possible energy states. By the *Pauli exclusion principle,* only two electrons can occupy any state, one with spin "up" and one with spin "down." As a consequence of this principle, at 0°K the electrons fill all the states up to a certain maximum energy level and none have energies above this. This is indicated by the plot shown in Fig. 12–3. The energy levels for a crystal of ordinary size are extremely close together and would not be visible as separate points on the curve. At higher temperatures some electrons are excited to higher energy levels. The distribution is altered as indicated by the dashed line in Fig. 12–3. The level at which the probability of occupation is $\frac{1}{2}$ is the *Fermi level* E_F. Following the work of Dirac and Fermi, the probability that any particular allowed state of energy E would be occupied at temperature T is given by

$$f(E) = \frac{1}{e^{(E-E_F)/kT} + 1} \tag{12-7}$$

and is called the *Fermi-Dirac distribution function;* E_F is of the order of a few electron volts; e.g., for Cu, Ag, and Au it is 7.04, 5.51, and 5.51 ev, respectively. In the Fermi-Dirac distribution at ordinary temperatures the distribution still falls rather abruptly, actually much more rapidly than indicated in Fig. 12–3, for the amount of energy an electron can acquire from a thermal source is only of the order of kT, which at room temperature amounts to 0.025 ev. Electrons in lower energy states cannot acquire these small energy increments because all states are already occupied to the extent permitted by the Pauli exclusion principle; only the electrons with energies near E_F can absorb heat and thereby contribute to the electronic specific heat. This explains why the electronic contribution to the specific heat of a metal is extremely small.

In theories of metals and alloys and in many problems of physics, it is of interest to know the number of states in any energy interval and the total number within the range up to E_F. The number of states may be quickly computed if one plots the states as points on a simple cubic space lattice having a lattice constant of unity (the three numbers n_x, n_y, n_z are used as the lattice coordinates of a state). Each cell of such a lattice has a unit volume and can be occupied by two electrons (of opposite spin) with energies proportional to the square of their distance $|n|$ from the origin. The number of states having E less than the maximum value E_F is then equal to the volume of a sphere of radius n_{max}. The volume of the sphere is $(4\pi/3)(n_{max})^3$. If there are N valence electrons in the crystal, these will fill $N/2$ cells, and hence the volume required to accommodate them is $N/2$. Equating this volume with the volume of the sphere and substituting the value of n_{max} in Eq. (12–6), one obtains

$$E_F = \frac{\hbar^2\pi^2}{2m}\left(\frac{3N}{\pi V}\right)^{2/3} \tag{12–8}$$

It is clear from this equation that the maximum energy of the electrons is a function of N/V, the number of electrons per unit volume, and not of the size or shape of the whole crystal.

The number of states per unit volume of the crystal which are contained in the interval between E and $E + dE$ defines a *density of states*, $N(E)$, and is of particular interest. It may be determined from the rate at which the volume of the sphere described above increases when n increases, and is given by

$$N(E) = \frac{(2m)^{3/2}}{4\pi^2\hbar^3} E^{1/2} \tag{12–9}$$

In a free-electron theory the density of states thus varies parabolically with E.

Sommerfeld's free-electron model provides a good explanation of some phenomena but not of several others. The theory accounts for Ohm's law and also for the law of Wiedemann and Franz (which states that the ratio of thermal to electrical conductivity at any particular temperature is a constant). It also accounts for electronic specific heats being very small, for the fact that electrons can tunnel through barriers, and in a qualitative way for various thermoelectric and galvanomagnetic effects (though quantitatively the theory is often unsatisfactory). On the other hand, the theory is seriously deficient in accounting for the difference in conductivity of different solids and, in fact, for a large number of experimental observations involving the motion of electrons in solids under various conditions of imposed electric and magnetic fields.

Electron motion in a lattice A great advance in the electron theory was made when the assumption that electrons are free in a solid was replaced by a more realistic one: that the potential V to which electrons are subjected varies periodically as a result of the presence of atoms arranged on a space lattice. The potential in a crystal varies with the periodicity of the space lattice.

The interaction of an electron with the periodic field of a crystal results in diffraction. The diffraction relations will first be discussed in terms of the momentum of an electron. As mentioned earlier, an electron has wave number k and momentum mv, with $k = mv/\hbar$.

Any electron will be reflected from a set of atomic planes in a crystal if Bragg's law, $n\lambda = 2d \sin\theta$, is satisfied. For an electron moving exactly along the x axis of a crystal, reflection from (100) planes, of spacing a, will occur when $n\lambda = 2a$, since $\theta = 90°$. Since the wavelength of the electron is $\lambda = 2\pi/k$, the reflecting condition can be written $2\pi n = 2ak \cos(90 - \theta)$, which in vector notation can be rewritten

$$\mathbf{k}\cdot\mathbf{x} = k_x = \pi n/a \tag{12–10}$$

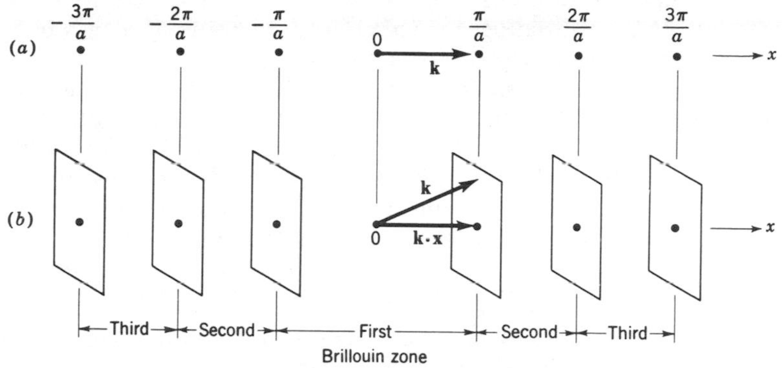

Fig. 12-4 (*a*) **Reciprocal lattice points corresponding to Bragg reflections.** (*b*) **Brillouin-zone planes in one direction.**

where **x** is a unit vector along the x axis, k_x is the component of **k** along this axis, and n is 1 for first-order reflection, 2 for second-order, and so on.

When **k** is considered as a vector with components k_x, k_y, k_z along the x, y, and z directions of a crystal, it may be said to define a **k** *space*. (This **k** space is actually a reciprocal space with a certain scale factor, and it differs from "momentum space" only by the scale factor.) If the vector **k** is plotted so that it extends out from the origin in **k** space along the x axis, it will be seen that the reflecting condition (Eq. 12–10) corresponds to the vector touching the points π/a, $2\pi/a$, $3\pi/a$, $\cdots$, which are arranged along the x axis (Fig. 12–4*a*); for any other direction of **k**, the value of k_x required for diffraction will be met if the vector ends on *planes* in **k** space through the points $n\pi/a$ and standing perpendicular to x (Fig. 12–4*b*). In general, any planes will reflect when k_d, the component normal to them, is given by $k_d = n\pi/d$, where d is the spacing of the planes.

How are the reflecting conditions, discussed above, related to the solutions of Schrödinger's equation? In Sommerfeld's free-electron model the solutions are wave functions of the form

$$\psi \sim e^{i\mathbf{k}\cdot\mathbf{r}} \qquad \text{i.e., } \psi = Ce^{i\mathbf{k}\cdot\mathbf{r}} \tag{12–11}$$

which is the equation for plane waves that are moving through space and carrying momentum as specified by **k**. Subsequent models treat the potential in a crystal as a periodically varying quantity. Bloch was able to show that the wave functions

$$\psi(\mathbf{k}) = U(\mathbf{r})e^{i\mathbf{k}\cdot\mathbf{r}} \tag{12–12}$$

are valid solutions of Schrödinger's equation when the potential energy of an electron is modulated by the periodicity within a crystal. In these *Bloch functions*, $U(\mathbf{r})$ is periodic with the periodicity of the lattice; in other words,

$$U(\mathbf{r}) = U(\mathbf{r} + \mathbf{R}) \tag{12–13}$$

where **R** is the identity period of the lattice in the direction of **k**. Diffraction occurs when conditions are such that $k_d = n\pi/d_{hkl}$; the corresponding solutions of the Schrödinger equation are not *traveling* waves but *standing* waves made up of waves traveling in opposite directions.

Consider a simple case with $n = 1$. A wave traveling along the $+x$ direction is reflected to become a wave moving in the $-x$ direction; this in turn is reflected into the $+x$ direction and so forth. The resultant is a stationary state made up of the two equal and oppositely moving parts $e^{i\pi x/a}$ and $e^{-i\pi x/a}$. We must consider both the sum and difference of these waves, however, for both the sum and the difference are solutions:

$$\psi_1 \sim \sin(\pi x/a) \sim (e^{i\pi x/a} - e^{-i\pi x/a})$$

$$\psi_2 \sim \cos(\pi x/a) \sim (e^{i\pi x/a} + e^{-i\pi x/a}) \tag{12–14}$$

The solutions ψ_1 and ψ_2 correspond to different values of energy even though they apply to the same value of **k**. Since the quantity ψ^2 indicates the density distribution, it will be seen that Eq. (12–14) represents two different stationary distributions. The electron charge distribution of one is maximized at atom centers, the other at midpoints between the atoms. The two have potential energies differing by a finite amount, and no solution exists with values of energy between these two; i.e., there is an "energy gap" at the Bragg reflecting condition.

In general, **k** does not have any of the values required for Bragg reflection and the waves moving in one direction are not matched by a Bragg-reflected component moving oppositely in a way that results in *standing* waves. Instead, in the general case, the result is a *traveling* wave, a wave with its energy modified by an oppositely moving component. As k increases from zero the oppositely moving component becomes increasingly important. At first (near $k = 0$) the resultant energy is that of a free electron, i.e., $E \sim k^2$, but with increasing k the energy falls *below* the free-electron value as indicated in Fig. 12–5, and the maximum deviation is reached at the Bragg reflection value. At this value a *forbidden energy range*, or an energy gap, exists. Further increase in k corresponds to energies above the energy gap; these lie *above* the free-electron parabola and gradually ap-

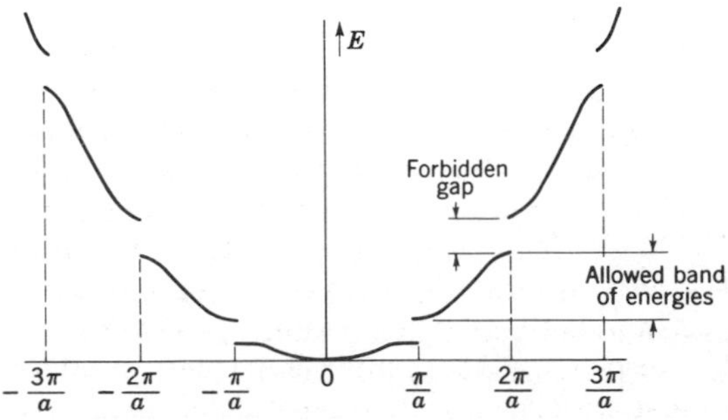

Fig. 12-5 Energy bands and forbidden energy gaps in one dimension.

proach the parabola as k becomes more divergent from the Bragg relation, until the next-higher-order reflection is approached.

A curious result of the interaction of the electron with the lattice is that the electron responds to an applied electric or magnetic field as if it had an effective mass m^* that is not, in general, the same as its ordinary mass m. The effective mass is different at different places on the curve of E vs. k of Fig. 12–5, in accordance with the formula

$$m^* = \frac{\hbar^2}{d^2E/dk^2} \tag{12–15}$$

Thus, m^* may be larger or smaller than m and is nearly equal to m only when $\mathbf{k}$ is remote from Bragg reflection values; m^* is, in general, anisotropic and may be negative. Formulas for several physical properties, including the behavior of an electron in applied fields, can be carried over from the free-electron case to the case of an electron in a periodic field by substituting m^* for m.

Brillouin zones The energy gaps that occur at certain values of the wave vector $\mathbf{k}$ have important consequences in the physics of crystals, as was pointed out by Brillouin. We saw in the preceding section that these energy gaps occur when the component of $\mathbf{k}$ along a certain direction in a lattice meets the requirement for reflection, which is that $k_d = n\pi/d_{hkl}$, which means that the wave vector in $\mathbf{k}$ space ends on one of the planes normal to $\mathbf{k}$ and at distances $n\pi/d_{hkl}$ from the origin. We now discuss the zones of allowed energy between these critical values of $\mathbf{k}$.

Consider first a one-dimensional case with periodicity a along the x direction. The range of k_x between $+\pi/a$ and $-\pi/a$ is the first "Brillouin zone"; the second Brillouin zone consists of the two segments $-2\pi/a$ to $-\pi/a$, and $+\pi/a$ to $+2\pi/a$. The pair of segments next farthest away from the origin constitutes the third zone and so on. Each zone contains a band of allowed energies.

For the case of periodicity in two dimensions, say the x and y directions normal to each other with periodicities a and b, the first Brillouin zone corresponds to k_x values as specified for the one-dimensional case plus analogous values for k_y, namely k_y between $-\pi/b$ and $+\pi/b$.

For the three-dimensional case, consider first a crystal with a simple cubic space lattice. The first Brillouin zone is a cube in $\mathbf{k}$ space enclosed between planes normal to the x, y, z axes, since each of these planes represents appropriate wave-vector components k_x, k_y, k_z for reflection from crystal planes with indices $\{100\}$, and since no other planes in $\mathbf{k}$ space lie closer to the origin than these (i.e., no planes in the crystal have greater spacings than $\{100\}$ planes). If we turn now to b.c.c. crystals, the first Brillouin zone is bounded by planes parallel to the crystal planes of greatest spacing, namely the $\{110\}$ planes, and is therefore a rhombic dodecahedron. The first Brillouin zone for f.c.c. crystals is determined in part by the fact that the $\{111\}$ planes are the most widely spaced; the octahedron formed by planes in $\mathbf{k}$ space parallel to the $\{111\}$ crystal planes would describe the zone if it were not for the fact that planes normal to the x, y, z axes,

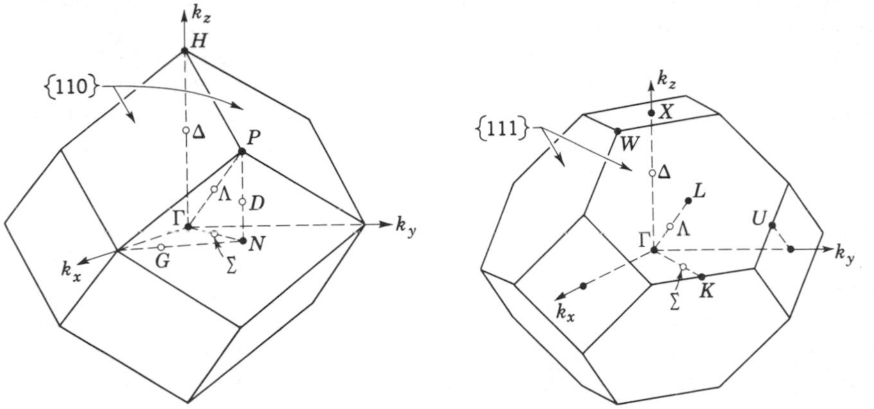

Fig. 12-6 The first Brillouin zones of the b.c.c. and f.c.c. structures showing the usual symmetry points and axes. The zone for the b.c.c. structure is a rhombic dodecahedron of (110) planes. The zone for the f.c.c. structure is a cubo-octahedron bounded by (200) and (111) planes. Centers of (111) planes (marked *L*) are nearer the origin than centers of (200) planes (marked *X*).

respectively, truncate the octahedron. The resulting first Brillouin zone is as shown in Fig. 12–6.

A simple way of determining the first Brillouin zone for any crystal is to use the reciprocal lattice of the crystal and draw lines from the origin to all the reciprocal lattice points that lie near the origin. Planes are then erected at the midpoints of these lines to enclose a solid figure which includes the origin. This solid is the first Brillouin zone. Another polyhedron is enclosed by planes drawn at midpoints on the lines to the reciprocal lattice points next removed from the origin, and these planes define the outer faces of the second Brillouin zone, and so on.

If the reciprocal lattice points are only those that represent appreciable reflecting power for electrons or x-rays, each face of the Brillouin zone will have an appreciable energy gap, for the width of the gap is a function of the electron distribution within the unit cell of the crystal just as is the structure factor in formulas for electron or x-ray diffraction. To have this reciprocal lattice plot be a plot in **k** space, the period along any direction should be considered as being $2\pi/d_{hkl}$ instead of $1/d_{hkl}$, which is the conventional scale used in diffraction and in other chapters of this book.

The occupation of energy levels Many properties of solids are strongly dependent on how the energy levels are occupied. At 0°K there are two electrons in each of the lowest levels. At higher temperatures, some electrons are excited to levels slightly above the Fermi level E_F, which is defined as the energy at which the probability of occupation is $\frac{1}{2}$. It is important to consider how the Fermi level compares with the energies corresponding to the faces of the first Brillouin zone. To do this we must consider how many quantum states exist in the first Brillouin zone.

Since $k = 2\pi/\lambda$, the possible values of **k** in a crystal of cubic shape are those with components $k_x = 2\pi n_x/L$, $k_y = 2\pi n_y/L$, and $k_z = 2\pi n_z/L$, where L is the edge length of the crystal; these correspond to wavelengths that are integral numbers of wavelengths in the x, y, z directions, respectively. In **k** space the allowed values of the x components are thus separated by $2\pi/L$ and the same is true for the y and z components. There is therefore a volume in **k** space of $(2\pi/L)^3 = 8\pi^3/V$ per allowed **k** value, where V is the volume of the crystal. For a simple cubic crystal the first Brillouin zone is a cube $2\pi/a$ on an edge and therefore its volume in **k** space is $8\pi^3/a^3$; thus, there is one allowed **k** value per unit cell in the crystal. Whatever the size or shape of the crystal the same relationship holds: if there are N unit cells in the crystal there will be N energy levels, each of which can accommodate two electrons, or a total of $2N$. If, then, a unit cell is chosen such that it has one valence electron per cell, the first zone will be only half filled. If there are two electrons per unit cell, there will be exactly enough states in the first zone to accommodate them, but it may be energetically more favorable for some of these with energies near the top of the filled levels to "spill over" into the next zone. If there are more than two electrons per unit cell, some must occupy states in one or more of the higher zones.

To explain in another way the situation that causes spilling over into higher zones when the first zone is not completely filled, we wish to refer to the energy gaps of Fig. 12–5. This one-dimensional plot applies to a certain direction in a crystal and thus to a certain periodicity. Other directions have different periodicities and therefore have gaps at different places along the axis of the abscissa. It may happen that the gaps differ so much in position that the second band of allowed energies overlaps the first, so that an electron can occupy a low level in the second band with less energy than a high level in the first. Band overlaps of this sort are by no means uncommon in metals.

To summarize, we have discussed valence electron wave functions in a periodic medium. Quantum states exist, each of which can be represented by a point in **k** space. The energy of electrons in some of these states is strongly influenced by the periodic potential in the crystal. States deviating most from free-electron energies are those most strongly diffracted; each of these has a specified value for its momentum component along the reflecting-plane normal, or in other words, the most strongly altered ones have **k** vectors ending on planes in **k** space that are boundaries of Brillouin zones. As electrons fill the quantum states, starting with the lowest energies and allowing two electrons per state, it is possible to accommodate two electrons per unit cell in the states of the first Brillouin zone, because there are as many states in the zone as there are primitive unit cells in the crystal. In some metals it is energetically more favorable for some electrons to spill over into the second zone (or perhaps into even higher zones) when there are band overlaps or small energy gaps at the Brillouin zone faces.

Conductors, insulators, and semiconductors The electrical conductivity of a solid is essentially determined by how completely the Brillouin zones are filled. In the absence of an externally applied electric field, there is no net drift of electrons in any direction, since any electron whose momentum corresponds to the vector **k** is matched by one with the vector −**k**. When a field is applied, however, this balance is upset and there is a net drift of electrons in the direction of the field, provided the field can actually accelerate some of the electrons, i.e., if some unoccupied states are available with energies just above the occupied levels. This net flux of electrons carries a current. In a *metal* there are always unfilled states at energy levels very slightly higher than the highest energy of the filled states and there is no difficulty in raising the electron energies by applying a field. But in an *insulator* there is an energy gap, just above the filled levels, large enough to prevent electrons from being raised to the next allowed energy level by the field. This situation occurs if a zone is completely filled and is separated from higher zones by an energy gap sufficiently large to prevent any spilling over of electrons into the next zones. An intermediate case occurs in *semiconductors*, which have conductivities several orders of magnitude smaller than metals, but higher than insulators.

Semiconductors always have small gaps between filled levels and empty levels, since thermal energies are sufficient to raise electrons to conducting levels. Materials such as Ge, grey Sn, InSb, InAs, PbSe, and PbTe, with gaps in the range of a few tenths of an electron volt, are semiconducting in the pure state and are called *intrinsic semiconductors*. *Extrinsic semiconductors*, on the other hand, require the addition of impurities to confer semiconducting properties. Impurities of higher valence add electrons when replacing atoms of lower valence and are known as *donors*. For example, P, As, and Sb are donors when added to Si, and by providing negative carriers they convert the Si into an *n-type semiconductor*. The impurities go into donor levels which lie slightly lower in energy than the empty levels of the conduction band, from which they are promoted to the conduction band by thermal energies. On the other hand, the addition of a few parts per million of a *lower* valence element than the Si atoms (e.g., B or Al) has the effect of providing conductivity by positive carriers, and converts the Si into what is known as a *p-type semiconductor*. This occurs because energy levels that are acceptors for electrons thermally excited from the filled levels are introduced by the impurity at energies just slightly above the filled levels of Si. Thus, the levels that are filled in pure Si contain holes when acceptor impurities have been added, and the holes act as positive carriers. The many types of imperfections discussed in Chap. 14 alter the path of conducting electrons and holes and thereby have an important effect on the resistivity. Imperfections also serve as traps for electrons and holes.

Metals do not have a gap just above the filled levels. Monovalent metals have enough valence electrons merely to fill half the states of the first zone. The conditions for good conductivity are therefore met. But in the divalent metals the presence of metallic conductivity comes from a different situation, since two electrons per atom are enough to fill the states in the first zone exactly. In these, the conductivity is conferred by zone overlap: some of the electrons are found in the second zone.

Energy bands Another approach to the problem of allowed and forbidden energies in solids is based on the energy levels of isolated atoms and the perturbations of these levels when a number of atoms are brought close to each other to form a solid. The interaction between neighboring atoms increases as they approach each other. The interaction is felt more strongly by the outer levels of an atom than by the inner ones. The result is a broadening of the sharp discrete energy levels of the individual atoms into bands of levels. For example, the 3*s* levels broaden into a band of the solid which can appropriately be called the 3*s* band. These allowed energy bands in a solid and the forbidden energy bands between them are the same ones discussed in connection with Brillouin zones. When the energy bands are calculated on the basis of perturbations of the energies of tightly bound electrons of an isolated atom, the method of calculation is known as the *tight-binding method*. It appears that this method cannot deal very satisfactorily with the large proportion of the volume of a metal *between* the ions, where the potential is nearly constant. A somewhat better approximation seems to be to treat the interstitial region as a region of constant potential, and among the means devised to solve Schrödinger's equation for this model of a metal is the *orthogonalized plane-wave* (OPW) *method*. This method takes a plane wave of the simple form $\exp(i\mathbf{k}\cdot\mathbf{r})$ that is suitable for the constant-potential region and combines it with a suitable mixture of wave functions that characterize the ion cores, the combining being done only for plane waves that are "orthogonal"[1] to the core-wave functions.[2] The effect of the orthogonality condition is to lessen the effective potential of the individual ions and give them a pseudopotential which does not have an overpowering effect on the motion of the electrons, so that electron motion closely resembles that of free electrons.[3]

The Fermi surface As mentioned earlier, it is convenient to indicate both the magnitude and the direction of the momentum of an electron in a crystal by plotting a vector in **k** space. The **k** vectors of electrons that have energies at the Fermi level terminate upon a surface in **k** space, the *Fermi surface*. At absolute zero this surface separates the occupied from the "empty" quantum states. At higher temperatures, even though there is a slight spread of vacant and occupied states near E_F (as shown in Fig. 12–3), the Fermi surface still essentially provides all the electron states that can play any significant part in the electrical and thermal conductivities and in the electronic specific heat.

For the case of free electrons the Fermi surface is a sphere. Probably no real metal exists with a Fermi surface that is truly spherical, although only a slight distortion from the spherical shape may be expected if E_F corresponds with states which lie well inside the first Brillouin zone. Sodium is believed to have a Fermi surface that is very nearly spherical. As discussed above, on approaching a zone boundary the electron energies depart from the parabolic relationship between E_k and k for free electrons and increase more and more slowly with k. In three-dimensional **k** space this is equivalent to the Fermi surface's "bulging" toward the boundaries and becoming

[1] A wave function ψ_1 is orthogonal to another, ψ_2, if the integral of $\psi_1\psi_2$ over space vanishes, which will be true if ψ_1 and ψ_2 have different numbers of nodes.

[2] C. Herring, *Phys. Rev.*, vol. 57, p. 1169, 1940.

[3] J. C. Phillips and L. Kleinman, *Phys. Rev.*, vol. 116, p. 287, 1959. W. Harrison, *Phys. Rev.*, vol. 118, p. 1190, 1960. M. H. Cohen and V. Heine, *Phys. Rev.*, vol. 122, p. 1821, 1961. M. H. Cohen in T. B. Massalski (ed.), "Alloying Behavior and Effects in Concentrated Solid Solutions," Gordon and Breach, New York, 1965. J. M. Ziman, *Advan. Phys.*, vol. 13, p. 89, 1964.

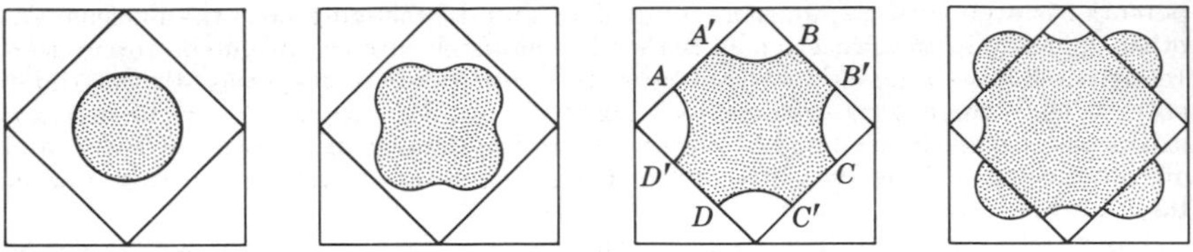

Fig. 12-7 Schematic illustrations of Fermi surfaces: (*a*) few electrons, (*b*) more electrons, (*c*) contact, (*d*) overlap.

distorted from a sphere. If there are enough electrons within the first zone, the Fermi surface may contact the zone boundaries, and if the band gaps are relatively small, it may overlap into the next zone. Hypothetical shapes of the Fermi surface in two dimensions are shown in Fig. 12–7.

A few properties of Fermi surfaces may be noted. At the points of contact the Fermi surface always meets the zone boundary at right angles along the line of intersection. If the Fermi surface is a sphere (Fig. 12–7*a*), the velocity of an electron is always parallel to its momentum vector. However, since the velocity is perpendicular to the energy surface, the **v** and **k** vectors may be no longer parallel (Fig. 12–7*b*) if the Fermi surface is nonspherical. The states along the lines of contact, such as AA', BB', CC', and DD' in Fig. 12–7*c*, have energies that are smaller than E_F and hence lie below the Fermi surface; only the points at the ends of these lines of contact (A, A', etc.) are at the Fermi surface.

Even in a monovalent metal the energy in certain directions in **k** space may be reduced by the energy gaps to such an extent that the Fermi surface is drawn into contact with some zone faces. This occurs, for example, in the noble metals, copper, silver, and gold, where the {111} zone faces are touched. This possibility had been conjectured for some years because of anomalies observed in certain transport properties, thermoelectric power, magnetoresistance, and Hall effects, but the first determination of the geometrical shape of the Fermi surface in copper occurred in 1957 when Pippard[1] published his studies of the "anomalous skin effect" and its interpretation.

The revised Bloch model The Bloch model (see page 315) until recently has served as a very successful basis for discussion of the motion of electrons in metals and alloys. Its main features may be characterized as follows[2]: Only the conduction electrons (identified with valence) are involved, and they move independently of one another. The motion is described by one-particle functions in a crystalline lattice with a periodic potential vaguely atomic in character. However, when experimental data on simple metals and alloys are fitted in the model, the required matrix elements of the crystal potential turn out to be small compared to the free-electron Fermi energy, which is a measure of the average kinetic energy of the electrons. This result implies the existence of only weak electron-atom interactions and appears to contradict the notion that the

[1] A. B. Pippard, *Phil. Trans. Roy. Soc.*, vol. A250, p. 325, 1957.

[2] See M. H. Cohen in T. B. Massalski (ed.), "Alloying Behavior and Effects in Concentrated Solid Solutions," Gordon and Breach, New York, 1965.

potential is atomic in character and therefore rapidly changing near the nucleus. Another difficulty in the Bloch model is that it completely neglects Coulomb interactions (electron-electron interactions) that must be comparable in magnitude to the kinetic energy. Thus, there is an apparent paradox that certain features necessary for the very existence of the Bloch model—electron-atom and electron–electron interactions—are not prominent in the original model. Recent theoretical advances appear to remove these obstacles.

The problem of reconciling possible strong electron electron interactions is removed by the realization that the Bloch model describes the motion of more complex entities, called *quasi particles*, introduced by Landau.[1] Quasi particles have an electron at the center, surrounded by a region of electron deficiency (correlation hole) and a further region containing electrons that have been pushed by the Coulomb repulsion away from the central electron and "flow around it much as water around a moving obstacle."[2] Reinterpretation of the Bloch model in terms of nearly independent quasi particles constituting a Fermi liquid, as compared with the original particles constituting a Fermi gas, removes much of the difficulty associated with the apparent neglecting of electron–electron correlations.

The problem of looking realistically at electron-atom interactions, thus considering in effect the influence of core electrons, has been tackled by introduction of a *pseudopotential*.[3] The core-electron wave functions are ignored to some extent and one deals instead with pseudowave functions which are not required to satisfy the orthogonality condition. Adding a repulsive potential has the effect of statistical exclusion of the valence electrons from the regions of space occupied by the core electrons. The pseudowave functions are then used as the original Bloch functions, and the crystal potential is represented by the pseudopotential.[4]

The mapping of Fermi surfaces The method of the *anomalous skin effect* involves applying a high-frequency (r.f.) field to a very pure single crystal of the metal at very low temperatures. This field induces currents in the metal which tend to prevent penetration of the field into the specimen, limiting it to a certain *skin depth*. At very low temperatures the electron mean free path is much longer than the skin depth. Hence, only those electrons that run very nearly parallel to the surface and within the "anomalous region" remain in the field long enough to receive appreciable energy from the field. The effective resistance under these conditions is a function of the radius of curvature of the Fermi surface in the region near the end of the velocity vector of these electrons. The anisotropy of the resistance, measured with crystals cut to have different crystallographic orientations, is used as a test of assumed models of the Fermi surface and has enabled physicists to propose reasonable shapes for the Fermi surfaces of some of the metals.

A perspective drawing, as proposed by Pippard, of the Fermi surface in copper enclosed within the first Brillouin zone is shown in Fig. 12–8. Following the study of copper, the details of the geometry and the presence of contact in all three noble metals, copper, silver, and gold, have been reported with the help of a number of elegant experimental techniques which are directly related to the Fermi surface.[5] Most of these techniques require the application of a strong magnetic field.

A method which has been particularly useful in revealing details of the Fermi surface

[1] L. Landau, *Soviet Phys. JETP (English Transl.)*, vol. 3, p. 920, 1956; vol. 5, p. 101, 1957; vol. 8, p. 70, 1959.

[2] Cohen, loc. cit.

[3] J. C. Phillips and L. Kleinman, *Phys. Rev.*, vol. 116, p. 287, 1959. See also the paper by J. M. Ziman, *Advan. Phys.*, vol. 13, p. 89, 1964.

[4] Note added in proof: A recent discussion of pseudopotentials will be found in an article by V. Heine in P. S. Rudman and J. Stringer (eds.), "Phase Stability in Metals and Alloys," McGraw-Hill, New York, 1966.

[5] See, for example, W. A. Harrison and M. B. Webb (eds.), "The Fermi Surface," Wiley, New York, 1960.

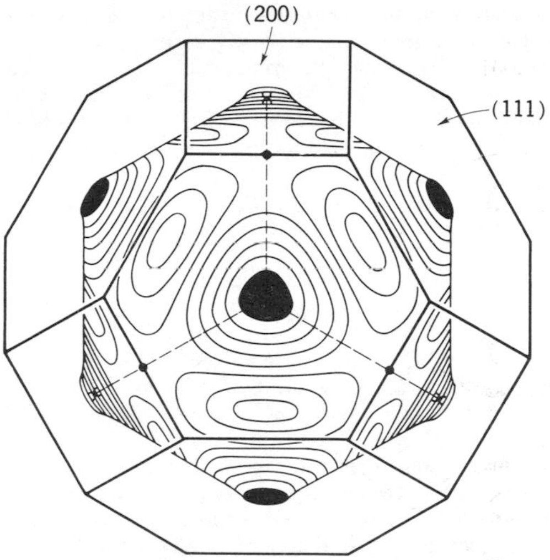

Fig. 12-8 Pippard's model of the Fermi surface in copper. (A. B. Pippard, Phil. Trans. Roy. Soc., vol. A250, p. 325, 1957.)

in many metals is associated with the *deHaas – van Alphen effect* in which a strong magnetic field H is applied to a single crystal of the metal at very low temperatures, and the magnetization (magnetic susceptibility χ) induced by the field is measured. The magnetic susceptibility is not constant but oscillates periodically with the inverse of the magnetic field. According to theory, the oscillations should occur at equal intervals P of the quantity $1/H$; the periods P are inversely proportional to A, the cross-sectional extremal (maximal or minimal) area of the Fermi surface normal to the direction of H. From the measurements of P for different directions of H, the corresponding value of A can be deduced. Direct information about the shape and size of the Fermi surface can thus be derived.

Under a strong magnetic field the **k** vector of an electron on the Fermi surface may be said to trace out an "*orbit*" at right angles to the applied field. This orbit gives directly the area of a cross section through the Fermi surface. If the Fermi surface is wholly contained within the first zone, as in Fig. 12–7*a* or *b*, the path of the electrons when the field is directed, for example, normal to the drawing, may be visualized as lying along the contours of the respective shaded cross-sectional areas. When the Fermi surface contacts the zone faces, as in Fig. 12–7*c*, the "orbit" is more difficult to visualize because of the multiple reflections which occur at the zone faces. Thus, for an anticlockwise orbit, an electron will move along the segment $A'B$, then undergo a "jump" to D', then move along the segment $D'A$, make a jump to C', follow along $C'D$, jump to B' . . . . Since the jumps are instantaneous, the orbit will consist of the four segments of solid line around the cross section of the Fermi surface. It is desirable, then, to connect the segments into a closed loop so as to visualize the complete orbit. This can be done if, instead of imagining the electron as jumping back by a whole reciprocal lattice vector each time it is reflected by a zone boundary, we translate the whole zone itself, at each boundary, and the Fermi surface with it. This procedure is the basis of the *repeated-zone scheme* in which individual portions of the Fermi surface in each zone are linked into a multiply connected surface. In a three-dimensional construction the polyhedral first zones are stacked together to fill **k** space with no voids, each zone face being shared with two polyhedra. This arrangement in **k** space is analogous to the cellular method of Wigner and Seitz, mentioned on page 307, in real space. Each polyhedron is centered on a reciprocal lattice point; if the crystal is f.c.c., the reciprocal lattice will be a b.c.c.

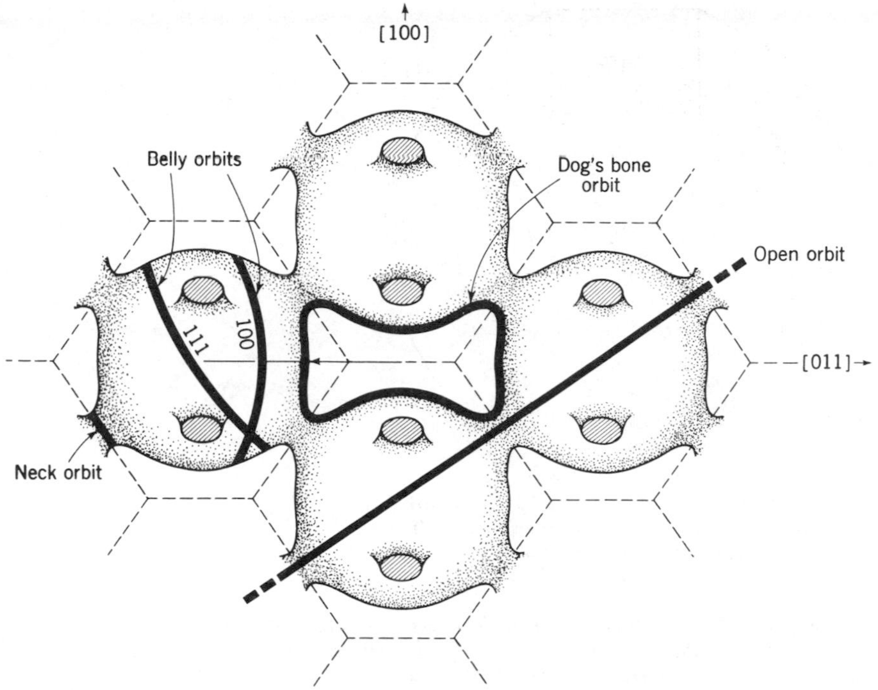

Fig. 12-9 A schematic (110) section through the multiply connected Fermi surface of copper, silver, or gold, illustrating the various possible orbits that can be described by electrons.

array of points and therefore will have a b.c.c. stacking of polyhedra. The Fermi surface, if drawn within each of these identical polyhedra and linked into a multiply connected surface as in Fig. 12–8, will be similar to that shown in perspective in Fig. 12–9, which represents a section through the multiple surface for gold. The connecting portions near the zone boundaries in Fig. 12–9 have been described as "necks" and the nearly spherical central portions as the "belly."[1] Under an applied magnetic field an electron at the Fermi surface can thus "move" around the so-called neck orbit or belly orbit, and these orbits enclose occupied states. Other orbits enclosing unoccupied states in **k** space are possible, such as the "dog's bone" orbit shown in the center of Fig. 12–9. It can be shown with the help of the multiply connected Fermi surfaces that sections can be made whose outer contours are not represented by any closed circuit at all, for example, in the case of sections along any row of necks and perpendicular to the plane of the drawing in Fig. 12–9. We may then speak of "open orbits." In a low magnetic field an electron only manages to complete a small portion of its orbit before it is scattered, but with high fields an electron may make many circuits before it is scattered; hence, it becomes important whether the circuit is an open or a closed orbit. The presence of an open orbit means that the Fermi surface must make contacts with the zone boundaries because only then can it be drawn as a multiply connected surface. The data from high-field magnetoresistance and Hall effect measurements can be interpreted, for example,

[1] D. Shoenberg, *Nature*, London, vol. 183, p. 171, 1959; *Phil. Mag.*, vol. 5, p. 105, 1960. See also Harrison and Webb, loc. cit.

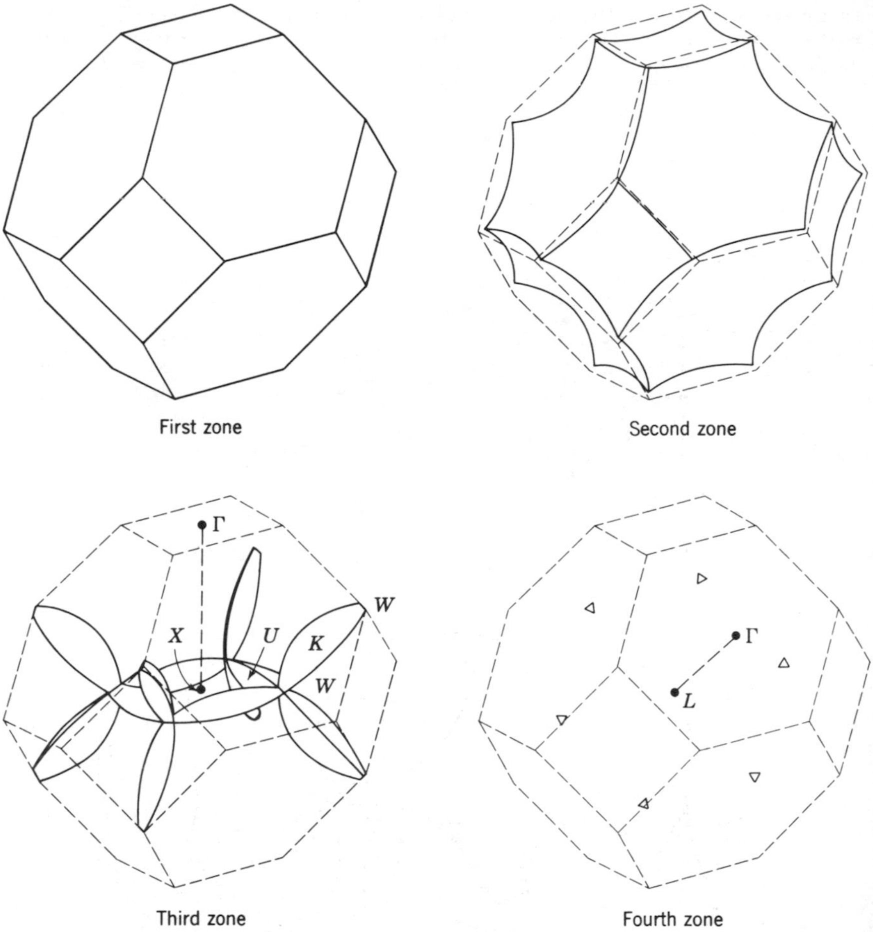

Fig. 12-10 Fermi surfaces of f.c.c. three-valent aluminum according to the single-OPW approximation. Regions with convex surfaces are occupied; those with concave surfaces are unoccupied. Note that some symmetry points shown in Fig. 12-6 have been translated to equivalent positions. (**After W. A. Harrison, Phys. Rev., vol. 118, p. 1190, 1960.**)

in terms of contacts along the [111] directions for copper and silver.[1] The Fermi surface may consist of a number of sheets in different zones, making its overall representation as a "solid body" difficult.

The scheme of picturing the Fermi surface as extending out from the origin in **k** space past the first and successive zone boundaries, as illustrated in Fig. 12–7, involves the so-called *extended-zone scheme*. There are advantages in considering the Fermi surface, alternatively, with a *reduced-zone scheme*, which makes use of the periodic properties of the **k** vector in **k** space. It is possible to translate the origin of coordinates

[1] J. R. Klauder and J. E. Kunzler, *J. Phys. Chem. Solids*, vol. 18, p. 256, 1961.

in **k** space by the amount of the period along any direction without changing any of the properties of the **k** vectors. This means that it is possible to replace any **k** vector that extends into one of the higher zones by a vector in the first zone and still have it represent the same energy state. To do this, a vector representing the periodicity in any direction in **k** space may be subtracted from the **k** vector. Suppose, then, that a Fermi surface overlaps slightly beyond the zone boundary that stands normal to the x axis at π/a. This bulge can be drawn as an *inward* bulge from the plane standing at $-\pi/a$, since it will then have been translated a full period of $2\pi/a$ to the left. If this

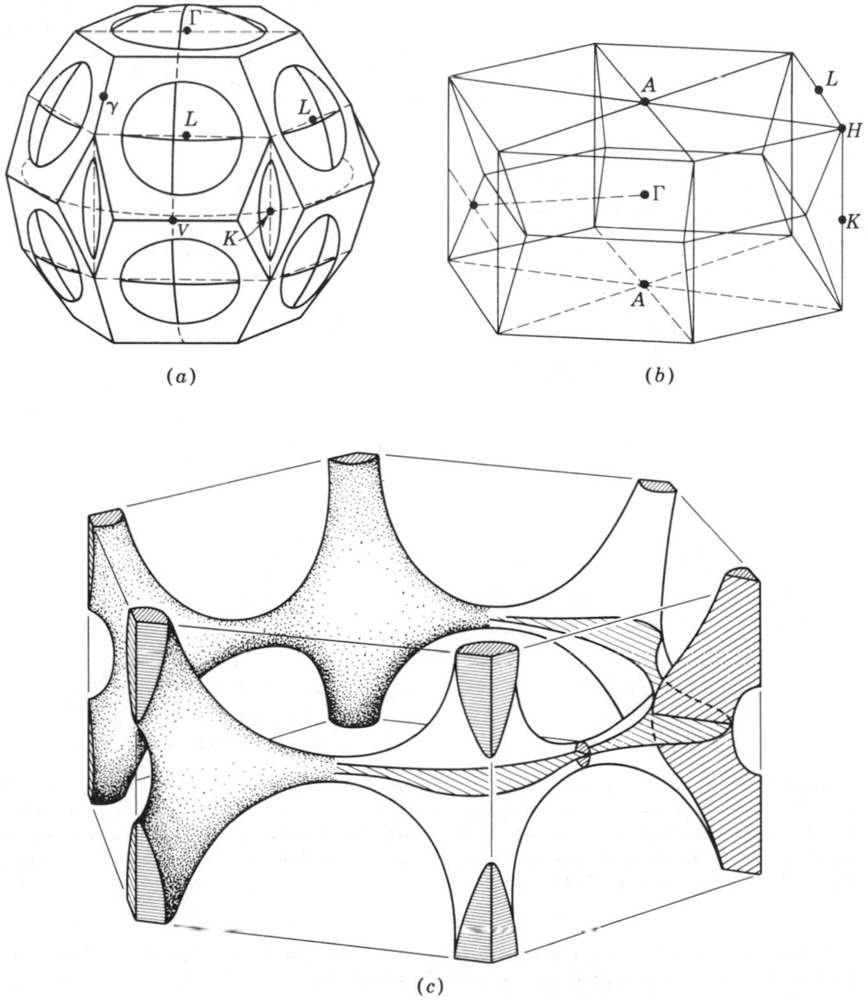

Fig. 12-11 Fermi surfaces in relation to the Brillouin zone for the c.p.h. structure. (*a*) The large (extended) zone for axial ratio c/a = 1.633 with Fermi surface of free electrons overlapping at symmetry points Γ, L, and K. (*b*) The reduced second zone with symmetry points indicated. (*c*) Holes in the combined first and second zones. The "monster" is partially sectioned; "caps" in the first band are cross-hatched. (See **D. F. Gibbons and L. M. Falicov, *Phil. Mag.*, vol. 8, p. 177, 1963.**)

construction is followed at each of the boundary faces of the first zone, all the Fermi surface found in the various segments of the second zone can be mapped *within* the polyhedron that is used for the first zone, and the same principle applies for the higher zones. An illustration of the reduced-zone scheme for trivalent aluminum is seen in Fig. 12–10.[1] For c.p.h. metals, such as Mg, Zn, Cd, or Be, the Fermi surface corresponding to two electrons per atom is usually drawn within the combined first and second zones, with overlaps into the third and fourth zones. In Fig. 12–11 hypothetical Fermi surfaces are shown for a divalent c.p.h. metal with a nearly ideal axial ratio, with the extended-zone scheme and the reduced-zone scheme. Fig. 12–11*b* represents holes in the first and second zones after remapping; it has been aptly described as a many-armed "monster."[2] It has been shown that this monster of holes may become multiply connected when spin-orbit coupling is taken into consideration, and this leads to open electronic orbits. The resulting Fermi surface provides an interpretation of the observed behavior of the magnetoresistance.[3]

Electronic structure of the noble metals Cu, Ag, and Au The calculated cohesive energies of the noble metals Cu, Ag, and Au are too low when compared with experimental values. Kambe,[4] in calculating these, considered only the contributions from electrons in the *s* band. Clearly, the neglect of the *d* band is a serious omission. It is believed that the extra cohesion observed experimentally must be the result of hybridization of the 3*d* band electrons with those in the 4*s* and 4*p* bands. Some of the *d* electrons participate in the bonding. In a sense, therefore, Cu, Ag, and Au seem to belong more with the transition elements than with the *B*-subgroup elements. The high melting points partially bear this out. However, although remarkably successful models have been proposed for the form of the Fermi surface in the noble metals, a successful calculation of the cohesive energies is not yet available.

The nature of the electronic states at the Fermi level in some prominent parts of the Brillouin zone is of interest in theories of alloying behavior. For example, in copper, at the point where the Fermi surface touches the {111} zone faces and forms necks, the electronic states are believed to be chiefly *p*-like in character, with *s*-like states falling outside the zone and into the higher-energy levels, making $E_p < E_s$. (This view is supported by experimental evidence based upon soft x-ray spectroscopy and the Knight-shift effect in the nuclear magnetic resonance,[5] as well as by recent theoretical calculations.[6]) Since in the *free atoms* of copper the energy of the 4*p* state is higher than that of the 4*s* state, on forming a *crystal* the energy of the *p* states must be lowered below that of the *s* states. The situation for silver and gold is less clear; possibly in gold crystals the energies at the centers of {111} zone faces are reversed with $E_p > E_s$.[7] These relationships

[1] W. A. Harrison, *Phys. Rev.*, vol. 116, p. 555, 1959; vol. 118, p. 1190, 1960.

[2] L. Falicov, *Phil. Trans. Roy. Soc.*, vol. A255, p. 55, 1962.

[3] M. H. Cohen and L. Falicov, *Phys. Rev. Letters*, vol. 5, p. 544, 1960.

[4] H. Kambe, *Phys. Rev.*, vol. 99, p. 419, 1955.

[5] See, for example, a review by M. H. Cohen and V. Heine, *Advan. Phys.*, vol. 7, p. 395, 1958.

[6] B. Segall, *Phys. Rev. Letters*, vol. 7, p. 154, 1961; *Phys. Rev.*, vol. 125, p. 109, 1962.

[7] Cohen and Heine, loc. cit.

are thought to influence alloying behavior and are discussed in a later section dealing with alloys.

Electronic structure of the transition metals The transition metals occupy three horizontal rows in the Periodic Table, with the alkali metals at the beginning of each row (or period) and the noble gases at the end (see groups A, Table 10 1). If n denotes the quantum number of valence electrons, the electronic energy levels of the free atoms change from the order $ns < (n-1)d < np$ in the alkali metals and *early transition elements* to the order $(n-1)d < ns < np$ in the *late transition elements* and the noble metals, with the result that along each row the order of the s and d states is gradually reversed. On forming crystalline solids, some of the degenerate atomic energy levels broaden into bands, but the same general effect is expected, with energies of s and d states again crossing over along each row and resulting in nearly the same energy in several intermediate elements. As a consequence of this there will be a tendency for some of the elements to exist in *hybrid sd states* (or mixed atomic orbitals). At the same time, on moving along a transition series from left to right, the p levels are lowered nearer to the s levels, so that when the *sd* hybridization decreases in late transition elements, the *sp* hybridization increases. There is general agreement that the *sd* and *spd* hybridization should give rise to strong cohesion somewhere in the middle of each period, the details depending on the electronic structure of each particular element. However, the electronic theories of the transition metals are very complicated and, as yet, incomplete.

Judging by the experimental data such as melting points, lattice spacings, compressibilities, thermal expansion, etc., in the first long period, cohesion appears to increase from group IA to group VA, with manganese showing an unexpectedly weak cohesion, and Fe, Co, and Ni showing stronger cohesion and little change. In the second and third long periods, cohesion increases up to group VIA, remains high in groups VIIA and VIIIA, and then falls. Also, on passing down along vertical columns between the three periods, there is a striking increase in bonding as evidenced by the higher melting points. According to Hume-Rothery,[1] the above differences between the three periods have been rather neglected in theoretical considerations.

Some of the early challenges to the possible theories of transition elements were provided by the data on magnetic properties of the elements Fe, Co, and Ni and their alloys in the first long period, as indicated by their *atomic magnetic moments*. If a ferromagnetic substance is subjected to a magnetic field, the saturation moment at absolute zero represents the greatest degree of magnetization that can be induced. These saturation moments are usually expressed in *Bohr magnetons* per atom (μ_B), and the accepted values for iron, cobalt, and nickel are 2.22, 1.71, and $0.61\mu_B$, respectively. These are not whole numbers, partly because of a contribution from the

[1] W. Hume-Rothery in P. A. Beck (ed.), "Electronic Structure and Alloy Chemistry of the Transition Elements," Interscience, New York, 1963.

orbital momentum of the electrons in the atom,[1] and partly because the atoms in the crystal may not all possess identical electronic structure, each with the same number of unpaired electrons in the $(3d)^{10}$ band, but rather may consist of atoms with different electronic states. This possibility has been developed into two alternative approaches. In the *Pauling hypothesis* the trends in certain observed properties, and in particular the magnetic saturation moments, were assumed to be the consequence of some of the *d* electrons entering into localized *atomic* (or nonbonding) *d* orbitals, described by localized wave functions. The remaining *d* electrons were regarded as forming hybrid bonding orbitals in association with the *s* and *p* electrons, in which electrons continuously exchange positions (resonate) and are responsible for metallic bonding. In this scheme a maximum of 2.44 electrons is considered to be in localized atomic *d* orbitals, and at most, six electrons contribute to bond formation.[2] Many authors consider the Pauling model unsatisfactory in that it gives no reason for the difference between the properties of the elements in the first and later long periods and in that the scheme does not agree with a number of physical properties[3] since it implies a constant number of bonding electrons per atom from group VI to group VIII*C*. A modification of the Pauling scheme proposed by Hume-Rothery, Irving, and Williams[4] takes into account certain experimental data as well as valences of the transition elements as observed in inorganic chemistry, but retains some idea of atomic orbitals.

In the alternative approach, electrons are treated collectively, with the whole assembly giving rise to a band or bands in which the number of unpaired spins per atom corresponds to the observed saturation moments but in which the magnetic electrons are not truly localized and could in principle contribute to the Fermi surface.[5] The necessity for making some electrons responsible for magnetism and others responsible for conduction and bonding led to proposals of two-band models in which $4p$ electrons

[1] This contribution is small but measurable. A. J. P. Mayer and G. Asch, *J. Appl. Phys.*, vol. 32, p. 312, 1961, give a value of $0.08\mu_B$ as orbital contribution in iron, leaving about 2.14 as contribution due to spin.

[2] These ideas are described in a number of papers by L. Pauling: *Phys. Rev.*, vol. 54, p. 889, 1938; *Proc. Roy. Soc. (London)*, vol. A196, p. 343, 1949; in American Society for Metals Symposium, "Theory of Alloy Phases," p. 220, Cleveland, Ohio, 1956. They have been criticized by: B. R. Coles and W. Hume-Rothery, *Advan. Phys.*, vol. 3, p. 149, 1954; B. R. Coles, *Advan. Phys.*, vol. 7, p. 40, 1958; W. Hume-Rothery in P. A. Beck (ed.), 1963, op. cit.

[3] See, for example, Coles and Hume-Rothery, loc. cit.

[4] W. Hume-Rothery, H. M. Irving, and R. G. J. Williams, *Proc. Roy. Soc. (London)*, vol. A208, p. 431, 1951.

[5] Various aspects of these theories are discussed by Mott: N. F. Mott and H. Jones, "The Theory of the Properties of Metals and Alloys," Clarendon, Oxford, 1936. N. F. Mott, *Phil. Mag.*, vol. 44, p. 187, 1953. N. F. Mott, *Rept. Progr. Phys.*, vol. 25, p. 218, 1962. N. F. Mott, *Advan. Phys.*, vol. 13, p. 345, 1964. The possibility of localized magnetic electrons in iron has been the subject of much controversy, and various theories have been proposed following x-ray scattering measurements, neutron diffraction, magnetic behavior in alloys of iron with transition and nontransition elements, etc. References to this work will be found in the recent papers by Mott.

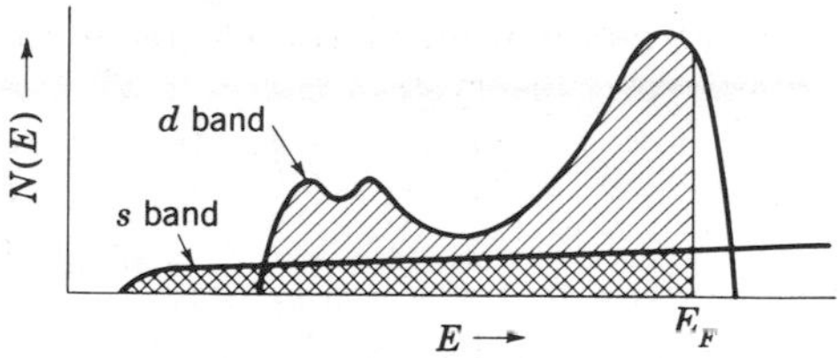

Fig. 12-12 A hypothetical two-band model of overlapping *d* and *s* bands.

are ignored and the *s* and *d* electrons are shown in two separate bands, as in Fig. 12–12, suggested for nickel. In order to produce ferromagnetism there must be a predominance of electrons with one spin; hence, the *d* band is subdivided into two hypothetical subbands, with "holes" in one subband being responsible for unpaired spins. The fact that there is a non-integral average difference between the numbers of electrons with positive and negative spins per atom is usually taken as evidence that wave functions that describe the magnetic electrons, which would be the 3*d* electrons in isolated atoms, overlap significantly between neighboring atoms and form a band. It is usually assumed that the 3*d* band contains its full quota (five per atom) of electrons with positive spin (that all the holes that give rise to the magnetic moment are on the negative side).[1] Some of the most direct evidence for the various proposed models comes from the study of alloys. It seems likely, for example, that in b.c.c. transition elements the density of states is roughly as in Fig. 12–13, and that the Fermi level for electrons with the spin-up and spin-down is as marked by the vertical lines.[2] It is clearly an approximate model since it makes no detailed provision for the hybridized states. However, the concept of separated *s* and *d* bands and a multiple *d* band is of historical interest since it has provided a stimulus to much experimental work in many directions. The number of subbands in the *d* band suggests, for example, that the distinction between atomic and bonding electrons may depend on direction in the lattice and on each particular subband. In one direction, high-energy *d* electrons may have some atomic character (antibonding) while low-energy electrons may have overlapping wave functions which give rise to bonding.

Stimulus toward the understanding of the electronic theories of the transition elements has been provided by the measurement of heat of solution of hydrogen, the hydrogen absorption, and the electronic specific heat, in a number of b.c.c. low-atomic-number transition elements and their alloys. Such measurements can give valuable information regarding the density of states in these metals, and the results obtained thus far suggest several maxima and minima with a particularly low value in the density of states at an electron concentration of about 5.7 and continuing low until chromium, for which group valence is 6. The electronic specific

[1] See, for example, Coles and Hume-Rothery, loc. cit.

[2] Mott, loc. cit., 1964.

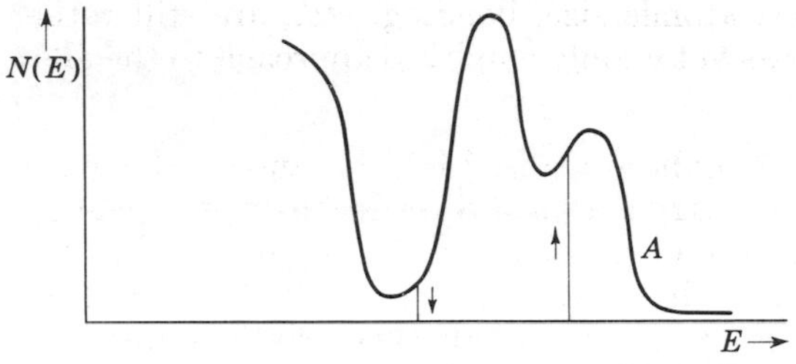

Fig. 12-13 Schematic density of states $N(E)$ for b.c.c. transition metals with spin-up and spin-down Fermi levels marked for iron. The point A marks the upper limit of the d band. (After **N. F. Mott, Advan. Phys., vol. 13, p. 335, 1964.**)

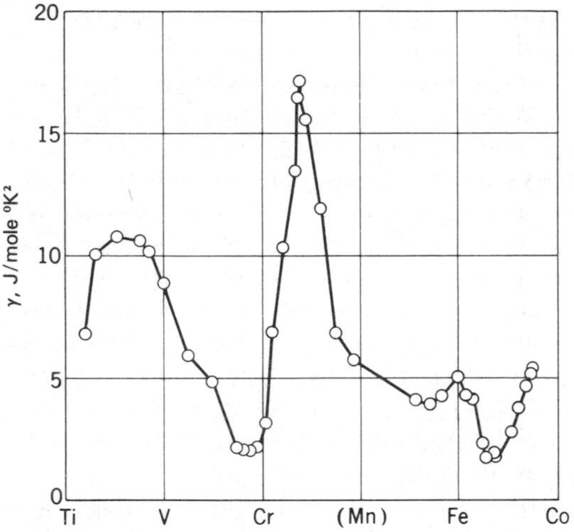

Fig. 12-14 Coefficient γ of electronic specific heat for b.c.c. Ti-V, V-Cr, Cr-Fe, and Fe-Co alloys. (**C. H. Cheng et al., Phys. Rev., vol. 126, p. 2030, 1962.**)

heat data for Ti-V, V-Cr, Cr-Fe, and Fe-Co alloys[1] are shown plotted together in Fig. 12–14. The general features suggest that a sharp peak occurs in the density of states for a Co-Fe alloy with approximately 19 percent Fe. It is from these and similar results and from recent theoretical calculations that the general band form as shown in Fig. 12–13 has been deduced. There is much similarity between the two curves, which gives encouragement to further work.

With the many uncertainties regarding the behavior of electrons in the transition elements and their alloys, it is not surprising that the notions

[1] C. H. Cheng, C. T. Wei, and P. A. Beck, *Phys. Rev.*, vol. 120, p. 426, 1960. C. H. Cheng et al., *Phys. Rev.*, vol. 126, p. 2030, 1962.

about such factors as valence, atomic size, bonding, etc., are still rather vague. This is reflected in the as yet mainly empirical approach to the alloy chemistry of these elements.

Charge oscillations It has long been realized that because of the wave-like nature of electrons and the sharp cutoff in their distribution in energy, there should be oscillations in the distribution of charge around a perturbation in a metal such as an impurity atom or a defect. Although there is a tendency for the electrons in the vicinity of the perturbation to adjust themselves so as to screen the disturbance perfectly and within a distance from the perturbation of the order of one-half the electron wavelength at the Fermi level, theoreticians have proposed that some oscillations in the charge density extend much beyond this range. Nuclear magnetic resonance measurements on alloys[1] have now confirmed these long-range oscillations, by showing *NMR* line shifts and broadening that could be accounted for by them.[2] Recently, some important implications of these oscillations for metallic crystals have been realized.[3]

The potential around a localized charge or perturbation varies with distance from the charge somewhat in the manner indicated in Fig. 12–15 (but dependent upon the geometry of the defect and the Fermi surface). In some alloys, for example in alloys of a monovalent metal such as copper, the potential around an impurity atom has the same sign at the nearest-neighbor position as at the center of the impurity itself. Consequently, substitutional impurity atoms tend to avoid being adjacent to each other, and tend, instead, toward short-range order (for example, in alpha-brass). The tendency for copper ions adjacent to a zinc atom to be repelled by it produces an expansion of the lattice on alloying with zinc. (Effects of this sort are also discussed in Chap. 13, page 358). In superlattices, it is believed that the oscillations of potential should account for the periodic arrangement of some antiphase boundaries in space.

In a polyvalent matrix, the larger wave number at the Fermi surface produces oscillations of smaller wavelength in the crystal, and as indicated in Fig. 12–15, the potential at nearest-neighbor distances is opposite in sign, slope, and curvature to what would occur in copper. Impurity atoms of the same sign of charge therefore attract; there should be a tendency to clustering and the forming of Guinier-Preston zones.

Predictions for transition elements require, in general, a better knowledge of their Fermi surfaces than is available in the early 1960s. But it already appears clear[4] that an interstitial impurity should be repelled by an adjacent

[1] N. Bloembergen and T. J. Rowland, *Acta Met.*, vol. 1, p. 731, 1953. T. J. Rowland, *Phys. Rev.*, vol. 119, p. 900, 1960.

[2] A. Blandin and E. Daniel, *J. Phys. Chem. Solids*, vol. 10, p. 126, 1959. W. Kohn and S. H. Vasco, *Phys. Rev.*, vol. 119, p. 912, 1960. A. Blandin and J. Friedel, *J. Phys. Radium*, vol. 21, 689, 1960.

[3] A. Blandin in T. B. Massalski (ed.), "Alloying Behavior and Effects in Concentrated Solid Solutions," Gordon and Breach, New York 1965. J. Friedel, *Trans. AIME*, vol. 230, p. 616, 1964 (a correction to an earlier formula is given in this review).

[4] J. Friedel, *Trans. AIME*, vol. 230, p. 616, 1964.

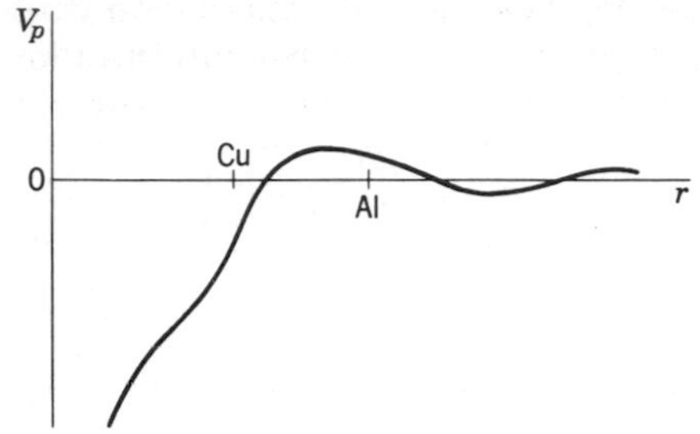

Fig. 12-15 Screened potential around a localized perturbation. (After J. Friedel, Trans. AIME, vol. 230 p. 616, 1964.)

substitutional atom of higher valence and attracted by a substitutional impurity of lower valence.

Superconductivity The importance of superconductivity to science and technology has steadily increased since 1911 when H. Kamerlingh Onnes discovered that electrical resistance drops to an immeasurably low value in mercury at low temperatures. In each of the many superconducting elements and compounds discovered since that date, it has been of importance to determine the transition temperature T_c below which the superconducting property exists, and the critical magnetic field for the material (the maximum value of the field in which superconductivity still remains). Much research has also centered around the perfect diamagnetism that accompanies superconductivity—the *Meissner effect,* which is the expulsion of the magnetic flux from a superconductor.

Interest in theoretical explanations of the phenomenon led, after many years, to the Bardeen-Cooper-Schrieffer (BCS) theory,[1] which successfully accounts for the occurrence of a second-order transition and explains in a very useful way the electrical and magnetic characteristics of the superconducting state. The BCS theory links the state to the presence of an energy gap for the excitation of the electrons from the superconducting ground state, in which electrons are *paired* (because of a screening charge that arises from electron-phonon interaction).

On the experimental side, research has also been stimulated by the finding of materials that remain superconducting at much higher temperatures than those first discovered, and materials that retain their superconductivity in the presence of magnetic fields of considerable strength.[2] The development of wires of Nb_3Sn and other alloys has revolutionized laboratory work involving high magnetic fields, because these wires retain their superconductivity in the presence of remarkably high magnetic fields. Solenoids wound from these wires become powerful magnets with zero dissipation of power when operated within the superconducting range of temperatures and magnetic fields. The fields can be made to be as high as 100 kG or perhaps even higher. Applications are being found for the new superconducting materials, also, in computers, accelerators, radiation de-

[1] J. Bardeen, L. N. Cooper, and J. R. Schrieffer, *Phys. Rev.*, vol. 106, p. 162, 1957; vol. 108, p. 1175, 1957. For reviews of theories, see references on page 337.

[2] J. E. Kunzler, *Rev. Modern Phys.*, vol. 33, p. 499, 1961.

tectors (bolometers), and other solid-state devices and experiments. Reviews of the field in general are appearing frequently[1]; it is appropriate to limit the present discussion to a few statements on crystal structure, electron concentration, and defects vs. superconductivity.

There is no direct and fundamental connection between the crystal-structure type and the superconducting property. Superconductivity is not confined to one or two crystal-structure types, and in fact, it is impossible to predict whether crystals of a given structure type will or will not be superconductors; some metals even in the amorphous state are superconductors.

Because of their possible applications—especially for constructing superconducting magnets—materials with high critical magnetic fields and high transition temperatures have been actively sought. Several important materials have an ordered A15 structure at room temperature; examples, with their transition temperatures T_c, are: Nb_3Sn (18.05°K), Nb_3Al (17.5°K), V_3Si (17.1°K), Nb_3Ga (14.5°K), Nb_3Au (11.5°K), Nb_3Pt (9.2°K), Mo_3Ir (8.8°K), and Mo_3Os (7.2°K). Solid solutions in general (perhaps invariably) have lower T_c values than the stoichiometric compounds or the pure elements. The A15 structure does not always remain stable in these materials down to 0°K, yet the change in crystal structure, when a change occurs, does not correlate with T_c—at least not in V_3Si.[2] The highest transition temperatures T_c among pure elements are reported as Tc, Nb, and Pb with T_c = 11.2, 9.46, and 7.23°K, respectively. Some elements become superconductors when deposited as thin films on cold substrates (below 10°K)[3]; others become superconductors when subjected to high pressures which induce structure changes.[4]

The atomic volume of the elements is related to the electronic states of metals and therefore to superconductivity. In Fig. 12–16, due to Roberts,[5]

[1] D. Shoenberg, "Superconductivity," Cambridge, New York, 1960. E. A. Lynton, "Superconductivity," Wiley, New York, 1962. J. Bardeen and J. R. Schrieffer in C. J. Gorter (ed.), "Progress in Low Temperature Physics," vol. 3, p. 170, North Holland Publishing, Amsterdam, and Interscience, New York, 1961. D. Douglass and L. Falicov in C. J. Gorter (ed.), "Progress in Low Temperature Physics," vol. 4, loc. cit., 1964. J. E. Kunzler, *Rev. Modern Phys.*, vol. 33, p. 499, 1961. J. Bardeen, *Rev. Modern Phys.*, vol. 34, p. 667 1962. B. T. Matthias, T. H. Geballe, and V. B. Compton, *Rev. Modern Phys.*, vol. 35, p. 1, 1963. A brief introduction and bibliography also appeared as a "Resource Letter" by D. M. Ginsberg in *Am. J. Phys.*, vol. 32, no. 2, p. 1, February, 1964.

[2] B. W. Batterman and C. S. Barrett, *Phys. Rev. Letters*, vol. 13, p. 390, 1965. Some V_3Si crystals transform from cubic to tetragonal with c/a increasing from 1.0000 to 1.0025 on cooling below temperatures in the 20 to 30°K range, the c/a ratio becoming constant at T_c; but the onset of superconductivity occurs even in crystals that do not transform from the cubic structure.

[3] W. Buckel, J. R. Dillinger (ed.), "Proceedings of the 5th International Conf. on Low Temperature Physics and Chemistry," p. 326, University of Wisconsin, Madison, Wisc., 1958.

[4] M. B. Brandt and N. I. Ginsberg, *Soviet Phys., JETP*, vol. 12, p. 1082, 1961.

[5] B. W. Roberts, *Prog. Cryog.*, vol. 4, p. 167, 1964.

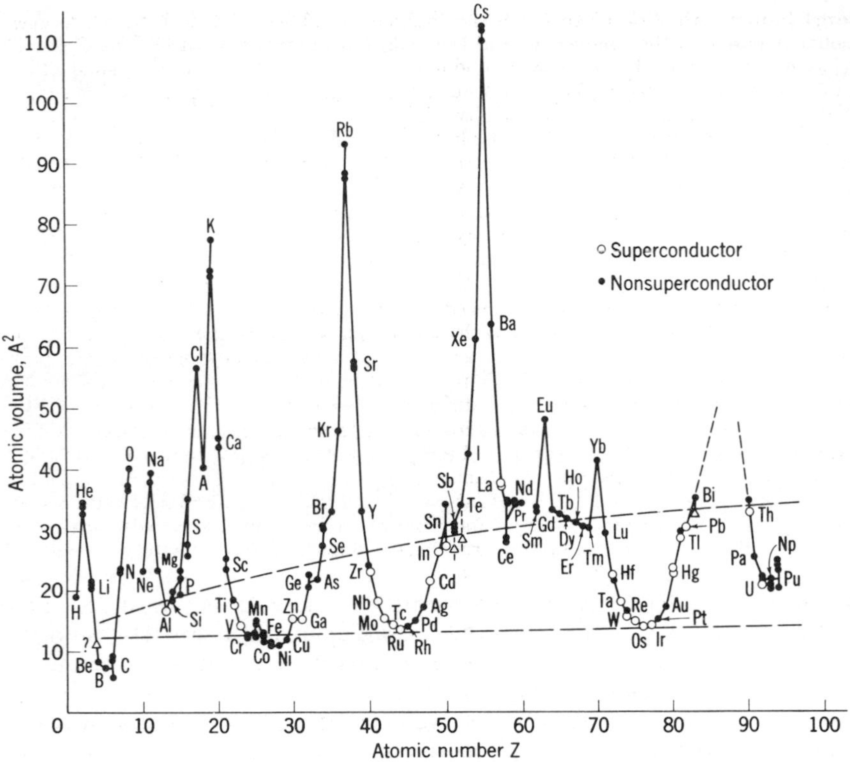

Fig. 12-16 Atomic volumes of the elements and the occurrence of superconductivity. (**After B. W. Roberts, Progr. Cryogen., vol. 4, p. 167, 1964.**)

the atomic volumes are plotted against the atomic number and the dashed lines show the approximate range of volumes within which superconductivity has been observed. Many elements which fall into this range have still not been identified as superconductors, but the possibility exists that their T_c temperatures are extremely low and may be very sensitive to impurity.[1]

There is an important correlation between the occurrence of superconductivity and the number of valence electrons, as stated in Matthias' rule. The number of valence electrons per atom, counting all electrons outside a filled inner shell, in any superconducting phase is "appreciably greater than one and less than ten."[2] Within this range high T_c values occur at certain values of electron concentration so that T_c is an "un-

[1] Ibid.

[2] B. T. Matthias, *Prog. Low Temp. Phys.*, vol. 2, p. 138, 1957. B. T. Matthias, T. H. Geballe, and V. B. Compton, *Rev. Modern Phys.*, vol. 35, p. 1, 1963.

dulating" function of valence concentration, and maximum T_c values occur near three, five, and seven electrons per atom.

It is becoming clear that one of the basic problems associated with tabulating data pertaining to superconducting properties is the sensitivity of the samples to prior metallurgical processing. Such factors as crystal perfection, purity, annealing times, etc., are often of utmost importance.

There is a phenomenological theory of superconductivity, the theory called "GLAG" (after the developers, Ginzburg, Landau, Abrikosov, and Gor'kov), which is more appropriate for practical applications than the BCS theory.[1] Two types of superconductors are differentiated by GLAG. Type I superconductors have complete exclusion of magnetic flux (i.e., complete "Meissner effect"). Type II superconductors have a "flux-line state" or "mixed state" when the metal is in a magnetic field of medium strength (between field strengths H_{c1} and H_{c2}); in this state the metal is superconducting everywhere except along the cores of cylindrical threads ("flux lines" or "vortex lines") which lie parallel to the applied field. The flux lines are surrounded by circulating electric supercurrents; each line constitutes a quantum of magnetic flux. These threads repel one another and take up regular or semiregular arrangements in the metal. Neutron diffraction in the range of diffraction angles near $2\theta = 20$ min has disclosed evidence of a lattice (currently believed to be hexagonal)[2] of the flux lines. The lines interact with, and may be pinned by, various imperfections in the metal (for example, dislocations[3]). The flux-line pinning is responsible for the retention of the superconductivity up to the critical field H_{c2}. The nature of the lattice of flux lines, its imperfections, and its interactions with inhomogeneities in the material form a very active new field of research of fundamental and practical importance in the structure of metals.

[1] For reviews, see: J. D. Livingston and H. W. Schadler, *Progr. Mat. Sci.*, vol. 12, p. 183, 1965; J. A. Catteral, *Met. Rev.*, vol. 11, p. 25, 1966; D. Dew-Hughes, *J. Mat. Sci. Eng.*, vol. 1, p. 2, 1966.

[2] D. Cribier et al., *J. Appl. Phys.*, vol. 37, p. 952, 1966; *Phys. Letters*, vol. 9, p. 106, 1964.

[3] E. Nembach, *Phys. Stat. Sol.*, vol. 15, in press, 1966.

13

Theories of metallic phases

The 1950s and 1960s are seeing a marked acceleration in our understanding of the relation of electron energies to the structure and stability of metallic phases. Qualitative and semiquantitative understanding seems to have been reached regarding a number of interesting problems in this field, and we can expect that more quantitative and more soundly based theories will soon follow the rapid expansion of experimental data. It is therefore appropriate in the present edition of this book to summarize at considerable length some of the empirical facts and current views relating to electron energies and Brillouin zones.

The theories of alloy phases cannot as yet be treated in a rigorous mathematical manner because the whole problem of the stability of different structures is immensely complicated; it involves calculation of small differences between large cohesion energies. The observed forms of the equilibrium diagrams of alloys can be interpreted in terms of the principle of minimum free energy and the "common-tangent" rule applied to mixtures of phases; but, although this principle can be successfully used as an illustration of phase stability, only a few energy calculations have been attempted from first principles.[1]

In the more complex alloy systems, it is clear that the understanding of the basic atomic and electronic factors that contribute to the free energy of alloy phases is still very limited; such factors as strain, geometrical configuration, complex electronic interaction, and various entropy factors—to mention only a few—obviously contribute to the free energy and must be correctly evaluated in a successful theory.

Considering metals and alloys as a whole, there are a number of empirical generalizations concerning metallic properties that should be emphasized. To a first approximation many properties of simple metals and alloys show a remarkable structural insensitivity. This applies, for example, to changes of cohesive energies upon solidification, ordering, allotropic changes, etc. The ratios of the boiling-point temperatures (representing dissociation

[1] The thermodynamical principles of equilibrium, and of alloy phase diagrams, are treated in a number of books and publications; see, for example: A. H. Cottrell, "Theoretical Structural Metallurgy," 2d ed., E. Arnold, London, 1955. Carl Wagner, "Thermodynamics of Alloys," Addison-Wesley, Cambridge, Mass., 1952. A. W. Lawson in American Society for Metals Symposium, "Thermodynamics in Physical Metallurgy," Cleveland, Ohio, 1949.

into atoms) to transformation temperatures (representing merely changes of agglomeration) are much higher in metallic structures than in molecular crystals. Many electronic properties that can be described by "one-electron" functions also show little dependence on structure. Thus, the increase of resistivity on melting is often no greater than a factor of 2 despite the usual implication that a long-range periodic lattice is needed for high conductivity. Knight-shift effects follow a similar trend.

Within the Periodic Table the properties of pure metals depend primarily on the column rather than on the row, which means that valence and electron concentration are more important than atomic numbers. Even when rows and columns are mixed, alloys with similar average valence often have similar properties. In dilute alloys the impurity atoms will tend to interact with one another as well as with the atoms of the matrix. On the whole, impurity atoms repel each other in monovalent matrices and attract in multivalent ones, producing a tendency to short-range order and clustering, respectively.

In this chapter we shall review some of the general rules as well as the more recent theories of alloy formation; the main emphasis will be on the factors that determine the crystal structure. The three major factors that are commonly considered in connection with the stability of alloy phases are: the *electron concentration*, related to valence and composition; the *atomic size*, related to radius ratios and space filling; and the *chemical affinity*, related to charge polarization, electronegativity effects, and stoichiometry. Each factor is at the moment only partially formulated.[1]

Electrochemical factor The influence of chemical affinity has already been emphasized when the formation of phases at stoichiometric ratios corresponding to the normal chemical valences of the participating elements was discussed in Chap. 10. For example, it is sometimes convenient to discuss the electrochemical factor in terms of the so-called *Zintl line*, which can be drawn vertically in the Periodic Table to divide elements that can form electropositive partners from those that form electronegative partners. The use of normal valence rules in predicting isotropic structures has found applications in the field of semiconductors and tetrahedral

[1] The progress in the understanding of alloying behavior has been summarized and discussed in a number of recent symposia and publications; see, for example, "Theory of Alloy Phases," American Society for Metals Symposium, Cleveland, Ohio, 1956. "The Physical Chemistry of Metallic Solutions and Intermetallic Compounds," National Physical Laboratory Symposium No. 9, Her Majesty's Stationery Office, London, 1959. T. B. Massalski and H. W. King, *Progr. Mater. Sci.*, vol. 10, p. 1, 1961. W. Hume-Rothery, *J. Inst. Met.*, vol. 90, p. 42, 1961. J. Friedel and A. Guinier (eds.), "Metallic Solid Solutions," W. A. Benjamin, New York, 1963. P. A. Beck (ed.), "Electronic Structure and Alloy Chemistry of the Transition Elements," John Wiley, New York, 1963. T. B. Massalski (ed.), "Alloying Behavior and Effects in Concentrated Solid Solutions," Gordon and Breach, New York, 1965.

Note added in proof: A comprehensive discussion of the progress and problems in the field of alloys will be found in a recent symposium: P. S. Rudman and J. Stringer (eds.), "Phase Stability in Metals and Alloys," McGraw-Hill, New York, 1966.

structures.[1] For the compounds of the MX type (sphalerite, wurtzite, and rock-salt structures) it has been shown empirically that the directional bonding of the tetrahedral type becomes weaker and is replaced by an ionic rock-salt-type bonding as the average quantum number of the valence electrons of M and X increases.[2]

Quantitative assessment of the electrochemical factor meets with difficulties. The basic definition relates differences in electronegativities to the bond energies, using a relationship similar to that for the case of ordering energy (see Chap. 11).

$$(\mathcal{E}_A - \mathcal{E}_B)^2 = V_{AB} - \tfrac{1}{2}(V_{AA} + V_{BB}) \qquad (13\text{–}1)$$

where $\mathcal{E}_A$ and $\mathcal{E}_B$ are electronegativities. However, while V_{AB} may be obtained from the heat of formation of an AB compound, the V_{AA} and V_{BB} values for metals are less easy to define. Hence, in practice, the electronegativities are assessed according to several measurable properties that are expected to be related to electronegativity, such as standard electrode potentials, the heat of formation of a halogen bond, surface work functions, and ionization potentials. The derived values of electronegativity frequently differ according to which scheme is employed.

Electron concentration in alloys Empirical studies have shown that in many alloy systems the *electron concentration* is the important parameter that influences such factors as the extent of primary solid solubility, the presence or absence of a particular structure, the range and stability of intermediate phases, the formation of long-period superlattices, trends in lattice spacings, the number of vacant sites in defect structures, the stacking-fault energy, and many others. The electron concentration is usually expressed as the number of valence electrons per unit cell of the structure (provided that all sites are occupied), or as the ratio of all valence electrons to the number of atoms, e/a. The use of e/a rather than atomic or weight percent composition as a parameter against which properties are compared almost never fails to bring about interesting correlations when applied to experimental data. However, the physical meaning of e/a is by no means clear; it is not certain whether or not, on alloying, all the valence electrons of a solute atom enter into the conduction band of the alloy,[3] since some of them may be retained to screen the positive charge on the nucleus. When a monovalent noble metal, say copper, is alloyed with a B-subgroup element, such as the trivalent gallium, some of the $s + p$ electrons of gallium may be expected to remain in *bound states* around the gallium ions, amount-

[1] For details, see E. Parthé in J. Westbrook (ed.), "Intermetallic Compounds," Wiley, New York, 1965. F. Laves in W. M. Mueller and M. Fay (eds.), "Advances in X-ray Analysis," vol. 6, Plenum, New York, 1963. E. Mooser and W. B. Pearson, *J. Chem. Phys.*, vol. 26, p. 893, 1957; "Progress in Semi-Conductors," vol. 5, p. 103, Haywood, London, 1960. C. H. L. Goodman, *J. Phys. Chem. Solids*, vol. 6, p. 305, 1958.

[2] Mooser and Pearson, loc. cit.

[3] N. F. Mott, *Progr. Metal Phys.*, vol. 3, p. 76, 1952; *Report Prog. Phys.*, vol. 25, p. 218, 1962. J. Friedel, *Advan. Phys.*, vol. 3, p. 446, 1954.

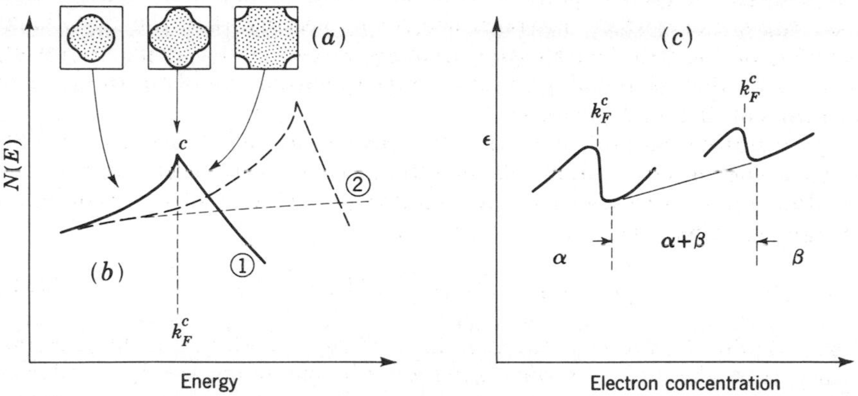

Fig. 13-1 Schematic relationship between (*a*) the shape of the Fermi surface and (*b*) the density of states. Diagram (*c*) illustrates the possible dependence of phase stability upon the total electronic energy, $\mathcal{E}$.

ing effectively to *incomplete ionization*. Hence, although the nominal value of e/a will be changed with additions of gallium to copper, the true disposition of the electrons in the solid solution is undoubtedly more complex than allowed for in the simple theories. According to Friedel, this does not conflict with the electron-concentration rules, because the bound states may be subtracted from the main conduction band.[1] Other problems are connected with the very definition of valence, as discussed in Chap. 12. Occasionally the average of the number of electrons outside the inert-gas core (i.e., $s + p + d$ electrons in the case of first-row elements) has been used to denote the electron concentration. This is sometimes known as the *average group number*, but even this value may be difficult to define. For example, the d electrons may be involved in the band structure of the noble metals and their alloys, due to hybridization with the s electrons as discussed earlier, and hence may give the appearance of a variable valence for the solvent.

Model of the band structure of an alloy The simple electronic model of an alloy formed between elements of different valence involves only the expansion or contraction of the Fermi surface on alloying as the electron concentration is increased or decreased. If the Fermi surface lies well within the Brillouin zone, the relationship between the energy E and the density of states $N(E)$ in the band is parabolic, but on approaching the zone boundaries the curve becomes modified as shown diagrammatically in Fig. 13–1*b*. The peak in the density of states corresponds to contact between the Fermi surface and the Brillouin zone, as shown diagrammatically in Fig. 13–1*a*. Following contact, the curve begins to fall as the corners of the zone are filled. The first successful model of an alloy band structure was proposed in the early 1930s by Jones and was subsequently developed in a

[1] See Friedel, loc. cit.

number of papers.[1] This model assumes a "rigid band" during alloying; i.e., the shape of the density-of-states curve for the pure solvent remains unaffected by alloying. The added or subtracted electrons merely change the position of the Fermi level E_F in the band. A particularly favorable energy condition is obtained if the available electrons fill the band only to the region for which the density-of-states curve (curve 1 in Fig. 13–1*b*) falls above the parabolic relationship for free electrons shown by the dashed line (curve 2). The total Fermi energy of the electrons, $\bar{E} = \int^{E_F} E\, N(E)\, dE$, when compared with the energy of the free electrons, $\bar{E}_0$, is lowered by this effect with the corresponding increase in stability. According to Jones,[2] the maximum difference $\Delta\bar{E} = \bar{E} - \bar{E}_0$, calculated from the density of states, should occur within the energy range lying to the right of the point C in the density of states, but this does not mean that a minimum in $\bar{E}$ will also occur in the same range of energy. However, a rather different approach, involving the long-range oscillations of electronic density around impurity atoms[3] (see Chap. 12, page 333), does suggest that a minimum occurs in this region in the total electronic energy $\mathcal{E}$, taking into account the kinetic energy of the electrons and the ion-electron, the ion-ion, and the electron-electron interaction energies. This is illustrated diagrammatically in Fig. 13–1*c*, in which the points $k_F{}^c$ indicate the points of contact between the Fermi surface and the Brillouin zone. When the zones of two different structures are contacted by the Fermi surface at different values of e/a, the favorable ranges of stability will be determined by the common-tangent principle, as illustrated in Fig. 13–1*c*. Theoretical arguments of this nature permit a qualitative interpretation of solid solubility ranges and of the stability of electron phases. However, a detailed quantitative theory is not yet available.

Fermi surfaces and Brillouin zones in alloys The concept of the Fermi surface in alloys has been used for many years as a means of discussing the enhanced stability of certain phases. Yet direct experimental determination of the Fermi surface in alloys, using the methods described for pure metals in Chap. 12, is practically impossible. The problem is one of greatly reduced *mean free path* of the electrons as a result of scattering by solute atoms. The mean free path is reduced from several thousand angstroms to perhaps 200 A or less. Hence, unless the extremal orbits of the Fermi surface (see Chap. 12) are very small, viz., around some small specific portions of the total surface, electrons in alloys will be scattered before completing a full orbit. Additions of a few percent of solute therefore usually will destroy the de Haas–van Alphen effect, the classic method for testing the possible "sharpness" of the Fermi surface and other topographic details. The situation is somewhat improved in alloys and com-

[1] H. Jones, *Proc. Roy. Soc.* (*London*), vol. A144, p. 225, 1934; vol. A147, p. 396, 1934; *Proc. Phys. Soc.* (*London*), vol. A49, p. 250, 1937; *Phil. Mag.*, vol. 43, p. 105, 1952.

[2] H. Jones, *J. Phys. Radium*, vol. 23, p. 637, 1962.

[3] A. Blandin in T. B. Massalski (ed.), "Alloying Behavior and Effects in Concentrated Solid Solutions," Gordon and Breach, New York, 1965.

pounds that are fully ordered, for then electron scattering is very considerably reduced, as indicated by the increased ratio of *residual resistivity* at low temperatures to the room-temperature resistivity. In such cases the de Haas–van Alphen oscillations have been reported in a number of stoichiometric structures, among them ordered beta-brass (Cu-Zn), SeTe, $MgZn_2$, AuSn, $AuGa_2$, etc.[1] The residual resistance ratio for such phases is about 1/20.

Recently, two experimental methods have been suggested that show some promise for the future. One involves the use of a pulsed magnetic field of very high intensity, exceeding 10^5 gauss and of some 10 msec in duration, coupled with the acoustic attenuation technique[2]; the other involves the use of the positron annihilation effect.

In the latter, positrons emitted from a radioactive isotope, such as Na^{22} or Cu^{64}, undergo collisions with electrons in a sample placed nearby. Occasionally the positrons are annihilated on collision and converted into two photons (γ rays) which are emitted instantaneously at approximately 180° to one another from the center of mass of the pair. A very slight departure from the 180° is directly proportional to the transverse component of the momentum of the pair. The momenta of the electrons involved in such collisions can thus, at least in principle, be worked out from the geometry and intensity of the emitted γ rays,[3] but the very high-angle collimation required is a disadvantage.

The experiments mentioned above and the rather substantial, although indirect, evidence based on x-ray diffraction, elastic constants, electronic specific heat, lattice spacings, phase stability, etc., suggest that the concepts of Fermi surfaces and Brillouin zones are applicable to alloys and that electron concentration has a striking and indisputable influence upon alloying behavior. One reason for this appears to be that the mean free path in most alloys is sufficiently long to make the moving electrons encounter a more or less uniform lattice potential and hence undergo reflections as in pure metals. Charge oscillations, related both to the geometry of the particular impurity or defect and to the form of the Fermi surface, must also be considered.[4]

Hume-Rothery rules As a result of studies by Hume-Rothery and his associates,[5] mainly of alloys based on copper and silver, certain general

[1] See articles by J. A. Rayne and W. B. Pearson in T. B. Massalski (ed.), "Alloying Behavior and Effects in Concentrated Solid Solutions," Gordon and Breach, New York, 1965. W. B. Pearson in P. S. Rudman and J. Stringer (eds.), "Phase Stability in Metals and Alloys," McGraw-Hill, New York, 1966.

[2] E. R. Dobbs in W. A. Harrison and M. B. Webb (eds.), "The Fermi Surface," p. 311, Wiley, New York, 1960.

[3] See, for example, a review by P. R. Wallace in F. Seitz and D. Turnbull (eds.), "Solid State Physics," vol. 10, Academic, New York, 1960. A. T. Stewart, *Phys. Rev.*, vol. 133, p. A1651, 1964.

[4] J. Friedel, *Trans. AIME*, vol. 230, p. 616, 1964.

[5] See W. Hume-Rothery and G. V. Raynor, "The Structure of Metals and Alloys," 4th ed., Institute of Metals, London, 1962.

rules have been postulated concerning the extent of primary solid solubility to be expected upon alloying. These rules refer to the difference between the relative atomic radii of the participating elements, their relative valences, and their electrochemical differences. The Hume-Rothery rules may be summarized as follows:

1. If the difference between the atomic radii of the elements forming an alloy exceeds about 14 to 15 percent, solid solubility is restricted. This is sometimes known as thc *15 percent rule*. Hume-Rothery and his associates took the *closest distance of approach of atoms* in the structure of each pure element as the measure of atomic size. If this value for a particular solute element lies outside the favorable size zone for the solvent, the *size factor* is said to be unfavorable, and as a rule, the primary solid solubility is restricted. Within the favorable zone the solubility should be extensive, but other factors, such as electrochemical interactions, may prevent this. In a sense, therefore, the 15 percent rule is a negative rule stressing the role of size differences mainly when they *restrict* alloy formation; and even in this aspect it is only partially successful. Theoretical justification for the 15 percent rule is provided by considerations of elastic strain energy in a solid solution (see below).

2. The likelihood of the formation of a particularly stable phase in an alloy system may be related to the electrochemical difference between the participating elements and is increased as one of the elements becomes more electronegative and the other becomes more electropositive. The above principle is then known as the *electronegative valence effect.* However, stable phases, such as Laves phases (see below), often limit solid solubility, although the increased stability in such cases appears to be due to size considerations rather than electronegativity.

3. Mutual solid solubility of two given elements is related to their respective valences, the amount of the solid solution in the element of lower valence being, as a rule, greater than vice versa. This general principle is known as the *relative valence effect.* It appears to be valid for copper, silver, or gold, which are monovalent, alloyed with the B-subgroup elements of the Periodic Table which possess valences greater than 1. When both partners are polyvalent, the relative valence rule appears to be less general.[1]

Attempts to test Hume-Rothery rules for a large number of elements as solvents are of interest.[2] One of the suggested methods is to plot primary solid solubilities as a function of two parameters at the same time.[3]

A large number of binary systems have been analyzed using a computer. If a 5 percent solid solubility is used as the criterion of an extensive solid solution, then the two-parameter analysis gives a correct prediction in some 76 percent of the 1500 cases studied.

[1] W. Hume-Rothery, "Atomic Theory for Students of Metallurgy," 3d ed., Institute of Metals, London, 1960.

[2] See, for example, J. T. Waber et al., *Trans. AIME*, vol. 227, p. 717, 1963.

[3] L. S. Darken and R. W. Gurry, "Physical Chemistry of Metals," p. 86, McGraw-Hill, New York, 1953.

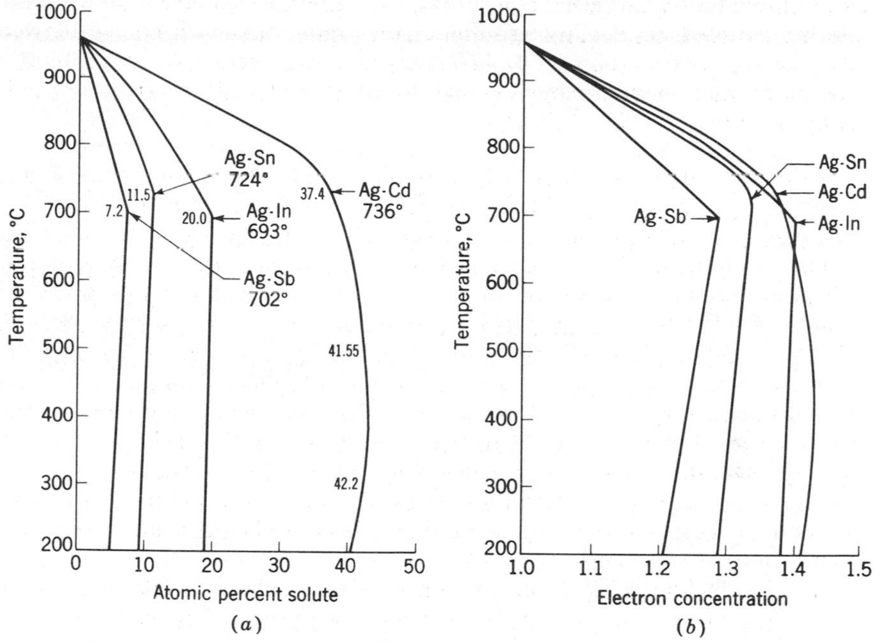

Fig. 13-2 Phase fields of primary solid solutions based on silver: (*a*) in terms of atomic composition and (*b*) in terms of *e/a*.

Electronic theories of primary solid solubility of alloys based on the noble metals A survey of the binary systems of copper, silver, and gold with the *B*-subgroup elements shows that the maximum extent of primary solid solubility may be correlated with an electron concentration value that does not change too much from system to system.[1] This becomes evident, for example, when pertinent portions of several related systems are superimposed, as in Fig. 13–2. The majority of silver-based primary solid solutions terminate within a fairly close range of *e/a* values near 1.4, whereas in copper-based alloys the *e/a* values show a wider scatter but are only a little less than 1.4 (see Fig. 10–15). In the case of gold alloys, the primary solid solubility is more restricted than in copper or silver, ranging between 1.2 and 1.3. Undoubtedly, the general extent of a primary solid solution must also be related to the stability of the intermediate phases that follow, and this fact should not be overlooked. For example, Fig. 10–15 shows that, apart from the systems Cu-In and Cu-Sn, the primary solutions followed by a cubic β phase reach somewhat higher values of *e/a* than those followed by a c.p.h. ζ phase. Nevertheless, although the correlation between the primary solubility and *e/a* does not lead to any unique value, it is quite striking when compared with a similar examination in terms of atomic composition (Fig. 13–2). Hence, it has been evident for a long time that

[1] W. Hume-Rothery, G. W. Mabbott, and K. M. Channel-Evans, *Phil. Trans. Roy. Soc.*, vol. A233, p. 1, 1934.

there must be an important link between primary solid solubility and electronic structure. Several attempts have been made to produce a satisfactory theory of the observed phenomenon; one in particular, notably the theory of Jones, is well known in metallurgical literature.

In his original paper, Jones[1] attempted to interpret the extent of primary solid solubility of Cu–Zn alloys (α phase) and the stability of the β phase in terms of the Brillouin zones and the Bloch wave functions of the conduction electrons. As mentioned previously, the *rigid band condition* was assumed. In addition, the *nearly free electron approximation* was used and extended from pure metals to random solid solutions. At a certain range of energy (and therefore e/a) the Brillouin zone of the α phase would possess a higher density of electronic states than the β phase, and thus "accommodate" the available electrons within lower total energy, making the total free energy lower for the α phase than for the β phase. Using identical values of the atomic volumes and energy discontinuities at the Brillouin zone faces for both α and β phases and making them equal to those of copper ($\Delta E \approx 4.1$ ev), Jones calculated the density-of-states curves for both phases. These curves are well known and are of considerable interest because the calculation from which they result represents an attempt at interpreting primary solid solubility in terms of energy relationships related to contact between the expanding Fermi surface and the Brillouin zone. However, if the calculated values of the density of states are plotted in terms of e/a, the peaks in $N(E)$ for contact with the $\{111\}$ zone faces in the α phase and with the $\{110\}$ faces in the β phase are found to occur at rather low e/a values, ≈ 1.0 and ≈ 1.23, respectively, and do not correspond with experimental data for the observed termination of the α phase ($e/a \approx 1.3$) or the range of stability of the β phase ($e/a \approx 1.5$). Thus, the correlation between peaks in the density of states and ranges of phase stability is poor and does not indicate a simple correspondence. On the other hand, the calculation correctly indicated a possible contact between the Fermi surface (clearly no longer spherical) and the $\{111\}$ zone faces in pure copper at one electron per atom, although this fact was not recognized until some twenty years later when Pippard measured the shape of the Fermi surface (see Chap. 12, page 324). Allowing for the existing contact, it is possible to interpret the termination of primary solid solubility by using the proposal, illustrated in Fig. 13–1, that greatest stability might be expected at some point lying to the right of the peak in the density of states. However, this approach no longer represents a simple band model. As pointed out by Hume-Rothery,[2] incorporation of Jones' original model into metallurgical literature has led to a good deal of confusion, mainly because of the parallel that could be drawn between the attempt by Jones to calculate the relative stability of two phases in terms of energies resulting from contact between nearly spherical Fermi surfaces and some assumed energy discontinuities, and attempts to calculate merely the electron concentration at which an inscribed Fermi sphere would contact a given set of zone faces (which by implication would have to possess zero gaps). Simple free-electron calculation shows that contact of a Fermi sphere with the Brillouin zone would occur in the α phase at 1.36 electrons per atom and in the β phase at 1.48 electrons per atom, and these values are strikingly close to observed phase stabilities. This, however, must now be regarded as somewhat fortuitous unless it can be shown that the band gaps in the zones of the α and β phase are very small, i.e., unless they are substantially reduced upon alloying from those in the pure noble metals. This calls for a variable behavior of the band gaps and the departure from a rigid band approach. There is some evidence that this may indeed be true in Cu–Zn α alloys.[3] Some twenty years after Jones' work,

[1] H. Jones, *Proc. Phys. Soc.* (*London*), vol. A49, p. 250, 1937.

[2] W. Hume-Rothery, *J. Inst. Met.*, vol. 90, p. 42, 1961; *Metallurgist*, vol. 3, p. 11, 1964.

[3] M. A. Biondi and J. A. Rayne, *Phys. Rev.*, vol. 115, p. 1522, 1959; also see the discussion of beta-brass on p. 349.

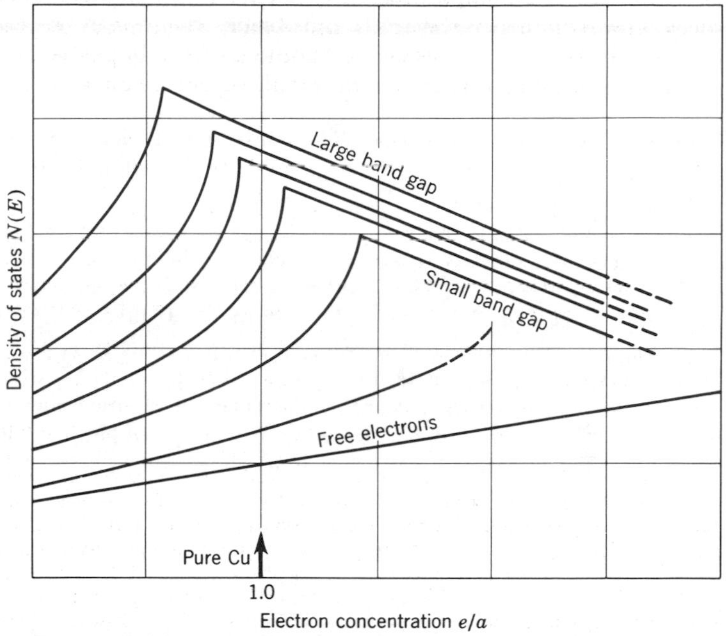

Fig. 13-3 Relationship between the density of states $N(E)$ and energy E (hence electron concentration e/a) for different values of the band gaps in the Brillouin zone. (After J. M. Ziman, *Advan. Phys.*, vol. 10, p. 1, 1961.)

Cohen and Heine[1] formally introduced the concept of "soft," or variable, band gaps in alloys.

The possible trend in the density of states corresponding to variable, and decreasing, band gaps may be inferred from the diagram shown in Fig. 13–3, which is due to Ziman. It corresponds to a succession of rigid band calculations for different values of the band gap. The net effect of decreasing the band gap is to decrease the density of states at any given e/a. Hence, the e/a at contact between the Fermi surface and the Brillouin zone is shifted from left to right as the band gaps decrease. If the band gaps become small, the nearly free electron model with a nearly spherical Fermi surface becomes plausible.

It must be emphasized that the above interpretation differs from that illustrated in Fig. 13–1*c* and related to the total electronic energy $\mathcal{E}$. Both interpretations connect the solid solubility to some details of the Fermi surface and the Brillouin zone. However, the calculation of $\mathcal{E}$ includes terms related to the long-range charge oscillations due to introduced solute atoms, while the band picture requires solute atoms to act mainly as donors of

[1] M. H. Cohen and V. Heine, *Advan. Phys.*, vol. 7, p. 395, 1958.

electrons to a common band. The fact that in the variable band models, solute atoms influence the band gaps brings the two interpretations closer together and emphasizes that contributions from both the solvent and the solute atoms are important in the electronic structure of the alloys.

Stability of electron phases The stability of electron phases β, γ, ζ, and ϵ, like the stability of the primary solid solutions, appears to be the consequence of interactions between the Fermi surface and the Brillouin zones of these structures, with emphasis on the influence of such interactions upon the density of states $N(E_F)$ at the Fermi surface.[1]

The Brillouin zone for the β phases with the b.c.c. structure is shown in Fig. 12–6. It is a rhombic dodecahedron of {110} faces which can hold two electrons per atom. The planes that define the energy zone (Jones' zone) of the γ structure belong to two sets, {330} and {411}, as illustrated in Fig. 13–4*a*. Together they result in 36 faces, and since these faces are all equidistant from the origin of **k** space, the shape of the zone approximates a sphere. The corresponding x-ray line, $h^2 + k^2 + l^2 = 18$, in a Debye-Scherrer pattern always shows high intensity, but due to the high multiplicity, the band gaps across the zone faces will not necessarily be large. In the original paper on the stability of the gamma-brass structure, Jones[2] has estimated band gaps in gamma-brass of the Cu-Zn system to be of the order of 1.46 to 1.93 ev. These values are of the same order of magnitude as those estimated for the zone of c.p.h. ϵ and ζ phases (see page 324). At the onset of contact between the undistorted Fermi surface of free electrons (i.e., a *Fermi sphere*) and the principal faces of the respective Brillouin zones for the β, γ, ζ, and ϵ phases, the zones are relatively full. This is particularly true for gamma-brass; this fact became the basis for the theoretical interpretation of the occurrence and stability of all electron phases in terms of nearly full Brillouin zones. The important values of e/a are: 1.48 for contact between the Fermi sphere and the zone for the beta-brass, 1.54 for contact between the Fermi sphere and the {300} and {411} faces of the zone of the gamma-brass structure, and 1.72 (depending upon the axial ratio) associated with the filling of the inner zone of the zeta- and epsilon-brasses. These e/a values bear similarity to the e/a ratios suggested for the electron "compounds" by Hume-Rothery (see Chap. 10, page 249)[3] and based upon chemical formulas [compare β-CuZn ($\frac{3}{2} = 1.5$), γ-Cu_5Zn_8 ($\frac{21}{13} = 1.62$), and ϵ-$CuZn_3$ ($\frac{7}{4} = 1.75$), with the values 1.48, 1.54, and 1.72 respectively], but it must be remembered that in both cases the actual values are derived from the particular model, electronic or chemical. The chemical formulas are now known to be inapplicable and the simple free-electron–Brillouin-zone models suffer from the limitation that for the e/a

[1] See the various articles by H. Jones listed on p. 342. Also, T. B. Massalski and H. W. King, Alloy Phases of the Noble Metals, *Progr. Mater. Sci.*, vol. 10, p. 1, 1961.

[2] H. Jones, *Proc. Roy. Soc.* (*London*), vol. A144, p. 255, 1934.

[3] See W. Hume-Rothery and G. V. Raynor, "The Structure of Metals and Alloys," 4th ed., Institute of Metals, London, 1962.

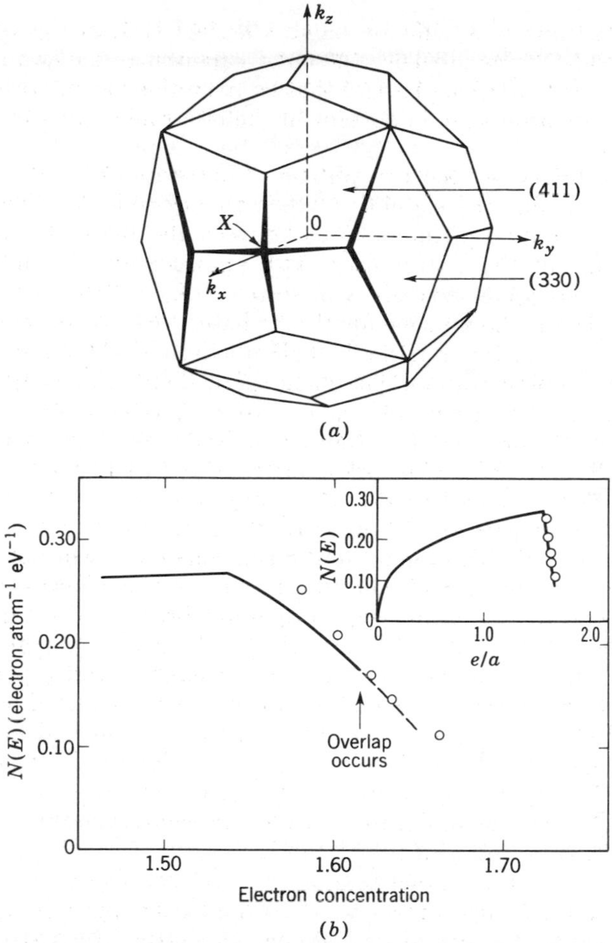

Fig. 13-4 Gamma-brass: (*a*) the Jones' zone and (*b*) the density of states derived from the electronic specific heat. (After B. W. Veal and J. A. Rayne, *Phys. Rev.*, vol. 132, p. 1617, 1963.)

values quoted above, the band gaps across the Brillouin zones would have to be zero or near zero, a fact not yet confirmed experimentally.

In the case of beta-brasses the remarkable agreement between the free-electron values $e/a = 1.48$ and the observed narrow range of actual values corresponding to the tips of the *V*-shaped fields (Fig. 10–16) is in accord with the possibility that band gaps are small. Additional support comes from the measurements of the de Haas–van Alphen effects in beta-brass[1]

[1] J. P. Jan, W. B. Pearson, and M. Springford, private communication, 1965, to be published.

and from band calculations,[1] both of which take into account the existence of the {100} Brillouin zone due to order.

The electronic specific heat measured within the range of the γ phase that follows beta-brass in the Cu-Zn system also suggests small band gaps and a nearly free electron behavior (see Fig. 13–4*b*).

The zone of the γ structure differs from the zones of the α and β structures in that it is not a true Brillouin zone, being *incomplete*,[2] and is sometimes called the *energy zone* or the *Jones' zone* because it has been selected from the x-ray reflections that correspond to the large structure factors, and not on the basis of geometrical symmetry considerations alone. No matter how large the energy gaps across the {411} faces, there are states outside these faces with energies no greater than the energies at interior points in other regions of the zone. Hence, at some value of e/a after the Fermi surface has contacted the {411} faces, electrons may move from the interior to the outside of the zone, not directly "across" the {411} faces, but to regions (such as that marked X in Fig. 13–4) near the points of intersections of four {411} faces. Thus, even though it may be calculated that the Jones' zone of the gamma-brass structure may contain 90 electrons per unit cell (1.73 e/a), the zone can never be exactly filled with electrons before states outside the zone begin to be occupied.

The zone commonly used for discussion of the c.p.h. phases is the "1–2 zone" illustrated in Fig. 12–11*a*, (page 327) for an ideally close-packed structure. The zone is bounded by 20 faces: 6 of the {10·0} type (the A faces), 2 of the {00·2} type (the B faces), and 12 of the {10·1} type (the C faces). The energy discontinuity vanishes across certain lines in the {00·1} faces[3] unless the structure is ordered, and hence these planes do not form part of the energy zone. However, the {00·1} faces, together with the {10·0} faces, may be used to obtain a *reduced zone* for the structure.

The number n of states per atom enclosed within the zone bounded by the three sets of faces is a function of the axial ratio and equals

$$n = 2 - \frac{3}{4}\left(\frac{a}{c}\right)^2\left[1 - \frac{1}{4}\left(\frac{a}{c}\right)^2\right] \tag{13–2}$$

Energy discontinuities also vanish along the intersections of planes of type A with other A planes, and A planes with C planes (see Fig. 13–5), and hence the energy zone is again incomplete in the sense that the inner portion without the "domes" (produced by intersections of the C faces) can never be filled completely without some electrons overlapping into the domes. For this reason Eq. (13–2) must be regarded as approximate. The values of n for ζ phases with ideal axial ratio and for ϵ phases with $c/a = 1.55$ are approximately equal to 1.745 and 1.721, respectively. For the outer zone bounded by the {00·2} and {10·1} faces, $n = 2$.

The Brillouin zone of the c.p.h. structure is of particular interest in the theory of alloy phases because interactions between the Fermi surface and the zone faces may alter the shape of the zone without affecting its volume and the basic symmetry. If the axial ratio is ideal (1.633), the {10·0} zone faces are closest to the origin of the k space, as indicated by the vector $\mathbf{k}_A$ in Fig. 13–5*b*, and are followed in order by the {00·2} faces (vector $\mathbf{k}_B$) and the {10·1} faces (vector $\mathbf{k}_C$). At values of the axial ratio corresponding to 1.50 or 1.73, the zone has a special symmetry since $\mathbf{k}_B = \mathbf{k}_C$ or $\mathbf{k}_B = \mathbf{k}_A$, respectively. It is significant that no known c.p.h. electron phases are stable in the region of these critical values of c/a.

[1] K. H. Johnson and H. Amar, *Phys. Rev.*, vol. 139, p. A760, 1965. K. P. Wang, H. Amar, and K. H. Johnson, *Bull. Am. Phys. Soc.*, vol. 11, p. 74, 1966.

[2] H. Jones, "The Theory of Brillouin Zones and Electronic States in Crystals," North Holland Publishing, Amsterdam, 1960.

[3] H. Jones, *Proc. Roy. Soc.* (*London*), vol. A147, p. 396, 1934.

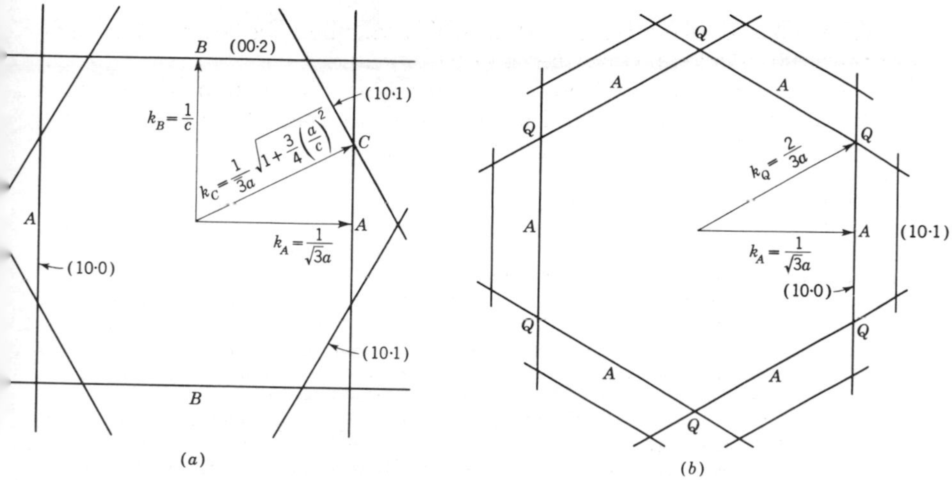

Fig. 13-5 Sections through the Brillouin zone for the c.p.h. structure: (*a*) vertical section and (*b*) horizontal section.

Stability of long-period superlattices As discussed in an earlier section (Chap. 11, page 279), the original discovery of the CuAuII structure was followed by reports of numerous other stable ordered structures which possess periodically spaced antiphase domains. It became obvious that the stability of this kind of ordered atomic arrangement cannot be interpreted in terms of a classical pair-interaction model because it would require postulating unusually long-range forces. The theories that have been proposed deal, instead, with the electronic structure of the crystal as a whole and are based upon the concept that the large regular unit cells and their periodic displacement result from a collective interaction between the electrons and the crystal lattice.[1]

One of the major indications that conduction electrons may be involved collectively is provided by the observed dependence of the long periods upon the electron concentration. This is observed both when the value of the superperiod M is plotted vs. e/a for a number of different systems and when it is plotted for a given binary long-period superlattice alloyed with additional elements of different valence, thereby changing e/a. An illustration of this effect is shown in Fig. 13–6.

Empirical models have been proposed by Schubert and his associates,

[1] J. C. Slater, *Phys. Rev.*, vol. 84, p. 179, 1951. J. F. Nicholas, *Proc. Phys. Soc.* (*London*), vol. A66, p. 201, 1953. K. Schubert, B. Kiefer, and M. Wilkens, *Z. Naturforsch.*, vol. 9a, p. 987, 1954. K. Schubert et al., *Z. Metallk.*, vol. 46, p. 692, 1955. K. Schubert, *Z. Metallk.*, vol. 46, p. 43, 1955; *Z. Naturforsch.*, vol. 14a, p. 650, 1959. H. Sato and R. S. Toth, *Phys. Rev.*, vol. 124, p. 1833, 1961; *J. Appl. Phys.*, vol. 33, p. 3250, 1962; *Phys. Rev.*, vol. 127, p. 469, 1962. A very detailed account of both the occurrence and theories of long-period superlattices is given by H. Sato and R. S. Toth in T. B. Massalski (ed.), "Alloying Behavior and Effects in Concentrated Solid Solutions," Gordon and Breach, New York, 1965.

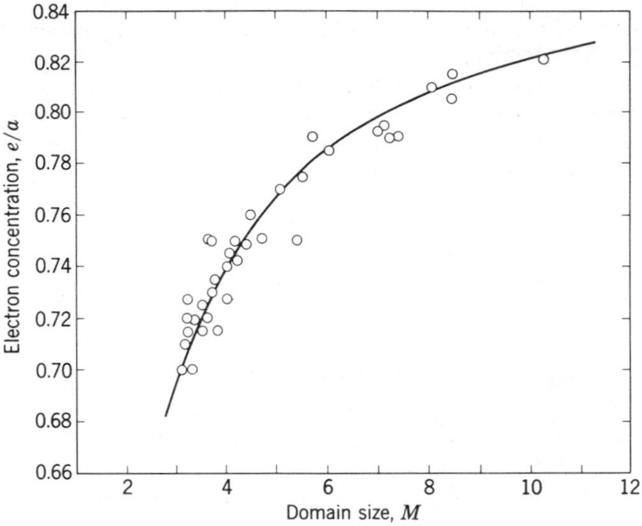

Fig. 13-6 Variation of the superperiod M with e/a in a number of long-period superlattices. (**After H. Sato and R. S. Toth, *Phys. Rev.*, vol. 127, p. 469, 1962.**)

based mainly upon periodic fluctuations in the electron-density distribution throughout the crystal structure. Models due to Slater, Nicholas, and Sato and Toth are based upon the interaction of the conduction electrons with the Brillouin zone of the superlattice along the lines of Jones' model, discussed earlier. The detailed theory of Sato and Toth appears to account for the stabilization of the long-period superlattices and also for additional features: the nature of lattice distortions which accompany their formation, the temperature and composition dependence of the periods, and the influence of the electron-concentration changes. It also deals with the balance between electron energies and antiphase boundary energies.

The description of the interaction between the Fermi surface and the Brillouin zone, an interaction responsible for the lowering of the energy, is mainly qualitative since the actual shape of the Fermi surface is unknown. The first Brillouin zone for the disordered f.c.c. lattice is bounded by the {111} and {200} discontinuity planes and has been described earlier (page 318). In Fig. 13–7 this zone is shown, together with additional discontinuity planes of the general form {001} and {110} which result when the AuCuI ordered tetragonal structure is formed. The volume of the larger zone is $4/a^3$ (a = lattice parameter) and that of the inner zone is $2/a^2c$; they can accommodate two and one electrons per atom, respectively. Since it is known from the recent experimental work and calculations that the Fermi surface is distorted and touches the {111} faces of the zone in both pure copper and pure gold,[1] it is likely that a similar situation exists in the CuAu alloy. Nevertheless, relatively little distortion has been reported in the [110] and [100] directions, where apparently the Fermi surface remains nearly spherical.[2] Therefore, when the {100} and {110} discontinuities appear due to the formation

[1] The details of the Fermi surface in these metals have been described in Chap. 12.

[2] See, for example, W. A. Harrison and M. B. Webb (eds.), "The Fermi Surface," Wiley, New York, 1960.

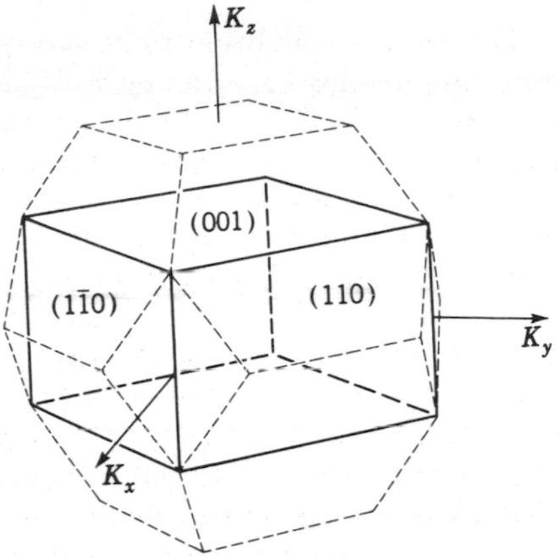

Fig. 13-7 The Brillouin zone for the ordered structure CuAuI.

of the CuAuI superlattice, the Fermi surface, which encloses one electron per atom, may be pictured as a distorted body bulging in the [111] direction and touching the {111} planes while being nearly spherical in the [001] and [110] directions. The free electron energies at the centers of the {100} and {110} faces are 2.4 and 4.8 ev, respectively, and the energy at the surface of a spherical Fermi surface enclosing one electron per atom is 6.5 ev. Therefore, with small band gaps, electrons might be expected to overlap substantially across the {100} zone faces and only slightly across the {110} faces although, of course, overlap is a sensitive function of the energy gaps. The possible small overlap across the {110} faces is shown diagrammatically in Fig. 13–8 as a two-dimensional section in a plane through the origin and parallel to the {001} plane of the Brillouin zone. If the {110} faces can be separated as shown diagrammatically in Fig. 13–8*b* and *c*, so that two slightly tilted discontinuities are formed in place of each original {110} face, it can be seen that a contour of a Fermi surface of a similar average radius, as in the case of Fig. 13–8*a*, can be contained within a set of four outer faces, while overlapping the remaining four inner faces (Fig. 13–8*b*); or alternatively, a Fermi surface that has a smaller average diameter than in the case of Fig. 13–8*a* and *b* can be contained completely within the four inner faces (Fig. 13–8*c*).

Both the possibilities shown in Fig. 13–8*b* and *c* prevent the Fermi surface from overlaps of the type shown in Fig. 13–8*a*, while stabilization presumably occurs due to contact. In addition, changes in the electron concentration on alloying, and therefore the

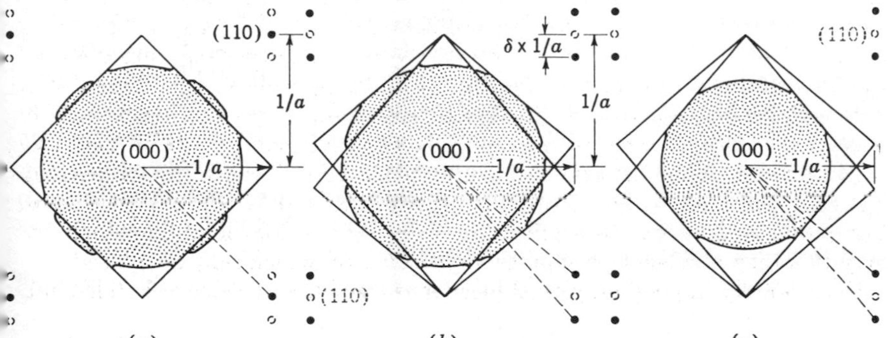

Fig. 13-8 Diagrammatic illustration of the possible Fermi surface contours in the Brillouin zones for the CuAu superlattice: (*a*) CuAuI; (*b*) and (*c*) CuAuII.

changes in the size of the Fermi surface, can be accommodated merely by a change in the separation of the tilted faces, thereby increasing the size of the Brillouin zone without necessitating additional overlaps. This separation is directly related to the long-period M of the superlattice. In the model shown in Fig. 13–8*b* and *c*, one-half of the separation of the superlattice spots, $\delta/a = 1/(2Ma)$.

The characteristic splitting as depicted in the model is obtained as a direct consequence of the superperiod in the *b* direction of the CuAuII superlattice and is confirmed by the existence of separated superlattice spots in the electron diffraction photographs, as represented by the reciprocal lattice diagram discussed earlier (see Fig. 11–12). Hence, it appears that the occurrence of the ordered structure with the long periods affects the geometry of a small portion of the Fermi surface in such a way that it permits a lowering of the total Fermi energy. Presumably, as may be inferred from Fig. 13–8, the local curvature of the Fermi surface and the increased contact with the Brillouin zone, enhanced by the splitting of the zone, contribute to an increase in the density of states. Further refinements of the above model necessitate also taking into consideration the effect of antiphase domain energy, which becomes particularly important as the period M decreases, or when electron concentration does not change on alloying.

Electron concentration as a general parameter in alloy structures The term "electron phase" or "electron compound" may be applied with some justification to phases other than those based on the noble metals. It is increasingly evident, for example, that the σ phase (see Chap. 10, page 266) possesses features of an electron phase, and so do also other variable-composition phases (μ, χ, P, R, etc.) based on the transition elements. Since the d electrons unquestionably contribute to e/a in these phases, and since the d bands are incompletely filled, the details of possible electronic interactions are bound to be complex and not necessarily restricted only to Brillouin-zone–Fermi-surface effects; one may expect some of the bonding forces to be highly directional, or the "d-band vacancies," rather than electrons, to play a role. The general importance of e/a has been clearly established by careful studies. Interesting correlations, which emphasize the influence of e/a, are particularly evident, for example, from studies of constitution of ternary systems, where many such phases frequently form narrow, elongated phase fields that appear to follow approximately constant contours of e/a rather than atomic composition, and which show characteristic shifts from system to system.[1]

Typical displacements of σ-phase fields in a number of binary systems are illustrated in Fig. 10–27. An example of the alignment along certain contours influenced by e/a is shown in Fig. 13–9. In four different ternary systems, Fe-Mo-Co, Fe-Mo-Ni, Ni-Mo-Co, and Mn-Mo-Co, the elongated μ, σ, P, δ, and R fields follow approximate contours of *electron vacancy concentration* calculated on the assumption of simple additivity of electron vacancy numbers in the d bands of the pure elements. An alternative consideration of such elongated phase fields in terms of packing arrangements of different-sized atoms is not possible.[2]

The limits of primary solid solutions of some transition metals, such as

[1] S. P. Rideout et al., *Trans. AIME*, vol. 191, p. 872, 1951. P. Greenfield and P. A. Beck, *Trans. AIME*, vol. 200, p. 252, 1954; vol. 206, p. 265, 1956.

[2] See M. V. Nevitt, in P. A. Beck (ed.), "Electronic Structure and Alloy Chemistry of the Transition Elements," John Wiley, New York, 1963.

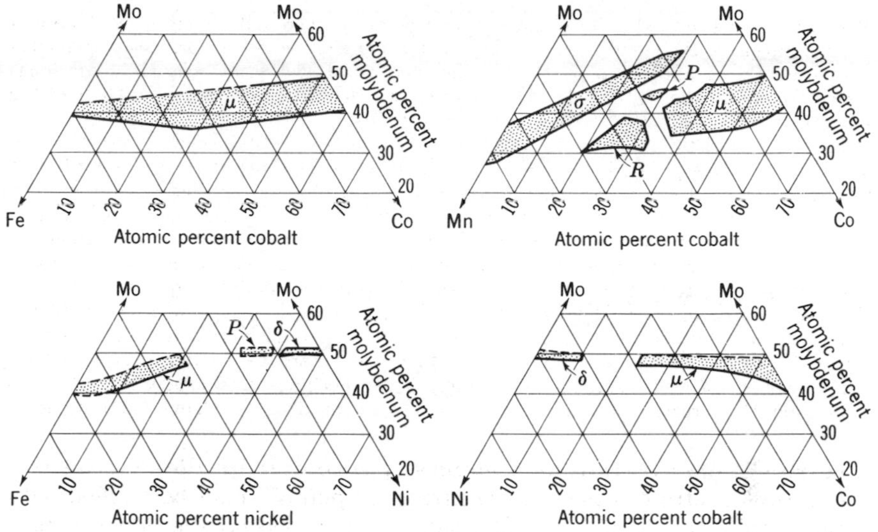

Fig. 13-9 Relationship between phase fields of the μ, σ, P, R, and δ phases and electron vacancy concentration. (Courtesy P. A. Beck.)

Fe, Co, and Ni, when plotted in terms of average group numbers of the alloys, occur at roughly the same average group number of about ≈ 7.7. This behavior resembles the trends in the alloys of the noble metals.[1] Manganese, as is often the case, does not seem to fit the general scheme.

Examples of dependence upon electronic factors are also provided by the behavior of essentially stoichiometric phases, such as the family of Ti_2Ni-type phases (A_2B "η carbides") which contain atomic sites of high coordination similar to the σ, μ, δ, P, and R phases mentioned above. The relative abundance of titanium-group elements as A partners and the frequent occurrence of cobalt-group elements as B partners in these phases appear to be electronic in nature.[2]

Additions of nontransition elements such as silicon and aluminum influence the stability of σ phases, apparently increasing the e/a ranges. The choices of particular partners in the transition-element phases that possess the cesium-chloride structures AB, the close-packed ordered structure AB_3, and the AB_5 structures indicate trends that are clearly dependent upon the position of the component elements in the Periodic Table.[3]

The influence of electronic factors is of particular significance in Laves phases in determining both the type of structure and the ranges of stability. This fact has been recognized for nearly thirty years. Dependence upon e/a is particularly striking in the magnesium alloys studied by Laves and Witte.[4] The three modifications, $MgCu_2$, $MgZn_2$, and $MgNi_2$, occur in several ternary systems, and their ranges of stability are clearly dependent on e/a. While it is undoubtedly true that the ratio of the radii constitutes an important parameter in stabilizing the Laves phases (see pages 258 and 377), which particular structure is chosen appears to be related to the density of states and

[1] W. Hume-Rothery, *Phil. Mag.*, vol. 6, p. 769, 1960.

[2] M. V. Nevitt and J. W. Downey, *Trans. AIME*, vol. 221, p. 1014, 1961.

[3] H. J. Wallbaum, *Naturwissenschaften*, vol. 31, p. 91, 1943. A. E. Dwight and P. A. Beck, *Trans. AIME*, vol. 215, p. 976, 1959. A. E. Dwight, *Trans. AIME*, vol. 215, p. 283, 1959; *Trans. ASM*, vol. 53, p. 477, 1961.

[4] F. Laves and H. Witte, *Metallwirtschaft*, vol. 15, p. 840, 1936.

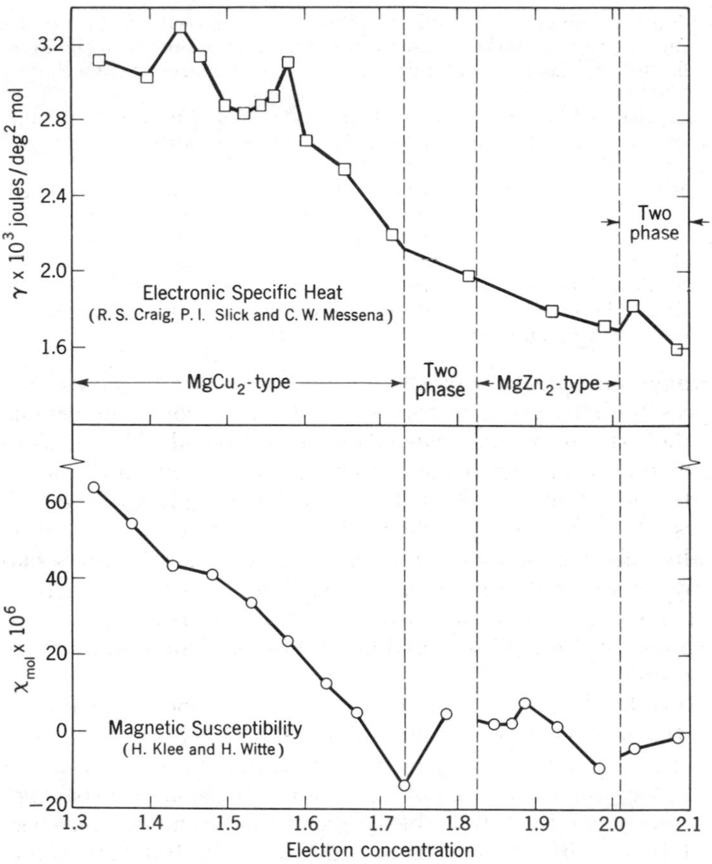

Fig. 13-10 Measurements of (*a*) electronic specific heat and (*b*) magnetic susceptibility in the pseudobinary system $MgCu_2$-$MgSi_2$ plotted as a function of *e/a*. (Courtesy R. S. Craig.)

the interactions between the Fermi surface and the Brillouin zone. Measurements of the changes in magnetic susceptibility,[1] hydrogen solubility,[2] and electronic specific heat[3] of several alloys within the pseudobinary sections, such as $MgCu_2$–$MgZn_2$, $MgNi_2$–$MgZn_2$, $MgCu_2$–$MgAl_2$, $MgZn_2$–MgAl, and $MgCu_2$–$MgSi_2$, support the hypothesis that phase boundaries on the electron-rich side of typical Laves phases are sometimes restricted following contact by the Fermi surface with an appropriate Brillouin zone, resulting in a decrease of the density of states. The trends in the electronic specific heat and in the magnetic susceptibility of the free electron gas, both of which are proportional to the density of states, are shown in Fig. 13–10 for the pseudobinary system, $MgCu_2$–$MgSi_2$. They clearly indicate that a low density of states is associated with

[1] H. Klee and H. Witte, *Z. Phys. Chem.*, vol. 202, p. 352, 1954.

[2] K. H. Liester and H. Witte, *Z. Phys. Chem.*, vol. 202, p. 321, 1954.

[3] R. S. Craig, P. I. Slick, and C. W. Massena, private communication, 1965, to be published. R. S. Craig and E. Klar, private communication, to be published. See also W. E. Wallace in P. S. Rudman and J. Stringer (eds.), "Phase Stability in Metals and Alloys," McGraw-Hill, New York, 1966.

phase boundaries and termination of solid solubility. The observed maxima in the density of states can be shown to be the consequence of interactions between the Fermi surface and specific Brillouin-zone discontinuities in a manner analogous to the electron phases.

The range of stability of Laves phases formed by 3*d*, 4*d*, and 5*d* transition elements with one another has an upper electron concentration limit. Even with favorable r_A/r_B ratio, Laves phases for which $e/a \geqq 8$ do not form.

Additions of silicon to binary Laves phases based on transition elements also suggest electronic effects. Silicon occupies the *B* positions in such structures, and an increase in its content makes the R_A/R_B ratio further removed from the value 1.225, which is characteristic of ideal packing. Yet, Laves phases are found in ternary systems such as V–Co–Si and V–Ni–Si and not in the binary systems V–Co and V–Ni for which the radius ratios are already borderline, 1.10 and 1.08, respectively. Clearly, the appearance of the ternary phases must be an electron concentration effect.[1]

Lattice spacings Introduction of solute atoms into the structure of a solvent produces both localized distortions and macroscopic distortions, involving the whole structure. As discussed on pages 295 and 370, a measure of the local distortions may be provided by changes that occur in the actual diameters of the participating atoms. However, such studies reveal only averaged effects; they are unable to show whether or not atoms possess or retain essentially spherical shape on alloying and what local effects may be produced by departures from randomness. Since a certain degree of deviation from randomness appears to be almost a universal rule in solid solutions, much remains to be discovered about atomic shapes and sizes on the microscopic scale.

Measurement of lattice spacings provides information about the average dimensions of the unit cell. Such measurements in solid solutions have contributed to the understanding of a number of factors that are of importance in the theory of alloy phases.[2] The change in the lattice spacing *a* of the cubic primary solid solutions based on the noble metals has been considered in terms of percentage distortions, $\Delta a/a \times 100$, per atomic percent of solute expressed as a function of valence and atomic number.[3] For a given common solvent, the importance of valence difference $(V_A - V_B)$ is demonstrated if the distortions are plotted as in Fig. 13–11. For a given row of elements in the Periodic Table, the percentage distortion is frequently of the form

$$\frac{\Delta a}{a} = k(V_A - V_B) + C \qquad (13\text{-}3)$$

where *k* and *C* are constants. However, the percentage distortions cannot be attributed solely to the net change in the number of conduction electrons,

[1] D. I. Bardos, K. P. Gupta, and P. A. Beck.. *Trans. AIME*, vol. 221, p. 1088, 1951.

[2] See, for example, the following reviews: T. B. Massalski, *Met. Rev.*, vol. 3, p. 45, 1958. W. B. Pearson, "A Handbook of Lattice Spacings and Structures of Metals and Alloys," Pergamon, New York, 1958. G. V. Raynor, *Trans. Faraday Soc.*, vol. 45, p. 698, 1949. W. Hume-Rothery and G. V. Raynor, "The Structure of Metals and Alloys," Institute of Metals, London, 1962.

[3] W. Hume-Rothery, "Atomic Theory for the Students of Metallurgy," Institute of Metals, London, 1960. E. A. Owen, *J. Inst. Met.*, vol. 73, p. 471, 1947.

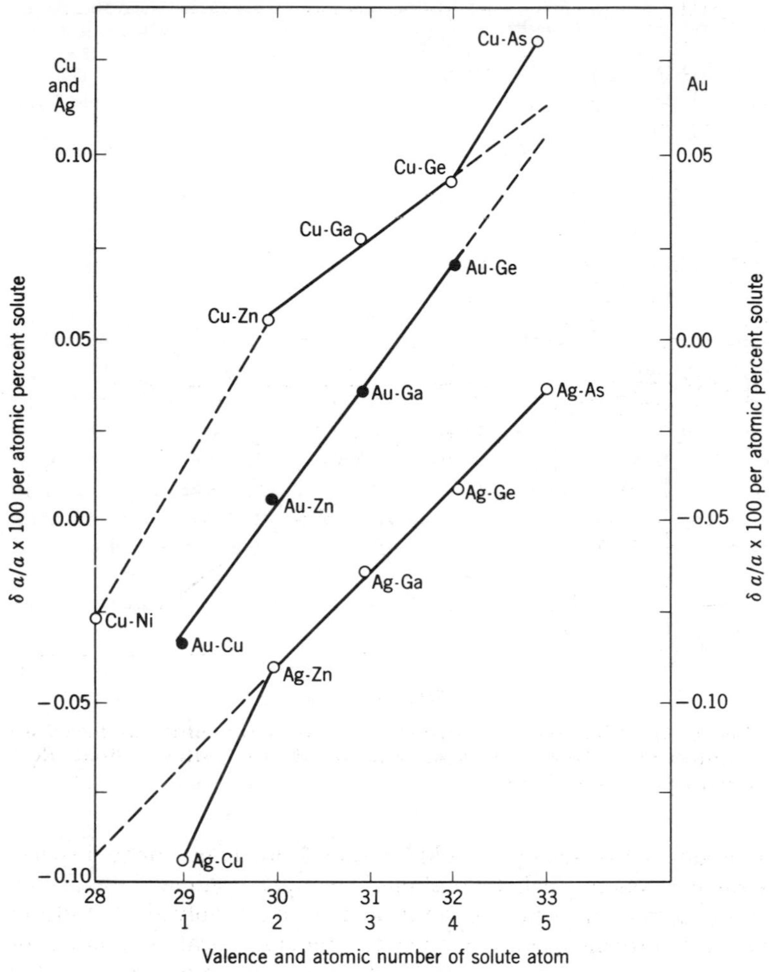

Fig. 13-11 Percentage distortions of lattice spacings in solid solutions based on copper, silver, and gold. (After E. A. Owen.)

for when plotted in terms of e/a the spacing changes do not superimpose. Nor is the above linear relationship valid in all cases. Other factors must also contribute, such as size differences.[1]

In addition to the noble metals, detailed trends of lattice spacings have been established for a number of other solvents such as Al, Mg, Sn, Pb, Ru, Pu, Ce, etc. The data for the aluminum-based systems are plotted in Fig. 13–12.

As mentioned above (page 333), the difference in valence between solvent

[1] Raynor, loc. cit.

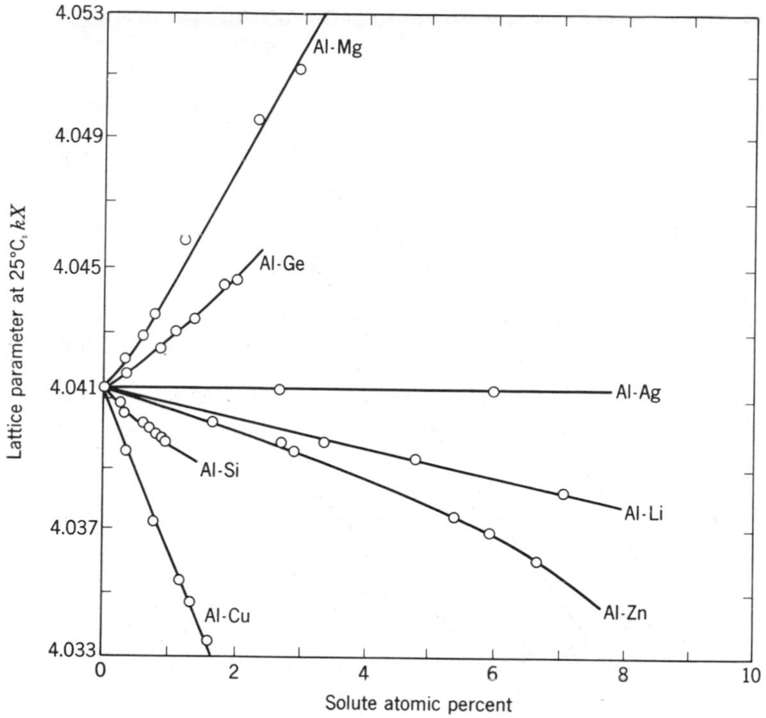

Fig. 13-12 Changes of lattice spacings in solid solutions based on aluminum. (After H. J. Axon and W. Hume-Rothery, *Proc. Roy. Soc.*, vol. A193, p. 1, 1948.)

and solute produces a tendency for solute atoms to aggregate or to disperse within the solvent matrix. Such fluctuations in composition may be expected to occur also in systems where several solutes are involved that differ sufficiently in electrochemical behavior to undergo local attraction or repulsion. For example, in the system Al-Mg-Si,[1] a minimum occurs in the lattice-spacing trend with composition corresponding to an electrochemical interaction between magnesium and silicon atoms. However, in other ternary systems, lattice spacings of ternary alloys are frequently a linear function of composition and may be calculated from binary data using empirical additive relationships. This has been demonstrated in the systems Cu-Al-In,[2] Cu-Zn-Ge, Cu-Ga-Zn,[3] and Cu-Zn-Si.[4] Such additive linear behavior suggests that in simple ternary solid solutions based on copper there is no appreciable solute-solute interaction, since copper atoms are

[1] R. B. Hill and H. J. Axon, *J. Inst. Met.*, vol. 83, p. 354, 1955.

[2] P. H. Sterling and G. V. Raynor, *J. Inst. Met.*, vol. 84, p. 57, 1955.

[3] B. B. Argent and D. W. Wakeman, *J. Inst. Met.*, vol. 85, p. 413, 1956.

[4] H. Pops, *Trans. AIME*, vol. 230, p. 267, 1964.

presumably able to screen the solute atoms. Similar behavior is also observed in silver-based systems such as Ag-Mg-Sb,[1] despite the fact that a strong electrochemical interaction might be expected between magnesium and antimony, similar to the interaction between magnesium and silicon observed in the system Al-Mg-Si. The absence of departure from linearity seems to suggest that the noble metals such as copper and silver are more effective in screening the solute elements from one another than are the "more open" metals such as Al.

Lattice spacings and zone overlaps Lattice-spacing trends with e/a in electron phases with c.p.h. structure provide interesting information about the electronic structure of such alloys. The changes of lattice spacings of the ϵ and η phases in the system Cu-Zn, and the curious fact that these two phases with the c.p.h. structure are separated by a two-phase field, were considered some thirty years ago by Jones[2] and interpreted in terms of electronic interactions related to Brillouin zone overlaps. Axial ratio changes in magnesium alloys were subsequently interpreted analogously,[3] and the general theory of the influence of electronic effects has more recently been extended to the whole group of c.p.h. electron phases ζ, ϵ, and η found in systems based on the noble metals.[4]

In all, some twenty-five binary systems possess at least one, but sometimes more, intermediate phases that are c.p.h. A hypothetical schematic phase diagram based on any of the three noble metals in which the primary solid solution α is followed by the typical phases with the c.p.h. structure, ζ, ϵ, and η, is shown in Fig. 13–13. Within the hexagonal phases the overall trends of observed decreases and increases of axial ratio with e/a are shown in Fig. 13–14.

The whole group of such phases, including the primary solid solutions, has close-packed structures, and the c.p.h. structures have c/a ratios approximately those for close-packed spheres. Thus, the $\{111\}_\alpha$ and $\{00\cdot2\}_\zeta$ planes shown diagrammatically within the unit cells in Fig. 13–13 are essentially equivalent, and the spacings between atoms within these planes can be compared across a given phase diagram.

In Fig. 13–15 the lattice spacing trends in the α, ζ, ϵ, and η phases of the system Ag-Cd are shown. One notices that between the α and ζ phases of the Ag-Cd system the a spacings begin an accelerated increase Δa above the trend established at lower values of e/a. Since to a first approximation the packing of atoms in the close-packed planes of these phases should be affected in about the same way by such factors as the size changes and the

[1] R. B. Hill and H. J. Axon, *J. Inst. Met.*, vol. 85, p. 109, 1956.

[2] H. Jones, *Proc. Roy. Soc.* (*London*), vol. A147, p. 396, 1934.

[3] W. Hume-Rothery and G. V. Raynor, *Proc. Roy. Soc.* (*London*), vol. A174, p. 471, 1940. G. V. Raynor, *Proc. Roy. Soc.* (*London*), vol. A174, p. 457, 1940; *Proc. Roy. Soc.* (*London*), vol. A180, p. 107, 1942.

[4] T. B. Massalski and H. W. King, *Progr. Mater. Sci.*, vol. 10, p. 1, 1961. T. B. Massalski, *J. Phys. Radium*, vol. 23, p. 647, 1962.

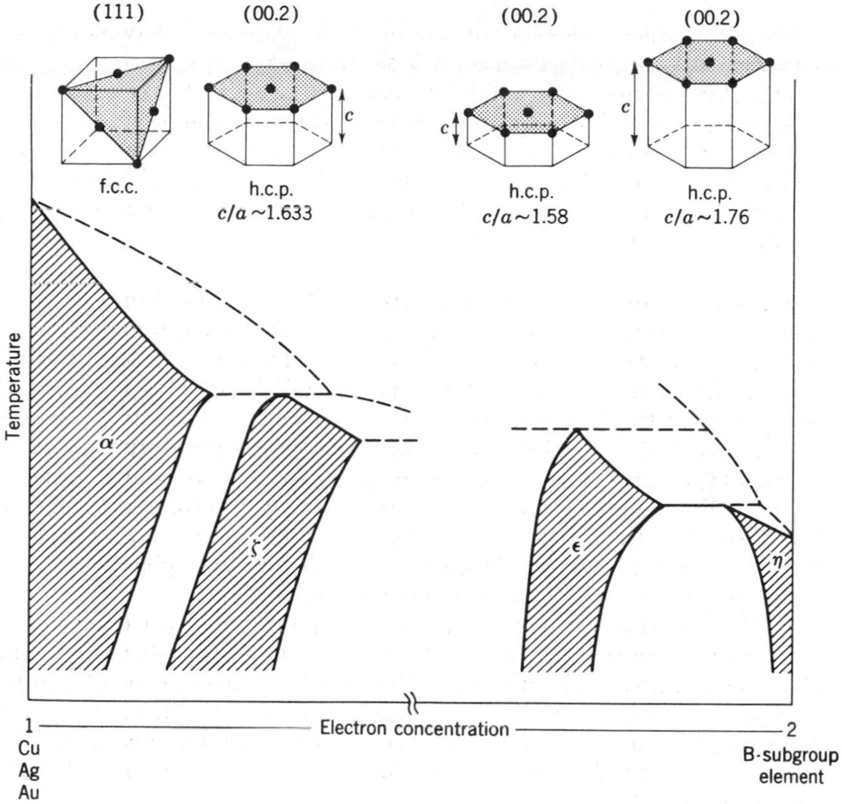

Fig. 13-13 A hypothetical phase diagram showing phase fields of close-packed α, ζ, ϵ, and η phases.

changes in the electrochemical affinity, the reason for the accelerated increase appears to be associated with the change of the crystal structure from cubic to hexagonal. Analogous increases Δa have been observed in all binary systems based on copper, silver, and gold. The accelerated expansion in the closest-packed planes occurs in the range of e/a values between 1.36 and 1.42, but seems to begin at a slightly different value in each system. When many systems are compared together, it is found that the Δa expansion begins at an almost constant value of e/a for any given solvent.

Lattice-spacing trends of a related kind are observed at higher values of e/a in the range of ϵ and η phases (see Fig. 13–15). The accelerated trend Δa established in the a spacings of the ζ phases, as mentioned previously, is simply continued throughout most of the homogeneity range of ζ phases. Thus, the same effect that is responsible for the Δa increase in the ζ phases is present and further continued in the ϵ phases, but in addition an accelerated increase in the c spacings Δc occurs at higher values of e/a. This is emphasized by the reversal of the axial-ratio trends, from a decrease to

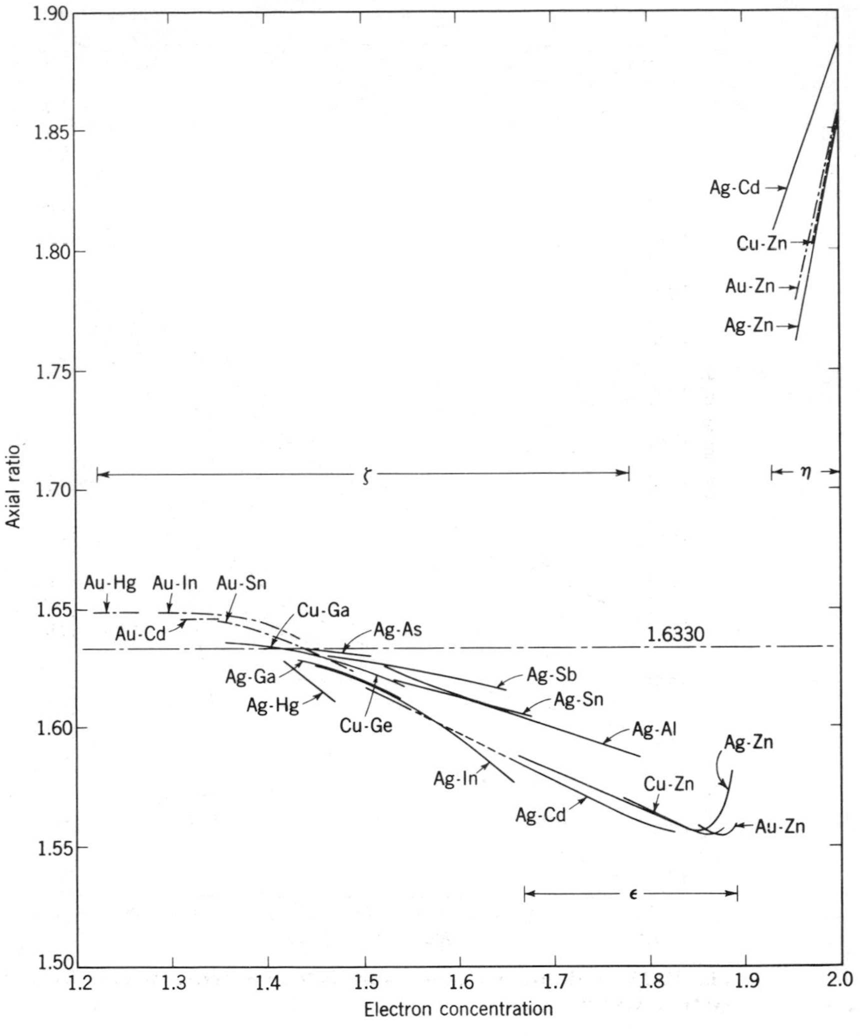

Fig. 13-14 Changes of axial ratio with electron concentration in c.p.h. electron phases.

an increase as shown in Fig. 13–14. The electron concentration value for the onset of Δc effects changes slightly from system to system, but it is in the vicinity of about 1.85 electrons per atom.

An interpretation of the above effects has been provided by consideration of the interactions between the Fermi surface and the Brillouin zone. Since the e/a values associated with the ranges of stability of the ϵ and η phases are larger than those calculated for the electron content of the inner Brillouin zone using Eq. (13–2), overlaps of electrons must exist across certain Brillouin-zone faces. Actually, interpretation of the lattice-spacing

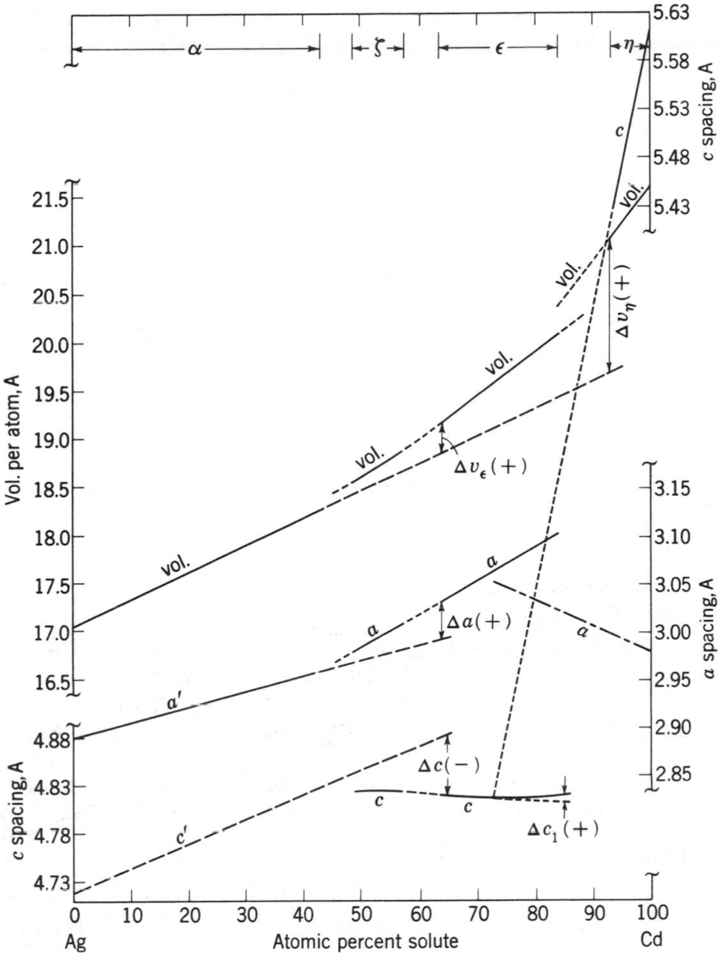

Fig. 13-15 Volume-per-atom and lattice-spacing trends in the system Ag-Cd.

trends suggests that overlaps exist also in the ζ phases and that the differences between various overlaps are responsible for the observed trends in the axial ratio and individual lattice spacings *c* and *a*.

According to the original theory[1] overlaps tend to contract or expand the zone in directions normal to the overlapped planes and thus tend to produce changes in the corresponding lattice parameters in the opposite direction in the real space. The above mechanism appears to be responsible for the observed accelerated expansion in the *a* and *c* spacings. The Δa effect occurs when the Fermi surface overlaps the {10·0} or *A* faces shown in the vertical section of the Brillouin zone in Fig. 13–5, and the Δc expansion occurs when the Fermi surface overlaps the horizontal {00·2} or *B* faces. Since

[1] H. Jones, *Proc. Roy. Soc.* (*London*), vol. A147, p. 396, 1934; *Phil. Mag.*, vol. 41, p. 663, 1950.

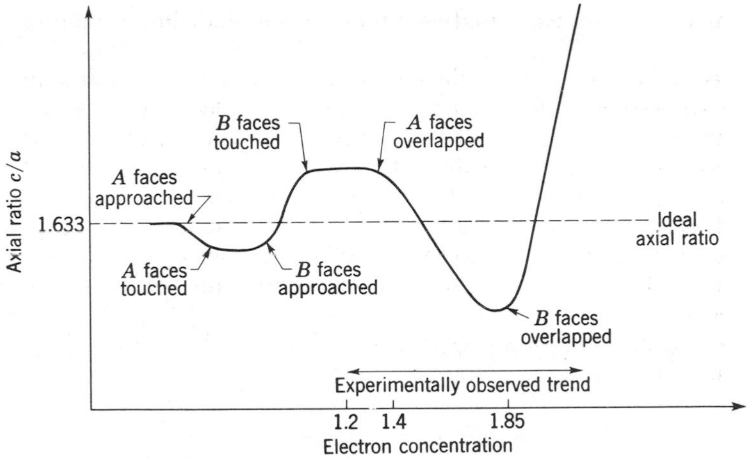

Fig. 13-16 A schematic dependence of c/a on e/a expected from theoretical considerations.

the {10·0} faces are nearer to the origin of the zone than are the {00·2} faces, it is natural that the Δa effect should occur at a lower value of electron concentration than the Δc effect. According to Goodenough[1] interaction between an approaching Fermi surface and a given set of Brillouin-zone faces should likewise tend to move the zone faces inwards in order to lower the energy. Hence, qualitatively the combined influence of approach and overlap effects will result in an overall trend in the axial ratio as that shown diagrammatically in Fig. 13–16. Comparison of this figure with Fig. 13–14 suggests that the initial higher-than-ideal axial ratio in the ζ phases is probably a result of an interaction between the Fermi surface and the B faces of the Brillouin zone. These faces correspond to the $\{111\}_\alpha$ faces in the Brillouin zone of the f.c.c. structure, where contact between the Fermi surface and the Brillouin zone already exists in the pure noble metal. A simple calculation shows that the band gaps are of the order of 1 ev or less. The analysis also indicates that the band gaps for c.p.h. phases decrease with the increase of solute content, emphasizing the possibility of a variable-band model.

Defect structures A phase in which some atom sites are vacant is a *defect structure*. It is believed that at least some phases of this kind can be understood in terms of electron concentration effects: these phases appear to tolerate vacant lattice sites rather than excess numbers of electrons in the unit cell, and when the concentration is changed, vacancies are produced in numbers that maintain an optimum electron concentration.[2]

Evidence for this has been found in a number of alloys with defect structures: CoAl, NiAl, CoGa, NiGa, etc. Evidence of omission of atoms from sites in a unit cell has also been observed in the study of some gamma-brasses,[3] and Al alloys,[4] in which no transition elements are involved,

[1] J. B. Goodenough, *Phys. Rev.*, vol. 89, p. 282, 1953.

[2] A. J. Bradley and A. Taylor, *Proc. Roy. Soc.* (*London*), vol. A159, p. 56, 1937. H. Lipson and A. Taylor, *Proc. Roy. Soc.* (*London*), vol. A173, p. 232, 1939.

[3] W. Hume-Rothery, J. O. Betterton, and J. Reynolds, *J. Inst. Metals*, vol. 80, p. 609, 1952.

[4] E. C. Ellwood, *Nature*, vol. 163, p. 772, 1948; *J. Inst. Met.*, vol. 80, p. 217, 1952.

so that the valence of the participating atoms is more definite than in others.

As mentioned in Chap. 10, tetrahedral structures in which normally every atom in the structure has four nearest neighbors located on a surrounding tetrahedron can form defect structures with vacant sites. It is possible to describe defect tetrahedral structures by assuming that all atoms still have tetrahedral orbitals, but some atoms have fewer than four neighbors. In this scheme every tetrahedral orbital that has not been used for bonding requires an additional electron with antiparallel spin to become a nonbonding orbital. Hence, e/a must be greater than 4 and related to the number of vacancies.

For a compound of composition $Q_nR_m \cdots O_p$ this may be expressed by the general equations[1]

$$\frac{e}{a} = \frac{4}{y} = 4$$

$$y = \frac{n + m + \cdots}{p} \tag{13-4}$$

where Q and R are elements forming the defect tetrahedral structure, O denotes the unoccupied structure sites, and n, m, and p are their atomic fractions.

Band structures of group V semimetals and the IV-VI semiconductors Recent theories and band calculations at the University of Chicago[2] have provided a better understanding than has hitherto been possible of the structures and conductivity properties of the group V semimetals As, Sb, and Bi and the IV-VI semiconducting compounds such as PbS. There is a taste of the future in these theories. Although the details are complex and not as yet fully published, and some of the principles used are rather sophisticated, the conclusions reached can be summarized quite simply.

Two structure types are involved: (1) The rhombohedral elements As, Sb, Bi, of *A*7 structure are semimetallic under normal conditions, i.e., they are electrical conductors at 0°K with equal small numbers of electrons and holes as carriers, due to a small overlap of valence and conduction bands. The compound GeTe also has the arsenic structure and is semimetallic. (2) The IV-VI compounds PbS, PbSe, PbTe, and SnTe have the cubic NaCl-type structure, which consists of two interpenetrating f.c.c. lattices (page 240); they are semiconductors, having a small band gap. If the difference between atoms is neglected, these IV-VI compounds can be visualized as simple cubic with one atom per cell. When viewed in this

[1] E. Parthé, *Z. Krist.*, vol. 119, p. 204, 1963.

[2] Morrel H. Cohen, L. M. Falicov, and S. Golin, *IBM J. Res. Develop.*, vol. 8, no. 3, p. 215, 1964. Morrel H. Cohen, *J. Phys. Radium*, vol. 23, p. 643, 1962; *Phys. Rev.*, vol. 121, p. 387, 1961; in T. B. Massalski (ed.), "Alloying Behavior and Effects in Concentrated Solid Solutions," Gordon and Breach, New York, 1965. L. M. Falicov and S. Golin, *Phys. Rev.*, vol. 137, p. A871 1965. S. Golin, to be published.

way, the difference between the kinds of structures 1 and 2 above can be stated very simply. The rhombohedral structures have interaxial angles α near 57°; the corresponding angles for the type-2 structures are 60° (the angles between face diagonals $\langle 110 \rangle$ of a cube). Both structures have atoms on two interpenetrating face-centered lattices, but while the NaCl structure has points of one lattice centered midway between the points of the other, the rhombohedral $A7$ structure has one lattice displaced from this position by an amount expressed by the parameter u (which is near 0.23 in each case); u is defined so that $u = 0.250$ corresponds to the undisplaced NaCl structure, and $u \neq 0.250$ signifies a displacement along the f.c.c. body diagonal, [111]; $2u$ is the distance between nearest neighbors in the [111] direction in terms of the body diagonal of the unit cell.

An average group number of 5 is favorable to a simple cubic structure, just as an average group number of 4 is favorable to a diamond cubic structure. There is ample reason to treat the band energy calculations of both the substances listed above in a unified way, and by so doing it becomes understandable that the cubic structures are stabilized by the chemical difference between group IV and VI elements. The simple cubic structure tends to be *unstable* against a shear which makes $\alpha \neq 60°$ and an internal displacement which makes $u \neq 0.250$.

The conventional chemical interpretation of group IV elements with the diamond structure and of III-V and II-VI compounds with the zinc blende and wurtzite structures is that the atoms of these crystals have an sp^3 electron configuration, with tetrahedral bonding resulting from the hybridized orbitals. Similarly, the group V and IV-VI structures under discussion are conventionally accounted for by a p^3 configuration. The s electrons play a negligible role in these, and bonding is due chiefly to three p bonds at right angles to each other; the coordination is octahedral.

Band-energy calculations have been carried out for these group V and IV-VI crystals with particular attention to degeneracies at symmetry points in the Brillouin zone, to the bonding vs. antibonding character of the wave functions, and to the effect of sublattice displacement on the Fourier coefficients of the crystal pseudopotential The conventional chemical interpretation is corroborated. The s bonding and s antibonding bands are both occupied; the three p bonding bands are also occupied, but the three p antibonding bands are unoccupied. Stabilization of the cubic structure of IV-VI compounds arises when the difference in potentials of the IV and VI elements is *large* enough; if the difference is *small* a compound has the rhombohedral structure and is semimetallic.

A displacement of the sublattices ($u \neq 0.250$) causes important changes in the Fourier coefficients: degeneracies are lifted, and configurations appropriate to semimetals or to insulators can arise. When spin-orbit coupling is taken into consideration, computations indicate that semimetallic properties and the rhombohedral $A7$ structure can be expected. Stabilization of the $A7$ structure is believed to arise chiefly from the internal displacement; the shear which reduces $\alpha = 60°$ to $\alpha \approx 57°$ plays only a secondary role.

The magnitude of the difference in the pseudopotential increases from GeTe to PbTe and is believed to be strong enough to stabilize the NaCl structure in PbTe but not in GeTe (which is rhombohedral below 700°K). The cubic-to-rhombohedral transformation characteristics in GeTe–SnTe alloys[1] are also in accord with this theory. The strength of the pseudopotential of Sn is greater than that of Ge, as judged by the atomic ionization potentials. Accordingly, increasing the Sn content of (SnGe)Te alloys increases the preference for the NaCl structure. (The transition temperature above which

[1] J. N. Bierly, L. Muldawer, and O. Beckman, *Acta Met.*, vol. 11, p. 447, 1963.

the NaCl structure is stable and below which the rhombohedral is stable is lowered from 700 to near 0°K.)

Pressure may also serve as a stabilizing factor for the NaCl structure. It was pointed out some years ago that this can be concluded theoretically from the magnitude of the two smallest interatomic distances, for the cubic structure is favored when these distances are within a critical magnitude. This has recently been corroborated by the fact that as pressure is increased, the ordinary form of Bi (BiI) has a steadily increasing rhombohedral angle, and at critical values of the pressure changes to the phase BiII, which is simple cubic ($\alpha = 60°$).[1] A similar change in α occurs for SbI under pressure, with α finally reaching 60° at 70 kb (simple cubic SbII).[2] In summary, with $u \neq 0.250$ and with α sufficiently different from 60°, the band structure is that of a semimetal (for example, As, Sb, Bi at atmospheric pressure); with α very close to 60° the situation is that of an insulator (as observed in BiI under pressure, and predicted for SbI under pressure). But with $u = 0.250$, the configuration of bands is that of a good metal, as is observed in the high-pressure phases BiII and SbII.

Atomic sizes in metals In order to consider the influence of atomic sizes on crystal structure, it is first necessary to define the "atomic size." A number of schemes exist for describing the magnitude of atomic sizes in pure elements. The concept of atomic radius, defined as half the mean bond length, was introduced originally for ionic salts by W. L. Bragg[3] and was extended to metals by Goldschmidt.[4] Since then, a number of excellent treatments and reviews have been published.[5]

A number of elements crystallize in different allotropic structures, and a study of atomic distances in such structures shows that the atomic radius diminishes with decreasing coordination number.[6] The empirical correction for relating atomic radii to a standard close-packed structure of coordination number 12 is as follows:

Coordination number (*CN*)	4	6	8	10	12	14	16
Required correction in percent	12	4	3	1.4	0	−1.2	−2.2

[1] R. Jaggi, Solid State Physics Conference, Bristol, England, 1965; Colloquium of Asso. Française de Cryst., Rennes, France, 1965.

[2] L. F. Vereshagin and S. S. Kabalkina, *Soviet Phys. JETP*, vol. 20, p. 274, 1965.

[3] W. L. Bragg, *Phil. Mag.*, vol. 40, p. 169, 1920.

[4] V. M. Goldschmidt, *Z. Phys. Chem.*, vol. 133, p. 397, 1928.

[5] The early history of the development of the concepts of metallic radii, ionic radii, covalent bond radii, and van der Waals' radii is given by L. Pauling, "The Nature of the Chemical Bond," 2d and subsequent eds., Cornell University, Ithaca, N.Y., 1946. A discussion of various atomic radii and the variability of atomic sizes is given by F. Laves in American Society for Metals Symposium, "Theory of Alloy Phases," Cleveland, Ohio, 1956. Difficulties in assessment of atomic sizes and their role in the structure of metals and alloys are considered by W. Hume-Rothery and G. V. Raynor in "The Structure of Metals and Alloys," 4th ed., Institute of Metals, London, 1962. Atomic volume and size correlations in alloys are discussed by H. W. King in T. B. Massalski (ed.), "Alloying Behavior and Effects in Concentrated Solid Solutions," Gordon and Breach, New York, 1965. Atomic volume correlations in allotropic changes and in some Laves phases are discussed by P. S. Rudman, *Trans. AIME*, vol. 233, pp. 864 and 872, 1965.

[6] Goldschmidt, loc. cit.

The above correlation indicates that the conservation of volume during structural changes is an important factor. Thus, for example, for the typical change f.c.c. ⇄ b.c.c., if a and r denote the lattice parameter and atomic radius respectively, then assuming that volume per atom, Ω, remains unchanged, one has $(a_{\text{f.c.c.}})^3/4 = (a_{\text{b.c.c.}})^3/2$, where $a_{\text{f.c.c.}} = \sqrt{2}(r_{\text{f.c.c.}})$ and $a_{\text{b.c.c.}} = 2/\sqrt{3}(r_{\text{b.c.c.}})$. Hence, $r_{\text{f.c.c.}} \approx 1.03\ r_{\text{b.c.c.}}$, which gives the 3 percent correction for these two structures. It is important to realize that it is the atomic radius of the *higher* coordination structure, $r_{\text{f.c.c.}}$, that becomes larger than the radius of the *lower* coordination structure $r_{\text{b.c.c.}}$ while the volume per atom remains constant, or nearly constant. This is evident, for example, from the plot for iron given in Fig. 10–5. Apart from special cases such as Sn, Ce, Pu, and perhaps a few others, the allotropic volume changes in the metallic elements are very small, usually less than 1 percent. Thus, the close-packed structures are on the whole no more dense than are the non-close-packed structures.[1]

In principle, all atomic radii can be recalculated to correspond with a particular coordination, and this constitutes the basis of many tabulations of atomic radii. Schemes that employ coordination corrections sometimes break down for complex or noncubic structures such as those of Ga, Sb, Se, Zn, Cd, Be, α-Mn, Pu, etc., where the basic coordination number is difficult to define since there may be several bonds of slightly differing lengths.

The *single-bond metallic radii* of Pauling are based upon the concept that a number of covalent bonds that represent the bonding valence, say e, resonate between nearest-neighbor atoms within a coordination shell of a given value, say CN. The bond number n is defined as e/CN and is related to the single-bond diameter d_1 by the semiempirical relation derived from studies of carbon bonds,

$$d_n = d_1 - 0.60 \log_{10} n \qquad (13\text{–}5)$$

where d_n is the closest distance of approach of atoms in the crystal. Half the value of d_1 gives the single-bond metallic radius. The above scheme has been used with success in many cases, but it also leads to many difficulties, where Eq. (13–5) clearly does not apply.[2]

The *closest distance of approach*, d, derived from lattice spacings, is traditionally used in connection with the 15 percent size-factor rule of Hume-Rothery. In many instances d is remarkably successful as the initial measure of atomic sizes if used in a qualitative manner. However, the r_{CN12} and $d/2$ values will naturally show a large difference in elements of low coordination number. Thus, for silicon $d/2 = 1.17$ and $r_{CN12} = 1.34$. Compared with copper, for which $d/2 = r_{CN12} = 1.28$, the radius of silicon may be taken as bigger or smaller, depending on which scheme is used.

Occasionally, a good empirical measure of an effective size of an element

[1] P. S. Rudman, *Trans. AIME*, vol. 233, p. 864, 1965.

[2] See, for example, W. Hume-Rothery and B. Coles, *Advan. Phys.*, vol. 3, p. 149, 1954. King, loc. cit.

in solid solution is obtained by extrapolation of a nearly linear trend in lattice spacings of a f.c.c. solvent in which this element, the solute, is dissolved, to the value corresponding to 100 percent solute. The obtained value of the atomic diameter is sometimes known as the *apparent atomic diameter* (AAD).[1] However, this approach is complex since, for example, silicon when dissolved in three-valent aluminum gives larger extrapolated AAD value than when dissolved in monovalent copper, showing that AAD values depend upon the nature of the solvent.

Recently the volume per atom, Ω, or the *radius derived from volume per atom*, $r_\Omega = (\frac{3}{4}\,\Omega/\pi)^{1/3}$, also known as the *Seitz radius*, has been used as a comparative measure of atomic sizes. The volume per atom is the volume of the unit cell divided by the number of atoms in the cell. It is one of the basic parameters in the equation of state, and hence it is related to the energy of the metal or the alloy; in metallic structures it is relatively independent of coordination and crystal structure, not only in pure elements but also in alloys.[2] Values are given in the last column of Table *A*-6 in the Appendix.

Atomic sizes in solid solutions As discussed in Chap. 11 (page 295), the static displacements of atoms from lattice sites in a solid solution, corresponding to changes in the individual atomic sizes, may be estimated from modulation in diffuse x-ray scattering, and from quasi temperature reduction in the Bragg reflections. In the former case the modulations of the diffuse x-rays diffracted by a solid solution are described by the coefficients α_i, related to the nature of local atomic order of atoms, and by the size-effect coefficients β_i, which are related to the differences in the sizes of the component atoms (see page 295). *A* and *B* atoms of dissimilar size can be packed together closer if they are displaced by small distances from the points of the regular crystal lattice. Hence, the atomic diameter *d*, averaged over many aggregates of atoms, may differ considerably from the atomic diameters d_{AA}, d_{BB}, and d_{AB}, which represent the average individual atomic sizes and interatomic bonds. These parameters, calculated from the size-effect coefficients, constitute the nearest approach to the assessment of actual atomic sizes in solid solutions.

Assessments based upon lattice spacings derived from the Debye-Scherrer method constitute a much further departure from reality. Thus, *d* corresponds to the average closest distance of approach of atoms in a solid solution ($d = 2r$), but it cannot be simply related to the atomic radii of either of the two metals. This is evident from a plot, as in Fig. 13–17, where the average parameters calculated from experimental data on diffuse scattering are shown for the complete range of solid solubility in the system Co-Pt[3]; a substantial relaxation of atomic sizes appears to occur on alloying.

[1] H. J. Axon and W. Hume-Rothery, *Proc. Roy. Soc.* (*London*), A193, p. 1, 1948.

[2] N. F. Mott, "Reports on Progress in Physics," vol. 25, p. 218, Institute of Physics, London, 1962. T. B. Massalski and H. W. King, *Progr. Mater. Sci.*, vol. 10, p. 1, 1961.

[3] P. S. Rudman and B. L. Averbach, *Acta Met.*, vol. 5, p. 65, 1957.

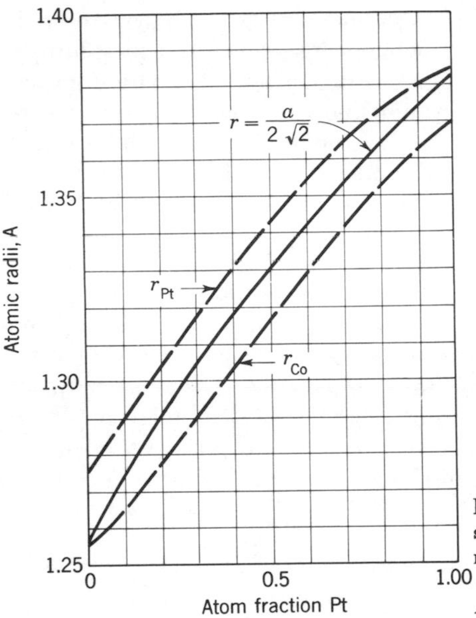

Fig. 13-17 Atomic radii in disordered solid solutions of Co-Pt, evaluated by means of diffuse x-ray scattering. (P. S. Rudman and B. L. Averbach, *Acta Met.*, vol. 5, p. 65, 1957.)

Size effects in solid solutions The influence of the ratio of the radii of the components upon the choice of crystal structure is discussed in Chap. 10. The effect of this ratio on solid solubility was recognized early by Hume-Rothery as a possible restrictive hindrance. The suggested maximum tolerable disparity between atomic diameters constitutes the 15 percent rule discussed on page 344. This size-factor rule[1] appears to be almost universally a necessary but not sufficient condition of substantial solid solubility. The basic assumption of Hume-Rothery's 15 percent rule is that the atomic diameter is given by the *closest distance of approach*, *d*, of atoms in each pure element. As mentioned before, both this criterion of atomic size and others provide a useful though limited guide to the possible prediction of restricted solid solubility. Support for the 15 percent rule has been obtained for a very large number of systems[2] where the rule is more than 90 percent effective in predicting limited solubility (less than 5 atomic percent). The estimates of the strain energy associated with substitutional alloying show a direct link between the limitation of primary solid solubility

[1] The size factor in a binary system *A-B* is defined as $SF_{AB} = (d_B - d_A)/100d_A$, and the "effective" size factor in a ternary system as

$$SF_{ABC} = \frac{(xSF_{AB} + ySF_{AC})}{(x + y)}$$

where d_A, d_B, and d_C are values of closest distance of approach and x and y are atomic percentages of components *B* and *C*. A possible role of the size factor based on *d* values is discussed by G. V. Raynor in American Society for Metals Symposium, "Theory of Alloy Phases," Cleveland, Ohio, 1956.

[2] J. T. Waber et al., *Trans. AIME*, vol. 227, p. 717, 1963.

and Hume-Rothery's 15 percent rule.[1] Nevertheless, there seems to be relatively little justification for using d as a quantitative measure of the atomic size for other comparative purposes, particularly when the two component metals have structures of different coordination.[2]

The volume size-factors Calculations based on simple elastic models make it possible to relate the strain energy in a solid solution to the atomic volume. The strain energy $E_s(c)$ may be expressed by the general equation of the form[3]

$$E_s(c) = A\mu \frac{1}{\Omega}\left(\frac{\partial\Omega}{\partial c}\right)^2 f(c) \tag{13-6}$$

where A is a numerical constant, μ is the shear modulus, Ω is the mean atomic volume, and c is the concentration. In many binary systems it is found that the mean volume per atom varies nearly linearly with composition over quite wide ranges of composition, and hence can be expressed by a general equation

$$\Omega = (1 - c)\Omega_0 + c\Omega_\alpha \tag{13-7}$$

where Ω_0 is the initial volume per atom and Ω_α is the extrapolated effective volume per atom of component B when in solid solution with component A within a phase denoted α. The fractional rate of change of volume per atom with composition is thus almost constant, and for dilute solutions

$$\frac{1}{\Omega}\frac{\partial\Omega}{\partial c} = \frac{\Omega_\alpha - \Omega_0}{\Omega_0} \tag{13-8}$$

The latter quantity is an expression of a *size factor in terms of volume*, Ω_{SF}. On substituting in Eq. (13-6), one finds that the strain energy of a solid solution is proportional to the square of the volume size-factor.

Volume size-factors for the primary solid solutions in binary alloys based on copper, silver, gold, aluminum, iron, and magnesium are given in Table 13-1. A comparison of the Ω_{SF} with known solubility limits in primary solid solutions based on the noble metals suggests an approximately 30 percent difference as a limiting tolerance.[4]

[1] L. S. Darken and R. W. Gurry, "Physical Chemistry of Metals," McGraw-Hill, New York, 1953. J. D. Eshelby, "Solid State Physics," vol. 3, p. 79, Academic, New York, 1956.

[2] W. Hume-Rothery and G. V. Raynor, "The Structure of Metals and Alloys," 3d ed., Institute of Metals, London, 1954. T. B. Massalski and H. W. King, *Progr. Mater. Sci.*, vol. 10, p. 1, Pergamon, London, 1961. H. W. King in T. B. Massalski (ed.), "Alloying Behavior and Effects in Concentrated Solid Solutions," Gordon and Breach, New York, 1965.

[3] Massalski and King, loc. cit.

[4] Note added in proof: Quantitative size-factors, defined in terms of the effective atomic volume of the solute, have been calculated for 469 substitutional solid solutions by H. W. King, *J. Mater. Sci.*, vol. 1, p. 79, 1966.

Table 13–1 Volume size-factors in binary alloys based on Cu, Ag, Au, Al, Fe, or Mg*

Solute	Solvent					
	Cu	Ag	Au	Al	Fe	Mg
Li				+3.2		−12.4
Cu	—	−26.9	−14.2	−18.9	+15.3	
Ag	+40.8		−0.5			−55.0
Au	+47.9	− 1.0	—			
Be	−27.2				−25.2	
Mg	+51.0	+ 7.7		+32.2		—
Zn	+19.9	−13.8	−14.5	− 5.9		−43.5
Cd	+59.3	+17.0	+13.6			−20.0
Hg		+21.7	+19.2			
Al	+20.3	− 9.1	−10.3	—	+11.7	−33.7
Ga	+24.6	− 5.0	− 4.4			
In	+76.3	+25.5	+26.2			− 0.7
Tl		+39.9				− 3.9
Si	+ 5.9			−12.5	− 6.6	
Ge	+25.0	+ 2.6	+ 2.7	+13.4		
Sn	+85.6	+33.1	+31.0			− 0.6
Pb		+56.6				+15.1
As	+39.8	+10.6				
Sb	+89.0	+45.4	+38.7			
Bi		+71.8				
Ti					+24.9	
Cr					+ 4.3	
Mo	+19.4				+20.2	
W					+27.3	
Mn	+31.0	− .35		−51.2	+ 4.3	
Fe	+ 6.2				—	
Co			−14.6		− 0.2	
Ni	− 8.0	−15.9			+ 4.5	
Pd	+27.9	−16.5	−14.1			
Pt	+31.3	−20.8	−13.1			

* H. W. King in T. B. Massalski (ed.), "Alloying Behavior and Effects in Concentrated Solid Solutions," Gordon and Breach, New York, 1965. See also H. W. King, *J. Mater. Sci.*, vol. 1, p. 79, 1966.

Deviations from Vegard's law The expected linear dependence of lattice spacings on composition, to follow a line joining the values for the pure elements, has come to be known as *Vegard's law*. In practice, approximately linear variation is often found within the limits of primary solid solubility based on one of the elements; but, when extrapolated to 100 percent of the solute, the obtained value rarely does coincide with the initial value assumed for the solute. Even when the two elements are mutually soluble, the lattice spacings as a rule follow a curve. One may thus speak of deviations from Vegard's law although, as postulated originally, the law was proposed as a linear relationship between molecular volumes of mutually soluble pairs of ionic salts (such as KCl-KBr, etc.),[1] and this law seems to be an exception rather than the rule in metallic solid solutions.

The trends in lattice spacings with composition in binary systems, and the deviations from an assumed Vegard's law, are in reasonable quantitative agreement with a number of calculations based on elastic models that employ experimentally established parameters for the pure metals, such as shear modulus, compressibility, Poisson's ratio, bulk modulus, pressure, the number of atoms per unit cell, and of course, the assumed initial values of the atomic sizes in the pure components. The various numerical expressions and their success for predicting deviations from Vegard's law have been recently reviewed.[2] Calculations of deviations from Vegard's law in terms of volume per atom for several alloys of Cu, Ag, Au, Al, Fe, and Mg[3] show that in all systems where smaller atoms are substituted for larger atoms, $\Omega_B < \Omega_A$, the difference between the observed and calculated deviations, $\Delta_{cal} - \Delta\Omega_{obs}$, is greater than 5 percent and that in general the agreement is poor. Perhaps this should not be surprising since the most important objection to the elastic analog of a solid solution is the fact that, on the atomic scale, the model is very unrealistic. Interactions between atoms involve electronic charges and potentials that are perhaps highly directional and nonuniform, and which are subject to considerable variation on alloying. Some progress has been made recently in the field of quantum-mechanical calculations of size effects that take into account various charge interactions and screening effects.[4]

Geometrical principles In addition to satisfying certain requirements related to electronic interactions and sizes, atoms must adjust to the properties of space when they arrange themselves into a regular crystalline structure.[5] It appears that certain conditions tend to be fulfilled when this occurs. A few general geometrical principles are given by F. Laves:

Space principle By far the largest number (58) of metallic elements tend to crystallize in close-packed structures. When atoms are indistinguishable from one another, the highest coordination number geometrically possible is 12 (f.c.c. or c.p.h. structure with $c/a = 1.633$), but higher co-

[1] L. Vegard, *Z. Physik*, vol. 5, p. 17, 1921; *Z. Krist.*, vol. 67, p. 239, 1928.

[2] K. A. Gschneidner and G. H. Vineyard, *J. Appl. Phys.*, vol. 33, p. 3444, 1962. Also see references of Table 13–1.

[3] King, loc. cit.

[4] A. Blandin in T. B. Massalski (ed.), "Alloying Behavior and Effects in Concentrated Solid Solutions," Gordon and Breach, New York, 1965.

[5] Several important contributions are of particular interest in this general area: F. Laves in American Society for Metals Symposium, "Theory of Alloy Phases," Cleveland, Ohio, 1956; in W. M. Mueller and M. Fay (eds.), "Advances in X-ray Analysis," vol. 6, p. 43, Plenum, New York, 1963. F. C. Frank and J. S. Kasper, *Acta Cryst.*, vol. 11, p. 184, 1958; vol. 11, p. 184, 1958; vol. 12, p. 483, 1959. G. E. R. Schultze, *Z. Krist.*, vol. 115, p. 261, 1961. E. Parthé, *Z. Krist.*, vol. 115, p. 52, 1961.

ordination values may be found in alloys between atoms of suitably different sizes, thus allowing for even tighter packing. This occurs perhaps even more frequently than has been realized thus far. The general tendency in such cases seems to be one of maximum filling of available space, i.e., *maximum density*.

Symmetry principle When a f.c.c. structure changes to a b.c.c. structure, the coordination, taken in the usual sense as the number of nearest neighbors, changes from 12 to 8. The next-largest number (33) of metallic elements have the b.c.c. structure despite the fact that sphere-packing arrangements are possible with coordinations equal to 11, 10, and 9. It can be shown that the resultant structures would, however, be less symmetrical than the b.c.c. structure. Hence, a *tendency towards a high symmetry* is another geometrical principle.

Connection principle Within a given structure, atoms may be imagined as an interconnecting array of spheres, or points, with links or bonds between them. There will always be some short links. If all the links except the shortest are dropped, those atoms that are still connected into units are said to form a *connection*. The connection is *homogeneous* if all atoms in it are of the same kind, and *heterogeneous* if they differ. The connections can be finite or infinite and they can extend in one, two, or three dimensions. One may thus speak of islands (I), chains (C), nets (N), or lattices (L). The geometrical connection principle expresses the general tendency of structures to form *connections of "high" dimensions*. The f.c.c. structure represents the case where all three geometrical principles are satisfied.

The geometrical principles may be expected to compete with other factors that determine crystal structures in alloys, such as e/a, the electrochemical difference, or the temperature; but in many cases they are found also to compete with one another. For example, the connection principle competes with the symmetry principle in AB-type phases. A choice of the NaCl-type structure offers a high symmetry but the NiAs type offers a better connection since it permits the metallic "ions" that fit into octahedral voids to approach each other more closely than in the cubic NaCl-type structure.[1]

The space-filling principle finds particular recognition in alloy phases with variable composition formed between transition elements, such as the σ phase and related phases mentioned on page 266 and in other phases that depend on size factors.[2] Taking only two types of atoms, A and B, and disregarding the chemical difference between them, one finds that groups of such atoms can be arranged into one of four types of polyhedra of the general formula AB_n in which nB atoms surround an A atom. The value of n can be 12, 14, 15, or 16, corresponding to a high coordination. The four "coordination polyhedra" or "Kasper polyhedra" are shown in Fig.

[1] Laves, loc. cit.

[2] J. S. Kasper in American Society for Metals Symposium, "Theory of Alloy Phases," Cleveland, Ohio, 1956.

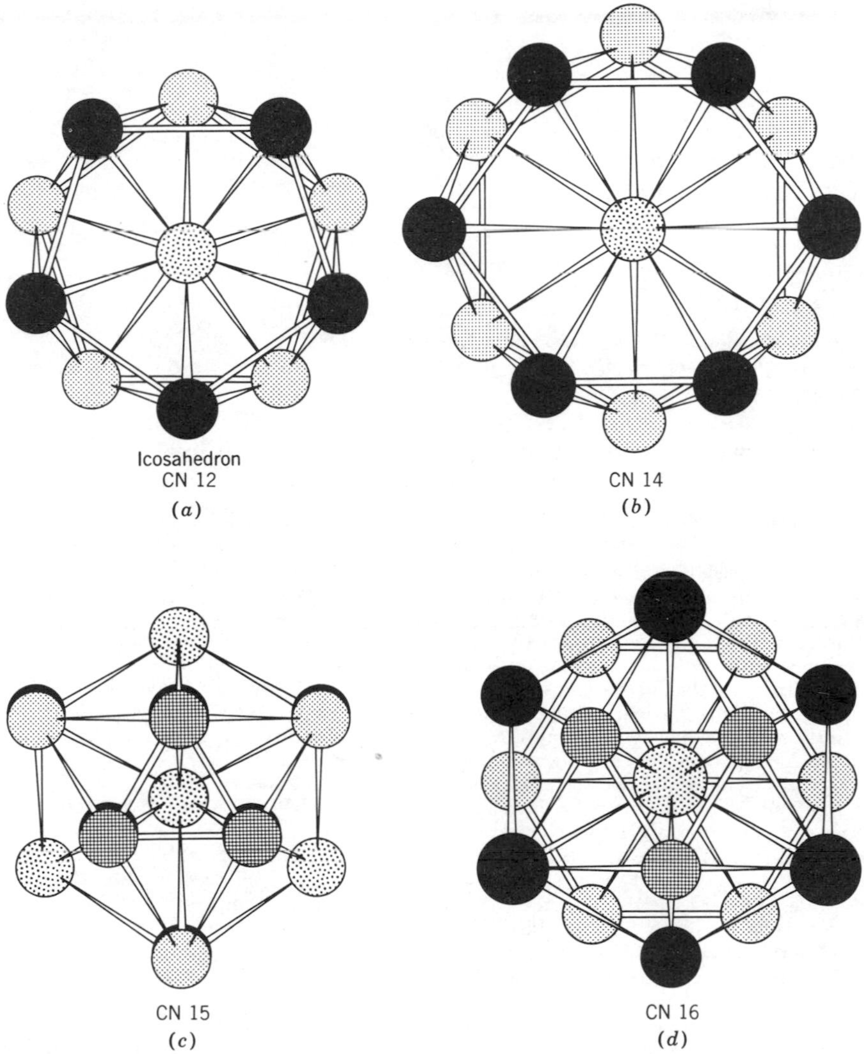

Fig. 13-18 Kasper coordination polyhedra. For *CN* 12, the two spheres above and below the central sphere, along the fivefold axis, are not shown. Similarly, two spheres above and below the central one of *CN* 14, along the sixfold axis, are not shown. For *CN* 16, one sphere below the central one is not shown. (After J. S. Kasper in "*Theory of Alloy Phases*," American Society for Metals, Cleveland, Ohio, 1956.)

13–18. They are formed by a network of *B* atoms consisting only of triangular faces with five or six triangles meeting at each corner. Such polyhedral structures can be constructed by placing an *A* or a *B* atom at the centers of the polyhedra. The four coordination structures shown in Fig. 13–18 are the only geometrically possible structures that are convex, are

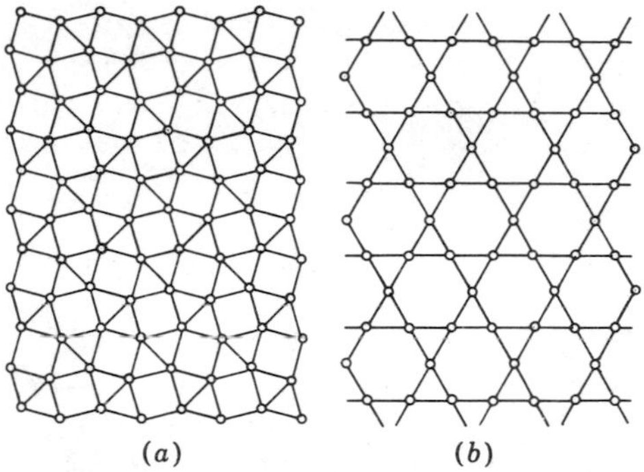

Fig. 13-19 Two-dimensional coordination (see text). (After F. C. Frank and J. S. Kasper.)

bounded only by triangles, and have corners at which at least five triangles meet.[1]

The geometry of the coordination polyhedra involves a modified concept of coordination. Consider spheres of equal size. The centers of *three* spheres in the closest approach will form a triangle, and the centers of *four* spheres will form a tetrahedron. If it is necessary to construct a coordination shell with triangular faces in which all links are equal, only tetrahedron, octahedron, and icosahedron (Fig. 13–18) are possible. However, if it is required further that all atoms in a coordination shell make equilateral triangles with the central atom, no solution is possible; there appears to be no space-filling structure that utilizes icosahedral coordination and provides twelvefold coordination of *all* atoms. Hence, in the coordination polyhedra observed in nature, a moderate variation from equality of interatomic distances is observed—a condition particularly suited for mixing atoms of different sizes or ones that modify their sizes and electronic structures by a localized transfer of electrons.

The *domain* of an atom is defined as the space in which all points are nearer to the center of that atom than to any other. The number of neighbors of the central atom is called the coordination number of that atom, and the set of neighbors is called the *coordination shell*. Together, they form a *coordination polyhedron*. In this description it is possible to regard the b.c.c. structure as having coordination number 14 and the c.p.h. structure as always having coordination number 12, no matter what is the value of the axial ratio. However, some caution is required in this use of coordination since the Goldschmidt coordination correction discussed on page 367 will, of course, no longer apply.

The concept of coordination polyhedra can be represented by two-dimensional basic networks, as shown in Fig. 13–19, which frequently occur in complex structures. The coplanar arrays can be derived from a triangle (Fig. 13–19*a*) or from a hexagon (Fig. 13–19*b*). However, since the coordination polyhedra are bounded by triangles, the whole crystal space is divided into irregular, i.e., somewhat deformed, polyhedra. As pointed out in Chap. 10, the ideal close-packed structures have octahedral or tetrahedral voids, or alternatively, they can be said to consist of octahedral and tetrahedral aggregates of spheres which touch one another at the middle of the edges. Hence, the density will fluctuate in the structure, being high in the tetrahedral regions and low in octahedral regions. By contrast, the density distribution in a space divided entirely into deformed tetrahedra will be more even and will permit greater density of packing. Various coordination polyhedra can be stacked together in space as shown in Fig. 13–20. Other examples of structures that can be represented as coordination structures are the cubic

[1] Frank and Kasper, loc. cit.

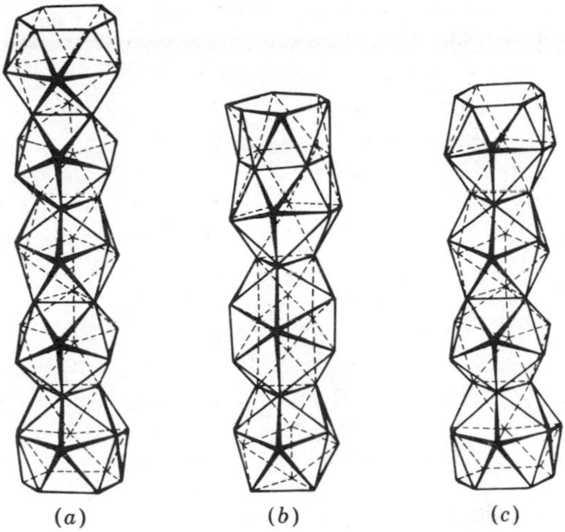

Fig. 13-20 Construction of complex structures from coordination polyhedra along the three-fold axes. (*a*) *R* phase, entire *c* repeat: *CN* 16, 12, 12, 12, 16. (*b*) μ phase, one-half *c* repeat: *CN* 12, 15, 16, 14, (*CN* 12 is shared with lower half, below.) (*c*) $MgNi_2$, entire repeat: *CN* 16, 12, 12, 16. (After Y. Komura, W. G. Sly, and D. P. Shoemaker, *Acta Cryst.*, vol. 13, p. 575, 1960.)

and hexagonal Laves phases, which have *CN*12 and *CN*16, and the Cr_3Si-type structures with *CN*12 and *CN*14.[1]

In addition to lattice spacings, a large number of independent interatomic distances can be measured in these complicated phases, ranging from 20 in the σ phase to 94 in the δ phase. Some attempts have been made to compute the lattice spacings and their variation with composition in σ phases from average atomic radii of the component atoms, assuming that the structure can be represented by a packing of spheres.[2] A reasonable agreement is obtained between calculated and observed values for the *a* spacings of the tetragonal unit cell, except in σ phases containing silicon. Silicon appears to exist in different electronic states in many structures, and hence it possesses a variable effective atomic radius.[3] The agreement for the *c* spacings is less good, suggesting that some atoms arranged in the *c* direction are probably considerably distorted from spherical

[1] Elucidation of space principles in complex structures is mostly due to Frank and Kasper, loc. cit., and to Shoemaker and his associates: G. Bergman and D. P. Shoemaker, *Acta Cryst.*, vol. 7, p. 857, 1954. D. P. Shoemaker, C. B. Shoemaker, and F. C. Wilson, *Acta Cryst.*, vol. 10, p. 1, 1957. Y. Komura, W. G. Sly, and D. P. Shoemaker, *Acta Cryst.*, vol. 13, p. 575, 1960. C. B. Shoemaker and D. P. Shoemaker, *Acta Cryst.*, vol. 16, p. 997, 1963; vol. 18, p. 37, 1965.

[2] H. P. Stüwe, *Trans. AIME*, vol. 215, p. 408, 1959.

[3] As pointed out earlier, p.368, the AAD values of Si depend on the nature of the solvent. For Si size-effects in complex structures, see, for example, Bertil Aronsson, *Arch. Kem.*, vol. 16, p. 379, 1960. M. V. Nevitt, *Trans. AIME*, vol. 212, p. 349, 1958. A. M. Bardos, D. I. Bardos, and P. A. Beck, to be published. The changes of atomic volume of Si, Ge, and Sn, when dissolved in different solvents, are discussed by T. Yoshioka and P. A. Beck, *Trans. AIME*, vol. 233, p. 1788, 1965.

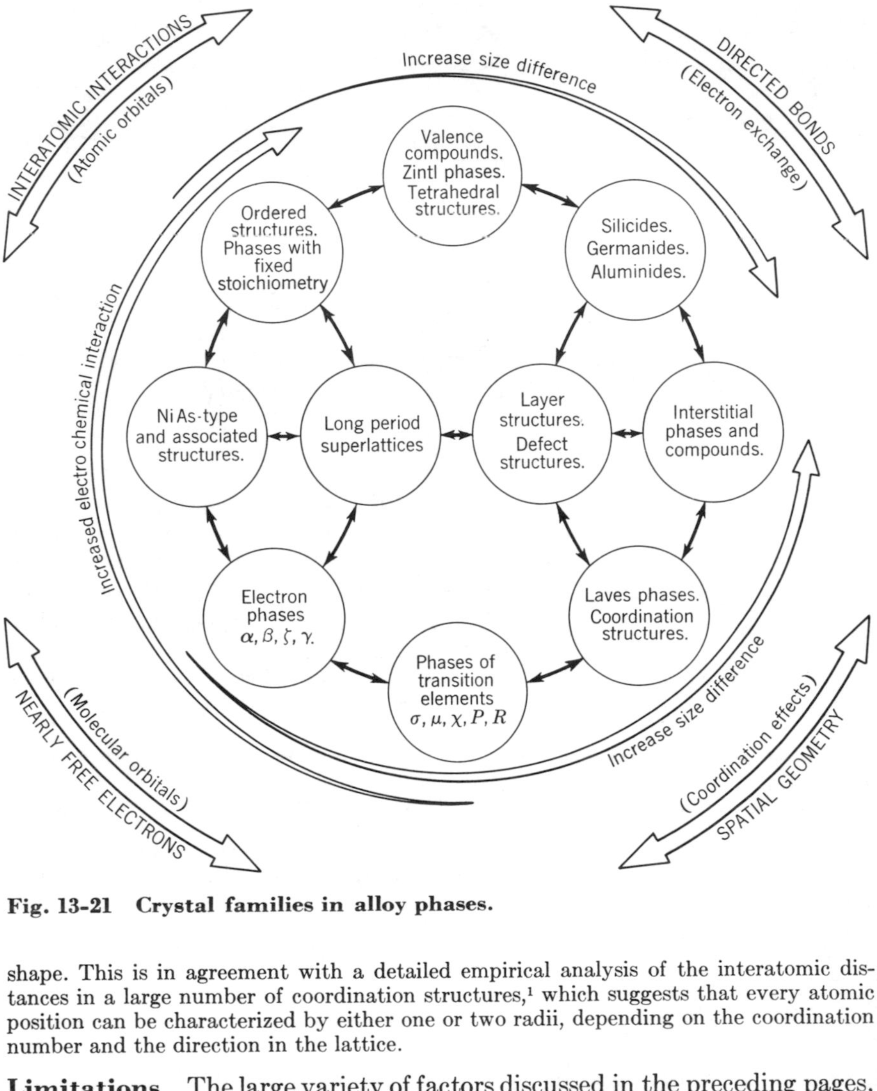

Fig. 13-21 Crystal families in alloy phases.

shape. This is in agreement with a detailed empirical analysis of the interatomic distances in a large number of coordination structures,[1] which suggests that every atomic position can be characterized by either one or two radii, depending on the coordination number and the direction in the lattice.

Limitations The large variety of factors discussed in the preceding pages, and their influence and interplay in determining the stability of metallic crystal structures, clearly indicate the unavoidable complexity of the whole field. With the present rate of advances in both the theories and the methods for computing small differences between crystal energies, a rigorous mathematical treatment of alloy structures should become possible in the not too distant future. At the moment, the relationship between the various factors and the more basic quantities, such as internal energy, entropy, or temperature, is mostly descriptive. For example, while many of the interpre-

[1] C. B. Shoemaker and D. P. Shoemaker, *Trans. AIME*, vol. 230, p. 486, 1964.

tations in terms of electronic energies are undoubtedly in the right direction, strictly speaking, comparison of electronic energies applies only at the absolute zero of temperature. Thus, it is with some astonishment that we find it possible to apply these same arguments to the stability of phases present well above the room temperature, where many additional contributions undoubtedly become important. Even in the case of primary solid solutions, such as the α phases of the noble metals, the maximum extent in terms of e/a is customarily associated with the "knee-shaped" phase fields, such as the α phase of copper-zinc which occurs at some 500°C. It is this type of maximum solubility, at different temperatures in each system, that correlates so well with e/a (Fig. 10–15). At lower temperatures most of the phase diagrams show an increasingly restricted solid solubility so that, for example, at room temperature many α phases would no longer fit the observed correlation, the situation becoming worse as the temperature approaches that of absolute zero. The reason for this is to be found in the increasing stability of the (usually ordered) intermediate phases which, by the common-tangent principle, restrict the primary solid solubility. Hence, for the purpose of comparison one must look for solid-solution effects at temperatures where the interference from other factors and neighboring phases is least. It is these discovered correlations that so frequently show a remarkable dependence upon e/a or size effects, despite the fact that they are observed many hundreds of degrees above absolute zero. That electronic effects, to be associated mainly with Fermi energy, are still effective at some 1000°C constitutes an almost astonishing proof of the powerful influence of electron concentration in the structure of metals and alloys.

Crystal families in alloy phases In order to emphasize the interplay among various factors discussed above that influence the choice of the crystal structure, it is sometimes convenient to summarize the gradually changing character of the influencing factors, and the resulting gradual change of the observed structural arrangements, in such a diagram as that shown in Fig. 13–21. Each major factor is accentuated in a given family of related structures. The diagram emphasizes the fact that no particular structure is solely dependent on one particular factor but rather that there is a continuous interplay with a changing emphasis on different factors.

14

Defects in crystals

The preceding chapters have been concerned with the structure of ideal crystals in which deviations from perfect periodicity are absent or can be neglected. In actual crystals, however, defects are always present and their nature and effects are often as important in understanding the properties of crystals as is the basic periodicity of the ideal crystal. As a result of an unparalleled activity in research in recent years, the existing knowledge of defects and their influence on many different mechanical and physical properties can now only be covered adequately by a series of books. In the space available here it is appropriate to limit the discussion to an introduction that underlines the crystallographic aspects of imperfections, provides background material for subsequent chapters, and directs the reader to important references for further details. Book-length reviews are available on defects and their effect on elastic and plastic deformation and other properties,[1] and shorter reviews are given in many symposium reports and journal articles. A noteworthy attempt has been made by a large committee to publish a critical appraisal of research in the science of materials, which includes statements regarding the as yet unresolved important problems in the several fields that comprise this science, including mechanical behavior, crystal growth, surface phenomena, diffusion, electrical and magnetic properties, and radiation effects—all fields in which crystal imperfection is important.[2]

Several types of imperfections must be considered in any reasonably complete listing for metallic crystals, and many more for nonmetallic sub-

[1] A. H. Cottrell, "Dislocations and Plastic Flow in Crystals," Clarendon, Oxford, 1953. J. Friedel, "Les Dislocations," Gauthier-Villars, Paris, 1956; English transl. and extended ed. published by Pergamon, New York, 1964. W. T. Read, Jr., "Dislocations in Crystals," McGraw-Hill, New York, 1953. E. Schmid and W. Boas, "Kristallplastizität, Springer-Verlag, Berlin, 1936; "Plasticity of Crystals" (transl.), F. A. Hughes, London, 1950. A. Seeger in S. Flügge (ed.), "Handbuch der Physik," vol. 7, pt. 1, pp. 383–665, Springer-Verlag, Berlin, 1955; pt. 2, pp. 1–210, 1955. W. Shockley et al. (eds.), "Imperfections in Nearly Perfect Crystals," Wiley, New York, 1952. C. Zener, "Elasticity and Anelasticity of Metals," University of Chicago, Chicago, 1948. "Impurities and Imperfections," American Society for Metals, Cleveland, Ohio, 1955. J. C. Fisher et al. (eds.), "Dislocations and Mechanical Properties of Crystals," Wiley, New York, 1956. "Vacancies and Other Point Defects in Metals and Alloys," Institute of Metals, London, 1958. "Lattice Defects in Quenched Metals," Plenum, New York, in press.

[2] L. Himmel, J. J. Harwood, and W. J. Harris, "Perspectives in Materials Research," Office of Naval Research, available from U.S. Government Printing Office, Washington, D.C., 1963.

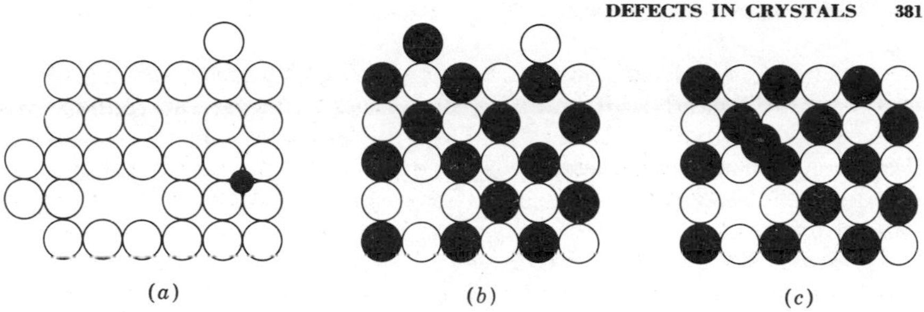

Fig. 14-1 Illustrating point defects in crystals: (*a*) Vacancies, di-vacancies, and interstitials. (*b*) Schottky defects in ionic crystals. (*c*) Frenkel defects in ionic crystals.

stances. These will be mentioned briefly before taking up the most important types in individual sections.

1. A *point defect* extends its influence only a few atom diameters beyond its lattice position. The point defects (Fig. 14–1) include vacancies (missing atoms), di-vacancies and tri-vacancies; interstitial atoms, di-interstitials and tri-interstitials; individual atoms of a species different from the surrounding matrix; and atoms out of place in an otherwise ordered superlattice.

An electron whose energy is within an otherwise unpopulated band of energies in a crystal may be considered to be a point defect, for its influence is local. The same is true of a "hole"—a missing electron in an almost-filled energy band. Furthermore, if the absorption of a photon in a nonmetallic crystal raises an electron to a higher state and the electron is bound tightly to the hole left by it in the filled state, the electron-hole pair may move together through the structure, constituting a point defect known as an *exciton*.[1] However, an exciton is not highly localized—it is spread over the volume of many atoms.

In nonmetallic crystals, the formation of a vacancy involves a local readjustment of charge in the surrounding crystal such that charge neutrality is maintained in the crystal as a whole. Thus, if in an ionic crystal there is a vacancy in a positive-ion site, charge neutrality may be achieved by creating a vacancy in a neighboring negative-ion site (the pair constitute a *Schottky defect*), or by having a positive ion in an interstitial position nearby (this pair constitutes a *Frenkel defect*, Fig. 14–1); or there may be pairs of interstitials of opposite charge.[2]

2. If a plane of atoms extends only partway through a crystal, the edge of such a plane is a defect in the form of a *line*, and is known as a *dislocation*.

[1] J. Frenkel, *Phys. Rev.*, vol. 37, pp. 17, 1276, 1931. R. Peierls, *Ann. Physik*, vol. 13, p. 905, 1932. J. C. Slater and W. Shockley, *Phys. Rev.*, vol. 50, p. 705, 1936. For a summary and other references, see C. Kittel, "Introduction to Solid State Physics," 2d ed., Wiley, New York, 1956.

[2] Summaries of imperfections, including those in nonmetallic crystals, will be found in: F. Seitz in W. Shockley et al. (eds.), "Imperfections in Nearly Perfect Crystals," Wiley, New York, 1952. H. G. van Bueren, "Imperfections in Crystals," North Holland Publishing, Amsterdam, and Interscience, New York, 1960.

Various types of dislocations are found in crystals (see later); they interact with each other and with point defects because they are surrounded by stress fields (and in ionic crystals they are also surrounded by electric fields). Another linear defect, believed to exist in radiation-damaged crystals, is the *crowdion*, discussed later in connection with radiation damage.

3. Since dislocations are frequently arrayed closely together on planes and curved surfaces within a crystal, it is often advantageous to think of such an array as a *two-dimensional defect*. Grain boundaries, subgrain boundaries, and twin boundaries are examples of such defects, as well as domain boundaries between differently oriented superlattice domains or differently oriented domains of a ferromagnetic or antiferromagnetic material. An error in the stacking sequence of atomic planes in a crystal, a *stacking fault*, is another type of two-dimensional defect.[1]

4. Carrying this classification one step further, the *three-dimensional defects* should be listed: clusters of point defects in an otherwise relatively perfect matrix, voids, cracks, bubbles, and particles of a different orientation or structure than the surrounding matrix.

Other defects of an extended nature are also present in crystals: *phonons* (the elastic vibrations or elastic waves that carry quantized amounts of energy equal to $h\nu$, where ν is their frequency), internal stress systems both microscopic and macroscopic in extent, certain dislocation configurations, and concentration gradients resulting from freezing or other causes.

Point defects Vacancies are formed when atoms or ions are ejected from their normal sites in crystals by interaction with phonons or by the combination of thermal energy with energy supplied through plastic deformation or irradiation. In metals the equilibrium ratio at T°K of the number of vacancies, n, to the number of atoms, N, is given by

$$\frac{n}{N} = Ae^{-E_f/kT}$$

where E_f is the net energy required for the formation of a vacancy, k is the Boltzmann's constant, and A is an entropy term of the order of unity in f.c.c. metals. The equilibrium number of vacancies (*Schottky* defects) thus increases exponentially with temperature. In metals, where E_f has values in the neighborhood of 1 ev, the concentration may reach 1 in 10^5 or 1 in 10^4 near the melting point.

In ionic crystals, paired vacancies of opposite sign exist in equilibrium concentrations given by

$$\frac{n}{N} = A'e^{-E_p/2kT}$$

where E_p is the energy of formation of a pair, which is estimated also to be near 1 ev (at least for the alkali halides). In ionic crystals, also, charge

[1] Stacking faults are mentioned in Chaps. 10, 11, and 16, and on p. 387.

neutrality can be maintained in the presence of vacancies by having interstitial ions. For example, in AgCl and AgBr the dominant point defects have been shown to be silver-ion vacancies and interstitial silver ions—a type of defect illustrated in Fig. 14–1 and known as a *Frenkel* type of defect.

The degree of association of vacancies into di-vacancies is difficult to estimate, but it must decrease with increasing temperature. There exists, also, a tendency for vacancies to be bound to impurity atoms. In many crystals the concentrations of vacancies present at high temperatures can be "quenched in" and retained to a considerable extent as a supersaturated concentration at room temperature. Supersaturation tends to be relieved by clustering into larger groups, and by diffusion to other sinks such as boundaries, subboundaries, dislocations, voids, and inclusions.

A vacancy, once formed, can move about from one lattice position to another by overcoming an activation barrier for motion that may be appreciably less than the energy of formation, and di-vacancies may move even more easily. Vacancy migration is the predominating mechanism of diffusion, though not necessarily the only one.[1]

Interstitial atoms in a crystal, like substitutional impurities, tend to locate themselves at positions such that their strain energy is at or near a minimum. The coordinates of such positions are discussed in Chap. 10, together with the kinds of atoms that are sufficiently small to form extensive interstitial solid solutions. Various interstitial positions that are exactly equivalent in an unstrained crystal may become distinguishable when a stress is applied. For example, a tensile stress along [001] of b.c.c. iron makes the space at coordinate position 0, 0, $\frac{1}{2}$ in the unit cell larger than the space at 0, $\frac{1}{2}$, 0. Interstitial carbon atoms in the smaller interstices tend to migrate to the larger ones in response to the stress. This migration can cause internal friction and alterations in the elastic modulus.

The concentration of interstitial atoms in a pure metal is believed to be lower than the concentration of vacancies, but high concentrations are attained when atoms of small radius are introduced into metal crystals, forming the interstitial solid solutions discussed in Chap. 10. Concentrations can be higher in crystals of rather open structure than in close-packed crystals.

It can be assumed that a tendency exists for interstitials to cluster together and to associate with certain substitutional impurity atoms. Thermal agitation tends to dissociate any such clustering or pairing, just as it tends to disorder a superlattice. Many defects serve as permanent sinks for interstitials; in particular, an interstitial may settle into a vacancy or a grain boundary.

Electrons in unfilled energy bands, and holes in nearly filled bands, tend to be attracted to structural defects. For example, the *F center* results from such attraction; it is a defect consisting of an electron bound to a negative-ion vacancy. It is common in alkali halide crystals. Several other kinds of defects, consisting of a combination of one or more electrons or

[1] A recent text covering the fundamentals of diffusion is by P. G. Shewmon, "Diffusion in Solids," McGraw-Hill, New York, 1963.

holes bound to some localized structural defect, have been identified which belong to the general class of *color centers* (they color crystals). There are many kinds of color centers, each with its special optical, electrical, and magnetic characteristics.[1]

Since defects interact with conduction electrons, the number of defects present in a crystal is of vital importance in semiconductor science and technology, and in experimental physics. Defects provide very effective *scattering centers* and *traps* for the electrons and thereby become an important factor limiting the mean free path of the electrons in many instances. For example, if a high-purity metal crystal is cooled to liquid-helium temperatures where there is little scattering of electrons by phonons, electrical resistivity becomes very small and the *residual resistivity* at these temperatures is chiefly dependent upon the content of imperfections.

Foreign atoms also have an important effect on electrical properties of semiconductors when they provide energy levels slightly above the highest levels of a filled band or slightly below the lowest level of an empty band. These impurities then become *acceptors* and *donors*, respectively, and contribute to the electrical conductivity.

Development of dislocation theory The concept of dislocations in crystals was developed by Orowan,[2] Polanyi,[3] and Taylor[4] for the purpose of accounting for the observed fact that single crystals have yield strengths 10^2 to 10^5 times smaller than would be expected if all atoms on a plane, the *slip plane*, glided simultaneously in a rigid array over atoms of the adjacent atomic plane. The low yield strengths of crystals became understandable by postulating that the displacement occurs locally along a shear front which moves across the slip plane—like a wrinkle that is moved across a rug on the floor. The shear front is a dislocation, a line imperfection which, as it moves through a perfect crystal, leaves material behind it that has been displaced. Peierls[5] showed that the cohesive forces across a slip plane should confine the shear front to a width of a few atomic spacings, though shear stresses from the dislocation would extend out much farther. The many powerful techniques subsequently developed for the direct observation of dislocations, several of which are presented in the following chapter, fully confirmed the early theories and stimulated an immense amount of research on dislocations and their effects. A wealth of information has been derived from actual observation on how dislocations move through crystals of various types, interact with each other, combine, split into partial dislocations, form networks, pile up, form tangles, curve around obstacles, form loops, originate or disappear at grain boundaries, and inter-

[1] For references in this field, see: F. Seitz in W. Shockley et al. (eds.), "Imperfections in Nearly Perfect Crystals," Wiley, New York, 1952. H. G. van Bueren, "Imperfections in Crystals," North Holland Publishing, Amsterdam, and Interscience, New York, 1960.

[2] E. Orowan, *Z. Physik*, vol. 89, 634, 1934.

[3] M. Polanyi, *Z. Physik*, vol. 89, p. 660, 1934.

[4] G. I. Taylor, *Proc. Roy. Soc. (London)*, vol. A145, p. 362, 1934.

[5] R. E. Peierls, *Proc. Phys. Soc. (London)*, vol. 52, p. 34, 1940.

act with other kinds of defects such as vacancies, subgrain boundaries, twin boundaries, grain boundaries, and inclusions. (See the references at the beginning of this chapter.)

Edge and screw dislocations The displacement that occurs when a dislocation passes a point is described by a vector, the *Burgers vector*, usually given the symbol **b**. The direction of **b** with respect to the dislocation line and the length of **b** with respect to the identity distance in the direction of **b** are the fundamental characteristics of a dislocation. Figure 14–2 illustrates an *edge dislocation*, one in which **b** is normal to the line of the dislocation. In this figure **b** lies in the plane of the drawing, which is normal to the dislocation. An incomplete plane of atoms ends on the dislocation; since this partial plane is *above* the slip plane in the upper row of drawings and *below* the slip plane in the lower row, the former may be termed a *positive dislocation*, the latter a *negative dislocation*. As shown in the figure, under an applied shear stress indicated by the vectors labeled τ, a positive edge dislocation moves to the right and a negative edge dislocation moves to the left, yet both result in the same displacement of the upper part of the crystal with respect to the lower, and upon emerging from the crystal both would produce the same step, or *slip line*, on the surface.

A *screw dislocation* has its Burgers vector *parallel* to the line of the dislocation. This type is illustrated in Fig. 14–3. Suppose one moves from atom to atom on a plane normal to the dislocation, such as the upper plane in the figure, and in so doing makes a circuit around the dislocation. This

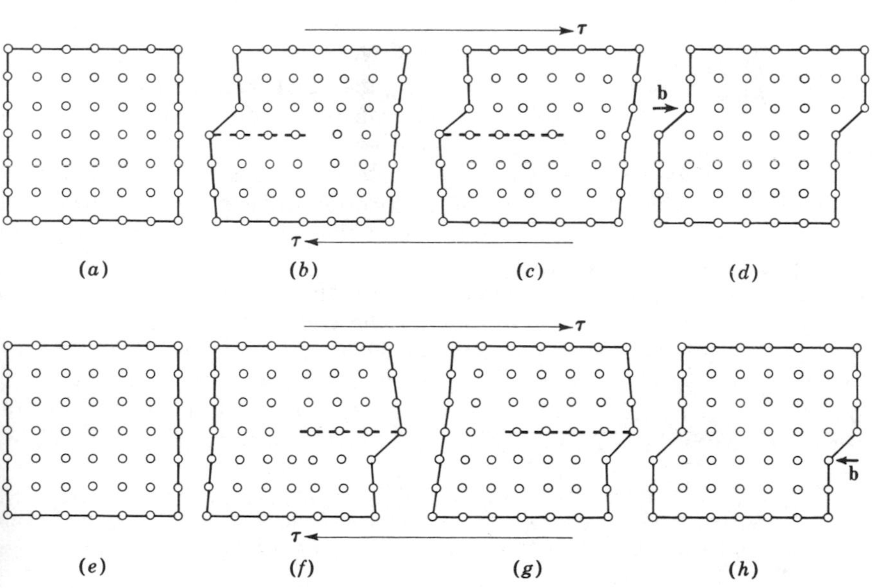

Fig. 14-2 Edge dislocations. In *b*, *c*, and *d*, a positive dislocation moves to the right; in *e*, *f*, *g* and *h*, a negative dislocation moves to the left. Burgers vectors are *b* and are normal to the dislocation line.

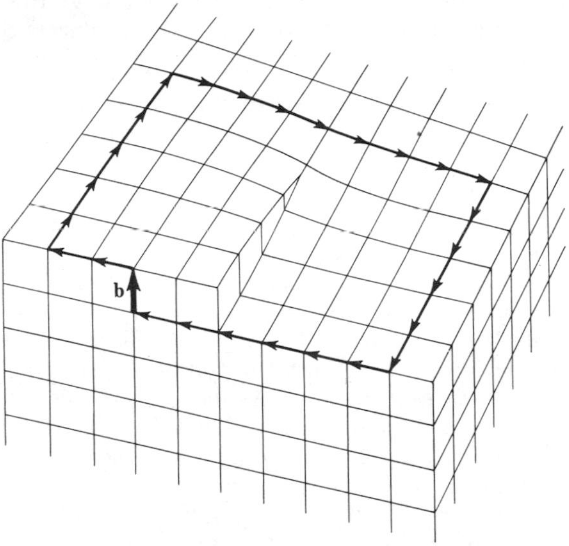

Fig. 14-3 A screw dislocation causes a distortion such that a Burgers circuit (line of arrows) fails to close by the amount indicated by the Burgers vector *b*, which is parallel to the dislocation.

Burgers circuit will be a spiral path which in one revolution advances by the distance represented by the Burgers vector **b**. (Similarly, a Burgers circuit around an edge dislocation fails to close by **b** for the edge dislocation, when compared with a circuit, otherwise identical, which encloses no dislocation.) The outer row of atoms on the plane indicated in Fig. 14–3 illustrates a circuit around a screw dislocation.

If a screw dislocation extends to the surface as in Fig. 14–3 and material is being deposited on this surface, the dislocation will continue as the crystal grows. Frank and his coworkers[1] proposed that the self-perpetuating step in the crystal surface at the point of emergence of a screw dislocation provide a place where growth can occur with almost arbitrarily low values of supersaturation or of undercooling. (The work of forming a nucleus of greater than critical size on a smooth face is thereby avoided.) This theory received confirmation when such points of emergence and the "growth spirals" radiating from them were recognized[2] on natural crystal faces of beryl ($Be_3Al_2Si_6O_{18}$), on long-chain paraffin crystals[3] ($C_{36}H_{74}$), on silicon

[1] F. C. Frank, *Discussions Faraday Soc.*, vol. 5, pp. 38, 67, 186, 1949. W. K. Burton, N. Cabrera, and F. C. Frank, *Nature*, vol. 163, p. 398, 1949; *Phil. Trans. Roy. Soc.*, vol. A243, p. 299, 1951. F. C. Frank, Advances in Physics, *Phil. Mag.*, Quart. Suppl., vol. 1, p. 91, 1952.

[2] L. J. Griffin, *Phil. Mag.*, vol. 41, p. 196, 1950.

[3] I. M. Dawson and V. Vand, *Nature*, vol. 167, p. 476, 1951; *Proc. Roy. Soc. (London)*, vol. A206, p. 555, 1951.

carbide,[1] and on magnesium, PbI_2, and CdI_2 crystals.[2] A segment of a curved dislocation may have some edge character and some screw character; the exact amount of each is determined by the Burgers-vector direction which may be anything between the pure screw and the pure edge.

Perfect and partial dislocations; stacking faults A dislocation in which the Burgers vector is an identity period in the lattice is spoken of as "complete," "whole," or "perfect." Passage of such a dislocation leaves the crystal unchanged in its atomic arrangement. A dislocation can be specified by giving its components along the three crystal axes. Thus, the Burgers vector for slip in a f.c.c. crystal has the length and direction of half a face diagonal and may be written $\mathbf{b} = (a_0/2)[110]$ or simply $\frac{1}{2}[110]$.

Because the strain energy of a dislocation is proportional to b^2, it is energetically favorable for dislocations in some crystals to form an "extended dislocation" by splitting into two "partial dislocations," each having smaller **b** than the whole dislocation. In f.c.c. metals it is thus common to find the complete dislocation $\frac{1}{2}[110]$ split into the partials $\frac{1}{6}[121]$ and $\frac{1}{6}[21\bar{1}]$, often referred to as *Shockley partials*, with a consequent reduction in shear strain energy.[3] A decomposition of a full dislocation into partials results in a separation of the partials. The area between the partials is a discontinuity in the stacking sequence of the atom layers, a stacking fault. In a f.c.c. crystal, the normal stacking sequence $ABCABC\ldots$ may be interrupted at a stacking fault and becomes $ABCA/CABCAB\ldots$. The interruption is indicated by the slant bar. The fault is therefore a local region with c.p.h. stacking. As will be seen from Fig. 14-4, slip on (111) planes in f.c.c. crystals is actually a succession of steps in zigzag fashion, alternately in directions $[1\bar{2}1]$ and $[\bar{2}11]$ (or on another pair of directions of this type).

The distance between two Shockley partials represents a balance between the action of two forces. The stress fields of the dislocations cause mutual repulsion, but the energy of the stacking fault that stretches between them causes attraction. Since the force on one partial dislocation arising from the stress field of the other can be calculated, the observed equilibrium width of the fault in the absence of other stress fields can be used to determine the actual stacking-fault energy of a crystal. The energy varies widely from one substance to another. The width of a stacking fault, however, may be smaller or much greater than this equilibrium value if other stress fields are also present.

A "Shockley partial" may be defined as a partial that has **b** lying in the fault plane. A *Frank partial*, on the other hand, has **b** nonparallel to the fault plane; it can only be formed by removing or inserting a part of a plane of atoms. Since a partial can glide only if its Burgers vector is in its fault

[1] A. R. Verma, *Phil. Mag.*, vol. 42, p. 1005, 1951. S. Amelinckx, *Nature*, vol. 167, p. 939, 1951.

[2] A. J. Forty, *Phil. Mag.*, vol. 43, pp. 72, 481, 1952; vol. 42, p. 670, 1951.

[3] R. D. Heidenreich and W. Shockley, "Report on Strength in Solids," p. 37, Physical Society, London, 1948.

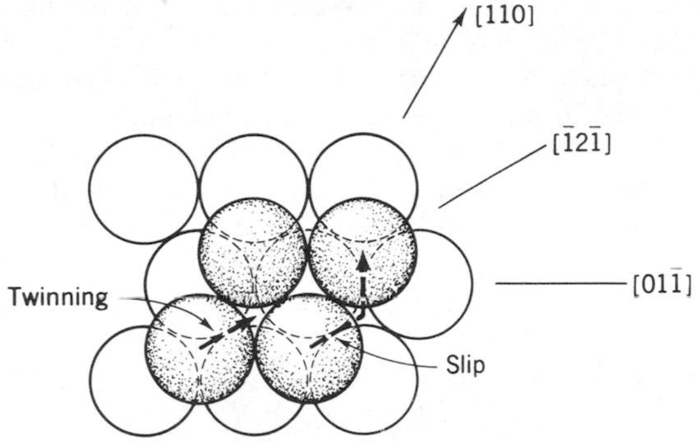

Fig. 14-4 Directions of movement of close-packed layers over each other during slip and during twinning in f.c.c. crystals.

plane, a Shockley partial can glide (it is "glissile"). Any dislocation with **b** parallel to *no* slip plane is "sessile" and can move only by a process involving mass transport—by acquiring or losing atoms at the edge of the incomplete atom layer.

The type of fault between two Shockley partials, which has been called an *intrinsic fault*, alters the f.c.c. stacking sequence in the manner indicated in Fig. 14–5*a*. It is equivalent to the removal of a close-packed layer of atoms. It is not the only type of fault possible, however. The result of inserting a layer, as indicated in Fig. 14–5*b*, has been called an *extrinsic fault;* this type is perhaps rare but has been found in cerium (see page 452).

If the crystal orientation on one side of a plane continues to be different from that on the other side in the sequence shown in Fig. 14–5*c*, the fault

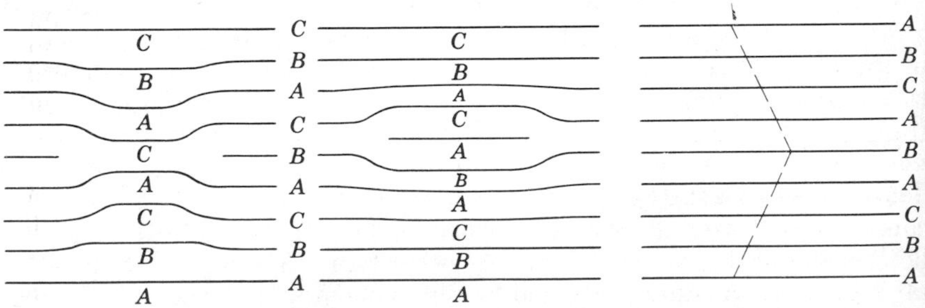

Fig. 14-5 Faults in the stacking sequence of f.c.c. crystals. The lines represent the edges of (111) planes. (*a*) Intrinsic fault, bounded by two Shockley partial dislocations. (*b*) Extrinsic fault, equivalent to an inserted plane. (*c*) Twin fault or growth fault. Stacking sequences are indicated by dashed lines and by sequence of letters.

is a *twin fault* or *growth fault*.[1] Twins that are extremely thin but more than two atom layers thick are often called *microtwins*.

In a superlattice, some directions have longer repeat distances in the ordered state than in the disordered. In such directions, **b** for the ordered state will be some multiple of **b** for the disordered. A perfect dislocation in the disordered crystal in this case will correspond to a partial dislocation in the ordered crystal, and the stacking fault will represent a break in the order.

Dislocation content of crystals Since the dislocation content of a crystal has an important influence on various properties, particularly on flow stress, electrical and thermal conductivities, and internal energy, countless researches are devoted to estimating the content of individual specimens. The content is defined in terms of the number of dislocations that intersect a random section through the crystal per square centimeter. Alternately it may be defined as the total length of dislocation line per cubic centimeter—a measure that usually gives a larger value since most dislocations are not normal to the random section.

Dislocation densities up to 10^{10} or more per square centimeter have been determined by transmission electron microscopy, but caution is necessary to be sure that the density present in a bulk sample is not lowered by losses while the specimen is being thinned in preparation for microscopy. Etch-pit methods are also used in counting dislocations—but etch-pit counting is reliable only if the etching technique is critically chosen and controlled to ensure that an etch pit forms at each point where a dislocation meets the surface. For the highest densities, above say 10^{10} dislocations per square centimeter, the microscopic and etch-pit methods fail, but estimates may be based on the broadening of diffraction lines. The range of densities that may be expected in metals may be judged from the following approximate figures:[2]

Carefully grown high-purity single crystals	0 to 10^3
Ordinary single crystals, annealed or unstrained	10^5 to 10^6
Polycrystalline specimens, annealed	10^7 to 10^8
Severely cold-worked specimens	10^{11} to 10^{12}

Thompson's notation for dislocations in f.c.c. crystals[3] It is often convenient to use a notation for dislocations in f.c.c. crystals that will be described with reference to Fig. 14–6. Consider a tetrahedron with corners at A, B, C, D, and with faces representing the $\{111\}$ planes of the crystal, which are the slip planes. The edges of the tetrahedron then are parallel to

[1] W. T. Read, "Dislocations in Crystals," McGraw-Hill, New York, 1953. Diffraction effects from these types are discussed on p. 459.

[2] D. McLean, "Mechanical Properties of Metals," Wiley, New York, 1962.

[3] N. Thompson, *Proc. Phys. Soc. (London)*, vol. B66, p. 481, 1953.

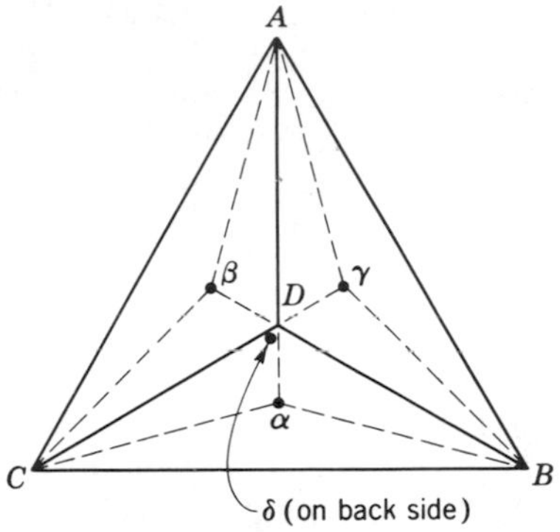

Fig. 14-6 Thompson's notation for dislocations in f.c.c. crystals, based on a tetrahedron.

the Burgers vectors of complete dislocations. Thus, AB represents a possible Burgers vector for a dislocation slipping on the plane ABD (and also for a dislocation slipping on the plane ABC). The dashed lines drawn to the face centers labeled with Greek letters then represent Burgers vectors for partial dislocations. For example, the vector AB may dissociate into the two partials $A\gamma$ and γB, since these two partials add up to AB by vector addition. The tetrahedron is useful in dealing with problems having to do with dislocations meeting at a point or combining with each other. Whenever two or more dislocations meet at a point, the sum of their Burgers vectors must equal zero, a requirement readily tested with the aid of the tetrahedron if proper attention is paid to the signs of the vectors (Thompson's article discusses signs in detail).

Nodes, stacking fault energy, cross slip A common configuration in crystals, in which three dislocations lying in the same slip plane come together at a node, will serve to illustrate Thompson's notation and some of the geometry of nodes. Let the slip plane be represented on the tetrahedron by the face containing the label δ; then the Burgers vectors of the three complete dislocations must be AB, BC, and CA, since they must add up vectorially to zero. These three will split up into the partials $A\delta + \delta B$, $B\delta + \delta C$, and $C\delta + \delta A$. The partials then arrange themselves in two ways depending on the cyclic arrangement of the three dislocation lines (with certain pairs canceling each other out). In Fig. 14–7, the two cyclic arrangements of complete dislocations and of their corresponding fault-plane and partial-dislocation arrangements are shown. These are commonly seen in electron microscope pictures of dislocations.

The *Lomer-Cottrell sessile dislocation* that has often been discussed in theories of work hardening may be illustrated as follows: Let a dislocation with vector DC lie in the slip plane represented by the face of the tetrahedron marked β; we may write this as $DC(\beta)$. Let this dislocation be intersected by one with vector CB that lies in slip plane δ. These combine to give a dislocation with vector DB. The two original dislocations are parallel to AC, and the result of the combination is a line that is also parallel to AC, which is the line of intersection of the two slip planes

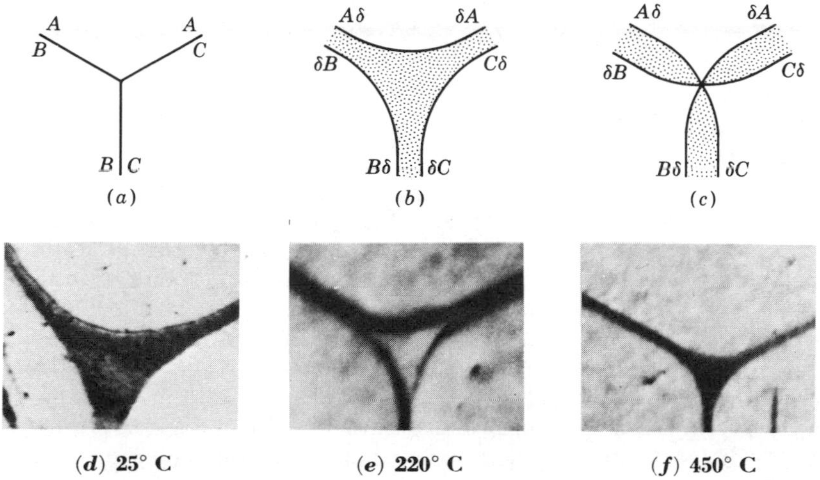

Fig. 14-7 Dislocation nodes. (*a*) Full dislocation with Thompson's notation; (*b*) and (*c*) partial dislocations; (*d*), (*e*), and (*f*) electron micrographs of nodes in Cu–7.33 percent Al. Node radius depends on stacking-fault energy and alters during annealing, provided solute impedance is overcome. (Courtesy P. R. Swann.)

involved. The reaction is then

$$DC(\beta) + CB(\delta) = DB(\beta, \delta)$$

This resultant dislocation has a vector that does not lie in either slip plane, and since it cannot glide in either plane it is sessile. Actually, the reaction involves partials and may be written

$$D\beta + \beta C + C\delta + \delta B = D\beta + \beta\delta + \delta B$$

where the vector $\beta\delta$ from point β to point δ on the tetrahedron is responsible for the sessile nature of the resultant dislocation. The appearance of the arrangement is illustrated in Fig. 14–8.

This *Lomer-Cottrell lock*,[1] as it is sometimes called, which is one of the types of interactions between dislocations that are slipping on intersecting slip planes, has been postulated as a barrier to further slip and a significant contributor to strain hardening. Its effectiveness as a barrier has been questioned, however, for it can be "undone from its ends."[2]

The configurations of Fig. 14–7 can be clearly seen in some electron micrographs, and from a measure of the curvature of the partials near the

[1] W. M. Lomer, *Phil. Mag.*, vol. 42, p. 1327, 1951. A. H. Cottrell, *Phil. Mag.*, vol. 43, p. 645, 1952.

[2] P. B. Hirsch in "Internal Stresses and Fatigue in Metals," p. 139, Elsevier, New York, 1959.

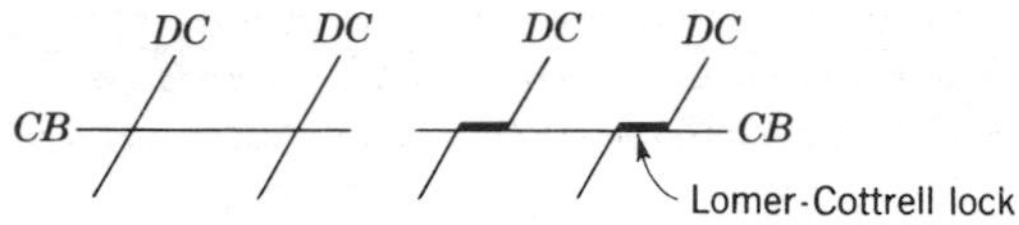

Fig. 14-8 Dislocation interactions resulting in "Lomer-Cottrell locks."

nodes (the node radius) the stacking-fault energy γ has been calculated.[1] From such measurements and from the widths of stacking-fault ribbons[2] it has become clear that γ is particularly low in stainless steels ($\approx$13 ergs per cm), also low in alpha-brass and some other Cu and Ag alloys, is intermediate in Cu, Ag, and Au (in the range between 20 and 90 ergs per cm), and high in Al (200) and Ni.[4] However, many of the early determinations are only approximate, and later, more elaborate methods of computing γ from the nodes[5] suggest that earlier methods that have been used extensively for metals and alloys overestimate γ by factors between 1 and 2. A variety of complications are involved in such determinations: elastic anisotropy, the nature of the dislocations and the stacking faults, the stress fields encountered, the possibility of solute segregation, the inequalities of curvatures and widths of the stacking faults, and the possible lack of equilibrium configurations when solutes impede the motion of dislocations. Therefore, most determinations must be considered as estimates.[6] Other methods for estimating γ have also been employed, in addition to the inferences that can be drawn from the stability of defects, such as dislocation nodes or tetrahedra produced by collapsed vacancies,[7] or from the relationship between the twin-boundary energy and γ (see page 406). For example, in deforming single crystals of f.c.c. metals, τ_3 is the stress at which thermally activated *cross slip* occurs; and since cross slip is related to the collapse under stress of the extended partial dislocations, a relation-

[1] M. J. Whelan, *Proc. Roy. Soc.*, vol. A249, p. 114, 1958. A. Howie and P. R. Swann, *Phil. Mag.*, vol. 6, p. 1251, 1961. M. H. Loretto, *Phil. Mag.*, vol. 10, p. 467, 1964. H. M. Loretto, L. M. Clarebrough, and R. L. Segall, *Phil. Mag.*, vol. 10, p. 731, 1964.

[2] S. Amelinckx and P. Delavignette in J. B. Newkirk and J. H. Wernick (eds.), "Direct Observation of Imperfections in Crystals," Interscience, New York, 1962; G. Thomas and J. Washburn (eds.), "Electron Microscopy and Strength of Crystals," Interscience, New York, 1963.

[3] A. Art et al., *Appl. Phys. Letters*, vol. 2, p. 40, 1963.

[4] P. B. Hirsch in G. M. Rassweiler and W. L. Grube (eds.), "Internal Stresses and Fatigue in Metals," Elsevier, New York, 1959. Additional values will be found in: P. R. Thornton and P. B. Hirsch, *Phil. Mag.*, vol. 3, p. 738, 1958; in ref. 2 above; and in T. Jøssang and J. P. Hirth, *Phil. Mag.*, vol. 13, p. 657, 1966. See also Chap. 16, p. 465.

[5] T. Jøssang et al., *Acta Met.*, vol. 13, p. 149, 1965.

[6] J. W. Christian and P. R. Swann in T. B. Massalski (ed.), "Alloying Effects in Concentrated Solid Solutions," Gordon and Breach, New York, 1965.

[7] J. Silcox and P. B. Hirsch, *Phil. Mag.*, vol. 4, p. 72, 1959. G. Czjzek, A. Seeger, and S. Mader, *Phys. Stat. Solidi*, vol. 2, p. 558, 1962. M. H. Loretto, L. M. Clarebrough, and R. L. Segall, *Phil. Mag.*, vol. 11, p. 459, 1965.

ship may be derived between τ_3 and γ.[1] Particularly low values are expected and found in the layer structures, such as graphite, talc, molybdenum sulfide, and aluminum nitride; γ in graphite is about 0.035 ergs per cm^2, for example.

In metals with high γ, the stacking-fault ribbons are narrow and more easily forced together by stresses. Since the ribbons must come together before a dislocation can leave its slip plane and move onto an intersecting slip plane, high values favor such cross slip. Because cross slip is a means by which a dislocation can work its way around an obstacle, the value of γ is important in understanding various mechanical properties of solids, including slip and creep resistance, strain hardening, and age hardening.

Cross slip is sometimes also involved in one of the common mechanisms for generating a succession of dislocations, the *Frank-Read source*. A segment of an edge dislocation moving on a (111) plane of a f.c.c. crystal may cross slip on $(\bar{1}11)$ until an obstacle is avoided, and then cross slip from $(\bar{1}11)$ back onto one of the (111) planes. The ends of the segment that has undergone two successive cross-slip processes are pinned, but the segment itself can bow out in response to stress (see Fig. 14–9). If it continues to bow out, it will wind around the pinning points, close on itself and form a dislocation loop within which there has been a displacement of magnitude **b**. This process leaves a dislocation segment stretching between the pinning points as at the beginning of the process; consequently, it can be repeated indefinitely. A Frank-Read source may thus throw out a succession of loops.

Loops Dislocation loops are frequently observed in metals and other crystals. If a loop is not smaller than about 50 A in diameter, it can be recognized unmistakably by electron microscopy, and under favorable circumstances it can be seen as a round or elliptical ring or as a polygon; its nature and probable origin can often be inferred (a vacancy loop is distinguished from an interstitial loop) and its influence on other dislocations observed.[2]

Some loops leave the stacking sequence of the atomic planes unchanged, but others alter the sequence so that the stacking inside the loop contains a fault in stacking sequence that is not found in the perfect crystal outside the loop. A stacking fault within the loop can be recognized by electron microscopy, as is discussed in connection with Fig. 15–15 on page 442.

A brief list follows of the ways a loop may be formed:

1. Slip may occur on a limited area of a slip plane; the boundary of the slipped region is a dislocation loop. A source of dislocations, such as a Frank-Read source, may send out a sequence of loops and become surrounded by concentric rings.

[1] See, for example, the articles by P. Haasen and by J. W. Christian and P. R. Swann in T. B. Massalski (ed.), "Alloying Behavior and Effects in Concentrated Solid Solutions," Gordon and Breach, New York, 1965.

[2] See also Chap. 15. For longer reviews of loops and other defects, see references at the beginning of this chapter and G. Thomas and J. Washburn (eds.), "Electron Microscopy and Strength of Crystals," Interscience, New York, 1963.

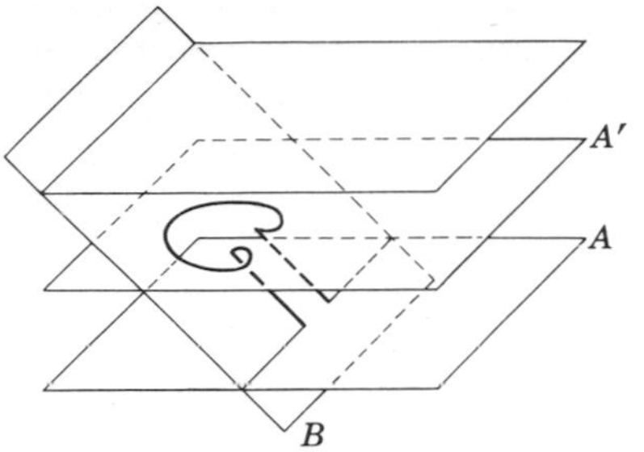

Fig. 14-9 A Frank-Read source. A dislocation on plane *A* cross slips on plane *B* and again on *A′*, where it bows out to become a Frank-Read source of dislocations.

2. A loop results if vacancies cluster together on an atomic plane, provided the void thus produced collapses.

3. Loops can be formed, analogously, by interstitial atoms, for if these cluster together in sufficient numbers they may form an extra plane inserted between the planes of the perfect crystal and surrounded by a loop.

4. Some loops form (at least in MgO) as follows:[1] A moving screw dislocation may develop a *jog*, a segment which lies at a high angle to the rest of the dislocation; **b** in the jog is unfavorable to slip, although the adjacent portions glide easily in response to the imposed stress. If the positive and negative segments on opposite sides of the jog are on slip planes very close together, the stress fields of the two may prevent them from passing each other and they will form a stable pair or "dipole." The dipole may become longer and longer as plastic flow continues, or it may pinch together, forming a narrow loop.

5. A moving dislocation may loop around an obstacle such as a small precipitate, come together after passing the obstacle and move on, leaving a dislocation that rings the obstacle.

6. Loops may be thrown out into the surrounding crystal from an embedded inclusion, in a sequence resembling a series of smoke rings following one another. This "punching out" of loops may arise, for example, from differential thermal expansion of the inclusion and the surrounding matrix.[2]

Jogs, helices, and tangles Important configurations of dislocations result from the process of "climb." Since an edge dislocation is the edge of a half plane of atoms, the movement of a vacancy to the dislocation amounts to the removal of an atom from the half plane, and puts an indentation in the edge of the half plane. A succession of such events results in a dislocation with many jogs in it. The dislocation "climbs" out of the slip planes on which it was originally located. Movement by climb is diffusion-controlled. It may occur either by the diffusion of vacancies to the dislocation or by the diffusion of interstitials (climb involving interstitials is in the direction opposite to climb caused by vacancies). A jog can move along the dislocation readily by the jumping of atoms from one position on the dislocation to a neighboring one. A jog can also move by slip, but this can occur easily only in the direction of **b**; if an applied stress forces it to move in any direction other than parallel to the Burgers vector, it must generate either vacancies or interstitials, and it consequently requires a much greater expenditure of energy. When of sufficient length, a jog be-

[1] J. Washburn et al., *Phil Mag.*, vol. 5, p. 991, 1960. G. Thomas and J. Washburn (eds.), "Electron Microscopy and Strength of Crystals," p. 30, Interscience, New York, 1963.

[2] D. A. Jones and J. W. Mitchell, *Phil. Mag.*, vol. 3, p. 1, 1958.

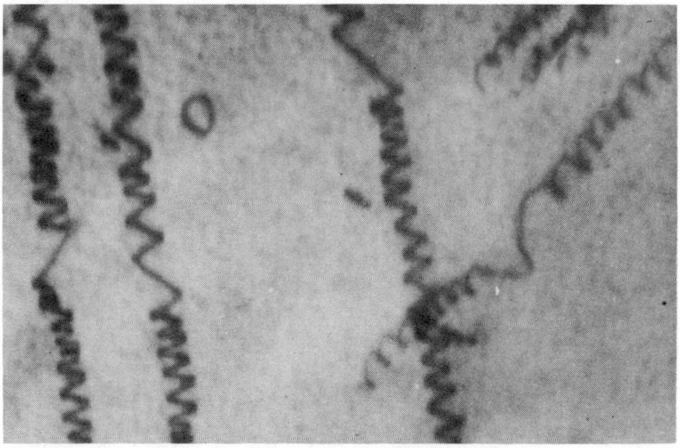

Fig. 14-10 Screw dislocations deformed into helices by climb. Al–4 percent Cu quenched from 540°C. ×60,000. (Courtesy G. Thomas and M. J. Whelan.)

comes a very effective and stable pinning point and is sometimes spoken of as a *superjog*.

The effect of diffusing point defects to or from a screw dislocation is to convert it by climb into a curved line or a helix, such as that illustrated in Fig. 14–10. An edge dislocation can also turn into a helix[1] by a flow of point defects along the core of the dislocation.

The helices can develop into tangles,[2] or irregular three-dimensional dislocation arrays, illustrated in Fig. 14–11. Many crystals develop during deformation into irregular cells having tangles as cell walls. Tangles do not occur at temperatures too low for vacancy migration, and interaction of point defects with dislocations must therefore be necessary; plastic deformation is also somehow involved, since tangles tend to form zones parallel to the active slip planes at high deformations. They occur in many different substances,[3] both b.c.c. and f.c.c. in structure, but apparently not in materials with low stacking-fault energy, such as alpha-brass or stainless steel. It has been proposed that tangles can be formed by the cooperative action of many processes primarily involving slip-induced point-defect precipitation.[4]

In deformed b.c.c. metals, tangles form in the early stages of deformation, and a cell structure develops with a low density of dislocations in the cell

[1] J. Weertman, *Trans. AIME*, vol. 227, p. 1439, 1963.

[2] R. deWitt, *Trans. AIME*, vol. 227, p. 1443, 1963.

[3] Observations and theories are covered by several investigators in *Phys. Soc. Japan*, vol. 18, supplement I, 1963; and in G. Thomas and J. Washburn (eds.), "Electron Microscopy and Strength of Crystals," Interscience, New York, 1963.

[4] The processes, given the term "mushrooming," have been proposed in the following: D. Kuhlmann-Wilsdorf, R. Maddin, and H. G. F. Wilsdorf in "Strengthening Mechanisms in Solids," p. 137, American Society for Metals, Cleveland, Ohio, 1962. H. G. F. Wilsdorf and D. Kuhlmann-Wilsdorf, *Phys. Rev. Letters*, vol. 3, p. 170, 1959.

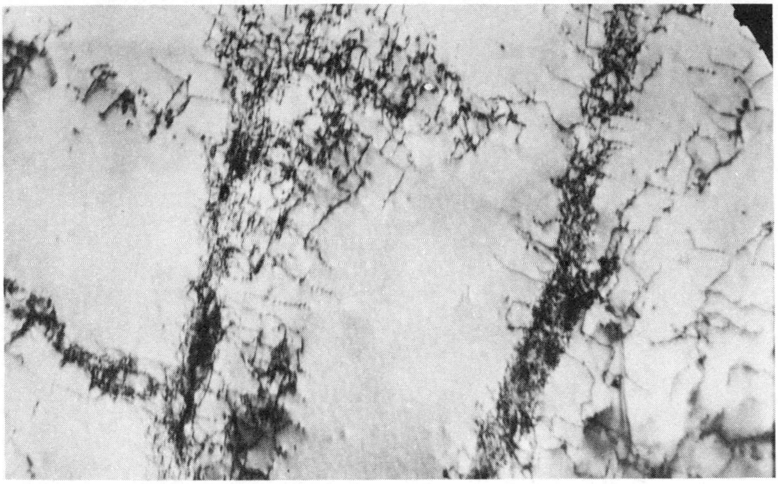

Fig. 14-11 Dislocation tangles at cell walls in polycrystalline iron strained 9 percent at 25°C. ×35,000. (Courtesy A. S. Keh.)

interiors and a high density in the cell walls. Long jogs and elongated loops are common.[1] Annealing transforms the cells into subgrains with better-defined boundaries. Semiregular networks are occasionally found during the early stages of deformation in Mo but not in Fe or Ta.[2]

Tangles serve both as sources of dislocations and as barriers to the movement of dislocations. They are also of importance in influencing the distribution of solutes and in the nucleation of precipitates.

Dislocation nets and small-angle boundaries Dislocations often arrange themselves in networks, some of which show great regularity. This was shown first by Mitchell and coworkers in crystals of the silver halides when the dislocations within the crystals were "decorated" by groups of silver atoms during the formation of an internal latent image.[3] Decoration methods for other crystals have also been developed; for example, Dash and others have used the principle that silicon is transparent to infrared light, so that when copper atoms are moved to dislocations in silicon and precipitated there, the dislocations can be viewed by infrared light. Although various other methods have also been used to see networks, the greatest advances have been made by using transmission electron microscopy.

It is found that a network may be two-dimensional or three-dimensional;

[1] A. S. Keh and S. Weissmann in G. Thomas and J. Washburn (eds.), "Electron Microscopy and Strength of Crystals," Interscience, New York, 1963.

[2] R. Benson, G. Thomas, and J. Washburn in J. B. Newkirk and J. H. Wernick (eds.), "Direct Observation of Imperfections in Crystals," Interscience, New York, 1962.

[3] J. W. Mitchell in J. C. Fisher et al. (eds.), "Dislocations and Mechanical Properties of Crystals," Wiley, New York, 1956. J. M. Hedges and J. W. Mitchell, *Phil. Mag.*, vol. 44, pp. 223, 357, 1953.

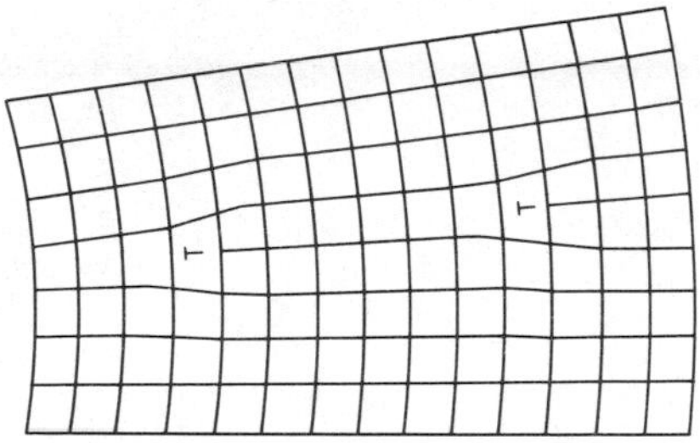

Fig. 14-12 The array of edge dislocations in a tilt boundary.

the individual segments of the dislocations join at nodes as in Fig. 14–7. At each node the Burgers vectors add vectorially to zero and the line tensions are balanced.

A *tilt boundary* is a simple type consisting of similar edge dislocations arranged parallel to each other as in Fig. 14–12. The misorientation across the boundary is merely a rotation about an axis parallel to the dislocations, and if these each have Burgers vectors of magnitude b it can be seen directly from Fig. 14–12 that the misorientation is simply

$$\theta = \frac{b}{s} \tag{14-1}$$

where s is the distance between the dislocations. Another type of small-angle boundary is a *twist boundary,* which is produced by a crossed grid of screw dislocations, and in which the misorientation is a twist about an axis normal to the boundary.

Even though there are five degrees of freedom that must be specified to describe fully a plane boundary (three in the relative orientation of the two subgrains or grains, and two in the orientation of the boundary), it is possible to specify it in terms of a dislocation model.[1] Boundaries may be either plane or curved; in fact, a subgrain can be completely surrounded by—wrapped in—a network and have an orientation differing from its surroundings. Figure 14–13 shows a well-developed network boundary.

Many details of the growth of grains and subgrains and the nature of the changes that take place during annealing have been reviewed in detail.[2] The strain energy of a dislocation assembly is a driving force that tends to produce and retain regular subboundaries; they form spontaneously during deformation or during annealing in many crystalline materials.

Substructures have been observed directly after deformation at room

[1] W. Shockley and W. T. Read, *Phys. Rev.*, vol. 75, p. 692, 1949; vol. 78, p. 275, 1950; also in Shockley et al. (eds.), "Imperfections in Nearly Perfect Crystals," Wiley, New York, 1952.

[2] P. Beck, *Phil. Mag.*, Suppl., vol. 3, p. 245, 1954. Earlier reviews include: W. G. Burgers in "Handbuch der Metallphysik," vol. 3, Springer, Berlin, 1941, and Edwards Brothers, Ann Arbor, Mich., 1944. C. S. Barrett, "Structure of Metals," 2d ed., McGraw-Hill, New York, 1952. R. W. Cahn in "Impurities and Imperfections," p. 41, American Society for Metals, Cleveland, Ohio, 1955.

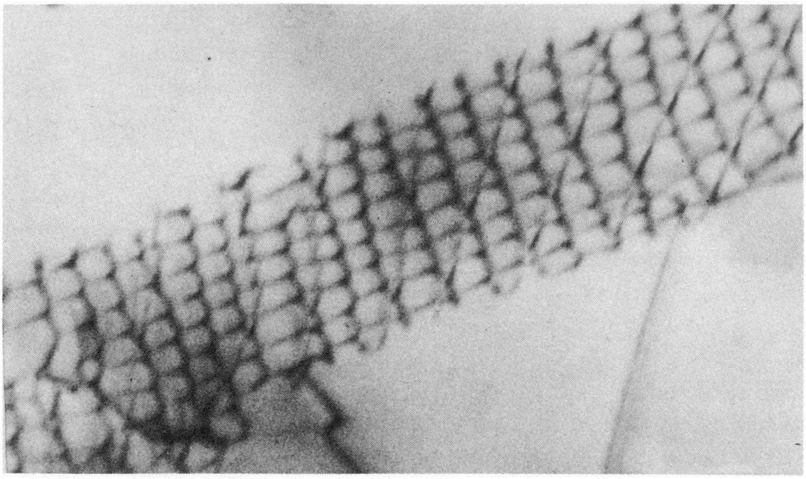

Fig. 14-13 A network of dislocations forming a subboundary. Iron strained 16 percent at 25°C, then annealed 16 hours at 550°C. ×40,000. (Courtesy A. S. Keh.)

temperature in Al,[1] Cu, Ni, and Fe[2] and even Mo,[3] and in Zn when deformed and studied by x-ray diffraction without heating above 78°K.[4] A simple type of dislocation wall known as a *kink* (Fig. 14–14) can be produced by deforming some crystals. The kink forms athermally and is stable during subsequent deformation.[5]

Thermally activated processes of glide and climb of dislocations also produce subboundaries. *Polygonization* is a term applied to such mechanisms since they subdivide crystals into polygonal subgrains. Annealing tends to develop sharp dislocation boundaries and to lower the density of dislocations in the interior of the subgrains.[6] Creep deformation produces similar results.[7] The lower energy of a polygonized grain may prevent absorption by more highly strained grains during *recrystallization*,[8] defined as the process of migration of high-angle boundaries to relieve strain energy. In fact, polygonization competes with and may entirely prevent recrystallization, especially in weakly deformed specimens. Substructures are also

[1] P. A. Beck and H. Hu, *Trans. AIME*, vol. 194, p. 83, 1952.

[2] P. Gay and A. Kelly, *Acta Cryst.*, vol. 6, p. 165, 1953.

[3] N. K. Chen and R. Maddin, *Trans. AIME*, vol. 197, p. 300, 1953.

[4] G. P. Conard II, B. L. Averbach, and M. Cohen, *Trans. AIME*, vol. 197, p. 1036, 1953.

[5] E. Orowan, *Nature*, vol. 149, p. 643, 1942. J. B. Hess and C. S. Barrett, *Trans. AIME*, vol. 185, p. 599, 1949. J. Washburn and E. R. Parker, *Trans. AIME*, vol. 194, p. 1076, 1952. D. W. Bainbridge et al., *Acta Met.*, vol. 2, p. 322, 1954.

[6] R. W. Cahn, *J. Inst. Metals*, vol. 76, p. 121, 1949. C. G. Dunn and F. W. Daniels, *Trans. AIME*, vol. 191, p. 147, 1951. E. C. W. Perryman, *Acta Met.*, vol. 2, p. 26, 1954.

[7] G. R. Wilms and W. A. Wood, *J. Inst. Metals*, vol. 75, p. 693, 1949. R. W. Cahn, I. J. Bear, and R. L. Ball, *J. Inst. Metals*, vol. 82, p. 481, 1954.

[8] P. Lacombe and A. Berghezan, *Metaux* (*Corrosion-Ind.*), vol. 25, p. 1, 1949.

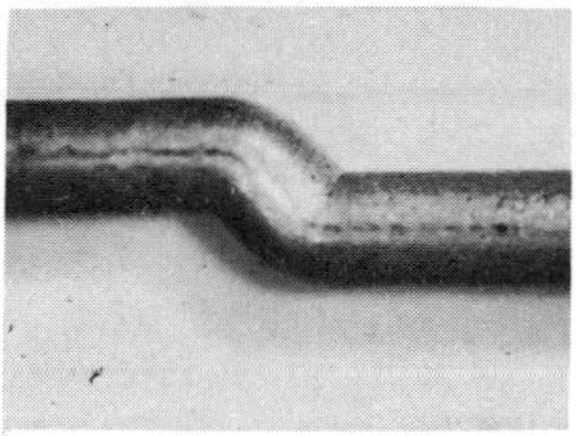

Fig. 14-14 A kink in a zinc crystal after compression.

generated by phase transformation. A metallographic appearance in ferrite, called *veining*, which results from transformation of austenite to ferrite, is an example[1]; other examples are found in the β phase in brass[2] and in alpha-uranium.[3]

The energy of a subboundary can be computed[4] from the attractive and repulsive forces of the dislocations comprising the boundary—provided the dislocations are not too close together. The energy in these small-angle boundaries (if θ is less than 45°) rises from 0 to E_m as θ increases from $\theta = 0$ to a maximum at $\theta = \theta_m$ according to the relation

$$\frac{E}{E_m} = \frac{\theta}{\theta_m}\left(1 - \ln \frac{\theta}{\theta_m}\right) \tag{14–2}$$

Experimental data on the orientation dependence of boundary energy has accumulated rapidly in recent years since the importance of boundary energies in controlling microstructures was explained.[5] Early data[6] include experimental determinations on tin, lead[7], silicon ferrite, and copper[8] and confirm the theory within an angular range up to θ_m (originally assessed[7] at $\approx 20°$ but later determined[8] as $\approx 6°$ in copper). If misorientations are increased beyond θ_m, cusps are found in the curve, where E drops to a sharp minimum, again confirming the theory. Cusps can be expected whenever the orientations are such that rows of atoms in the two grains come into registry across the boundary at regular intervals of a few interatomic distances, forming coherent or semicoherent interfaces.

Imperfections resulting from solidification The growth of a crystal from the melt is a frequent source of imperfections.[9] A common mac-

[1] A. Hultgren and B. Herrlander, *Trans. AIME*, vol. 172, p. 493, 1947.

[2] R. J. Davis, R. Pearce, and W. Hume-Rothery, *Acta Cryst.*, vol. 5, p. 36, 1952.

[3] R. W. Cahn, *Acta Met.*, vol. 1, p. 176, 1953.

[4] W. Shockley and W. T. Read, vol. 75, p. 692, 1949; vol. 78, p. 275, 1950; also in W. Shockley et al. (eds.), "Imperfections in Nearly Perfect Crystals," Wiley, New York, 1952.

[5] Cyril S. Smith in W. Shockley et al. (eds.), "Imperfections in Nearly Perfect Crystals," Wiley, New York, 1952; *Trans. Am. Soc. Metals*, vol. 45, p. 533, 1953; *Trans. AIME*, vol. 175, p. 15, 1948.

[6] Summarized by J. C. Fisher and C. G. Dunn in W. Shockley et al. (eds.), "Imperfections in Nearly Perfect Crystals," Wiley, New York, 1952.

[7] B. Chalmers, *Proc. Roy. Soc.* (*London*), vol. A196, p. 64, 1949. K. T. Aust and B. Chalmers, *Proc. Roy. Soc.* (*London*), vol. A201, p. 210, 1950.

[8] N. A. Gjostein and F. N. Rhines, *Acta Met.*, vol. 7, p. 319, 1959; vol. 8, p. 263, 1960.

[9] For reviews, see: U. M. Martius in "Progress in Metal Physics," vol. 5, p. 279, Interscience, New York, 1954. D. T. J. Hurle, "Mechanisms of Growth of Metal Single Crystals from the Melt," Pergamon, New York, 1962.

roscopic defect is *lineage structure* or *striations*, a substructure consisting of columnar regions differing 0.2 to 5° from neighboring ones. Striations are not removed by long annealing, but can be lessened by controlling the rate of growth from the melt.[1] Another defect structure resulting from freezing is a system of parallel subgrains, microscopic in size and prismatic in form, referred to as *fine lines* or *corrugations*. This structure arises from a cellular structure of the solid-liquid interface. The cell centers project into the melt slightly farther than the hexagon-shaped cell boundaries. The cells increase in size with decreasing growth rates and disappear at slow rates, at least in tin, leaving only striations.[2] A steep thermal gradient in the liquid near the interface suppresses the corrugated structure, for it is caused by the phenomenon of *constitutional supercooling*, a supercooled region in the melt near the advancing interface with the solid. The impurity content of the liquid adjacent to the interface is raised by the rejection of impurities from the interface, and if the phase diagram is such that increased impurity content corresponds to an increased liquidus temperature, the actual temperature in the liquid for a small distance ahead of the interface may be below the liquidus temperature (i.e., the liquid here is supercooled). A flat interface is then unstable, and cells (or in extreme cases, dendrites) develop. Crystallographic features of the structure and their dependence on rates, purity gradients, etc., have been extensively studied,[3] as well as the uneven solute distributions that result.

The methods of growing crystals with minimum numbers of defects include careful *pulling* of a crystal upward from the surface of the melt (with suitable rotation and temperature control), and *zone melting*—moving a molten zone along a rod.[4]

Dislocation interactions Since energy is required for one dislocation to cut through another and produce a jog in each, a major hindrance to slip is encountered if a moving dislocation has to "cut through a forest" of many dislocations that are located so as to intersect the slip plane of the moving dislocation. As mentioned earlier, the movement of jogs themselves is also restricted, and can occur only in certain directions without the production of point defects. Another process that requires extra energy is the pinching together of an extended dislocation, yet this must be forced to occur if the dislocation is to cross slip onto another slip plane or to acquire a jog.[5]

[1] A. J. Goss and S. Weintraub, *Proc. Phys. Soc.* (*London*), vol. B65, p. 561, 1952. E. Teghtsoonian and B. Chalmers, *Can. J. Phys.*, vol. 29, p. 370, 1951; vol. 30, p. 388, 1952.

[2] J. W. Rutter and B. Chalmers, *Can. J. Phys.*, vol. 31, p. 15, 1953.

[3] See summaries by Hurle, *loc. cit.*, and the various contributors to "Liquid Metals and Solidification," American Society for Metals, Cleveland, Ohio, 1958. Texts with treatments of the subject include: R. E. Reed-Hill, "Physical Metallurgy Principles," Van Nostrand, New York, 1964. B. Chalmers, "Physical Metallurgy," Wiley, New York, 1959.

[4] W. G. Pfann, "Zone Melting," Wiley, New York, 1958.

[5] A. Seeger in J. C. Fisher et al. (eds.), "Dislocations and Mechanical Properties of Crystals," Wiley, New York, 1956.

The force exerted on a dislocation per unit length by a shear stress is given by $\mathbf{F} = \boldsymbol{\tau}\cdot\mathbf{b}$, where $\boldsymbol{\tau}\cdot\mathbf{b}$ is the component of shear stress along the direction of **b** in the glide plane. Each dislocation sets up a stress field and therefore interacts with other dislocations in its neighborhood.[1] A dislocation in a stress field may be pinned at certain points, in which case it will bow out between the pinning points, until the tendency to increase the curvature is balanced by the line tension of the dislocation (thus altering the elastic modulus from the value that would be obtained in a dislocation-free sample). If the applied stress is so high that it overbalances the line tension, the dislocation may continue to move, winding around the pinning points and becoming a Frank-Read source.

Point defects are strongly attracted by the stress fields of some dislocations, particularly the edge type. Vacancies and small dissolved atoms, for example, tend to remain in the regions around an edge dislocation where the stresses are compressive; interstitials and large solute atoms prefer regions of tension. Thermal agitation acts contrary to this segregation tendency and produces a more or less diffuse distribution. The result is an "atmosphere" in neighboring regions, appropriately called a *Cottrell atmosphere*, which is bound to the dislocation whenever the temperature is not so high that the clustering is dispersed, or whenever the dislocation is not moving so fast that the atmosphere cannot keep up with it.[2] If such atmospheres have collected they can cause the *yield-point phenomenon*: the dislocations can move only very slowly, dragging the atmosphere along, until the stress is increased to a value sufficient to tear the dislocations away from their atmospheres. As a result, a yield point, or sudden drop in load-bearing capacity, occurs. The only dislocations free from atmospheres are freshly formed or fast-moving ones. In addition to this Cottrell locking, solid solutions may be more or less strengthened by segregation of the solute at stacking faults (*Suzuki segregation*).[3] A similar effect of impurities segregating in grain boundaries and hindering the motion of the boundaries has been suggested in a theory of the recrystallization process proposed by Lücke and Detert.[4] However, measurements of grain-boundary mobility and recrystallization rates[5] indicate the importance of other factors

[1] Specification of the stress field will be found in practically every text on dislocations or on mechanical properties. For example: A. H. Cottrell, "Dislocations and Plastic Flow in Crystals," Oxford, New York, 1953, 1961. D. McLean, "Mechanical Properties of Metals," Wiley, New York, 1962. Detailed calculations of the effects of the stresses appear frequently in the technical journals.

[2] A. H. Cottrell, "Dislocations and Plastic Flow in Crystals," Oxford Press, New York, 1953, 1961. D. McLean, "Mechanical Properties of Metals," Wiley, New York, 1962.

[3] H. Suzuki, *Sci. Rep. Res. Inst. Tohoku Univ.*, vol. 4A, no. 5, p. 455, 1952; also in "Dislocations and Mechanical Properties of Crystals," p. 361, Wiley, New York, 1957. Further discussion of flow stress will be found in recent texts, such as G. E. Dieter, "Mechanical Metallurgy," McGraw-Hill, New York, 1961.

[4] K. Lücke and K. Detert, *Acta Met.*, vol. 5, p. 628, 1957.

[5] K. T. Aust and J. W. Rutter, *Trans. AIME*, vol. 215, p. 119, 1959; vol. 218, p. 682, 1960. W. C. Leslie, F. J. Plecity, and F. W. Aul, *Trans. AIME*, vol. 221, p. 982, 1961.

not treated in the Lücke-Detert theory—for example, inhibition of boundary motion by clusters of solute atoms at imperfections in the unrecrystallized matrix.[1]

Dislocations may stream out in succession from a source and pile up at an obstacle, such as a grain boundary, until their combined stress field either stops the source or starts sources in the neighboring grain or in the boundary, or until a crack is formed by the tensile stresses across a cleavage plane.[2]

Grain boundaries and grain shapes High-angle boundaries cannot be usefully analyzed in terms of their dislocations. They are thin regions—only a very few atom diameters thick—with higher energies than subboundaries. They exert strong tensions that are responsible for grain growth.[3] A few fundamentals may be summarized as follows:

The three-dimensional polygons giving the shape of grains in polycrystalline metals (or beer foam or soap froth)—with either straight or curved sides—obey certain topological principles. In an infinite array of grains (or a finite sample with suitably chosen periphery) the average grain has exactly 6 sides. Grains with n sides, which may be represented by the symbol P_n, must conform in number, in such an array, to the relation

$$4P_2 + 3P_3 + 2P_4 + 1P_5 + 0P_6 - 1P_7 - 2P_8 \cdots - (n - 6)P_n = 0$$

Therefore, if there is a single 7-sided grain anywhere in the array there must be a 5-sided one. A 9-sided grain must be accompanied by a 3-sided one, etc. The films in a soap froth meet each other at angles of 120°, and the same tendency is prominent in metals, although crystallinity here introduces some variation in the surface tension of different boundaries. In a soap froth the regularity that would correspond to minimum film area is absent (though the area does not differ greatly from the minimum). Polygons that are not 6-sided exist; they have curved rather than planar surfaces so that junctions remain at 120°, and the curvature introduces varying pressures in the individual bubbles. The cells with fewer than 6 sides have convex faces; those with more than 6, concave.

Polycrystalline metals are similarly irregular, and here also similarly curved surfaces occur. Each surface tends to move toward its center of

[1] There is also, curiously, correlation between the "recrystallization temperature" of binary iron alloys and the number of *d*-shell electrons in the solute atoms, noted by E. P. Abramson II and B. S. Blakeney, Jr., *Trans. AIME*, vol. 218, p. 1101, 1960.

[2] C. Zener, *Trans. Am. Soc. Metals*, vol. 40A, p. 3, 1948. J. S. Koehler, *Phys. Rev.*, vol. 85, p. 480, 1952. A. N. Stroh, *Proc. Roy. Soc.* (*London*), vol. A223, p. 404, 1954; *Advan. Phys.*, vol. 6, p. 418, 1957. Fracture reviews include: B. L. Averbach et al. (eds.), "Fracture," Wiley, New York, 1959. D. C. Drucker and J. J. Gilman (eds.), "Fracture of Solids," Wiley, New York, 1963.

[3] Cyril S. Smith has written comprehensive treatments of this field in *Trans. AIME*, vol. 175, p. 15, 1948; in "Metal Interfaces," p. 65, American Society for Metals, Cleveland, Ohio, 1952; in W. Shockley et al. (eds.), "Imperfections in Nearly Perfect Crystals," Wiley, New York, 1952; in *Trans. Am. Soc. Metals*, vol. 45, p. 533, 1953; in *Sci. Am.*, p. 58, January 1954; and in *Met. Rev.*, vol. 9, p. 1, 1964.

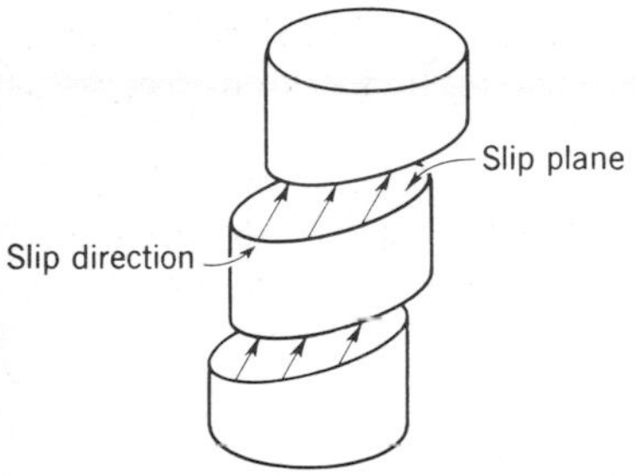

Fig. 14-15 Deformation of a crystal by slip.

curvature and thereby to reduce the energy of its boundary. At temperatures high enough to permit diffusion and grain boundary motion, the grains with 5 sides tend to become 4-sided, and these in turn tend to become 3-sided and then to disappear. Grains with more than 6 sides tend to have concave sides and on the average will grow. An ideally stable configuration is never reached. Slowing and cessation of growth are due to other causes—sometimes to boundaries being pinned by inclusions. The general subject of quantitative metallography, including the measurement of grain sizes and matters concerning grain shapes, has received much attention.[1]

The crystallography of slip, twinning, and cleavage Turning now to more macroscopic aspects of deformation, we summarize briefly the main crystallographic features of slip, twinning, and cleavage. When plastic deformation of a crystal occurs by slip, it is usually found that lamellae glide over each other on crystallographic *slip planes* (Fig. 14–15) of low indices and in a *slip direction* of low indices (normally the direction of closest-packed rows of atoms, in metallic elements). Much research has been devoted to these crystallographic features.[2]

In f.c.c. metals the predominating slip plane is the closest-packed plane (111). Other planes may become active at elevated temperatures. In b.c.c. and tetragonal metals, slip can occur on several planes, most of which are among those most densely packed with atoms and most widely spaced. Many metals alter their slip planes when deformed at unusually high or low temperatures, when solutes are added, or when high strain rates are involved; the data are so extensive and so dependent upon the exact conditions involved in each experiment that an exhaustive summary cannot be given here. Listed in Table 14–1 is merely a selection of data that might be considered to be some of the most important facts.

Although the slip planes that are active may vary with conditions, depending on which ones happen to have the lowest resistance to slip, the

[1] E. E. Underwood, "Quantitative Metallography," Addison-Wesley, Cambridge, Mass., 1965.

[2] Summaries: E. Schmid and W. Boas, "Kristallplastizität," Springer, Berlin, 1935; in English transl., Hughes, London, 1950. H. J. Gough, Edgar Marburg Lecture, *ASTM Proc.*, vol. 33, pt. 2, p. 3, 1933. Report of a Conference on Internal Strains in Solids, *Proc. Phys. Soc.* (*London*), vol. 52, 1940. R. Maddin and N. K. Chen, *Progr. Metal Phys.*, vol. 5, p. 53, 1954.

Table 14–1 Slip systems in common crystals[a]

Structure	Examples	Slip plane, slip direction
F.c.c. ($A1$)	Cu, Ag, Au, Ni, Al	(111) [110]
	$A1$ at elevated temperatures	(100) $[10\bar{1}]$
B.c.c. ($A2$)	α-Fe	(110), (112), (123) $[11\bar{1}]$
	W, Mo, Na at 0.08 to 0.24 T_m	(112)[b]
	Mo, Na, β-CuZn, at 0.26 to 0.50 T_m	(110)[b]
	Na, K, at 0.80 T_m	(123)[b]
	Nb	(101)
C.p.h. ($A3$)	Cd, Be, Te	$(0001)[11\bar{2}0]$
	Zn	$(0001)[11\bar{2}0]$, $(11\bar{2}2)$ $[11\bar{2}\bar{3}]$
	Be, Re, Zr	$(10\bar{1}0)[11\bar{2}0]$[c]
	Mg[d]	$(0001)[11\bar{2}0]$, $(11\bar{2}2)[10\bar{1}0]$, $(10\bar{1}1)[11\bar{2}0]$
	Ti, Zr, Hf	$(10\bar{1}0)[11\bar{2}0]$, $(10\bar{1}1)[11\bar{2}0]$, $(0001)[11\bar{2}0]$, $(11\bar{2}2)$[e]
CsCl type ($B1$)	LiTl, MgTl, AuZn[f]	(110)[100]
	AgMg[f]	(321)[111]
	β-CuZn	(110)[111]
Orthorhombic ($A20$)	α-U	(010)[100]; cross slip on (011), (013) (110)
Tetragonal	β-U[g]	(110)[001]
Tetragonal ($A5$)	β (white) Sn	(110)[001], (100)[001], (100)[011?], $(10\bar{1})[101]$, (121)[101]
Rhombohedral ($A7$)	Bi	$(111)[10\bar{1}]$
	Hg	$(11\bar{1})[\bar{1}10]$, $(11\bar{1})[011]$[h]
NaCl type ($B1$)	NaCl, AgCl[f]	$(110)[1\bar{1}0]$, (100)

[a] Except as noted otherwise, data are from: Review by R. Maddin and N. K. Chen in "Progress in Metal Physics," vol. 5, p. 53, Interscience, New York, 1954. R. W. K. Honeycomb in "Progress in Materials Science," vol. 9, Pergamon, New York, 1961. C. S. Barrett, "Structure of Metals," 2d ed., McGraw-Hill, New York, 1953.
[b] T_m is the melting temperature in these determinations by E. N. Da C. Andrade and L. C. Tsien. See also A. T. Churchman, *Proc. Roy. Soc. (London)*, vol. 226A, p. 216, 1954.
[c] In single crystals of c.p.h. Cu-Ge ζ phase, $\{10\bar{1}0\}$ slip has been observed in addition to {0001} slip. P. H. Thornton, *Phil. Mag.*, vol. 11, p. 71, 1965.
[d] R. E. Reed-Hill and W. D. Robertson, *Trans. AIME*, vol. 212, p. 256, 1958.
[e] $(11\bar{2}2)$ in Hf reported by R. E. Reed-Hill, private communication, 1964.
[f] A. H. Cottrell in R. Grammel (ed.), "Deformation and Flow in Solids," p. 33, Springer, Berlin, 1956.
[g] A. N. Holden, *Acta Cryst.*, vol. 5, p. 182, 1952.
[h] J. G. Rider and F. Heckscher, *Phil. Mag.*, vol. 13, p. 687, 1966.

slip directions are not so fickle. Among the equivalent planes of given indices $\{hkl\}$, and the equivalent directions $\langle uvw \rangle$, the ones that become active under a given applied stress are the one or two subjected to the highest resolved shear stress.[1] In iron, where (110), (112), and (123) all

[1] If a force F is applied to a single-crystal bar having a cross section A, as in Fig. 14–16, and a slip plane lies at angle ϕ to the cross-section plane (i.e., $90 - \phi$ to the direction of

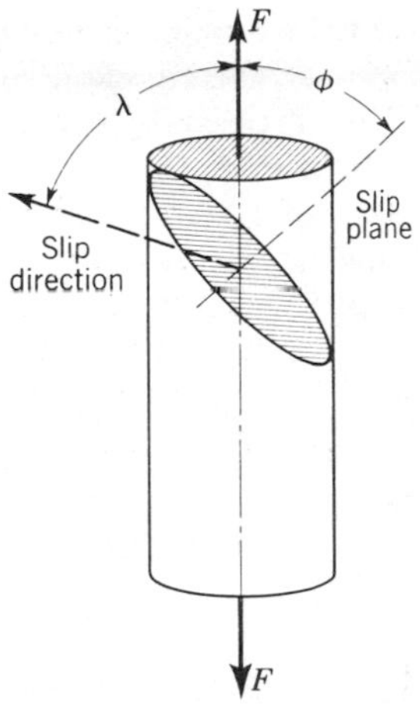

Fig. 14-16 Coordinates for calculating resolved shear stresses.

function as slip planes, the direction of slip is always the close-packed direction [111], which is common to all three sets of planes.[2] Slip in iron is characteristically on a corrugated surface rather than on a plane. In ionic lattices of the sodium-chloride type, slip is also along the lines of greatest atomic density [1$\bar{1}$0], although the slip plane (110) is not the plane of greatest density.[3]

A plane of slip and a direction of slip lying in that plane constitute a *slip system*. Face-centered cubic metals, having 4 (111) planes and 3 [101] directions in each, possess 12 slip systems; body-centered iron has 4 [111] directions, around each of which are arranged 12 slip planes having the slip direction as their zone axis, thus giving 48 slip systems. The spacing between slip lines, the displacement on each, the clustering of lines into slip bands of appreciable width, and the velocity of dislocations forming them are much studied.

The reorientation of a single crystal resulting from slip under uniaxial strain can be simply stated. Disregarding the local distortions at the ends of a rod-shaped specimen, slip on a single plane rotates the active slip plane and the active slip direction toward the tensile axis during elongation in tension. (On a stereographic projection the tensile axis moves along a great circle toward the slip direction.) When rotation increases the resolved

---

F), the area of the slip plane will be $A/\cos\phi$ and the stress on the plane will be $F/(A\cos\phi)$; the component of this stress in the slip direction will be this stress multiplied by $\cos\lambda$, where λ is the angle between the direction of F and the direction of slip. Thus, the resolved shear stress tending to produce slip will be

$$\tau = \frac{F}{A}\cos\phi\cos\lambda$$

[2] H. J. Gough, *Proc. Roy. Soc. (London)*, vol. A118, p. 498, 1928.

[3] An extensive discussion of the geometry of slip and its relation to atomic configuration, particularly in minerals, has been published by M. J. Buerger, *Am. Mineralogist*, vol. 15, pp. 45, 174, 226, 1930.

shear stress on a second slip plane to a value sufficient to make it also become active, both slip simultaneously or alternately thereafter in *double slip* or possibly later in *multiple slip* involving still more slip systems. Rotation is then modified accordingly. If a single crystal is deformed in compression, on the other hand, the direction of rotation of the lattice is such as to bring the pole of the slip plane that is active into parallelism with the axis of compression.[1]

Twins Crystals are said to be twinned if they are composed of two or more portions that have certain specific orientations with respect to each other, known as *twin orientations*. Two adjacent crystals are *reflection twins* if their lattices are mirror images of each other across a lattice plane, called the *twinning plane*. This plane must be parallel to the same lattice plane in both crystals, but obviously, it is not a mirror plane that would simply perpetuate the lattice of one portion directly into the other. Although the *lattice rows* in the pair of twinned crystals will be symmetrically oriented, not all *atom positions* need be mirror images across the twinning plane; and in some crystals they are not. Crystals are *rotation twins* if a two-, three-, four-, or sixfold rotation of one crystal about a *twinning axis* produces the orientation of the other. The rotation axis lies either in the twinning plane or normal to it and is not a symmetry element of the lattice of the individual crystals. Twinned crystals can be related in orientation by yet another macroscopic symmetry element, namely, a center of symmetry—provided the crystal structure of an individual crystal does not have a center of symmetry.

In the more symmetrical crystals, such as the b.c.c. and the c.p.h., the specifications are met for both rotation and reflection twins. Such twins are sometimes designated *compound twins*. Thus, f.c.c. crystals have a (111) twinning plane and also a [111] twinning axis; b.c.c. crystals have a(112) twinning plane and also a [112] twinning axis. A description of a twinned crystal that is a rotation twin with respect to another portion of a crystal does not imply that the lattice rotation was achieved by mechanical means. In many minerals large numbers of thin twins, the phenomenon known as *polysynthetic twinning*, may occur.

Paired crystals joined along a twin plane are said to be *contact twins*, but not all boundaries between twinned regions are along twin planes even for reflection twins. Although the crystal structures in the two individuals of a twinned pair are identical, there may be distortion in the structure within a few atom diameters of the interface between the pair; the interface may have a low or a high interface energy depending on the amount of this distortion and on whether or not the coordination across the interface of the structure is appreciably altered. The interface energy of a twin plane of contact twins constitutes a *coherent boundary*, and its energy is generally much lower than the energy of the other interfaces. In many varieties of twins a special lattice (*coincidence lattice*) exists, con-

[1] Simple formulas give the amount of rotation accompanying a given overall strain; see: E. Schmid and W. Boas, "Kristallplastizität," Springer, Berlin, 1935; in English transl., Hughes, London, 1950. G. E. Dieter, "Mechanical Metallurgy," p. 102, McGraw-Hill, New York, 1961.

sisting of certain lattice points of both individual crystals, which continues without disturbance across the twin interface.

Twins may be formed by deformation, by growth during crystallization from the liquid or the vapor state, by growth during annealing (i.e., by recrystallization or by grain-growth processes), or by the movement of boundaries between different solid phases (e.g., during phase transformation).

Deformation twins An important mechanism of plastic flow in many metals is deformation twinning, consisting of shearing movements of the atomic planes over one another. The shearing appears to be homogeneous under the microscope, for it results in a uniform tilting of the surface in the twinned region; but in many crystals not every atom shifts a distance proportional to its distance from the twinning plane. The *lattice points* undergo homogeneous shear but in addition there is a "shuffle" of some of the *atom positions* accompanying the shear, without which the crystal structure would have been altered by the shearing. This mode of deformation increases in prominence when temperatures are low or strain rates are high. Detailed treatments of various subjects in the field will be found in the proceedings of a conference.[1]

A typical mode of development of deformation twins is illustrated in Fig 14–17, which shows the metallographic appearance of polycrystalline zinc at successive stages of a tensile test. Thin lamellae first appear very quickly, then grow slowly in width as deformation proceeds. The lenticular shape is characteristic not only of many deformation twins but also of many martensitic transformation products, in which there is a similar change of shape of a portion of a crystal.

Iron and alloyed ferrites are twinned by impact at room temperature and by slower deformation at lower temperatures to form the narrow lamellae known as *Neumann bands*, shown in Fig. 14–18. These have small but readily visible widths, unlike slip lines, and appear prominently after polishing and etching. A small fraction of the volume is disoriented by twinning in b.c.c. crystals, and in c.p.h. crystals almost any fraction can be converted to twins. Some bismuth and antimony crystals, which are rhombohedral in structure, appear to deform entirely by twinning and without visible slip lines,[2] but soft, ductile ones also deform by slip.[3]

Twinning is a major factor in the deformation of c.p.h. metals whenever the applied stresses are directed so as to convert substantial portions into a twin orientation. The reoriented material may provide slip planes that are favorably oriented for further deformation by slip.[4]

[1] R. E. Reed-Hill, H. C. Rogers, and J. P. Hirth (eds.), "Deformation Twinning," Gordon and Breach, New York, 1964.

[2] H. J. Gough and H. L. Cox, *Proc. Roy. Soc.* (*London*), vol. A127, p. 431, 1930; *J. Inst. Metals*, vol. 48, p. 227, 1932.

[3] W. F. Berg and L. Sander, *Nature*, vol. 136, p. 915, 1935.

[4] C. H. Mathewson and A. J. Phillips, *Trans. AIME*, vol. 74, p. 143, 1927. G. Edmunds and M. L. Fuller, *Trans. AIME*, vol. 99, p. 175, 1932. E. Schmid and G. Wassermann, *Z. Physik*, vol. 48, p. 370, 1928. H. Mark, M. Polanyi, and E. Schmid, *Z. Physik*, vol. 12, p. 58, 1922.

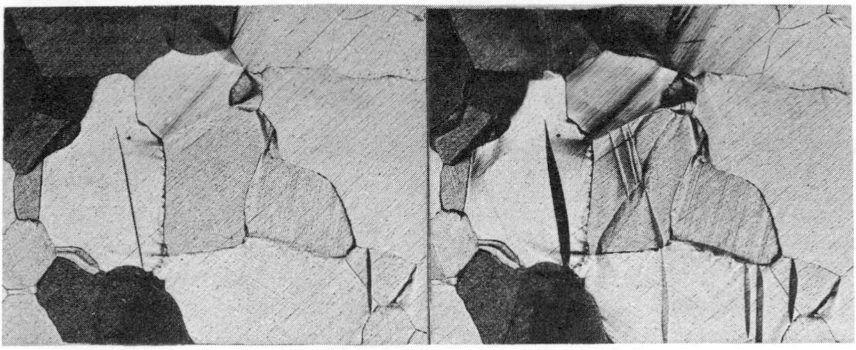

Fig. 14-17 Growth of deformation twins during tensile straining of polycrystalline zinc. (Courtesy J. E. Burke.)

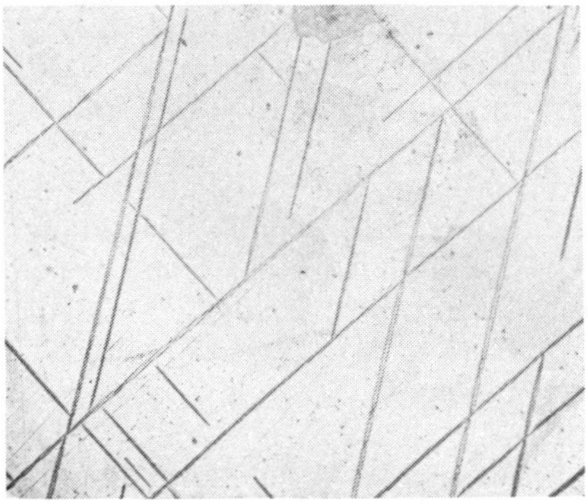

Fig. 14-18 Deformation twins in a crystal of ferrite. Polished and etched. ×100.

For many years deformation twinning was believed to be absent in f.c.c. metals although Samans in 1934 had reported evidence of it in oscillating crystal patterns of severely deformed alpha-brass.[1] But in 1957, Blewitt, Coltman, and Redman[2] obtained unmistakable x-ray and metallographic evidence of twinning in copper at liquid helium temperatures, and deformation twins were later found in many other f.c.c. metals and alloys[3] after low-temperature deformation and in copper after explosive loading[4] at ordinary temperatures. Suzuki and Barrett found that silver crystals could

[1] C. H. Samans, *J. Inst. Metals*, vol. 55, p. 209, 1934.

[2] T. H. Blewitt, R. R. Coltman, and J. K. Redman, *J. Appl. Phys.*, vol. 28, p. 651, 1957.

[3] Some examples are: H. Suzuki and C. S. Barrett, *Acta Met.*, vol. 6, p. 156, 1958 (Ag, Ag-Au, Au). P. Haasen, *Phil. Mag.*, vol. 3, p. 384, 1958 (Ni). P. R. Thornton and T. E. Mitchell, *Phil. Mag.*, vol. 7, p. 361, 1962 (alpha-brass). P. Haasen and A. King, *Z. Metallk.*, vol. 51, p. 722, 1960 (Cu-Ge, Cu-Ga).

[4] C. S. Smith, *Trans. AIME*, vol. 214, p. 574, 1958. C. S. Smith and C. M. Fowler in "Response of Metals to High Velocity Deformation," p. 309, Interscience, New York, 1961. R. J. De Angelis and J. B. Cohen in R. E. Reed-Hill, H. C. Rogers, and J. P. Hirth (eds.), "Deformation Twinning," p. 430, Gordon and Breach, New York, 1964.

twin at 0°C even in slow tensile tests, whereas progressively lower temperatures were required for Ag-Au solid solution crystals as the gold content was increased, reaching temperatures in the 78°K range for crystals of gold. The twins, of macroscopic width, contained many individual twin lamellae.

Twinning during deformation is usually accompanied by a sharp click or series of clicks (the "cry" of tin is an example). The time required to form a twin in polycrystalline tin has been found to be of the order of 30 μsec, with a fine structure present having a period of 3 μsec.[1] In single crystals of some metals (e.g., Sn[2] and Zn[3]) a definite expenditure of energy per unit volume twinned is observed in experiments on twinning by impact. The strains accompanying heating and cooling of polycrystalline metals having anisotropic thermal expansion will often cause twinning (e.g., in Zn, Cd, and Sn,[4] and also in α–U,[5] which is particularly subject to twinning). In a number of instances it has been observed that reversing the stresses that caused a twin will cause untwinning.

There is now some quantitative evidence that there is a resolved shear stress criterion for twinning, at least in some materials,[6] although not all experimenters have concluded so. The stress required for nucleation is presumably higher than that for propagation whenever there is an abrupt drop in load with the onset of twinning, but when there is no accompanying load drop the twins can be seen to widen gradually. In these cases propagation requires higher stresses than nucleation.[7] That the twinning stress decreases with decreasing temperature is suggested by some experiments, but the variation is small and difficult to establish with certainty. However, recent experiments[8] clearly indicate a correlation of twinning stress with stacking-fault energy in copper alloys (Cu–Zn, Cu–Al, and Cu–Ge).

Processes that cause tangled dislocations at cell walls tend to inhibit twinning; thus, columbium (b.c.c.), deformed at room temperature to produce a cell structure, resists twinning during subsequent deformation at low temperatures. Interstitial impurities likewise impede twinning, perhaps by controlling the stress concentrations reached in dislocation pileups.[9]

Theories of the mechanism of deformation twinning Many ingenious models have been devised to provide the dislocations that must move through a crystal to generate a deformation twin. A partial dislocation

[1] W. P. Mason, H. J. McSkimin, and W. Shockley, *Phys. Rev.*, vol. 73, p. 1213, 1948.

[2] B. Chalmers, *Proc. Phys. Soc. (London)*, vol. 47, p. 733, 1953.

[3] E. I. Salkovitz, *Phys. Rev.*, vol. 85, p. 1046, 1952.

[4] W. Boas and R. W. K. Honeycombe, *Proc. Roy. Soc. (London)*, vol. A186, p. 57, 1946; vol. A188, p. 427, 1947.

[5] R. W. Cahn, *Acta Met.*, vol. 1, p. 49, 1953.

[6] See, for examples: E. O. Hall, "Twinning and Diffusionless Transformations," Butterworth, London, 1954. B. Thompson and D. J. Millard, *Phil. Mag.*, vol. 43, p. 421, 1952. J. J. Gilman, *Nature (London)*, vol. 169, p. 149, 1952. J. J. Gilman and T. A. Read, *J. Metals*, vol. 4, p. 875, 1952. E. A. Gyndyn and V. E. Startsev, *J. Exp. Theoret. Phys.*, vol. 20, p. 738, 1950. R. King, *Nature (London)*, vol. 169, p. 543, 1952. C. S. Barrett and H. Suzuki, *Acta Met.*, vol. 6, p. 156, 1958. P. B. Price, *Proc. Roy. Soc. (London)*, vol. A260, p. 251, 1961.

[7] J. A. Venables in R. E. Reed-Hill, H. C. Rogers, and J. P. Hirth (eds.), "Deformation Twinning," Gordon and Breach, New York, 1964.

[8] P. R. Thornton and T. E. Mitchell, *Phil. Mag.*, vol. 7, p. 361, 1962.

[9] J. O. Stiegler, C. K. H. DuBose, and C. J. McHargue, *Acta Met.*, vol. 12, p. 263, 1964. C. J. McHargue, *Trans. AIME*, vol. 224, p. 234, 1962.

must be nucleated on every lattice plane, or a single dislocation must somehow move from one plane to the next as first proposed by Cottrell and Bilby in their *pole mechanism.*[1] In the pole mechanism, a *pole dislocation* is assumed to exist which has a component of the Burgers vector **b** normal to the (112) plane of a b.c.c. crystal. A partial dislocation on the (112) plane rotates around this pole with one end tied to it and climbs one layer per resolution. By doing so it produces twinning in successive layers, thus generating a homogeneous twin. A number of mechanisms have been proposed that are more or less closely related to this first model,[2] but not all are based on a pole model.[3] At least two of the recent transmission electron microscope studies of deformation twinning have led to the conclusion that twins grow by repeated nucleation mainly at stress concentrators or obstacles to dislocation motion where dislocations pile up during slip.[4,5] Price[5] has observed deformation twinning in whiskers of tin that were completely lacking in prior dislocations and concluded that a pole mechanism could not be involved. Whatever the mechanism, in many types of crystals it involves, in addition to shear, a "shuffle" consisting of local rearrangement of some of the atom positions.[6]

The reasons for one twinning mode (twin system) to operate in preference to another are presumed to include one or more of the following principles: (1) the homogeneous shear is minimized (thus minimizing the accommodation slip or accommodation twinning in the twin and its surroundings); (2) the shuffle is avoided or at least minimized and confined to a small fraction of the atoms; (3) coordination and interatomic distances at the interface are nearly normal (thus minimizing the interface energy); (4) order in a superlattice is undisturbed; (5) resolved shear stress reaches the critical value.[7] That the disruption of long-range order tends to suppress

[1] A. H. Cottrell and B. A. Bilby, *Phil. Mag.*, vol. 42, p. 573, 1951.

[2] N. Thompson and D. J. Millard, *Phil. Mag.*, vol. 43, p. 422, 1952 (a model for Cd). P. B. Hirsch, A. Kelly, and J. W. Menter, *Proc. Phys. Soc.* (*London*), vol. B68, p. 1138, 1955. D. G. Westlake, *Acta Met.*, vol. 9, p. 327, 1961 (a model for Zr; see also comments by R. E. Reed-Hill, W. A. Slippy, Jr., and L. J. Buteau, *Trans. AIME*, vol. 227, p. 977, 1963). H. Suzuki and C. S. Barrett, *Acta Met.*, vol. 6, p. 156, 1958 (for f.c.c.). J. A. Venables, *Phil. Mag.*, vol. 6, p. 379, 1961.

[3] J. B. Cohen and J. Weertman, *Acta Met.*, vol. 11, pp. 996, 1368, 1963.

[4] J. T. Fourie, F. Weinberg, and F. W. C. Boswell, *Acta. Met.*, vol. 8, p. 851, 1960.

[5] P. B. Price, *Proc. Roy. Soc.* (*London*), vol. A260, p. 251, 1961.

[6] This was noted in various examples: C. S. Barrett in "Cold Working of Metals," American Society for Metals, Cleveland, Ohio, 1948 (in Mg). R. W. Cahn, *Acta Met.*, vol. 1, p. 49, 1953 (in α-U). F. Laves, *Naturwissenschaften,* vol. 30, p. 546, 1952 (in feldspars). R. W. Cahn, *Phil. Mag.*, Suppl., vol. 3, p. 363, 1954. H. S. Rosenbaum in R. E. Reed-Hill, H. C. Rogers, and J. P. Hirth (eds.), "Deformation Twinning," p. 43, Gordon and Breach, New York, 1964 (c.p.h. metals).

[7] M. A. Jaswon and D. B. Dove, *Acta Cryst.*, vol. 9, p. 621, 1956; vol. 10, p. 14, 1957; vol. 13, p. 232, 1960. H. Kiho, *J. Phys. Soc. Japan*, vol. 9, p. 739, 1954; vol. 13, p. 269, 1958. A. G. Crocker, *J. Iron Steel Inst.*, vol. 198, p. 167, 1961; *Acta Met.*, vol. 10, p. 113, 1962; *Phil Mag.*, vol. 7, p. 1901, 1962. See also reviews cited under the section on crystallography of twins.

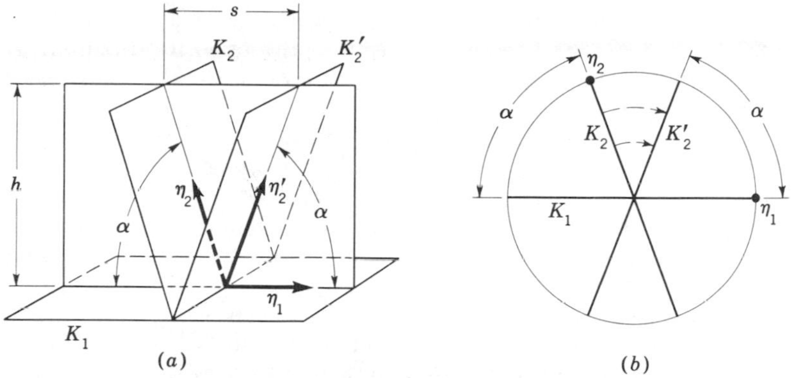

Fig. 14-19 Crystallographic twinning elements. Shear of magnitude s/h during deformation twinning causes plane K_2 to shear over to position K_2'; K_1 remains invariant.

twinning, as originally suggested by Laves, now has been verified experimentally in certain cases, though not in all.[1]

Shear in deformation twinning Figure 14–19 illustrates two important reference planes to be considered in twinning. The shear associated with twinning leaves two lattice planes undistorted—i.e., all distances and angles in these planes are left unchanged by the twinning shear. One plane is K_1, the shear plane, which contains the shear direction η_1. The other is K_2 before twinning, K_2' after twinning. As indicated in the sketch and the corresponding stereographic projection in Fig. 14–19, the angle between the two undistorted planes, α, is unchanged by the twinning and is related to the magnitude of the shear strain. If a point at a height h above K_1 is moved a distance s by the twinning shear so that the shear strain is s/h, it follows that $s/h = 2 \tan (90 - \alpha)$.

The planes K_1 and K_2 divide the stereographic plot of Fig. 14–19 into four segments. Any crystal direction in the upper left segment between K_1 and K_2 is *shortened* by the shear. Any direction in K_2 is undistorted, and any direction in the segment at the upper right is *lengthened.* If a tension test causes a crystal to twin, the length l_0 in the twinned volume will be elongated to the length $l = l_0(1 + 2S \sin \chi \cos \lambda + S^2 \sin^2 \chi)^{1/2}$, where χ and λ are the angles from the direction of l to K_1 and to η_1, respectively, and S is the shear strain.[2]

The homogeneous shear varies with the axial ratio of hexagonal crystals from about 0.186 for metals with c/a near 1.585 (Be, Ti, Zr, Hf, Ru) and

[1] F. Laves, *Naturwissenschaften*, vol. 39, p. 546, 1952. R. W. Cahn and J. A. Coll, *Acta Met.*, vol. 9, p. 138, 1961. R. H. Richman in Reed-Hill, Rogers, and Hirth (eds.), op. cit., p. 236.

[2] E. Schmid and G. Wassermann, "Plasticity of Crystals," Hughes, London, 1950; *Z. Physik*, vol. 48, p. 370, 1928. E. O. Hall, "Twinning and Diffusionless Transformations," Butterworth, London, 1954.

0.131 for Mg with $c/a = 1.624$ ($\cong$ Co, Cr, La, Ce, Pr, Er, Ca) to zero for $c/a = 1.732$; it becomes a shear in the opposite direction for ratios above this, reaching 0.143 and 0.175 for Zn and Cd of $c/a = 1.856$ and 1.886, respectively.

The magnitude of this homogeneous shear governs the maximum amount of deformation that can be provided by twinning alone. For example, the elongation of a zinc crystal when completely converted to a twin can amount to 7.39 percent at most. The direction of shear determines whether or not a stress applied in a given direction will cause twinning. Thus, compression parallel to the basal plane of magnesium, beryllium, or titanium will cause twinning because twinning leads to shortening of the axis of compression. A rolled (or rolled and recrystallized) plate of any of these metals is twinned by edgewise compression, since the grains in the plate lie with their basal planes parallel to the plane of the plate. Bending such a plate will twin the metal on the inside of the bend. (Unbending will also untwin this region.)[1] On the other hand, compression *normal* to a plate of rolled zinc or cadmium tends to twin it.

The crystallographic elements of twins The well-defined crystallographic features of twins have been reviewed in detail by Cahn[2] for both metallic and nonmetallic structures; other reviews have been published for metals[3] and for minerals.[4] The formal crystallographic requirements for twinning in the various crystal systems have been tabulated by Donnay and Donnay.[5] The twinning modes have been determined in a great variety of crystals.[6]

Refer again to Fig. 14–19 and consider the plane that lies normal to the twinning plane K_1 and that contains the vector η_1. This is the shear plane. Along the intersection of this shear plane with K_2 is an important vector η_2, which is at angle α from K_1. Twinning causes the movement of η_2 to η_2' and leaves η_2 at angle α from K_1. Now η_2 is the *only* direction in K_2 that makes the same angle before and after twinning with *any* direction in K_1. If we assume that K_1 is a rational plane and η_2 is a rational direction in the

[1] C. S. Barrett and C. T. Haller, *Trans. AIME*, vol. 171, p. 246, 1947.

[2] R. W. Cahn, *Phil. Mag.*, Suppl., vol. 3, p. 363, 1954.

[3] Hall, *loc. cit.* R. Clark and G. B. Craig in "Progress in Metal Physics," vol. 3, Pergamon, London, and Interscience, New York, 1952. C. G. Mathewson, *Trans. AIME*, vol. 78, p. 7, 1928. M. V. Klassen-Neklyudova, "Mechanical Twinning of Crystals," Consultants Bureau, New York, 1964 (transl. from Russian ed. of 1960). Reed-Hill, Rogers and Hirth (eds.), *loc. cit.*

[4] H. Tertsch, "Die Festigkeitserscheinunger," Springer, Berlin, 1949.

[5] J. D. H. Donnay and G. Donnay in "International Tables for X-ray Crystallography," vol. 2, p. 101, Kynoch, Birmingham, England, 1959.

[6] Some stereographic projection manipulations of value in studying shears and orientation relationships are given in Chap. 2; selected-area electron diffraction is a powerful tool recently developed (Chap. 15); for classical methods of studying twins, see the previously cited books by E. Schmid and W. Boas and by E. O. Hall, and various texts on mineralogy. For x-ray methods, see R. J. DeAngelis and J. B. Cohen in Reed-Hill, Rogers, and Hirth (eds.), op. cit, p. 430.

lattice (i.e., their indices are rational numbers), it is then possible for them to both be rational after the twinning shear. When this criterion is met and the other twinning elements K_2 and η_1 are irrational, the twin is said to be of type I. If, on the other hand, η_1 and K_2 are the only rational elements, the twin is of type II. In many of the more symmetrical crystals, *all* four elements K_1, K_2, η_1, and η_2 are rational (these are called *twins of compound type*). Crocker has pointed out[1] that if a twin again twins, the *doubly* twinned region should have crystallographic twinning elements equal to those of a *singly* twinned region in which the four single twinning elements are *all irrational*. However, the predictions of such a theory include a range of uncertainty in resulting reorientations because accommodation by slip may occur in the twin, in the untwinned matrix, or partly in one and partly in the other, in unpredictable fashion.

The reorientations resulting from twinning can be geometrically determined from the twinning elements. In the cubic metals, twins on (111) and on (112) both yield the same new orientations: there are a total of 4 new first-order twin orientations; if each of these is twinned, there are a total of 12 new second-order twin orientations.[2] In more general terms, in any twin of the common type I, the lattice reorientation is equivalent to a 180° rotation about the pole of the twinning plane K_1; in any twin of type II, a type encountered only with lower-symmetry crystals, the reorientation is equivalent to a 180° rotation about η_1.

Regardless of the type, twinning leaves the unit cell unchanged in size, shape, and structure, but altered in orientation; any alteration in *structure* by a twinlike deformation is properly classed as a transformation rather than a twinning (but the transformation mechanism may involve some twinning, as discussed in the chapter on transformations).

The interface between twin and matrix may be parallel or nearly parallel to K_1, as it is in deformation twins, or portions of the interface may be a plane of higher indices (and higher energy), as they are twins formed in metals during annealing.

Coherent twin boundaries have surface energies of the order of 2 to 25 percent of the energy of an ordinary, high-angle grain boundary. These relative energies are basic to theories of the origin of *annealing twins*. In one model, annealing twins are nucleated at the time when a grain in a polycrystalline metal increases its number of sides and when, by inserting a twin in the corner that is becoming the additional side, the total boundary energy of the piece is lessened.[3]

There is a close relation between the geometry of twin boundaries and stacking faults. This is particularly striking in f.c.c. metals. Neglecting the interaction energy between adjacent twin boundaries, the twin-boundary energy is half the stacking-fault energy.

[1] A. G. Crocker, *Phil. Mag.*, vol. 7, p. 1901, 1962.

[2] First order twins have one (111) plane, three (110) planes, three (112) planes, and many others in common with the parent crystal. A. B. Greninger, *Trans. AIME*, vol. 120, p. 293, 1936. C. G. Dunn, *Trans. AIME*, vol. 161, p. 90, 1945. C. H. Mathewson, *Trans. AIME*, vol. 78, p. 7, 1928 (plot for zinc). Matrices (Chap. 1) are useful for twinning problems.

[3] J. E. Burke and D. Turnbull in "Progress in Metal Physics," vol. 3, p. 220, 1952, Interscience, New York. R. L. Fullman, *J. Appl. Phys.*, vol. 22, p. 1350, 1951. K. T. Aust and J. W. Rutter, *Trans. AIME*, vol. 218, p. 1023, 1960.

Hence, the method of estimating twin-boundary energies from dihedral angles formed between twin boundaries and other grain boundaries[1] provides a means of estimating the stacking-fault energy.

Electron microscope pictures[2] of an Fe–78 percent Ni alloy show that annealing twins are nucleated at grain boundaries; it is suggested that the nucleus of a twin is a stacking fault, or possibly a thin packet of stacking faults, that "literally explodes" into the recrystallized grain from a migrating grain boundary during the intense release of energy at the boundary. The nucleus acts as a mechanism for this energy release. Once the first sheet of a twin is formed, contiguous sheets can form relatively easily.

The edge of the twin probably consists of all three of the possible (a/6) ⟨112⟩-type partial dislocations on the twinning plane, with the net Burgers vector of the boundary being zero. There is thus no net shear strain when the twin grows edgewise.

Observed twinning elements In b.c.c. metals, the twinning plane K_1 is nearly always (112) and the direction of shear η_1 is always $[11\bar{1}]$ (or equivalent plane and direction). In f.c.c. metals, the mode is always $(111)[11\bar{2}]$. In the rhombohedral structures (As, Sb, and Bi), the mode is always $(110)[00\bar{1}]$, and in germanium (diamond cubic structure), it is $(111)[11\bar{2}]$. Twinning modes in hexagonal, tetragonal, and orthorhombic metals are more complicated, and they vary from metal to metal and also between a metal and its alloys. New modes are continually being found, and various rare modes are not yet widely accepted as firmly established, either because of insufficient or ambiguous data, or because the investigator has not given proper attention to the possibility of double twinning. A summary of the more firmly established twinning elements is presented in Table 14–2.

Cleavage The commonly accepted cleavage planes in some common crystals are listed in Table 14–3. Magnesium does not appear in the table because confusion as to its true cleavage planes has apparently been caused by the presence of deformation twins which fracture internally.[3] Habit planes for fracture have appeared in some tests to be $\{11\bar{2}4\}$ and in others[4] to be $\{10\bar{1}1\}$.

Low temperatures, high strain rates, and stress concentrations tend to favor cleavage by raising flow stress. There is considerable evidence that the requirement for opening a cleavage crack in a brittle crystal involves the component of stress normal to a cleavage plane; for cleavage this component must reach a critical value.[5] The initiation of cleavage is a

[1] R. Fullman, *J. Appl. Phys.*, vol. 22, p. 448, 1951. M. C. Inman and H. R. Tipler, *Met. Rev.*, vol. 8, p. 105, 1963. G. F. Bolling and J. Winegard, *J. Inst. Metals*, vol. 86, p. 492, 1957.

[2] S. Dash and N. Brown, *Acta Met.*, vol. 11, p. 1067, 1963.

[3] R. E. Reed-Hill and W. D. Robertson, *Acta Met.*, vol. 5, p. 728, 1957.

[4] J. G. Byrne in R. E. Reed-Hill, H. C. Rogers, and J. P. Hirth (eds.), "Deformation Twinning," p. 397, Gordon and Breach, New York, 1964. H. Asada and H. Yoshinaga, *J. Japan Inst. Met.*, vol. 23, p. 649, 1959.

[5] E. Schmid and W. Boas, "Kristallplastizität," Springer, Berlin, 1935; in English transl. Hughes, London, 1950, *loc. cit.*; M. Georgieff and E. Schmid, *Z. Physik*, vol. 36, p. 759, 1926. This law was probably first stated by L. Sohnke, *Poggendorf's Ann.*, vol. 137, p. 177, 1869, as applied to rock salt.

Table 14–2 Principal twinning elements for metals[a]

Crystal structure	Twinning plane, K_1	Twinning direction, η_1	Second undistorted plane, K_2	Direction η_2	Shear[b]
B.c.c.	$\{112\}$	$\langle 11\bar{1}\rangle$	$\{11\bar{2}\}$	$\langle 111\rangle$	0.707
F.c.c.	$\{111\}$	$\langle 11\bar{2}\rangle$	$\{11\bar{1}\}$	$\langle 112\rangle$	0.707
Rhombohedral (As, Sb, Bi)	$\{110\}$	$\langle 00\bar{1}\rangle$	$\{001\}$	$\langle 110\rangle$	[b]
All c.p.h.	$\{10\bar{1}2\}$	$\langle \bar{1}011\rangle$	$\{\bar{1}012\}$	$\langle 10\bar{1}1\rangle$	[c]
Some c.p.h.[d]	$\{11\bar{2}1\}$	$\langle 11\bar{2}\bar{6}\rangle$	$\{0001\}$	$\langle 11\bar{2}0\rangle$	
	$\{11\bar{2}2\}$	$\langle 11\bar{2}\bar{3}\rangle$	$\{11\bar{2}\bar{4}\}$	$\langle 22\bar{4}3\rangle$	
	$\{11\bar{2}3\}$	—	—	—	
Hg[e]	$\{135\}$	$\langle \bar{1}21\rangle$	$\{\bar{1}11\}$	$\langle 0\bar{1}1\rangle$	
Diamond cubic (Ge)	$\{111\}$	$\langle 11\bar{2}\rangle$	$\{11\bar{1}\}$	$\langle 112\rangle$	0.707
Tetragonal (In, β-Sn)	$\{301\}$	$\langle \bar{1}03\rangle$	$\{\bar{1}01\}$	$\langle 101\rangle$	
	$\{101\}$	$\langle 10\bar{1}\rangle$	$\{\bar{1}01\}$	$\langle 101\rangle$	
Orthorhombic[f] α-U	$\{130\}$	$\langle 3\bar{1}0\rangle$	$\{1\bar{1}0\}$	$\langle 110\rangle$	0.299
	$\{172\}$[e]	$\langle 312\rangle$	$\{112\}$	[e]	0.228
	$\{112\}$	[e]	$\{172\}$[e]	$\langle 312\rangle$	0.228
	$\{121\}$	[e]	[e]	$\langle 311\rangle$	0.329
	$\{176\}$[e]	$\langle 5\bar{1}2\rangle$	$\{1\bar{1}1\}$	$\langle 123\rangle$[e]	0.214

[a] Data chosen mainly from E. O. Hall, "Twinning and Diffusionless Transformations,' Butterworth, London, 1954; and from R. E. Reed-Hill, H. C. Rogers, and J. P. Hirth (eds.), "Deformation Twinning," Gordon and Breach, New York, 1964, with other data as noted.

[b] The rhombohedral shear is $(2 \cos \alpha)/\sin (\alpha/2)$ where α is the rhombohedral angle; for Sb the shear is 0.34, for Bi, 0.11.

[c] Shears, S, for axial ratio c/a and twinning of mode $\{10\bar{1}2\} = K_1$ are

	c/a	S		c/a	S
Cd	1.886	0.175	Zr	1.592	−0.169
Zn	1.856	0.143	Ti	1.587	−0.175
Mg	1.623	−0.131	Be	1.568	−0.186

The general law is $S = \{(c/a)^2 - 3\}/(c/a)\sqrt{3}$

[d] In addition to $\{10\bar{1}2\}$ twinning seen in all c.p.h. metals, various modes are reported. Those listed are for Zr. Ti, Mg, and Yt are similar to these, in general; but for Ti, see E. A. Anderson, D. C. Jillson, and S. R. Dunbar, *Trans. AIME*, vol. 197, p. 1191, 1953; also D. G. Westlake in Reed-Hill, Rogers, and Hirth (eds.), op. cit., p. 29; and H. Kiho, *J. Phys. Soc. Japan*, vol. 13, p. 269, 1958. For Mg, see R. E. Reed-Hill, *Trans. AIME*, vol. 218, p. 554, 1960; and R. E. Reed-Hill and W. D. Robertson, *Acta Met.*, vol. 5, p. 717, 1957. For Co, see K. G. Davis and E. Teghtsoonian, *Acta Met.*, vol. 10, p. 1189, 1962. In contrast to pure metals, $\{10\bar{1}2\}$ twins have not been observed in the deformation of c.p.h. alloy phases such as, for example, the Cu-Ge or Ag-Sn ζ phases; here the only twins found are $\{10\bar{1}1\}$ and $\{11\bar{2}1\}$. P. H. Thornton, *Acta Met.*, vol. 13, p. 611, 1965; vol. 14, p. 444, 1966.

[e] Irrational indices. For Hg see A. G. Crocker et al., *Phil. Mag.*, in press, 1966.

[f] Data from: R. W. Cahn, *Acta Met.*, vol. 1, p. 49, 1953. L. T. Lloyd and H. H. Chiswick, *Trans. AIME*, vol. 203, p. 1206, 1955.

Table 14–3 Cleavage planes in common types of crystals[a]

Structure	Examples	Cleavage plane
B.c.c. (*A*2)	Fe, W	(001)
C.p.h. (*A*3)	Cd, Zn, Be	(0001)
	ζ-Cu-Ge	$(10\bar{1}1)$[b]
Rhombohedral (*A*7)	As, Sb	(111), (110), $(11\bar{1})$
	Bi	(111), $(11\bar{1})$
Hexagonal (*A*8)	Te, Se	$(10\bar{1}0)$
NaCl (*B*1)	NaCl, LiF, MgO, KCl	(100)
Fluorite type (*C*1)	CaF_2	(111)
Zinc blende	ZnS, InS	(110)
Diamond (*A*4)	Ge, Si, C	(111)
Hexagonal graphite (*A*9)	C	(0001)

[a] Data are from "Progress in Metal Physics," vol. 5, Interscience, New York, 1954; and B. L. Averbach et al. (eds.), "Fracture," Wiley, New York, 1959.
[b] In c.p.h. single crystals of ζ-Cu-Ge phase, formation of $\{10\bar{1}1\}$ twins accompanied by {0001} slip causes cleavage along $\{10\bar{1}1\}$. P. H. Thornton, *Acta Met.*, vol. 13, p. 611, 1965.

complex matter, however, involving the interaction of groups of dislocations with barriers, the interaction of twins with barriers, or the presence of brittle constituents. The critical stress is probably, in general, a function of the amount of prior plastic strain and the degree of recovery from the plastic strain.

Cleavage is sometimes controlled by the twinning process: deformation twins may cause the stress concentrations that nucleate cracks; twins may provide the lattice orientations most appropriate for cracking; and parting may occur at twin boundaries. On the other hand, the occurrence of twinning may *hinder* crack nucleation and crack propagation, by providing plastic deformation that relieves stress concentrations and also by providing a barrier to crack propagation. In some b.c.c. materials that exhibit low-temperature brittleness, the onset of deformation twinning does not coincide with the increase in brittleness; in others the abrupt shear stress concentrated at the edge of a twin does cause crack nucleation.[1] There is a strong orientation dependence of the ductile-brittle transition temperature in alpha-iron single crystals[2]: all orientations of the tensile axis falling within 30 to 39° from [001] result in brittle behavior in tensile tests at −196°C. The particular deformation-twin intersections that could produce appreciable normal stress across a {100} cleavage plane in a b.c.c. crystal during a tensile test have been listed.[3] Such pairs occur in crystals with the tensile axis in the [001] direction or falling within a zone of directions extending from $[\bar{1}12]$ to $[\bar{1}11]$. Twin pairs have been discussed also in terms of *emissary dislocations*,[4] which are glissile dislocations emitted from the tip of a twin and which may interact with emissary disloca-

[1] B. L. Averbach et al. (eds.), "Fracture," Wiley, New York, 1959. R. Priestner in Reed-Hill, Rogers, and Hirth (eds.), op. cit., p. 321.

[2] N. P. Allen, B. E. Hopkins, and J. E. McLennan, *Proc. Roy. Soc.* (*London*), vol. A234, p. 221, 1956. W. D. Biggs and P. L. Pratt, *Acta Met.*, vol. 6, p. 694, 1958. B. Edmondson, *Proc. Roy. Soc.* (*London*), vol. A264, p. 176, 1961. J. J. Cox, G. T. Horne, and R. F. Mehl, *Trans. AIME*, vol. 49, p. 118, 1957.

[3] Priestner, loc. cit.

[4] A. W. Sleeswyk, *Acta Met.*, vol. 10, pp. 705, 803, 1962.

tions from another twin. Emissary dislocations may pile up also against other barriers and thus nucleate cleavage. The crack-nucleating propensity of twins is directly related to their thickness. Various mechanisms by which cracks may be nucleated have been proposed and criticized.[1]

A few general rules[2] about cleavage in nonmetallic lattices may be mentioned here. It is always found that layer lattices (such as graphite, MoS_2, $CdCl_2$, mica, $CaSO_4 \cdot 2H_2O$, and CdI_2) cleave parallel to the layers. Zn, Cd, As, Sb, and Bi can also be classed in this group (although other cleavages are sometimes found in addition). Cleavages never break up radicals in ionic crystals or molecular complexes in homopolar crystals, but within the limits imposed by these two restrictions the cleavage planes are those most widely spaced. In ionic crystals which have no radicals, cleavage occurs so as to expose planes of anions, where such planes exist. Cubic crystals of the chemical formula AX cleave on $\{100\}$ planes.

Dislocations and stress-corrosion cracking The dislocation distribution in deformed alloys is closely correlated with the mode of failure by stress-corrosion cracking.[3] Deformed alloys with a cellular arrangement of dislocation tangles, such as illustrated in Fig. 14–11, have a superior resistance to transgranular stress-corrosion cracking. On the other hand, alloys susceptible to this cracking characteristically have dislocations grouped on planes as either pileups or planar networks, such as that of Fig. 14–13, or more irregularly, as a result of extensive slip concentrated in a thin slip band. However, the presence of planar dislocation groups does not necessarily imply susceptibility to transgranular cracking.

The preferential grouping of dislocations on planes presumably requires the absence of extensive cross slip. A small or even a negligible tendency for cross slip is known to be characteristic of metals and alloys of low stacking-fault energy γ, and of alloys with short-range or long-range order. Such alloys are, in fact, found to be among those susceptible to transgranular cracking under stress-corrosion conditions.

Perhaps the active sites for the nucleation of stress-corrosion cracks are local regions where the motion of dislocations has caused such local disorder as to provide the conditions for localized chemical attack. Antiphase boundaries in highly ordered superlattices may also be chemically active sites in either deformed or annealed specimens—provided they form a connected network. Lack of a connected network of antiphase boundaries in materials such as beta-brass could account for the resistance to stress-corrosion that has been observed in such materials.

[1] Some reviews of the subject: B. L. Averbach et al. (eds.), "Fracture," Wiley, New York, 1959. D. C. Drucker and J. J. Gilman (eds.), "Fracture of Solids," Interscience, New York, 1963. R. E. Reed-Hill, H. C. Rogers, and J. P. Hirth (eds.), "Deformation Twinning," Gordon and Breach, New York, 1964. F. P. Bullen, *Trans. AIME*, vol. 227, p. 1069, 1963 (Zn crystals).

[2] W. A. Wooster, "Crystal Physics," Cambridge University, New York, 1938.

[3] P. R. Swann, *Corrosion*, vol. 19, pp. 102t–112t, March, 1963.

15

X-ray and electron microscopy

Many useful techniques have been developed in which crystals are photographed by processes involving the absorption or diffraction of x-rays or electrons. It is appropriate in this book to present the fundamentals of these techniques, especially those that depend directly on the principles of diffraction.

Direct photography of crystals by x-rays or electrons supplies needed information about the nature of inhomogeneities in crystals. The radiographic and diffraction methods are useful at magnifications below about 100 diameters; the electron microscope is useful at higher magnifications, up to about 150,000 diameters.

X-RAY METHODS

For a treatment of *microradiography* the reader is referred to the extensive reviews that have appeared elsewhere.[1] Modern microradiography includes both *contact* microradiography (with the film in contact with the specimen) and *projection* microradiography (with the specimen close to a point source of x-rays and the photographic film at some distance, thereby recording a magnified shadow image). With the sharply focused electron beams now available, a resolution of the order of 0.1 μ can be obtained. Since the absorption of x-rays in a specimen depends on the atomic number of the elements present, the variation of intensity from point to point in a radiograph reveals the variation in composition as well as the variation in mass per unit area in the specimen. Both absorption spectroscopy and fluorescent-emission spectroscopy are now employed in the field of microchemical analysis.

Electron probe microanalysis has rapidly advanced to the status of a major laboratory tool since its introduction by Castaing and Guinier.[2] The

[1] V. E. Cosslett, *Met. Rev.*, vol. 5, p. 225, 1960. J. J. Trillat, *Met. Rev.*, vol. 1, p. 3, 1956. V. E. Cosslett and W. C. Nixon, "X-ray Microscopy," Cambridge University, New York, 1960. V. E. Cosslett, A. Engström, and H. H. Pattee (eds.), "X-ray Microscopy and Microradiography," Academic, New York, and Academic, London, 1957. "X-ray Microscopy and X-ray Microanalysis," Proceedings of Stockholm Symposium, 1959, Elsevier, Amsterdam, 1960. B. Chalmers and A. G. Quarrell, "The Physical Examination of Metals," 2d ed., Arnold, London, 1960. W. C. Nixon, *Research*, vol. 8, p. 473, 1955.

[2] R. Castaing and A. Guinier, Proceedings of a Conference on Electron Microscopy at Delft, 1949, Delft, 1950. R. Castaing, National Bureau of Standards Symposium, 1951, U. S. National Bureau of Standards Circular 527, p. 305, 1954. L. S. Birks, "Electron Probe Microanalysis," Interscience, New York, 1963. R. Castaing in "Advances in Electronics and Electron Physics," vol. 13, p. 317, Academic, New York, 1960.

method employs a focused electron beam which is moved over the surface of the specimen to excite characteristic x-ray radiation. An analysis of the emitted radiation by x-ray spectrometry provides information about the chemical composition of the irradiated small area of the specimen; the resolution is a few cubic microns of material at each position of the electron beam. In the most advanced forms of electron probe apparatus the emitted radiation of a chosen wavelength is received by a counter and displayed on a viewing screen by a spot which scans the screen in synchronism with the scanning of the specimen by the focused electron beam. Alternatively, the screen displays an image that represents the intensity of the electrons scattered by the specimen. In either method the resolution obtainable is in the micron range, and the method is a nondestructive one, except for a slight surface burning and carbon deposition.

The most common applications include the studies of precipitates, segregates and inclusions, and the determination of solid solubilities, as well as the identification of phases and investigations of short-range diffusion and composition gradients.

Focused x-rays can be used to form micrographs. By using the diffraction of x-rays from a cylindrically bent crystal, it is possible to form a focused image of the characteristic radiation of a given wavelength emitted from a sample,[1] but resolution is generally poor owing to aberration and crystal imperfections. Better resolution—of the order of 1 μ—has been predicted and actually reached by focusing x-rays with curved mirrors.[2] The rays are incident on the mirrors at angles within the critical angle for the mirror material, which is less than 1° for most substances. The problem of producing mirrors of sufficiently accurate shape has thus far been an extreme hindrance to the development of such microscopes.

X-ray topographs (reflection images) Methods for obtaining images of a crystalline sample by using a diffracted beam of x-rays can reveal imperfections in crystals, twins, subgrain boundaries, strains, segregations, and transformation products. Table 15–1 summarizes the most used methods. These and others are discussed in the following sections. The use of topographs (sometimes called *x-ray reflection micrographs*) is steadily increasing, for they frequently save an experimenter many days of frustrating work by aiding him to select specimens that are free from trouble-causing imperfections of certain types.

The simplest, low-magnification methods Ordinary Laue spots made with general radiation from an ordinary x-ray tube can reveal by

[1] G. Gouy, *Ann. Phys.*, ser. 9, vol. 5, p. 241, 1926. H. H. Johann, *Z. Physik*, vol. 69, p. 185, 1931. Y. Cauchois, *Ann. Phys.*, ser. 11, vol. 1, p. 215, 1934; *Rev. Opt.*, vol. 29, p. 151, 1950. L. v. Hámos, *Am. Mineralogist*, vol. 23, p. 215, 1938; *J. Sci. Instr.*, vol. 15, p. 87, 1938.

[2] P. Kirkpatrik and A. V. Baez, *J. Opt. Soc. Am.*, vol. 38, p. 766, 1948. E. Prince, *J. Appl. Phys.*, vol. 21, p. 698, 1950. M. Montel, *Opt. Acta*, vol. 1, p. 117, 1954. J. Dyson, *Proc. Phys. Soc. (London)*, vol. 65, p. 580, 1952. See also recent symposia volumes on microscopy.

Table 15–1 Most-used x-ray topographic methods

Common Name	Nature	Characteristics; uses; approximate range of useful enlargement of images
Back-reflection Laue	Large Laue spots	Simple; shows subgrains, subboundary angles, twins and distortion within the (small) irradiated area. ×1 to ×5.
Schulz	Laue images, point-focus x-ray tube	Simple; as above but covering entire surface. ×1 to ×10.
Berg-Barrett	$K\alpha$ reflected image, line-focus tube	Simple; for subgrains, twins, imperfections, dislocations over entire surface of a crystal. ×1 to ×100.
Lang: section topograph	Stationary crystal and film, $K\alpha$ or $K\alpha_1$ transmitted image, point-focus tube	Uses special camera; for subgrain boundaries, dislocations, faults, within a section of a crystal. ×1 to ×300.
Lang: projection topograph	As above, but crystal and film shifted	As above, for surveying entire interior; limited to lower dislocation densities. ×1 to ×300.
Bonse	Uses crystal monochromatized $K\alpha$	Uses 2-crystal apparatus; for strain field of individual dislocations and lattice tilts as small as 0.1 sec. ×1 to ×50.
Guinier-Tennevin	Focused transmission Laue image	Uses point focus tube; semitopographic; for lattice tilts as small as 10 sec. ×1 to ×5.
Borrmann	Anomalous transmission in highly perfect crystal	See Fig. 15–7; dislocations and other imperfections cast shadows. ×1 to ×50.

their internal structure major distortions, subgrains, twins, and inclusions. In general, however, only a small area of the crystal is irradiated in either transmission or back-reflection Laue patterns.[1]

Schulz[2] showed that it was useful to irradiate a considerable area on the surface of a crystal, then with the surface-reflection arrangement to record the reflected Laue pattern on a film several centimeters from the crystal. Parts of the crystal that differ slightly in orientation from others send out diffracted beams at an angle to the others and produce fragmented Laue spots. Orientation differences at a subgrain boundary can thus be de-

[1] Some early use was made of the method, however, with transmission through quartz crystals: C. S. Barrett, *Phys. Rev.*, vol. 38, p. 832, 1931. S. Nishikawa, Y. Sakisaka, and I. Sumoto, *Phys. Rev.*, vol. 31, p. 1078, 1931. C. S. Barrett and C. E. Howe, *Phys. Rev.*, vol 39, p. 889, 1932.

[2] L. G. Schulz, *Trans. AIME*, vol. 200, p. 1082, 1954.

termined from the images, though the *Schulz photographs* lack high resolution for details.

Berg[1] produced interesting images of the cleavage surface of rock salt by using a source of x-rays distant from the crystal and a photographic film quite close to the crystal and parallel to the surface. The geometrical arrangements for polychromatic and for monochromatic radiation are indicated in Fig. 15–1 by the upper and lower drawings, respectively. With either of these arrangements, each point on the surface of a perfect crystal sends a diffracted ray a short distance to a single point on the film placed parallel to the surface, so as to form an image or topograph of the surface.

Another topographic arrangement, which utilizes characteristic radiation diverging from a pinhole or small focal spot, consists of a crystal and a film mounted together on a carriage that oscillates a few degrees, the range of oscillation serving in the same way as the slit source in Fig. 15–1 to provide for a Bragg reflection of the various points on the specimen surface.[2]

Berg-Barrett method A convenient modification of the technique originated by Berg, which has frequently been called the *Berg-Barrett method*, has been rather widely used. A crystal is mounted on a tilting stage or goniometer and brought into reflecting position for $K\alpha$ radiation from a standard x-ray tube by watching for the reflection on a fluorescent screen, or by detecting it in a counter. The screen is then replaced by a fine-grained film or plate placed (usually) very close to the crystal, as indicated in Fig. 15–2. The crystal is stationary during the exposure. The conditions for sharp detail in the reflected image are[3] (1) a minimum distance between crystal and film—preferably under a millimeter, (2) a large distance from x-ray tube to crystal, or a tube oriented so that the focal spot is foreshortened into a narrow line corresponding to the lower drawing of Fig. 15–1, (3) an x-ray tube with a target preferably of low atomic number such as Cr, Fe, Co, or Cu to minimize penetration into the crystal, (4) a specimen surface undistorted by sawing, abrasion, or mechanical polishing, (5) a film or plate with fine-grained emulsion, which is shielded from the direct radiation. Under these conditions, fine-grained dental x-ray film and lantern-slide emulsions require exposure times of a few minutes and permit enlargements of 10 or 20 diameters; high-resolution spectroscopic plates or nuclear plates, with a resolving power of 1000 lines per millimeter, require several hours of exposure but produce images that can be enlarged optically to useful magnifications of about 100 diameters. If a block of polycrystalline material is used instead of a single crystal, occasional grains may find themselves in reflecting position at any arbitrary setting of the block, and the reflected rays leaving the surface will yield images of these; the reflected beams most nearly perpendicular to the photographic plate yield the sharpest and least distorted images.

[1] W. Berg, *Z. Krist.*, vol. 89, p. 286, 1934; *Naturwissenschaften*, vol. 19, p. 391, 1931.

[2] N. Wooster and W. A. Wooster, *Nature*, vol. 155, p. 786, 1945. C. S. Barrett, *Trans. AIME*, vol. 161, p. 15, 1945.

[3] C. S. Barrett, *Trans. AIME*, vol. 161, p. 15, 1945.

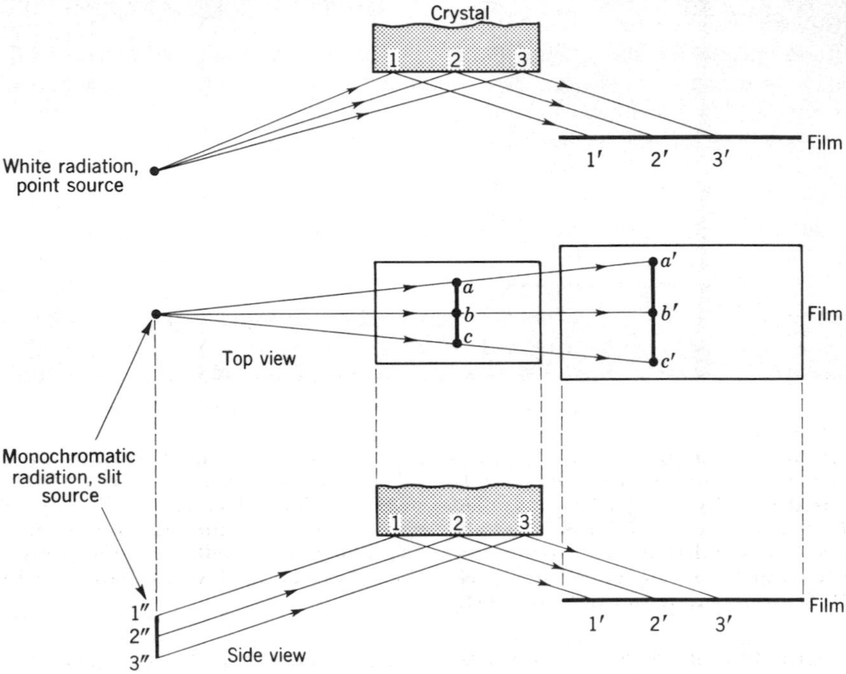

Fig. 15-1 Berg's two arrangements for x-ray topographs. Points on crystal surface, 1, 2, 3, *a*, *b*, *c*, reflect to corresponding points on the film: 1′, 2′, 3′, *a*′, *b*′, *c*′. The upper drawing is for general radiation, the lower for characteristic.

Examples of x-ray reflection micrographs (topographs) made by the reflection of characteristic radiation from a standard type of x-ray tube are reproduced in Fig. 15–3.

Variations in intensity in the x-ray image can be caused by variations either in the orientation or spacing of the reflecting planes, or by variations

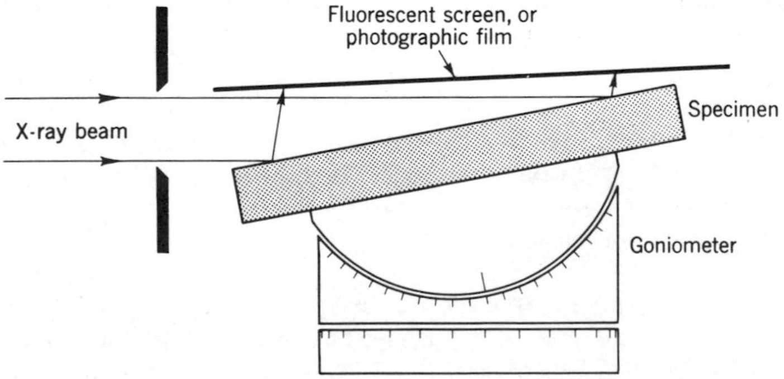

Fig. 15-2 A crystal (or polycrystalline specimen) mounted on a goniometer in position for Berg-Barrett photographs.

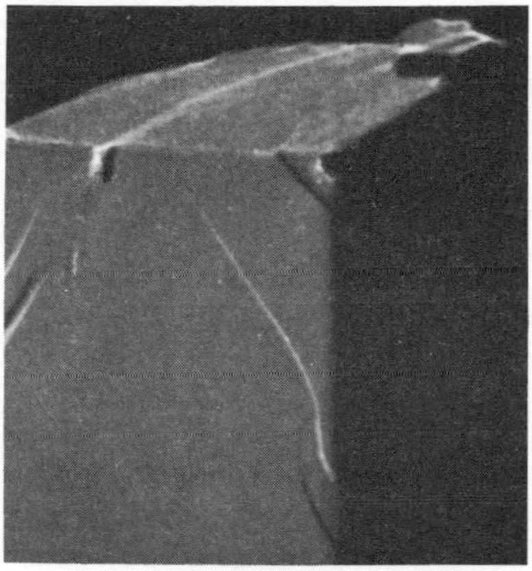

(*a*)

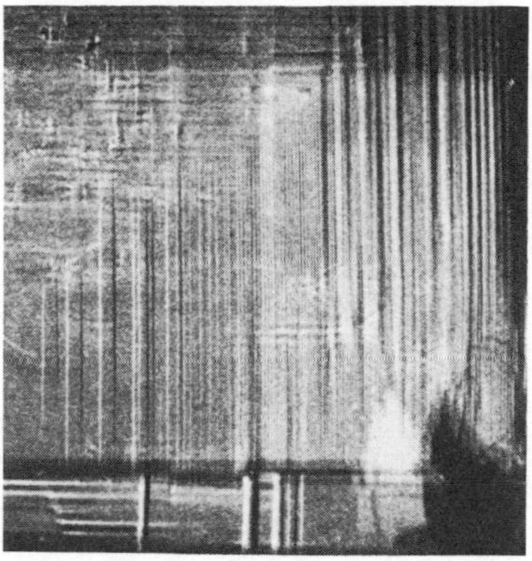

(*b*)

Fig. 15-3 X-ray topographs by the Berg-Barrett method. (*a*) Zirconium crystal with subboundaries, reflection from front and top faces. The crystal appeared visually to be a flawless block. Lantern-slide plate, $\times\sim 8$. (*b*) Crystal of V_3Si after cooling to 4.2°K. The cubic crystal of *A*15 structure has transformed to a tetragonal structure with $c/a = 1.0025$ and twin lamellae have appeared, tilted tenths of a degree from each other. Fine-grained x-ray film. $\times\sim 15$. (B. W. Batterman and C. S. Barrett.)

in extinction of the rays in the crystal. Extinction effects become relatively more important when intense reflections are used and when crystal perfection is high, since extinction arises from a shading of underlying layers of a crystal due to reflection of a portion of the incoming rays by the overlying layers. With weak reflections, on the other hand, contrast effects should be caused chiefly by orientation or spacing variations in the sample, due to portions of the sample being partially or fully out of reflecting position as compared with neighboring material.

Even relatively low-resolution x-ray topographs, such as those of Fig. 15–3, can reveal twins, included grains, deformation bands, lineage structure resulting from the freezing process, subgrains, slip lines, and effects from segregation and from local strain. Higher-resolution images clearly reveal individual dislocations[1] if the dislocation density is not too high, and if the Burgers vector does not lie in the reflecting plane. Just as in electron microscope pictures (see later), the contrast increases with the product $\mathbf{g}\cdot\mathbf{b}$, where $\mathbf{g}$ is the diffraction vector (normal to the diffracting plane) and $\mathbf{b}$ is the Burgers vector.

Improvement in resolution may be obtained with small crystals such as whiskers if a narrow (0.1-mm) x-ray beam is used and only the $K\alpha_1$ radiation is diffracted, as shown by Webb,[2] who obtained better than 5-μ resolution with high-resolution spectroscopic plates, and 5- to 10-μ resolution with nuclear emulsions 25 to 50 μ thick requiring exposures of only a few minutes with Mo $K\alpha$ or Ag $K\alpha$ radiation. Thinner nuclear emulsions are recommended for use with radiations of longer wavelength. For maximum

[1] J. B. Newkirk, *Phys. Rev.*, vol. 110, p. 1465, 1958; *Trans. AIME*, vol. 215, p. 483, 1959. M. Yoshimatsu and K. Kohra, *J. Phys. Soc. Japan*, vol. 15, p. 1760, 1960.

[2] W. W. Webb, *J. Appl. Phys.*, vol. 31, p. 194, 1960.

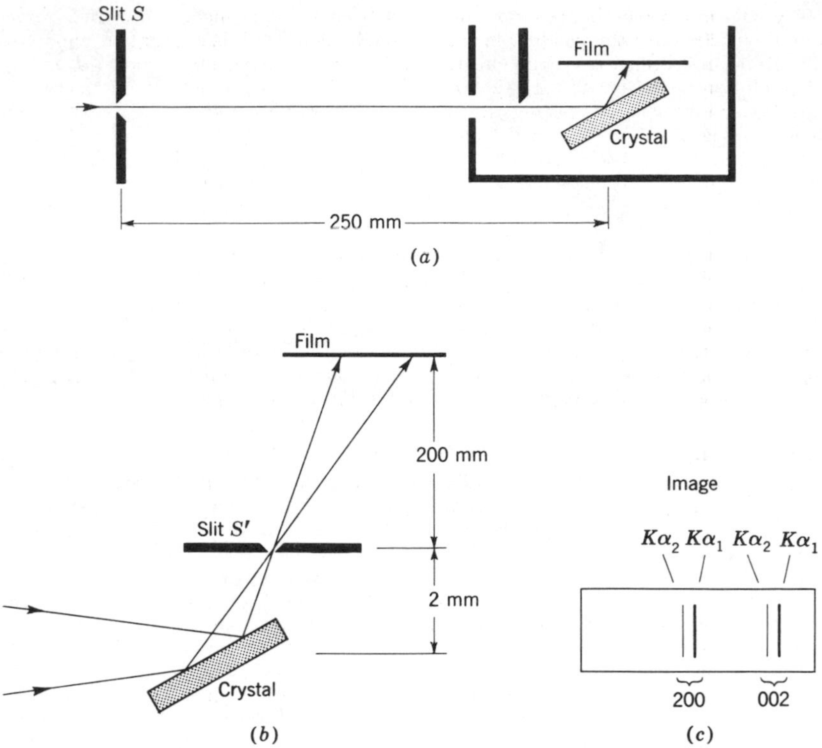

Fig. 15-4 Modifications of Berg-Barrett method. (*a*) Narrow slit *S* positioned for detecting slight disorientations. (*b*) Converging rays and narrow slit *S*′ for detecting disorientations and changes in interplanar spacings. (*c*) *K*α doublets displaced by spacing differences between (200) and (002) planes in $BaTiO_3$, in camera of type *b*. (After P. C. Bousquet et al.)

sharpness in registering dislocations, it is not desirable to collimate the incident beam to the extent that gives maximum sensitivity to small strains, since the strain field extends out many tens of microns from a dislocation; therefore, some of the simpler experimental arrangements serve well. Surface details such as scratches, steps, ridges, pits, and irregular oxide layers also appear in the images and require attention if they are to be differentiated from internal defects.

Additional refinements of the Berg–Barrett method have been developed[1] by using the arrangements indicated in Fig. 15–4. These have been applied to relatively imperfect crystals in which image contrast due to variation in extinction is absent and all contrast is due to orientation and spacing variations. In the arrangement of Fig. 15–4*a*, the narrow slit *S* (about 0.1 by 2 mm) collimates the beam so that the angle of incidence of rays on the crystal is determined for each point on the crystal with a precision of about 1.5 sec—with respect to rotation about an axis *normal* to the plane of incidence of the beam (the plane of the drawing). The camera is thus sensitive to these orientations in the crystal, but is increasingly insensitive to disorientations consisting of rotations about axes that deviate more and more from this position. This variation in sensitivity can aid in interpreting the nature of the disorientations in a specimen if photographs are made with different orientations of the specimen.

The arrangement in Fig. 15–4*b* permits deduction of additional information. A narrow

[1] P. C. Bousquet et al., *Acta Cryst.*, vol. 16, p. 989, 1963.

slit (say 0.03 mm wide) is placed about 2 mm from the specimen, and a fine-grained film at about 200 mm; the incident beam, although collimated, is convergent enough to permit Bragg reflections from the substructure elements being investigated. Point-to-point resolution in the image with the arrangement in Fig. 15–4*b* is low, but sensitivity to variations in interplanar spacing d is high, with the result that a pattern such as that sketched in Fig. 15–4*c* is obtained from barium titanate, in which alternate lamellae reflect 002 and 200 beams with d_{002} and d_{200} differing by 1 percent. If orientation differences are about axes normal to the plane of incidence, the film records the difference in Bragg angles plus or minus the differences in orientation between the reflecting planes of the different lamellae.

Berg–Barrett photographs can also be made with the reflected beam transmitted through a specimen, if absorption in the specimen is not too great; but resolution is likely to be lower than with reflection geometry.

If a succession of Berg–Barrett images is made with the photographic film at different distances from the specimen, the diffracted beams can be traced back to the originating grains or subgrains and can be combined with information obtained from double-crystal spectrometer rocking curves to determine misorientations of subgrains.[1]

Bonse method A modification developed by Bonse[2] involves making the incoming beam to the sample highly collimated by reflecting it from a high-perfection germanium-crystal monochromator. The intensity distribution near the image of individual dislocations is discussed in detail by Bonse.[3]

With a 444 reflection from the monochromator and again from the sample, and with germanium crystals containing no more than about 1000 dislocations per cm^2, one- or two-day exposures yield pictures in which the strain field around a dislocation in germanium can be seen out to distances of 50 μ from the core. At this distance the rotation of the lattice is no more than 0.1 or 0.2 sec of arc. The setting of the crystals known as the $(+, -)$ setting is found to be more sensitive to small strains than the setting with the specimen reflecting to the opposite side of the monochromatic beam, the $(+, +)$ setting.

Renninger[4] has made use of a modification of this double-crystal technique in which the second crystal is intentionally tilted by some minutes of arc in such a way that its reflecting-plane normal is no longer in the plane defined by the incident and reflected beam from the first (i.e., monochromator) crystal. The second crystal then does not reflect as a whole; only a narrow strip reflects. A succession of pictures, with a slightly different tilt for each, then produces a "zebra" pattern of stripes if they are superimposed on a single film, and imperfections in the crystal appear as curves and discontinuities in the zebra stripes.

Lang method A method developed by Lang[5] consists of irradiating a crystal with a slit-collimated beam of monochromatic x-rays from a point-

[1] S. Weissmann, *J. Appl. Phys.*, vol. 27, pp. 389, 1335, 1956. J. Intrater and S. Weissmann, *Acta Cryst.*, vol. 7, p. 729, 1954. Y. Nakayama, S. Weissmann, and T. Imura in J. B. Newkirk and J. H. Wernick (eds.), "Direct Observation of Imperfection in Crystals," Interscience, New York, 1962.

[2] U. Bonse, *Z. Physik*, vol. 153, p. 278, 1958. U. Bonse and E. Kappler, *Z. Naturforsch.*, vol. 13a, p. 348, 1958. U. Bonse in J. B. Newkirk and J. H. Wernick (eds.), "Direct Observation of Imperfections in Crystals," p. 431, Interscience, New York, 1962. See also K. Kohra in Newkirk and Wernick, op. cit., p. 461.

[3] Bonse, loc. cit.

[4] M. Renninger in "Crystallography and Crystal Perfection," Academic, New York, p. 145; *Phys. Letters*, the Netherlands, vol. 1, p. 104, 1962.

[5] A. Lang, *Acta Met.*, vol. 5, p. 359, 1957; *Acta Cryst.*, vol. 12, p. 249, 1959; *J. Appl. Phys.*, vol. 30, p. 1748, 1959. Suitable cameras are now marketed.

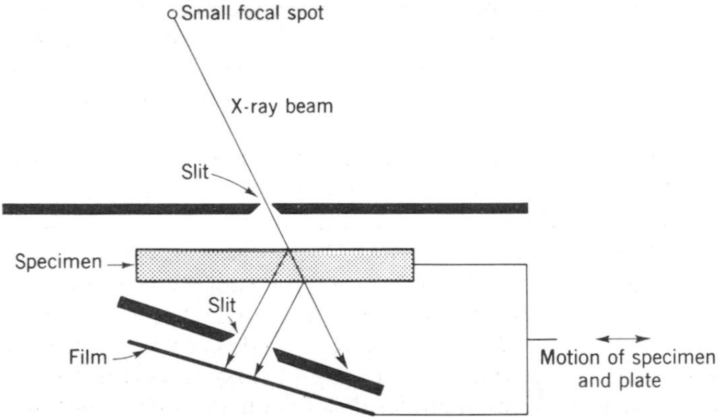

Fig. 15-5 Arrangement for the Lang method. Specimen and film can be translated together. Slits are normal to the plane of the drawing.

focus tube and photographing the reflected beam that is transmitted through the crystal. The geometry of the arrangement is indicated in Fig. 15–5. The diffracted beam of $K\alpha_1 + K\alpha_2$ radiation (or for highest resolution, $K\alpha_1$ only) reaches the film through a slit, while the direct beam is stopped by the absorbing screen. If the film is normal to the diffracted beam, a thick emulsion can be used, such as a nuclear emulsion as thick as 100 μ together with Ag $K\alpha$ or Mo $K\alpha$ radiation, and crystals can be as thick as a millimeter or so if they are reasonably transparent to these rays.

With the specimen and film held in a stationary position, a photograph is obtained of a thin section of the crystal (the section irradiated by the direct beam); the photograph is appropriately called a *section topograph*. On the other hand, the crystal and film may be mounted on a carriage and moved back and forth during the exposure with a constant-velocity translation mechanism, recording the Bragg reflection in a photograph known as a *projection topograph*. This photograph is a projection that shows defects throughout the volume of the specimen that is moved through the

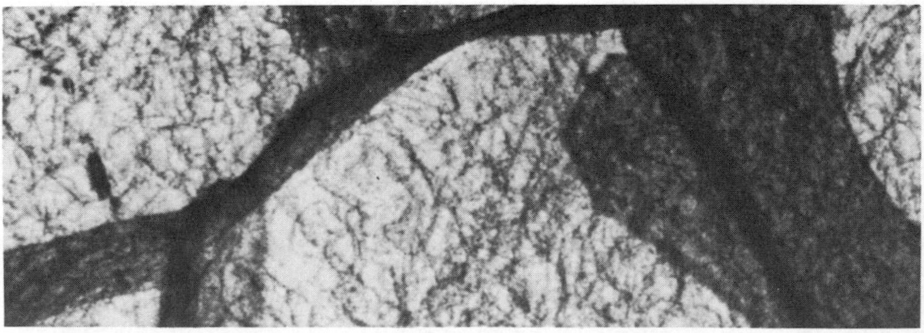

Fig. 15-6 Projection topograph by the Lang method. Dislocations and subboundaries in an undeformed LiF crystal. ×25. (Courtesy A. R. Lang)

direct beam, and is thus the sum of many section topographs. It serves as a general survey of the imperfection distribution, but is capable of registering individual dislocations only if there is a relatively low dislocation density. For high dislocation densities the section topograph is more suitable.

One of Lang's projection topographs is reproduced in Fig. 15–6. Not only individual dislocations but also subboundaries can be clearly seen throughout the interior of the crystal, which is a {100} plate of LiF about 0.5 mm thick containing a random network of dislocations and dense walls of dislocations (not individually resolved) that formed subboundaries across which disorientations were 10 to 20 sec of arc. If two pictures are made with two different reflections, *hkl* and $\bar{h}\bar{k}\bar{l}$, and viewed stereoscopically, it is possible to determine orientation in space and Burgers vector for all segments of dislocation lines, provided the dislocation density is not too high.

Lang's x-ray micrographs show fringes similar to those seen in electron microscope pictures of thin foils—"thickness contours"—which are due to the intensity variation of the wave field in the crystal.[1] The fringes are accounted for by applying the dynamical theory of diffraction—but in a somewhat more complex fashion than is needed for electron diffraction. It is found that stacking faults and twin boundaries also produce fringe patterns, as they do in electron microscope pictures, but it cannot readily be determined from the topographs whether a given fringe pattern is caused by one stacking fault or by a number of them superimposed.[2]

Borrmann method The anomalous transmission of x-rays, often called the *Borrmann effect* or the *Campbell-Borrmann effect,* is also used as a basis for x-ray diffraction topography. When a single crystal of high perfection is placed in reflecting position in a collimated monochromatic beam, as indicated in Fig. 15–7, the transmitted and diffracted beams experience abnormally low absorption and emerge from the crystal with anomalously high intensity.[3] (As indicated in Fig. 15–7, they also emerge in a displaced position.) The absorption in the crystal is greatly reduced because a stationary-wave system is set up in the crystal as a result of interference between incident and reflected waves, and the nodes of this standing-wave system occur at the atomic sites. The displaced position of emergence of the beams is a result of the fact that the direction of energy flow in the crystal is along the reflecting planes. Both the transmitted beam T and the reflected beam R, if photographed with a fine-grained emulsion, are capable of revealing dislocations in the crystal as well as other internal variations in perfection due to segregation or precipitation. Since the strain field of a dislocation destroys the perfection that is needed for anomolous

[1] N. Kato and A. R. Lang, *Acta Cryst.*, vol. 12, p. 787, 1959; these are designated by Kato and Lang as "Pendellösung fringes," following P. P. Ewald's discussion of the phenomenon in *Ann. Phys.*, vol. 49, p. 117, 1916.

[2] K. Kohra and M. Yoshimatsu, *J. Phys. Soc. Japan*, vol. 17, p. 1041, 1962.

[3] G. Borrmann, *Z. Physik*, vol. 127, p. 297, 1950. H. N. Campbell, *J. Appl. Phys.*, vol. 22, p. 1139, 1951; *Acta Cryst.*, vol. 4, p. 180, 1951.

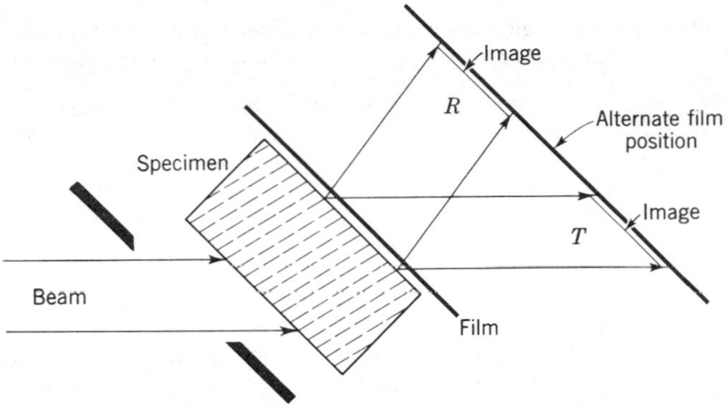

Fig. 15-7 Arrangements for the Borrmann method. The reflecting planes in the crystal are indicated by dashed lines.

transmission, the dislocation casts a shadow and appears as a dark line in the transmitted beam image T and also in the diffracted beam image R.[1] The alternate film position indicated in Fig. 15–7 yields images with much lower resolution than is obtained with the emulsion placed against the back of the crystal.

General characteristics of topographic methods The incident beams and even the diffracted beams in most of these x-ray methods are very intense. It is therefore important to take all necessary precautions against the danger of x-ray exposure, especially of the hands and face. The need for caution is particularly important if a specimen is being adjusted when the x-ray beam is on.

X-ray topographs do not reach the extreme resolution that can be obtained with electron microscopes, but they do offer certain unique advantages. When the dislocation density in a sample is not too high, the dislocations can be photographed without the sample being thinned to an extremely thin foil; the danger of altering the dislocation array by thinning is thus avoided. Nonuniform distribution of impurities and effects due to segregation and precipitation can be seen (even when they are distributed on a scale too large for electron microscopy), but the small-scale effects encountered in early-stage precipitation are only accessible with electron microscopy. Exposure times are measured in hours for high-resolution work, but with the faster emulsions used for low-power work, a few minutes are sufficient.

Other methods for photographing substructure A semitopographic method of studying mosaic structure in crystals is the *Guinier–Tennevin method,*[2] in

[1] Additional references to discussions of this effect will be found in "Direct Observation of Defects in Crystals," pp. 47, 409, 509, Interscience, New York, 1962.

[2] A. Guinier and J. Tennevin, *Acta Cryst.*, vol. 2, p. 133, 1949; "Progress in Metal Physics," vol. 2, p. 177, Pergamon, New York, 1950; "Imperfections in Nearly Perfect Crystals," p. 402, Wiley, New York, 1952.

which Laue spots in transmission photographs are made sensitive to slight misorientations in the specimen. A narrow focal spot (effective width, 0.04 mm) and a large crystal-to-film distance (25 to 100 cm) permit the registration of substructure-boundary disorientation that is less than 10 sec of arc. In this method the distances from source to specimen and from specimen to film are chosen so that the diffracted rays for a certain set of planes are brought to a focus at the film (the sheet specimen is tangent to the focusing circle, and the film is also located on this circle, or is tangent to it at the focus of the reflected beam). The focusing action has the result that the photograph is not topographic in the plane of incidence (the plane of the reflection circle), but some topographic character (with low resolution) is found in the direction normal to the plane.

The misorientations of a crystal can also be photographed semitopographically with monochromatized radiation by placing a specimen at the focus of a curved crystal monochromator and the film about a meter from the specimen either in the reflection or the transmission geometry. This is the so-called *Lambot method*.[1]

A sensitive means of studying small misorientations in crystals is based on *divergent x-ray beam patterns*, including those in the back-reflection region.[2] Characteristic x-rays diverging from a point source, when reflected from a perfect crystal, produce a diffraction pattern of curves (Kossel lines) nearly elliptical in shape and very sharply delineated.[3] If a small-angle boundary crosses the reflecting region of a specimen, the material on the two sides of the boundary produces patterns slightly differing in orientation, and the Kossel curves on the film will consequently show abrupt breaks and displacements.[4,5] With the use of a special x-ray tube in which the electron beam is sent down a capillary tube and focused magnetically at the end of it, back-reflection patterns have been obtained that could reveal small-angle boundaries with misorientations down to about 2 min of arc.

If a Laue pattern is made with a fine-focus tube and with the geometry of the Schulz method, some of the Laue spots will be crossed by lines made by the divergent characteristic radiation from the x-ray tube. These also can show displacements caused by small-angle boundaries in the specimen where misorientation is only a few minutes of arc.[6]

Hirsch[7] has reviewed mosaic crystals and methods of studying these, including Debye-Scherrer photographs made with very small-diameter beams, the *microbeam technique*.

In the 1960s it has become possible to construct *x-ray interferometers* by using the principle that a beam is split into two coherent beams in anomalous transmission through a crystal.[8] One of the two beams is absorbed in the crystal, the other is transmitted through it and emerges as two coherent diverging beams.

Experiment has shown that it is possible to reflect one or both of the diverging beams in a second crystal, thereby converting them into converging beams which set up moiré fringes. The beam-splitting crystal, the second reflecting crystal, and a third crystal used as an analyzer can be made of silicon of high perfection. When properly oriented, the analyzer permits detection of lateral shifts of about 2 A, rotation of 10^{-2} sec, or d spacing changes of about 1 part in 10^8. The sensitivity of the interferometers to phase

[1] H. Lambot, L. Vassamillet, and J. Dejace, *Acta Met.*, vol. 1, p. 711, 1953.

[2] T. Imura, S. Weissmann, and J. J. Slade, *Acta Cryst.*, vol. 15, p. 786, 1962.

[3] These are also discussed on p. 97.

[4] D. R. Schwarzenberger, *Phil. Mag.*, vol. 4, p. 1242, 1959.

[5] P. E. Lighty et al., *J. Appl. Phys.*, vol. 14, p. 2233, 1963.

[6] T. Fujiwara, *Mem. Defense Acad. Math. Phys. Chem. Eng.* (*Yokosuka, Japan*), vol. 2, p. 127, 1963.

[7] P. B. Hirsch, "Progress in Metal Physics," vol. 6, p. 236, Pergamon, New York, 1956.

[8] U. Bonse and M. Hart, *Appl. Phys. Letters*, vol. 6, p. 155, 1965; vol. 7, p. 99, 1965; *Z. Physik*, vol. 188, p. 154, 1965; vol. 189, pp. 161, 259, 1966. M. Renninger has also investigated the subject: *Phys. Letters* (the Netherlands), vol. 1, p. 103, 1962.

shifts, thickness variations, index of refraction values, and elastic strains from applied stresses or from defects suggests various potential uses. Topographs can be made by placing a film behind the analyzer crystal.

ELECTRON MICROSCOPY

Transmission electron microscopy has had widespread use in recent years, owing to the availability of convenient instruments with high resolving power, great advances in specimen-preparation techniques, and rapid developments in the interpretation of electron micrographs. There are now many ingenious techniques and countless applications to the science of materials; a general knowledge of the more important principles and applications is needed by all well-informed workers in the fields of physical chemistry, physics, metallurgy, materials, and the geophysical sciences. For those requiring more detailed treatments than can be included here, various reviews and symposia are available.[1]

Equipment Figure 15–8 is a diagram that illustrates the lens arrangement in a typical high-resolution instrument employing magnetic lenses. Electrons are emitted from a heated tungsten filament and accelerated by a potential usually between 40 and 100 kv. The electron beam is directed successively through condenser lenses, an aperture, a specimen holder, an objective lens, another aperture, an intermediate lens, another aperture, and a projector lens, and onto a fluorescent screen, a photographic plate, or a film. A set of vacuum pumps maintains a pressure below about 10^{-5} mm Hg during operation. Air locks are usually provided to permit insertion of the specimen and rapid return to operating pressures. Adjustments are provided for alignment of the apertures and to permit electron diffraction patterns to be made of a selected area of the specimen. Attachments are frequently installed to permit tilting of the specimen, and some work is also done with attachments that permit heating or cooling the specimen, or that permit stressing the specimen during observation.

Most modern instruments when properly cleaned and adjusted give resolutions of 10 to 30 A, and under the best conditions 3- to 10-A resolution

[1] G. Thomas, "Transmission Electron Microscopy of Metals," Wiley, New York, 1962. C. E. Hall, "Introduction to Electron Microscopy," McGraw-Hill, New York, 1953. F. S. Sjöstrand and J. Rhodin (eds.), "Electron Microscopy," Academic, New York, 1956. K. Lark-Horovitz, "Methods of Experimental Physics, vol. 6, Part A, Preparation, Structure, Mechanical and Thermal Properties," Academic, New York, 1959. "Proceedings of the Third International Conference on Electron Microscopy," Royal Microscopy Society, London, 1956. "Proceedings of the Fourth International Conference on Electron Microscopy, 1958," Springer-Verlag, Berlin, 1960. G. Thomas and J. Washburn (eds), "Electron Microscopy and Strength of Crystals," Interscience, Wiley, New York, 1963. "Proceedings of The European Regional Conference on Electron Microscopy," vol. 1, De Nederlandse Vereniging voor Electronenmicroscopie, Delft, 1961. D. Kay, "Techniques for Electron Microscopy," Blackwell, Oxford, 1961. "Proceedings of the Fifth International Congress for Electron Microscopy," vol. 1, Academic, New York, 1962. J. W. Menter, *Advan. Phys.*, vol. 7, p. 299, 1958. R. D. Heidenreich, "Fundamentals of Transmission Electron Microscopy," Interscience, New York, 1964. Subjects outlined in this chapter are treated in detail in a book which appeared after this chapter was in proof: P. B. Hirsch, A. Howie, R. B. Nicholson, D. W. Pashley, and M. J. Whelan, "Electron Microscopy of Thin Crystals," Butterworths, London, 1965.

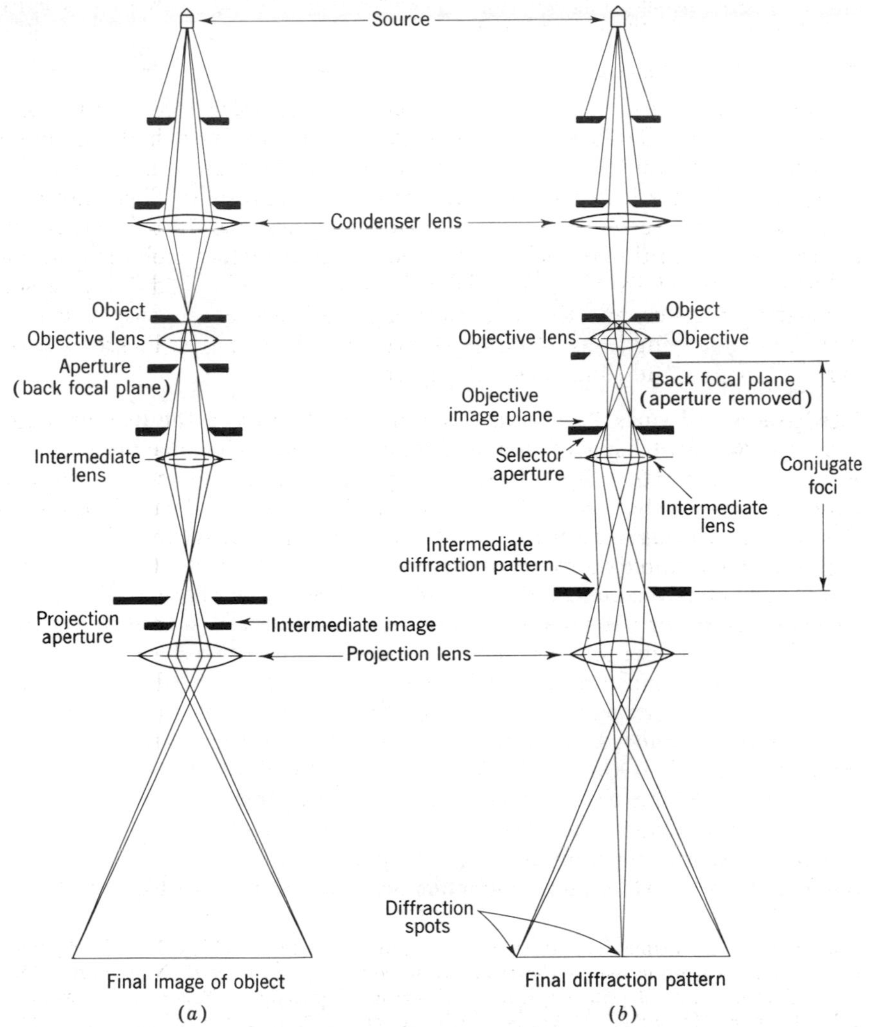

Fig. 15-8 Arrangement of lenses and diaphragms in a modern high-power electron microscope. (*a*) Operating as a three-state electron microscope. (*b*) Operating to produce a selected-area diffraction pattern. (**After R. D. Heidenreich, "Fundamentals of Transmission Electron Microscopy," Interscience, New York, 1964.**)

is achieved, which means that magnifications of ×10,000, ×20,000, and ×40,000 diameters are common. The magnification at any setting of the operating voltage and the lens currents of the microscope can be calibrated by using a fine mesh screen or a grating of known spacing as the specimen to be magnified; for example, a replica of an optical grating. Small spherical particles of latex have also been used (but are subject to diameter changes when carbon is deposited on them during the observing period). Since the image rotates as the lens currents are changed, the rotation is also calibrated—for example, by comparing microscopic images of, say, a long flat crystallite of MoO_3 with the diffraction pattern of the same crystal at the same current and voltage settings (illustrated in Fig. 22–1).

The ability to make both photographs and diffraction patterns in the same instrument is extremely useful, particularly when the possibilities of *selected-area diffraction* are exploited. By using an intermediate aperture of small diameter and moving it to a selected area of the image, it is possible to photograph the position of the aperture on the image and then to record the diffraction pattern of the electrons transmitted through this part of the specimen. The orientation of the diffraction pattern with respect to the image of the crystal can also be determined (provided image rotation has been calibrated), so that the various details in the image can be directly related to the orientation of the crystal lattice. Selected-area diffraction is therefore a powerful research tool for identifying and studying the habit and orientation of microscopic crystallites, twins, included grains, subgrains, precipitates, and transformation products.

Orientation determinations Owing to the short wavelength of the electrons in the range of voltages above 40 kv, the Ewald sphere of reflection has a very large radius; consequently, it deviates only slightly from a plane.[1] Frequently, because of this, many reflections can occur simultaneously—particularly when the reciprocal lattice points are somewhat extended rather than sharp. The diffraction pattern then is a direct image of a considerable portion of a plane in reciprocal space. It is then relatively easy to assign indices to the spots and to determine the orientation of the diffracting crystal. Often this is done by inspection—for example, by comparing the pattern with the appearance of various planes in a three-dimensional model of the reciprocal lattice. A tentative solution can be tested by identifying individual spots by their distance from the central spot.[2] The patterns for several orientations of a f.c.c. crystal are sketched in Fig. 15–9.[3]

In indexing spots it is helpful to have previously determined values of the *camera constant* for the various magnifications of the instrument. The camera constant is the wavelength of the electrons used times the effective length of the camera from specimen to film for the lens current used. The camera constant divided by the distance from the central spot to a dif-

[1] The Ewald sphere is discussed on p. 88, and electron diffraction is treated more fully in a later chapter.

[2] Aberration in the objective lens may be sufficient to cause some distortion of the pattern, and the electrons passing near the edges of the limiting aperture are most subject to such aberration. The pattern may also be somewhat confused by diffracted rays from portions of the specimen just *outside* the area covered by the limiting aperture. See A. W. Agar and R. Phillips, *Brit. J. Appl. Phys.*, vol. 11, pp. 185, 504, 1960.

[3] Plots of the patterns for many orientations of f.c.c., b.c.c., c.p.h., NaCl, and CsCl structures are shown in the book by Hirsch et al., *loc. cit.*, 1965. Kikuchi line patterns are also plotted.

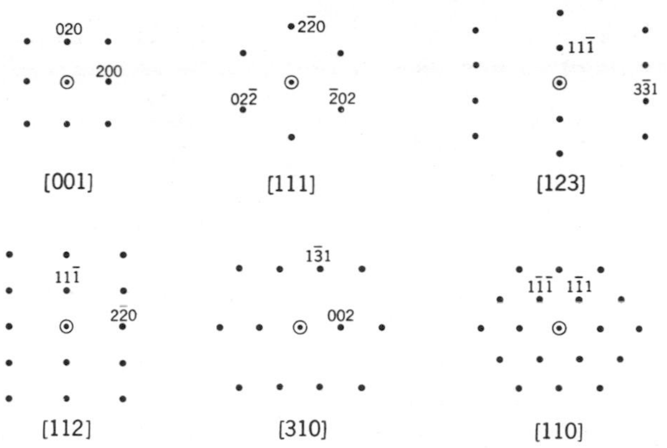

Fig. 15-9 Electron diffraction patterns observed with the direct beam along certain [*hkl*] directions of f.c.c. crystals.

fraction spot gives directly the interplanar spacing d of the reflecting plane. This follows from the reciprocal lattice relationships (see Chap. 4); the distance from the origin in reciprocal space to the reflecting point is proportional to $1/d$, and the corresponding distance from the central spot on the film to the diffraction spot is proportional to $\sin \theta$, which for small angles is approximately equal to θ. Sometimes also the *ratios* of distances from the central spot are used in identifying pairs of spots, since these are equal to the ratios of $1/d$ values. Frequent use is made of stereographic projections of indexed spots or zones. From a knowledge of crystal orientation, the thickness of a foil specimen can be determined.[1]

Replicas A specimen too thick to be penetrated by an electron beam can be examined indirectly in an electron microscope by preparing a replica of the specimen surface. Several different methods of replication have been developed by which the metallographic details of a specimen's surface can be transferred to a replica thin enough to serve in electron transmission microscopy. With suitable techniques, resolution approaching 20 A for the surface details can be achieved, which represents a great advance over the highest attainable resolution in optical microscopy, though it is inferior to the limits reached in direct transmission through thin foil specimens. Microscopes for replica work do not need to be as elaborate as those for direct-transmission work at the highest powers and consequently can be made in less expensive models which are easier to maintain and operate. The purpose of the present discussion is to introduce and characterize the important techniques briefly, giving references to detailed descriptions of the individual techniques that have appeared in the more readily available publications. A concise but well-documented listing of each technique is given in several references cited on page 430 (for example, "Methods of

[1] See p. 445.

Experimental Physics, vol. 6, Part A" and "Transmission Electron Microscopy of Metals").

Plastic replicas are simply and quickly made; consequently, they have been extensively employed for many years. Replication can be effected without destroying the sample; the resolution, however, is not better than 100 or 200 A. A solution of a plastic such as Formvar, Parlodion, or Collodion is used. (These are trade names, respectively, for polyvinyl formal dissolved in dioxane, cellulose dissolved in ethyl ether and ethanol, and cellulose nitrate dissolved in amyl acetate.) The solution is flowed over the surface in a manner that leaves a film a few hundred angstroms thick when dried. If the surface has been metallographically polished, etched, washed, and dried, previously, the film then follows the contours of the surface and carries an image of the metallographic details in the surface. Stripping and mounting the replica is a rather delicate task.[1]

Since the contrast obtained with plastic replicas is low, it is often desirable to evaporate a heavy metal such as gold, platinum, or chromium onto the replica. If the evaporated metal arrives at the replica at an oblique angle, the shadow cast by the condensing metal greatly enhances the contrast in the micrographs. Several two-stage plastic replication methods have also been devised.[2]

Carbon replicas have largely replaced plastic replicas in many laboratories because of the better resolution that can be achieved with them.[3] After depositing a film about 200 A thick on a metallorgaphically prepared surface, the replica can be scratched into small squares, removed by electropolishing the sample, washed, perhaps shadowed, and mounted on an electron microscope grid. It is sometimes possible to etch a multiphase metallic specimen through a deposited carbon film in such a way that the matrix is dissolved but some of the precipitate particles (e.g., carbides in steels) are left unattacked and attached to the replica. The extracted particles can then be identified by electron diffraction or by x-ray or chemical means.[4]

Oxide replicas consisting of the thin oxide layer formed on the specimen itself have been used, particularly anodic films on aluminum and aluminum-rich alloys, for disclosing slip lines on the surface and precipitated particles.[5]

Silica replicas are made by depositing in vacuum a thin layer of SiO_2 or SiO either directly on the specimen itself[6] or on a polystyrene replica.

Transmission microscopy with thin foils When only replica methods were available, the resolution obtainable was limited by the replica. Far

[1] V. J. Schaefer and D. Harker, *J. Appl. Phys.*, vol. 13, p. 427, 1942. V. J. Schaefer, *Phys. Rev.*, vol. 62, p. 495, 1942. C. M. Schwartz, A. E. Austin, and P. M. Weber, *J. Appl. Phys.*, vol. 20, p. 202, 1949. A. E. Austin and C. M. Schwartz, *J. Appl. Phys.*, vol. 22, p. 847, 1951. I. W. Fischbein, *J. Appl. Phys.*, vol. 21, p. 1199, 1950; *Proc. Am. Soc. Testing Mater.*, vol. 50, pp. 444, 489, 1950; vol. 52, p. 543, 1952; vol. 57, p. 452, 1957.

[2] One is described by R. D. Heidenreich and V. G. Peck, *J. Appl. Phys.*, vol. 14, p. 23, 1943.

[3] D. E. Bradley, *J. Inst. Metals*, vol. 83, p. 35, 1954–1955. E. Smith and J. Nutting, in "Proceedings of The Third International Conference on Electron Microscopy, 1954," p. 206, Royal Microscopy Society, London, 1956. J. Diehl, S. Mader, and A. Seeger, *Z. Metallk.*, vol. 46, p. 650, 1955. E. Smith and J. Nutting, *Brit. J. Appl. Phys.*, vol. 7, p. 214, 1956; *J. Iron Steel Inst.*, vol. 187, p. 314, 1959.

[4] R. M. Fisher, *J. Appl. Phys.*, vol. 24, p. 113, 1953; American Society for Testing Materials, Special Tech. Publ. No. 155, p. 49, 1954.

[5] M. S. Hunter and F. Keller, *Am. Soc. Testing Mater.*, Spec. Tech. Publ. no. 155, 1953. See also: G. Thomas, "Transmission Electron Microscopy of Metals," Wiley, New York, 1962. K. Lark-Horowitz and V. A. Johnson, "Solid State Physics," vol. 6, Part A, Academic, New York, 1959.

[6] H. Wilsdorf and D. Kuhlmann-Wilsdorf, *Z. Angew. Phys.*, vol. 4, pp. 361, 409, 1952.

greater resolution is now obtained by transmission through thin foils of the specimen itself. Various methods of specimen preparation are employed.

Samples thin enough for direct-transmission microscopy can be prepared by vacuum deposition. When deposited on a suitable substrate, the film can be stripped from the substrate or freed by dissolving the substrate; if the substrate is a single crystal, the deposit may grow epitaxially and be a single crystal.[1] In some instances solid solutions and compounds can also be prepared in this way. It is also possible first to deposit one substance and then to superimpose a second one of slightly different lattice constant. When the two layers condense in parallel orientation (which more frequently occurs if the substrate is hot), a transmission micrograph shows a moiré pattern and provides a direct image, enormously enlarged, of crystal planes and defects.[2] A major disadvantage of vapor-deposited films and of films deposited electrolytically or from solution is the fact that their defect structure and many of their properties differ markedly from those of bulk samples. Various special techniques have occasionally been employed, including those involving extremely rapid cooling of molten specimens (see page 161). Solid samples can be cut into sufficiently thin sections by the use of *microtomes*,[3] but severe deformation of metallic samples prepared in this way must be expected.[4]

Some materials can be cleaved suitably. By far the most useful methods, however, involve the chemical and electrolytic thinning of bulk specimens. Many specific thinning preparations and techniques are used,[5] and the references on page 430 include several. Even when specimens are thinned from bulk materials, however, the final specimen may differ in important ways from the material in bulk, and the danger of drawing false inferences about the bulk material is always present. A very common trouble of this type is the loss of dislocations and the rearrangement of the residual dislocations during thinning.

Image contrast Since an electron microscope image is formed by the electrons emerging from the back side of the specimen and traveling in a direction such that they pass through the objective aperture, the contrast in the image is dependent on the factors governing the number of these emerging electrons at each point on the back surface. Variations in the number will occur from point to point over the specimen if the beams that

[1] A summary of work in the field of epitaxy is given by D. W. Pashley, *Advan. Phys.*, vol. 5, p. 173, 1956.

[2] G. A. Bassett, J. W. Menter, and D. W. Pashley, *Proc. Roy. Soc. (London)*, vol. A246, p. 345, 1958.

[3] H. B. Haanstra, *Philips Tech. Rev.*, vol. 17, p. 178, 1955. H. Fernandez-Moran, *Ind. Diamond Rev.*, vol. 16, p. 128, 1956. L. Reimer, *Z. Metallk.*, vol. 50, p. 37, 1959.

[4] A. Phillips in "Direct Observations of Imperfections in Crystals," Interscience, New York, 1962.

[5] G. Thomas, "Transmission Electron Microscopy of Metals," Wiley, New York, 1962. W. J. McG. Tegart, "The Electrolytic and Chemical Polishing of Metals in Research and Industry," Pergamon, Oxford, and Macmillan, New York, 1956. D. Kay, "Techniques for Electron Microscopy," Blackwell, Oxford, 1961. P. B. Hirsch et al., *loc. cit.*, 1965.

pass through have encountered different thicknesses or densities, or different average atomic concentrations. Electrons are removed from the direct beam by both elastic and inelastic scattering processes (inelastic processes involve change of velocity as well as change of direction; the amount of the inelastic scattering is approximately $1/Z$ times the amount of elastic scattering, where Z is the atomic number). More electrons emerge to form an image as the electron accelerating voltage becomes higher, the specimen becomes thinner, and the atomic number of the atoms in the specimen becomes lower. The amount of the scattering is such that specimens must be very thin for transmission work—under 1000 A for most materials. The elastically scattered electrons from a crystalline sample are the diffracted electrons; hence, the contrast effects produced by these can only be understood by considering what diffracted beams of high intensity can occur in each instance.

A *bright-field image* is produced when the *direct beam*, having passed through the specimen and the objective lens, is permitted to pass through a small objective aperture as indicated in Fig. 15–10, and is enlarged by the subsequent lens or lenses. The diffracted beams from the specimen do *not* pass the aperture; hence, the diversion of some electrons from the direct beam into one or more strong diffracted beams results in lowered intensity in the bright-field image. The stronger the chief diffracted beam, the greater the loss in direct-beam intensity. Strongly diffracting areas of a specimen appear dark against the bright background of the image.

A *dark-field image* is produced, as indicated in Fig. 15–10, if one strong diffracted beam is allowed to pass through the aperture (which is adjustable), but the direct beam and other diffracted beams are not. Then a few strongly reflecting areas located within a nonreflecting matrix will appear in the image as bright spots on a dark background. In some applications a comparison of bright-field and dark-field images removes ambiguities of interpretation.

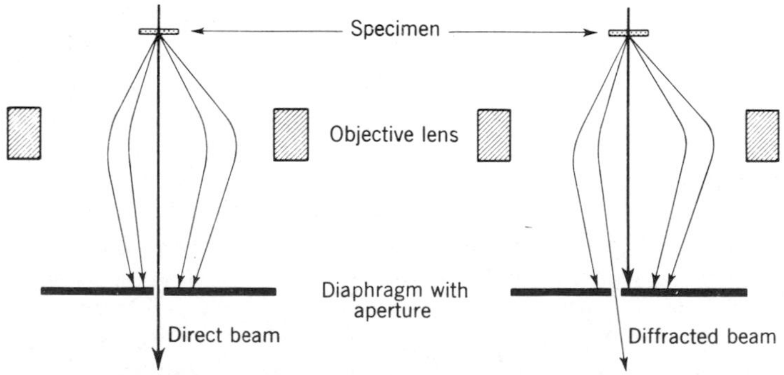

Fig. 15-10 Position of aperture with respect to the diffraction pattern for bright-field and dark-field images.

The kinematic theory of image contrast The fundamental principles of diffraction of x-rays (Chap. 4) carry over with little modification to the diffraction of electrons. The theory of diffraction contrast in electron microscope images can therefore be presented quite briefly here, provided the treatment is based on the kinematic theory of diffraction rather than on the more rigorous and much more complex dynamic theory (which involves the interaction of direct and diffracted beams). The kinematic theory applies to the great majority of the problems encountered in practice, for it holds well except when the specimen is oriented exactly in a reflecting position or is unusually thick. The simple kinematic theory involves the following assumptions:

1. Only one diffracted beam of appreciable intensity occurs, and the intensity of this is very small in comparison with the intensity of the direct beam (the crystal is nearly but not exactly in the Bragg reflecting position).
2. One component of the diffracted beam arises from a single column of unit cells extending from the top to the bottom of the specimen along the path of the direct beam, and the total diffracted beam is the sum of all such components added together.
3. The specimen is so thin that absorption and rescattering effects can be neglected.

Consider, then, the diffracted beam from a single column of unit cells. The amplitude of the beam coherently scattered from this column will be

$$A_D = \sum_j F_j \exp 2\pi i \phi_j \qquad (15\text{-}1)$$

where F_j is the scattering factor for the unit cell located at position $\mathbf{r}_j$ in the column and ϕ_j is the phase of the electron wave from this unit cell. The phase angle ϕ_j is determined by the unit vectors representing the incident beam and the diffracted beam, as mentioned in Chap. 4 (page 87), vectors here designated as $\mathbf{k}_o$ and $\mathbf{k}$, respectively[1]; namely,

$$\phi_j = \frac{\mathbf{k} - \mathbf{k}_o}{\lambda} \cdot \mathbf{r}_j = (\mathbf{g} + \mathbf{s}) \cdot \mathbf{r}_j \qquad (15\text{-}2)$$

where λ is the wavelength of the electrons and $\mathbf{s}$ is a vector in reciprocal space representing the deviation of a certain reciprocal lattice point $\mathbf{g}$ from the sphere of reflection, as in Fig. 15–11. (The reciprocal lattice point at position $\mathbf{g}$ is near the reflection sphere but by assumption 1 it cannot lie exactly on the sphere; hence, $\mathbf{s}$ is small but nonzero.) The origin of coordinates in the crystal can be chosen so that the vector $\mathbf{r}_j$ is a lattice vector for all values of j; then

$$\mathbf{g} \cdot \mathbf{r}_j = n \qquad (15\text{-}3)$$

[1] In the reciprocal lattice discussion of p. 86 the symbols used were $\mathbf{S}_o$ and $\mathbf{S}$. Changes in nomenclature here are desirable to conform to the custom in electron microscopic publications.

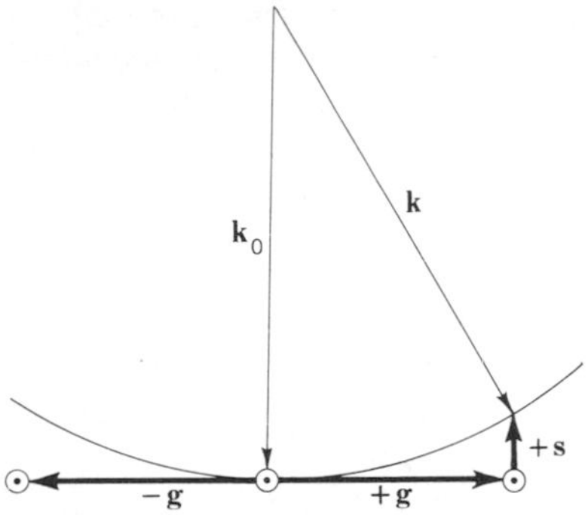

Fig. 15-11 Position of Ewald sphere and vectors in reciprocal space.

where n is always an integer for each cell in the column. Equation (15–1) therefore reduces to

$$A_D \sim \sum_j \exp\,(2\pi i\mathbf{s}\cdot\mathbf{r}_j) = \int_{-t/2}^{+t/2} \exp\,(2\pi i\mathbf{s}\cdot\mathbf{r})dr$$

$$= \frac{\sin \pi ts}{\pi s} \tag{15–4}$$

when all unit cells scatter with equal amplitude, where t is the thickness of the specimen in the direction of the direct beam. The intensity of the diffracted beam, I_D, thus varies periodically as thickness is increased, for

$$I_D \sim \frac{\sin^2 \pi ts}{(\pi s)^2} \tag{15–5}$$

and it varies periodically, also, as s is increased. One period of oscillation of intensity occurs as the crystal thickness increases to the *extinction distance* t_o', where

$$t_o' = s^{-1} \tag{15–6}$$

(This equation does not hold when s approaches zero.) The total intensity incident on the specimen, I_I, is equal to the sum of the diffracted intensity I_D and the transmitted intensity I_T:

$$I_I = I_D + I_T \tag{15–7}$$

Therefore, the transmitted beam also oscillates in intensity, but drops to *minimum* intensity and produces minimum intensity in the bright-field image when the diffracted beam reaches *maximum* intensity, and vice versa. In a sample of uniform composition and density, variations in the brightness of the image from point to point can be caused by variations in either s or t. As a consequence, electron microscope images frequently contain *extinction contours*; in bright-field images they are broad dark bands which are contours of constant s (*inclination contours*) or of constant t (*thickness contours*).

It follows from the principles outlined above that contrast in micrographs containing subgrains, twins, distorted areas, dislocations, and other details depends strongly on specimen orientation. A slight tilting of a specimen may enhance or reduce the visibility of a given feature. In general, contrast is high whenever there is a strong Bragg reflection, i.e., at positions lying near an extinction contour; thus, the operator looking for fine detail should gradually tilt the specimen so as to cause extinction contours to sweep across the specimen and watch the resulting changes in the image.

The types of periodic bands formed at internal discontinuities such as subgrain boundaries and stacking faults are mentioned below, but before taking up these it should be mentioned that the value of t_o in a given experiment should be known if contrast is to be fully understood. If the specimen thickness in the direction of the beam is less than t_o, no extinction contours can occur. The extinction distance can be calculated for any given set of conditions[1] and for metallic specimens will be found in the range 100 to 1000 A for the most important cases. Only for light elements, superlattice diffraction, and large spacings of the reflecting planes is the extinction distance likely to be greater than the specimen thickness.

Amplitude-phase diagrams The detailed consideration of contrast effects is aided by the construction of amplitude-phase diagrams, which display graphically the result of adding vectors that represent the amplitude and phases of the waves scattered by each of the unit cells of a column, as required by Eq. (15–1).[2]

Let the scattering amplitude from each unit cell in the column be set equal to unity and let the phase difference between successive cells in the column be ϕ. The dependence of ϕ on the magnitude s of the vector that expresses the deviation of the reflection sphere from a reciprocal lattice point is given by $\phi = 2\pi s$. The addition of the waves scattered by the column is then represented by the addition of equal-length vectors, each successive vector lying at an angle ϕ from the preceding one. The vectors are chords of a circle, as indicated in Fig. 15–12*a*. Since the number N of cells in the column is very large—equal to t/a where t is the length of the column and a is the dimension of the unit cell—the total length of arc is approximately equal to N, the total angle subtended by the vectors is $N\phi$, and the radius of the circle is $R = N/N\phi = 1/\phi$. The length of the resultant vector A is given by the relation $A/2 = R \sin (N\phi/2)$, from which Eq. (15–4) follows directly. For any given value of s, the diffracted beam builds up to a maximum as the direct beam penetrates deeper into the crystal, i.e., as t increases; the maximum corresponds to the resultant vector becoming a diameter of the circle. The diffracted beam amplitude then decreases and passes through zero when the vectors make a complete revolution of the circle, which occurs at depth t_o'. The sinusoidal variation continues as depth increases further. A typical example is illustrated in

[1] R. D. Heidenreich, *J. Appl. Phys.*, vol. 20, p. 993, 1949. M. J. Whelan and P. B. Hirsch, *Phil. Mag.*, vol. 2, pp. 1121, 1303, 1957. P. B. Hirsch, A. Howie, and M. J. Whelan, *Phil. Trans. Roy. Soc.*, vol. A252, p. 499, 1960.

[2] Hirsch, Howie, and Whelan, loc. cit. A related plot is mentioned in the discussion of x-ray diffraction on p. 82 and in A. J. C. Wilson, "X-ray Optics," Methuen, London, 1949, 1960.

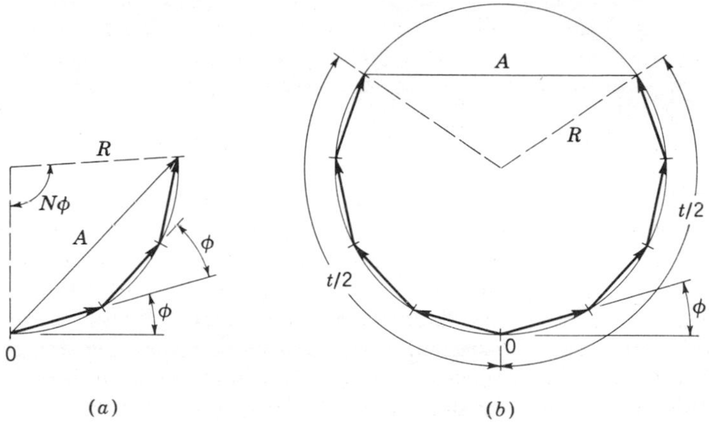

Fig. 15-12 Amplitude-phase diagrams. (*a*) Origin at the end of a column of *N* cells. (*b*) Origin at the center of a foil of thickness *t*.

Fig. 15–12*b*, with the midpoint of the sample placed at the bottom of the circle.

In an imperfect crystal, the phase of diffracted electrons is no longer given by Eq. (15–2) but by the expression

$$\phi_j = (\mathbf{g} + \mathbf{s}) \cdot (\mathbf{r}_j + \mathbf{R}_j) \tag{15–8}$$

where $\mathbf{R}_j$ represents the displacement of the jth cell from its position in the undistorted perfect crystal. Since $\mathbf{s}\cdot\mathbf{R}_j$ and $\mathbf{g}\cdot\mathbf{r}_j$ are nearly zero, the resultant amplitude is altered from the relationship of Eq. (15–4) and becomes

$$A_D \sim \sum_j \exp\,(2\pi i\mathbf{g} \cdot \mathbf{R}_j)\,\exp\,(2\pi i\mathbf{s} \cdot \mathbf{r}_j) \tag{15–9}$$

Images of stacking faults If a vector **R** describes the displacement of the exit side of a foil with respect to the entrance side, due to the presence of a stacking fault, the effect on the amplitude A_D in Eq. (15–9) will be to introduce an additional phase angle α of magnitude given by

$$\alpha = 2\pi\mathbf{g} \cdot \mathbf{R} \tag{15–10}$$

Consider, now, the influence of the stacking fault on the diffracted intensity. Let the line *SQT* in Fig. 15–13 represent a stacking fault in a f.c.c. crystal. Then **R** will be equal to the Burgers vector of a partial dislocation, a vector of type $(a/6)[121]$, if the fault is of the intrinsic type (such as would be caused by the passage of a single Shockley partial dislocation). This fault will introduce a phase difference $\alpha = (\pi/3)(h + 2k + l)$ between the waves diffracted from the material on one side of the fault plane with respect to the waves diffracted by the material on the other

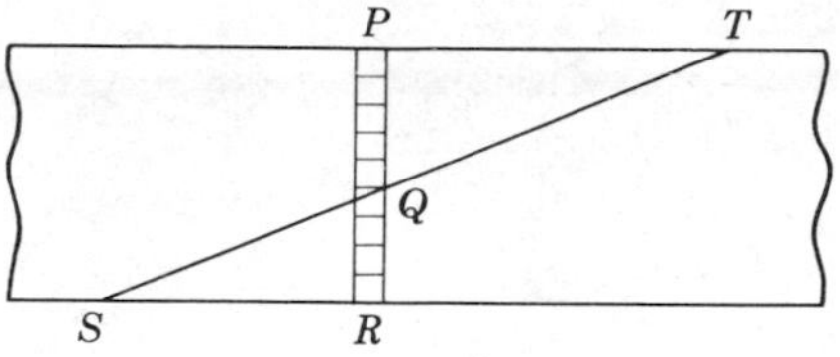

Fig. 15-13 A column of cells, *PQR*, intersecting a stacking fault *SQT* in a foil.

side, for waves producing the *hkl* reflection. The plot in Fig. 15–14 represents this situation for the particular column of unit cells that intersects the fault plane at point Q. On the circle at the point corresponding to Q a phase angle $\alpha = 120°$ is plotted and the portion of the column in the crystal below Q is represented by points along a new circle, identical in radius to the first, drawn as indicated by QB. The resultant amplitude is then given by the vector AB drawn from the point representing the top of the crystal, A, to the point representing the bottom, B. If one considers columns at increasing distances from the column PR, it will be obvious that the abrupt change of phase will occur at different depths in the crystal and at different points on the first circle of the plot; the result will be a sinusoidal variation in the intensity of the image of the fault. The striped appearance of a fault plane, accounted for in this way, is a very common feature of the micrographs of metallic materials of low stacking-fault energy, where stacking faults are wide (see, for example, Fig. 15–15).

The principles just discussed are applicable not only when a given displacement is produced by slip, but also when it results from the removal (or insertion) of a plane or partial plane of atoms, and when it concerns an antiphase boundary in a superlattice; the contrast effects associated with a dislocation are amenable to similar treatment. Detailed treatments of contrast for stacking faults have been published for the case of no

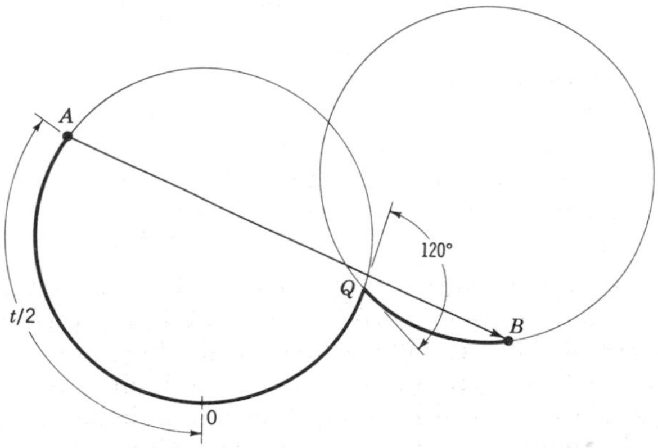

Fig. 15-14 Amplitude-phase diagram corresponding to Fig. 15-13.

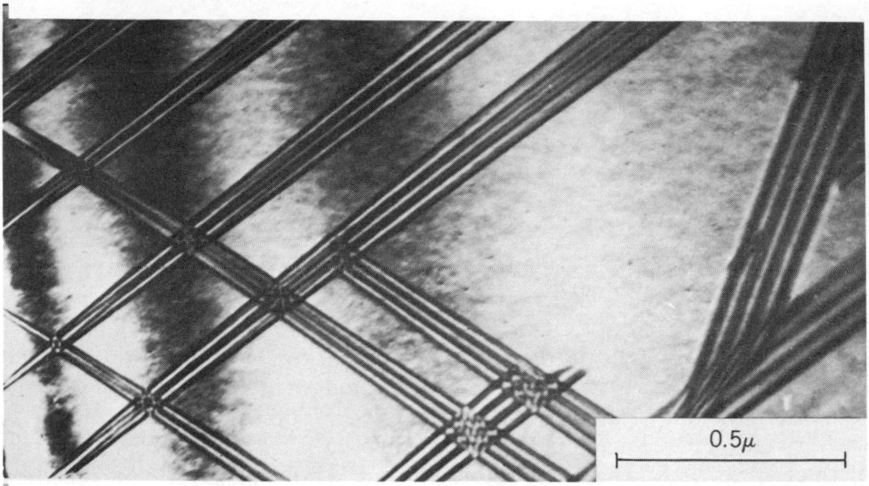

Fig. 15-15 Striated images of stacking faults, and broad thickness contours ("fringes," vertical on the micrograph) in an electron micrograph of a wedge-shaped Cu–10 percent Ge alloy crystal. Crystal thickness just to the right of center is four extinction distances; successive broad white bands to the left of this are, respectively, three, two, and one extinction distances in thickness. (Courtesy J. W. Christian and P. R. Swann.)

absorption[1] and with anomalous absorption considered[2,3]; when suitable information is obtained from images it is possible to determine the sense of the displacement vector at a stacking fault, and to distinguish extrinsic from intrinsic faults and vacancy loops from interstitial loops.[3,4] Absorption in the specimen not only tends to reduce fringe contrast in the middle region of a foil, but more importantly, it destroys the symmetry of the fringes about the foil midpoint, thereby making it possible, if the foil is thick enough, to recognize which edge of the image of a fault corresponds to the top of the foil and which to the bottom. Positive values of α can be distinguished from negative, and as a consequence intrinsic faults can be distinguished from extrinsic.

[1] M. J. Whelan and P. B. Hirsch, *Phil. Mag.*, vol. 2, pp. 1121, 1303, 1957. P. B. Hirsch, A. Howie, and M. J. Whelan, *Phil. Trans. Roy. Soc.*, vol. A252, p. 499, 1960.

[2] H. Hashimoto, A. Howie, and M. J. Whelan, "Proceedings of the European Regional Conference on Electron Microscopy," vol. 1, p. 194, de Nederlandse Vereniging voor Electronmicroskopie, Delft, 1960.

[3] H. Hashimoto, A. Howie, and M. J. Whelan, *Phil. Mag.*, vol. 5, p. 967, 1960; *Proc. Roy. Soc. (London)*, vol. A269, p. 80, 1962. A. Howie, *Rev. Met.*, vol. 6, p. 467, 1961. P. B. Hirsch et al., *loc. cit.*, 1965.

[4] A. Art, R. Gevers, and S. Amelinckx, *Phys. Stat. Solids*, vol. 3, p. 697, 1963. Other contributions to the field will be found in *Phil. Mag.*, vol. 7, 1962, and *Phys. Stat. Solids*, vol. 3, 1963. A relatively nonmathematical summary is given by A. Howie in J. B. Newkirk and J. H. Wernick (eds.), "Direct Observations of Imperfections in Crystals," p. 269, Interscience, New York, 1962. A simplified but detailed summary which cites over 140 references and contains many suggestions and precautions for observing defects is being prepared by S. Amelinckx; see also the book by P. B. Hirsch et al., *loc. cit.*, 1965.

The magnitude of the displacements and α values encountered at the domain boundaries of superlattices have been worked out[1] for the common types of superlattices.

Images of dislocations and dislocation loops Certain dislocations are invisible in electron micrographs. If atoms are displaced by vectors lying exactly parallel to the atomic plane that is providing the contrast-producing reflection, the phase of the reflection will be uninfluenced by the displacement ($\mathbf{g} \cdot \mathbf{R} = 0$, where $\mathbf{R}$ is the displacement) as in Eq. (15–10). This can occur with screw dislocations whenever the diffraction vector $\mathbf{g}$ is normal to the Burgers vector $\mathbf{b}$ (i.e., $\mathbf{g} \cdot \mathbf{b} = 0$). Consequently, reflection from planes parallel to the axis of a screw dislocation produces no contrast, and such a dislocation is invisible. Lack of contrast also occurs with edge dislocations, for $\mathbf{g} \cdot \mathbf{b} = 0$ when reflection is from a plane lying normal to the axis of the dislocation. Also, there will be only a slight warping of planes that lie parallel to the slip plane of an edge dislocation; hence, only low contrast can be expected when these reflect. Planes parallel to the extra half plane can provide the greatest contrast, and as these are bent *toward* the reflecting orientation on one side of the dislocation and *away from* it on the other side, the former will reflect more strongly than the latter and will cause a darker line on that side of the dislocation in the bright-field image.

Finding an orientation for which a dislocation becomes invisible is a good technique for determining the direction of the Burgers vector of the dislocation. When a tilt of the specimen is found that extinguishes the contrast of a dislocation, a selected-area diffraction pattern is made of it (without changing the specimen orientation, preferably using an area not too close to an extinction contour). Then, if one diffraction spot is found that is more intense than the others, this spot is responsible for contrast, and its diffraction vector $\mathbf{g}$ is normal to the Burgers vectors of dislocations that are lacking contrast. If it is found that a certain dislocation lacks contrast for two different diffraction vectors $\mathbf{g}_1$ and $\mathbf{g}_2$, the Burgers vector is normal to both, i.e., is along $\mathbf{g}_1 \times \mathbf{g}_2$. Caution is needed in this test, however, if partial dislocations are present, since dislocations may be invisible because the Burgers vector is too small.

Dislocation loops such as those of Fig. 15–16 can be caused by the clustering of vacancies or of interstitials; these two types of loops can be distinguished[2] (unless they are too small). The nature of a loop may be determined by a test which involves moving an extinction contour across a loop by tilting the foil [thereby changing the sign of the vector $\mathbf{s}$ (Fig. 15–11) without changing $\mathbf{g}$] and making both diffraction patterns and micrographs with the different tilts. From the apparent widths of a loop in the two micrographs the nature of the loop can then be ascertained. Another method involves moving a Kikuchi line through a diffraction spot. Kikuchi lines in a diffraction pattern are due to Bragg reflection of electrons that have been scattered inelastically by the specimen.[3] Since these electrons spread out in all directions from the point of impact of the direct beam, the locus of the reflected electrons is a very wide cone with an opening angle of $90 - \theta$ (θ is the Bragg angle). The cone intersects the photographic plate

[1] M. J. Marcinkowski in G. Thomas and J. Washburn (eds.), "Electron Microscopy and the Strength of Crystals," Interscience, New York, 1963.

[2] G. W. Groves and M. J. Whelan, *Phil. Mag.*, vol. 7, p. 1603, 1962. J. W. Edington and R. E. Smallman, *Australian J. Metals*, vol. 8, p. 8, 1963. Further references and details of operational procedures are given by Amelinckx, *loc. cit.*, and by P. B. Hirsch et al., *loc. cit.*, 1965.

[3] Kikuchi lines are also discussed on p. 599.

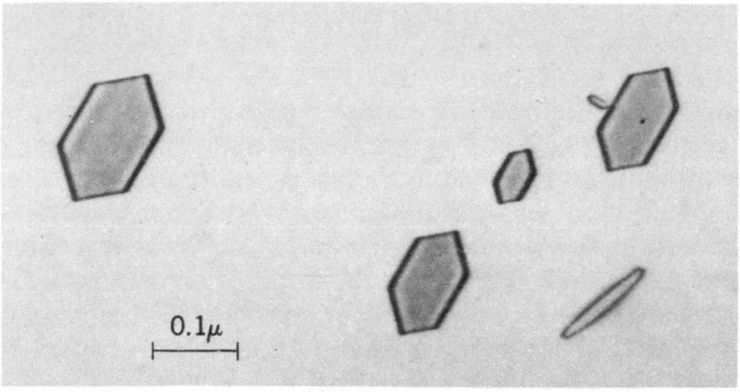

Fig. 15-16 Electron micrograph of dislocation loops in high purity aluminum quenched from 650°C, aged at room temperature. Beam along [100]. (Courtesy R. M. J. Cotterill.)

to make a slightly curved line which passes through a diffraction spot if the orientation is exactly right for reflection, but otherwise it is slightly displaced from the spot. The displacement of the Kikuchi line is *toward* the direct beam if the reciprocal lattice point for the reflecting plane is *inside* the Ewald reflecting sphere, which may be considered as the case for which **s** is positive, and *away from* the direct beam for negative values of **s**.

The sense of a Burgers vector may be determined by similar operations—by comparing photographs taken with different signs of **g** and the same sign of **s**. The position of a dislocation is between the images of the dislocation in the two photographs, and the displacement of the image from the actual position of the dislocation is in the direction to be expected from the tilting of the atomic planes in the neighborhood of the dislocation.[1]

In low-stacking-fault-energy hexagonal metals and in graphite, the Burgers vector for a vacancy loop is inclined with respect to the basal plane, whereas for an interstitial loop it lies normal to the basal plane. In these crystals the two types of loops can therefore be distinguished merely by determining the Burgers-vector direction.[2] In f.c.c. metals, vacancies as well as interstitials are believed to cluster into sheets on {111} planes, which may collapse into loops. A collapsed void contains a stacking fault of the intrinsic type, which can be eliminated by passage of a partial dislocation over the fault plane; the loop then becomes a perfect prismatic dislocation. The resulting loop is then capable of glide along a cylinder parallel to the Burgers vector. On the other hand, an interstitial loop contains an extrinsic fault, and two partials must sweep over it if it is to be removed—one in a glide plane just above the fault plane and the other in the plane just below it. These characteristics aid in distinguishing the loop types when they are used in connection with theoretical estimates of the energy required for unfaulting the loops, and with observations of the heat treatments required to initiate unfaulting.[3]

In gold, quenched from 950 to 1000°C and aged at temperatures below about 140°C,[4] vacancies in supersaturated concentrations cluster and somehow produce an unusual type of defect, the small tetrahedra of Fig. 15-17. Each tetrahedron is bounded by

[1] G. W. Groves and M. J. Whelan, *Phil. Mag.*, vol. 7, p. 1603, 1962. R. Siems, P. Delavignette, and S. Amelinckx, *Phys. Stat. Solidi*, vol. 2, p. 421, 1962.

[2] S. Amelinckx and P. Delavignette, *Phys. Rev. Letters*, vol. 5, p. 50, 1960. G. Williamson and C. Baker, *Phil. Mag.*, vol. 6, p. 313, 1961.

[3] R. M. J. Cotterill and R. L. Segall, *Phil. Mag.*, vol. 8, p. 1105, 1963. A. Eikum and G. Thomas, *Acta Met.*, vol. 12, p. 537, 1964.

[4] T. Mori and M. Meshii, *Acta Met.*, vol. 12, p. 104, 1964.

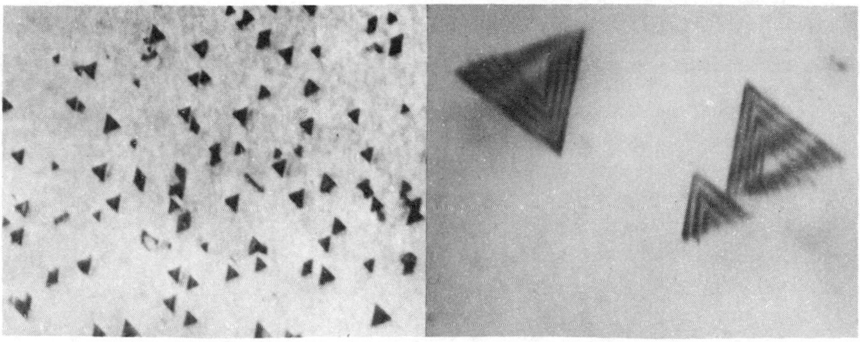

(*a*) (*b*)

Fig. 15-17 Tetrahedra in 99.999 percent pure gold, quenched from 1000°C and aged. (*a*) Quenched and aged at 120°C for 20 min. ×100,000. (T. Mori and M. Meshii.) (*b*) Quenched and aged 1 hour at 100°C; beam along [112]. ×160,000. (Courtesy R. M. J. Cotterill.)

stacking faults on {111} planes with edges composed of "stair-rod" dislocations having $(a/6)[110]$-type Burgers vectors.[1]

Coherent precipitates A precipitate platelet that causes an outward lattice displacement in its surroundings (as from hydrostatic pressure) is similar in its effects to an interstitial loop, and a precipitate platelet that causes inward displacement is similar to a vacancy loop. The contrast effects for these, as seen edge on, have been worked out.[2]

Measures of foil thickness The periodic variations in intensity along a fault plane serve as a means of determining the thickness of a foil; extinction contours at a foil edge can also be used for thickness determination. A knowledge of foil thickness is needed in determining the number of defects per unit volume.

Dislocations moving in a foil are found to leave traces at the two surfaces of the foil, which appear in the image of the foil; the traces are seen only for a few seconds with some metals, but for much longer times with others. The traces can be accounted for by assuming that they are dislocations left behind in the interface between the metal and the oxide film on the surfaces.[3] If slip is on a known plane and the crystal orientation has been determined, these traces on the two surfaces serve to indicate the thickness of the foil, for they outline the projection of the slip plane on the plane of the foil. The striated appearance of a fault plane serves the same purpose. From the known tilt of the slip plane or the fault plane in the foil and the projected width of the plane, it is easy to calculate the thickness of the foil.

[1] J. Silcox and P. B. Hirsch, *Phil. Mag.*, vol. 4, p. 72, 1959.

[2] M. E. Ashby and L. M. Brown, in "Proceedings of the International Conference on Electron Microscopy," Philadelphia, Pa., 1962, Academic, New York, 1962.

[3] A. Howie and M. J. Whelan, *Proc. Roy. Soc. (London)*, vol. A267, p. 206, 1962.

"Supervoltage" electron microscopy The use of very high voltages in electron microscopy has begun to be important in the 1960s and can be expected to increase in the future. The primary advantage of using voltages above 100 kv is the increased thickness that can be penetrated: at 800 or 1000 kv Hashimoto[1] computes that the useful specimen penetration should be increased threefold over that for 100 kv. Actual micrographs taken at 1000 kv by Dupouy and Perrier[2] appear much sharper than expected, which suggests that an improvement in image quality (in some applications and at least for defects near the exit surface) may be an additional advantage of importance—possibly arising from changes in the inelastic scattering of the electrons in the higher-voltage range. This feature is expected to aid in studies of precipitation and phase changes, cold-worked materials, and faulted structures, but the relativistic increase in mass of the electron cannot be neglected and should prevent improved domain contrast in magnetic materials.[3]

[1] H. Hashimoto, *J. Appl. Phys.*, vol. 35, p. 277, 1964. The energy dependence of extinction contour fringes and image resolution is discussed.

[2] G. Dupouy and F. Perrier, *J. Microscop.*, vol. 1, p. 167, 1962.

[3] R. M. Fisher, private communication, 1965.

16

Diffraction from imperfect and cold-worked metals

Studies of the nature of imperfections introduced into crystals during growth and during plastic deformation have continued for many decades and are still continuing, for the complexities are many. Since the fundamental types of defects (Chap. 14) and some direct methods for observing defects (Chap. 15) have been outlined, it is appropriate here to present the general principles of the diffraction methods that are used in these studies.

Orientation spread; asterism The first x-ray method to be widely used with deformed metals was the Laue method. As mentioned in Chap. 5, plastic and elastic bending of crystal planes results in elongation of Laue spots ("asterism"). Analysis of the streaks provides an understanding of the range of orientation present in the portion of the specimen that formed the diffraction pattern. If this spread is a rotation around a single axis, for example, this fact can be deduced from the pattern. Interesting results have been obtained in this way, even though methods employing characteristic radiation have largely superseded the Laue method, which uses general radiation.

A major portion of the orientation range in a plastically deformed crystal in many experiments is now known to result from deformation bands that are visible microscopically.[1] There are microbands as well, and in many metals cells develop during deformation (see Chap. 14). The resulting reorientations are of interest in studies of deformation textures (Chap. 20). On the other hand, if slip occurs in a single crystal on a single slip plane (i.e., "easy glide" deformation), little or no asterism results until this type of flow is succeeded by slip on conjugate slip planes or by a more turbulent flow. Extensive reviews of research on slip and deformation bands and their effects on reorientation, asterism, residual dislocation densities, and distributions have been published.[2] The range of orientations found in a deformed crystal containing deformation bands or kink bands consists not

[1] C. S. Barrett, *Trans. AIME*, vol. 135, p. 296, 1939; vol. 137, p. 128, 1940. C. S. Barrett and L. H. Levenson, *Trans. AIME*, vol. 135, p. 327, 1939; vol. 137, p. 112, 1940. P. Gay and R. W. K. Honeycombe, *Proc. Phys. Soc.* (*London*), vol. A64, p. 844, 1951.

[2] R. Maddin and N. K. Chen, *Progr. Metal Phys.*, vol. 5, p. 53, 1954. P. B. Hirsch, *Progr. Metal Phys.*, vol. 6, p. 236, 1956 (for asterism, see p. 296). A review of work up to 1950 appeared in C. S. Barrett, "Structure of Metals," 2d ed., McGraw-Hill, New York, 1952.

only of differing orientations across band boundaries, but also of a range of orientation within each band. Some of the substructure within deformed crystals may be judged from x-ray patterns, x-ray topographs (see Chap. 15), and optical metallography, but some is on a scale so small that it can be seen properly only on transmission electron micrographs. In some experiments the substructure has become polygonized by the time it is examined, in which case the regions between the walls of dislocations may be large enough to be seen by ordinary x-ray or optical means, or they may be visible only with electron microscopy.

The distortion of individual grains is seen not only by asterism in Laue photographs but also by the circumferential widening of spots on Debye rings. It is shown better if the spots are registered on a film that is mounted on a carriage which oscillates, together with the specimen, through an angular range of a few degrees; the spots then can broaden in *all* directions, which indicates that the distortion of individual grains is complex.[1] As deformation increases, the individual spots steadily decrease in sharpness, and overlapping of spots increases, which leads to an almost uniform blackening around a Debye ring. Further deformation may lead to an intensification of some portions of various rings, signifying that preferred orientation has developed.

The use of x-ray microbeam techniques, employing beams about 35 μ in diameter, has yielded the most detailed knowledge of the curvatures resulting from plastic deformation that could be obtained prior to the use of the electron microscope. The group at Cambridge University that used microbeam cameras showed that deformation of polycrystalline samples of many metals produces a cellular structure in which dislocations are concentrated in cell walls separating regions of lower dislocation density.[2] With increasing deformation, the size of the cells decreases, they become more distorted, and the angular misorientations increase.

Reciprocal lattices of deformed and imperfect crystals The distortions discussed above are concerned with orientation spreads, not with changes in interplanar spacings; therefore, the reciprocal lattice representations of these distortions do not involve changing the distances of reciprocal lattice points from the origin, but only the spread of each point in reciprocal space over the surface of the sphere on which it lies. As explained in Chap. 4, the vector $\mathbf{r}^*$ in reciprocal space, from 000 to a point *hkl*, is equal in magnitude to the reciprocal of the spacing of the (*hkl*) plane, i.e., to $1/d_{hkl}$, and extends in the direction of the normal to (*hkl*). A spread in orientation around a single axis in the crystal corresponds, on the sphere centered at 000 and passing through a reciprocal lattice point, to an elongation of the point into an arc of a circle concentric with the axis of ro-

[1] C. S. Barrett, *Metals and Alloys*, vol. 8, p. 13, 1937.

[2] P. B. Hirsch, *Progr. Metal Phys.*, vol. 6, p. 236, 1956, gives a summary (see p. 304). P. Gay, P. B. Hirsch, and A. Kelly, *Acta Cryst.*, vol. 7, p. 41, 1954. P. B. Hirsch, *Acta Cryst.*, vol. 5, pp. 168, 172, 1952. P. B. Hirsch and J. N. Kellar, *Acta Cryst.*, vol. 5, p. 162, 1952. P. Gay and A. Kelly, *Acta Cryst.*, vol. 6, pp. 165, 172, 1953. A. Kelly, *Acta Cryst.*, vol. 7, p. 554, 1954.

tation; a point *on* the axis of rotation is not elongated. With more complex distortions each point is enlarged into an area on the sphere.

With some imperfections and distortions, discussed later in this chapter, the reciprocal lattice points may also spread in other directions following changes of their distance to the origin of reciprocal space; i.e., Debye rings and diffraction peaks may broaden into a range of Bragg angles, and individual spots may spread in one direction or in many directions. Stacking faults make a contribution to the broadening of spots, as do also inhomogeneous residual stresses and small crystallite sizes.

Inhomogeneous residual strains A uniformly stressed body has interplanar spacings altered to values that can be measured by the displacement of Debye rings of polycrystalline or powder patterns or by the change of Bragg angles of single crystals. The altered diffraction angles yield data from which the state of stress in the irradiated area of the specimen can be computed (see Chap. 17). However, the stresses thus computed are *macroscopic*, either those homogeneous on a macroscopic scale, or the macroscopic average of stresses that vary on a microscopic scale, *microstresses*. Atomic displacements may even vary on a scale of only tens of angstroms. They may arise from inhomogeneous plastic flow connected with the development of slip lamellae, deformation bands, or deformation twins, or from inhomogeneous flow near precipitates, inclusions, twins, subboundaries, grain boundaries, cracks, voids, and notches. Even the anisotropic thermal contraction of differently oriented grains in a polycrystalline metal or the unequal contraction of adjacent phases in an alloy can introduce microstresses. The inhomogeneities in the internal stress pattern cause a range of lattice spacings, consequently a range of Bragg angles in diffraction, and therefore a broadening of Debye rings in powder diffraction patterns.[1]

To see the effect of inhomogeneous strains on the reciprocal lattice, consider a cubic crystal subjected to a tensile stress along one axis.[2] The reciprocal lattice will be compressed in the direction along which the crystal is stretched, because of the reciprocal relationship. Because the crystal contracts laterally, the reciprocal lattice will expand in directions normal to the tensile stress. If the stress varies in magnitude throughout the crystal but is constant in direction, the points will be spread into lines, each point on a line corresponding to the position of a reciprocal point from a portion of the crystal that is under a certain stress. Along any row in the reciprocal lattice that extends out from the origin, the length of the lines will be proportional to their distance from the origin. If the stresses vary in direction as well as in magnitude, each point will elongate into bundles of nonparallel lines that fill a small volume surrounding it. This volume will

[1] Some prefer to say that it is the *strain* that is inhomogeneous, since it is strain, not stress, that is directly computed from a Bragg angle, and it is the amplitude of the fluctuations of residual strain that is inferred from the broadening. The term "stress," however, seems to imply somewhat more clearly the elastic, nonequilibrium nature of the distortions.

[2] This discussion follows H. Lipson in "Symposium on Internal Stresses in Metals and Alloys," p. 35, Institute of Metals, London, 1948.

not be spherical, in general, because even if the average stresses in the various directions were equal, the value of Young's modulus varies with orientation in the crystal. Even with volumes of irregular shape at each reciprocal lattice point, however, it will be true that the sizes of these volumes along any row extending radially from the origin will be proportional to their distance from the origin. Since $r^* = 1/d = (2 \sin \theta)/\lambda$, the spread in r^* at any point, δr^*, is proportional to r^* for the points in any such row. Therefore, if $\delta(2 \sin \theta)/\lambda$ is proportional to $(2 \sin \theta)/\lambda$, then $(\cos \theta \; \delta\theta)/\lambda$ is proportional to $(\sin \theta)/\lambda$, and finally, $\delta\theta$ is proportional to $\tan \theta$ and independent of λ. In actual crystals strain broadening is generally associated with particle-size broadening and often also with stacking-fault broadening.

Much of the recent work on broadening is based on Fourier analysis of the intensity data. While specific questions, such as whether one type of fault or another is predominant, can be answered by simplified tests, it is preferable in the analysis of diffraction patterns to use all data that can be obtained from a precise determination of line profiles of several lines, to ignore no known causes of broadening, to make as few prior assumptions as possible in the analysis, and to use Fourier series for the analysis. Detailed treatments have been published by Guinier[1] and by Warren.[2] Warren's work contains not only a thorough explanation of the Fourier method but also a discussion of experimental procedures. A section on Fourier analysis of particle-size broadening is given in Chap. 7 (page 156) and a general discussion of the fault problem is presented below. The fundamental principles of Fourier-transform theory as applied to crystallographic problems in general have been summarized by Lipson and Taylor.[3]

Small coherent domains Imperfections left in a crystal from growth or plastic deformation effectively reduce the size of the regions that diffract as coherent crystallites. Consider the reciprocal lattice for crystals or coherent domains within crystals that are extremely thin in one dimension. Each reciprocal lattice point is elongated into a line, because of lack of resolving power (relaxation of one of the Laue conditions). All lines are of equal length and are normal to the plane of the crystal platelet. If more than one dimension is small, the points become enlarged into three-dimensional volumes, but again are of equal size. For points along a given row, δr^* is a constant, so that $(2 \cos \theta \; \delta\theta)/\lambda$ is a constant and $\delta\theta$ is proportional to $\lambda/\cos \theta$. Thus, line broadening from small crystallite size follows a different law from that produced by stresses: crystallite-size broadening in reciprocal space is the same in different orders of reflection of a plane, but strain broadening varies in the different orders. In Chap. 7 an analysis of line broadening in terms of crystallite size is presented which assumes no

[1] A. Guinier in "Théorie et Technique de la Radiocristallographie," Dunod, Paris, 1956; "X-Ray Diffraction in Crystals, Imperfect Crystals and Amorphous Bodies," Freeman, San Francisco, 1963.

[2] B. E. Warren, *Progr. Metal Phys.*, vol. 8, p. 147, 1959.

[3] H. Lipson and C. A. Taylor, "Fourier Transforms and X-ray Diffraction," G. Bell, London, 1958.

strain broadening. In a following section, a Fourier method for separating the two is given.

Stacking faults Faults in the stacking sequence of atomic planes (Chaps. 10, 14, and 18) produce characteristic alterations of the reciprocal lattice. If faults occur on a single set of planes, such as the basal planes of hexagonal cobalt, certain reciprocal lattice points will be elongated normal to the faulted planes, and others will be left sharp; the broadening is thus unlike the crystallite-size broadening, which affects *all* points. The unaltered points correspond to reflecting planes that have structure factors unaltered by the fault displacements. Thus, in c.p.h. crystals, $00 \cdot l$ points remain sharp for any value of l, and $hk \cdot l$ points remain sharp when $h - k = 3n$, where n is any integer, including zero.

The various points that are elongated may belong to more than one class with regard to the effect of the faults upon their reflecting power, but all points in a given class are elongated equally, by an amount that increases with the density of the faults. Therefore, as in crystallite-size broadening, points along a given radial row in the reciprocal lattice that belong to the same class will have constant δr^* and, as in crystallite-size broadening, will have $\delta\theta$ proportional to $\lambda/\cos\theta$. If faulting occurs on *several* sets of planes, such as the $\{111\}$ planes of f.c.c. crystals, several "spikes" may extend outward from a reciprocal lattice point. Although stacking-fault broadening for a given set of planes increases with $\lambda/\cos\theta$, as does crystallite-size broadening, an analysis of different hkl reflections would yield different apparent crystallite sizes.

The distribution of intensity along a spike contains information regarding the density of faults (i.e., the stacking-fault probability), the distribution of the faults, and the nature of the faults. Highly developed treatments are now available that give the intensity relationships to be expected in powder patterns for different fault types and also give methods that aim at inferring from the observed intensity distributions the fault types and the faulting probability for each type.[1]

Plastic deformation may produce deformation (or intrinsic) faults (see page 388) on $\{111\}$ planes of a f.c.c. metal, changing the sequence $ABCABC$ to $ABCA/CAB$, or twin faults (growth faults) which change the normal sequence to the reverse sequence, $ABCA/CBA$. Both of these are illustrated diagrammatically in Fig. 16–1 by a change in direction of the stacking pattern. Deformation faults in f.c.c. metals produce a shift in the position of certain peaks in a powder diffraction pattern, whereas twin faults cause line profiles to become asymmetric. Deformation-fault probabilities are usually designated by a parameter α (viz., $1/\alpha$ is the mean distance between deformation-fault planes) and twin-fault probabilities by β. The extrinsic fault, with probability coefficient α', alters the sequence to $ABCA/C/BABC$ (Fig. 16–1c), with the layer $/C/$, in effect, a plane

[1] A detailed discussion of the methods for studying the stacking fault probability and the obtained results is given by J. W. Christian and P. R. Swann in T. B. Massalski (ed.), "Alloying Behavior and Effects in Concentrated Solid Solutions," Gordon and Breach, New York, 1965.

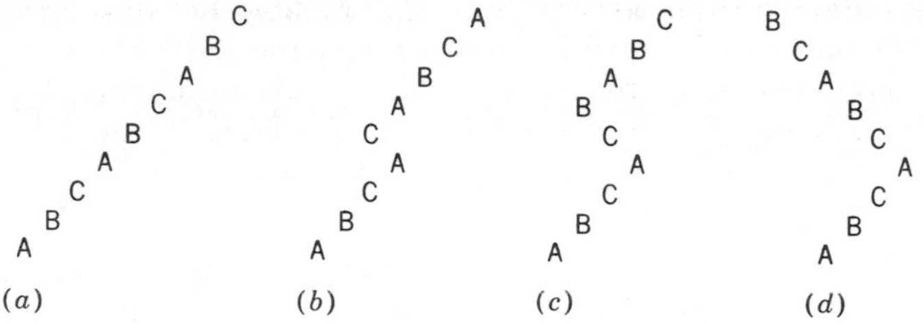

Fig. 16-1 Stacking faults in the f.c.c. lattice represented as sequence of (111) planes. (*a*) Normal f.c.c. lattice. (*b*) Deformation (intrinsic) fault. (*c*) Double deformation (extrinsic) fault. (*d*) Twin fault.

inserted into the regular sequence. This type has also been called a *double deformation fault*. Its effect on the reciprocal lattice and diffraction patterns differs from the effects of other faults.[1] In f.c.c. crystals, extrinsic faults cause both peak shifts and peak asymmetry in the x-ray powder patterns, and the directions of shifts are exactly equal and opposite to those expected from intrinsic faults. Hence, if α' is present but ignored in the evaluation of peak shifts, the obtained α coefficient will be less than the true value. It appears that extrinsic faults may be produced in some of the rare earths, e.g., cerium.[2]

Many of the theories of diffraction from faulted structures are properly applied only to randomly spaced faults (i.e., the probability of a fault between two neighboring atom layers is assumed to be independent of the position of all other faults). But there is evidence that nonrandomness occurs in some specimens with a given type of fault,[3] which implies that more than one parameter is needed in the analysis of such cases. In some cases this nonuniformity may consist simply of densely packed groups of randomly spaced faults separated by relatively fault-free regions, in which case there would be merely a superposition of unbroadened and broadened reflections. A more serious problem to analyze would be one in which the density distribution of faults followed a complex pattern related to the deformation process. This may be expected, for example, in filings, and its contribution must not be overlooked (see page 464). Many treatments also assume that faults occur on only one set of equivalent planes (e.g., one set of $\{111\}$ planes of f.c.c. crystals), whereas cold-worked grains of a polycrystalline sample undoubtedly have faults distributed on *all* sets. The

[1] C. A. Johnson, *Acta Cryst.*, vol. 16, p. 490, 1963. B. E. Warren, *J. Appl. Phys.*, vol. 34, p. 1973, 1963. See also Christian and Swann, loc. cit.

[2] H. M. Otte and H. Chessin in J. E. Hilliard and J. B. Cohen (eds.), "Local Atomic Arrangements Studied by X-Ray Diffraction," Gordon and Breach, New York, in press.

[3] C. S. Barrett, *Trans. AIME*, vol. 188, p. 123, 1950. J. Singer and G. Gashurov, *Acta Cryst.*, vol. 16, p. 601, 1963. H. M. Otte, *Acta Met.*, vol. 2, p. 349, 1954; vol. 5, p. 614, 1957. Faults occur in a regular or ordered sequence in Cu-Al martensite (H. Warlimont and M. Wilkens, *Acta Met.*, vol. 11, p. 1099, 1963).

fault probabilities obtained from powder data can probably be considered to be the sum of the individual probabilities for the separate sets of equivalent planes to a good enough approximation.[1] Most analyses assume that spacings across the faulted plane are the normal ones, unchanged, and most tests support this assumption.[2] Experiments on niobium and a stainless-steel b.c.c. martensite, however, disclosed peak shifts that were ascribed to spacing changes.[3]

As mentioned previously (Chap. 14, page 392), early work has established that the stacking-fault energy varies in magnitude amongst the various f.c.c. metals and alloys. Practically no faults are produced by deformation in Al and only a few in Ni, but faults are more profuse following deformation of Ag, Au, Cu, and Pb (the probability of faults decreasing more or less in that order). Examples of stacking disorders have been found in a steadily expanding list, starting with the early work on cobalt[4] and followed by a large number of investigations of filed or deformed f.c.c. metals and alloys that have relatively low stacking-fault energies.

Stacking disorders have also been found in the micas,[5] graphite,[6] and many other layer structures, and in various solid rare gases and their solid solutions with f.c.c. or c.p.h. structures and van der Waals binding.[7] They may occur in crystals as grown, but are far more common after deformation or after a phase transformation of the martensitic type (see page 351, Chap. 18). Plastic deformation is particularly effective in producing faults in f.c.c. solid solutions at temperatures and compositions for which c.p.h. and f.c.c. phases have nearly the same free energy,[8] where stacking-fault energies are extremely low. Severe deformation under these conditions may even produce an amorphous state, with crystallite sizes no bigger than 10 A or so, in van der Waals crystals[9] and in metals when subjected to extreme shearing strains by Bridgeman's method.[10]

Fourier analysis of particle-size and strain broadening Fourier analysis can now be used for interpreting the precise shape (profile) of

[1] B. E. Warren and E. P. Warekois, *Acta Met.*, vol. 3, p. 473, 1955.

[2] C. N. J. Wagner, *Arch. Eisenhüttenw.*, vol. 29, p. 489, 1958 (Fe). O. J. Guentert and B. E. Warren, *J. Appl. Phys.*, vol. 29, p. 40, 1958 (beta-brass).

[3] C. N. J. Wagner, A. S. Tetelman, and H. M. Otte, *J. Appl. Phys.*, vol. 33, p. 3080, 1962 (Ta, Nb, stainless-steel martensite).

[4] O. S. Edwards and H. Lipson, *Proc. Roy. Soc. (London)*, vol. A180, p. 268, 1942. T. R. Anantharaman and J. W. Christian, *Acta Cryst.*, vol. 9, p. 479, 1956.

[5] S. B. Hendricks, *Phys. Rev.*, vol. 57, p. 448, 1940.

[6] B. E. Warren, *Phys. Rev.*, vol. 59, p. 693, 1941.

[7] L. Meyer, C. S. Barrett, and P. Haasen, *J. Chem. Phys.*, vol. 40, p. 2744, 1964. C. S. Barrett and L. Meyer, *J. Chem. Phys.*, vol. 41, p. 107, 1965.

[8] C. S. Barrett in "Imperfections in Nearly Perfect Crystals," Wiley, New York, 1952, p. 97. C. S. Barrett and M. A. Barrett, *Phys. Rev.*, vol. 81, p. 311, 1951.

[9] Barrett and Meyer, loc. cit. (Ar-N_2).

[10] P. W. Bridgeman, *J. Appl. Phys.*, vol. 8, p. 328, 1937.

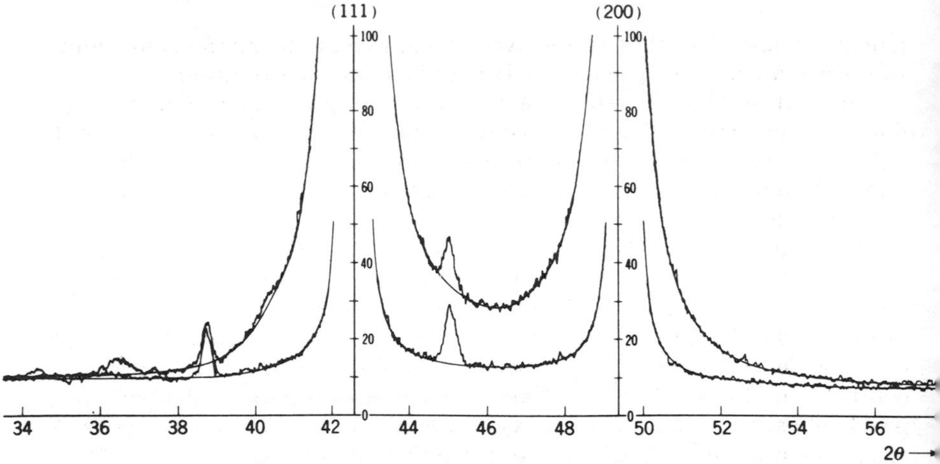

Fig. 16-2 Recordings at room temperature of the (111) and (200) reflections from cold-worked and annealed filings of 70–30 alpha-brass using crystal monochromated Cu $K\alpha$ radiation. Small peaks at 38.8 and 45° are from traces of LiF put on the sample surfaces for 2θ calibration purposes. The small peak at $2\theta = 36.5°$ has the position of the strongest reflection in Cu_2O and ZnO. (After B. E. Warren, *Progr. Metal Phys.*, vol. 8, p. 147, 1959.)

diffraction peaks in terms of microstrains, particle size, and stacking faults, and can be applied when more than one of these causes is active.[1]

The first step is to subtract from each measured intensity value a correction for background intensity. This may involve some uncertainty, for a powder diffraction peak of a cold-worked metal often has tails that extend many degrees from the center of the peak, as will be seen from Fig. 16–2, which shows the 111 and 200 reflections from cold-worked 70–30 alpha-brass with crystal-monochromated radiation, compared with the curve for the same filings after annealing. A curve for the annealed sample aids in estimating the background intensity. The example shown in Fig. 16–2 illustrates the need for caution in judging background intensities, since the long tails and the overlapping of two tails can easily lead to choosing too high a background for cold-worked samples. If peaks are very broad, it may be desirable to first correct the diffracted intensity values by the Lorentz-polarization factor, the absorption factor, and the atomic scattering factor f. It may then be advisable to obtain the profile of the $K\alpha_1$ peak uninfluenced by overlap with the $K\alpha_2$ peak—unless the doublet separation is negligible compared with the broadening, or unless a precision

[1] B. E. Warren, *Prog. Metal Phys.*, vol. 8, p. 147, 1959. B. E. Warren and B. L. Averbach, *J. Appl. Phys.*, vol. 21, p. 595, 1950. B. E. Warren and E. P. Warekois, *Acta Met.*, vol. 3, p. 473, 1955. C. N. J. Wagner, *Acta Met.*, vol. 5, p. 427, 1957. J. B. Cohen and C. N. J. Wagner, *J. Appl. Phys.*, vol. 33, p. 2073, 1962. C. R. Houska and B. L. Averbach, *Acta Cryst.*, vol. 11, p. 139, 1958. B. E. Warren, *Acta Met.*, vol. 11, p. 995, 1963. J. B. Cohen, "Diffraction Methods in Materials Science," Macmillan, New York, 1965. A somewhat different Fourier analysis that yields rather similar results has been developed by J. W. Christian and J. Spreadborough, *Proc. Phys. Soc.* (*London*), vol. B70, p. 1151, 1957.

monochromator has already removed the $K\alpha_2$. Various methods for doing this are available (see page 463). It may be done on a computer.

The next step is to correct for instrumental broadening. This may be done by the method of Stokes,[1] as mentioned in Chap. 7 (page 156) with either Bevers-Lipson strips[2] and a desk calculator, or by a computer program.[3] Removing instrumental broadening is an unfolding operation that involves the Fourier coefficients representing the profile for a standard sample that lacks broadening from distortion, particle size, or faults. For the standard, a sample is annealed enough to sharpen lines but not enough to produce grain growth that could reduce intensities by extinction. When proper background corrections are made, Warren finds that cold-working does not lessen the integrated intensity of the reflections from a metal, contrary to the conclusions of some earlier investigators. The corrected profile of each reflection is then expressed by its separate Fourier series. A brief presentation of the Fourier-series approach follows.

A general distortion is considered in which the position of a unit cell $m_1m_2m_3$ is given by the vector $\mathbf{R}$. In terms of the unit cell vectors $\mathbf{a}_1$, $\mathbf{a}_2$, and $\mathbf{a}_3$,

$$\mathbf{R}_{m_1m_2m_3} = m_1\mathbf{a}_1 + m_2\mathbf{a}_2 + m_3\mathbf{a}_3 + \boldsymbol{\delta}_{m_1m_2m_3} \qquad (16\text{–}1)$$

The displacement vector $\boldsymbol{\delta}_m = X_m\mathbf{a}_1 + Y_m\mathbf{a}_2 + Z_m\mathbf{a}_3$ is different, in general, for each cell $m = m_1m_2m_3$. Let the direction of the primary and diffracted beams be represented by the unit vectors $\mathbf{S}_0$ and $\mathbf{S}$, so that their difference gives

$$\mathbf{S}_0 - \mathbf{S} = \lambda(h_1\mathbf{b}_1 + h_2\mathbf{b}_2 + h_3\mathbf{b}_3) \qquad (16\text{–}2)$$

where $\mathbf{b}_1$, $\mathbf{b}_2$, and $\mathbf{b}_3$ are the vectors of the reciprocal lattice and h_1, h_2, h_3 are continuous variables. The intensity from the crystal is then related to the displacement at *pairs* of cells—cells at $\mathbf{R}_{m_1m_2m_3}$ and $\mathbf{R}_{m_1'm_2'm_3'}$—by the relation

$$I(h_1h_2h_3) = F^2 \sum_{m_1}\sum_{m_2}\sum_{m_3}\sum_{m_1'}\sum_{m_2'}\sum_{m_3'} e^{(2\pi i/\lambda)(\mathbf{S}-\mathbf{S}_0)\cdot(\mathbf{R}_m - \mathbf{R}_{m'})} \qquad (16\text{–}3)$$

This distribution of intensity in reciprocal space is then integrated throughout reciprocal space so as to give the power in the corrected plot of intensity distribution vs. 2θ in a powder diffraction line. For convenience, crystal axes $\mathbf{a}_1$, $\mathbf{a}_2$, and $\mathbf{a}_3$ are chosen so that each reflection is of the type $00l$. In this way, the contributions to the intensity are thought of as arising from columns of cells in the crystal, each column being parallel to $\mathbf{a}_3$ (normal to the reflecting plane). The summation with respect to m_3 and $m_{3'}$ is carried out for all pairs in a given column, and the other summations

[1] A. R. Stokes, *Proc. Phys. Soc.* (*London*), vol. 61, p. 382, 1948.

[2] C. A. Bevers, *Acta Cryst.*, vol. 5, p. 670, 1952.

[3] R. J. De Angelis and L. H. Schwartz, *Acta Cryst.*, vol. 16, p. 705, 1963 (program for IBM 650 and 709 computers). E. N. Aqua, to be published. Other programs are also cited by these authors.

are to add the contributions from the different columns. Following Warren's notation, let N_3 be the average number of cells in a column of a coherently diffracting domain, averaged over all diffracting domains in the sample, and let $n = m_3 - m_{3'}$. Consider an arbitrary displacement, varying with position m and having a Z component equal to $Z(m_3)$ and $Z(m_{3'})$ at m_3 and $m_{3'}$, respectively; let $Z_n = Z(m_3) - Z(m_{3'})$. Let $N_1N_2N_3 = N$ be the average number of cells per domain, and let $N_n(m_1m_2)$ be the number of cells with an nth neighbor in the same column. Then the distribution of diffracted power per unit length of the diffraction line, $P'(2\theta)$, is

$$P'_{2\theta} = K(\theta)N \sum_{-\infty}^{+\infty}{}_n \{A_n \cos 2\pi nh_3 + B_n \sin 2\pi nh_3\} \tag{16–4}$$

where
$$A_n = \frac{N_n}{N_3} \langle \cos 2\pi lZ_n \rangle$$

$$B_n = \frac{N_n}{N_3} \langle \sin 2\pi lZ_n \rangle$$

$$K(\theta) = \frac{G(u + b)}{|\,\mathbf{b}_3\,| \sin^2\theta}$$

In this expression, G contains the θ-dependent functions that are due to the variation of the atomic scattering factor f, and to the Lorentz-polarization factor; it has the form $f^2(1 + \cos^2 2\theta)/\sin^2 \theta$. It also contains a numerical factor to take account of the number of components, b, of a given peak that are affected by stacking faults, and the number u that are unaffected. *Component* is defined as a reflection from one set of parallel planes of a crystal, for example 111, as distinct from, say $11\bar{1}$. The Fourier coefficients A_n and B_n are therefore concerned with the *corrected* intensities. The averages $\langle \cos 2\pi lZ_n \rangle$ and $\langle \sin 2\pi lZ_n \rangle$ are averages over pairs that are nth neighbors in all columns in the sample. If there are equal probabilities for positive and negative Z_n values in the sample, the sine coefficients are zero and the peaks will be symmetrical; otherwise, there will be peak displacements and peak asymmetries. It is convenient to take an origin θ_0 near a peak center and to let h_3 be replaced in Eq. (16–4) by $h_3 - l = 2a_3(\sin \theta - \sin \theta_0)/\lambda$.

In the Warren-Averbach method for separating the particle-size and distortion coefficients, each coefficient A_n is taken as the product of a particle-size coefficient $A_n{}^S = N_n/N_3$ and a distortion coefficient $A_n{}^D = \langle \cos 2\pi lZ_n \rangle$. Since $A_n{}^S$ is independent of the order of reflection, l, whereas $A_n{}^D$ is a function of l and equal to unity when $l = 0$, measurements for a series of orders of reflections, $00l$, such as 001, 002, 003, etc., provide coefficients that can be written in the form

$$\ln A_n(l) = \ln A_n{}^S + \ln A_n{}^D(l). \tag{16–5}$$

The experimentally determined values of A_n must be normalized to unity for $n = 0$. If $\ln A_n(l)$ is plotted against l^2 for a given value of n, an extrapolation to $l = 0$ then gives directly $\ln A_n^S$. For small values of l and small n, such that Z_n is small, Eq. (16–5) can be put in the form

$$\ln A_n(l) = \ln A_n^S - 2\pi^2 l^2 Z_n^2 \qquad (16\text{–}6)$$

which shows that the extrapolation function will be linear if $\ln A_n(l)$ is plotted against l^2. Therefore, a plot vs. l^2 rather than l should be used. (Extrapolation can then be done if only two reflections are available, e.g., 200 and 400 in f.c.c. powders—600 is overlapped by another peak.) The curves are plotted for a series of values of lengths L, where $L = |\mathbf{a}_3|\, n$. The mean-square strain averaged over the length L along a column and averaged over all regions in the sample is $\langle \epsilon_L^2 \rangle = (\Delta L/L)^2$.

Small $\langle \epsilon^2 \rangle$ should give linear curves of $A_n(l)$ vs. l^2, and large strains should also be linear if they have a Gaussian distribution (which has been found in several cases). But if the strains fall off with increasing L more slowly than Gaussian, say as $\exp[-a\,|\epsilon|]$, the curves should be concave upwards, and if the strains fall off more rapidly than Gaussian, say as $\exp[-a^3\,|\epsilon^3|]$, the curves should be concave downwards, i.e., should slope more steeply, at higher values of l^2.[1] It is usually found that the root-mean-square strains $\langle \epsilon_L^2 \rangle^{1/2}$ plotted vs. L show a rapid increase at small values of L, indicating that the distortions are inhomogeneous and also possibly are greater in the smaller domains.

For cubic crystals of unit cell size a_0, with $h_0^2 = h^2 + k^2 + l^2$ and $a_3/l = d_{hkl} = a_0/h_0^2$, Eq. (16–6) becomes

$$\ln A_L(h_0) = \ln A_L^S - 2\pi^2 h_0^2 \langle (\Delta L)^2 \rangle / a_0^2 \qquad (16\text{–}7)$$

A series of experimentally determined curves for different lengths L in the sample are shown in Fig. 16–3.[2] These were obtained from cold-worked filings of thoriated tungsten in the following way. Coefficients A_l were obtained by the Stokes method from Fourier coefficients $A_l(\text{cw})$ of the cold-worked sample and $A_l(\text{ann})$ of the annealed sample by the Stokes relation $A_l = A_l(\text{cw})/A_l(\text{ann})$, which is valid if the asymmetry is small. Using $L = a_3 n$, curves for each of 8 hkl reflections with h_0^2 ranging from 2 to 16 were plotted vs. L, and from these curves at chosen values of L the $\ln A_L$ vs. h_0^2 curves of Fig. 16–3 were plotted. For this metal and also for aluminum filed under liquid nitrogen and measured at $-160°$C, it was found[3] that points for the different hkl reflections fall on a single curve for constant L, because of the isotropy of these metals and the absence of appreciable numbers of stacking faults.

The intercepts at $h_0 = 0$ in Fig. 16–3 are not zero; on the contrary, they indicate that the particle-size broadening is rather large. If the intercepts of such A_L^S curves are plotted vs. L, the initial slope gives the mean

[1] B. E. Warren, *Acta Met.*, vol. 11, p. 995, 1963. These microstrains include the macrostrain, if the origin for the Fourier analysis is chosen to be the same as that for the annealed peak—but not if the origin is that for the cold-worked peak (J. B. Cohen, private communication).

[2] M. McKeehan and B. E. Warren, *J. Appl. Phys.*, vol. 24, p. 52, 1953.

[3] C. N. J. Wagner, *Acta Met.*, vol. 5, p. 477, 1957.

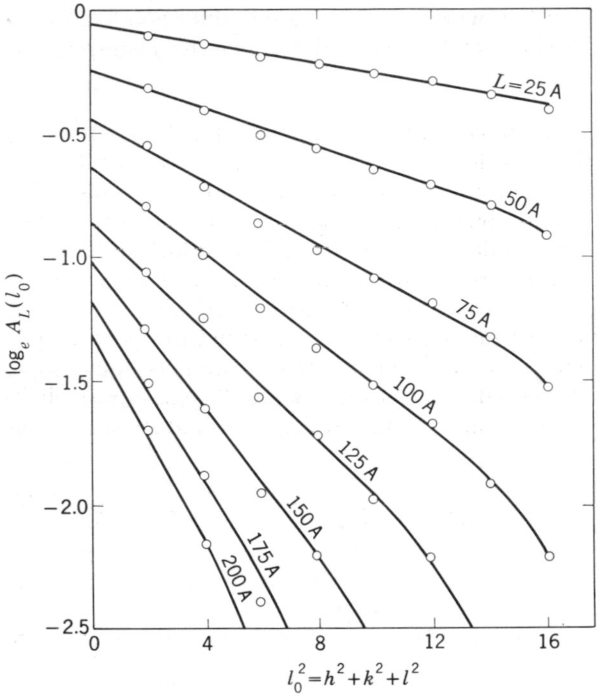

Fig. 16-3 Plot of ln $A_L(l_0)$ vs. $l_0^2 = h^2 + k^2 + l^2$ for cold-worked thoriated tungsten filings. (M. McKeehan and B. E. Warren, *J. Appl. Phys.*, vol. 24, p. 52, 1953).

volumn length $\bar{L}$.[1] For the example given in Fig. 16–3, it was found that $\bar{L}$ = 200 A, which is to be interpreted as the average effective size of the coherently diffracting domains. The distribution function of the column lengths is given[2] by a plot of the second derivative of $A_L{}^S$ vs. L.

Thus, from plots such as Fig. 16–3 the intercepts at $h_0{}^2 = 0$ give particle-size coefficients from which the mean size of coherently diffracting domains and the size-distribution function can be determined. The initial slopes give the mean-square strain.

In the experiment of Fig. 16–3 the strain broadening gave values of $\langle \epsilon_L{}^2 \rangle^{1/2}$ varying from 0.004 at L = 25 A to 0.0015 at L = 200 A, but these values were judged to be uncertain because the strain broadening was small compared to the particle-size broadening. Unlike the situation in tungsten, which has isotropic elastic constants, the strains in cold-worked

[1] B. E. Warren and B. L. Averbach, *J. Appl. Phys.*, vol. 21, p. 595, 1950. M. F. Bertaut, *Compt. Rend.*, vol. 228, p. 492, 1949.

[2] In many experiments the curve of $A_L{}^S$ vs. L is found to bend downward near $L = 0$; known as the *hook effect*, this indicates the presence of errors that should be corrected for. A correction has been applied by extrapolating the linear part of the curve and then renormalizing to make the new (extrapolated) intercept equal to unity. B. Y. Pines and A. F. Sirenko, *Soviet Phys. Cryst.*, vol. 7, no. 1, p. 15, 1962.

anisotropic materials will differ with crystallographic direction, and only orders of a particular (hkl) can then be used to determine the corresponding strain function.

Analysis of stacking-fault broadening We now turn to a more detailed consideration of the broadening from faults. Since faults interrupt the coherent pattern of a diffracting crystal, faults produce broadening that is similar to particle-size broadening in that it is independent of order of reflection for the particular reflections that are affected; but fault broadening differs from particle-size broadening in that some reflections remain unaltered by faults and others are broadened unsymmetrically.

Theoretical treatments of diffraction from faulted structures now include analyses for all three typical metallic structures as well as some more complex structures. In f.c.c. structures, deformation faults cause peak shifts and symmetrical peak broadening in the diffraction patterns; double

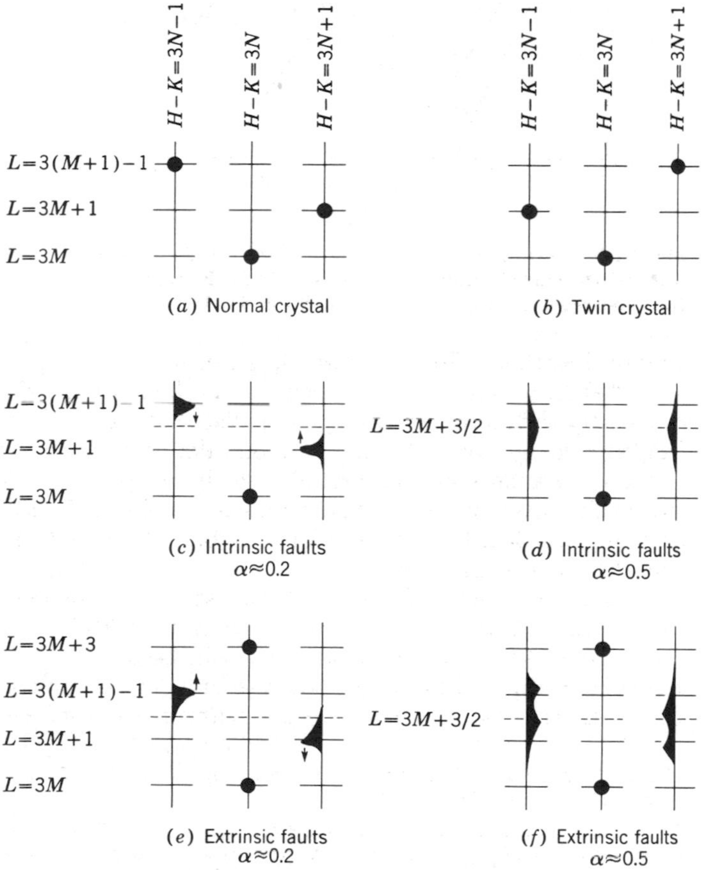

Fig. 16-4 Diagrammatical maps of intensity distribution in reciprocal space for crystals with stacking faults on HK·L planes.

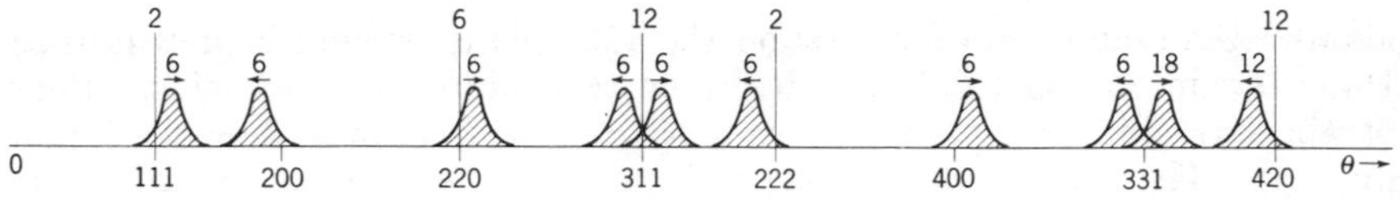

Fig. 16-5 The influence of deformation faulting on the x-ray powder pattern; the vertical lines represent the sharp components and the shaded areas, the diffuse components. Marks just above the *hkl* indices indicate positions of the reflections from unfaulted lattice. (After **M. S. Paterson, *J. Appl. Phys.*, vol. 23, p. 805, 1952.**)

deformation faults (extrinsic) cause both nonsymmetrical broadening and peak shifts; and twin faults produce essentially only a nonsymmetrical peak broadening.[1] The problem of separating the effects due to residual stresses and particle-size broadening from those due to stacking faults requires analysis of some diffraction peaks. To obtain trustworthy results in such an analysis is difficult; an investigator should have a thorough knowledge of both the theoretical and the experimental limitations.

For f.c.c. metals the theoretical prediction regarding peak shifts and peak broadening may be conveniently summarized with the aid of plots of the reciprocal lattice.[2] For this purpose the axes of the reciprocal lattice may be chosen so that $\mathbf{c}^*$ is normal to the plane of faulting. In the case of the f.c.c. structure the reflections then have indices $HK \cdot L$ based on hexagonal axes with a_1 and a_2 lying in the close-packed plane. Points in each row parallel to $\mathbf{c}^*$ have H and K indices belonging to one of the three classes $H - K = 3N$, $3N + 1$, and $3N - 1$, and each point in a given row is distinguished by $L = 3M$, $3M + 1$, and $3M - 1$, where N and M are integers. Then all the diffraction peaks can be described with reference to a diagram, as in Fig. 16–4. The black circles in Fig. 16–4*a* represent reflections from a perfect f.c.c. crystal, and those in Fig. 16–4*b* represent reflections from its twin. When deformation faults are introduced, reflections with $H - K = 3N$ and $L = 3M$ remain sharp, but other reflections broaden and their maxima shift. This is shown diagrammatically in the successive sketches in Fig. 16–4 for the different types of faults and for different values of the faulting parameter. Paterson showed that deformation faults cause the types of shifts of powder diffraction peaks that are illustrated schematically in Fig. 16–5. Subsequently, various detailed analyses were developed, which will be found in the references cited in this chapter.

In b.c.c. metals, faults should occur preferentially in $\{112\}$ planes; Warren has shown that a measurement of asymmetry should yield values of β, although the effects should be smaller than with f.c.c. metals. If the center of gravity of peaks near $2\theta = 90°$ can be determined to $\pm 0.01°$ 2θ,

[1] Parameters used to denote stacking-fault probabilities are α for intrinsic (deformation) faults, β for twin (growth) faults, and α' for extrinsic (double deformation) faults.

[2] M. S. Patterson, *J. Appl. Phys.*, vol. 23, p. 805, 1952. C. A. Johnson, *Acta Cryst.*, vol. 16, p. 490, 1963.

about 1 twin fault in every 1800 on all {112} planes should be detectable.[1] However, in careful tests[2] no detectable twin faults were found in filings of Fe, Nb, Ta, or beta-brass, even when the filing was done in liquid nitrogen. There is no possibility of determining α in b.c.c. metals from *shifts* of diffraction peaks of randomly oriented powders because all shifts average to zero for any α. If there is a change in the interplanar spacing of {112} planes where there is a stacking fault, as would be expected from the atom positions across a fault plane, there should be a change in the lattice constant from the normal lattice constant. This change has been detected and related to α in b.c.c. martensite (Fe–30 weight percent Ni) but not in an alloy of Fe–16 percent Cr–25 percent Ni when these were filed in liquid nitrogen.[3]

In c.p.h. materials there are no peak displacements and no peak asymmetries resulting from either deformation faults or growth faults, but there is peak broadening. The early x-ray work was based on the assumption of only one kind of fault, the growth fault, in hexagonal cobalt,[4] but interpretations requiring a mixture of growth and deformation faults have been published subsequently.[5] Values of α and β may be obtained, at least in principle, from measurements of the integral breadths of various reflections, from Fourier coefficients giving the shape of a particular reflection, or from measurements of integrated intensities. Both types of faults produce symmetrical broadening of the peaks for which $H - K = 3N \pm 1$, where N is any integer (see Table 16–1), but the broadening is different for lines that are "odd" or "even" with respect to the number L in the index $HK \cdot L$, based on the usual hexagonal cell. For growth faults, the diffraction breadth of the even lines is three times that of the odd lines, and the intensity of all lines decreases as β increases. For the deformation faults, however, even and odd lines retain the same breadth but the decrease in the intensity of the even lines is more rapid than that of the odd lines as α increases. Instrumental broadening and broadening due to effects other than faults can be eliminated by using lines in the diffraction pattern with $H - K = 3N$ as standards.

Details of methods for determining probabilities for the various types of stacking faults are given in the references cited below, and a critical discussion of some results and some important difficulties is given in the subsequent paragraphs.

General treatments A. J. C. Wilson, "X-Ray Optics," Methuen, London, 1949; *Proc. Roy. Soc. (London)*, vol. A180, p. 277, 1942; *Acta Cryst.*, vol. 2, p. 245, 1949. W. H.

[1] See also C. N. J. Wagner, A. S. Tetelman, and H. M. Otte, *J. Appl. Phys.*, vol. 33, p. 3080, 1962.

[2] DeAngelis and Cohen, loc. cit., page 408, ref. 4.

[3] A. J. Goldman and C. N. J. Wagner, *Acta Met.*, vol. 11, p. 405, 1963.

[4] A. J. C. Wilson, *Proc. Roy. Soc. (London)*, vol. A180, p. 277, 1942.

[5] T. R. Anantharaman and J. W. Christian, *Acta Cryst.*, vol. 9, p. 479, 1956. C. R. Houska and B. L. Averbach, *Acta Cryst.*, vol. 11, p. 139, 1958.

Table 16–1 Diffraction effects from faults (of a single type) in c.p.h. crystals

Type of fault	Fault parameter	Diffraction effects for $H - K = 3N \pm 1$
Growth	$0 \leq \beta \leq 0.5$	Symmetrical broadening of all peaks. Integrated intensity of even peaks, I_{even}, decreases to zero at $\beta = 0.5$. Ratio of breadths, $B_{\text{odd}}/B_{\text{even}} = 1/3$ for all β. Ratio of intensities, $I_{\text{odd}}/I_{\text{even}} = 3, 3.1, 4.2$ and ∞ for $\beta = 0, 0.2, 0.4$, and 0.5.
	$0.5 < \beta < 1$	Only odd lines present. Peaks remain symmetrical and broad.
Deformation	$0 \leq \alpha \leq 0.5$	Symmetrical broadening of all peaks. Ratio of breadths, $B_{\text{odd}}/B_{\text{even}} = 1$ for all α. Ratio of intensities, $I_{\text{odd}}/I_{\text{even}} = 3, 5.5, 35.3$, and ∞ for $\alpha = 0, 0.2, 0.4$, and 0.5.
	$\alpha > 0.5$	Breadths decrease again, and intensity transferred back from odd to even peaks.
	$\alpha = 1.0$	Perfect c.p.h. structure.

For further details, see: J. W. Christian and P. R. Swann, Ref. 1, p. 451, and other references in the text.

Zachariasen, *Phys. Rev.*, vol. 71, p. 715, 1947; *Acta Cryst.*, vol. 1, p. 277, 1948; "Theory of Diffraction of X-rays in Crystals," Wiley, New York, 1945. H. Jagodzinski, *Acta Cryst.*, vol. 2, pp. 201, 208, 298, 1949. J. Mering, *Acta Cryst.*, vol. 2, p. 371, 1949. R. Gevers, *Acta Cryst.*, vol. 7, p. 337, 1954. B. E. Warren and B. L. Averbach, *J. Appl. Phys.*, vol. 21, p. 595, 1950. B. E. Warren, *Progr. Metal Phys.*, vol. 8, p. 147, 1959.

F.c.c. and c.p.h. powder patterns M. S. Paterson, *J. Appl. Phys.*, vol. 23, p. 805, 1952. J. W. Christian, *Acta Cryst.*, vol. 7, p. 415, 1954. B. E. Warren and E. P. Warekois, *Acta Met.*, vol. 3, p. 473, 1955. C. N. J. Wagner, *Acta Met.*, vol. 5, pp. 427, 477, 1957. J. W. Christian and J. Spreadborough, *Proc. Phys. Soc. (London)*, vol. B70, p. 1151, 1957. C. R. Houska and B. L. Averbach, *Acta Cryst.*, vol. 11, p. 139, 1958. B. E. Warren, *Progr. Metal Phys.*, vol. 8, p. 147, 1959; *Australian J. Phys.*, vol. 13, p. 384, 1960. J. B. Cohen and C. N. J. Wagner, *J. Appl. Phys.*, vol. 33, p. 2073, 1962. C. A. Johnson, *Acta Cryst.*, vol. 16, p. 499, 1963 (extrinsic vs. intrinsic faults). C. N. J. Wagner, A. S. Tetelman, and H. M. Otte, *J. Appl. Phys.*, vol. 33, p. 3080, 1962 (spacing changes at fault planes). B. T. M. Willis, *Proc. Roy. Soc. (London)*, vol. A248, p. 183, 1958; *Acta Cryst.*, vol. 12, p. 683, 1959 (effect of Suzuki segregation). D. O. Welch and H. M. Otte, *Advan. X-Ray Anal.*, vol. 6, p. 96, 1963. H. M. Otte and D. O. Welch, *Phil. Mag.*, vol. 9, p. 299, 1964. C. N. J. Wagner, J. P. Boisseau, and E. N. Aqua, *J. Metals*, vol. 16, p. 753, 1964 (effect of macrostrains in faulted structures).

B.c.c. powder patterns P. B. Hirsch and H. M. Otte, *Acta Cryst.*, vol. 10, p. 447, 1957. O. J. Guentert and B. E. Warren, *J. Appl. Phys.*, vol. 29, p. 40, 1958. C. N. J. Wagner, A. S. Tetelman, and H. M. Otte, *J. Appl. Phys.*, vol. 33, p. 3080, 1962 (spacing changes at fault planes). R. J. DeAngelis and J. B. Cohen in R. E. Reed-Hill, J. P. Hirth, and H. Rogers (eds.), "Deformation Twinning," Gordon and Breach, New York, 1965.

In principle, the deformation-fault probability α in f.c.c. samples can be determined in various ways. One of the best is by measuring the opposite displacements of 111 and 200 peaks. With such measurements the errors due to incorrect placement of the specimen with respect to the diffractometer axis are minimized, and also peak shifts due to changes in lattice constants by cold-work are partially compensated for. This method relates α to the relative displacements of the peaks by a simple formula. If the separation $2\theta_{200} - 2\theta_{111}$ of the 200 and 111 peaks of an annealed sample is altered by the amount $\Delta(2\theta_{200} - 2\theta_{111})$ by cold-work and if the diffraction angles are expressed in degrees, α is given by the relation

$$\Delta(2\theta_{200} - 2\theta_{111}) = -\frac{45\sqrt{3}}{\pi^2}(\tan\theta_{200} + \tan\theta_{111})\alpha \qquad (16\text{–}8)$$

In practice, the difficulties encountered in interpreting the measured x-ray peak shifts are experimental as well as theoretical. Step-by-step counting procedures should be used, for peaks are broad and are usually superimposed on a background of considerable intensity. Because of statistical fluctuations, one should not take the position of a peak to be simply the 2θ value with the largest number of counts. Various methods are used with symmetrical peaks, including (1) extrapolating the line of peak midpoints to the point of intersection with the peak and (2) fitting a parabola to the peak (see Chap. 17).

However, symmetrical peaks seldom exist. Causes of asymmetry are: (1) $K\alpha_1$-$K\alpha_2$ doublet, (2) nonflat background, (3) vertical divergence in the x-ray beam, (4) Lorentz-polarization factor, (5) asymmetry in the deformation-intensity distribution function, (6) twin faults, (7) extrinsic faults, (8) segregation at stacking faults, and (9) apparent asymmetry due to incorrect choice of the center of a peak (which serves as the origin in a Fourier analysis).

Separation of the pattern due to $K\alpha_1$ from that due to $K\alpha_2$ is done in various ways. The Rachinger method[1] assumes that there exists a well-defined background level which can be judged from the appearance of the pattern and that a point can be found where the scattering can be assumed to be from only one component of the doublet. This method is not applicable in the case of broad and overlapping peaks. In the Papoulis method,[2] it is assumed that the diffraction profile of $K\alpha_1$ or $K\alpha_2$ is independently symmetrical, a condition not satisfied if some of the causes of asymmetry listed above are involved. The Stokes method[3] has been used most frequently in these and related problems (see Chap. 7, page 156); it is a Fourier method applicable to both symmetrical and unsymmetrical peaks. One normally assumes that the angular dispersion is constant throughout the range of 2θ of each peak that is analyzed. An analytical method of separating one of the $K\alpha$ components from the combined $K\alpha_1 + K\alpha_2$ pattern published by Keating[4] can be applied to the whole or to any part of the diffraction pattern. It assumes only that the ratio of intensities of $K\alpha_1$ and $K\alpha_2$ is known and that their wavelength-intensity profiles are identical (or nearly identical).

Considering the experimental difficulties in stacking-fault work, it is not likely that the positions of peak maxima are meaningful to better than 0.05° in 2θ. The reproducibility of the centroid positions is better, perhaps of the order of 0.01° in 2θ. A major uncertainty is always that of properly judging the background intensity, a problem which becomes more acute as stacking-fault densities increase and as solute content increases. A truncation procedure[5] has been proposed for lessening the effects of this uncertainty.

A principle used in several studies is based on the deviation of apparent a_0 values for individual reflections from a straight line on a Nelson-Riley plot (with cylindrical sample) or a plot vs. the function $\cos\theta\cot\theta$ (with a flat sample, in a diffractometer).[6]

A method of determining the twin-fault probability β which minimizes the difficulties of choosing the correct background intensity is to measure the difference between the

[1] W. A. Rachinger, *J. Sci. Instr.*, vol. 25, p. 254, 1948.

[2] A. Papoulis, *Rev. Sci. Instr.*, vol. 26, p. 423, 1955.

[3] A. R. Stokes, *Proc. Phys. Soc.* (*London*), vol. B61, p. 382, 1948.

[4] D. T. Keating, *Rev. Sci. Instr.*, vol. 30, p. 725, 1959.

[5] E. R. Pike and A. J. C. Wilson, *Brit. J. Appl. Phys.*, vol. 10, p. 57, 1959.

[6] The $\cos\theta\cot\theta$ extrapolation function is used by R. P. I. Adler and C. N. J. Wagner, *J. Appl. Phys.*, vol. 33, p. 3451, 1962.

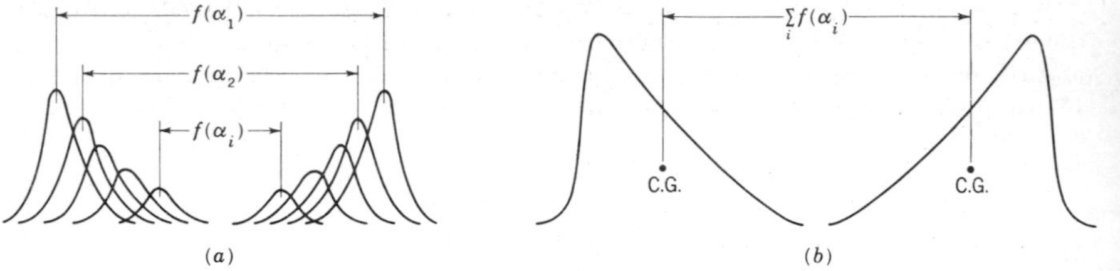

Fig. 16-6 Diagrammatic representation of peak asymmetry due to a particular density distribution of deformation faults. (*a*) Individual components with densities, $\alpha_1 \cdots \alpha_i$. (*b*) Asymmetric peaks resulting from summation of components.

centers of gravity and the *peaks* in the diffraction pattern of f.c.c. metals and alloys, for the displacement of the center of gravity from the positions of the peaks is a function of β.[1] This method of evaluating twin faulting has been successful in the study of electrodeposited films when deformation faulting is absent. However, the problem of evaluating mixed faulting in cold-worked samples, such as metal filings, is more difficult.

Approximate theoretical treatments, applicable to small values of parameters α, β, α', have been worked out and used.[2] Warren points out that when all three types of fault are present it should be very difficult to derive reliable values of the three faulting parameters. Probability evaluation for a sample with mixed faulting is also of questionable validity because the values of α, α', and β may not be uniform throughout the sample, but may differ appreciably from one to another of the coherently diffracting domains.[3] In such a case the contribution from the intrinsic deformation faults alone could produce peak asymmetry in addition to asymmetry due to twin faulting or extrinsic faulting. This is illustrated diagrammatically in Fig. 16–6. Under these conditions, at best only an approximate analysis of the different parameters α, β, and α' seems possible. This constitutes a particularly serious handicap of the x-ray method if *filings* are used, since experiment has shown that there is a tendency for the smaller particles in the filings to have a higher fault probability than the larger particles.[4] Similar difficulties from nonuniform stacking-fault probabilities can also be expected in cold-worked *polycrystalline samples*. A further complication in *alloys* arises from the possibility of segregation of solute atoms to stacking faults, which would also affect the diffraction pattern.

The influence of the mode of deformation on the extent of faulting may be studied if x-ray measurements are made on bulk samples rather than on filings. Since most physical and mechanical properties are determined on bulk samples, and since thinned bulk samples can be examined by transmission electron microscopy, the possibility of comparing x-ray and microscopic examinations of the same material is attractive, but presents certain difficulties. Stacking-fault probabilities have been determined in polycrystalline alloy tensile specimens, using a *method of slopes*, and have been compared with results of electron microscope studies of samples thinned from the bulk specimens.[5]

[1] J. B. Cohen and C. N. J. Wagner, *J. Appl. Phys.*, vol. 33, p. 2073, 1962. R. J. DeAngelis and J. B. Cohen in R. E. Reed-Hill, J. P. Hirth, and H. Rogers (eds.), "Deformation Twinning," Gordon and Breach, New York, 1965.

[2] C. N. J. Wagner, *Acta Met.*, vol. 5, p. 427, 1957. B. E. Warren, *Progr. Metal Phys.*, vol. 8, p. 147, 1959; *J. Appl. Phys.*, vol. 32, p. 2428, 1961; *Australian J. Phys.*, vol. 13, p. 384, 1960; *J. Appl. Phys.*, vol. 34, p. 1973, 1963. DeAngelis and Cohen, loc. cit.

[3] M. Wilkens, *Z. Physik*, vol. 158, p. 483, 1960; *Phys. Stat. Solidi*, vol. 2, p. 692, 1962. D. O. Welch and H. M. Otte, *Advan. X-Ray Anal.*, vol. 6, p. 96, 1964.

[4] T. R. Anantharaman, *Acta Met.*, vol. 9, p. 903, 1961.

[5] H. M. Otte, D. O. Welch, and G. F. Bolling, *Phil. Mag.*, vol. 8, p. 345, 1963. H. M. Otte and D. O. Welch, *Phil. Mag.*, vol. 9, p. 299, 1964.

Although the two methods indicated about the same stacking-fault energies, a microscopic examination of the thinned samples indicated a much lower stacking-fault density in these. From this experiment and various other experiments it is clear that there are complexities involved—such as the changes that may accompany thinning—and that an experimenter in the field should acquaint himself with the various pitfalls discovered in prior investigations.

Relationship between stacking-fault probability α and stacking-fault energy γ
By considering the strain fields around dislocations and using suitable approximations the stacking-fault probability α may be related to the stacking-fault energy γ. Since α expresses the fraction of slip planes that are faulted, it is proportional to $\rho\eta$ where ρ is the dislocation density of those dislocations that have split into partials and η is the fault width. From the quantity $\rho\eta$ the relative value of γ can be obtained only by making suitable assumptions both about η and ρ. If one assumes similar dislocation density for the filed noble metals, and an equilibrium value of η, the x-ray results indicate that $\gamma_{Ag}:\gamma_{Cu}:\gamma_{Au}$ are related as $1:3\frac{1}{4}:4$,[1] in general agreement with the nodal method and stacking-fault tetrahedra.[2]

In the field of f.c.c. alloys, a number of techniques have been used to evaluate γ, involving x-rays, plastic deformation of single crystals (τ_3), electron microscopy of thin films (nodes and tetrahedra), and density of annealing twins. These methods have shown that a pronounced increase in the stacking-fault density, or a lowering of stacking-fault energy, usually occurs with increasing solute concentration in the alloys based on Cu, Ag, and Au, although there are discrepancies between individual methods.

A basic reason why the results obtained thus far by different methods show a rather wide spread is that all "methods" of determining γ are in fact models describing some process in which γ is involved and will give wrong results if the model is wrong. Even in a relatively direct method, such as the measurement of the radii or width of nodes (see Chap. 14, page 391), the derived values of γ depend on a correct and complete theory of node formation. Work at the Cavendish Laboratory[3] indicates that the values of γ based on the nodal method published prior to 1964 are too low, some by a factor of ≈ 2.3. This factor varies with the character (edge or screw) of the node, and it is related to the line tension of the dislocations that are involved; other difficulties are mentioned in Chap. 14.

[1] L. F. Vassamillet, *J. Appl. Phys.*, vol. 32, p. 778, 1961. L. F. Vassamillet and T. B. Massalski, *J. Appl. Phys.*, vol. 34, p. 3402, 1963.

[2] H. M. Loretto, L. M. Clarebrough, and R. L. Segall, *Phil. Mag.*, vol. 10, p. 731, 1964; vol. 11, p. 459, 1965.

[3] L. M. Brown, *Phil. Mag.*, vol. 10, p. 441, 1964.

17

Stress measurement by x-rays

The Bragg angle for the reflection of x-rays from a set of crystal planes is sensitive to all factors that influence the interplanar spacing of the reflecting planes. Since stresses within the elastic range can alter the spacing of reflecting planes, d, enough to change the angle θ in the Bragg law by an easily measurable amount, the magnitude of the stresses that are altering the normal spacing of the planes can be deduced from the observed 2θ angles.

Characteristics of the x-ray methods Only *elastic* strains, not plastic strains, are indicated by the changes in 2θ. This fact is useful in computing residual stresses from the measurements. There are also other important characteristics of the method. With one technique, strains can be determined by comparing interplanar spacings in the stressed and unstressed states of the specimen. But in an alternate technique, strains can be determined from appropriate measurements on the stressed state only. It is therefore possible to study residual stresses in an object without cutting it up for the purpose of reducing stresses to zero. Steep stress gradients and highly localized stresses can be studied by x-rays if the beam of rays is allowed to strike only a small area on a specimen.

Although the method has been used from time to time for many years, renewed interest was initiated by the discovery that with diffractometer techniques it is possible to measure stresses in very hard steels.[1] There is interest, also, in x-ray determinations of the approximate magnitude of microstrains that are distributed on a microscopic scale throughout metals of high hardness. The microstrains are estimated from the broadening of x-ray reflections. This chapter is concerned chiefly with the macroscopic rather than the microscopic strains, i.e., with strains that represent some weighted mean of the microstrains and that can be studied by strain gauges or, in some materials, by photoelastic measurements.

Not all characteristics of the method are advantageous. The penetration of the x-rays into most metals is only slight—of the order of a few thousandths of an inch—and therefore the method reveals only strains at and very near the surface. Consequently, the stresses involved are always biaxial stresses because the stress normal to the surface is always zero at a free surface. If triaxial stresses deeper within an object are to be studied,

[1] A. L. Christenson (ed.), "Measurement of Stress by X-ray," Society of Automotive Engineers Information Rept. Tr–182.

special techniques are employed by which their magnitude can be inferred from a series of measurements made on the strains at surfaces exposed by cutting, grinding, or etching. Since the method as applied to ordinary polycrystalline metals is essentially of the powder type discussed in Chaps. 4 and 7, the presence of large grains in a sample tends to cause a troublesome "graininess" in the diffraction pattern which must be minimized by employing a device to oscillate the specimen or film. Difficulties are also encountered if the diffracted rays are broad or weak, as they may be from cold-worked or high-hardness metals. Precision of the order of ±3000 psi can be readily achieved with annealed steel; but with the broad lines of a hardened steel, approaching this precision is much more difficult.

In pioneer work with the method[1] only uniaxial stresses or the sum of the two principal stresses in the plane of the surface were determined. Later it was realized that the principal stresses could be *individually* determined if properly directed beams were used,[2] and also that x-raying an object in the unstressed state was not actually necessary,[3] since the unstressed value of the interplanar spacing could be computed from two suitable measurements on the stressed material. Some fundamentals of stress and strain in isotropic elastic solids are briefly reviewed below, followed by explanations of some of the more convenient x-ray methods.

Elastic stress-strain relations Strain ϵ is defined by the relation

$$\epsilon = \frac{\Delta l}{l} \tag{17-1}$$

where Δl is the change in length of a line in a stressed body having the original length l. If this is produced by a stress σ (a force divided by the area over which it is applied) and if the stress acts in a single direction, Hooke's law states that the strain will be proportional to the stress:

$$\epsilon = \frac{\sigma}{E} \tag{17-2}$$

The constant of proportionality, E, is Young's modulus. Consider a rectangular coordinate system with tension σ_x applied in the X direction, and assume that the body is isotropic. That is, assume that the elastic properties are the same in all directions through the material. There will then be a contraction in all directions at right angles to X, and if strains parallel to Y and Z are ϵ_y and ϵ_z, respectively, then

$$-\epsilon_y = -\epsilon_z = \nu\epsilon_x = \frac{\nu\sigma_x}{E} \tag{17-3}$$

where ν is Poisson's ratio and negative signs denote contraction.

[1] H. H. Lester and R. H. Aborn, *Army Ordnance*, vol. 6, pp. 120, 200, 283, 364, 1925–1926. G. Sachs and J. Weerts, *Z. Physik*, vol. 64, p. 344, 1930.

[2] C. S. Barrett and M. Gensamer, *Phys. Rev.*, vol. 45, p. 563, 1934 (abstract); *Physics*, vol. 7, p. 1, 1936. R. Glocker and E. Osswald, *Z. Tech. Physik*, vol. 16, p. 237, 1935.

[3] F. Gisen, R. Glocker, and E. Osswald, *Z. Tech. Physik*, vol. 17, p. 145, 1936.

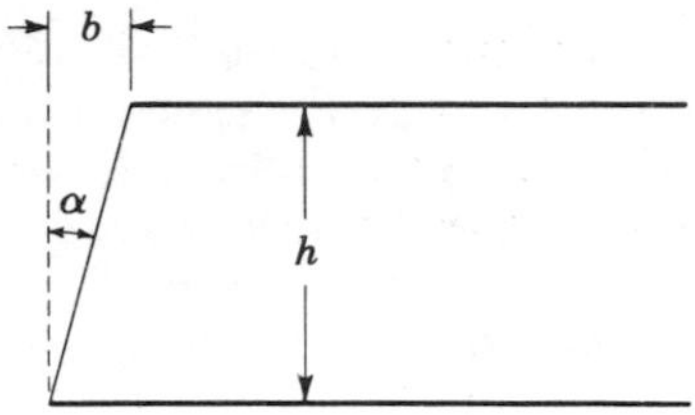

Fig. 17-1 Diagram illustrating shear strain.

In addition to the strains produced by stresses normal to a surface, the *normal strains* mentioned above, there are *shear strains*. The magnitude of a shear strain, γ, is defined by the lateral displacement of a plane relative to another parallel plane per unit distance separating the planes. Referring to Fig. 17–1,

$$\gamma = \frac{b}{h} = \tan \alpha$$

The proportionality between shear strain and the shear stress τ that produces it may be written

$$\gamma = \frac{\tau}{G} \tag{17–4}$$

where the constant G is the modulus of elasticity in shear, the torsion modulus. In a rectangular coordinate system, shear stresses and shear strains require subscripts to designate their directions; a shear stress acting on a plane perpendicular to the X direction and along the Y direction is written τ_{xy}, and a shear strain on a plane perpendicular to X and acting in the direction of Y is written γ_{xy}.

If an infinitesimal parallelepiped, say a cube, is inscribed in the stressed body and the cube edges are taken as coordinate axes, there will be in general three components of stress acting on each face, as in Fig. 17–2, but in ordinary conditions of equilibrium many of these are equal (for example, $\tau_{yz} = \tau_{zy}$ and $\sigma_x = \sigma_{-x}$). In fact, no more than six of these components of stress are required in order to specify completely the state of stress at a point in an isotropic solid: σ_x, σ_y, σ_z, τ_{xy}, τ_{yz}, and τ_{zx}.

A simplification results if the coordinate axes of Fig. 17–2 are directed in such a way that the shear stresses on all faces are zero. This is always possible, regardless of the complexity of the stress system. The stresses normal to the cube surfaces are then the *principal stresses* σ_1, σ_2, and σ_3, and these are related to the principal strains, ϵ_1, ϵ_2, and ϵ_3, in isotropic bodies by the equations

$$\begin{aligned} \epsilon_1 &= \frac{1}{E}[\sigma_1 - \nu(\sigma_2 + \sigma_3)] \\ \epsilon_2 &= \frac{1}{E}[\sigma_2 - \nu(\sigma_1 + \sigma_3)] \\ \epsilon_3 &= \frac{1}{E}[\sigma_3 - \nu(\sigma_1 + \sigma_2)] \end{aligned} \tag{17–5}$$

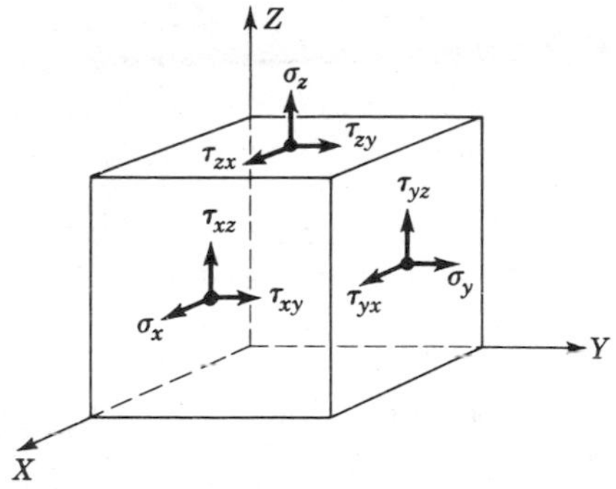

Fig. 17-2 Shear stresses τ, and normal stresses, σ, on an element of volume.

Equations (17–2) and (17–3) are special cases of these relations. Individual grains of a polycrystalline material are usually anisotropic (E varies with crystallographic direction), but if there is no preferred orientation of the grains the material may still follow closely the laws for an isotropic medium.

Method for sum of principal stresses ($\sigma_1 + \sigma_2$) A photographic x-ray method of stress analysis suitable for determining uniaxial stresses and the sum of the principal stresses in the plane of the surface requires one photograph of the stressed specimen and another of the same material in the stress-free state.[1] If we assume that the material is elastically isotropic and is subjected to principal stresses σ_1 and σ_2 lying in the plane of the surface, the normal strain perpendicular to the surface will be $\epsilon_3 = -(\sigma_1 + \sigma_2)\nu/E$, where ν is Poisson's ratio and E is Young's modulus. (At a free surface the *stress* normal to the surface is zero; thus $\sigma_3 = 0$.) For simple tension, $\sigma_2 = \sigma_3 = 0$ and the sign of σ_1 is positive; this produces a strain normal to the surface which is a contraction (indicated by the negative sign). The spacing of the atomic planes lying parallel to the surface will thereby be altered from d_0 (the unstressed value) to $d_\perp$, so that $\epsilon_\perp = (d_\perp - d_0)/d_0$. Therefore, measurements of $d_\perp$ and d_0 give the sum of the principal stresses,

$$\sigma_1 + \sigma_2 = -\frac{E}{\nu}\left(\frac{d_\perp - d_0}{d_0}\right) \tag{17–6}$$

Strain measurements by x-rays require precision technique, which means that back-reflection cameras must be employed. (These cameras have been discussed in Chap. 7, page 125.) Figure 17–3 illustrates a back-reflection camera set for determining stresses by this method. An x-ray beam passes through a pinhole placed at the focusing position, goes through a hole in the film, and strikes the specimen perpendicularly, reflecting from planes that are nearly parallel to the surface (N_1 and N_2 are the reflecting-plane normals). Using cobalt $K\alpha$ radiation for an iron or steel specimen, reflection will occur from (310) planes with $\theta = 80°37.5'$, and the strain measured along the (310) plane normals will closely approximate the desired strain $\epsilon_\perp$ and can be used in the above equations. Copper $K\alpha$ radiation is suitable

[1] G. Sachs and J. Weerts, *Z. Physik*, vol. 64, p. 344, 1930. F. Wever and H. Möller, *Arch. Eisenhüttenw.*, vol. 5, p. 215, 1931–1932.

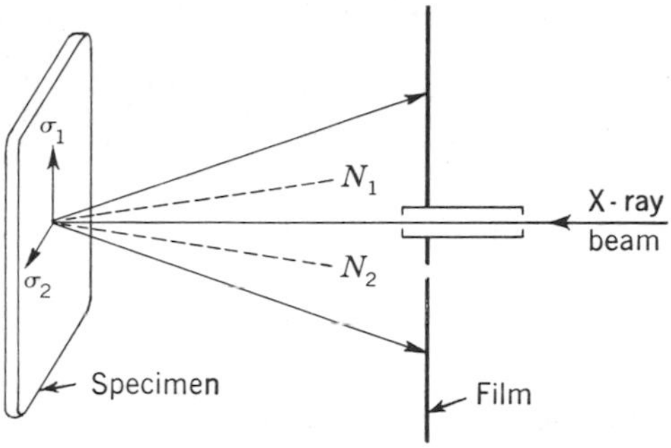

Fig. 17-3 Back-reflection camera set up for stress determination with beam normal to surface. Reflecting-plane normals are N_1 and N_2.

for aluminum and duralumin since the (511) reflection occurs at about 81°. Cobalt reflects from copper at $\theta = 81°46.5'$ (400 reflection) and from brass (68 percent copper) at 75°30′. Nickel reflects from cartridge brass (331) planes at 79°, and iron reflects from magnesium (105) planes at 83°.

The best accuracy to be expected of this method is about $\pm 1 \times 10^{-4}$ A for the spacing of atomic planes, corresponding to ± 2 kg per sq mm or about 3000 psi in stress for iron and steel specimens.[1] Owing to the fact that the reflecting planes are not exactly parallel to the surface, Eq. 17–6 is only an approximate expression; the computed stress $(\sigma_1 + \sigma_2)$ will be on the average about 7 percent too small.[2] The film is usually rotated about its center during the exposure to smooth out spottiness in the lines.

The distance between the film and the specimen surface can be determined by placing a powdered calibrating substance on the specimen. The lattice spacing of this substance is known, and one can therefore compute the specimen-to-film distance from a measurement of a Debye ring produced by it. Annealed gold or silver powder is suitable for iron, aluminum, duralumin, and brass samples.[3]

The distance from specimen to film can also be measured directly, permitting exposure times only about half as long as those when calibration substances are used. The distance can be measured by inside micrometers or can be set at a predetermined value, say 5.00 cm, by a metal pointer attached to the hub of the camera.[4] A feeler gauge slipped between the end of the pointer and the specimen enables one to determine when the camera is adjusted to the proper distance. The pointer is then removed

[1] H. Möller and J. Barbers, *Mitt. Kaiser-Wilhelm-Inst. Eisenforsch., Düsseldorf*, vol. 16, p. 21, 1934. H. Möller, *Mitt. Kaiser-Wilhelm-Inst. Eisenforsch., Düsseldorf*, vol. 21, p. 295, 1939.

[2] R. Glocker, "Materialprüfung mit Röntgenstrahlen," Springer, Berlin, 1936, 1958.

[3] The Au 333 reflection with copper $K\alpha_1$ is at $\theta = 78°56'$ and with cobalt $K\alpha_1$ the 420 reflection is at $\theta = 78°46'$; a_0 for Au is 4.0788 A; a_0 for Ag is 4.0857 A, at 25°C.

[4] D. E. Thomas, *J. Sci. Instr.*, vol. 18, p. 135, 1941.

during the exposure. A suitably inscribed circle on the film, concentric with the axis of rotation of the film, facilitates accurate measurement. Thomas states that film shrinkage errors amount to about 1 part in 84,000 in lattice-constant measurement and may be neglected.

At a standard distance the displacement of the diffracted line corresponds directly to a certain stress; for example, for a distance which gives a gold ring 50.0 mm in diameter, a shift of $\frac{1}{10}$ mm in the position of one side of the 310 line from iron corresponds to $(\sigma_1 + \sigma_2) = 9.2$ kg per sq mm (13,000 psi), assuming $E = 21{,}000$ kg per sq mm (30×10^6 psi) and $\nu = 0.28$. With aluminum, assuming $E = 7200$ kg per sq mm (10.3×10^6 psi) and $\nu = 0.34$, the same shift would indicate $(\sigma_1 + \sigma_2) = 2.4$ kg per sq mm (3420 psi).

Since only $(\sigma_1 + \sigma_2)$ is determined, this method gives a limited view of the stress situation, and in fact the method cannot detect torsional stresses in the surface since these have $\sigma_1 = -\sigma_2$. The unstressed reading can be obtained by removing the load, or in an internally stressed piece it sometimes can be had by cutting out a small piece of the specimen with saw cuts or with a hollow drill. Stress-relief annealing can also be resorted to, but at some risk of dissolving or precipitating some elements from the diffracting matrix, which would change the unstressed value of the lattice spacings.

Equations for the ellipsoid of strain and of stress The fundamental equations on which all procedures for stress analysis are based will be presented before other procedures. Under homogeneous elastic deformation a spherical element of volume in an isotropic solid is deformed into an ellipsoid. The normal strain ϵ in any chosen direction is given by the approximate equation for the ellipsoid of strain

$$\epsilon = a_1^2\epsilon_1 + a_2^2\epsilon_2 + a_3^2\epsilon_3 \tag{17–7}$$

where ϵ_1, ϵ_2, and ϵ_3 are the principal strains (Fig. 17–4) and a_1, a_2, and a_3 are the direction cosines of the chosen direction with respect to the directions of the principal strains.[1] In terms of the coordinates ϕ, ψ of Fig. 17–4, the direction cosines may be written

$$\begin{aligned} a_1 &= \sin\psi\cos\phi \\ a_2 &= \sin\psi\sin\phi \\ a_3 &= \cos\psi = \sqrt{1 - \sin^2\psi} \end{aligned} \tag{17–8}$$

If the direction cosines are substituted in (17–7) together with the values of the principal stresses ϵ_1, ϵ_2, and ϵ_3 from (17–5) and if we set $\sigma_3 = 0$ (since the stress normal to a free surface is zero), then Eq. (17–7) may be written

$$\epsilon - \epsilon_3 = \frac{1 + \nu}{E}(\sigma_1 \cos^2\phi + \sigma_2 \sin^2\phi)\sin^2\chi \tag{17–9}$$

[1] A derivation of elastic-theory formulas will be found in S. P. Timoshenko, "Theory of Elasticity," McGraw-Hill, New York, 1934.

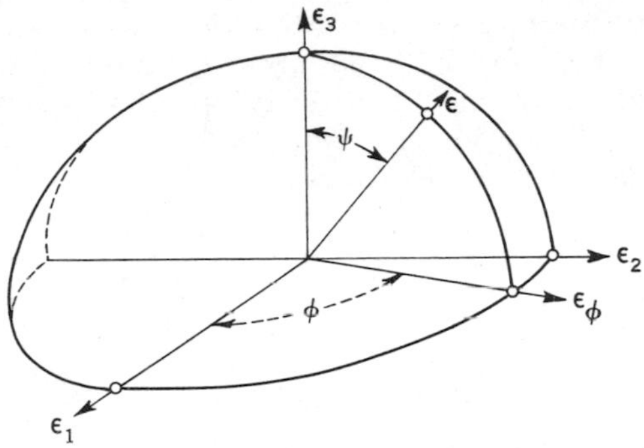

Fig. 17-4 The ellipsoid of strain. Principal strains are ϵ_1, ϵ_2, and ϵ_3.

Now the approximate equation for the stress ellipsoid is

$$\sigma = a_1^2\sigma_1 + a_2^2\sigma_2 + a_3^2\sigma_3 \tag{17–10}$$

where a_1, a_2, and a_3 are the direction cosines of the stress σ with respect to the principal axes of stress. The stress parallel to the surface at ϕ degrees from axis number 1 is

$$\sigma_\phi = \sigma_1 \cos^2 \phi + \sigma_2 \sin^2 \phi \tag{17–11}$$

Substitution of (17–11) in (17–9) leads to the relation

$$\sigma_\phi = (\epsilon - \epsilon_3) \cdot \frac{E}{1 + \nu} \cdot \frac{1}{\sin^2 \psi} \tag{17–12}$$

Let d_0 be the spacing of atomic planes in the unstressed condition, $d_\perp$ the spacing in the stressed metal perpendicular to the surface, and d_ψ the spacing in the direction specified by ψ, ϕ; then

$$\epsilon - \epsilon_3 = \frac{d_\psi - d_0}{d_0} - \frac{d_\perp - d_0}{d_0} = \frac{d_\psi - d_\perp}{d_0} \tag{17–13}$$

To a close approximation this may be written

$$\epsilon - \epsilon_3 = \frac{d_\psi - d_\perp}{d_\perp} \tag{17–14}$$

and Eq. (17–12) may be reduced to the convenient form

$$\sigma_\phi = \frac{d_\psi - d_\perp}{d_\perp} \cdot \frac{E}{1 + \nu} \cdot \frac{1}{\sin^2 \psi} \tag{17–15}$$

Two-exposure method for σ_ϕ The component of stress σ_ϕ in any desired direction in the surface can be determined from two exposures, one with the beam normal to the surface giving $d_\perp$, and one with the beam inclined in the plane of the normal and the component σ_ϕ, giving d_ψ in Eq. (17–15).[1]

[1] F. Gisen, R. Glocker, and E. Osswald, *Z. Tech. Physik*, vol. 17, p. 145, 1936.

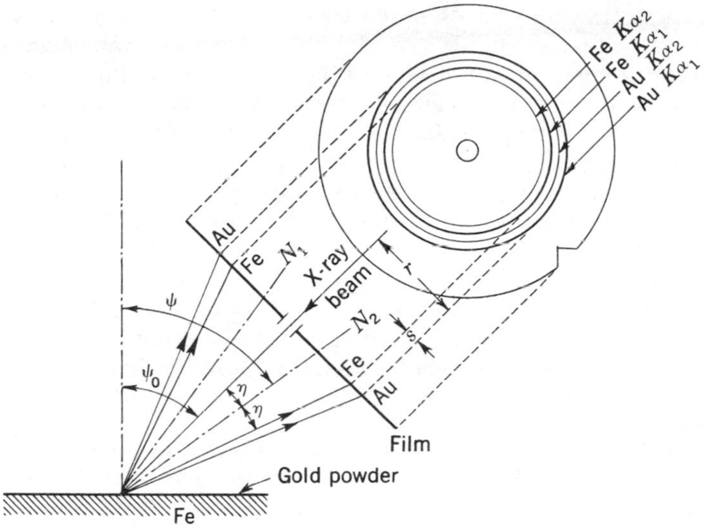

Fig. 17-5 Back-reflection camera inclined at $\psi_0 = 45°$, and type of film obtained with cobalt radiation, iron specimen, and gold calibrating powder on surface.

No exposure of the material in the unstressed state is necessary. The usual practice is to make the inclined exposure with the beam 45° from the perpendicular. Diffraction rings of a comparison substance (usually gold, silver, or brass) on the surface of the specimen may be used to determine the effective distance from the irradiated spot to the film, as illustrated in Fig. 17–5 where r is the radius of a gold calibrating ring and $r - s$ the radius of a specimen ring. Measurement of the distance s between the rings on the side nearest the specimen gives a calculation of d at the angle $\psi = \psi_0 + \eta$, which is different from the value of d on the other side of the rings at $\psi = \psi_0 - \eta$. Both measurements may be computed independently in Eq. (17–15), or the more sensitive one, $(\psi_0 + \eta)$, may be used.

For metals that are isotropic or nearly so, the value of E is independent of ψ. For others this is not always true; variations may be detected by making a series of exposures with different ψ and constant ϕ, and comparing the computed values of σ_ϕ. Christenson and Rowland[1] point out that from a practical point of view it is unnecessary that E be constant for the successful use of the method in stress measurement. If the spacings d are always determined at two specific ψ angles, such as 0 and 45°, the difference in d values will always be proportional to the stress despite any difference in E that may exist at those ψ angles; thus,

$$\sigma_\phi = K(d_{\psi_2} - d_{\psi_1}) \cong K'(2\theta_{\psi_2} - 2\theta_{\psi_1}) \qquad (17\text{–}16)$$

[1] A. L. Christenson and E. S. Rowland, *Trans. Am. Soc. Metals*, vol. 45, p. 638, 1953.

where the constant of proportionality K can be calibrated by tests with known stresses applied to the particular material being studied. Direct measurement of the distance from specimen to film is possible if sufficient care is taken and if a circle is put on the film concentric with the axis.

A convenient procedure is to calculate a relationship between stress and the ring displacement rather than to apply Eq. (17–15) to every reading. For this purpose one finds the multiplying factor that will bring the diameter of the calibration ring ($2r$) to some standard value, say 50 mm; the measured s is multiplied by this factor, then compared with a prepared table or chart relating s to σ_ϕ for the given experimental conditions.

Visual measurement of the films gives about the same accuracy as measurement with a recording microphotometer, viz., 1×10^{-4} A in atomic spacings for photographs with sharp lines. This corresponds to about 1.7 kg per sq mm (2400 psi) in stress with a steel specimen when $\psi_0 = 45°$.

Correction for oscillation of film (two-exposure method) The spottiness of rings from large-grained material should be removed by rotating the film in perpendicular exposures, but this is not advisable in inclined exposures where the rings deviate more from the circular. However, oscillation back and forth through a limited range of angles may be used to smooth out the lines. This decreases the average angle of inclination of the planes reflecting to the lower side of the film from $\psi_0 + \eta$ to $\psi_0 + \eta - \Delta\psi$, while on the upper side of the film $\psi_0 - \eta$ is increased to $\psi_0 - \eta + \Delta\psi$; the azimuthal position ϕ also changes through a range. The errors in stress computations decrease with the range of oscillation but can be kept small enough to be neglected if oscillation is limited to ±30° or less.[1]

Single-exposure method for σ_ϕ A stress component may be determined from a single inclined exposure if the diffraction rings are measured at two positions on their circumference.[2] The beam is inclined toward the direction in which the component σ_ϕ is to be measured; that is, the beam lies in the plane containing the normal and the component to be measured, thus at the angle ϕ, and ψ_0 from the normal. The ring from the specimen is measured with reference to the ring from the calibration substance, both at the top of the ring where $\psi_1 = \psi_0 - \eta$ and at the bottom where $\psi_2 = \psi_0 + \eta$, giving spacings d_1 and d_2, respectively. Equation (17–12) is written for each set of measurements, giving two simultaneous equations

[1] The value of $\Delta\psi$ is small—for iron specimens with cobalt radiation $\Delta\psi$ is only 0.1° for a range of oscillation $\delta = \pm 10°$, 1.3° for $\delta = \pm 30°$, and 2.8° for $\delta = \pm 45°$. Oscillation of the film also alters ϕ through a range $\pm\Delta\phi$. In a photograph of iron with $\psi_0 = 45°$, $\pm\Delta\phi = 2.0°$ for $\delta = \pm 10°$ for $\delta = \pm 10°$, 5.9° for $\delta = \pm 30°$, and 8.5 for $\delta = \pm 45°$, provided that the diffraction lines are measured on the lower side where $\psi = \psi_0 + \eta$. On the upper side the range $\pm\Delta\phi$ is larger, being 2.8, 7.9, and 10.8° for $\delta = \pm 10$, 30, and 45°, respectively. (R. Glocker, B. Hess, and O. Schaaber, *Z. Tech. Physik*, vol. 19, p. 194, 1938).

[2] R. Glocker, B. Hess, and O. Schaaber, *Z. Tech. Physik*, vol. 19, p. 194, 1938. An alternate method of computation is given by D. E. Thomas, *J. Sci. Instr.*, vol. 18, p. 135, 1941.

whose solution is

$$\sigma_\phi = \left(\frac{E}{1+\nu}\right)\left(\frac{d_1 - d_2}{d_0}\right)\frac{1}{\sin 2\psi_0 \sin 2\eta} \tag{17–17}$$

For the incident beam at $\psi_0 = 45°$, this reduces to

$$\sigma_\phi = \left(\frac{E}{1+\nu}\right)\left(\frac{d_1 - d_2}{d_0}\right)\frac{1}{\sin 2\eta} \tag{17–18}$$

In this formula, d_0 need not be determined accurately on each individual specimen, for the normal lattice constant for unstressed metal may be used (2.8610 for steel). This is a rapid method but is less accurate than the method employing a perpendicular exposure in addition to the inclined one, for an error of $\pm 1 \times 10^{-4}$ A in d in Eq. (17–18) introduces an error of ± 3.5 kg per sq mm (5000 psi) for steel specimens, twice the error of the two-exposure method.

Methods for determining σ_1 and σ_2 when their directions are known When the directions of the principal stresses are known (for example, in a cylinder containing quenching stresses), Eq. (17–15) can be applied to give the principal stresses σ_1 and σ_2. For this purpose the perpendicular exposure is combined with an inclined exposure in which the beam is tipped toward σ_1 and also with an exposure in which the beam is tipped toward σ_2. Thus, three exposures, all of them on the stressed material, serve to determine completely the surface stresses.

If d_0 is obtained by an exposure of the specimen in the stress-free state, then two inclined exposures of the stressed material will serve, one giving the spacing d_ψ at the angle ψ from the normal and at the azimuth ϕ, the other giving d_ψ' at the same angle from the normal and at azimuth $\phi + 90°$.* Equations (17–2), (17–3), and (17–7) for these conditions lead to the relations

$$\left.\begin{aligned} \sigma_1 + \sigma_2 &= \left(\frac{d_\psi + d_\psi' - 2d_0}{d_0}\right)\frac{E}{(1+\nu)\sin^2\psi - 2\nu} \\ \sigma_1 - \sigma_2 &= \left(\frac{d_\psi - d_\psi'}{d_0}\right)\frac{E}{(1+\nu)\sin^2\psi} \end{aligned}\right\} \tag{17–19}$$

The sum of these equations gives σ_1 while the difference gives σ_2.

If one uses the single inclined exposure method for σ_ϕ, with the beam inclined toward σ_1 and again with the beam tipped toward σ_2, the two principal stresses may be determined with only two exposures.[1] However, this shorter method, based on Eq. (17–17) or Eq. (17–18), has lower accuracy than three-exposure methods.

The various equations of this chapter can be solved to give the unstressed lattice spacing d_0 from three or more measurements of the stressed state. For example, a perpendicular exposure combined with two inclined exposures at azimuths ϕ and $\phi + 90°$ will give d_0 by simultaneous equations of types (17–6) and (17–15).

Method giving magnitude and direction of principal stresses Both the magnitudes and directions of σ_1 and σ_2 can be determined if the components of stress are computed for three directions in the surface. The maximum accuracy can be obtained if one uses a perpendicular exposure paired with three inclined exposures at azimuth angles ϕ, $\phi + 60°$, and $\phi - 60°$.† An error of $\pm 1 \times 10^{-4}$ A in spacing then corresponds to

* C. S. Barrett and M. Gensamer, *Physics*, vol. 7, p. 1, 1936.

[1] R. Glocker, B. Hess, and O. Schaaber, *Z. Tech. Physik*, vol. 19, p. 194, 1938.

† H. Möller, *Mitt. Kaiser-Wilhelm-Inst. Eisenforsch.*, *Düsseldorf*, vol. 21, p. 295, 1939.

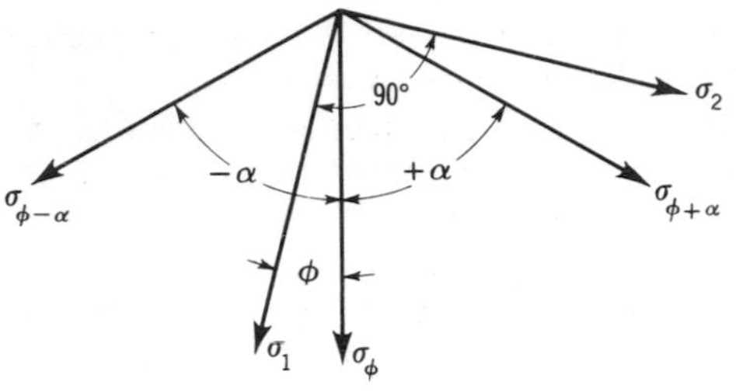

Fig. 17-6 Arrangement of principal stresses (σ_1, σ_2) and stress components in the specimen surface.

± 2.3 kg per sq mm (± 3300 psi) in the principal stresses in steel and an error in their direction that is given by the relation $\pm 65°/(\sigma_1 - \sigma_2)$ where the stresses σ_1 and σ_2 are given in kilograms per square millimeter. These figures are computed for x-ray beams at an angle of inclination of 45° and for stress directions that would give minimum accuracy.

The stress components in the plane of the surface at angles ϕ, $\phi + \alpha$, and $\phi - \alpha$ (Fig. 17–6) are given by the relations

$$\left.\begin{aligned}\sigma_\phi &= \tfrac{1}{2}(\sigma_1 + \sigma_2) + \tfrac{1}{2}(\sigma_1 - \sigma_2)\cos 2\phi \\ \sigma_{\phi+\alpha} &= \tfrac{1}{2}(\sigma_1 + \sigma_2) + \tfrac{1}{2}(\sigma_1 - \sigma_2)\cos 2(\phi + \alpha) \\ \sigma_{\phi-\alpha} &= \tfrac{1}{2}(\sigma_1 + \sigma_2) + \tfrac{1}{2}(\sigma_1 - \sigma_2)\cos 2(\phi - \alpha)\end{aligned}\right\} \qquad (17\text{–}20)$$

Solving for σ_1, σ_2, and ϕ with $\alpha = 60°$, we have

$$\sigma_1 = \tfrac{1}{3}[\sigma_\phi + \sigma_{\phi-60} + \sigma_{\phi+60} + \sqrt{(2\sigma_\phi - \sigma_{\phi-60} - \sigma_{\phi+60})^2 + 3(\sigma_{\phi-60} - \sigma_{\phi+60})^2}]$$

$$\sigma_2 = \tfrac{1}{3}[\sigma_\phi + \sigma_{\phi-60} + \sigma_{\phi+60} - \sqrt{(2\sigma_\phi - \sigma_{\phi-60} - \sigma_{\phi+60})^2 + 3(\sigma_{\phi-60} - \sigma_{\phi+60})^2}]$$

$$\tan 2\phi = \frac{\sqrt{3}(\sigma_{\phi-60} - \sigma_{\phi+60})}{2\sigma_\phi - \sigma_{\phi-60} - \sigma_{\phi+60}} \qquad (17\text{–}21)$$

The components of stress and σ_ϕ, $\sigma_{\phi+60}$, and $\sigma_{\phi-60}$, are determined by inclined beam exposures at azimuth angles ϕ, $\phi + 60$, and $\phi - 60$, each exposure being combined with an exposure perpendicular to the surface, using Eq. (17–15). Alternatively, this four-exposure technique may be shortened by merely using three inclined exposures and Eq. (17–17), but the accuracy is then cut to about half. The simultaneous equations relating stress components to the principal stresses can be solved graphically if desired.[1]

Values of the elastic constants; anisotropy The elastic constants that are measured mechanically do not necessarily apply accurately to x-ray determinations of stress. Each grain is anisotropic, and the strain is measured always along a certain crystallographic direction (e.g., along [310] for iron). Therefore, the grains that reflect have only certain orientations with respect to the axes of stress, and the effective values of E and ν in these orientations may differ from the overall average orientations, the latter being measured in a mechanical test. The effective values of E and ν are also influenced by the fact that each grain is surrounded by a polycrystalline mass. There are complex interactions between a grain and its surroundings; while these have been studied theoretically, it seems

[1] W. R. Osgood and R. G. Strum, *J. Res. Natl. Bur. Stand.*, vol. 10, p. 685, 1933 (three components). W. R. Osgood, *J. Res. Natl. Bur. Stand.*, vol. 15, p. 579, 1935 (four components). A. H. Stang and M. Greenspan, *J. Res. Natl. Bur. Stand.*, vol. 19, p. 437, 1937 (four components). "Handbook of Experimental Stress Analysis," Wiley, New York, 1950.

obvious that an x-ray determination of E and ν is preferable to any theoretically derived values, or to mechanically measured values.

The x-ray values should be determined for the particular conditions under which they are to be used, for the constants depend upon the indices of the reflecting planes, the wavelength of x-rays used, and perhaps on grain size or other microstructural variables.[1] For this purpose it is recommended that the constant in the much-used Eq. (17–16) be determined in the same apparatus to be used for stress analysis, with a calibrating specimen that is as similar as possible metallurgically, and by using the data from strain gauges attached to the specimen when the specimen is subjected to a series of loads, rather than by stresses *computed* from known loads. A bar or slotted ring with wire strain gauges mounted near the spot to be x-rayed is subjected to stresses, and the gauge readings serve to determine the constant K' to be used with the x-ray readings in Eq. (17–16) or the value $E/(1 + \nu)$ in Eq. (17–15). Apparatus alignment should be tested by seeing that 2θ does not vary with ψ when a sample that is free from macrostresses is x-rayed with different angles ψ. The calibrating stresses should be kept within the elastic range. An actual calibration curve is reproduced in Fig. 17–7.

When a diffractometer is used (see below) it is also well to be certain that the profiles of the reflections from the standard sample at different values of ψ are the same—after the intensity data have been corrected by all angle-dependent correction factors and normalized to the same peak height. The constants E and ν can be evaluated individually by using at least three ψ angles and plotting the d spacings against the applied stresses.

Although there is general agreement that the precautions discussed above should be taken, it is by no means certain that the effective x-ray values will always differ from published mechanical values. In fact, calibration experiments on steels so often lead to the values $E = 30 \times 10^6$ psi and $\nu = 0.29$ that these values can well be assumed for steels whenever high accuracy is not needed.

Uncertainties in the interpretation of x-ray data can be expected whenever an object has been subjected to plastic flow; the stresses deduced by x-rays then often differ from mechanically determined stresses. The reasons appear to be complex (see below, page 483). Additional work on elastic constants and anisotropy as determined by x-rays in iron and steel has been published recently.[2]

Recent experiments indicate that the presence of stacking faults can introduce errors in stress determinations for materials of relatively low stacking-fault energy. Stacking faults of the deformation type, intrinsic faults, which result from a displacement at one atomic layer, cause peak

[1] "Measurement of Stress by X-ray," Society of Automotive Engineers Information Rept. TR–182. V. Hauk, *Z. Metallk.*, vol. 36, p. 120, 1944. E. Macherauch and P. Müller, *Arch. Eisenhüttenw.*, vol. 29, p. 257, 1958.

[2] G. B. Greenough, *J. Iron Steel Inst.*, vol. 180, p. 233, 1955. E. Macherauch and P. Müller, *Arch. Eisenhüttenw.*, vol. 29, p. 257, 1958. G. B. Greenough, *Progr. Metal Phys.*, vol. 3, p. 176, 1952.

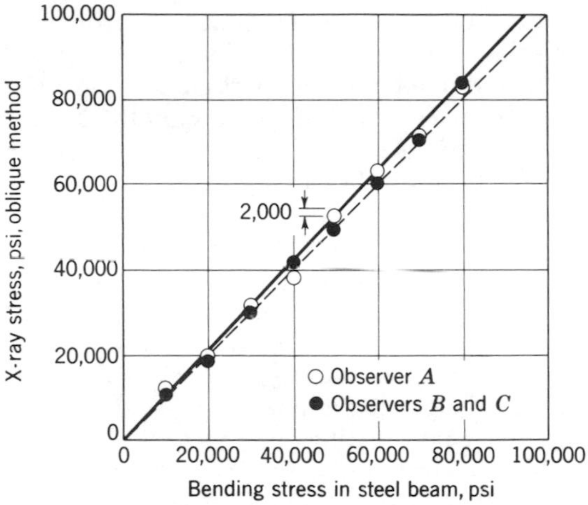

Fig. 17-7 Comparison of stresses determined by x-rays (two-exposure method) and stresses computed from the curvature of a bent beam. Annealed mild steel. (J. T. Norton and B. M. Loring, *Welding J.*, Res. Supplement, 1941.)

shifts and symmetrical broadening of peaks,[1] whereas those of the double-deformation type, extrinsic faults, cause both unsymmetrical broadening and peak shifts.[2] Stacking-fault probabilities have been determined from Nelson-Riley plots by a comparison of the slopes of lines drawn through first- and second-order reflections of different planes,[3] and when this method was applied to silicon bronze (Cu–6.6 atomic percent Si–1.2 atomic percent Mn) it was found that stacking-fault probabilities increased with increasing applied stress. The resulting shifts in peaks yielded computed stresses that were 25 percent or more too low unless the shifts resulting from the stacking faults were removed.[4] Errors in stresses computed from x-ray data were negligible, however, in similar experiments on brass (Cu–30 percent Zn) which has a higher stacking-fault energy.

Equipment for photographic stress measurement Convenient x-ray equipment for stress determination with photographic films consists of a portable x-ray tube mounted in a rayproof and shockproof housing (the smaller the better) and supported on an adjustable stand, as, for example, in Fig. 17–8. The tube may be connected to the portable power source by

[1] C. N. J. Wagner, *Acta Met.*, vol. 5, p. 427, 1957.

[2] C. A. Johnson, *Acta Cryst.*, vol. 16, p. 490, 1963. B. E. Warren, *J. Appl. Phys.*, vol. 34, p. 1973, 1963.

[3] H. M. Otte and D. O. Welch, *Phil. Mag.*, vol. 9, p. 299, 1964. C. N. J. Wagner, A. S. Tetelman, and H. M. Otte, *J. Appl. Phys.*, vol. 33, p. 3080, 1962.

[4] A. L. Esquivel and H. M. Otte in S. Locke and H. M. Otte (eds.), "Materials Science Research," vol. 2, Plenum, New York, 1964.

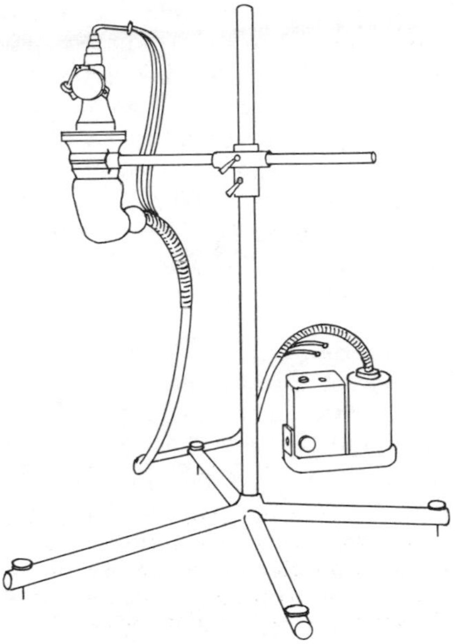

Fig. 17-8 Portable x-ray stress-measuring apparatus.

an insulated cable. A back-reflection camera is essential and should be mounted on the x-ray tube itself, somewhat in the fashion sketched in Fig. 17–9. The film is mounted on a small film holder that can be rotated or oscillated in its own plane to remove spottiness from the rings, with the beam passing through an adjustable pinhole system at its center. The smallest pinhole should be placed at the focusing position, viz., on the

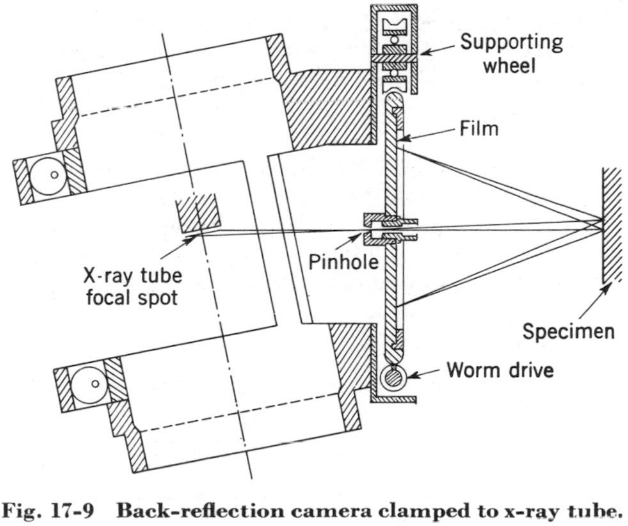

Fig. 17-9 Back-reflection camera clamped to x-ray tube.

circumference of a circle that passes through the Debye ring on the film and is tangent to the irradiated spot on the specimen. With coarse-grained specimens it may be necessary to oscillate the specimen about an axis normal to the x-ray beam, in addition to oscillating the film. Greater intensities may be had if slits (not too long) are used instead of pinholes; and if the slits are oriented to be parallel to the tangent of the Debye ring at the place it is measured, the distortion of the ring caused by the length of the slits need not be excessive. Slits 0.030 in. wide have been found to be satisfactory for normal incidence work, and 0.020 in. wide for 45° incidence.[1]

Stress measurement with diffractometer equipment An increasing number of x-ray stress measurements are being made with counter diffractometers. The general principles are the same as for the film techniques mentioned above. The specimen surface is mounted in the parafocusing position whenever possible, in order that the diffraction peaks will be as sharp as possible. Various counters have been used but the advantages lie in a choice of counters and counting circuits that will yield good line intensities over the background radiation (see Chap. 3).

Counter diffractometers were applied to stress measurement in high-hardness steels by Christenson and Rowland[2] with greater success than would have been expected in view of the great width of the diffraction lines given by such materials. It was found that chromium radiation filtered with 0.001-in.-thick vanadium foil in the *diffracted* beam greatly improved the contrast between diffraction peaks and background radiation, yet gave a martensite 211 reflection and an austenite 220 reflection at angles of about 115 and 128 degrees 2θ, respectively, which were found to be satisfactory. It is now recognized that film cameras and diffractometers are competitive for stress measurement when the diffraction lines are sharp, but that diffractometers are unquestionably superior when lines are broad. It is not unusual to determine stresses from hardened steels by the use of peaks having a breadth of 8 to 12° 2θ at half height. With broadening of this magnitude the profile of the diffuse diffraction line (and therefore its apparent 2θ position) is influenced by the variation of the absorption factor and the Lorentz-polarization factor with angle. To apply angle-dependent correction factors, it is almost necessary to use a diffraction peak plotted from diffractometer measurements (either a strip-chart record of rate meter readings or a plot of scaler readings). With a diffractometer, also, the use of a counter filled with argon for Cu $K\alpha$ or Cr $K\alpha$ radiation improves the peak-to-background ratio, and further improvement is possible if a proportional or a scintillation counter is used with pulse-height discriminating circuits.

Good diffraction peaks are only obtained when all slits are designed and placed properly. It is particularly important when radiation strikes the specimen at an inclined angle, such as the usual $\psi = 45°$, to have the counter slit at the best focusing position. With a flat sample the diffracted rays do not come to an exact focus, but good parafocusing can

[1] G. A. Hawkes, *Brit. J. Appl. Phys.*, vol. 8, p. 229, 1957.

[2] A. L. Christenson and E. S. Rowland, *Trans. Am. Soc. Metals*, vol. 45, p. 638, 1953.

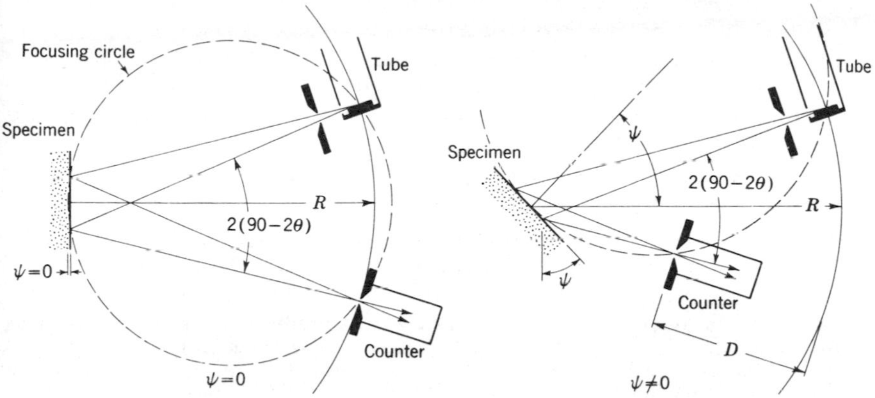

Fig. 17-10 Best position of counter slit of a diffractometer when diffuse lines are measured.

be obtained (see Chap. 7) if the counter slit is moved to a position on the focusing circle. To obtain parafocusing with *inclined* incidence of the rays, it is necessary to move the slit at the counter closer to the specimen than the slit at the x-ray tube (or the line focus of the tube), as indicated in Fig. 17–10. If the tube is at a distance R from the irradiated spot on the specimen, then the counter slit should be moved a distance D from its normal position (at distance R from the specimen), where D varies according to the relation

$$\frac{D}{R} = 1 - \frac{\cos [\psi + (90 - \theta)]}{\cos [\psi - (90 - \theta)]} \tag{17–22}$$

Of the several alternative methods that have been used in different laboratories to measure the positions of diffuse peaks, the most serviceable ones seem to be those in which the curve of measured intensities vs. 2θ is corrected by angle-dependent factors and then fitted in its upper regions with a parabola. Five data points are used for the parabola fitting in one procedure,[1] three in another.[2] In fitting a parabola with a vertical axis to three points on a recorded peak, the points may be equally spaced along the x axis, the axis of 2θ. Let h be the x coordinate of the vertex and x_1 be the position of the first data point; then

$$h = x_1 - \frac{c}{2} \frac{(3a + b)}{(a + b)} \tag{17–23}$$

where c is the interval in the x direction between data points, and a and b are differences in the vertical coordinate (y) between the middle data point and the data points on either side of it, as indicated in Fig. 17–11. The results obtained on diffuse peaks by this procedure have been found to correlate well with strain-gauge results and with tests in which samples with internal stresses were dissected.[3] The results are reproducible to within 3000 to 4000 psi even with diffuse reflections from martensitic steels.

A twin-counter apparatus for stress measurement is in use in Japan.[4] The counters

[1] R. E. Ogilvie, "Stress Measurement with X-ray Spectrometer," Thesis, Massachusetts Institute of Technology, Cambridge, Mass., 1952.

[2] D. P. Koistinen and R. E. Marburger, *Trans. Am. Soc. Metals.*, vol. 51, p. 537, 1959. R. E. Marburger and D. P. Koistinen in G. M. Rasseweiller and W. L. Grube (eds.), "Internal Stresses and Fatigue in Metals," p. 98, Elsevier, New York, 1959.

[3] Koistinen and Marburger, loc. cit.

[4] K. Kojima and T. Tamaru, *Mem. Inst. Sci. Ind. Res. Osaka Univ.*, vol. 18, p. 147, 1960. K. Kojima et al., Third International Conference on Nondestructive Testing, 1960.

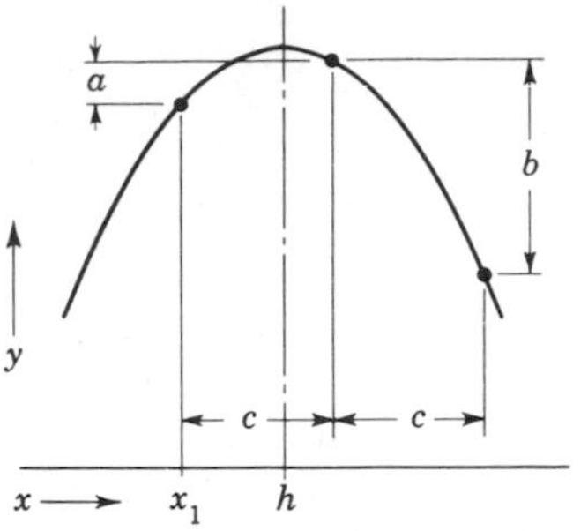

Fig. 17-11 Parabola with axis at $x = h$ fitted at three points to a diffraction peak.

measure a pair of intensities on a diffraction peak at positions differing by a constant angular separation that amounts to a fraction of the half-width of a peak. The counters are moved together through the diffraction peak, and the *difference* in the count rate in the two counters is fed into the rate meter and recorded. The angular setting of the pair that corresponds to the peak intensity of the diffracted beam is determined from a curve plotted at 5-min intervals, and is the point where this curve indicates zero difference in intensity.

The use of differential counters minimizes errors arising from variations in x-ray tube voltages, such as might be encountered in portable apparatus lacking adequate voltage stabilizing circuits. To make the angular separation $\Delta 2\theta$ of the two counters small enough, they are mounted at the *same* value of 2θ, but the slit in front of one is displaced in a direction normal to its length by an amount $\Delta 2\theta$ from the slit in front of the other. The incident beam is directed at the specimen through a divergent Soller slit from an x-ray tube slit that is located on the focusing circle through the specimen surface and the receiving slits. The metal sheets in the Soller slit are parallel to the length of the counter slits and the x-ray tube slit. Apparently the technique is only used on specimens that give peaks so sharp that angle-dependent correction factors can safely be ignored.

X-ray beam penetration In specimens that have steep stress gradients, the observed 2θ values are not characteristic of the surface but represent a weighted average for the volume effectively penetrated by the x-ray beam, which may extend distances below the surface of the order of a few thousandths of an inch with aluminum-base materials and Co $K\alpha$ radiation, or a few ten-thousandths with Cr $K\alpha$ radiation on steel. The shallow penetration of the rays requires that attention be given to the preparation of the surface. A machined surface of steel, for example, may contain stresses that differ from the stress in underlying layers by as much as 60,000 psi,[1] unless the disturbed layer is removed by chemical or electrolytic means.

Even rubbing a surface with a rubber ink eraser that contained abrasive particles has been known to create stresses in excess of 10,000 psi. In a steel sample, the thickness of the layer that contributes two-thirds of the diffracted intensity of Cr $K\alpha$ radiation varies with ψ:

ψ	Depth in inches for $2\theta = 128°$	Depth in inches for $2\theta = 156°$
0	0.000217	0.000236
45	0.000105	0.000159

[1] F. Wever and H. Möller, *Mitt. Kaiser-Wilhelm-Inst. Eisenforsch., Düsseldorf,* vol. 18, p. 27, 1936.

When steep stress gradients are being studied, the shallow penetration can be an advantage (for example, when the stresses resulting from surface abrasion or shot blasting are being studied). It is rare that stress varies markedly in the depths penetrated with steel samples having properly prepared surfaces, but corrections can be applied if necessary.[1]

Stress vs. depth below surface X-rays cannot directly reveal stresses throughout the interior of an object, because of their shallow penetration. However, they can be used together with dissecting methods to deduce such stress distributions in ways similar to the sectioning methods used with ordinary strain gauges. Thus, the longitudinal and transverse stresses throughout a plate have been determined by cutting out rectangular blocks and then slicing layers from them, with x-ray measurements being made after each cut.[2] By repeatedly removing thin concentric shells from a cylinder and x-raying the new surfaces cut and electropolished, the distribution of stresses that existed in the original cylinder can be deduced (if rotationally symmetrical).[3]

Practical applications; the need for caution The many applications that have been made of x-ray stress analysis are now too numerous to list. They include residual stresses in castings, forgings and welds, heat-treated steel objects such as ball bearings and bearing races, machined and shot-peened surfaces, and fatigued specimens. An example of a stress determination in a hardened steel below a ground surface by x-rays and by mechanical means[4] is given in Fig. 17–12. Since the stress gradients were high, correction for penetration was used in this case. Stress relief as a result of annealing can readily be followed.

X-rays do not necessarily measure a volume-average stress in the irradiated volume. Much attention has been given to the fact that when a bar is stretched plastically in tension and then unloaded, the observed shift in diffraction peaks indicates the presence of residual macroscopic compressive stresses at or near the surface,[5] although these are not indicated by mechanical tests.[6] Thus, removal of a layer from one surface does not produce curvature in the remainder of the bar.

Various factors are believed to contribute to this effect, and these may differ in relative importance in different materials.[7] Surface grains may have a lower yield point than

[1] "Measurement of Stress by X-ray," Society of Automotive Engineers Information Rept. TR–182.

[2] D. Rosenthal and J. T. Norton, *Welding J.*, Res. Suppl., vol. 24, pp. 295-s, 307-s, 1945.

[3] Measurement of Stress by X-ray," loc. cit.

[4] Ibid. Mechanical data are from H. R. Letner, *Trans. Am. Soc. Mech. Eng.*, vol. 77, p. 1089, 1955.

[5] F. Wever and B. Pfarr, *Mitt. Kaiser-Wilhelm-Inst. Eisenforsch, Düsseldorf.*, vol. 15, p. 137, 1933. S L. Smith and W. A. Wood, *Proc. Roy. Soc.*, vol. A178, p. 93, 1941; vol. A181, p. 72, 1942. C. J. Newton and H. C. Vacher, *Trans. AIME*, vol. 203, p. 1193, 1955. M. J. Donachie, Jr., and J. T. Norton, *Trans. Met. Soc. AIME*, vol. 221, p. 962, 1961.

[6] Donachie, Jr., and Norton, loc. cit. S. L. Smith and W. A. Wood, *Proc. Roy. Soc.*, vol. A182, p. 404, 1944. N. Davidenkov and M. Timofeeva, *J. Tech. Phys., U.S.S.R.*, vol. 16, p. 283, 1946.

[7] R. I. Garrod and G. A. Hawkes, *Brit. J. Appl. Phys.*, vol. 14, p. 422, 1963.

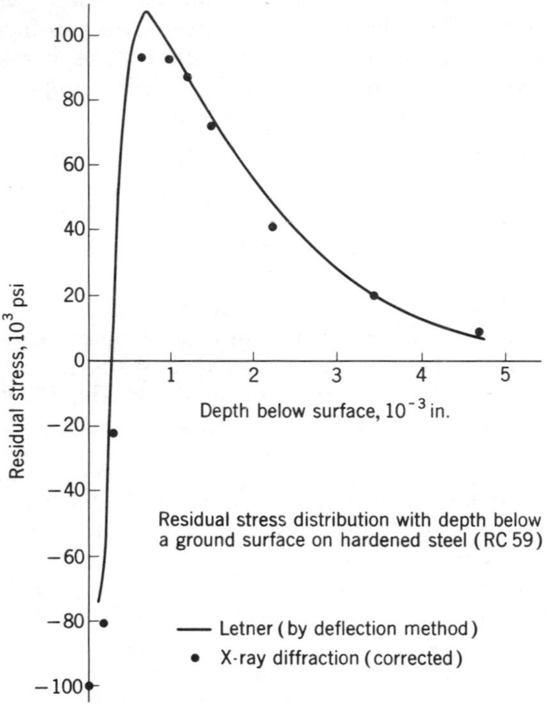

Fig. 17-12 Comparison of stresses measured by x-ray and by mechanical methods. X-ray results have been corrected for x-ray penetration.

interior grains and may thus be elongated plastically under loads that are still carried elastically by the grains in the interior. Grains or portions of subgrains that have been least distorted during plastic flow (or contain the lowest density of dislocations) may be the most important in determining the *apparent* position of a diffraction peak, and the more diffuse contribution from the other peaks may be easily ignored in evaluating peak positions. Portions of a grain near a boundary or subboundary may thus differ from interior portions.[1] However, this characteristic of the x-ray method does not render it useless for practical applications. There is need for caution, however, and the use of data from several different wavelengths has been recommended, on the basis that the average of these is presumably closer to the true macroscopic stress than can be obtained by using a single wavelength.

Broadened reflections and inhomogeneous strains Metals in the cold-worked state and metals hardened by martensitic transformations or aging produce broadened peaks. Analysis of the broadening by Fourier methods is discussed on page 453; some practical applications of breadth measurements can be made, however, without elaborate analysis. With

[1] Ibid.; Garrod and Hawkes refer to and discuss the various theories of the effect; earlier reviews appeared in the second edition of this book; by G. B. Greenough in "Progress in Metal Physics," vol. 3, p. 176, Interscience, New York, 1952; by B. D. Cullity, *Trans. AIME*, vol. 227, p. 356, 1963; and by Donachie and Norton, loc. cit.

only a simple index of the amount of broadening of martensite lines in steel, Marburger and Koistinen have found that there is a direct correlation between the broadening and the surface hardness (Rockwell C hardness between 50 and 62).[1] The index used was the breadth at half height of the parabola fitted to the diffractometer record of the diffraction peak. With I as the intensity maximum of a martensite reflection, the breadth at half height is $2c[I/(a + b)]^{1/2}$ in terms of the quantities of Eq. (17–23). As a more convenient index, Koistinen and Marburger used the square of the half-height breadth with data points at 1° intervals, so that the index was then merely $I(a + b)$.

The broadening of reflections by microscopically distributed stresses in an object that has been plastically deformed is believed to be of different types in different metals.[2] In some, the cell walls that contain high densities of dislocations are wider than in others, and perhaps make a significant contribution to the diffraction peaks.

[1] D. P. Koistinen and R. E. Marburger in G. M. Rassweiller and W. L. Grube (eds.), "Internal Stresses and Fatigue in Metals," p. 98, Elsevier, New York, 1959.

[2] R. I. Garrod and R. A. Coyle, *Trans. AIME*, vol. 230, p. 519, 1964.

18

Phase transformations in the solid state

From the wealth of information that has been accumulated on transformations in the solid state, it is possible to classify various types and to recognize some general characteristics and typical behaviors in each, despite the many conflicting results and conclusions reached at various times by different observers. An understanding of the structural and crystallographic features of several *typical* transformations is a prerequisite for successful research in this field. A knowledge of the specific crystallographic features for a reaction in any *particular* substance is a prerequisite for a deeper insight into the atomic mechanisms involved, the kinetics of the reaction and any effects of the reaction on the physical, mechanical, and chemical properties of the substance.

As a first approximation it is customary to divide phase transformations in metallic systems into two main classes, *nucleation-and-growth transformations* and *martensitic transformations*, according to the assumed mechanism involved in the transformation process. The former can, if required, occur as *isothermal transformations* in the sense that the process of transformation can be completed at a constant temperature with thermal activation and diffusion playing an important role. Martensitic transformations, on the contrary, occur usually only while the temperature is changing and are dependent only slightly, if at all, on thermal agitation; they have been accordingly designated as *athermal transformations*. Although there is some dissatisfaction with this terminology, it has become widely accepted.[1] However, it has become desirable to base the classification of martensitic transformations not on the athermal characteristic, but instead, as we point out later, on the characteristic that there is a *change of shape* of each region that transforms.

Over the last two decades a very large literature on the whole subject of solid-state transformations has been published, including some excellent review articles, symposia, and books. Several general and detailed aspects of transformations are discussed in various articles in R. Smoluchowski, J. E. Mayer, and W. A. Weyl (eds.), "Phase Transformations in Solids," Wiley, New York, 1951. G. V. Kurdjumov, *Dokl. Akad. Nauk. SSSR*, vol. 60, p. 1543, 1948. J. S. Bowles and C. S. Barrett, *Progr. Metal Phys.*, vol. 3, p. 1, 1953. Institute of Metals Symposium, "The Mechanism of Phase Transformations in Metals," Monograph and Rept. ser. no. 18, London, 1956. J. W. Christian in R. W. Cahn (ed.), "Physical Metallurgy," North Holland Publishing, Amsterdam,

[1] One can object that martensitic transformations also "nucleate and grow." Discussions of terminology will be found following the papers published in "The Mechanism of Phase Transformations in Metals," Institute of Metals, London, 1956.

1965. J. W. Christian, "The Theory of Transformations in Metals and Alloys," Pergamon, New York, 1965.

An interesting review of progress and problems in solid-state transformations is given in L. Himmel, J. J. Harwood, and W. J. Harris (eds.), "Perspectives in Materials Research," Office of Naval Research, Washington, D.C., 1961. Transformations occurring mainly by diffusional processes are discussed in: American Society for Metals Symposium, "Precipitation from Solid Solution," Cleveland, Ohio, 1957. American Institute of Mining, Metallurgical, and Petroleum Engineers Symposium, V. F. Zackay and H. I. Aaronson (eds.), "Decomposition of Austenite by Diffusional Process," Interscience, New York, 1962. "Physical Properties of Martensite and Bainite," Iron and Steel Institute Special Report No. 93, London, 1965.

Crystallographic and geometrical aspects of the martensite transformation are considered by: F. C. Frank, *Acta Met.*, vol. 1, p. 15, 1953. M. A. Jaswon, *Research*, vol. 11, p. 315, 1958. J. K. Mackenzie, *J. Australian Inst. Met.*, vol. 5, p. 90, 1960. B. A. Bilby and J. W. Christian, *J. Iron Steel Inst.*, vol. 197, p. 122, 1961. C. M. Wayman, "Introduction to the Crystallography of Martensitic Transformations," Macmillan, New York, 1964.

Problems of nucleation and kinetics in solid-state transformations are discussed by: J. H. Hollomon and D. Turnbull, *Progr. Metal Phys.*, vol. 4., p. 333, 1953. M. Cohen, *Trans. AIME*, vol. 212, p. 171, 1958. H. Knapp and U. Dehlinger, *Acta Met.*, vol. 4, p. 289, 1956. Thermodynamics of transformations have been reviewed by: L. Kaufman and M. Cohen, *Progr. Metal Phys.*, vol. 7, p. 115, 1958. G. V. Kurdjumov, *J. Iron Steel Inst.*, vol. 195, p. 26, 1960. J. W. Christian, "The Theory of Transformations in Metals and Alloys," Pergamon, New York, 1965.

Precipitation hardening has been reviewed by: H. K. Hardy and T. J. Heal, *Progr. Metal Phys.*, vol. 5, p. 143, 1954. A. Kelly and R. B. Nicholson, *Progr. Mater. Sci.*, vol. 10, p. 149, 1963.

Certain relationships between solid-state transformations and microstructure are discussed by: C. S. Smith, *Trans. ASM*, vol. 45, p. 533, 1953; *Met. Rev.*, vol. 9, p. 1, 1964. J. B. Newkirk in American Society for Metals, "Precipitation from Solid Solution," Symposium, Cleveland, Ohio, 1957. H. I. Aaronson in V. F. Zackay and H. I. Aaronson (eds.), "Decomposition of Austenite by a Diffusional Process," Interscience, New York, 1962. A. H. Geisler in R. Smoluchowski, J. E. Mayer, and W. A. Weyl (eds.), "Phase Transformations in Solids," Wiley, New York, 1951.

Articles describing recent investigations of precipitation hardening alloys and of transformations by means of transmission electron microscopy will be found in G. Thomas and J. Washburn (eds.), "Electron Microscopy and Strength of Crystals," Wiley, New York, 1963.

General characteristics of martensitic transformations The distinguishing characteristics of martensitic transformations as they were known in the 1940s were summarized by Troiano and Greninger.[1] The principal characteristics were—and in general still are—that diffusion is not required, that platelike volumes transform with great speed while the temperature is falling, that almost none form while the temperature is held constant, that the transformation is aided by cold-work, that an element of volume in the transforming material changes its shape, and that the product has definite crystallographic habit and lattice orientation relationship with respect to the parent phase. Because of these common characteristics the terms *diffusionless transformation* and *shear transformation* were often applied to it.

Although these are still recognized as important features of most martensitic transformations, various exceptions have now been discovered that

[1] A. R. Troiano and A. B. Greninger, *Metal Progr.*, vol. 50, p. 303, 1946.

make it advisable to discard most of these characteristics as criteria for this type of transformation and to retain only the criterion of a *change of shape*, such as can cause a tilting of the surface of a specimen.

As examples of the classification difficulties, one can cite certain martensites that form isothermally at a low rate[1] and some nucleation-and-growth transformations that proceed with surprising rapidity.[2] In the case of *thermoelastic martensite*[3] the growth velocity is strictly controlled by the rate of change of temperature, or of stress; yet the slow growth does not imply that thermal activation energy is required. Some martensites occur both rapidly and slowly. In In-Tl alloys a thin plate of the martensitic phase forms very rapidly at first across a crystal of the parent phase, and then thickens at a slow observable rate,[4] while in Cu-Zn alloys[5] an initially slow growth of martensitic plates is followed by "bursts" at very high velocity. The *bainite transformation* in steels forms by a slow shearlike transformation that has the features of martensite,[6] yet it cannot proceed unless the driving force for the transformation is increased by an accompanying precipitation of carbides in the new phase, which mechanism therefore controls the rate of growth and is related to the diffusion of carbon.[7] There are other transformations in which diffusion undoubtedly controls the process, but which nevertheless conform to the basic geometry of a martensitic change. Precipitation of CuBe from α solid solution[8] and the ordering reaction in CuAu[9] provide examples. There are also other precipitation reactions that appear to have typically martensitic characteristics.[10] Thus, it became increasingly difficult to differentiate between martensitic and nucleation-and-growth transformations when special cases were considered, unless the *change of shape* is taken as the criterion.[11]

[1] E. S. Machlin and M. Cohen, *Trans. AIME*, vol. 194, p. 489, 1952. R. E. Cech and J. H. Hollomon, *Trans. AIME*, vol. 197, p. 685, 1953. A. N. Holden, *Acta Met.*, vol. 1, p. 617, 1953.

[2] T. B. Massalski, *Acta Met.*, vol. 6, p. 243, 1958. A. Gilbert and W. S. Owen, *Acta Met.*, vol. 10, p. 45, 1962. M. J. Bibby and J. G. Parr, *J. Iron Steel Inst.*, vol. 202, p. 100, 1964.

[3] R. W. Cahn, *Nuovo Cimento* (Suppl.), vol. 10, p. 350, 1953. This name is, again, rather unfortunate since it does not convey any precise meaning. A discussion of thermoelastic martensites is given on p. 529.

[4] J. S. Bowles, C. S. Barrett, and L. Guttman, *Trans. AIME*, vol. 188, p. 1478, 1950. Z. S. Basinski and J. W. Christian, *Acta Met.*, vol. 2, p. 148, 1954.

[5] H. Pops and T. B. Massalski, *Trans. AIME*, vol. 230, p. 1662, 1964.

[6] G. V. Smith and R. F. Mehl, *Trans. AIME*, vol. 150, p. 211, 1942. B. A. Bilby and J. W. Christian in "The Mechanism of Phase Transformations in Metals," Institute of Metals, London, 1956.

[7] T. Ko and S. A. Cottrell, *J. Iron Steel Inst.*, vol. 172, p. 307, 1952; vol. 173, p. 224, 1953. T. Ko, *Acta Met.*, vol. 2, p. 75, 1954.

[8] J. S. Bowles and W. J. McG. Tegart, *Acta Met.*, vol. 3, p. 590, 1955.

[9] J. S. Bowles and R. Smith, *Acta Met.*, vol. 8, p. 405, 1960.

[10] H. M. Otte and T. B. Massalski, *Acta Met.*, vol. 6, p. 494, 1958.

[11] D. Hull, *Bull. Inst. Met.*, vol. 2, p. 134, 1954. B. A. Bilby and J. W. Christian, *J. Iron Steel Inst.*, vol. 197, p. 122, 1961 (see also, however, discussion on p. 172 and J. W. Christian in V. F. Zackay and H. I. Aaronson (eds.), "Decomposition of Austenite by the Diffusional Process," Interscience, New York, 1962.

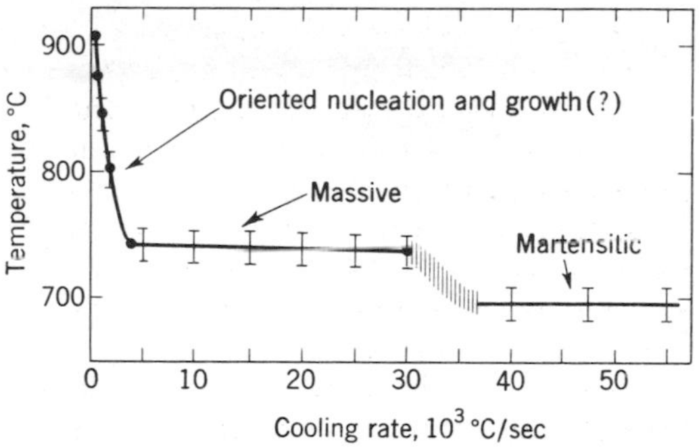

Fig. 18-1 The possible modes of $\gamma \rightarrow \alpha$ transformation in pure iron on cooling. (**Data from M. J. Bibby and J. G. Parr, *J. Iron. Steel Inst.*, vol. 202, p. 100, 1964.**)

Polymorphic changes As already discussed in Chap. 10, polymorphic changes involve alteration of structure but not of composition; they are numerous among the metallic elements and particularly prolific among certain heavy elements such as plutonium or uranium. Not only temperature changes but also pressure changes, shock waves, and plastic deformation can induce new structural forms.

It is possible for a polymorphic change to occur by a nucleation-and-growth process under one set of conditions and by a martensitic process under a different set. The purity, the grain size of the starting material, and the rate of quenching all play an important part.

The existence of different transformation mechanisms for the $\alpha \rightleftharpoons \gamma$ change in pure iron is a case in point and has been demonstrated experimentally.[1] In Fig. 18–1 a diagrammatic representation of the transformation temperature and the possible mechanism as a function of the rate of cooling is given, following mainly the experimental work of Bibby and Parr. With slow cooling rates the $\gamma \rightarrow \alpha$ transformation occurs near the equilibrium temperature, ~910°C, and under these conditions it presumably involves oriented nucleation and growth of the α phase by a diffusional process. At rates between 5000 and 30,000°C per sec, the transformation temperature is depressed further below 750°C, and it appears to be of the massive type (see page 507), and becomes martensitic, with typical surface rumpling and needlelike appearance when quenching rates reach 4×10^4 to 5×10^4 °C per sec. Slower quenching rates than these can produce martensite if the iron is less pure.[2]

[1] W. S. Owen and A. Gilbert, *J. Iron Steel Inst.*, vol. 196, p. 142, 1960. A Gilbert and W. S. Owen, *Acta Met.*, vol. 10, p. 45, 1962. M. J. Bibby and J. Gordon Parr, *J. Iron Steel Inst.*, vol. 202, p. 100, 1964.

[2] A. R. Entwistle in "The Mechanism of Phase Transformation in Metals," Institute of Metals, London, 1956. K. P. Singh and J. G. Parr, *Acta Met.*, vol. 9, p. 1073, 1961. C. M. Wayman and C. J. Altstetter, *Acta Met.*, vol. 10, p. 992, 1962.

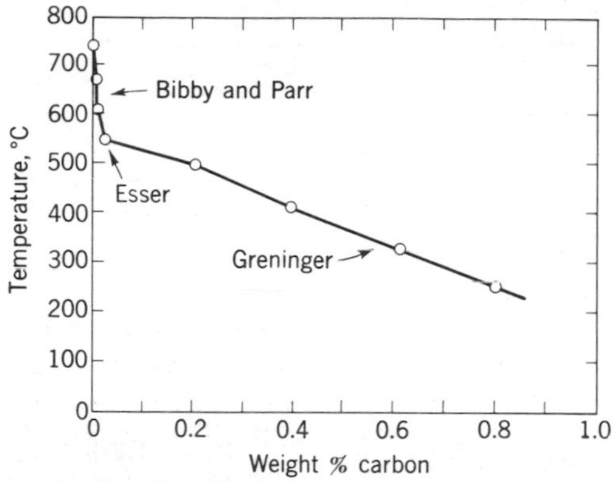

Fig. 18-2 The temperature of martensitic transformation, M_S, in the iron-carbon system.

Introduction of impurities (particularly carbon) into iron depresses the M_S temperature at first very rapidly and then at a slower rate (see Fig. 18–2); but it also tends to depress the temperature of the massive transformation below that of the martensitic. The existence of the initial steep decrease of M_S has only been realized recently.[1] With total impurity contents exceeding ~0.6 weight percent for interstitial solid solutions and several percent for substitutional solutions, a change occurs in the morphology and in the habit of the resultant martensite in many alloy systems.[2] In the Fe-Ni system the polymorphic change can be suppressed altogether below the room temperature when more than 30 atomic percent of nickel is added to iron.

When polymorphic changes occur by a diffusional process, their speed depends on the purity of the sample, the transformation temperature, and other factors such as, for example, exceptionally large volume changes, as in tin.

The phase change from the b.c.c. (white, β) to diamond-type (gray, α) tin takes place by a typical nucleation-and-growth mechanism. It begins at about room temperature (~18°C) but reaches its maximum velocity well below this temperature, near the temperature of "dry ice" (−78°C). The specific volume of gray tin is about 20 percent larger than that of the white tin; hence, the transformation involves a large strain-energy contribution and is accompanied by deformation and cracking, mainly of the gray form, which is more brittle than the white form. During growth, regions of the gray tin advance from a few nuclei in the form of gray pustules visible on the surface; the process is sometimes given the picturesque

[1] W. D. Swanson and J. Gordon Parr, *J. Iron Steel Inst.*, vol. 202, p. 1295, 1964. R. B. G. Yeo, *Trans. AIME*, vol. 224, p. 1222, 1962. A. Gilbert and W. S. Owen, *Acta Met.*, vol. 10, p. 45, 1962. Bibby and Parr, loc. cit.

[2] W. S. Owen, E. A. Wilson, and T. Bell, in V. F. Zackay (ed.), "High Strength Materials," Wiley, New York, 1965.

name "tin pest" or "tin disease." The transformation may be hastened by "inoculation" with gray tin, i.e., by implanting a few particles of gray tin on the surface of an untransformed white tin specimen. Excessive cracking of the gray form may be reduced by alloying.[1]

Transformations between f.c.c. and c.p.h. structures are common. The manner of stacking layers to form the f.c.c. and c.p.h. structures results in very similar environments for the atoms—in fact, with similarly spaced layers, both the 12 nearest neighbors and the 6 second-nearest neighbors remain at the same distance from a given atom, and only third-nearest and more remote ones change. This transformation has been reported in calcium, scandium, lanthanum, and cerium, which are hexagonal at low temperatures and f.c.c. at high, and in cobalt, which has a similar relationship: the stable structure at room temperature is c.p.h., but this transforms to f.c.c. at temperatures above 400°C. The transformation involves the shearing of close-packed atomic planes in such a way as to change the stacking sequence from *ABCABC* in f.c.c. to *ABABAB* in c.p.h. (as discussed in Chap. 10). There is a contraction of interatomic spacings normal to the close-packed planes, and expansion within the planes produces a c/a ratio that is slightly less than ideal. The characteristics of the transformation suggest that it occurs by a martensitic change,[2] the nucleus being essentially a stacking fault.[3] The nature of faulting, the fault density, and the role of faults in the transformation have been discussed by a number of investigators.[4] This transformation is also observed in crystals with van der Waals binding, such as argon and mixtures of argon with nitrogen, oxygen, or CO at low temperatures,[5] which are stable in the c.p.h. structure above a temperature range of stability for the f.c.c. structure.

Iron, titanium, zirconium, thallium, lithium, and sodium transform from the b.c.c. structure to the close-packed structure that is stable at lower temperatures. The b.c.c. structure is mechanically less stable, provides opportunity for greater amplitudes in the thermal vibrations of the atoms than the c.p.h., and would therefore be expected to have larger entropy S in the equation $F = U - TS$, where F is the free energy, U is the internal energy, and T is the temperature. It is thought that, because the entropy

[1] See, for example, E. O. Hall in "The Mechanism of Phase Transformations in Metals," Institute of Metals, London, 1956. P. H. Van Lent, *Acta Met.*, vol. 10, p. 1089, 1962.

[2] A. Troiano and J. Tokich, *Trans. AIME*, vol. 175, p. 728, 1948. F. Sebilleau and H. Bibring in "The Mechanism of Phase Transformation in Metals," op. cit., p. 209. M. A. Gedwill, C. J. Altstetter, and C. M. Wayman, *Trans. AIME*, vol. 230, p. 453, 1964.

[3] A. Seeger, *Z. Metallk.*, vol. 44, p. 247, 1953; vol. 47, p. 653, 1956.

[4] See, for example: M. A. Jaswon, *Research*, vol. 11, p. 315, 1958. C. R. Houska, B. L. Averbach, and M. Cohen, *Acta Met.*, vol. 8, p. 81, 1960.

[5] The results of experiments are presented in a series of papers by C. S. Barrett and L. Meyer, beginning with argon-nitrogen mixtures, *J. Chem. Phys.*, vol. 41, p. 1078, 1964; vol. 42, p. 107, 1965. Even pure argon can exist in a metastable c.p.h. form which martensitically changes to f.c.c.: L. Meyer, C. S. Barrett, and P. Haasen, *J. Chem. Phys.*, vol. 40, p. 2744, 1964.

term is larger for the body-centered structure, this form becomes more stable with respect to the close-packed form as the temperature is raised.[1] The martensitic nature of the b.c.c. $\rightleftharpoons$ c.p.h. transformations is well established. All are subject to an approximate orientation relationship: $(101)_{\text{b.c.c.}} \| (0001)_{\text{c.p.h.}}$ and $[11\bar{1}]_{\text{b.c.c.}} \| [11\bar{2}0]_{\text{c.p.h.}}$ sometimes known as the *Burgers relation* following a study of the zirconium transformation. In titanium the deviations from the Burgers relation are of the order of $\frac{1}{2}$ to 1° and the habit plane (see page 496) is close to {8, 9, 12}.[2] Of the b.c.c. alkali metals, only Li and Na transform, not K, Rb, or Cs.[3] Some beta-brasses also transform at cryogenic temperatures.[4] Lithium transforms as a result of cold work at liquid nitrogen temperature into a faulted f.c.c. structure. The amount of transformation increases with further lowering of the temperature of cold working, and the sample also will transform spontaneously when cooled below an M_S temperature of about 72°K. The phase formed by simple cooling is c.p.h. The spontaneously formed martensitic phase in sodium is likewise c.p.h. with the M_S temperature of about 36°K. The c.p.h. phase that forms on simple cooling in sodium, unlike lithium, is not changed to f.c.c. as a result of cold work; but it contains numerous stacking faults.

Not only temperature and cold work, but also pressure, and pressure combined with plastic flow, can induce phase transformations in metals, as the extensive experiments of Bridgman, and more recently of others, have shown.

Morphology of phase transformations A precipitation reaction is usually initiated by cooling or quenching an alloy from a temperature at which its structure is single-phase solid solution to a temperature where the phase is supersaturated. Slow cooling and low-temperature annealing tend to promote precipitation, while fast cooling tends to suppress precipitation. Solute atoms clustering in the matrix and forming very small dispersed particles of precipitate are a very important cause of hardening (age hardening), which usually reaches a peak and begins to decline before the particles become large enough to be seen in an optical microscope.

The detailed physical location and distribution of different phases in an alloy can in many cases be determined solely from the study of polished and etched sections through a specimen. However, as already mentioned in

[1] C. Zener, "Elasticity and Anelasticity of Metals," University of Chicago, Chicago, 1952. This reasoning, in fact, led to the search for the lithium transformation: C. S. Barrett, *Phys. Rev.*, vol. 72, p. 245, 1947. The b.c.c. form of iron is stable below the range of stability of the f.c.c.; divergent suggestions have been made as to the reason for this, and theoretical opinion cannot be said to have crystallized as yet; however, the close-packed form (c.p.h.) can be induced in iron by pressure and shock (see Chap. 10, pp. 231 and 233).

[2] A. J. Williams, R. W. Cahn, and C. S. Barrett, *Acta Met.*, vol. 2, p. 117, 1954.

[3] C. S. Barrett, *J. Inst. Metals*, vol. 84, p. 1653, 1955–1956.

[4] See C. S. Barrett, *Trans. ASM*, vol. 49, p. 53, 1957; *Acta Cryst.*, vol. 9, p. 621, 1956. L. L. Isaacs and H. Pops, *Trans. AIME*, vol. 230, p. 266, 1964.

Chap. 2, information derived from a one-surface analysis must be treated with caution when interpreted in terms of a three-dimensional pattern. When the manner of distribution of one phase within another shows distinctive regularity, it is common to speak of a *habit* or *habit planes* as the crystallographic family of planes of the matrix along which the minor phase is physically distributed. Such habits may be *rational*, involving crystallographic planes of the matrix with relatively simple indices such as {111}, {110}, or {001}, or they may be *irrational* if the planes cannot be described with simple indices, as for example, the habit planes near {225} of martensites in the high-carbon steels and the habit of martensite formed from beta-brass which is near $\{2, 12, 11\}_\beta$. Habit planes can usually be identified from an analysis of polished surfaces, but a precise *crystallographic orientation relationship* between the *lattices* of the two phases may sometimes exist even if there is no clearly distinguishable habit. The morphology of the constituents resulting from a transformation and the properties of the resulting assemblage are strongly influenced by the nature of the boundaries between the various phases.[1]

Incoherent boundaries have disordered, isotropic boundary structures. Diffusion occurs rapidly both along them and across them, and their mobility is high except when they encounter inclusions. These are the typical high-energy boundaries of a randomly oriented polycrystalline metal. They are the type that moves in recrystallization and grain growth, and the type found at the edges of recrystallized regions in *discontinuous precipitation*, the class of reaction in which precipitation is concentrated in certain regions.

Coherent boundaries represent a maximum in the continuity of the lattices adjoining the boundaries, and contain few dislocations or none. The boundaries of this type have low surface energy, but the surrounding material may be strained for considerable distances. Their movement is believed to be arrested by the accumulation of volume strain energy. When precipitation from solid solution develops these boundaries, dendritic branching onto crystallographically equivalent habit planes may be frequent

Partially coherent boundaries differ from the fully coherent type by the presence of arrangements of dislocations. The lattices on both sides in the vicinity of a boundary are strained. Boundary migration takes place mainly by the movement of dislocations normal to the boundary. This type, as well as the fully coherent type, is frequently formed in the early stages of precipitation from solid solution.

When precipitation proceeds simultaneously throughout the parent phase (i.e., in *continuous precipitation*), the nuclei tend to form first on dislocations. This tendency is used to "decorate" dislocations to make them

[1] C. S. Smith, *Trans. AIME*, vol. 175, p. 15, 1948; *Trans. ASM*, vol. 45, p. 533, 1953; *Met. Rev.*, vol. 9, no. 33, p. 1, 1964. Some fundamentals are discussed in Chap. 14. An earlier review was published by R. F. Mehl and L. K. Jetter in "Symposium on Age Hardening of Metals," American Society for Metals, Cleveland, Ohio, 1940.

Fig. 18-3 Octahedral pattern of Widmanstätten precipitation in the San Francisco Mountains meteorite. The surface heating during flight through the atmosphere has produced a partial obliteration of the octahedral pattern in the peripheral zone. ×2. (Courtesy E. P. Henderson, Museum of Natural History, Washington, D. C.)

visible in crystals transparent to visible or infrared light, and is strikingly evident in transmission electron micrographs.[1]

The oldest known examples of continuous decomposition are found in meteorites that exhibit the *Widmanstätten structure* (Fig. 18–3), usually visible to the naked eye. The beautifully regular markings on a polished and etched surface of an iron-nickel meteorite were first discovered by Count Alois von Widmanstätten early in the nineteenth century. The study of meteorites has occupied an important place in the development of physical metallurgy and, more recently, in space science.[2] It now comprises a very large literature.[3]

Networks of parallel lamellae similar to the meteoritic structure have been observed in artificially prepared Fe-Ni alloys and in a great number of

[1] Numerous examples will be found in: "Dislocations and Mechanical Properties of Crystals," Wiley, New York, 1957. J. M. Hedges and J. W. Mitchell, *Phil. Mag.*, vol. 44, p. 223, 1953. S. Amelinckx, "The Direct Observation of Dislocations, Solid State Physics Supplement 6," Academic, New York, 1964.

[2] The record in meteorites is directly linked to their past history; it reflects the thermal, mechanical, and environmental conditioning that these objects received and therefore clearly contains important information about the origin and processes that have occurred in some parts of the solar system. Meteorites thus constitute "a poor man's space probe" (E. Anders).

[3] See, for example, articles quoted in: E. Anders, "Origin, Age, and Composition of Meteorites," *Space Sci. Rev.*, vol. 3, p. 583, 1964. C. E. Moore (ed.), "Research on Meteorites," Wiley, New York, 1962. J. I. Goldstein and R. E. Ogilvie, *Geochim. Cosmochim. Acta*, vol. 29, p. 893, 1965.

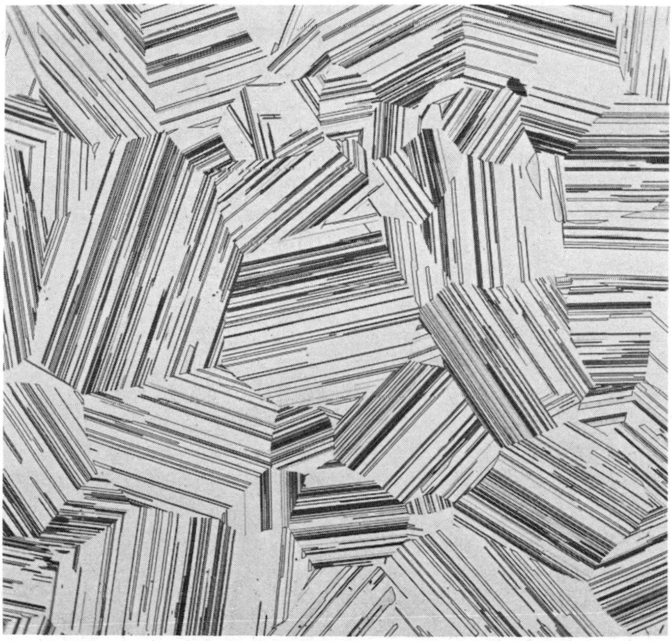

Fig. 18-4 Widmanstätten precipitation of a c.p.h. (ζ) phase from a f.c.c. (α) phase in Cu-Si alloys. Precipitation occurs on the octahedral $\{111\}$ planes. The interfaces are highly coherent. A simple crystallographic correspondence is maintained: $(0001)_\zeta \| (111)_\alpha$.

other alloys in which precipitates form from a supersaturated solid solution, and have come to be known as Widmanstätten structures. In the majority of cases the precipitating crystals are thin plates on matrix planes of low indices, as in Fig. 18–4; occasionally, they form on planes of high indices or on irrational planes, and in some instances they form needles, geometrical shapes, rosettes, irregular particles instead of plates, or coherent duplex structures. The orientation of the precipitate lattice is definitely related to the orientation of the matrix lattice. The research on such Widmanstätten relationships in many different systems has grown so rapidly that a tabulation of the results seems no longer feasible here.[1]

Lattice relationships in Widmanstätten structures Young[2] first showed by x-rays that the individual lamellae of octahedrite meteorites were single crystals of definite orientation. The high temperature f.c.c.

[1] Many crystallographic relationships in precipitation reactions are listed by: C. S. Barrett, "The Structure of Metals," 2d ed., p. 548, McGraw-Hill, 1952. J. B. Newkirk in "Precipitation from Solid Solutions," American Society for Metals Symposium, p. 70, 1959.

[2] J. Young, *Proc. Roy. Soc.* (*London*), vol. A112, p. 630, 1926.

γ phase (the taenite) of the Fe-Ni meteoritic iron (roughly 8 percent nickel) precipitates the b.c.c. α-phase crystals (the kamacite) as plates along the octahedral planes of the γ matrix. Within the α plates a (110) plane is parallel to the (111) plane of union with the matrix. The crystallography and habit of some other Widmanstätten patterns have similar simple relationships.[1]

One of the first Widmanstätten structures of microscopic dimensions to receive a thorough analysis is found in Al-Ag alloys. Each plate of the precipitate, c.p.h. γ', forms along an octahedral plane of the f.c.c. matrix; the basal plane of the precipitate lies parallel to the octahedral plane of the matrix, similarly to the Cu-Si case illustrated in Fig. 18–4. Furthermore, the closest-packed rows of atoms in matrix and precipitate are parallel; the lattices are coherent. The γ' is a transitional phase; the basal plane of γ' maintains complete registry with the (111) plane of the matrix until the plate has grown to a thickness of many hundreds of angstroms.[2] Finally, presumably, the stresses tending to tear the particle from registry become intolerable, and the transition to the stable structure occurs. The stable phase γ differs from the transition phase only in its dimensions a and c. It has been suggested that the transition phase is merely a stressed form of the stable phase and that the stresses arise from the coherence between precipitate and matrix. Assuming that biaxial stresses in the precipitate strain it into registry with the matrix, Mott and Nabarro[3] computed the thickness to which the precipitate would grow before the strain energy becomes too large. Further details of stages in Al-Ag precipitation are discussed later.

Plane of precipitation and shape of precipitates Upon what plane of the matrix will a plate of precipitate form, i.e., what will be the habit plane? Observation shows that this plane is not solely determined by the crystal structure of the matrix; thus, it is not necessarily the plane of greatest atomic density, or the one containing the closest-packed row of atoms, or a cleavage, slip, or twinning plane. The plane chosen is the result of an interaction between the matrix and precipitate structures. A rule that is found to apply to most alloys is that the atomic planes facing each other across the matrix-precipitate interface have very similar atom patterns and atomic spacings, a minimum mismatch. This is presumably

[1] R. F. Mehl and C. S. Barrett, *Trans. AIME*, vol. 93, p. 78, 1931 (Al-Ag, Cu-Si). R. F. Mehl, C. S. Barrett, and D. W. Smith, *Trans. AIME*, vol. 105, p. 215, 1933. R. F. Mehl, C. S. Barrett, and H. S. Jerabek, *Trans. AIME*, vol. 113, p. 211, 1934 (Fe-N, Fe-P). R. F. Mehl, C. S. Barrett, and F. N. Rhines, *Trans. AIME*, vol. 99, p. 203, 1932 (Al-Cu, Al-Mg_2Si). R. F. Mehl and O. T. Marzke, *Trans. AIME*, vol. 93, p. 123, 1931 (Cu-Zn, Cu-Al).

[2] C. S. Barrett and A. H. Geisler, *J. Appl. Phys.*, vol. 11, p. 733, 1940. C. S. Barrett, A. H. Geisler, and R. F. Mehl, *Trans. AIME*, vol. 143, p. 134, 1941.

[3] N. F. Mott and F. R. N. Nabarro, *Proc. Phys. Soc.* (*London*), vol. 52, p. 86, 1940. F. R. N. Nabarro, *Proc. Phys. Soc.*, (*London*), vol. 52, p. 90, 1940; *Proc. Roy. Soc.* (London), vol. A175, p. 519, 1940.

a manifestation of a more fundamental rule that *interface energy is a minimum.*

A Widmanstätten precipitate with a boundary that over most of its area is coherent and of low energy may have portions that are noncoherent and of high mobility. The shape changes during growth are thought to be influenced by the rapid growth of the mobile parts and the sessile character of semicoherent parts.[1]

In order for a precipitation reaction to proceed, the reduction in the volume-free energy must be greater than the sum of the interface energy and the strain energy. When a precipitate particle is very small, the interface energy may dominate the reaction and may lead to the formation of a coherent transition phase, as in the case of many precipitation hardening alloys. Theories have been developed also that relate precipitate morphology to strain-energy minimization.[2]

Discontinuous precipitation and eutectoid reactions These transformations are nucleated mainly at grain boundaries or imperfections. Growth involves duplex regions, consisting of alternate particles or lamellae of different structure and composition, advancing in colonies into a single-phase parent matrix. It seems quite likely that in some cases the nuclei of the growing regions are themselves duplex.[3]

The mean composition of the product of a *discontinuous precipitation* is equal to the composition of the parent. As the duplex regions ("cells" or "colonies") grow edgewise, the individual plates maintain an approximately constant spacing. However, the variation of the growth rate and the interlamellar spacing is usually a function of temperature and undercooling.[4] Branching occurs in preference to repeated nucleation. This effect is particularly evident in the pearlite structures (see Fig. 18–5) of steels and numerous ferrous and nonferrous systems. Recent metallographic work[5] has shown that in pearlite, the cementite and ferrite act as equal partners and either one can be the so-called *active nucleus.* According to Hillert, the ferrite and cementite can have any orientation relationship with respect to the matrix austenite, except certain combinations leading

[1] H. I. Aaronson in "Decomposition of Austenite by Diffusional Processes," p. 387, Interscience, New York, 1962. Aaronson explains in this way the irregularities ("ruggedness") that develop in austenite-cementite boundaries in samples of plain-carbon hypereutectoid steels after long holding near 775°C.

[2] F. R. N. Nabarro, *Proc. Phys. Soc. (London)*, vol. 52, p. 90, 1940; *Proc. Roy. Soc.* (London), vol. A175, p. 519, 1940, N. F. Mott and F. R. N. Nabarro, *Proc. Phys. Soc. (London)*, vol. 52, p. 86, 1940. F. Laszlo, *J. Iron Steel Inst.*, vol. 147, p. 173, 1943; vol. 148, p. 137, 1943; vol. 150, p. 183, 1944; vol. 152, p. 207, 1945; vol. 164, p. 5, 1950.

[3] See C. S. Smith in "Decomposition of Austenite by Diffusional Processes," loc. cit., p. 238.

[4] See, for example, the study of the growth rate in the austenitic Fe 30-Ni 6-Ti alloy: G. R. Speich, *Trans. AIME*, vol. 227, p. 754, 1963.

[5] M. Hillert in "Decomposition of Austenite by Diffusional Processes," loc. cit. (references and an evaluation of earlier work are included).

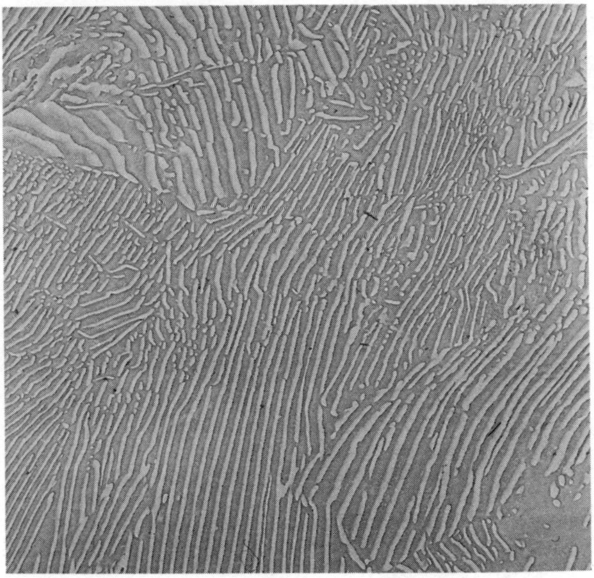

Fig. 18-5 The structure of pearlite formed by an eutectoid decomposition of a simple carbon steel. ×15,000. (Courtesy United States Steel Research Laboratory.)

to coherent interfaces that would make cooperative growth impossible. Electron microscopy and electron diffraction investigations[1] seem to suggest, on the other hand, that orientation relationships of cementite and ferrite with respect to austenite do play a role in the formation of pearlitic structures and that stacking faults and imperfection phenomena are important. The suggested orientation relationships are as follows:

$$(111)_\gamma \,||\, (110)_\alpha \,||\, (001)_c, \qquad [110]_\gamma \,||\, [111]_\alpha \,||\, [010]_c$$

and between ferrite and cementite

$$(011)_\alpha \,||\, (001)_c, \qquad [\bar{1}00]_\alpha \,||\, [100]_c \quad \text{and} \quad [0\bar{1}1]_\alpha \,||\, [010]_c.^2$$

Orientation relationships obtained when cementite precipitates from ferrite or from martensite seem to be different from the above (see page 514).

Microstructures and reactions in eutectoid reactions have been reviewed recently[3]; these include the pearlite, bainite, and martensite reactions in

[1] L. S. Darken and R. M. Fisher in "Decomposition of Austenite by Diffusional Processes," loc. cit., p. 249.

[2] A. J. Baker, P. M. Kelly, and J. Nutting in G. Thomas and J. Washburn (eds.), "Electron Microscopy and Strength of Crystals," Interscience, New York, 1962.

[3] C. W. Spencer and D. J. Mack in "Decomposition of Austenite by Diffusional Processes," loc. cit., p. 549. See also R. F. Mehl and C. A. Dube in "The Phase Transformations in Solids," Wiley, New York, 1951.

steels, and similar reactions, together with the massive type, in other alloys. These are discussed in later sections.

Changes occurring during aging The hardening of alloys by aging is of great industrial importance and unusual scientific interest. The changes of physical properties, which are exceedingly varied and complex, have been the subject of so many investigations that it would not be possible to summarize them adequately here. The present discussion is limited to the more striking crystallographic features of the subject.[1]

Merica, Waltenberg, and Scott[2] were the first to point out that for an alloy to age-harden it must have a decreasing solubility with decreasing temperature, so that when the alloy is given a solution heat treatment and then a quench to a lower temperature, it will be supersaturated with respect to one or more dissolved elements and will undergo structural changes.

In many alloys the precipitation process consists of the sequential reactions: *supersaturated solid solution* → *one or more transition states* → *saturated solution* + *equilibrium precipitate*. The initial and final states correspond to the phases of the equilibrium phase diagram. The transition states are often less easily identified; examples of these are discussed later. The successive stages may overlap so that two or more exist simultaneously.

Slip bands and grain boundaries are favored regions for rapid precipitation in some alloys. In the regions of rapid precipitation the lattice parameter may alter to that of the depleted matrix before the diffraction lines from other regions register any parameter change.[3] Examples are found in the alloys of beryllium in copper[4] and zinc in aluminum (Fig. 18-6). Electron microscopy has recently added much information on aging.[5] As research continues to reveal unsuspected complexities in the aging of various alloys, it has become evident that a sequence of events that occurs in one alloy does not necessarily hold in another; there is evidence, also, that some alloys go through different stages during low-temperature and high-temperature aging and that the behavior may be different in thin sheets from behavior in the bulk of the material. Each of these various stages has its own specific effect on physical and mechanical properties,

[1] In addition to the references mentioned on p. 487, the reader is referred to the following reviews for extensive bibliographies and summaries: American Society for Metals, "Symposium on the Age-hardening of Metals," Cleveland, Ohio, 1940. W. L. Fink, *J. Appl. Phys.*, vol. 13, p. 75, 1942. A. H. Geisler in "Phase Transformations in Solids," p. 387, Wiley, New York, 1951. G. C. Smith in "Progress in Metal Physics," vol. 1, p. 163, Interscience, New York, 1949. Several articles in American Society for Metals Symposium, "Precipitation from Solid Solution," Cleveland, Ohio, 1959.

[2] P. D. Merica, R. G. Waltenberg, and H. Scott, *Trans. AIME*, vol. 64, p. 41, 1920.

[3] W. L. Fink and D. W. Smith, *Trans. AIME*, vol. 122, p. 284, 1936.

[4] A. G. Guy, C. S. Barrett, and R. F. Mehl, *Trans. AIME*, vol. 175, p. 216, 1948.

[5] A. Saulnier and R. Syre, *Rev. Met.*, vol. 49, p. 1, 1952. G. Thomas and J. Nutting, *J. Inst. Met.*, vol. 82, p. 610, 1954; in "The Mechanism of Phase Transformations in Metals," Institute of Metals, London, 1956. See also the review article by A. Kelly and R. B. Nicholson, *Progr. Mater. Sci.*, vol. 10, p. 149, 1963.

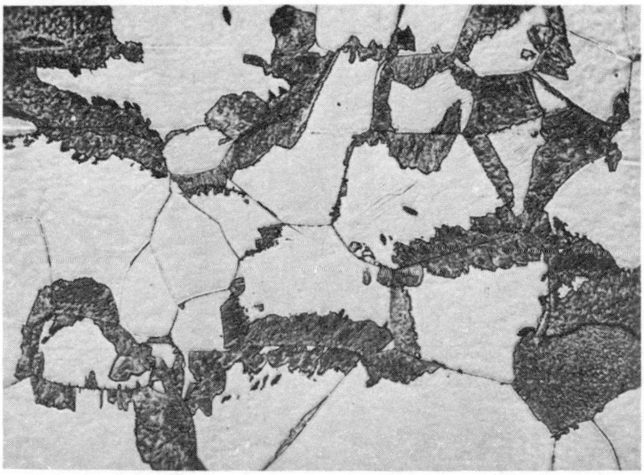

Fig. 18-6 Discontinuous precipitation from the Al-rich solid solution of the Al-Zn system. ×150.

either through the simple effect of a dispersed phase or through an interaction with the matrix.

A most important contribution to hardening is made when a metastable phase is coherent with the matrix in a way that produces severe coherence strains.[1] Strains arising merely from the difference in specific volume between matrix and precipitate are of interest but probably are minor causes of hardening compared with those arising from coherence. The release of coherence strains brings softening (overaging).

Structure changes during aging in Al–Cu, Al–Ag, and Cu–Be Some details of thoroughly investigated structure changes during aging should be mentioned as examples of the complexities that may be encountered. Aluminum-rich Al-Cu alloys have been perhaps the most carefully studied.[2] Copper atoms segregate on the cube planes of the aluminum matrix forming *Guinier Preston* (G.P.) *zones*; then the tetragonal phase θ'' (which has sometimes been called G.P.II) forms, which is coherent with the matrix and ordered; next a cubic (CaF_2-type) phase θ' appears which is also coherent with the matrix; and finally the equilibrium precipitate θ ($CuAl_2$), with a tetragonal crystal structure, replaces all transitional

[1] R. F. Mehl and L. K. Jetter, "Symposium on the Age-hardening of Metals," American Society for Metals, Cleveland, Ohio, 1940. J. S. Bowles and C. S. Barrett, "Progress in Metal Physics," vol. 3, Interscience, New York, 1952. Cyril S. Smith, *Trans. AIME*, vol. 175, p. 15, 1948. Kelly and Nicholson, loc. cit.

[2] A. Guinier, *Nature*, vol. 142, p. 569, 1938; *J. Phys. Radium*, ser. 8, vol. 3, p. 124, 1942; *Acta Cryst.*, vol. 5, p. 121, 1952; *Compt. Rend.*, vol. 231, p. 655, 1950. G. D. Preston, *Nature*, vol. 142, p. 570, 1938. V. Gerold, *Z. Metallk.*, vol. 45, pp. 593, 599, 1954. J. M. Silcock, T. J. Heal, and H. K. Hardy, *J. Inst. Met.*, vol. 82, p. 239, 1953–1954. Reviews have been published by: A. H. Geisler in R. Smoluchowski, J. E. Mayer and W. A. Weyl (eds.), "Phase Transformations in Solids," p. 387, Wiley, New York, 1951. R. B. Nicholson, G. Thomas, and J. Nutting, *J. Inst. Metals*. vol. 87, p. 429, 1959. A. Guinier, in "Advances in Solid State Physics," vol. 9, p. 293, Academic, New York, N.Y., 1959.

structures. The proposed transformation sequence is thus: G.P. zones $\rightarrow \theta'' \rightarrow \theta' \rightarrow \theta(CuAl_2)$. In the same alloy, slow cooling seems to generate only the stable phase. (In commercial alloys of the duralumin type, which contain magnesium and silicon, there is also a precipitation of platelets of Mg_2Si.) A different sequence has been observed in Al-Ag, where spherical regions rich in Ag atoms have an ordered superlattice arrangement before any new precipitate structure appears.[1] (Further examples are given in Table 18–1, introduced later.)

In Cu-rich Cu-Be alloys, diffuse x-ray diffraction patterns from single crystals indicate that in the absence of recrystallization, the precipitate first appears as beryllium-rich thin plates parallel to {100} planes of the copper solid-solutions matrix.[2] Various x-ray diffraction observations can be accounted for by three different precipitate structures.[3]

X-ray diffraction in the early stages of aging The existence of transitional states in the early stages of the precipitation process during age-hardening has been proved by x-ray and electron diffraction experiments on the systems Al-Cu,[4] Al-Ag,[5] and many others. Before transition structures become fully developed, however, x-ray diffraction effects can be obtained from regions where the precipitation process is beginning. As mentioned earlier, this was discovered in Al-Cu alloys by Guinier and his coworkers[6] and independently by Preston,[7] who found *streaks* on Laue photographs such as would be caused by two-dimensional diffraction gratings located on {100} planes of the matrix. These local concentrations of copper atoms are called G.P. zones. They appear to be platelets of copper a few atoms in thickness and a few hundred angstroms in diameter; the dimensions increase as aging is continued and eventually reach a value that gives three-dimensional diffraction from θ''. At a later stage the lattice transforms to the stable phase (θ).

Similar streaks in x-ray photographs have been observed and studied in detail in other age-hardening alloys. In aluminum-rich aluminum-silver alloys, Geisler and Hill[8] found evidence of a stage that precedes the platelet stage, during aging at room temperature. This stage produces the diffraction

[1] A. Guinier, *J. Phys. Radium*, ser. 8, vol. 3, p. 124, 1942. C. B. Walker and A. Guinier, *Compt. Rend.*, vol. 234, p. 2379, 1952.

[2] A. Guinier and P. Jacquet, *Compt. Rend.*, vol. 217, p. 22, 1943; *Rev. Met.*, vol. 41, p. 1, 1944. A. G. Guy, C. S. Barrett, and R. F. Mehl, *Trans. AIME*, vol. 175, p. 216, 1948.

[3] A. H. Geisler, J. H. Mallery, and F. E. Steigert, *Trans. AIME*, vol. 194, p. 307, 1952.

[4] G. Wassermann and J. Weerts, *Metallwirtschaft*, vol. 14, p. 605, 1935. W. L. Fink and D. W. Smith, *Trans. AIME*, vol. 122, p. 284, 1936; vol. 137, p. 95, 1940. G. D. Preston, *Proc. Roy. Soc.*, (*London*), vol. A167, p. 526, 1938; *Phil. Mag.*, vol. 26, p. 855, 1938.

[5] C. S. Barrett, A. H. Geisler, and R. F. Mehl, *Trans. AIME*, vol. 143, p. 134, 1941.

[6] A. Guinier, *Compt. Rend.*, vol. 204, p. 1115, 1937; vol. 206, p. 1641, 1938; *Nature*, vol. 142, p. 669, 1938. J. Calvet, P. Jacquet, and A. Guinier, *J. Inst. Metals*, vol. 6, p. 177, 1939.

[7] G. D. Preston, *Nature*, vol. 142, p. 570, 1938; *Phil. Mag.*, vol. 26, p. 855, 1938; *Proc. Roy. Soc.* (*London*), vol. A167, p. 526, 1938; vol. 52, p. 77, 1940.

[8] A. H. Geisler and J. K. Hill, *Acta Cryst.*, vol. 1, p. 238, 1948.

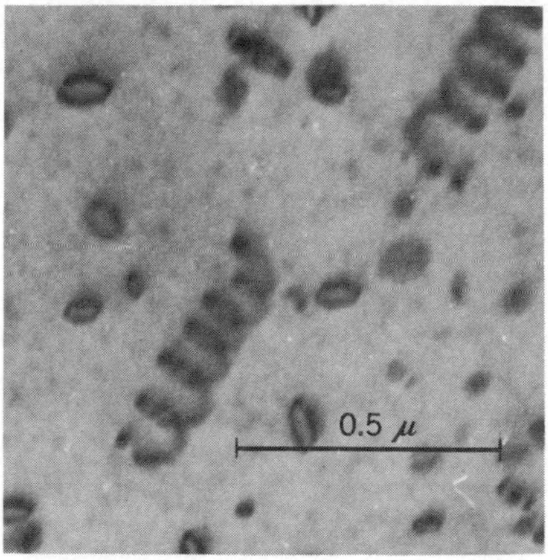

Fig. 18-7 Electron micrograph of a thin foil of Al–4 percent Ag alloy quenched from 550°C. Showing helical dislocations and prismatic dislocation loops. (A. Kelly and R. B. Nicholson, *Progr. Mater. Sci.*, vol. 10, p. 151, 1963.)

streaks which correspond to planes in reciprocal space, and which are interpreted as diffraction from a one-dimensional grating in the form of a "stringlet" of precipitate. The proposed stringlets lie along the close-packed rows ⟨110⟩ of the matrix, and are estimated to have dimensions roughly 10 by 20 by 100 A. As aging continues, the diffraction effects change; the reciprocal lattice at the next stage contains rods, which may be accounted for by two-dimensional gratings consisting of platelets on the close-packed planes of the matrix. At a later stage, the rods coalesce into spots, as if the platelets thickened into three-dimensional crystals.

In deducing the changes, Guinier and his coworkers rely heavily on a comparison of the vicinity of reciprocal lattice point 000 with others, since size effects due to precipitates, superlattices, or enriched zones should be the same around 000 as around all other points, whereas effects due to stacking disorders and various distortions should not be found around 000. Stringlet and platelet effects in Al-Ag are ascribed not to particles of precipitate but to superlattice domains and to stacking disorders.

Fig. 18-8 Electron micrograph of a thin foil of Al–1.7 percent Cu alloy quenched from 540°C and aged 16 hours at 130°C. Showing disk-shaped G.P. zones. (R. B. Nicholson and J. Nutting, *Phil. Mag.*, vol. 3, p. 531, 1958.)

In the systems Al-Mg and Al-Mg_2Si[1] the precipitate involves new atom sites since the streaks correspond to reciprocal lattice rods that do not pass through the reciprocal lattice points of the matrix.

Electron microscope studies of age-hardening alloys During early stages of aging it is possible to see with transmission electron microscopy[2] that the solute atoms aggregate to form small clusters (G.P. zones) and eventually precipitate while, at the same time, vacancies are annihilated at grain boundaries, surfaces, and dislocations. Large supersaturations of vacancies produce dislocation loops (prismatic, helical, and other), and the resultant dislocation substructure has an important effect on the observed precipitation sequence.[3]

Rapid quenching produces excess vacancies, and these are found to precipitate mainly as prismatic dislocation loops in dilute alloys and as helical dislocations in concentrated alloys. Both effects can be seen in an Al–4.4 percent Ag alloy quenched from 550°C (Fig. 18–7). Subsequently, they serve as centers for solute segregation and nucleation of precipitates. Figure 18–8 shows an electron micrograph of a thin foil of an Al–1.7 percent Cu alloy during early stages of aging at 130°C, following quenching from 540°C. After 16 hours the structure contains many disk-shaped G. P. zones lying on a plane perpendicular to the plane of the foil. The zones are about 3 to 4 A wide. After aging for one day at 130°C, θ'' plates, about 20 A in thickness and 400 A in diameter, become evident. Since both the G. P. zones and the θ'' precipitates should remain coherent with the matrix, the dark diffraction contrasts sometimes observed around them are interpreted as elastic strains in the matrix. Electron micrographs of Al–4.4 percent Ag alloys prepared from aged bulk material have confirmed that the G. P. zones are spherical in this system, with zone diameters consistent with the x-ray results.[4] On prolonged aging, the γ' precipitates are nucleated at helical dislocations and to a much smaller extent on Frank sessile dislocation loops. The γ' precipitates grow rapidly from each helix to cover most of the specimen in a typical Widmanstätten pattern. The precipitates are heavily faulted in the early stages of growth but become perfect as they grow larger. The suggested sequence of precipitation in Al-Ag alloys is as follows: Spherical silver-rich G. P. zones (possibly ordered) $\rightarrow$ silver-rich stacking faults at helical dislocations $\rightarrow$ γ' rods at helical dislocations $\rightarrow$ γ' plates containing stacking faults $\rightarrow$ perfect γ' (possibly ordered) $\rightarrow$ γ.[5] This sequence illustrates the considerable complexity encountered in a typical age-hardening system. Electron microscope investigations on a number of other systems have been reviewed recently,[6] and a tabulation of results as summarized by Thomas is given in Table 18–1.[7]

[1] A. H. Geisler and J. K. Hill, *Acta Cryst.*, vol. 1, p. 238, 1948.

[2] R. B. Nicholson and J. Nutting, *Phil. Mag.*, vol. 3, p. 531, 1958. R. B. Nicholson, G. Thomas, and J. Nutting, *J. Inst. Metals*, vol. 87, p. 429, 1959.

[3] G. Thomas and M. J. Wheelan, *Phil. Mag.*, vol. 4, p. 511, 1959. R. B. Nicholson and J. Nutting, *Acta Met.*, vol. 9, p. 332, 1961. K. H. Westmacott et al., *Phil. Mag.*, vol. 6, p. 929, 1961. A. L. Davies and J. E. Nicholson, *J. Inst. Metals*, vol. 93, p. 109, 1964–1965.

[4] R. B. Nicholson and J. Nutting, *Acta Met.*, vol. 9, p. 332, 1961.

[5] Ibid.

[6] A. Kelly and R. B. Nicholson, *Progr. Mater. Sci.*, vol. 10, p. 151, 1963. Also see articles by G. Thomas, by R. B. Nicholson, and by M. M. Baker, P. M. Kelly, and J. Nutting in G. Thomas and J. Washburn (eds.), "Electron Microscopy and Strength of Crystals," Interscience, New York, 1963.

[7] This table is a modification of one published in Thomas and Washburn (eds.), loc. cit.

Table 18–1 Summary of structures observed by transmission electron microscopy (Courtesy G. Thomas)

Alloy	Structure and shape	Habit in the matrix	Diffraction effects	Coherence, strain, and distribution
Al–4%Cu	Zones, Cu discs	Form on $\{100\}_m$	Streaks $\|\|\langle 100\rangle$	Coherent; S; Ho
	θ'' plates	Form on $\{100\}_m$	Spots with streaks	Coherent; S: Ho
	θ' plates	$(100)_{\theta'} \|\|(100)_m$	Spot pattern	Semicoherent; N: Ho + Het
Al–16%Ag	Zones, Ag spheres			Coherent; N; Ho
	γ', ordered	Form on $\{111\}_m$; faulted	Spots; streaks at first	Semicoherent; N; Het
Al–Mg	No zones seen			
	β' rods	Length $\|\|\langle 100\rangle_m$	Spot pattern	Semicoherent (?); N; Ho + Het
Al–Si	Zones (?), Si spheres			Coherent (?); N; Ho
	Pure Si plates	On $\{100\}$, $\{111\}$, $\{112\}$	Spot pattern	Semicoherent; N; Ho
Al–Zn	Pure Zn plates	Form on $\{111\}$ (?)		Ho
Al–Mg_2Si	Zones, Mg_2Si needles	Length $\|\|\langle 100\rangle_m$	Streaks $\|\|\langle 100\rangle$, superlattice	Coherent; N; Ho
	Rods, ordered f.c.c.	Length $\|\|\langle 100\rangle_m$	Spot pattern	N; Ho
	Mg_2Si plates, f.c.c.	On $\{100\}$, $[110]\|\|[100]_m$	Spot pattern	Incoherent (?); N; Ho
Al–Zn–Mg	Zones, spheres			Coherent; N; Ho
	$MgZn_2$ plates	Form on $\{111\}_m$	Spot pattern	Semicoherent; N; Ho + Het
Cu–Be	Zones, discs, of CuBe (?) ordered	Form on $\{100\}_m$	Streaks $\|\|\langle 100\rangle$ Superlattice	Coherent; S; Ho
Fe–Mo	Mo discs (?)	Form on $\{100\}_m$	Streaks $\|\|\langle 100\rangle$	Coherent; S (?); Het (?)
Fe–Cu	Cu spheres			*N*; Het
Fe–Au	Au discs, faulted	Form on $\{100\}_m$	Spots; streaks	Semicoherent; Het
Fe–C	Fe_3C in pearlite	$(111)_\gamma \|\|(110)_\alpha \|\|(001)_c$	Streaked spots	Semicoherent; *N*

S = strain fields observed
N = no strain fields observed
Ho = homogeneous nucleation
Het = heterogeneous nucleation (i.e., at imperfections and discontinuities)

The structure of "side-band" alloys There are striking examples of unusual diffraction effects in certain ternary alloys. A phase in the Fe-Ni-Al system that has the CsCl-type structure for Ni and Al atoms, with Fe atoms replacing these at random, may be made to precipitate a disordered b.c.c. phase richer in iron, the matrix becoming poorer in iron but retaining its order.[1] When the precipitate and matrix phases are present in about equal amounts, they take the form of lamellae parallel to {100} planes of the matrix and are coherent on these planes. There appears to be a wave pattern of varying iron content, with a wavelength of the order of a micron after annealing for 16 days at 850°C. A wavelike precipitation is also found in the alloy Cu_4FeNi_3.[2]

At high temperatures this alloy consists of a f.c.c. structure, which transforms at lower temperatures into copper-rich and copper-poor phases, both f.c.c., of different parameter. The precipitation sequence in these alloys after quenching from the single phase region was proposed[3] as: high-temperature f.c.c. matrix → "modulated" f.c.c. matrix → alternate lamellae of two tetragonal phases → f.c.c. equilibrium phases. Hence, before the final precipitation is complete, the alloy passes through a stage in which the diffraction lines become flanked by slightly diffuse side bands. Daniel and Lipson account for the presence of side bands by assuming that there is a sinusoidal variation in lattice spacing of (100) planes with a wavelength in the range 100 to 5000 A, increasing with annealing time. This stage is followed by a stage in which two metastable tetragonal phases coexist before the equilibrium cubic phases finally appear.

Hargreaves,[4] in a further study, concluded that the diffraction effects can be interpreted in terms of tetragonal phases that exist in lamellar form and retain complete coherence with the matrix. Geisler and Newkirk[5] found an analagous situation in the Cu-Ni-Co system where the equilibrium precipitate and the depleted matrix are both face-centered in structure but have lattice parameters smaller and larger, respectively, than the parent phase. Another example of a side-band structure is found in Au-Pt alloys.[6]

A number of recent investigations involving electron microscope and x-ray studies support the modulation model.[7] Very convincing support is

[1] A. J. Bradley and A. Taylor, "Physics in Industry—Magnetism," Institute of Physics, London, 1938. A. Taylor, "X-ray Metallography," Wiley, New York, 1961.

[2] V. Daniel and H. Lipson, *Proc. Roy. Soc. (London)*, vol. A181, p. 368, 1943; vol. A182, p. 378, 1944. The analysis of these patterns is covered in: R. W. James, "Optical Principles of the Diffraction of X-rays," G. Bell, London, 1948. A. J. C. Wilson, "X-ray Optics," Methuen, London, 1949.

[3] H. K. Hardy and T. J. Heal, *Progr. Metal Phys.*, vol. 5, p. 143, 1954.

[4] M. E. Hargreaves, *Acta Cryst.*, vol. 2, p. 259, 1949; vol. 4, p. 301, 1951.

[5] A. H. Geisler and J. B. Newkirk, *Trans. AIME*, vol. 180, p. 101, 1949.

[6] T. J. Tiedema, J. Bouman, and W. G. Burgers, *Acta Met.*, vol. 5, p. 310, 1957. L. J. Van Der Toorn, *Acta Met.*, vol. 8, p. 715, 1960.

[7] E. Biedermann and E. Kneller, *Z. Metallk.*, vol. 47, p. 289, 1956. Tiedema, Bouman, and Burgers, loc. cit.

Fig. 18-9 Microstructure ascribed to superposition of two orthogonal sets of plane waves of composition fluctuation in the spinodal decomposition of a solid solution. Alloy "Triconal XX" aged in a magnetic field. ×37,500. (Courtesy F. E. Luborsky.)

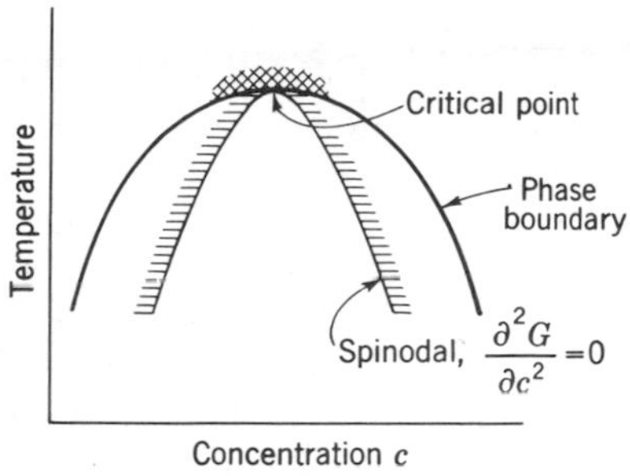

Fig. 18-10 Critical point, phase boundary, and spinodal boundary in a phase diagram. (After J. W. Cahn.)

also provided by the theoretical work on the development of modulations, known as *spinodal decomposition* (see below). The periodically arrayed precipitates of Fig. 18–9 are believed to have initiated in spinodal decomposition.

Spinodal decomposition In general a solid solution may be considered to become unstable on cooling with respect to two types of possible changes: (1) large in degree but small in extent and (2) infinitesimal in degree but large in extent.[1] The first type is exemplified by the formation of an infinitesimal particle of a new material with properties approaching those of final precipitate. The second type is exemplified by a small composition fluctuation spread over a large volume. A solid solution is stable with respect to such fluctuations only if the chemical potential of each component increases with increasing density of that component. For a two-component system, this condition is equivalent to $(\partial^2 G/\partial c^2)_{T,P} \geq 0$, where G is the Gibbs free energy per mole of solution and c is the concentration. On a binary phase diagram the boundary of the unstable region is thus defined by the locus of $(\partial^2 G/\partial c^2)_{P,T} = 0$ and is called the spinodal. An illustration of this boundary is given in Fig. 18–10 for the case of a critical point in a phase diagram. The mechanism of nucleation approaches the mechanism of spinoda decomposition continuously as the spinodal is approached.[2] In Fig. 18–10 the darkened area indicates a region of large fluctuation leading to easy nucleation and the cross-hatched area corresponds to the region of easy coherent large-scale fluctuations which cannot lead to coherent nucleation. True spinodal decomposition should occur everywhere within a sample, and the amplitude of composition fluctuations should grow continuously until a metastable equilibrium is reached with "a preferential amplification of certain wavelength components."[3] Experimental investigations of spinodal decompositions show that these criteria are satisfied.

Massive transformations The polymorphic $\gamma \rightarrow \alpha$ change in iron occurs under certain conditions as a *massive* type, mentioned on page 489, in which atoms are rapidly transferred from one structure to another across a high-energy interface. Since only a few jumps may be sufficient to relocate each atom from the parent structure into a new structure, there is no long-range diffusion; the boundary can move very rapidly.

Some solid solutions can transform in a similar way on heating or cooling, to give a new single-phase structure without a change of composition. Thus, a massive transformation in an alloy system resembles a polymorphic

[1] J. W. Gibbs, "Collected Works," Yale University, New Haven, Conn., 1948.

[2] J. W. Cahn and J. E. Hilliard, *J. Chem. Phys.*, vol. 31, p. 688, 1959. M. Hillert, *Acta Met.*, vol. 9, p. 525, 1961.

[3] J. W. Cahn, *Acta Met.*, vol. 9, p. 795, 1961; vol. 10, p. 179, 1962; vol. 10, p. 907, 1962.

change. Reactions of this type exhibit nucleation-and-growth characteristics and must be thermally activated.

The name *massive transformation* was first used by Greninger[1] to describe a patchy (massive) appearance of the microstructure obtained after quenching the high-temperature b.c.c. β phase in the Cu-Al system. Recent work[2] has shown that massive transformations occur at critical composition ranges in a number of alloy systems of copper, silver, iron, zirconium, and probably many others. The crystals of the new phase are nucleated at grain boundaries or imperfections and, at least in the extreme cases, are incoherent with the surrounding matrix.

Metallographic work has shown that in the decomposition of the high-temperature b.c.c. β phase in the Cu-Ga system the growing crystals of the massive c.p.h. phase frequently cross the prior grain boundaries of the parent β phase.[3] Massive transformations have been observed in pure iron and in iron alloys, particularly in Fe-Ni.[4] While the final product of such transformation is usually the equilibrium phase, corresponding to a lower temperature, the actual reaction may well occur at much higher temperatures at which slower cooling would produce an equilibrium mixture of two phases or a bainitic structure. Microstructures resulting from a massive transformation in Cu-Zn alloys and in relatively pure iron are shown in Fig. 18–11.

Massive transformations require certain critical conditions of supersaturation and quenching rates. The "driving force" must be high enough to permit nucleation and growth of mostly incoherent crystals, and the quenching rate must be high enough to suppress the possible equilibrium decomposition or ordering, but not so high that only a martensitic reaction can occur. Kinetically, a massive transformation, which is thermally activated, is intermediate between an equilibrium reaction and a possible martensitic transformation.

Morphology of quenched alloys indicates that sometimes planar facets develop inside and between various massive crystals. Such effects may be seen, for example, on the photomicrograph shown in Fig. 18–11, corresponding to an $\sim$38.40 atomic percent zinc alloy in the Cu-Zn system. The transformation in the Cu-Zn system is incomplete, which permits study of the crystallographic relationships and habits. Planar interfaces of the massive phase have been found to be parallel to matrix planes of simple

[1] A. B. Greninger, *Trans. AIME*, vol. 133, p. 204, 1939.

[2] T. B. Massalski, *Acta Met.*, vol. 6, p. 243, 1958. D. Hull and R. D. Garwood, "Mechanism of Phase Transformations in Metals," Institute of Metals, London, 1956. C. W. Spencer and D. J. Mack in V. F. Zackay and H. T. Aaronson (eds.), "Decomposition of Austenite by Diffusional Processes," Interscience, New York, 1962. W. S. Owen and A. Gilbert, *J. Iron Steel Inst.*, vol. 196, p. 142, 1960. H. Pops and T. B. Massalski, *Acta Met.*, vol. 13, p. 1021, 1965. J. E. Kittl and T. B. Massalski, to be published.

[3] Massalski, loc. cit.

[4] A. Gilbert and W. S. Owen, *Acta Met.*, vol. 10, p. 45, 1962. W. D. Swanson and J. G. Parr, *J. Iron Steel Inst.*, vol. 202, p. 104, 1964. G. R. Speich and P. R. Swann, *J. Iron Steel Inst.*, vol. 203, p. 480, 1965. R. R. G. Yeo, *Trans. ASM*, vol. 57, p. 48, 1964. R. F. Hehemann and R. H. Goodenow, *Trans. AIME*, vol. 233, p. 1777, 1965.

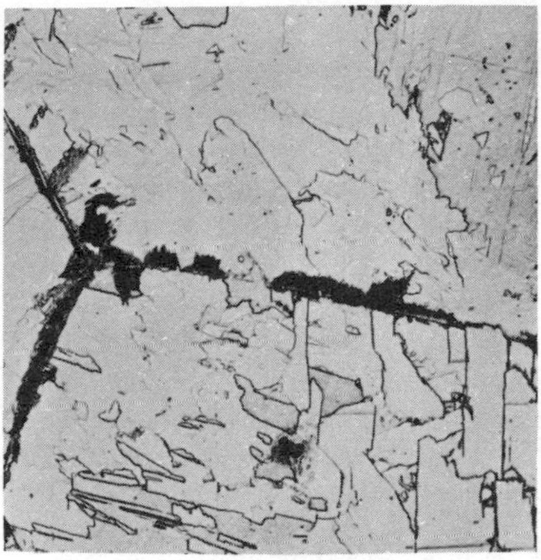

(*a*)

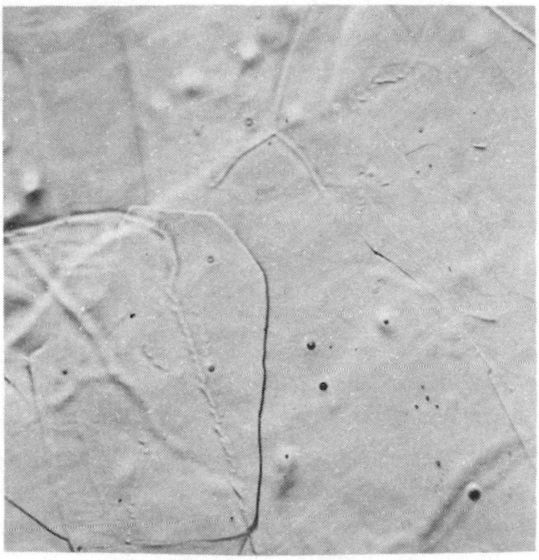

(*b*)

Fig. 18-11 (*a*) **Microstructure resulting from a massive transformation $\beta \rightarrow \alpha_{massive}$ in a Cu–38.4 weight percent Zn alloy quenched into iced water from the β phase field. Some high temperature precipitation of the equilibrium α phase has occurred at grain boundaries prior to the massive transformation. Polished and etched. ×100. (Courtesy J. E. Kittl.)** (*b*) **Microstructure resulting from a massive transformation in pure iron (0.0079 weight percent C, 0.002 weight percent N). Unpolished. ×500. (Courtesy R. F. Hehemann.)**

indices such as $\{100\}_\beta$, $\{110\}_\beta$, or $\{112\}_\beta$ and sometimes also coincide with the martensitic habit planes $\{155\}_\beta$ to $\{166\}_\beta$.[1] In such cases, lattice orientation relationships may exist.

Cinematographic study of Cu-Ga alloys at high temperatures has shown that the growth of the massive phase in some Cu-Ga alloys occurs by the propagation of irregular as well as nearly planar interfaces.[2] The numerous fine striations in some massive grains have been shown to correspond to a duplex structure of h.c.p. and f.c.c. platelets of identical composition.[3]

Order-disorder transformations in CuAu As pointed out in Chap. 11 (page 305), the evidence that is now available suggests that most order-disorder transformations are of the first order thermodynamically. The ordered phase that first appears at the ordering temperature possesses a finite degree of long-range order, and its lattice has different lattice constants than the lattice of the disordered phase. The transformation process, in addition to being a nucleation-and-growth reaction, sometimes possesses features that resemble a martensitic transformation. The superlattices that form from the f.c.c. solutions of an approximate composition AB are of particular interest. The f.c.c. solid solutions CuAu, CoPt, InMg, FePt,

[1] D. Hull and R. D. Garwood in "The Mechanism of Phase Transformations in Metals," p. 219, Institute of Metals, London, 1956.

[2] J. E. Kittl and T. B. Massalski, to be published.

[3] K. H. G. Ashbee, I. F. Vassamillet and T. B. Massalski, *Acta Met.*, in press.

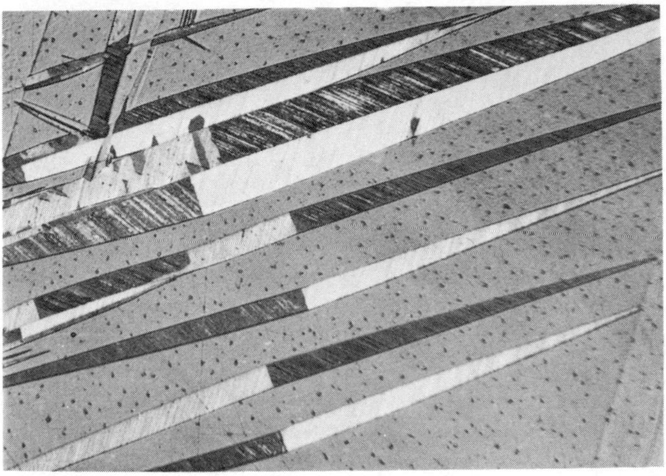

Fig. 18-12 Apparent martensitic formation of the CuAu II phase from the cubic CuAu. (R. Smith and J. S. Bowles, *Acta. Met.*, vol. 8, p. 405, 1960.)

FePd, MnNi, and NiPt all form the same type of the f.c.t. superlattice structure at room temperature, the $L1_0$ type characteristic of CuAuI (Fig. 11–4).[1] There is usually also a temperature interval in which the long-period superlattice is stable, of the general type CuAuII (Fig. 11–10) that is orthorhombic and is produced from the tetragonal structure by the introduction of shifts occurring periodically in the direction of the long period.

Isothermal transformation studies of CuAu[2] have established that the long-period superlattice forms from the disordered cubic phase by a process of slow nucleation and growth. However, the orthorhombic phase grows in the form of pyramidal plates. The formation of the pyramidal plates is accompanied by the production of surface relief effects in the manner of martensitic transformations.[3] A typical metallographic appearance of such plates is shown in Fig. 18–12. Each plate exhibits a system of fine parallel subbands which are twins in the orthorhombic lattice on the {101} plane.

Structures in carbon steels and the decomposition of austenite The various phases encountered in steels and their crystal structures have always been of major importance in studies of transformations in the solid state as well as of tremendous importance in industrial metallurgy. The

[1] J. S. Bowles and A. S. Malin, *Australian J. Metals*, vol. 5, p. 131, 1960.

[2] G. C. Kuczynski, R. F. Hochman, and M. Doyama, *J. Appl. Phys.*, vol. 26, p. 871, 1955.

[3] R. Smith and J. S. Bowles, *Acta Met.*, vol. 8, p. 405, 1960.

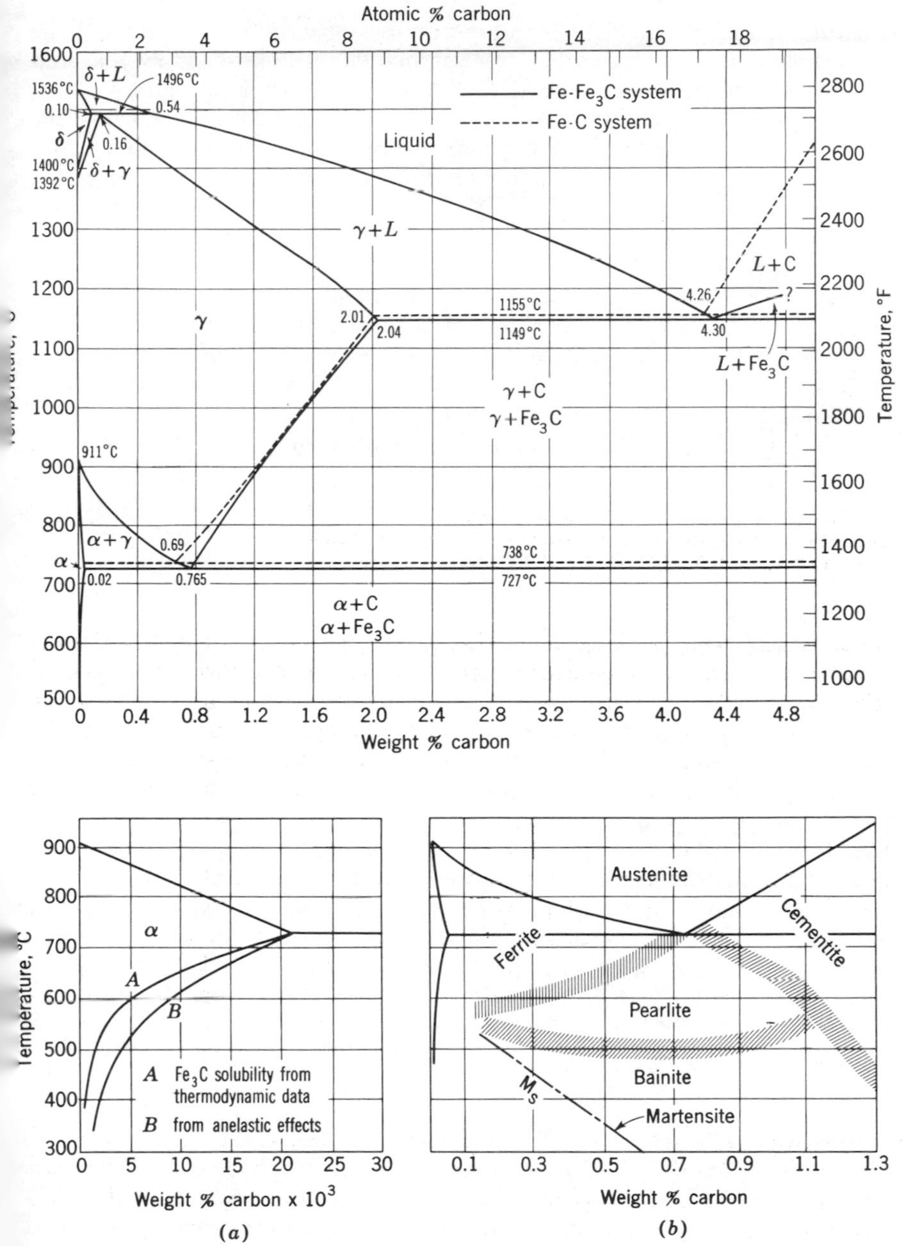

Fig. 18-13 The constitution diagram of Fe-C alloys in the region of eutectoid decomposition. (Courtesy J. S. Kirkaldy.) Below: (*a*) Details of the α-phase boundary, (*b*) Temperature-composition regions in which the ferrite, cementite, pearlite, bainite, and martensite reactions are dominant.

constitution diagram of Fig. 18–13 presents recent data for the limits of the phase fields both for the ordinary (but metastable) $Fe-Fe_3C$ system and for the equilibrium Fe-C system (i.e., the system after decomposition of the cementite to graphite).[1] We have inserted the M_S line showing the temperatures for the start of the martensite reaction on quenching, and have also indicated the regions in which the pearlite and bainite reactions are dominant.

Austenite, an interstitial solid solution of carbon in f.c.c. (γ) iron, has carbon atoms in interstitial positions at the center of the f.c.c. unit cell of iron atoms (at $\frac{1}{2}\frac{1}{2}\frac{1}{2}$) and also at the midpoints of the cell edges ($\frac{1}{2}00$, $0\frac{1}{2}0$, $00\frac{1}{2}$). These interstices, which are structurally equivalent and are the largest interstices in the lattice, provide more positions than can be filled by carbon atoms even in saturated austenite (2.0 weight percent C) (see also Chap. 10, page 236). That these largest interstices are the ones actually occupied was first indicated by precision measurements of the intensities of x-ray diffraction lines.[2]

Ferrite, the solid solution of carbon in b.c.c. (α) iron, has a maximum solubility of 0.025 weight percent C at the eutectoid temperature. The fact that ferrite dissolves far less carbon than does austenite is accounted for by the smaller size of the interstices in ferrite (see Chap. 10, page 237). The octahedral interstices, such as $00\frac{1}{2}$ and equivalent ones, are occupied by these and other interstitial solute atoms.[3]

The separation of pro-eutectoid ferrite from austenite is a nucleation-and-growth process. Under certain conditions it may take place almost entirely at grain boundaries or at least originate at grain boundaries. Alternatively, ferrite may form as a Widmanstätten pattern along the {111} planes of the austenite grains, analogously to the segregation pattern in meteorites (page 494). The morphology of growth patterns of ferrite crystals have been related to the structure of austenite grain boundaries and other imperfections that may serve as initial nuclei.[4] When ferrite takes the form of thin plates growing inside the austenite grains, the

[1] This diagram is based on one in M. Hansen and K. Anderko, "Constitution of Binary Alloys," McGraw-Hill, New York, 1958, with amendments assembled by J. S. Kirkaldy and G. R. Purdy (private communication) from researches published from 1959 to 1962. The amendments include [in figure (*a*)] curves from R. P. Smith, *Trans. AIME*, vol. 224, p. 105, 1962, which show the unexplained discrepancy for the limit of α-Fe solubility from thermodynamic data and from anelastic measurements.

[2] N. J. Petch, *J. Iron Steel Inst.* (London), vol. 145, p. 111, 1942.

[3] Internal friction proves this for most b.c.c. transition metals. For a review of this and other relaxation phenomena, see W. P. Mason, "Physical Acoustics III," Academic, New York, 1965. The position of oxygen atoms in silicon (diamond structure) is different: due to the covalent nature of the binding, the oxygen forms a noncolinear Si-O-Si configuration with a nearest-neighbor pair of Si atoms, and as there are six equivalent oxygen sites clustered around each Si-Si pair, barrier hopping from one site to another is believed to occur, with free rotation probable at 1000°C.

[4] See, for example, H. I. Aaronson in "The Mechanism of Phase Transformations in Metals," Institute of Metals, London, Monograph and Rept. Ser., no. 18, 1956; and in V. F. Zackay and H. I. Aaronson (eds.), "The Decomposition of Austenite by the Diffusional Process," Interscience, New York, 1962.

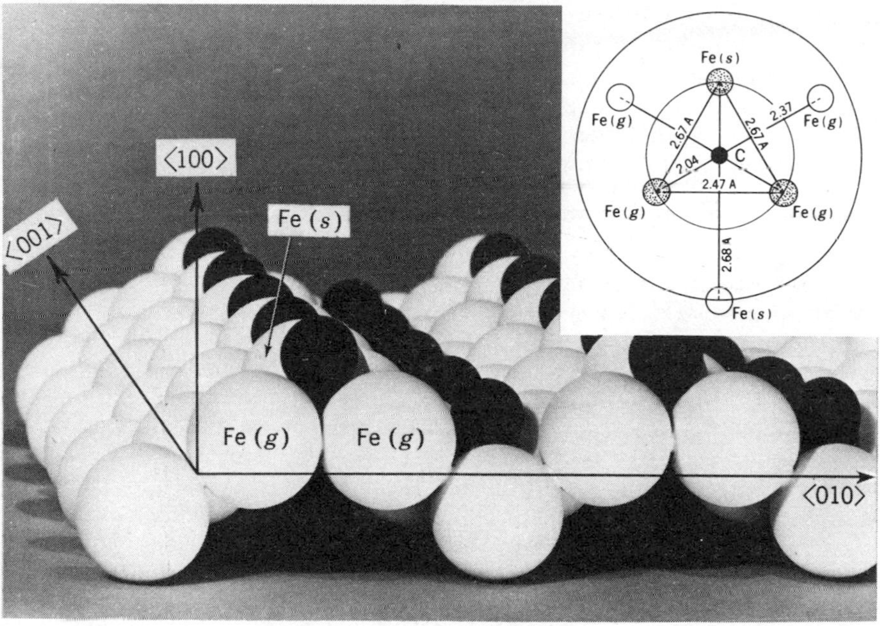

Fig. 18-14 A regular pleated layer of cementite structure. Insert: Stereogram of nearest- and next-nearest-neighbor Fe atoms around a C atom. Fe(g) and Fe(s) correspond to iron atoms in general and special positions, respectively. Shaded circles indicate two iron atoms, one above and one below carbon. (Courtesy G. A. Jeffrey.)

closest-packed planes and the close-packed directions of both structures are mutually parallel: $(111)_\gamma \| (110)_\alpha$ and $[110]_\gamma \| [111]_\alpha$, as in typical Widmanstätten precipitation.

Cementite, Fe_3C, has an orthorhombic unit cell containing 12 Fe and 4 C atoms. Lipson and Petch's values[1] for the lattice constants are $a = 4.5235$, $b = 5.0888$, and $c = 6.7431$ (when converted to angstroms). The carbon atoms are interstitially located in a nearly close-packed structure of iron atoms. As may be seen from the insert in Fig. 18–14, some iron atoms may be considered to be in general positions Fe(g) and some in special positions Fe(s) at a slightly larger distance from a given carbon atom. The lattice constants are not always the same, but vary with the temperature at which the cementite is brought to equilibrium with austenite. The phase is ferromagnetic below 210°C, and the Fe atoms can be replaced by Mn, Cr, Mo, W, and V. A way of representing the cementite lattice is shown in Fig. 18–14, in which the iron atoms are shown as corrugated layers with the carbon atoms fitting into the corrugations as indi-

[1] H. Lipson and N. J. Petch, *J. Iron Steel Inst.* (*London*), vol. 142, p. 95P, 1940. N. J. Petch, *J. Iron Steel Inst.* (*London*), vol. 149, p. 143P, 1944. For other determinations, see W. Hume-Rothery, G. V. Raynor, and A. T. Little, *J. Iron Steel Inst.* (*London*), vol. 145, p. 143, 1942. The space group is V_h^{16}-*Pbnm*.

cated.[1] Only small atomic shifts are necessary to translate the iron-atom positions in ferrite to those in cementite or in the structure of *epsilon-carbide* that forms with additions of manganese. The close-packed cubic [111] rows of iron atoms in ferrite can become "zig-zag" rows in the cementite [010] direction and in epsilon-carbide,[2] with only small zig-zag displacements of the iron atoms, a relationship that is in accord with, and accounts for, the observed lattice orientation relationship $\{112\}_\alpha \| (001)_c$ with $\langle 111 \rangle_\alpha \| [010]_c$ and the less defined relationship $\{125\}_\alpha \| (001)_c$ with $\langle 311 \rangle_\alpha \| [010]_c$.[3]

As mentioned earlier, in a steel of composition near the eutectoid (0.80 weight percent C), slow cooling produces a lamellar structure of ferrite and cementite known as *pearlite* (Fig. 18–5). The lamellae become thinner and more closely spaced as the cooling rate is increased or when less opportunity is provided for diffusion of carbon. The crystallographic relationships between cementite and ferrite in pearlite appear to be different from those given above (see page 498).

Martensite is a metastable phase that forms from austenite when a specimen is cooled at a rate exceeding a critical rate (depending on composition and metallurgical history) that is sufficient to suppress the formation of pearlite and other diffusion-controlled reactions. The martensite crystals are lenticular plates that form on certain planes of the austenite with high velocity, as discussed later. In iron-carbon steels the martensite structure is b.c.t., with an axial ratio that depends upon composition (Fig. 18–15).[4] The carbon atoms presumably lie in the $00\frac{1}{2}$ and $\frac{1}{2}\frac{1}{2}0$ distorted octahedral interstices, thereby distorting the structure from b.c.c. and leading to the observed dependence of the axial ratio on carbon content.[5] The carbon atoms occupy only a fraction of these sites, however. These carbon positions are occupied automatically if the carbon atoms are randomly distributed in the austenite, which then transforms to martensite by a specified homogeneous deformation (see page 521). The curves of a and c can be extrapolated linearly to the a parameter for carbon-free alpha-iron, and this is taken by many to indicate that martensite is tetragonal even with very low concentrations of carbon.[6] A recent investigation[7] has shown tetrag-

[1] E. J. Fasiska and G. A. Jeffrey, *Acta Cryst.*, vol. 19, p. 463, 1965.

[2] K. W. Andrews, *Acta Met.*, vol. 11, p. 939, 1963; vol. 12, p. 921, 1964.

[3] Ibid. When cementite precipitates from martensite, the reported orientation relationships are $(211)_\alpha \| (001)_c$ with $[0\bar{1}1]_\alpha \| [100]_c$ and $[1\bar{1}\bar{1}]_\alpha \| [010]_c$. (P. M. Kelly and J. Nutting, *J. Iron Steel Inst.*, vol. 197, p. 199, 1961. W. Pitsch and A. Schrader, *Arch. Eisenhüttenw.*, vol. 29, p. 485, 1958.) For details of additional orientation relationship studies, see W. Pitsch, *Arch. Eisenhüttenw.*, vol. 34, pp. 381 and 641, 1963. W. Pitsch, *Acta Met.*, vol. 10, pp. 791 and 897, 1962.

[4] The data in this figure have been collected by C. S. Roberts, *Trans. AIME*, vol. 197, p. 203, 1953.

[5] N. J. Petch, *J. Iron Steel Inst.*, vol. 147, p. 221, 1943.

[6] See, for example, the following articles: G. V. Kurdjumov, *J. Iron Steel Inst.*, vol. 195, p. 26, 1960. M. Cohen, *Trans. AIME*, vol. 224, p. 638, 1962. These authors propose that the cubic structure is observed in dilute interstitial alloys only because autotempering, with the rejection of interstitials, occurs during the quench or at room temperature.

[7] P. G. Winchell and M. Cohen, *Trans. ASM*, vol. 55, p. 347, 1962.

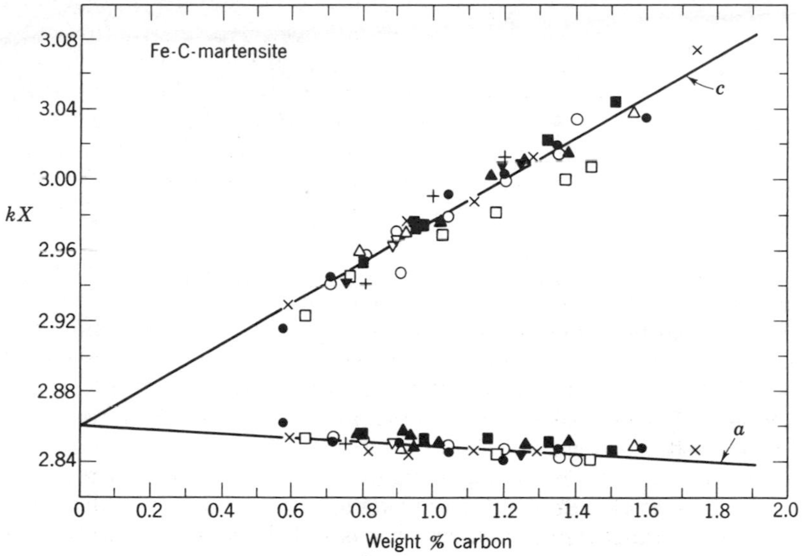

Fig. 18-15 Variation of lattice constants *a* and *c* of martensite with carbon content in plain carbon steels. (**After C. S. Roberts.**)

onality even with less than 0.05 weight percent C. Another view, however, is that tetragonality should disappear under certain conditions[1]; and that the disappearance of tetragonality accounts for observed changes in the morphology of Fe-C and Fe-N martensites.[2]

Bainite is the name given to the structure or structures that form on isothermal transformation or slow cooling at temperatures below those for pearlite formation, or below the so-called "knee" in the *T-T-T* reaction curves. At reaction temperatures just below those that produce the finest pearlite, bainite has the appearance of a feather or a comb (upper bainite). At lower temperatures, freshly formed bainite is dark-etching (low bainite), unlike fresh martensite, and is an aggregate of ferrite (b.c.c.) and finely divided carbide particles. Ferrite in bainite exhibits the same relief effects on polished surfaces as does martensite, in contrast to pro-eutectoid ferrite and pearlite, which do not exhibit such effects (see Fig. 18–16). However, as indicated by Fig. 18–16, bainite forms slowly, unlike martensite. The orientation relationship between bainitic ferrite and austenite may differ from that between martensite and austenite in the same steel.[3]

[1] C. Zener, *Trans. AIME*, vol. 167, p. 550, 1946, concluded on theoretical grounds that tetragonality should disappear when the composition and temperature of a sample puts it above a critical temperature for superlattice ordering, which may be at room temperature for some carbon concentrations between 2.5 atomic percent C and pure iron.

[2] W. S. Owen, A. E. Wilson, and T. Bell in V. F. Zackey (ed.), "High Strength Materials," Wiley, New York, 1965. The term *massive martensite* is used for packets of laths or fine plates on {111}, but this term tends to confuse the transformation with the massive transformation that occurs by diffusional nucleation and growth.

[3] W. Hoffmann and G. Schumacher, *Arch. Eisenhüttenw.*, vol. 26, p. 99, 1955.

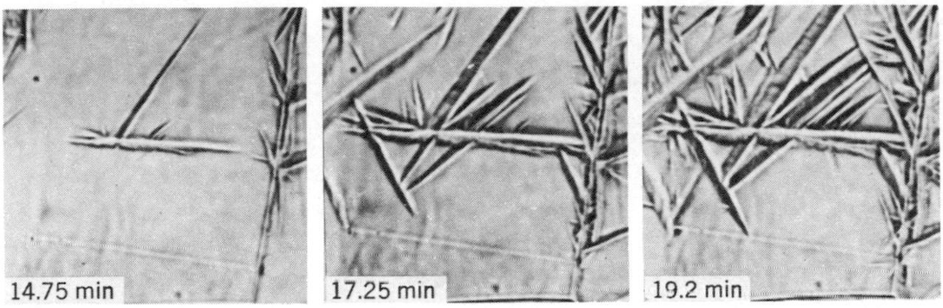

Fig. 18-16 A sequence of hot-stage micrographs showing the formation of bainite at 350°C in 3.32 percent Cr, 0.66 percent C steel. ×500. (Courtesy G. R. Speich.)

It is generally believed that the bainite transformation proceeds by the formation at an advancing interface of a supersaturated ferrite,[1] from which fine carbon particles precipitate subsequently.[2] It has been shown, in addition, that the rate of growth in thickness of the bainitic plates differs from the rate of edgewise (lengthwise) growth. According to Speich and Cohen,[3] the growth in thickness is controlled by the diffusion of carbon through the ferritic matrix and along the bainite-austenite interface, whereas the edgewise growth is controlled by the rate of carbon diffusion in austenite, near the advancing edge. Supersaturated ferrite may thus be not exactly of the austenite composition and may exist only transiently at the interface, but it is at this stage presumably that the crystallographic relationships and the habit become established. The observed ranges of habit planes for bainite formed at different temperatures are summarized in Fig. 18–17.[4] The behavior of lower bainite can be described in part in terms of the phenomenological martensitic theories, discussed later, but not the behavior of upper bainite.[5]

In alloy steels a constituent with acicular form may appear—near 500°C in the upper bainite range of temperatures—that is known as acicular ferrite,[6] or the *X* constituent.[7] It is prominent in steels rich in elements that have strong carbide-forming tendencies; these alloying elements are Cb, Ti, V, W, Mo, Cr, and Mn.

In summary, in a steel of a eutectoid composition the general tendency

[1] E. S. Davenport and E. C. Bain, *Trans. AIME*, vol. 90, p. 117, 1930. J. R. Vilella, G. E. Guellich, and E. C. Bain, *Trans. ASM*, vol. 24, p. 225, 1936. J. M. Robertson, *J. Iron Steel Inst.*, vol. 119, p. 391, 1929 (I). Further references and discussion will be found in C. Zener, *Trans. AIME*, vol. 167, p. 550, 1946; in a discussion thereof by A. Hultgren, *Trans. ASM*, vol. 39, p. 915, 1947; and in a review article by R. F. Hehemann and A. R. Troiano, *Metal Progr.*, vol. 70, p. 97, 1956.

[2] For details, see references to T. Ko given on p. 488.

[3] G. R. Speich and M. Cohen, *Trans. AIME*, vol. 218, p. 1050, 1960.

[4] J. S. Bowles and N. F. Kennon, *J. Australian Inst. Met.*, vol. 5, p. 81, 1960.

[5] Ibid.

[6] A. Hultgren, *Trans. ASM*, vol. 39, p. 915, 1947, and references given therein.

[7] E. S. Davenport, *Trans. ASM*, vol. 27, p. 837, 1939.

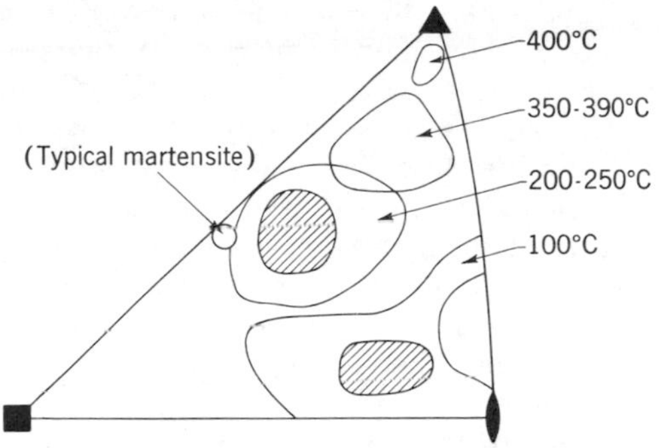

Fig. 18-17 Habit planes observed in bainite. Shaded areas include 80 percent of the observations made at 100°C and 200 to 250°C. (After J. S. Bowles and N. F. Kennon.)

for transformations, as the temperature is lowered, may be described as follows: coarse pearlite → fine pearlite → upper bainite → lower bainite → martensite (see Fig. 18–13). At other compositions pro-eutectoid separation of ferrite or cementite will occur first with additional complications to the resulting morphology. The structures of alloy steels, cast iron, and carbides are described in detail by Hume-Rothery.[1]

Martensitic shape change, morphology, and kinetics In a martensitic transformation, since a new crystalline phase grows out of the parent phase by the propagation of a semicoherent interface, there must be an ordered, cooperative, and rapid transfer of atoms from one structure to another across the interface as it moves. Clearly, such an interface must be composed of dislocations whose movement permits the martensitic product to form rapidly at any temperature. In the case of f.c.c. $\rightleftharpoons$ c.p.h. transformations, as in cobalt, the proposal has been made that the propagation mechanism involves partial dislocations rotating around a screw dislocation so as to generate a stacking fault along an advancing helix. The mechanism is known as the *pole* or *mill mechanism*.[2] Kinetics of martensitic transformations and the nature of hardening in martensitic steels have been discussed recently in detail.[3] If the nucleation is independent of time, as appears to be the case in most martensites, the transformation proceeds mainly while the temperature is changing—hence, the well-known *athermal* behavior of most martensites. However, nucleation may occur also iso-

[1] W. Hume-Rothery, "The Structures of Alloys of Iron, an Elementary Introduction," Pergamon, Oxford, 1966.

[2] A. Seeger, *Z. Metallk.*, vol. 47, p. 653, 1956. C. R. Houska, B. L. Averbach, and M. Cohen, *Acta Met.*, vol. 8, p. 81, 1960. A similar theory for twinning has as yet found no experimental verification in electron microscope pictures.

[3] L. Kaufman and M. Cohen, *Progr. Metal Phys.*, vol. 7, p. 165, 1958. M. Cohen, *Trans. AIME*, vol. 224, p. 638, 1962; vol. 212, p. 171, 1958.

thermally and the martensitic plates can grow—both rapidly[1] and slowly.[2] Since the growth of the martensitic plates is accompanied by shape changes, some *tilting* or *relief effects* usually become visible on flat surfaces.[3] The distortion of a polished surface is shown in Fig. 18–18. Optical examination with an interferometer shows that strained regions often extend from each martensite crystal all the way to each of its neighbors.[4] An important feature of the distortion is that the scratches remain continuous across the austenite-martensite interface.[5] This shows that a high degree of coherence is retained across the interface as it moves during the transformation; retention of coherence can occur only if the matrix is pulled along during the distortion.

The transformation in any single region may proceed at an extremely rapid rate. For example, individual plates have been reported to form in about 10^{-7} sec in Fe-Ni-C steels.[6] In some martensitic reactions, as in Au-Cd,[7] In-Tl,[8] or cobalt,[9] a single crystal may be transformed by the passage of a single-plane interface separating the two phases, in which case the matrix constraints are minimized. In general, however, the martensitic product is surrounded by the untransformed matrix and the constraints imposed on it produce the typical lenticular or acicular shape, although in some steels, needles or laths have also been reported.[10] The moving interface is stopped when it encounters a grain boundary, a pre-

[1] F. Förster and E. Scheil, *Z. Metallk.*, vol. 32, p. 165, 1940. R. F. Bunshah and R. F. Mehl, *Trans. AIME*, vol. 197, p. 1251, 1953.

[2] A. N. Holden, *Acta Met.*, vol. 1, p. 617, 1953. Also see laetr section on "thermoelastic martensites."

[3] In some cases, as in cobalt, the displacive features caused by the transformation may be so small that they escape detection; shears in cobalt are only about 100 to 300 A thick: S. Takeuchi and T. Honma, *Sci. Rept. Res. Inst. Tohoku Univ.*, ser. A, vol. 9, pp. 492, 508, 1957. In the titanium transformation, accommodation slip occurs in areas adjacent to the martensitic plates, and if it happens to be parallel to the plates, it tends to diminish, or compensate the tilting effect: A. J. Williams, R. W. Cahn, and C. S. Barrett, *Acta Met.*, vol. 2, p. 117, 1954. Surface upheavals can, at least in principle, result from volume changes in transformations that are not martensitic, with accommodation slip occurring in either the parent phase, the product phase, or both: see J. W. Christian in V. F. Zackay and H. I. Aaronson (eds.), "Decomposition of Austenite by Diffusional Processes," Interscience, New York, 1962.

[4] C. S. Barrett in "Imperfections in Nearly Perfect Crystals," Wiley, New York, 1952, p. 97.

[5] J. S. Bowles, *Acta Cryst.*, vol. 4, p. 162, 1951. M. Cohen, E. S. Machlin, and V. G. Paranjpe, *Trans. ASM*, vol. 42A, p. 242, 1950

[6] R. F. Bunshah and R. F. Mehl, *Trans. AIME*, vol. 197, p. 1251, 1953.

[7] L. C. Chang and T. A. Read, *Trans. AIME*, vol. 189, p. 47, 1951.

[8] Z. S. Basinski and J. W. Christian, *Acta Met.*, vol. 2, p. 148, 1954.

[9] V. J. Kehrer and H. Leidheiser, *J. Chem. Phys.*, vol. 21, p. 570, 1953.

[10] A. B. Greninger and A. R. Troiano, *Trans. AIME*, vol. 140, p. 307, 1940. P. Dörnen and W. Hoffmann, *Arch. Eisenhüttenw.*, vol. 30, p. 627, 1959. See also: B. A. Bilby and J. W. Christian, "The Mechanism of Phase Transformations in Metals," Institute of Metals, London, 1956. G. R. Speich and P. R. Swann, United States Steel Rept. no. 1170, to be published.

Fig. 18-18 Martensite crystals in a large grain of partially transformed Fe–30 percent Ni alloy. Polished and scratched before transformation, to show distortion produced by the transformation. Unetched. ×150.

viously formed martensitic plate, or another lattice disturbance that serves as a barrier. Further transformation must then occur by nucleation of new plates. The martensitic plates in many alloys contain a fine structure of slip bands or twins as a direct result of the transformation mechanism.[1] In high-carbon steels, and also in Fe-Ni-C steels, the fine twinning occurs on the usual $\{112\}$ twinning mode of the b.c.c. structure which corresponds to the $\{110\}_\gamma$ plane of the parent austenite lattice. The twin bands are between 100 and 400 A thick, and the particular twinning modes that would be expected to be active have been indicated theoretically[2] and in some cases have been determined experimentally. Electron microscopy and electron diffraction have been particularly effective in revealing the dislocations, orientation relationships, twin faults, deformation faults, and twins of various martensites.[3]

[1] P. M. Kelly and J. Nutting, *J. Iron Steel Inst.*, vol. 197, p. 199, 1961; *Proc. Roy. Soc. (London)*, vol. A259, p. 45, 1960. A. J. Baker, P. M. Kelly, and J. Nutting in G. Thomas and J. Washburn (eds.), "Electron Microscopy and Strength of Crystals," Interscience, New York, 1963. Z. Nishiyama and K. Shimizu, *Acta Met.*, vol. 7, p. 432, 1959.

[2] A. G. Crocker, *Acta Met.*, vol. 10, p. 113, 1962.

[3] W. Pitsch, *Arch. Eisenhüttenw.*, vol. 30, p. 503, 1959. The transformation substructure of Fe–18 percent Cr–12 percent Ni alloys has been studied by J. Dash and H. M. Otte, *Acta Met.*, vol. 11, p. 1169, 1963, and of Fe-Ni alloys by G. R. Speich and P. R. Swann, United States Steel Rept. no. 1170, to be published. The Japanese have been especially active in this field; see, for example: The review by Z. Nishiyama and K. Shimizu, *J. Electronmicroscopy*, vol. 12, p. 28, 1963; the work by Nishiyama's group reported in *Mem. Sci. Ind. Res. Osaka Univ.*, vol. 13, p. 1, 1956; vol. 16, pp. 73, 87, 1959; vol. 18, p. 71, 1961; vol. 20, p. 47, 1963; vol. 21, pp. 41, 51, 1964; *Acta Met.*, vol. 9, pp. 620, 980, 1961.

Crystallographic aspects of martensite transformations Consideration of the experimental data has encouraged the development of theories of martensite crystallography to explain, for any particular transformation, the orientation relationship, the habit plane, the shape deformation, and the nature of the fine structure within the martensitic plates.[1]

The first attempt to describe the shifts of atoms during the martensite transformation was made by Bain,[2] who proposed that the transformation merely involves a compression of the *c* axis of the austenite unit cell and an expansion of *a* and *b*. This mechanism could lead to the proper b.c.t. structure and the proper c/a ratio (in fact, austenite may be described as a b.c.t. structure with an axial ratio of $\sqrt{2}$). Bain also proposed that the interstitially dissolved carbon atoms prevent the axial ratio from going completely to unity, a view that is now fully substantiated.

The correspondence between individual atom positions before and after transformation, which was suggested by Bain, has been retained in all subsequently proposed mechanisms; however, it is now clear that the simple mechanism he proposed does not account for several observed facts. A sequence of improvements in theory then followed. The orientation relationships determined for steels are shown in Fig. 18–19, together with appropriate references.

Kurdjumov and Sachs,[3] from their determination of the orientation relationship between austenite and martensite in 1.4 percent carbon steel, namely, the *K-S* relationship $(111)_\gamma \| (110)_\alpha$, $[1\bar{1}0]_\gamma \| [1\bar{1}1]_\alpha$, proposed a theory that martensite is formed by the two consecutive shears $(111)_\gamma[\bar{1}\bar{1}2]_\gamma$ and $(1\bar{1}2)_M[\bar{1}11]_M$. Nishiyama[4] observed an orientation relationship for martensite in a 70–30 iron-nickel alloy that differed from the *K-S* relationship by a rotation of 5°16′; namely, the *N* relationship $[111]_\gamma \| [110]_\alpha$, $[\bar{2}11]_\gamma \| [1\bar{1}0]_\alpha$. Nishiyama suggested that the transformation resembled twinning in that a single shear on $(111)_\gamma[\bar{1}\bar{1}2]_\gamma$ was involved. This shear (of 19°28′) fails to produce the desired structure; in fact, it is merely the first shear of the *K-S* mechanism. Greninger and Troiano's work on steel containing 0.8 percent C and 22 percent Ni[5] has suggested that the martensite habit plane was approximately $(259)_\gamma$. A shear corresponding to this habit, when applied to the austenite lattice, produces from one of the $\{110\}$ austenite planes a plane identical with the (112) martensite plane;

[1] The crystallographic aspects of martensite are discussed in a number of review articles: A. B. Greninger and A. R. Troiano, *Trans. AIME*, vol. 185, p. 590, 1949. J. S. Bowles and C. S. Barrett, *Progr. Metal Phys.*, vol. 3, p. 1, 1952. Bilby and Christian, loc. cit. B. A. Bilby and J. W. Christian, *J. Iron Steel Inst.*, vol. 197, p. 122, 1961. M. A. Jaswon, *Research*, vol. 11, p. 315, 1958. J. K. Mackenzie, *J. Australian Inst. Metals*, vol. 5, p. 90, 1960. A summary of the orientation relationships in steels is given by H. M. Otte, *Acta Met.*, vol. 8, p. 892, 1960.

[2] E. C. Bain, *Trans. AIME*, vol. 70, p. 25, 1924.

[3] G. Kurdjumov and G. Sachs, *Z. Physik*, vol. 64, p. 325, 1930.

[4] Z. Nishiyama, *Sci. Rept. Tohoku Univ.*, ser. 1, vol. 23, p. 638, 1934–1935.

[5] A. B. Greninger and A. R. Troiano, *Trans. AIME*, vol. 145, p. 291, 1941; vol. 185, p. 590, 1949.

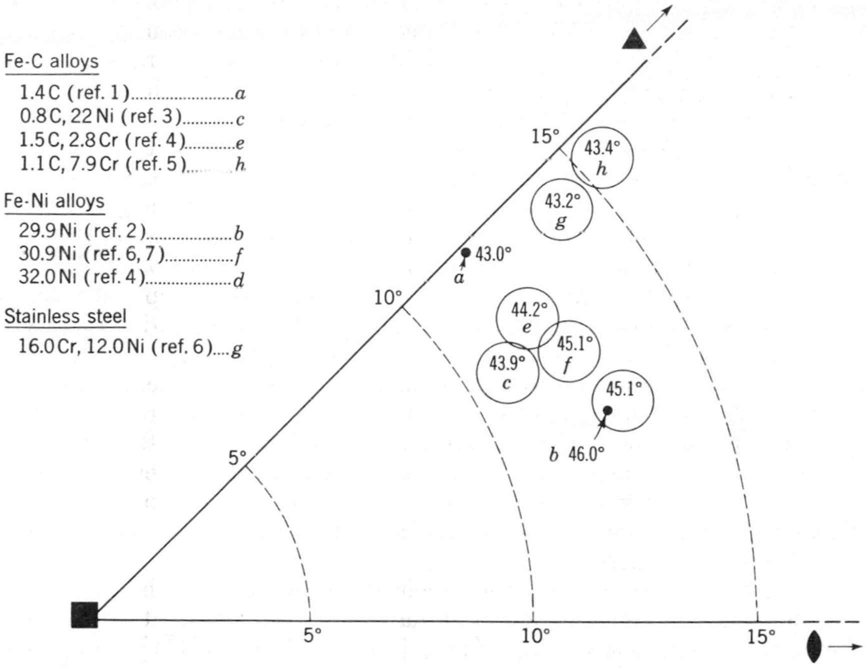

Fig. 18-19 Summary of orientation relationships in iron alloys. Diagram shows the location of the axis of rotation and the number gives the amount of rotation (in degrees) required to bring the austenite axes into coincidence with the martensite axes. Size of circle indicates probable experimental error in location of axis, except for *a* and *b* where error is unknown. Experimental error in angle of rotation is ±15°. (Courtesy H. M. Otte.) ([1]G. V. Kurdjumov and G. Sachs, *Z. Phys.*, vol. 64, p. 325, 1930. [2]Z. Nishiyama, *Sci. Rept. Tohoku Univ.*, vol. 23, p. 637, 1934. [3]A. B. Greninger and A. R. Troiano, *Trans. AIME*, vol. 145, p. 291, 1941; vol. 185, p. 590, 1949. [4]H. M. Otte, *Acta Met.*, vol. 8, p. 892, 1960. [5]C. M. Wayman, J. E. Hanajee, and T. A. Read, *Acta Met.*, vol. 9, p. 391, 1961. [6]J. F. Breedis and W. D. Robertson, *Acta Met.*, vol. 10, p. 1077, 1962. [7]J. F. Breedis and C. M. Wayman, *AIME*, vol. 224, p. 1128, 1962.)

the martensite lattice can then be produced by a second shear on $(112)_M[11\bar{1}]_M$. Greninger and Troiano therefore proposed a two-stage transformation: a homogeneous shear on the habit plane, producing relief effects (Fig. 18–18), and a different shear on the martensite twinning elements, i.e., the $\{112\}_M$ planes, in agreement with parallel markings sometimes observed on these planes. Further improvements, identifying the habit plane as the plane of least distortion and rotation, have followed subsequently (see below).

Experimental studies of the crystallographic features continue. A variety of martensitic morphologies have been studied recently in the carbon-free *marageing steels,* as composition and cooling rates were varied. In iron-nickel martensites, for example, in the 0 to 33 atomic percent Ni range three distinct morphological structures can be identified in quenched specimens.[1] Alloys containing less than 6 percent Ni frequently transform by a massive transformation. Martensitic structures at higher Ni contents change from packets of fine plates or needles, or laths[2] with a rational habit, such as $\{111\}_\gamma$, to the "acicular" martensite with irrational habit. The change in morphology occurs as nickel is increased, and is accompanied by changes in dislocation substructure and in the orientation and habit from $\{111\}_\gamma$ to $\{259\}_\gamma$ with no change in the symmetry of the (cubic) martensite but with the martensite becoming internally twinned. In binary interstitial steels containing carbon or nitrogen, the change in morphology occurs with small additions of these elements; the orientation is apparently unchanged, but the habit changes from $\{111\}_\gamma$ to $\{225\}_\gamma$ or $\{259\}_\gamma$. The accompanying slip in the martensite is replaced by twinning, and the lattice symmetry changes from cubic to tetragonal. However, bainite produced from plain carbon austenite of eutectoid composition has its ferrite in the N relationship between 450 and 350°C and in the K-S relationship at 250°C, similarly to martensite,[3] while the habit plane changes continuously throughout this temperature range (see Fig. 18–17).

Modern phenomenological theories of martensite formation Theoretical developments of the 1950s finally achieved a remarkable success in the task of interrelating the crystallographic features of martensite formation,[4] although research continues on some of the details. The principle of employing (when necessary) two strains, one causing the observed relief effects and the other completing the change of the lattice, has proved to be a very effective basis for the theories.[5]

The distortion that causes the change of shape of an individual martensite plate is a homogeneous strain; every point in the transforming region moves in the same direction and moves a distance proportional to its distance from an invariant plane of reference. It is thus an *invariant-plane strain* in that the reference plane (and all planes parallel to it) remain unrotated by it and also remain undistorted, except that in the Bowles-Mackenzie

[1] See, for example, W. S. Owen, E. A. Wilson, and T. Bell in V. F. Zackay (ed.), "High Strength Materials," Wiley, New York, 1965. G. R. Speich and P. R. Swann, *J. Iron Steel Inst.*, vol. 203, p. 480, 1965. R. Yeo, *Trans. ASM*, vol. 57, p. 48, 1964.

[2] Called "massive martensite" by Owen et al.; see p. 515.

[3] G. V. Smith and R. F. Mehl, *Trans. AIME*, vol. 150, p. 211, 1942.

[4] A nonmathematical summary is given by J. K. Mackenzie in *J. Australian Inst. Metals*, vol. 5, p. 90, 1960. Detailed summaries are given by: B. A. Bilby and J. W. Christian in "The Mechanism of Phase Transformations in Solids," Institute of Metals, London, 1956, p. 121. J. W. Christian, *J. Inst. Metals*, vol. 84, pp. 386, 394, 1955–1956. C. M. Wayman, "Introduction to the Crystallography of Martensitic Transformations," Macmillan, New York, 1964.

[5] D. S. Lieberman, M. S. Wechsler, and T. A. Read, *Trans. AIME*, vol. 197, p. 1503, 1953: *J. Appl. Phys.*, vol. 28, p. 532, 1957. R. Bullough and B. A. Bilby, *Proc. Phys. Soc. (London)*, vol. 69, p. 1276, 1956. D. S. Lieberman, *Acta Met.*, vol. 6, p. 680, 1958. B. A. Bilby and F. C. Frank, *Acta Met.*, vol. 8, p. 239, 1960. J. S. Bowles and J. K. Mackenzie, *Acta Met.*, vol. 2, pp. 138, 224, 1954. J. K. Mackenzie and J. S. Bowles, *Acta Met.*, vol. 5, p. 137, 1957.

(B-M) theory a small dilatation is permitted when necessary, while this dilatation is not assumed in the Wechsler-Lieberman-Read (W-L-R) theory. A "second" or "complementary" strain must be combined with the "first" in most instances to generate the final crystal structure from the intermediate state generated by the first strain. The second strain does not further alter the macroscopic shape change. It consists of slip on fairly regularly spaced slip planes, or of simple homogeneous twinning shears in small regions such that the shears proceed in opposite directions in neighboring regions to produce twinned lamellae.

The theories are *phenomenological*; they are concerned with the initial and final states and do not specify the path followed by atoms undergoing the transformation. Therefore, if the strains are called first and second, as they sometimes have been, it should not be implied that one necessarily precedes the other, for they may be combined and may merely represent the total change resolved into convenient components. Since the first alters the shape and the lattice, it may be called the *lattice strain* or the *shape strain* in contrast to the second, the *lattice-invariant strain*, which can counteract the shape change of the first, as indicated in Fig. 18–20, without altering the crystal structure.

The input data for these theories are the lattice constants (or axial ratios) for the two lattices, the correspondence between individual atom positions before and after transformation, and an assumed plane and direction for the second distortion. The quantities predicted by the theories are the habit plane, the lattice orientation relationship, and the direction

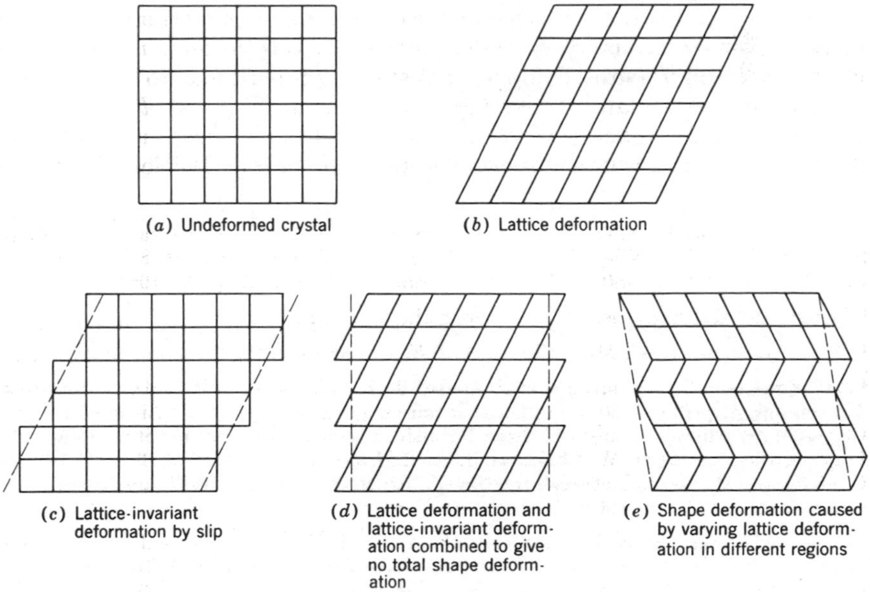
(a) Undeformed crystal
(b) Lattice deformation
(c) Lattice-invariant deformation by slip
(d) Lattice deformation and lattice-invariant deformation combined to give no total shape deformation
(e) Shape deformation caused by varying lattice deformation in different regions

Fig. 18-20 Types of deformation in martensite transformation. (After J. K. Mackenzie.)

and magnitude of the surface upheaval or shape deformation. In the W-L-R theory the lattice-invariant shear system is varied as a parameter, maintaining the requirement that the interface remain undistorted (have no dilatation), while the B-M theory uses a particular shear system and a dilatation parameter. These two approaches yield almost identical results. Still a third approach, that of Bullough and Bilby,[1] assumes the moving interface in a transformation to be an array of glissile dislocations.[2] With a digital computer this theory permits rapid calculation of orientation relationships. For example, over 100 different habit planes have been calculated for steels, assuming over 350 different modes of inhomogeneous deformation combined with the Bain strain.[3] The computer programs facilitate further systematic investigations of this type. The matrix-algebra approach, as opposed to the graphical method, has been employed profitably in this field.[4]

In phenomenological analyses of martensite crystallography the pure strain **P** describes, apart from a possible rotation, the structural change that converts a selected unit cell of the parent into one of the product. An unknown rotation **R**, combined with **P**, produces the total lattice deformation (structural change) that relates the two cells in their correct orientation relationship. The condition of an invariant-plane strain along the habit plane is not satisfied, however, unless some simple shape deformation **S** is introduced in combination with **RP**. Hence, in matrix notation, the total transformation distortion **F** is given by **F** = **SRP**. The deformation **S** may be affected by slip or twinning, as illustrated in Fig. 18–20. By assuming some information about **S**, and postulating that **F** is an invariant plane strain, the above relationship may be solved to determine **F**, **R**, and **S** completely. This procedure is the basis of all phenomenological theories, which, with minor differences, are essentially equivalent.[5] The small differences lie in the assumptions about **S**. The original theory of Wechsler, Lieberman, and Read[6] assumed the transformation interface to be undistorted and unrotated, whereas Bowles and Mackenzie[7] relax the condition of invariant plane strain and assume that lines in the interface are unrotated but may need an adjustment in length, involving a dilatation param-

[1] R. Bullough and B. A. Bilby, *Proc. Phys. Soc. (London)*, vol. B69, p. 1276, 1956.

[2] B. A. Bilby, R. Bullough, and E. Smith, *Proc. Roy. Soc. (London)*, vol. A231, p. 263, 1955.

[3] A. G. Crocker and B. A. Bilby, *Acta Met.*, vol. 9, p. 678, 1961. Results of the calculations are deposited at Sheffield University Library: A. G. Crocker, "Numerical Results on Martensitic Crystallography," vol. 1, 1961.

[4] See, for example, M. A. Jaswon and J. A. Wheeler, *Acta Cryst.*, vol. 1, p. 216, 1948; M. S. Wechsler, D. S. Lieberman, and T. A. Read, *Trans. AIME*, vol. 197, p. 1503, 1953; or B. A. Bilby and J. W. Christian, *J. Iron S eel Inst.*, vol. 197, p. 122, 1961.

[5] J. W. Christian, *J. Inst. Metals*, vol. 84, pp. 386 and 394, 1955–1956. An account of the above approach will also be found in an article by J. W. Christian in R. W. Cahn (ed.), "Physical Metallurgy," North Holland Publishing, Amsterdam, 1965. A comprehensive development of the subject is also given in C. M. Wayman, "Introduction to the Crystallography of Martensitic Transformations," Macmillan, New York, 1964, together with an account of the matrix algebra used.

[6] Wechsler, Lieberman, and Read, *loc. cit.* Also, H. M. Otte, *Trans. AIME*, vol. 218, p. 342, 1960. D. S. Lieberman, *Acta Met.*, vol. 6, p. 680, 1958. R. Bullough and B. A. Bilby, *Proc. Phys. Soc. (London)*, vol. 69, p. 1276, 1956. B. A. Bilby and F. C. Frank, *Acta Met.*, vol. 8, p. 239, 1960.

[7] See references given on p. 522.

eter, and that the second shear is mainly twinning.[1] Later modifications of the W-L-R theory also introduced dilatation parameters.[2]

The phenomenological theories have been very successful in accurately describing the main features of martensitic transformations in many systems, both ferrous and nonferrous, and they apparently have more far-reaching implications than has been initially supposed since it is possible to use them to account for crystallographic features of some order-disorder and precipitation reactions.[3]

With the progress of time, both matrix methods[4] and graphical methods[5] have been applied to the crystallography of a large number of different transformations.

Attempts have been made to improve the W-L-R theory for steels by making calculations that assume two complementary shears, in each of which the shear direction is fixed and the shear plane is rotated continuously about the shear direction, or the shear plane is fixed and the shear direction is rotated. The results obtained on a digital electronic computer for this problem suggest that to account for the $(225)_\gamma$ habit plane, either a dilatation of 1.5 percent as proposed by the B-M theory is required, or the shears must involve irrational and multiple-slip elements.[6]

Deficiencies of modern theories We should not leave the impression that the theories discussed above are adequate in all respects. None of the phenomenological theories alone can explain why the operating strain mechanisms are what they are, or what the physical meaning of the dilatation parameter is. The dilatation is usually assumed to be isotropic, but

[1] In high-carbon steels, the thin foil electron microscopy experiments now definitely point to twinning inside the martensitic plates as the mode of the second shear (A. J. Baker, P. M. Kelly, and J. Nutting in G. Thomas and J. Washburn (eds.), "Electron Microscopy and Strength of Crystals," Interscience, New York, 1963), but no internal twinning has been observed in low-carbon steels (P. M. Kelly and J. Nutting, *Proc. Roy. Soc. (London)*, vol. A259, p. 45, 1960; *J. Iron Steel Inst.*, vol. 197, p. 199, 1961); or in stainless steels (J. Dash and H. M. Otte, *Acta Met.*, vol. 11, p. 1169, 1963).

[2] M. S. Wechsler and H. M. Otte, *Acta Met.*, vol. 9, p. 117, 1961. H. M. Otte, *Acta Cryts.*, vol. 16, p. 8, 1963.

[3] R. Smith and J. S. Bowles, *Acta Met.*, vol. 8, p. 7, 1960. H. M. Otte and T. B. Massalski, *Acta Met.*, vol. 6, p. 7, 1958. J. S. Bowles and W. J. McG. Tegart, *Acta Met.*, vol. 3, p. 590, 1955.

[4] Examples of these are given in papers by D. S. Lieberman, M. S. Wechsler, and T. A. Read, *J. Appl. Phys.*, vol. 26, p. 473, 1955. H. M. Otte and T. A. Read, *Trans. AIME*, vol. 209, p. 412, 1957. J. S. Bowles and J. K. Mackenzie, *Acta Met.*, vol. 2, p. 129, 138 and 224, 1954. M. S. Wechsler, *Acta Met.*, vol. 7, p. 793, 1959. R. Bullough and B. A. Bilby, *Proc. Phys. Soc. (London)*, vol. B69, p. 1276, 1956. A. G. Crocker and B. A. Bilby, *Acta Met.*, vol. 9, pp. 678 and 992, 1961. B. A. Bilby and F. C. Frank, *Acta Met.*, vol. 8, p. 239, 1960. A. G. Crocker, *J. Iron Steel Inst.*, vol. 198, p. 167, 1961. K. A. Johnson and C. M. Wayman, *Acta Cryst.*, vol. 16, p. 480, 1963.

[5] M. S. Wechsler, T. A. Read, and D. S. Lieberman, *Trans. AIME*, vol. 218, p. 202, 1960. D. S. Lieberman, T. A. Read, and M. S. Wechsler, *J. Appl. Phys.*, vol. 28, p. 532, 1957. D. S. Lieberman, *Acta Met.*, vol. 6, p. 680, 1958 (stereographic method).

[6] A. G. Crocker and B. A. Bilby, *Acta Met.*, vol. 9, p. 678, 1961.

the suggestion has been made that to assume anisotropic distortions might prove useful.[1] Rational shear directions and planes have usually been assumed, but recent work has explored some irrational modes, and whether the modes should be rational or not has not been answered from first principles. No one has yet proven that any principle of minimum operative shears, minimum-interface strain energy, or minimum atomic shuffling must be valid, however intuitively reasonable such principles might appear. None of the theories account for the variation of any of the crystallographic features with composition or temperature.

In many respects, theoretical developments have been hampered by a lack of sufficiently precise experimental data, such as information on internal inhomogeneity in the product and precision information on *individual* plates of the martensitic phase rather than averages of many plates with different orientations. Sometimes the pattern of inhomogeneous shape deformations can be predicted from the crystallographic features of the transformation,[2] but much hinges upon the knowledge of precise data. Crystallographic features of a single transformation process that results in a single plate or domain of the martensitic phase are needed but are seldom available. In most cases, only "average" or "typical" values are reported both for the lattice orientation relationship and for the habit plane. A summary of the data on orientation relationships in steels is given in Fig. 18–19.

The meaning of scatter in orientation or in habit is often ambiguous. For example, when a martensite plate contains a "midrib," it is not clear whether the midrib represents the locus of the start of a transformation and its true habit plane or the meeting plane of two martensitic plates that may not have the same habit planes (the contact plane then causing a scatter in apparent habit).[3]

Characteristics of some nonferrous martensites Several nonferrous alloys change during cooling from f.c.c. structure to f.c.t. structure by a martensitic mechanism. Some are characterized by rather small shape deformations. The transformation that occurs in an alloy of indium containing 18.5 to 20.7 atomic percent thallium has been studied in particular detail.[4] The transformation produces a rather abrupt change of axial ratio from 1.0 to 1.020 at M_S, which increases gradually to 1.038 at room temperature. Accompanying this is a tilting of crystallographic lamellae on a metallographic surface; etching and proper lighting reveal subbands within the main bands, as shown in Fig. 18–21. Analysis shows that both the main bands and the subbands are traces of {110} planes. This diffusionless transformation is fully accounted for by two consecutive shears on two different {110} planes at 60° to each other, the shear directions being

[1] K. A. Johnson and C. M. Wayman, *Acta Cryst.*, vol. 16, p. 480, 1963. Work by J. K. Mackenzie, discussed in C. M. Wayman, "Introduction to the Crystallography of Martensite Transformations," Macmillan, New York, 1964.

[2] D. S. Lieberman, *Acta Met.*, vol. 6, p. 680, 1958. H. M. Otte, *Acta Met.*, vol. 8, p. 892, 1960.

[3] J. F. Breedis and C. M. Wayman, *Trans. AIME*, vol. 224, p. 1128, 1962. P. M. Kelly and J. Nutting, *J. Iron Steel Inst.*, vol. 197, p. 199, 1961; Kelly in discussion.

[4] J. S. Bowles, C. S. Barrett, and L. Guttman, *Trans. AIME*, vol. 188, p. 1478, 1950. L. Guttman, *Trans. AIME*, vol. 188, p. 1472, 1950. Z. S. Basinski and J. W. Christian, *Acta Met.*, vol. 1, p. 759, 1953; vol. 2, p. 101, 1954.

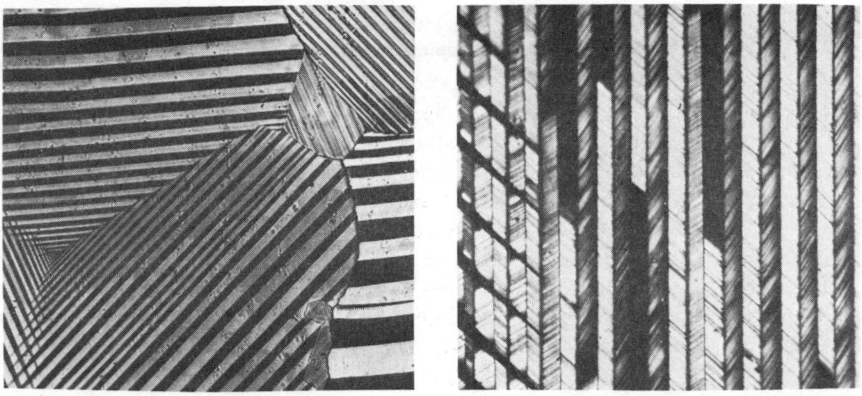

(a) (b)

Fig. 18-21 Microstructure of In-Tl alloy after transformation from cubic to tetragonal with $c/a = 1.03$. (a) ×75; (b) in polarized light, ×250. The main bands are ascribed to the first shear being alternately $(101)[\bar{1}01]$ and $(101)[10\bar{1}]$; the subbands are ascribed to the second shear being in opposite directions in adjoining subbands.

⟨110⟩ directions. The first shear amounts to one-third of the shear that would produce a twin in the tetragonal crystal on a {101} plane. This shear transforms the parent cubic phase into a structure that can be described as being one-third of the way between two tetragonal {101} twins. The second shear, operating in one sense along a ⟨110⟩ direction, produces the tetragonal structure, and operating in the opposite sense produces its twin. When it occurs in one sense, this second shear is equal in magnitude to the first; in the opposite sense, it is twice the magnitude of the first. The magnitude of the shears increases as cooling continues below M_S.

In Fig. 18–21 the main bands are regions in which the first shear occurs alternately in one sense and in the opposite sense. The subbands within the main bands are traces of the planes on which the second shear occurs. This shear corresponds to a twin deformation, and each set of subbands consists of alternating twins. There is satisfactory agreement between the measured orientation relationships and the predictions of the double-shear theory. Basinski and Christian showed that the transformation can be made to occur as a single interface moves across a single crystal. When it does so, the interface is at the edge of a region of homogeneous shear that causes a tilt of the atomic planes by 2ϵ, the shear angle for the first shear, $2\epsilon = (2/3)[(c/a) - 1]$, as required by the double-shear mechanism. The alternate twin lamellae of the f.c.t. structure can be converted to a single crystal by applying a suitable stress.

Similar transformations from cubic to tetragonal occur in other solid solutions, e.g., In-Cd, Cu-Mn, Cr-Mn, and Au-Mn.[1] There is a close similarity, also, to the microstructure of barium-titanate ($BaTiO_3$) after transformation from cubic to tetragonal.[2] An exactly similar reorientation accompanies the cubic (A-15 type, often called *beta-tungsten type*) to a tetragonal structure by a reversible martensitic transformation in the superconducting compound V_3Si below about 30°K.[3] Although the tetragonality in V_3Si is always very small (on the order of $c/a = 1.0025$ at 4.6°K), the orientation re-

[1] J. S. Bowles, C. S. Barrett, and L. Guttman, *Trans. AIME*, vol. 188, p. 1478, 1950. J. H. Smith and P. Gaunt, *Acta Met.*, vol. 9, p. 819, 1961.

[2] B. Matthias and A. von Hippel, *Phys. Rev.*, vol. 73, p. 1378, 1948. P. W. Forsbergh, Jr., *Phys. Rev.*, vol. 76, p. 1187, 1949.

[3] B. W. Batterman and C. S. Barrett, *Phys. Rev. Letters*, vol. 13, p. 390, 1964.

lationships are accurately accounted for by the double-shear theory. There is similarity, also, to the cubic-tetragonal transformation accompanying the onset of ordering in the alloys CoPt,[1] FePt,[2] AuCu,[3] and Ni_4W,[4] though the exact degree to which the mechanisms of these transformations resemble the In-Tl mechanism is not always clear.

The transformation in Au-Cd alloys has been carefully studied.[5] There is an interesting rubberlike stress-strain behavior in the transformed material, accounted for by the stress-induced movement of twin boundaries. The ordered b.c.c. (CsCl type) β phase transforms to an ordered orthorhombic β' for compositions near 47.5 atomic percent Cd, but with 50.0 atomic percent Cd the product is tetragonal β'', and with suitable quenching, still a different structure seems to result.

The transformation from b.c.c. to c.p.h. in lithium[6] has a $\{441\}_{b.c.c.}$ habit plane; this and related transformations in Ti with an {8, 9, 12} habit plane[7] in Ti-Mo alloys[8] and in Ti-Mn alloys[9] with {344} and {334} habits, respectively, have been treated with the B-M theory.[10]

Martensitic phases form from the majority of the b.c.c. β phases based on the noble metals Cu, Ag, and Au. The β phases formed by these metals with divalent Zn, Cd, Mg, and Hg occur in the region corresponding to 50–50 atomic percent of solvent and solute. They order rapidly into the CsCl structure during slow or even fast cooling. In the systems with Zn and Cd all ordered β' phases transform martensitically at cryogenic temperatures.[11] The β phases formed by the noble metals with trivalent (Al, Ga, Tl, In) and quadrivalent (Sn, Pb, Si) elements have a lesser tendency to form ordered structures on cooling, and when ordering does take place it is of the $FeAl_3$ type, the Heusler-alloy type, or a more complex type. Accordingly, the martensitic transformations in these phases involve several complexities. Transformations in Cu-Al, Cu-Sn, and Cu-Ga systems, and in ternary systems such as Cu-Zn-Ga, have received particular attention.[12]

Experimental observations show that the high-temperature b.c.c. phases (β) decompose

[1] J. B. Newkirk et al., *Trans. AIME*, vol. 188, p. 1249, 1950.

[2] H. Lipson, D. Shoenberg, and G. V. Stupart, *J. Inst. Metals*, vol. 67, p. 333, 1941.

[3] D. Harker, *Trans. ASM*, vol. 32, p. 210, 1944. J. L. Haughton and R. J. M. Payne, *J. Inst. Metals*, vol. 46, p. 457, 1931.

[4] E. Epremian and D. Harker, *Trans. AIME*, vol. 185, p. 267, 1949.

[5] L. C. Chang, *Acta Cryst.*, vol. 4, p. 320, 1951. L. C. Chang and T. A. Read, *Trans. AIME*, vol. 191, p. 47, 1951. D. S. Lieberman, M. S. Wechsler, and T. A. Read, *J. Appl. Phys.*, vol. 26, p. 473, 1955. M. S. Wechsler and T. A. Read, *J. Appl. Phys.*, vol. 27, p. 194, 1956. H. K. Birnbaum, *Trans. AIME*, vol. 215, p. 786, 1959.

[6] C. S. Barrett and O. R. Trautz, *Trans. AIME*, vol. 175, p. 579, 1948. C. S. Barrett in "Phase Transformations in Solids," chap. 13, p. 343, Wiley, New York, 1951. J. S. Bowles, *Trans. AIME*, vol. 191, p. 44, 1951.

[7] A. J. Williams, R. W. Cahn, and C. S. Barrett, *Acta Met.*, vol. 2, p. 117, 1954.

[8] S. Wenig and E. S. Machlin, *Trans. AIME*, vol. 200, p. 1280, 1954.

[9] Y. C. Lin and J. Margolin, *Trans. AIME*, vol. 197, p. 667, 1953.

[10] J. K. Mackenzie and J. S. Bowles, *Acta Met.*, vol. 5, p. 137, 1957.

[11] β'-AgZn does not transform spontaneously, according to H. Pops (private communication).

[12] G. V. Kurdjumov, *J. Tech. Phys.* (*USSR*), vol. 18, p. 999, 1948 (transl. no. 2300, Henry Brutcher, Altadena, Calif.). A. B. Greninger, *Trans. AIME*, vol. 133, p. 204, 1939. N. Nakanishi, *Trans. Japan Inst. Metals*, vol. 2, p. 85, 1961. T. B. Massalski and G. V. Raynor, *J. Inst. Metals*, vol. 82, p. 539, 1953. P. R. Swann and H. Warlimont, *Acta Met.*, vol. 11, p. 511, 1963. T. Saburi and C. M. Wayman, *Trans. AIME*, vol. 233, p. 1375, 1965; H. Warlimont in "Physical Properties of Martensite and Bainite," *Iron Steel Inst.*, Special Report No. 93, London, 1965. L. Delaey and H. Warlimont, *Phys. Stat. Solidi*, vol. 8, p. K121, 1965; *Z. Metallk.*, vol. 36, p. 437, 1965.

during quenching to form martensites which have basically either f.c.c. (β' and β_1') or c.p.h. (γ') structures. Frequently, an ordering reaction $\beta \rightarrow \beta_1$ occurs prior to the martensitic change. Earlier workers, using mainly x-ray methods, interpreted the β' and β_1' structures as distorted c.p.h.[1] or more complex[2]; but the more recent investigations with electron microscopes and electron diffraction[3] have shown that these structures are merely faulted to such an extent that certain diffraction maxima are displaced almost to hexagonal positions. The density of faults appears to be independent of composition and related to the nature of the martensitic transformation, i.e., faulting occurs as the second shear, or the lattice-invariant deformation, necessary to maintain an undistorted interface between the disordered parent β and martensitic β' (faulted f.c.c.) structure or the ordered parent β_1 and martensitic β_1' (faulted tetragonal ordered) structure. In the case of the c.p.h. γ' martensite (orthorhombic ordered), the lattice-invariant deformation occurs by twinning, as in steels. It is possible to calculate the extent of faulting or twinning in the martensites using the W-L-R or the B-M theory, and good agreement with experiment is obtained.

The "burst" phenomenon As may be expected, during a martensitic transformation the plates or needles of the new phase appear along several of the possible variants of the given habit plane. However, under certain conditions a group of plates of the new phase may appear nearly simultaneously in one portion of a specimen. Recent work on Fe-Ni single crystals containing 31.7 percent nickel has shown that the burst transformation is associated with an autocatalysis that has its origin in the mechanical coupling between certain variants of the habit plane during a single burst.[4]

Burst-type transformations also occur in nonferrous alloys. In the Cu-Zn system a burst-type transformation that occurs during cooling of the β phase at cryogenic temperatures is preceded by thermoelastic behavior (see page 530) that involves slow growth of martensitic plates along only one variant of the {2, 11, 12} habit plane.[5]

Thermoelastic and superelastic martensites Among the many martensitic transformations known, there are a few in which the process is almost perfectly reversible in that a reversal of the temperature change will cause an immediate reversal of the transformation; the martensite is said to be *thermally elastic* or *thermoelastic*, for it must involve a negligible amount of slip. Similarly, some stress-induced transformations are reversible, i.e., *mechanically elastic*, and are responsible for a rubberlike behavior, or *superelasticity*. A migration of twin boundaries in a fully transformed material can also provide rubberlike behavior (see previous section, page 528). Compared with the best-known spring materials, such as phosphor-bronze and beryllium-copper, which can withstand elastic strains of the order of 0.5 percent, a stress-induced, reversible, martensitic transformation in a Cu-Al-Ni alloy can cause a specimen to spring back from a strain

[1] Greninger, *loc. cit.*

[2] Nakanishi, *loc. cit.*

[3] See, for example, Swann and Warlimont, *loc. cit.*, and Warlimont, *loc. cit.*

[4] E. S. Machlin and M. Cohen, *Trans. AIME*, vol. 191, pp. 744, 1019, 1951. J. C. Bokros and E. R. Parker, *Acta Met.*, vol. 11, p. 1291, 1963.

[5] H. Pops and T. B. Massalski, *Trans. AIME*, vol. 230, p. 1662, 1964.

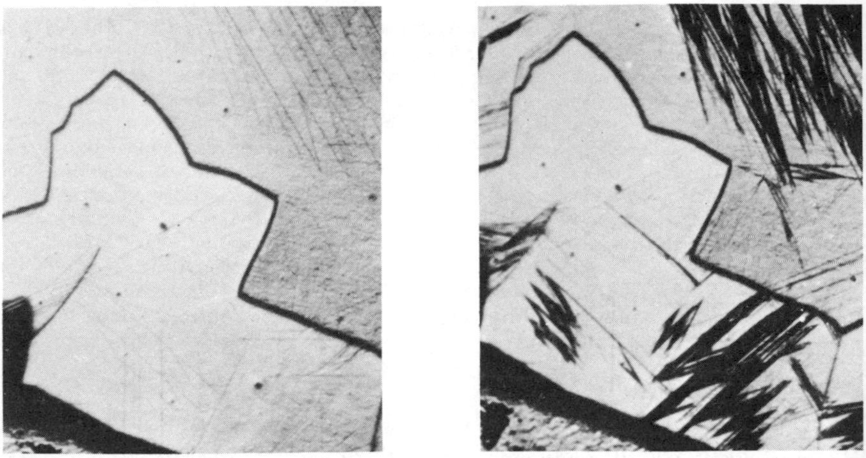

(*a*) (*b*)

Fig. 18-22 Mode of growth of martensite structures in a Cu–39.93 percent Zn alloy. Micrographs taken from a 16-mm film sequence. ×16.5. (*a*) Micrograph at −133°C shows needles of thermoelastic martensite; (*b*) Micrograph at −135°C shows burst-type martensite. (H. Pops and T. B. Massalski.)

eight times greater than that.[1] To date, a number of thermoelastic and superelastic transformations have been reported.[2]

In order that such large reversible changes can occur, the material must be in the form of a single crystal or must consist of unusually large grains. Furthermore, it appears that the parent matrix should be relatively stiff if accommodation strains due to a thermoelastic transformation are to remain elastic. Under such conditions, the velocity of the transformation interface is controlled by the rate of change of temperature or of stress, and will be as slow as the change of these two parameters.

As mentioned previously (page 488), in the Cu-Zn system the β phase, which has the ordered CsCl structure, transforms both thermoelastically and by bursts (see page 529). This is illustrated in Fig. 18–22 by means of photomicrographs taken from a cinematographic sequence.

Nucleation in some transformations may be intimately connected with

[1] W. A. Rachinger, *J. Australian Inst. Metals,* vol. 5, p. 114, 1960.

[2] G. V. Kurdjumov and L. G. Khandros, *Dokl. Akad. Nauk SSSR,* vol. 66, p. 211, 1949 (Cu-Ni-Al). L. C. Chang and T. A. Read, *Trans. AIME,* vol. 191, p. 47, 1951 (Au-Cd). J. E. Reynolds and M. B. Bever, *Trans. AIME,* vol. 194, p. 1065, 1952 (Cu-Zn). M. W. Burkart and T. A. Read, *Trans. AIME,* vol. 197, p. 1516, 1953 (In-Tl). Z. S. Basinski and J. W. Christian, *Acta Met.,* vol. 2, p. 148, 1954 (In-Tl). C. W. Chen, *Trans. AIME,* vol. 209, p. 1202, 1957. G. V. Kurdjumov, V. Lobodynk, and L. Khandros, *Soviet Phys. Cryst.,* vol. 6, p. 165, 1961 (Cu-Ni-Al). M. J. Duggin and W. A. Rachinger, *Acta Met.,* vol. 12, pp. 529 (Cu-Ni-Al) and 1015 (Au-Cu-Zn). H. Pops and T. B. Massalski, *Trans. AIME,* vol. 230, p. 1662, 1964 (Cu-Zn). For some comments on the theory of the thermoelastic effect, see R. W. Cahn, *Nuovo Cimento,* Suppl., vol. 10, p. 350, 1953.

partial dislocations in the parent phase. It has been proposed that transformations from the b.c.c. phase to close-packed phases, such as those that occur in Li, Na, Zr, FeTi, CuZn, Cu-Al, and Li-Mg, may start as dissociated partial dislocations on $\{110\}_{b.c.c.}$ planes.[1]

One of the resulting stacking faults resembles two-sequent (*ABAB*) planes of the close-packed or nearly close-packed product structure in some ternary thermoelastic martensites formed in Heusler-alloy ordered structures, where only slight alterations of the interatomic spacings are required.[2] With decreasing temperature, the stacking faults become extensive and their occurrence on alternate (110) planes of the matrix would result in a transformation whose major distortion would be a [110] displacement of alternate $(1\bar{1}0)$ planes, as proposed by Zener.[3] Various nucleating singularities have been proposed in the extensive discussions of nucleation.[4]

Transformations induced by strain Since every martensitic transformation involves a distortion that has a homogeneous component, it would be expected that all could be aided by externally applied stresses. This is found to be the case in general, and perhaps invariably. Plastic strain induces transformation at temperatures well above the martensite start temperature, M_S, in Fe-Ni,[5] Cu-Zn,[6] Li,[7] Au-Cd, and doubtless many other alloys. In Li,[8] Cu-Al,[9] and Fe-Mg alloys,[10] one martensitic phase transforms to another with cold work. The tendency for plastic strain to produce a transformation diminishes as the temperature is raised above M_S, and above a temperature designated as M_d the strain is no longer effective.[11] In a sample that has been cold worked *above* M_d, the temperature M_S is found to be lower than in an annealed sample.

It appears that cold working *below* M_d acts somewhat in the way thermal agitation does to aid atoms in crossing energy barriers from metastable states into more stable states. Cold work converts the martensitic phase in

[1] J. B. Cohen, R. Hinton, K. Lay, and S. Sass, *Acta Met.*, vol. 10, p. 894, 1962.

[2] M. J. Duggin and W. A. Rachinger, *Acta Met.*, vol. 12, pp. 529 (Cu-Ni-Al), 1015 (Au-Cu-Zn), 1964.

[3] C. Zener, *Phys. Rev.*, vol. 71, p. 846, 1947.

[4] L. Kaufman and M. Cohen in "The Mechanism of Phase Transformations in Metals," p. 187, Institute of Metals, London, 1956. R. E. Cech and D. Turnbull, *J. Metals*, vol. 8 (i.e., *Trans. AIME*, vol. 206), p. 124, 1956.

[5] E. Scheil, *Z. Anorg. Allgem. Chem.*, vol. 207, p. 21, 1932. A. W. McReynolds, *J. Appl. Phys.*, vol. 17, p. 823, 1946.

[6] A. B. Greninger and G. Mooradian, *Trans. AIME*, vol. 128, p. 337, 1938. T. B. Massalski and C. S. Barrett, *Trans. AIME*, vol. 209, p. 455, 1957.

[7] C. S. Barrett, *Phys. Rev.*, vol. 72, p. 245, 1947.

[8] L. C. Chang and T. A. Read, *Trans. AIME*, vol. 189, p. 47, 1951.

[9] A. B. Greninger, *Trans. AIME*, vol. 133, p. 204, 1939.

[10] A. R. Troiano and F. T. McGuire, *Trans. ASM*, vol. 31, p. 340, 1943.

[11] C. S. Barrett and O. R. Trautz, *Trans. AIME*, vol. 175, p. 579, 1948. A. W. McReynolds, *J. Appl. Phys.*, vol. 17, p. 823, 1946.

Li and in Li-Mg alloys, which appear to be highly faulted c.p.h., into a f.c.c. phase (probably also faulted). In the Cu-Si system, cold-work produces a phase, γ, that has not been seen in thermally treated samples, and that is metastable with respect to the equilibrium phase γ.[1] In the Co-Ni and Ar-N_2 systems, strain-induced transformations c.p.h., f.c.c., and f.c.c. $\rightleftharpoons$ c.p.h. have been used to locate the temperature-composition line along which the f.c.c. and c.p.h. phases have equal free energy.[2] Phases in which stacking faults are produced by plastic deformation may be classed as strain-induced metastable states; the Cu-Si system offers examples, as do also the Ag-Sn, the Cu-Ga, and probably also the Ag-Sb systems.[3] Profuse faulting occurs also in common stainless steels.[4]

Orientations in other reactions To the list of reactions in which new phases are generated whose orientations are crystallographically related to phases already present must be added *peritectic reactions*,[5] *oxidation reactions*,[6] and alloy layers produced by *diffusion*.[7] The orienting of *overgrowths* by substrates has been shown to depend upon a sufficient similarity in atomic positions in the matching planes[8]; if the atomic patterns are similar, as in alkali halides, the lattice parameters may differ as much as 25 or 30 percent without causing the deposited crystal to become randomly oriented.[9] When the habit of crystals is altered by impurities in the liquids from which they crystallize, it is probable that oriented adsorbed layers of the *impurities* are responsible.[10] Oriented electrodeposits are common; these are discussed in Chap. 19.

Orientation relationships are found in all types of transformations. When the two phases are of similar structure and approximately equal lattice dimensions, it can be predicted that the new will have an orientation identical with the old. When the reaction is at a *surface*, the pattern of

[1] C. S. Barrett, *Trans. AIME*, vol. 188, p. 123, 1950.

[2] J. B. Hess and C. S. Barrett, *Trans. AIME*, vol. 194, p. 645, 1952 (Co-Ni). C. S. Barrett and L. Meyer, *J. Chem. Phys.*, vol. 42, p. 107, 1965.

[3] C. S. Barrett and Marjorie A. Barrett, *Phys. Rev.*, vol. 81, p. 311, 1951. C. S. Barrett, "Imperfections in Almost Perfect Crystals," Wiley, New York, 1952. J. E. Kittl and T. B. Massalski, *J. Inst. Metals*, vol. 93, p. 182, 1965.

[4] H. M. Otte, *Acta Met.*, vol. 2, p. 349, 1954; *Acta Met.*, vol. 5, p. 614, 1957.

[5] A. B. Greninger, *Trans. AIME*, vol. 124, p. 379, 1937.

[6] H. C. H. Carpenter and C. F. Elam, *J. Iron Steel Inst.*, vol. 105, p. 83, 1922 (Fe-O). R. F. Mehl and E. L. McCandless, *Trans. AIME*, vol. 125, p. 531, 1937 (Fe-O). R. F. Mehl, E. L. McCandless, and F. N. Rhines, *Nature*, vol. 134, p. 1009, 1934 (Cu_2O). K. H. Moore, *Ann. Physik*, vol. 33, p. 133, 1938 (Cu_2O).

[7] S. Woo, C. S. Barrett, and R. F. Mehl, *Trans. AIME*, vol. 156, p. 100, 1944.

[8] L. Roger, *Compt. Rend.*, vol. 182, p. 326; vol. 179, p. 2050, 1925. C. A. Sloat and A. W. C. Menzies, *J. Phys. Chem.*, vol. 35, p. 2005, 1931.

[9] L. G. Schulz, *Acta Cryst.*, vol. 5, p. 130, 1952.

[10] P. Gaubert, *Compt. Rend.*, vol. 143, p. 776, 1906; vol. 167, p. 491, 1918; vol. 180, p. 378, 1925. C. W. Bunn, *Proc. Roy. Soc. (London)*, vol. A141, p. 567, 1933.

atoms in the two interfacial planes will be similar provided that similar planes exist in the phases. This is the essence of epitaxial growth.[1]

Solid state transformations in general It is instructive to consider transformations from a different standpoint, as Buerger has done,[2] on the basis of the changes in crystal structure that occur, and to look at examples more complex than the simple ones encountered among the metals. *Displacive transformations* occur when a space network of atoms is systematically distorted without disrupting the atomic linkage of the net; this type includes the martensitic transformations. In this type, the high-temperature form normally has a more open structure, a more symmetrical arrangement of atoms, a greater amplitude of thermal vibrations, and consequently a greater entropy than the low-temperature form into which it "collapses" on cooling. For example, cubic barium titanate collapses into the tetragonal form on cooling, with the formation of lamellae having low-energy interfaces and twin (or near-twin) orientation relationship to each other. Transformation rates are normally very high. *Reconstructive transformations* involve disruption of the space network and construction of a new network. (The new pattern may have the same coordination as the former in regard to the shortest primary interatomic distances and merely be altered in secondary coordination.) This type is more common in compounds of low coordination than in those of high coordination, for the possible alternate arrangements are more numerous with low coordination. Since low coordination is a requirement for the occurrence of a glassy phase,[3] polymorphism is common in substances that can be prepared in a glassy state. SiO_2 is an example; it exists in many polymorphs. In some reconstructive transformations, not only the secondary but also the primary coordination is changed, as in $CaCO_3$ when it changes from calcite to aragonite, and even the type of bonds may be changed, as in tin and in the transformation from diamond to graphite.

First-order vs. second-order transformations Superlattice transformations have been discussed in Chap. 11. Most are first-order reconstructive transformations—they involve a discontinuous change of the lattice, accompanied by discontinuities in entropy, energy, volume, and physical properties in general. Some, however, are second-order, with thermodynamic quantities changing continuously and only their derivatives changing discontinuously; these transitions do not involve a latent heat

[1] See, for example: J. Friedel, "Dislocations," Interscience, New York, 1964. J. P. Hirth and G. M. Pound, *Progr. Mater. Sci.*, vol. 11, p. 69, 1963.

[2] M. J. Buerger in "Phase Transformations in Solids," Wiley, New York, 1951.

[3] W. H. Zachariasen, *J. Am. Chem. Soc.*, vol. 54, p. 3841, 1932. This statement applies to solidification at normal rates of cooling; very fast cooling, deposits from the vapor onto cold substrates, and severe cold work can produce a glassy state in some substances of high coordination number. See pp. 3 and 161; also, C. S. Barrett and L. Meyer, *J. Chem. Phys.*, vol. 42, p. 107, 1965.

but they do cause an abrupt change in specific heat, coefficient of thermal expansion, and compressibility (see also p. 303).

Landau and Lifschitz developed a comprehensive theory of second-order transformations[1] which covers not only superlattice transformations but also ferromagnetic, antiferromagnetic, and ferroelectric transitions. They proved that only certain changes in symmetry or in the space lattice of a crystal are involved if a transition is to have second-order characteristics. These changes are necessary but not sufficient conditions for this type of transition; symmetry changes in first-order transitions, on the other hand, have no such restrictions on symmetry or lattice changes, and indeed the symmetry elements may remain unchanged. The theory proves that the necessary change in symmetry at a second-order transition is (1) a reduction, by half, of the rotational or translational symmetry elements during the transition to the lower symmetry state, (2) a doubling of the magnitude of certain translation vectors of the lattice, (3) a tripling of some vectors in hexagonal lattices, or (4) a quadrupling of some vectors in body-centered lattices.

The theory is of interest not only for magnetic and ferroelectric transitions, but also for many complex types of structure alterations of the superlattice type,[2] as well as for a curious transition in V_3Si, which has second-order characteristics.[3]

Anderson and Blount,[4] with a derivation related to that of Landau and Lifschitz, conclude that if a transformation such as this is actually of the second-order type, it must involve a symmetry change of a type other than a simple homogeneous strain from cubic to tetragonal. Some internal change must exist in the unit cell in addition to the changes of unit cell size and shape, but this is as yet undiscovered in V_3Si.

[1] L. Landau, *Phys. Z. Sowjet.*, vol. 11, p. 545, 1937. E. Lifschitz, *J. Phys.*, vol. 6, pp. 61, 251, 1942, in English. L. Landau and E. Lifschitz, "Statistical Physics," English transl. by E. Peierls and R. F. Peierls, Addison-Wesley, Cambridge, Mass., 1958.

[2] For example, recent work by E. F. Bertaut and collaborators in Grenoble, France, concerns many ordered arrangements of vacancies and transition-metal atoms in the NiAs-type structures of the sulfides, tellurides, and selenides of Ti, V, Cr, Mn, Co, Ni, and in the spinels.

[3] B. W. Batterman and C. S. Barrett, *Phys. Rev. Letters*, vol. 13, p. 90, 1964. J. E. Kunzler and coworkers, private communication, 1965. (See also page 527.)

[4] P. W. Anderson and E. I. Blount, *Phys. Rev. Letters*, vol. 14, p. 217, 1965.

19

Orientations in castings and deposits

Metals and alloys frozen in a mold frequently exhibit preferred lattice orientations of the grains in some regions, although the outer surface and in some cases also the center may have grains oriented at random. The preferred orientation is usually most highly developed in a *columnar zone*, where the grains are elongated in the direction of heat flow, normal to the mold wall. Columnar grains cover the entire cross section of a copper ingot in Fig. 19–1. A crystallographic axis tends to be parallel or nearly parallel to the longitudinal axis of columnar grains; all azimuthal positions around this axis are expected to be equally probable unless there is a considerable temperature gradient parallel to the mold wall. The columnar grains are those with orientations most suitable to rapid growth. They are the ones that have survived in competition with slower ones as freezing has progressed.

The outer skin of the casting usually contains rapidly frozen small grains, more nearly equiaxed, and this zone can be expected to have either a different texture or a randomness of grain orientations. In an alloy, freezing in this zone tends to enrich the neighboring liquid in one of its constituents until *constitutional supercooling* is great enough to produce columnar growth and dendritic growth (see Chap. 14, page 400). If the constitutionally supercooled zones from the opposite walls of growing grains overlap in the center part of an ingot, the liquid may become supercooled so much that many nuclei start growing throughout the remaining liquid. These grow into equiaxed grains of random orientation. (A central zone of this type did not occur in the casting of Fig. 19–1.)

The chilled surfaces of aluminum and of $(\alpha + \beta)$ brass die castings are random in texture, yet in zinc and cadmium the outer skin seems to have (0001) planes parallel to the surface, and an intermediate layer with $(10\bar{1}1)$ parallel to the surface has sometimes been found between this skin and the columnar zone.[1]

Throughout the freezing process, the shape and texture of the grains depend on whether they form under conditions that favor the growth of existing grains or the nucleation of new ones. Conditions during freezing also govern whether the liquid-solid interface is smooth, terraced by platelets which are growing edgewise, cellular, or dendritic. Dendritic growth is favored by high solute concentrations, high freezing speeds, high temper-

[1] G. Edmunds, *Trans. AIME*, vol. 161, p. 114, 1945.

Fig. 19-1 Columnar grains in cast copper. Etched cross section of an ingot.

ature gradients, and high degrees of supercooling (see page 400). The chilled outer skin may be frozen dendritically, yet if its nuclei are randomly nucleated, the grains in the skin will have no preferred orientation. A gradient along the mold wall favors equiaxed grains, and a gradient normal to the wall favors columnar grains. Large temperature intervals between liquidus and solidus favor long columnar grains.[1]

Columnar grain orientations Body-centered cubic metals always have their columnar grains with [100] normal to the mold wall, parallel to the axis of the columnar grains, and parallel to the direction of heat flow.[2] This has been observed in ingot iron, silicon-ferrite, ferritic chrome steel, ferritic Fe-Ni-Al, beta-brass, molybdenum, and chromium. The longitudinal direction of dendrite arms is also [100].[3]

Face-centered cubic metals also usually have columnar grains with [100] normal to the mold wall. This texture has been found with Al, Cu, Ni, Ag, Au; with various alloys of Al, Cu, Ni, and Pb; and with austenitic stainless steel.[4] However, there are exceptions.[5] Zone-refined lead was found to have a [111] direction parallel to the heat flow in the columnar zone and to have a (111) plane parallel to the surface in the chilled zone.[6] The addition of as little as 0.0005 weight percent silver caused the [111] texture in the columnar zone to be replaced by the [100] texture, as shown by the

[1] L. Northcott, *J. Inst. Metals*, vol. 65, p. 173, 1939; vol. 72, p. 283, 1946.

[2] G. Wassermann and J. Grewen, "Texturen metallischer Werkstoffe," pp. 120, 415, Springer, Berlin, 1962

[3] D. Walton and B. Chalmers, *Trans. AIME*, vol. 215, p. 447, 1959.

[4] Wassermann and Grewen, loc. cit.

[5] Walton and Chambers, loc. cit.

[6] G. F. Bolling, J. J. Kramer, and W. A. Tiller, *Trans. AIME*, vol. 227, p. 47, 1963. A. Rosenberg and W. A. Tiller, *Acta Met.*, vol. 5, p. 565, 1957.

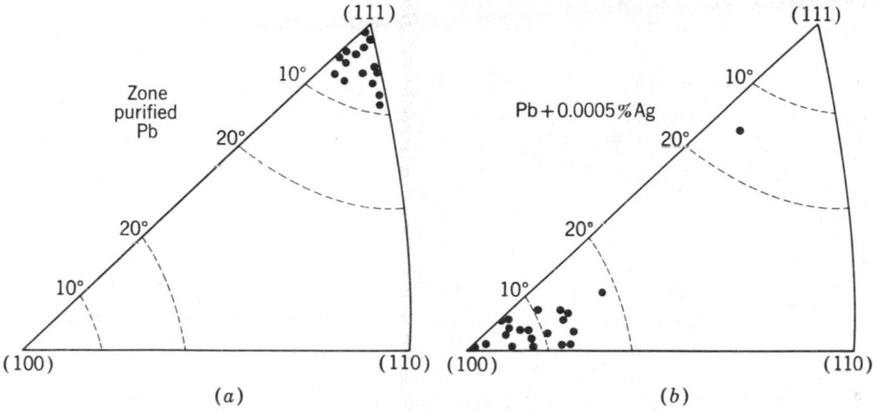

Fig. 19-2 Stereographic projection of orientations of columnar grains in lead after unidirectional solidification. (*a*) Zone-refined lead; grains have [111] within 10° of the direction of heat flow. (*b*) Pb + 0.0005 weight percent Ag; grains have [100] within 15° of the direction of heat flow. (**A. Rosenberg and W. A. Tiller.**)

stereographic plots of Fig. 19–2. The constitutional supercooling that resulted from this addition also changed the mode of freezing.

Some points of view on columnar growth have been included in a monograph by Ruddle,[1] and also in a discussion of the fact that in ingots of type 304 stainless steel (18 percent Cr, 9 percent Ni), a structure has been found that is entirely columnar, whereas type 430 (17 percent Cr) can have chill, columnar, and equiaxed zones.[2]

Close-packed hexagonal metals usually have (0001) parallel to the mold wall in the chilled zone. Cadmium and zinc have $[10\bar{1}0]$ normal to the mold wall in the columnar zone,[3] but magnesium is different.[4]

There have been some studies of binary eutectics after freezing. In Zn-Cd (17 percent zinc), each of the two phases has the texture it has when it is the only phase present, but in Al-Si (13 percent silicon), only the predominating phase is oriented (with [100] normal to the mold wall); the minor phase is randomly oriented.[5] Characteristically, there are crystallographic relationships between the two phases in a colony of a eutectic, i.e., an epitaxial relationship.[6] Silicon and germanium (diamond structure)

[1] R. W. Ruddle, "The Solidification of Castings," p. 49, Institute of Metals, London, 1957.

[2] F. C. Langenberg, J. K. McCauley, and M. D. Dias, *J. Metals*, vol. 15, p. 311, 1963.

[3] G. Edmunds, *Trans. AIME*, vol. 143, p. 183, 1941; vol. 161, p. 114, 1945.

[4] F. C. Nix and E. Schmid, *Z. Metallk.*, vol. 21, p. 286, 1929. G. Edmunds, *Trans. AIME*, vol. 143, p. 183, 1941.

[5] Nix and Schmid, loc. cit.

[6] W. Straumanis and N. Brakes, *Z. Physik Chem.*, vol. B38, p. 140, 1937. E. C. Ellwood and K. Q. Bagley, *J. Inst. Metals*, vol. 76, p. 631, 1949.

have many twins, which influence their cast texture,[1] and the (111) planes tend to be aligned in the direction of heat flow.[2]

Orientations in electrodeposits The nature of the orientation in electrodeposited metal is of considerable technical importance and has been the object of much investigation with the microscope and with x-ray and electron diffraction. The subject is complex, for the deposit is affected by the nature of the electrolyte, its hydrogen-ion concentration, the presence of addition agents, the temperature, the current density, the nature and condition of the base metal, stirring, rubbing, etc. There are also changes in texture as the deposited layer thickens, some of which have been ascribed directly to the thickness and others to changing conditions of temperature or concentration in the electrolyte.

Under some plating conditions the orientation of the base metal is copied by the deposit.[3] In fact, microscopic investigation of a cross section through deposit and base metal shows that the grains in the deposit are frequently continuations of the grains in the base metal.[4] The conditions most favorable to the continuation of base-metal grains into the electrodeposit are the following: clean, freshly etched surfaces, small current densities, electrolytes without colloidal additions, and similarity in structure between base metal and deposit. When both metals are cubic, continuation is found only if the base metal has a parameter approximately equal to that of the deposit (2.4 percent smaller to 12.5 percent greater).[5] Exceptions to these rules have been observed, however. It is possible to have continuation when the two metals have different structures, presumably because of related atomic patterns at the interface, as in Widmanstätten structures. It is also possible to have continuation across a thin interposed layer of a second substance[6] (perhaps through connecting pores or through orientations imparted to the layer). The continuity may occur only for certain grains of the base metal, as in rapidly applied industrial deposits, or it may be observed for all grains, for example, when the base metal itself is an electrodeposit.

The influence of the base metal does not extend very far into the de-

[1] W. C. Ellis and R. C. Treuting, *Trans. AIME*, vol. 191, p. 53, 1951. W. C. Ellis, *Trans. AIME*, vol. 188, p. 886, 1950; vol. 200, p. 291, 1954. E. I. Salkovitz and F. W. v. Batchelder, *Trans. AIME*, vol. 194, p. 165, 1952.

[2] E. Billig, *J. Inst. Metals*, vol. 83, p. 53, 1954.

[3] W. A. Wood, *Proc. Phys. Soc. (London)*, vol. 43, p. 138, 1931. W. Cochrane, *Proc. Phys. Soc. (London)*, vol. 48, p. 723, 1936. G. P. Thomson, *Proc. Roy. Soc. (London)*, vol. A133, p. 1, 1931.

[4] A. K. Huntington, *Trans. Faraday Soc.*, vol. 1, p. 324, 1905. A. K. Graham, *Trans. Am. Electrochem. Soc.*, vol. 44, p. 427, 1923. A. W. Hothersall, *Trans. Faraday Soc.*, vol. 31, p. 1242, 1935. Various studies of the effects of substrates will be found in the publications of G. I. Finch and coworkers in *Trans. Faraday Soc.*, 1935–1947.

[5] A. W. Hothersall, *Trans. Faraday Soc.*, vol. 31, p. 1242, 1935. Nonparallel but well-oriented deposits of iron on gold are accounted for by matching at the interface: G. I. Finch and C. H. Sun, *Trans. Faraday Soc.*, vol. 32, p. 852, 1936.

[6] A. M. Portevin and M. Cymboliste, *Trans. Faraday Soc.*, vol. 31, p. 1211, 1935.

posited metal, and the outer layers are free to take up their characteristic orientation. Under some conditions the underlying layers are randomly oriented and succeeding layers become progressively more sharply oriented for some distance.[1] Current density and electrolyte composition may be such that the entire deposit is random,[2] or varying degrees of preferred orientation can be produced.[3] Not only the degree of orientation but even the nature of the texture can be altered by the plating conditions.[4] In every case where a deposit is free to assume its own texture the orientations are such that a crystal axis stands perpendicular to the surface or parallel to the direction of current flow, and there is rotational symmetry around this axis. Thus, the deposit has a fiber texture (or a double fiber texture).

The conditions controlling the textures of electrodeposits have been studied and reviewed extensively.[5,6] A list of observed textures[6] will serve to indicate the types of textures that may be expected with a given metal, but it is impossible in limited space to specify all the conditions that produce each texture and the effects of subsequent annealing. It is probable, also, that the diffraction data were inadequate in some cases for a full description of the textures.

Silver deposits have been reported with fiber textures [111], [110], [211], [111] + [100], [111] + [211], [111] + [211] + $[10\bar{1}0]_{hex}$; copper with fiber textures [110], [100], [111], [110] + [100], [111] + [100]; nickel with [100], [110], [211], [311], [210], [111], [221], $[10\bar{1}0]_{hex}$, and 10 different mixtures of these single textures; lead with [111], [211]; gold with [110], [111], [111] + [100]; iron with [111], [110], [211]; chromium with [111], [100], [111] + [100], $[0001]_{hex}$; tin with [111], [100]; zinc with $[11\bar{2}0]$ + $[11\bar{2}2]$, [0001] within 75° of the fiber axis, [0001], $[11\bar{2}4]$, $[10\bar{1}1]$, $[11\bar{2}2]$, $[11\bar{2}0]$; cobalt with $[11\bar{2}0]$, $[110]_{f.c.c.}$; cadmium with $[11\bar{2}2]$; bismuth with [211], [100], $[1\bar{1}0]$; and antimony with [100].

Obviously, generalizations are difficult and of limited scope in this complex field, although attempts have been made.[7] Finch, Wilman, and Yang have proposed two general groups of deposits. In one, the deposited grains with the densest-packed atomic rows in the densest-packed planes perpendicular to the surface of the cathode tend to grow progressively outwards. This is said to be favored by low bath temperatures and concen-

[1] W. G. Burgers, *Philips Tech. Rev.*, vol. 1, p. 95, 1936.

[2] R. Glocker and E. Kaupp, *Z. Physik*, vol. 24, p. 121, 1924.

[3] W. A. Wood, *Phil. Mag.*, vol. 20, p. 964, 1935.

[4] R. Glocker and E. Kaupp, *Z. Physik*, vol. 24, p. 121, 1924. R. Bozorth, *Phys. Rev.*, vol. 26, p. 390, 1925.

[5] H. Fisher, numerous papers in *Z. Metall.*, and *Z. Electrochem.*, 1948–1955.

[6] G. Wassermann and J. Grewen, "Texturen metallischer Werkstoffe," pp. 127–147, Springer, Berlin, 1962.

[7] G. I. Finch, H. Wilman, and L. Yang, *Trans. Faraday Soc.*, Discussions of Electrode Processes, vol. 1, p. 144, 1947. B. C. Banerjee and A. Goswami, *J. Electrochem. Soc.*, vol. 106, pp. 20, 590, 1959; *J. Sci. Ind. Res.*, vol. 14B, p. 322, 1955; vol. 16B, p. 144, 1957. A. Goswami, *J. Sci. Ind. Res.*, vol. 15B, p. 322, 1956; vol. 16B, p. 315, 1957.

trations and by hydrogen evolution (e.g., Au, Ag, Cu, Ni, with [110] textures; Bi with $[1\bar{1}0]$; Fe with [111]; Co and Zn with $[11\bar{2}0]$). In the second group the densest-packed planes lie parallel to the surface of the substrate. This orientation is favored by high bath temperatures and no hydrogen evolution; atomic mobility on the surface of the deposit is higher for this group. The examples cited for the second group are the [111] textures in Au, Ag, and Cu, and [110] in Fe. Both groups are supposed to be produced only when hydrogen ions are not adsorbed on the face of the cathode. There are many deposits that fall in neither group. Brittleness, hardness, and brightness sometimes correlate with texture, but texture is not the only parameter to be considered in attempting to understand the properties of deposits of a given metal.

Evaporated and sputtered metal films Preferred orientations are found in many films produced by condensations from the vapor and by sputtering. As with electrodeposits, the presence or absence of a texture and its type depend upon many variables, including the nature and temperature of the surface on which the condensation occurs.

Deposits from the vapor phase onto noncrystalline substrates tend to be random in orientation when very thin, provided the evaporation is carried out with gases excluded, and provided the substrate is at a low temperature. The presence of a gas and the use of a hot substrate tend to produce a deposit with a single or duplex fiber texture. When a stream of vapor strikes a surface obliquely, the fiber axis tends to be inclined both to the stream and to the condensing surface.

Among the textures to be expected in vapor-deposited films are the following: aluminum and silver, [111], [100], [110]; copper, nickel, and palladium, [111]; gold, [110], [111]; iron, [111]; molybdenum, [110]; cadmium and zinc, [0001]; bismuth, [111], [110].[1]

Since thin films have become of increasing commercial importance in the electronics industry, the amount of research done on them has increased enormously. For example, there are some 700 titles listed for a single year (1963) in *Physics Abstracts* under the heading "Films, solid." Electron diffraction studies of deposits are discussed in Chap. 22.

1 Wassermann and Grewen, loc. cit.

20

Preferred orientations resulting from cold work

Most polycrystalline materials contain grains with lattice orientations that are not random but instead are clustered to some degree about a particular orientation or about a set of orientations. Bodies in which the grains are oriented nonrandomly are said to have a *preferred orientation* or *texture*, which can be detected and specified by the methods of Chap. 9.

The mechanical and thermal history of a specimen determines the nature of the texture that is developed; this chapter is concerned with textures developed by plastic flow at temperatures too low for simultaneous annealing to occur, i.e., textures developed by cold work. Although the statements in this book are concerned chiefly with metals and alloys, preferred orientations are also developed in many other materials, including both organic and inorganic compounds, natural and synthetic fibrous materials, mineralogical specimens, etc., and many of the principles discussed here also apply to these substances.

The importance of textures Preferred orientations have received much study because of the important effect they have on the properties of commercial products. A fine-grained metal specimen in which the grains have random lattice orientations will possess identical properties in all directions (provided that there are no elongated inclusions, segregations, or boundaries), but a specimen with a preferred orientation will have directional, or anisotropic, properties, which may be desirable or undesirable, depending upon the intended use of the material.

An example in which anisotropy is very important and is much desired is the steel sheet used for electrical transformer cores. A high permeability in the direction of the applied magnetic field is wanted, and may be obtained by rolling and annealing treatments designed to orient as many grains as possible with their easily magnetized directions in one direction in the sheet. Since the b.c.c. crystals of steel are most easily magnetized in [100] directions, much attention is given to procedures that will maximize the number of grains having [100] in the rolling direction, or (100) in the rolling plane.

An example in which anisotropy is often undesirable is the steel sheet used for deep drawing. Unequal mechanical properties caused by a specific texture in a sheet may cause difficulties or waste in certain deep-drawing operations. Thus, a cup drawn from such a sheet may have an uneven rim, or "ears." On the other hand, anisotropy may be put to advantage if the drawing operation itself is unsymmetrical and if the sheet can be turned to a position that makes favorable use of the directionality.

The principles governing texture development are of scientific as well as commercial interest. It should be possible to predict the textures and even the approach to final textures from first principles, and various attempts have been made to do this, but only limited success has been achieved, owing largely to the complexity of the problem and the assumptions that have to be made as a substitute for quantitative data that have not been obtained.

General principles of texture development Plastic flow causes a reorientation of the lattice of individual grains of a polycrystalline material and tends to develop a texture or preferred orientation of the lattice in the grains as well as a preferred change of shape of the grains. The progress of reorientation is gradual; the orientation change proceeds as plastic flow continues, until a texture is reached that is stable against indefinitely continued flow of a given type. For example, x-ray diffraction may disclose evidence of a partially developed texture when a sheet of metal is reduced in thickness by perhaps one-half, but the final rolling texture may not be reached until the thickness is reduced by 90 percent or even more. The nature of the stable deformation texture and the manner in which it is approached are characteristic of the material and of the nature of the flow throughout the deformation process (the magnitude of the three principal strains at all points within the specimen and at successive times in the process). The texture is also influenced by the temperature of the specimen during the deformation—especially if the temperature is high enough to permit recrystallization to occur during deformation.

Some general statements regarding textures deserve mention here: (1) The presence of elongated or flattened grains does not always imply a certain texture or even the presence of any texture at all; and the presence of equiaxed grains does not imply a random orientation. (2) From a detailed knowledge of the texture of a metallurgical or mineralogical specimen, it may be possible to deduce facts about its mechanical and thermal history. (3) By a controlled schedule of working and annealing, it is possible in certain cases to minimize texture formation or to develop a desired texture, but it is usually difficult to produce true randomness. (4) The plastic strains near the surface of a specimen may differ from those in the interior—especially in rolling and wire drawing—and may produce textures that vary with depth below the surface. (5) The symmetry of the preferred orientation tends to match the symmetry of the principal strains, but perturbing facts may be involved that destroy the equality of these two symmetries (for example, a partial retention of an unsymmetrical initial texture). (6) A given change of shape may be produced by intermediate deformations of many kinds, and these may have an effect on the final texture, so that textures cannot be uniquely predicted from the observed overall shape-change alone. (7) The orientation of a single crystal during rolling, swaging, wire drawing, compression, or tensile elongation does not necessarily proceed along a path identical with that for a similarly oriented grain of a polycrystalline matrix. (8) Preferred orientations may be of a simple type that can be adequately specified by one or more "ideal textures," or they may be so complex as to require description by pole figures or inverse pole figures, as mentioned in Chap. 9. (9) Anisotropy of mechanical properties in some specimens may be influenced as much (or more) by elongated or flattened inclusions, voids, fissures, or grain boundaries as by the preferred orientations of the grains. Even the presence of anisotropically distributed residual stresses may introduce directionality in some mechanical properties.

Summaries and discussion of the results of research will be found in the book by Wassermann and Grewen.[1] We are concerned here with occasional experiments that may be of special interest, and with general principles that appear to be valid.[2]

Textures of polycrystalline wires and rods The textures of wires are frequently termed *fiber textures* because they resemble the arrangement of oriented molecules and crystallites in fibrous materials, such as stretched rubber, various natural and synthetic polymers, and fibrous minerals. Simple wire textures consist of orientations that have a particular crystallographic direction parallel to the axis of the wire and other directions distributed with equal probability around the wire axis, giving "rotational" or "cylindrical" symmetry to the whole.

Many types of fiber textures are found that have greater complexity, however. Deviations from the simple ideal fiber texture with one crystallographic fiber axis $[uvw]$ parallel to the wire axis may have the nature of: (1) a scatter in orientation about an ideal $[uvw]$ fiber texture, (2) a mixture of randomly oriented grains (or subgrains) distributed in some way among the grains of a fiber texture, (3) a composite of two (or more) ideal fiber textures, $[u_1v_1w_1]$ and $[u_2v_2w_2]$, designated as a *double*, *duplex*, or in general a *multiple fiber texture*, (4) a *layered fiber texture* in which the outer layers are textured differently than the inner ones, perhaps owing to frictional forces at the surface during the wire-forming process, (5) a *spiral fiber texture* in which an assembly of fiber axes is inclined at a definite angle to the wire axis and distributed around the wire axis with rotational symmetry.[3]

A special type of spiral texture consists of fiber axes lying at 90° to the wire axis. Depending on the directions of flow during the wire-forming process, there may or may not be a plane of symmetry normal to the wire axis.[4]

Face-centered cubic metals and alloys have wire textures that are par-

[1] G. Wassermann and J. Grewen, "Texturen metallischer Werkstoffe," 2d. ed., Springer, Berlin, 1962 (808 pages).

[2] Note added in proof: I. L. Dillamore and W. T. Roberts, *Met. Rev.*, vol. 10, p. 271, 1965, have critically reviewed textures in wrought and annealed metals and current theories of these in 109 pages; reviews and new concepts are also presented in the 1965 ASM Symposium, "Recrystallization, Grain Growth, and Textures," American Society for Metals, Cleveland, in press.

[3] K. Weissenberg, *Z. Physik*, vol. 8, p. 20, 1921.

[4] Further discussion of the specification of fiber textures and their symmetry is given in R. J. Roe and W. R. Krigbaum, *J. Chem. Phys.*, vol. 40, p. 2608, 1964.

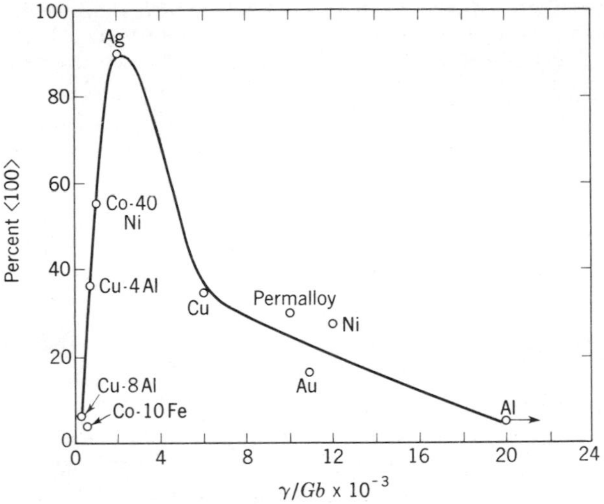

Fig. 20-1 Percent of [001] component in [111] + [001] duplex fiber textures of f.c.c. wires reduced 99 percent by cold drawing vs. the stacking-fault-energy parameter γ/Gb. Both high and low stacking-fault energies favor the [111] component. (A. T. English and G. Chin.)

ticularly sensitive to many variables. The textures are usually found to be duplex [111] + [100] fiber textures with the amount of material ascribed to each of the ideal orientations varying greatly from one material to another, as shown by Fig. 20–1 (which is discussed below), and even with a given material after deformation at different temperatures or after forming by cold-drawing, swaging, extruding, or tensile elongation. The texture may be influenced greatly, also, by the orientations existing at the start of a wire-forming operation, even though the reduction in area of the wire is extreme, at least if these starting orientations are symmetrical ones.[1] In general, however, it is probable that very high reductions will reduce most starting orientations to nearly identical final textures.

Whenever a f.c.c. metal is subject to deformation twinning at the deformation temperature, it becomes of importance to consider whether twinning is prominent at small reductions or only after considerable deformation, for twinning can cause reorientation of material, and can therefore cause changes in texture as deformation proceeds.

Stacking-fault energy γ appears to be a fundamental variable controlling the relative amounts of [111] and [100] components,[2] but the correlation between γ and texture is not simple, as will be seen from Fig. 20–1 for

[1] A. Freda and B. D. Cullity, *Trans. AIME*, vol. 215, p. 530, 1959, found that an initial [100] texture in a copper rod persisted throughout a reduction in area of 98.4 percent.

[2] N. Brown, *Trans. AIME*, vol. 221, p. 236, 1961. Some argue that stacking-fault frequency rather than energy may actually be the controlling factor, but there is *usually* a close correlation between these two parameters.

f.c.c. wires after 99 percent cold reduction in area.[1] The abscissa is a parameter used in theories of cross slip, γ/Gb, where G is the shear modulus and b is the Burgers vector, chosen because cross slip is thought to be responsible for reorientation into one of the final positions, [111]. Mechanical twinning can cause reorientation out of [111] into [511], which is near [100]; perhaps, therefore, the amount of material that appears to be in the [100] position is actually the amount of material that has undergone first-order mechanical twinning. If so, the variation of this twinning as a function of γ requires explanation, and can only be speculated about at present.[2] It is curious that the trends seen in Fig. 20–1 do not seem to correlate well with the trends in rolling textures of alloys with similar compositions.[3]

Textures in high-purity materials may be different from those in materials of ordinary purity. This is to be expected if partial or complete recrystallization occurs during deformation (which is the more likely the higher the purity); it may also result from altered deformation mechanisms.

Body-centered cubic metals invariably have a simple [110] fiber texture after cold drawing. The list investigated in various laboratories includes Fe, Mo, Nb, Ta, V, W, and Fe–4.6 percent Si, and beta-brass. There appear to be no minor texture components and no alterations by solid-solution formation, at least in alloyed ferrites.[4]

The *hexagonal* metal magnesium has a simple $[10\bar{1}0]$ fiber texture if the forming has been done at low temperatures.[5] (The $[10\bar{1}0]$ direction is midway between two close-packed atom rows of the c.p.h. structure.[6]) At working temperatures above 450°C the fiber axis is $[2\bar{1}\bar{1}0]$; the reason for this shift is unknown. Zinc wires have [0001] about 70° from the wire axis (i.e., a "spiral texture") after severe drawing; after smaller reductions [0001] is parallel to the wire axis.[7] Wires of zirconium[8] and titanium[9]

[1] A. T. English and G. Chin, *Acta. Met.*, vol. 13, p. 1013, 1965.

[2] Relevant possibilities are discussed by: P. R. Thornton and T. E. Mitchell, *Phil. Mag.*, vol. 7, p. 361, 1962. J. A. Venables, *J. Phys. Chem. Solids*, vol. 25, pp. 685, 693, 1964. J. F. W. Bishop, *J. Mech. Phys. Solids*, vol. 3, p. 130, 1954. E. A. Calnan, *Acta. Met.*, vol. 2, p. 865, 1954. H. Hu, R. S. Cline, and S. R. Goodman in the 1965 ASM Symposium, "Recrystallization, Grain Growth, and Textures," American Society for Metals, Cleveland, in press, suggest that the extensive faulting occurring in f.c.c. specimens of low γ makes them effectively an analogue of c.p.h. metals in that a maximum number of c.p. planes tend to approach the wire axis, this being accomplished in the f.c.c. metals by acquiring a [100] texture.

[3] See: R. E. Smallman, *J. Inst. Metals*, vol. 84, p. 10, 1955–1956. F. Haessner, *Z. Metallk.*, vol. 54, p. 79, 1963; vol. 53, p. 403, 1962.

[4] C. S. Barrett and L. H. Levenson, *Trans. AIME*, vol. 135, p. 327, 1939.

[5] L. G. Morell and J. D. Hanawalt, *J. Appl. Phys.*, vol. 3, p. 161, 1932.

[6] The close-packed rows are slip directions of the form $\langle 2\bar{1}\bar{1}0 \rangle$.

[7] F. Wolbank, *Z. Metallk.*, vol. 31, p. 249, 1939.

[8] W. G. Burgers, J. D. Fast, and F. M. Jacobs, *Z. Metallk.*, vol. 29, p. 410, 1937.

[9] I. Gokyu, H. Suzuki, and R. Horiuchi, *Nippon Kinzoku Gakkaishi*, vol. 18, p. 201, 1954 (*Met. Abstr.*, vol. 22, p. 196, 1954–1955). K. Sagel and U. Zwicker, *Z. Metallk.*, vol. 46, p. 835, 1955. C. J. McHargue and J. P. Hammond, *Trans. AIME*, vol. 197, p. 57, 1953.

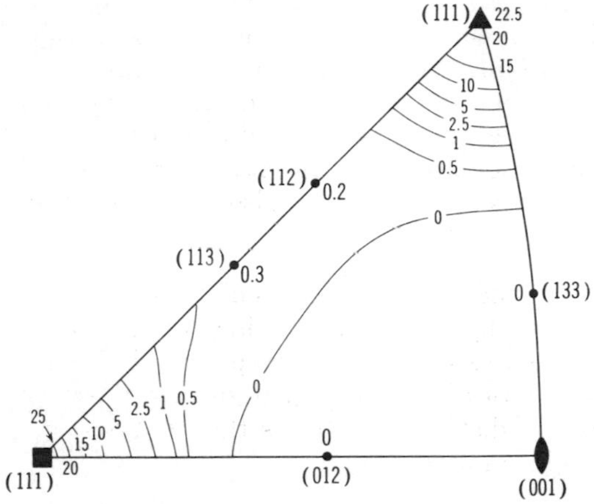

Fig. 20-2 Inverse pole figure of aluminum rod extruded slowly at 450°F. Fiber-axis density of unity corresponds to randomness. The [001] and [111] components contain 22 and 78 percent of the material, respectively. (**C. J. McHargue, L. K. Jetter, and J. C. Ogle.**)

have [10$\bar{1}$0] fiber textures. Twinning is responsible for removing material from a position in which the basal plane is parallel to the wire axis (the change in orientation amounts to 85°55′ when a single crystal of zinc is twinned).

Extruded rods often have approximately the same textures as drawn wires. This is true of magnesium[1] and sometimes of cubic metals, but is not true of zinc, perhaps because of varying amounts of twinning.[2] Inhomogeneities from surface to center in extruded rods may be due in part to the different thermal history of the parts. Aluminum extruded slowly at 450°F (232°C) has a texture shown by the inverse pole figure of Fig. 20-2.[3] The distribution of grains between [111] and [001] components varies with extrusion speeds and temperatures, and the amount of [001] material increases linearly with temperature at a given speed. This [001] component has been ascribed principally to recrystallization during the deformation as has been suggested also for the [001] component of other f.c.c. metals,[4] but electron microscopy has shown that the crystallites in the [001] orientation are apparently *not* recrystallized after 77°K extrusion.[5]

[1] L. G. Morrell and J. D. Hanawalt, *J. Appl. Phys.*, vol. 3, p. 161, 1932. E. Schiebold and G. Siebel, *Z. Physik*, vol. 69, p. 458, 1931.

[2] H. Unkel, *Metallwirtschaft*, vol. 21, p. 531, 1942. J. Grewen and G. Wassermann, *Z. Metallk.*, vol. 45, p. 498, 1954.

[3] C. J. McHargue, L. K. Jetter, and J. C. Ogle, *Trans. AIME*, vol. 215, p. 831, 1959.

[4] W. R. Hibbard, *J. Inst. Metals*, vol. 77, p. 581, 1950 (Al, Ag, Pb). W. A. Backofen, *Trans. AIME*, vol. 180, p. 250, 1951 (Cu).

[5] R. A. Vandermeer and C. J. McHargue, *Trans. AIME*, vol. 230, p. 667, 1964.

Wires formed by *rolling* and by *swaging* have the same textures at their centers as those formed by drawing through a die.[1] The *outer layers*, however, with almost any type of forming operation, contain orientations not found in the center.[2] The inclination of the fiber axes in the outer layers of a wire may reach 8 to 10° in some cubic metals and differs with different forming processes such as unidirectional vs. reversed passes through a die, drawing vs. swaging, etc. With hexagonal metals, also, there is a zonal structure.[3]

Textures of polyphase wires When the two phases of a binary eutectic are both deformed about equally by the wire-drawing operation, each phase takes up the texture it would exhibit in a single-phase wire.[4] On the other hand, if one phase is practically undeformed during drawing it disturbs the flow in the other and distorts or hinders the development of the texture, or it may even produce randomness.

Compression textures *Face-centered cubic* metals after uniaxial compression have a fiber texture that is most simply described as a [110] texture; i.e., a face diagonal is parallel to the axis of compression and normal to the plane of compression. However, this description is only a first approximation to a rather complex orientation; in the case of aluminum, half the crystallites are more than 10° from this orientation, regardless of the amount of compression.[5] Although there is a marked concentration near this [110] orientation, all possible orientations are found except those having [111] within about 20° of the compression axis. The range of orientations found can be presented best on a unit triangle of a standard stereographic projection, an inverse pole figure, as in Fig. 20-3. Each point on this plot shows the orientation of the compression axis in an area about 1 mm in diameter on an inner surface of a compression block. The orientations were determined by an optical goniometer after a compression of 84 percent. The distribution determined by x-rays after *compression rolling*—rolling with many passes, each pass in a different direction—is shown in Fig. 20-4 and illustrates again the absence of orientations around [111], the near absence around [100], the concentration around [110], and the spread from [110] to [311].

The compression texture of 70–30 brass is different, as shown in Fig.

[1] J. T. Norton and R. E. Hiller, *Trans. AIME*, vol. 99, p. 190, 1932. J. Thewlis, *Phil. Mag.*, vol. 10, p. 953, 1930. C. S. Barrett and L. H. Levenson, *Trans. AIME*, vol. 135, p. 327, 1939.

[2] E. Schmid and G. Wassermann, *Z. Metallk.*, vol. 19, p. 325, 1927; *Z. Physik*, vol. 42, p. 779, 1927.

[3] E. Schmid and G. Wassermann, *Naturwissenschaften*, vol. 17, p. 312, 1929. W. G. Burgers and F. M. Jacobs, *Metallwirtschaft*, vol. 14, p. 285, 1935. E. Schiebold and G. Siebel, *Z. Physik*, vol. 69, p. 458, 1931.

[4] G. Wassermann and J. Grewen, "Texturen metallischer Werkstoffe," p. 171, Springer, Berlin, 1962. R. M. Brick in "Cold Working of Metals," American Society for Metals, Cleveland, Ohio, 1949.

[5] C. S. Barrett and L. H. Levenson, *Trans. AIME*, vol. 137, p. 112, 1940.

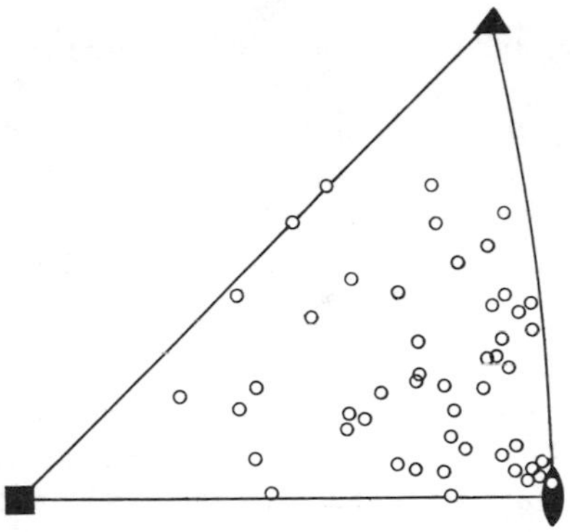

Fig. 20-3 Orientations on inner faces of aluminum blocks after 84 percent compression. Optical determination; the orientations of the compression axis of randomly chosen areas, about 1 mm in diameter, are indicated by small circles in the standard stereographic projection triangle bounded by the directions ■ [001], ▲ [111], and ⬮ [011]. Note absence around [100] and [111] and wide, unsymmetrical scatter around [110].

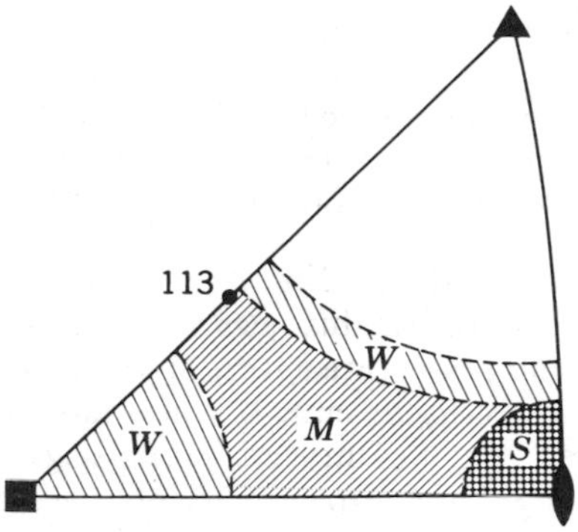

Fig. 20-4 Orientations in aluminum compressed 98 percent by compression rolling. Shaded areas on the projection indicate concentrations of orientations, deduced from x-ray line intensities (*S*, strong; *M*, medium; *W*, weak). Copper and nickel give similar plots.

20-5. No crystallites of alpha-brass have [100] directions near the axis of compression, while a fair number have [111] in this region; the concentration around the [110] position is still predominant, and the range from [110] to [311] still persists. A transition from an aluminum type to a brass type, at least in textures of cold-rolled sheet, can be caused by dissolving various elements in solid solution (see the discussion of rolling textures).

Body-centered cubic iron has a compression texture with [111] as major component plus a weaker component, [100].[1] The hexagonal metal magnesium (and Dowmetal) has a compression texture with the hexagonal axis parallel to the compression direction.[2] Compressing a rolled magnesium sheet parallel to the plane of the sheet causes twinning such that the (0001) planes that were parallel to the plane of the sheet become perpendicular to the compression direction(s).[3] By combining such compressions in two or more directions with slow hot-rolling at temperatures high enough to

[1] A. Onon, *Mem. Coll. Eng., Kyushu Imp. Univ.*, vol. 3, p. 267, 1923–1925. C. S. Barrett *Trans. AIME*, vol. 135, p. 296, 1939.

[2] S. Tsuboi, *Mem. Coll. Sci., Kyoto Imp. Univ.*, vol. A11, p. 375, 1928. L. G. Morrel and J. D. Hanawalt, *J. Appl. Phys.*, vol. 3, p. 161, 1932.

[3] C. S. Barrett and C. T. Haller, *Trans. AIME*, vol. 171, p. 246, 1947.

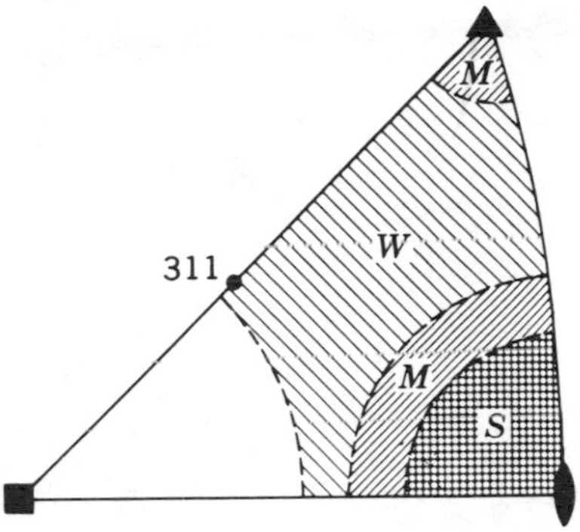

Fig. 20-5 Orientations in 70–30 brass compressed 97 percent by compression rolling (x-ray results).

suppress twinning ($\simeq$300°C), it is possible to produce a sheet with random orientation (though without appreciably increasing the tensile strength or ductility). The relatively small density of poles immediately around the [0001] orientation in rolled titanium and hafnium can be accounted for by twinning, which could reorient grains of that orientation.[1]

Theories of tension and compression textures Not merely one or two but several slip systems must be active in a grain during deformation—or at least in parts of a grain—because the grains are required to fit together without voids after the deformation. Early attempts to account for deformation textures undertook to treat the stability of various final orientations and differed chiefly in the number of slip systems that were assumed to operate in these final orientations. (The lattice rotations resulting from the operation of one slip system are cancelled by the operation of others when these final orientations are reached.) More recent theories, while retaining the idea that certain symmetrical orientations would be stable by such joint action of several slip systems, have emphasized geometrical principles that would rotate the lattice into stable positions, and some have attempted to account for the differences in the textures of different metals of the same crystal structure. The criterion for judging the stability of a position has been refined: it is now realized that it should be one to which a grain returns if it has been removed from the position by a temporary fluctuation in the state of flow.

Below, Taylor's theory is discussed first, then how it compares with studies of inhomogeneous flow; and last, more recent theories of fiber textures are discussed.

Taylor's theory of textures G. I. Taylor has worked out a theory of deformation textures that has a more rigorous basis than any previous one.[2] To permit any desired change of shape—such as will let grains fit together

[1] D. N. Williams and D. S. Eppelsheimer, *Trans. AIME*, vol. 194, p. 615, 1952 (Ti). D. S. Eppelsheimer and D. S. Gould, *Nature*, vol. 177, p. 241, 1956 (Hf).

[2] G. I. Taylor, *J. Inst. Metals*, vol. 62, p. 307, 1938; "Stephen Timoshenko 60th Anniversary Volume," p. 218, Macmillan, New York, 1938.

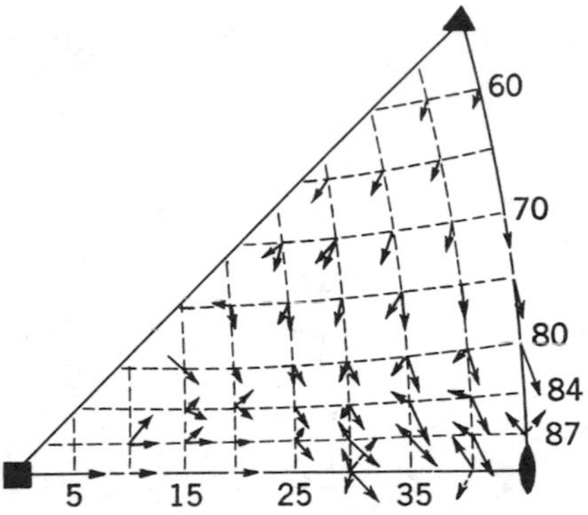

Fig. 20-6 Taylor's calculated rotations of crystal axes in grains of a f.c.c. polycrystalline aggregate during compression of 2.37 percent.

after deformation and will produce the same change of shape in the grains as in the aggregate as a whole—there must be at least five slip systems operating.[1] The *principle of least work* governs the choice of systems that must operate. This principle states that the *minimum* number will function which can produce the required change in shape. This number is five, except for special orientations. Furthermore, that group of five will be chosen for which the total work of deformation will be less than for any other group. Taylor computed the work that would be required on every group of five slip systems that could be chosen from the 12 possible systems in aluminum. There are 792 groups to be computed or ruled out by symmetry considerations of one kind or another, and Taylor considered all these for each of 44 different orientations of aluminum single crystals. Having computed what groups of systems would operate at each orientation, he predicted in what direction the lattice would rotate. Some ambiguity arose when two groups gave the same minimum values of total work; a grain that has an orientation in which this occurs will tend to rotate in two or more directions. Taylor remarked that in this case the direction of rotation can be a result of any combination of the two groups and thus can be any direction within a considerable range.

Taylor's computed rotations for grains of f.c.c. metals during compression are plotted in Fig. 20-6.[2] The rotation of the axis of compression is indicated by vectors originating at the orientation considered and extending in the direction of the calculated rotation. The length of the arrows corresponds to the rotation that should accompany a compression of 2.37 percent. Double arrows indicate equally favored sets of five slip systems; any direction that lies between them could equally well occur. The predicted rotations of Taylor have been tested[3] in individual grains of aluminum with an optical goniometer; the results are plotted in Fig. 20-7. In this figure the tails of the arrows are the initial orientations, the dots on

[1] R. v. Mises, *Z. Angew. Math. Mech.*, vol 8, p. 161, 1928.

[2] This figure is derived from the published chart that applies to tension, which is reproduced in Fig. 20-8.

[3] C. S. Barrett and L. H. Levenson, *Trans. AIME*, vol. 137, p. 112, 1940.

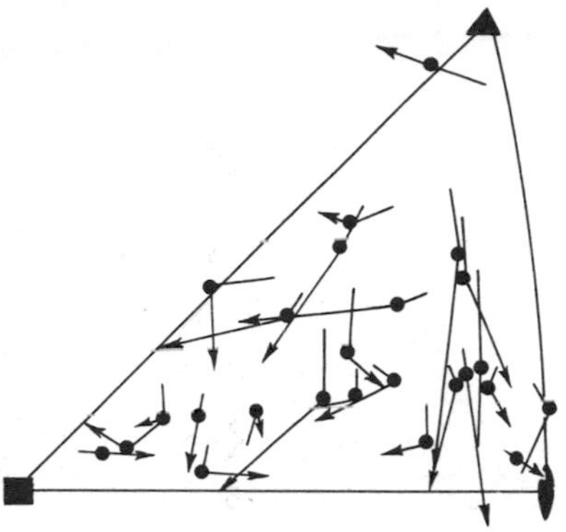

Fig. 20-7 Observed rotations of crystal axes in grains of aluminum during compression. Arrows connect initial orientations with those after 11 percent (dots) and 31 percent (arrowheads) compression.

the arrows are the orientations after 11 percent compression, and the heads of the arrows are those after 31 percent compression. When Figs. 20–6 and 20–7 are compared, it is found that about half the grains rotate as predicted by the theory, about a third do not, and the rest are uncertain.[1]

The effects of inhomogeneous flow and deformation bands on textures It is possible to explain the discrepancies between Figs. 20–6 and 20–7 by considering that nonuniform constraints on all sides of each grain affect its strain and result in an irregular strain throughout the material. That this occurs cannot be questioned, for it is the origin of the "orange-peel" surface sometimes encountered in deep-drawing operations, and it can readily be seen if a pair of polycrystalline blocks are placed together and compressed. After deformation the inner surfaces are roughened, clearly showing that each grain deforms in a manner influenced by the flow of its neighbors and not in the simple homogeneous manner that can be computed from the change of shape of the blocks. Consequently, the rotation in an individual grain is not solely a function of the orientation of the grain. The theoretical predictions, on the other hand, were based on the assumption that every grain changes its shape exactly as does the whole. If one considers Fig. 20–6 as merely indicating the general trends of lattice rotations, it is seen that the trends are downward and to the right, in qualitative agreement with the observed rotations. This accounts for the observed compression texture [110]. The regions in which the direction of rotation is uncertain correspond to the lower intensity regions of the pole figure (Fig. 20–4).

The corresponding plot for tensional deformation is reproduced in Fig. 20–8. The arrows here are the reverse of those for the compression case, and their trend predicts a tension texture with [111] and [100] components, in agreement with experimental data. Neither Taylor's theory nor the ones that preceded it predicted or discussed the deformation bands that form during plastic flow.

A micrographic and x-ray study of the structure of iron after compression has shown how the individual grains contribute to the preferred orien-

[1] For recent experiments on single crystals, see Ref. 2, p. 543.

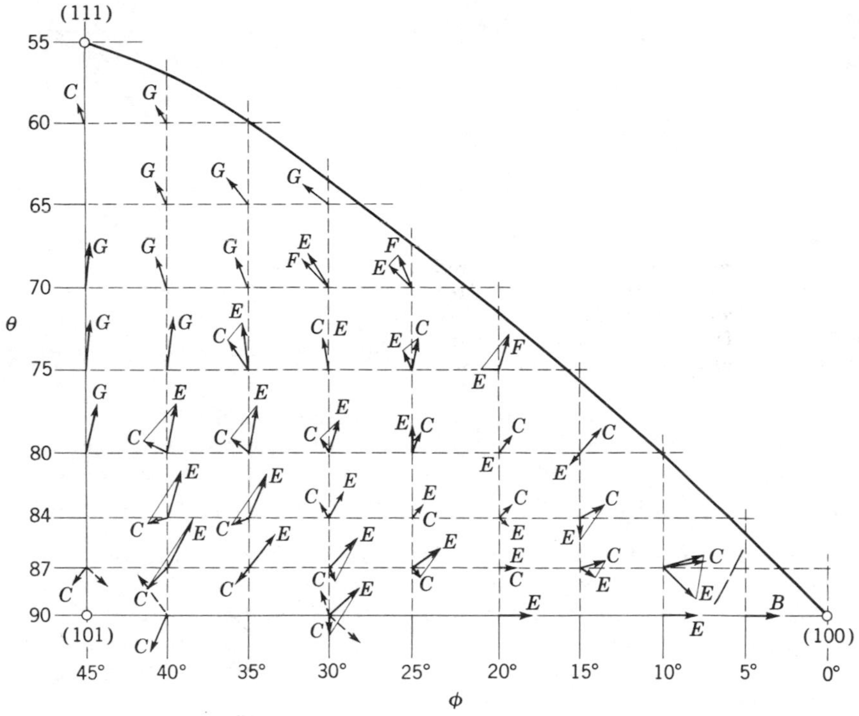

Fig. 20-8 Taylor's calculated rotations for f.c.c. polycrystalline grains during elongation of 2.37 percent. θ and ϕ are angles of the specimen axes from [101] and [100], respectively.

tations of the aggregate.[1] The rotation of a grain is conditioned by its original orientation as follows: If it initially has [100] nearly parallel to the axis of compression, it will seek the [100] position and retain it. If it initially has [111] near the axis of compression, it will seek and retain the [111] position. But if the initial orientation is removed from either of these two stable positions, the grain will subdivide into deformation bands, and alternate bands will rotate to the [111] and [100] positions (Fig. 20–9). X-ray measurements showing the rotation of the alternate deformation bands in a crystal of iron are plotted in Fig. 20–10, where the circular regions indicate the positions of the compression axis after various compressive deformations up to 72 percent reduction in thickness, at which point the final texture was reached.[1]

Individual grains of aluminum appear to undergo a more haphazard distortion during compression than do grains of iron.[2] Typical grains contain a spread in orientation as high as 10° after a 10 percent reduction in thickness and a maximum of about 50° after 60 percent reduction. The orien-

[1] C. S. Barrett, *Trans. AIME*, vol. 135, p. 296, 1939.

[2] C. S. Barrett and L. H. Levenson, *Trans. AIME*, vol. 137, p. 112, 1940.

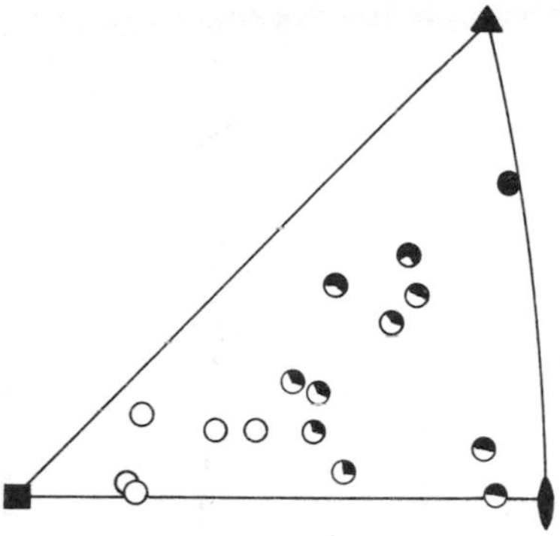

Fig. 20-9 Projection showing relative prominence of [100] texture (white segments of circles) and [111] texture (black segments) after compression, as a function of initial orientation of iron crystals.

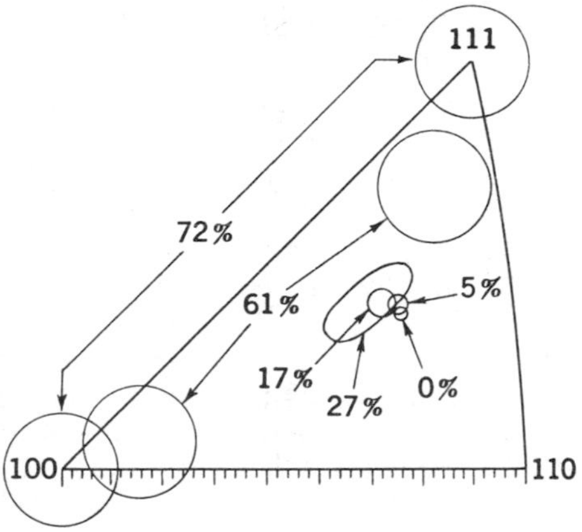

Fig. 20-10 Rotation of a crystal of iron during successive stages of compression up to 72 percent.

tation dependence of the banding tendency is plotted in Fig. 20–11, where filled and partly filled circles indicate the initial orientations of grains which subsequently formed bands or irregularities in orientation, and open circles indicate the initial orientations of those which did not.

During drawing, swaging, or elongating in tension, individual grains of iron also take up increasingly wide ranges of orientation.[1] Deformation bands form on crystallographic planes of the form [100] and [111], contribute to the scatter in orientation, and sometimes produce a complete [110] fiber texture in a single grain.

The crystallography of deformation bands, including *kink bands* and *bands of secondary slip*, has been summarized elsewhere.[2]

Later theories of the development of fiber textures Calnan and Clews[3] were concerned with the question of how the five slip systems

[1] C. S. Barrett and L. H. Levenson, *Trans. AIME*, vol. 135, p. 327, 1939.

[2] R. Maddin and N. K. Chen in B. Chalmers (ed.), "Progress in Metal Physics," vol. 5, Interscience, New York, 1954.

[3] E. A. Calnan and C. J. B. Clews, *Phil Mag.*, vol. 41, p. 1085, 1950. E. A. Calnan, *Acta Met.*, vol. 2, p. 865, 1954.

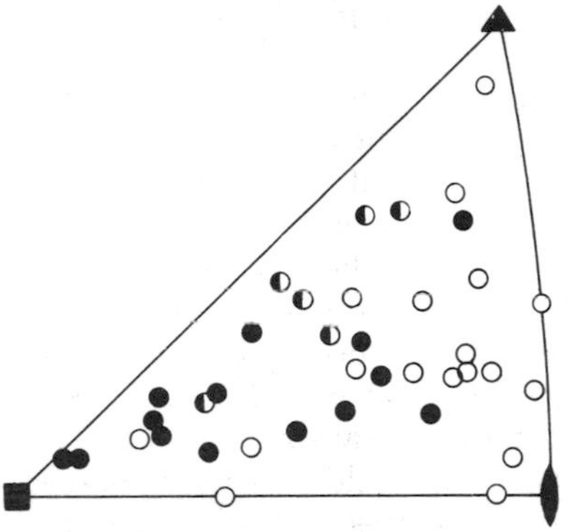

Fig. 20-11 Dependence of banding tendency, during compression, on initial orientation of grains of aluminum. (●) Grains forming bands. (○) No bands. (◐) Uncertain, irregular orientations.

required in Taylor's theory could come into operation simultaneously when, in general, only one or two initially reach the critical resolved shear stress for slip as the applied load is increased. They proposed that the problem be treated, for tensile deformation, on the basis of an effective tensile stress T_e, acting on a grain in a direction that is different, in general, from the direction of the applied tensile stress T_a; and that the effective tensile direction rotates by single slip, then by double slip until it reaches a direction that is symmetrical enough to provide equal resolved shear stresses on the necessary number of slip systems and to thereby halt further rotation. The constraints imposed by the surrounding grains together with the applied stress form a complex stress system, which admittedly is oversimplified in this treatment. The model nevertheless leads to definite predictions of the directions of rotation of grains and to stable orientations that are in reasonable agreement with observations.

Ōyane and Kojima[1] have proposed a version of the same type of theory to account for tension textures and compression textures in f.c.c. metals. Calnan and Clews have also applied their model to b.c.c. metals with {110}, {112}, and {123} slip[2] and to hexagonal metals,[3] assuming basal slip and $\{10\bar{1}2\}$ twinning. The model can account for the tension, compression, and cold-rolling textures of titanium if suitable assumptions are made regarding slip and twinning planes and their relative critical shear stresses.[4] In tension, the predicted rotation in titanium is to the stable $[10\bar{1}0]$ position, whereas in compression, $\{11\bar{2}2\}$ twinning prevents slip rotation from reaching the stable [0001] position and causes a spread in orientation preferentially along the $[0001]$–$[11\bar{2}0]$ edge of the unit triangle, from the position where slip and twinning are equally probable out toward $[11\bar{2}0]$. (Observing the inner edge of this spread is proposed as a means of determining the ratio of the critical shear stresses for slip and twinning

[1] M. Ōyane and K. Kojima, *Mem. Inst. Sci. Ind. Res.*, vol. 12, p. 15, 1955.

[2] E. A. Calnan and C. J. B. Clews, *Phil. Mag.*, vol. 42, p. 616, 1951.

[3] Calnan and Clews, op. cit., p. 919.

[4] D. N. Williams and D. S. Eppelsheimer, *J. Inst. Metals*, vol. 81, p. 553, 1952–1953.

as deformation proceeds.) The model has also been applied to orthorhombic metals.[1]

Attempts to understand the fact that different metals of the same f.c.c. structure have different texture have been based on the known (or assumed) differences in deformation mechanisms, and the effects of these on reorientation trends that are observed with single crystals. Bishop and Calnan independently[2] developed models in which the relative hardening of active and latent slip systems would account for varying amounts of "overshoot" (i.e., rotation with single slip past the orientation where slip on the second system should begin), and thereby would make texture differences understandable. Smallman's[3] extensive examination of textures of f.c.c. metals and alloys of various compositions yielded results in accord with the conclusion that these differences in textures might be attributed to differential hardening of active and latent slip systems. He pointed to a correlation between texture type and the tendency of a metal to show a yield point in a tensile test (during yield-point elongation, a single crystal tends to exhibit "easy glide" and overshoot). He also suggested that stacking-fault tendencies may play a role (as has more recently been indicated by the correlation with stacking-fault energy, Fig. 20–1); he mentioned also the unknown possible role of short-range order, but did not consider deformation twinning.

The specification of rolling textures Textures in rolled (or extruded) sheet can be described in an approximate way by choosing one or two "ideal orientations" involving the plane (hkl) that lies parallel to the plane of the sheet and the direction $[uvw]$ that lies in the rolling direction. Although the orientations found in greatest frequency can be specified in this way, i.e., by writing $(hkl)[uvw]$, the details of the scatter about these ideal orientations must be specified by listing a considerable number of less prominent orientations that are somewhat arbitrarily chosen. These may be of no fundamental significance, so that, when ideal textures are used to calculate the anisotropic distribution of elastic moduli or magnetic properties, they may lead to inaccurate predictions. A less artificial and more quantitative and accurate description is to plot a pole figure which gives the density distribution of one—or preferably two or more—sets of poles $[uvw]$ on a map (see Chap. 9).

Rolling textures of f.c.c. metals Many early determinations of textures in cold-rolled f.c.c. metals showed that they could be very roughly described as having (110) parallel to the rolling plane (RP) and $[1\bar{1}2]$ parallel to the rolling direction (RD); but with the increasing use of diffractometers and the development of theories of textures, more accurate descriptions have been introduced. The textures of aluminum, copper,

[1] E. A. Calnan and C. J. B. Clews, *Phil. Mag.*, vol. 43, p. 93, 1952.

[2] F. W. Bishop, *J. Mech. Phys. Solids*, vol. 3, pp. 130, 259, 1954–1955. E. A. Calnan, *Acta Met.*, vol. 2, p. 865, 1954.

[3] R. E. Smallman, *J. Inst. Metals*, vol. 83, p. 408, 1954–1955; vol. 84, p. 10, 1955–1956.

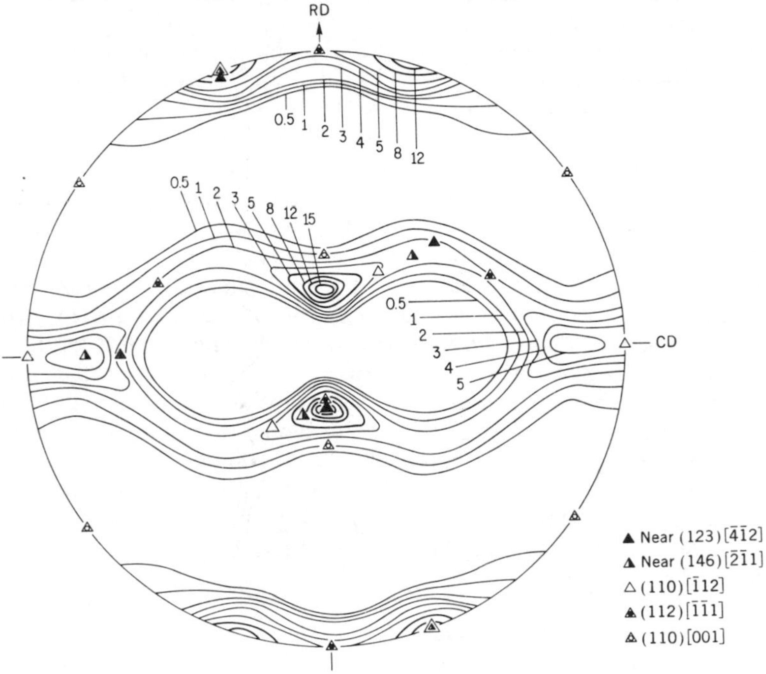

Fig. 20-12 (111) **pole figure of electrolytic copper rolled 96.6 percent at 25°C. Density of unity corresponds to randomness.** (**H. Hu and S. R.Goodman.**)

nickel, gold, iron-nickel containing 30 weight percent Ni, and thorium are very similar; the pole figures typical of these are shown in Figs. 20–12 and 20–13.[1] The texture may be approximated by the ideal texture $(123)[41\bar{2}]$,[2] or better by the textures $(123)[41\bar{2}] + (146)[21\bar{1}]$, as may be seen from the pole figure. Clearly, a simple $(110)[1\bar{1}2]$ description is inadequate.

There are metals, however, that have textures well described by $(110)[1\bar{1}2]$; for example, copper-zinc alloys of composition near 30 weight percent zinc. Pole figures for commercial 70–30 brass, for example, are shown in Figs. 20–14 and 20–15.[3] Materials with this brass type of rolling texture include[4] silver (if rolled near or below room temperature) and various

[1] H. Hu and S. R. Goodman, *Trans. AIME*, vol. 227, p. 627, 1963.

[2] Originally stated to be (123) $[1\bar{2}1]$, but the $[1\bar{2}1]$ referred to the direction in the sheet transverse to RD, which is not now common practice. Various other ideal positions have been used, such as (135) $[53\bar{3}]$ and (7, 12, 22) $[8, 4, \bar{5}]$ and (5, 6, 16) $[10, 13, \bar{8}]$; see G. Wassermann and J. Grewen, "Texturen metallischer Werkstoffe," Springer, Berlin, 1962. A criticism of (123) $[41\bar{2}]$ has been made by E. R. W. Jones and E. A. Fell, *Acta. Met.*, vol. 5, p. 689, 1957.

[3] H. Hu, P. R. Sperry, and P. A. Beck, *Trans. AIME*, vol. 194, p. 76, 1952.

[4] R. E. Smallman, *J. Inst. Metals*, vol. 84, p. 10, 1955–1956.

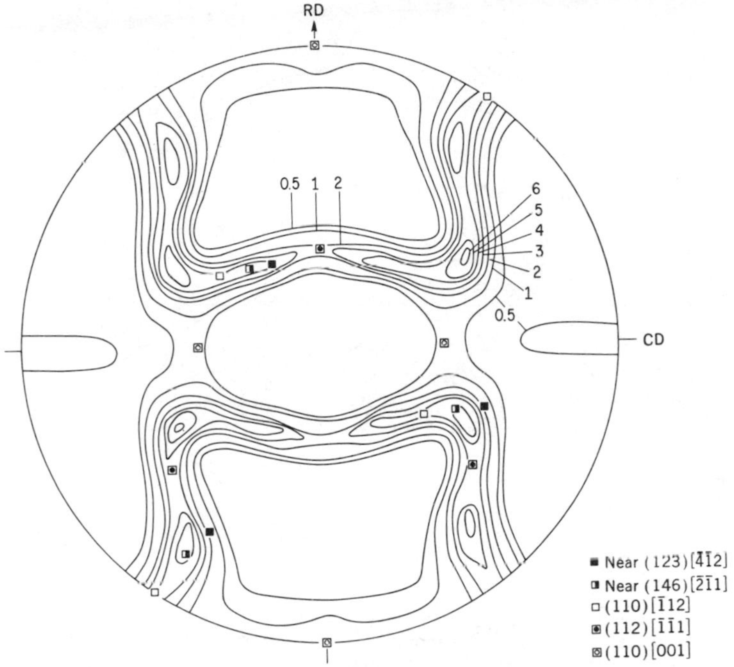

Fig. 20-13 (200) pole figure corresponding to Fig. 20-14.

substitutional solid solutions. Rolling at low temperatures tends to make a texture transition occur at a lower solute concentration (for example Cu–5 percent Zn becomes brass-type if rolled at −183°C, whereas the transition at room temperature is near Cu–11 percent Zn). The brass-type texture, although characteristic of silver when rolled at room temperature, is replaced by a copper type if rolled at 200°C (for the texture component corresponding to *unrecrystallized* material in the sheet).[1] H. Hu et al. find a correlation of silver textures with the stacking-fault frequency indicated by x-ray diffraction peak shifts for silver rolled at different temperatures. In silver rolled at room temperature, the brass-type texture is obtained, associated with high values of stacking-fault frequencies or high dislocation densities, or both. This texture changes gradually to the copper type as the rolling temperature is raised and as stacking-fault frequencies or dislocation densities decrease. In an analogous way, 18–8 stainless steel (type 304*L* or 304) has an austenite deformation texture that changes gradually from the brass-type to the copper-type as the temperature of rolling increases from 200 to 800°C (and as the stacking-fault frequency decreases).[2]

[1] H. Hu and R. S. Cline, *J. Appl. Phys.*, vol. 32, p. 760, 1961. H. Hu et al., *J. Appl. Phys.*, vol. 32, p. 1392, 1961.

[2] S. R. Goodman and H. Hu, *Trans. AIME*, vol. 230, p. 1413, 1964.

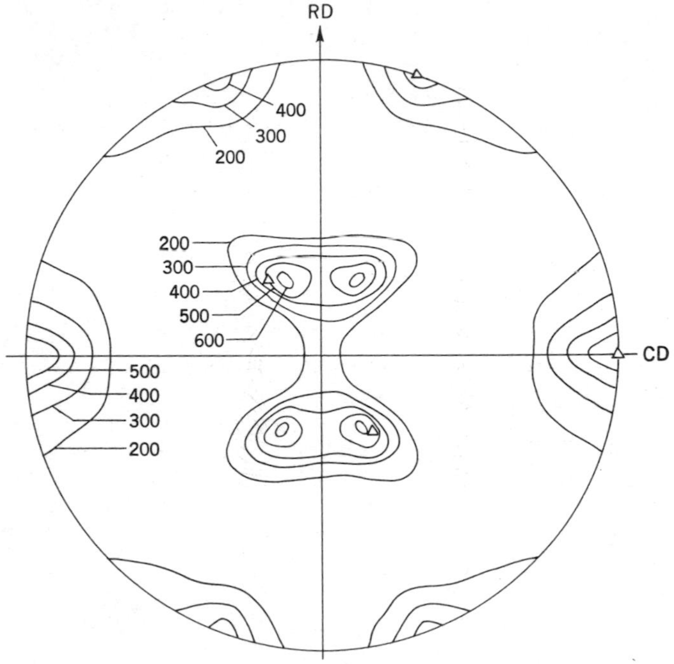

Fig. 20-14 (111) pole figure for the "inside texture" of 95 percent rolled strip of commercial 70–30 brass. Specimen 0.002 in. thick, from center portion of 0.024-in.-thick rolled strip; △ (110)[1$\bar{1}$2]. (H. Hu, P. R. Sperry, and P. A. Beck.)

A similar transition also occurs in cold-rolling copper. The ordinary copper-type obtained near room temperature changes gradually to the brass-type when rolling temperatures are reduced through the range -80 to -200°C, with a corresponding increase in stacking-fault frequency in the rolled metal.[1] Stacking-fault energies are known to increase with increasing temperature in electron microscope tests of some f.c.c. materials, which makes cross slip easier at the higher temperatures.[2]

Studies have shown the role of deformation bands in developing a f.c.c. rolling texture.[3] After high reductions each grain, and even each band, has a mean orientation within the more intense regions of the polycrystalline pole figure. Deformation bands, not deformation twins, constitute the principal mechanism of fragmentation by which complementary orientations are produced in the grains of most cubic metals.

Body-centered cubic rolling textures The numerous determinations of the texture of cold-rolled iron and steel have been in satisfactory agree-

[1] H. Hu and S. R. Goodman, *Trans. AIME*, vol. 227, p. 627, 1963.

[2] P. R. Swann and J. Nutting, *J. Inst. Metals*, vol. 90, p. 133, 1961–1962.

[3] C. S. Barrett and F. W. Steadman, *Trans. AIME*, vol. 147, p. 57, 1942.

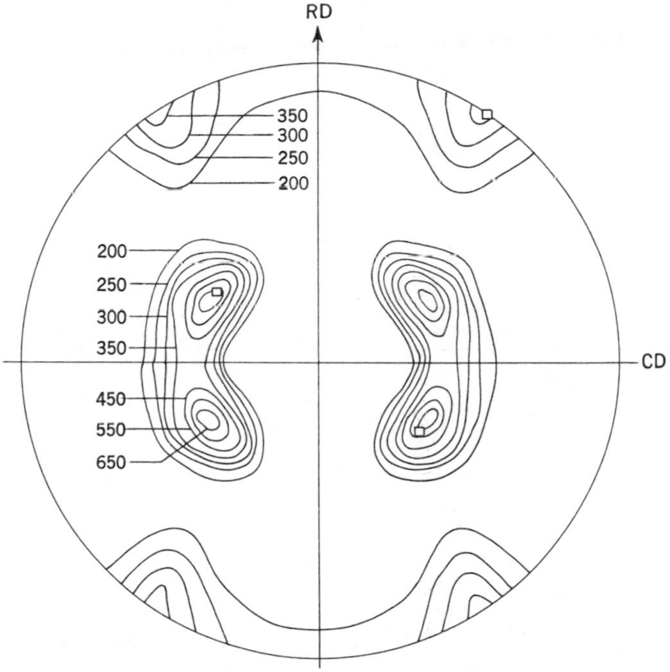

Fig. 20-15 (200) pole figure for the rolling texture of commercial 70–30 brass strip. Same specimen used for Fig. 20-14; □ (110)[1$\bar{1}$2]. (H. Hu, P. R. Sperry, and P. A. Beck.)

ment on the principal features of the texture (with the exception of a few conclusions that apparently have been based on insufficient data). The texture is chiefly one in which [110] directions of the grains lie along the direction of rolling—with a deviation of a few degrees—and (001) planes lie in the plane of the rolled sheet, with a deviation from this position chiefly about the rolling direction as an axis. The deviation includes orientations rotated away from the (001)[110] position various amounts up to 45 or 55° each way, and depends upon the amount of reduction.[1] The deviation about the direction in the sheet transverse to the rolling direction varies under different conditions from about 6 to 20°, decreases with increasing reductions, and in the surface layers decreases with increasing roll diameter. The crystallites decrease in number with increasing deviation from the ideal orientation.

Certain of the less prominent orientations were described by Kurdjumov and Sachs with the ideal orientations (112)[1$\bar{1}$0] and (111)[11$\bar{2}$].[2] Steels

[1] C. B. Post, *Trans. ASM*, vol. 24, p. 679, 1936. D. L. McLachlan and W. P. Davey, *Trans. ASM*, vol. 25, p. 1084, 1937. M. Gensamer and R. F. Mehl, *Trans. AIME*, vol. 120, p. 277, 1936.

[2] G. Kurdjumov and G. Sachs, *Z. Physik*, vol. 62, p. 592, 1930.

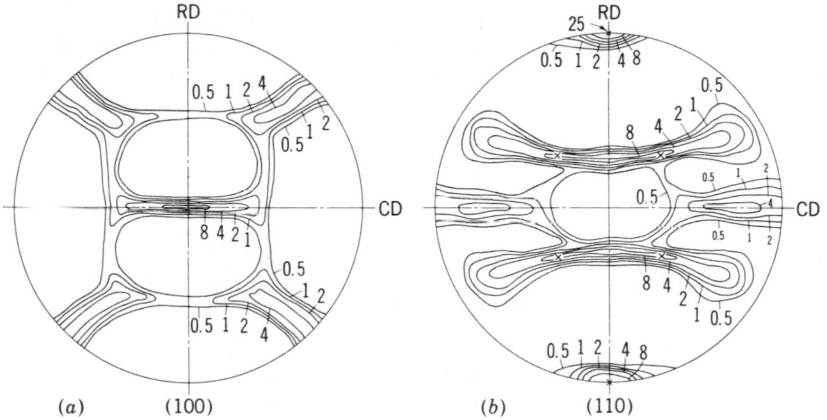

Fig. 20-16 Pole figures for cold-rolled molybdenum, 0.01-mm foil thickness. Diffractometer measurements referred to randomness as unity. (**A. Segmüller and G. Wassermann.**)

of higher carbon content tend to have less pronounced textures. Steels containing silicon in quantities to be of importance to the electrical industry for magnetic purposes have been extensively studied,[1] and in general they differ only slightly in cold-rolling texture from unalloyed iron. The textures are essentially independent of roll diameter, reduction per pass, rolling speed, and direct or reversed passes.[2] The application of tension to the strip as it passes through the rolls is without effect on the establishment or final degree of preferred orientation[3] (unless perhaps it lowers the relative amount of unoriented background material).

Molybdenum has a pronounced cold-rolling texture shown in Fig. 20–16,[4] which closely resembles iron; tantalum,[5] vanadium,[6] titanium with 18 weight percent niobium (b.c.c.),[7] beta-brass,[8] and tungsten[9] are also similar.

When the rotations of individual grains are followed during deformations simulating the change of shape of grains during rolling of a polycrystalline specimen, some crystals maintain a relatively sharp single orientation,

[1] C. S. Barrett, G. Ansel, and R. F. Mehl, *Trans. AIME*, vol. 125, p. 516, 1937. J. R. Brown, *J. Appl. Phys.*, vol. 29, p. 359, 1958. H. Möller and H. Stäblein, *Arch. Eisenhüttenw.*, vol. 29, p. 377, 1958.

[2] W. A. Sisson, *Metals Alloys*, vol. 4, p. 193, 1933.

[3] C. B. Post, *Trans. ASM*, vol. 24, p. 679, 1936 A. Hayes and R. S. Burns, *Trans. ASM*, vol. 25, p. 129, 1937. J. K. Wood, Jr., *Trans. ASM*, vol. 39, p. 725, 1947.

[4] A. Segmüller and G. Wassermann, *Freiberger Forschungsh.*, vol. B38, p. 38, 1960. C. J. McHargue and J. P. Hammond, *Trans. AIME*, vol. 194, p. 745, 1952.

[5] Segmüller and Wassermann, loc. cit.

[6] McHargue and Hammond, loc. cit.

[7] J. H. Keeler, *Trans. AIME*, vol. 206, p. 122, 1956.

[8] W. Iweronowa and G. Schdanow, *Tech. Phys. USSR*, vol. 1, p. 64, 1934.

[9] J. W. Pugh, *Trans. AIME*, vol. 212, p. 537, 1958.

others rotate inhomogeneously into two or more orientations, and some fragment into a texture resembling much of the entire pole figure.[1]

Cross-rolling textures differ from straight-rolling textures. When the amounts of reduction in the two perpendicular directions are roughly equal, the cross-rolling texture of b.c.c. metals is approximately the superposition of the two pole figures for straight rolling in the two directions.[2] Cross-rolled f.c.c. metals also have a more isotropic distribution of poles after cross rolling than after straight rolling.[3]

Close-packed hexagonal rolling textures Rolling should tend to rotate the slip plane of c.p.h. metals into the plane of the rolled sheet, and in accord with this tendency the predominating texture is one in which the basal plane, which is the predominant slip plane, lies in or near the rolling plane, although modifications are found. This simple texture is most pronounced in metals with an axial ratio near that for the close packing of spheres ($c/a = 1.633$), as in magnesium,[4] zirconium,[5] and hexagonal cobalt,[6] with $c/a = 1.624$, 1.589, and 1.624, respectively. There is a tendency for the $[11\bar{2}0]$ direction to align with the rolling direction, as might be expected from the fact that this is the slip direction.

Zinc and cadmium (with axial ratios $c/a = 1.856$ and 1.885, respectively) have rolling textures in which little material has the basal plane in the plane of the sheet. The hexagonal axis is found most frequently inclined 20 to 25° toward the rolling direction.[7] The texture on the surface differs, in general, from that in the interior,[8] and the texture can be modified by alloying, which alters the amount of twinning.[9] The texture of magnesium is also altered by alloying, especially by the presence of calcium,[10] which causes the central area to separate into two maxima.

The hexagonal metals zirconium, hafnium, titanium, and beryllium, with low c/a ratios (all less than 1.633), form rolling textures somewhat

[1] C. S. Barrett and L. H. Levenson, *Trans. AIME*, vol. 145, p. 281, 1941. C. G. Dunn and P. K. Koh, *Trans. AIME*, vol. 206, p. 1017, 1956. J. L. Walter and W. R. Hibbard, *Trans. AIME*, vol. 212, p. 731, 1958. H. Hu and R. S. Cline, *Trans. AIME*, vol. 224, p. 784, 1962. See also Ref. 2, p. 543.

[2] F. Haessner and H. Weik, *Arch. Eisenhüttenw.*, vol. 27, p. 153, 1956. H. Hu, *J. Metals*, vol. 9, p. 1164, 1957 (Fe). M. Semchyshen and G. A. Timmons, *Trans. AIME*, vol. 194, p. 279, 1952 (Mo). J. F. H. Custers and J. C. Riemersma, *Physica*, vol. 12, p. 195, 1946 (Mo).

[3] A. Merlini and P. A. Beck, *Acta Met.*, vol. 1, p. 598, 1953. J. T. Michalak and W. R. Hibbard, *Trans. AIME*, vol. 209, p. 101, 1957.

[4] V. Caglioti and G. Sachs, *Metallwirtschaft*, vol. 11, p. 1, 1932. J. C. McDonald, *Phys. Rev.*, vol. 52, p. 886, 1937. P. W. Bakarian, *Trans. AIME*, vol. 147, p. 266, 1942.

[5] W. G. Burgers and F. M. Jacobs, *Metallwirtschaft*, vol. 14, p. 285, 1935.

[6] G. Wassermann, *Metallwirtschaft*, vol. 11, p. 61, 1932.

[7] Caglioti and Sachs, loc. cit. M. A. Valouch, *Metallwirtschaft*, vol. 11, p. 165, 1932.

[8] A. Hargreaves, *J. Inst. Metals*, vol. 71, p. 73, 1945.

[9] M. L. Fuller and G. Edmunds, *Trans. AIME*, vol. 111, p. 146, 1934. E. Schmid, *Z. Metallk.*, vol. 31, p. 125, 1939. F. Wolbank, *Z. Metallk.*, vol. 31, p. 249, 1939.

[10] P. W. Bakarian, *Trans. AIME*, vol. 147, p. 266, 1942.

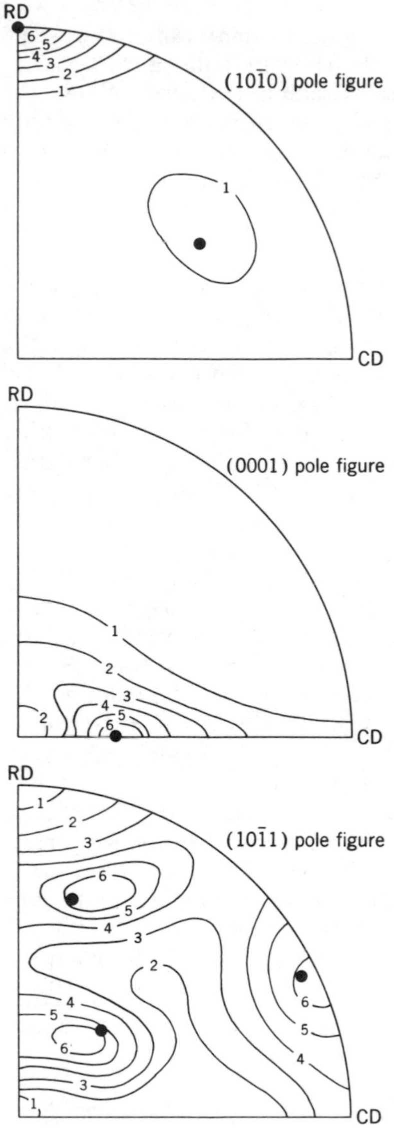

Fig. 20-17 Pole figures for cold-rolled titanium. Points (●) are for (0001) [10$\bar{1}$0] rotated ±30° around RD axis. (D. N. Williams and D. S. Eppelsheimer.)

different from those discussed above, as might be expected from the tendency of these crystals to slip on other than (0001) planes, and from the fact that the modes of twinning depend on c/a ratios. The deviation from a simple (0001)[10$\bar{1}$0] texture is toward the *transverse* direction in the pole figure of these metals instead of toward the rolling direction as in zinc and cadmium.[1] In zirconium, the tilt of basal planes away from the plane of the sheet and around the rolling direction as an axis averages 30 to 40° in each direction.

[1] J. H. Keeler, W. R. Hibbard, and B. F. Decker, *Trans. AIME*, vol. 197, p. 932, 1953.

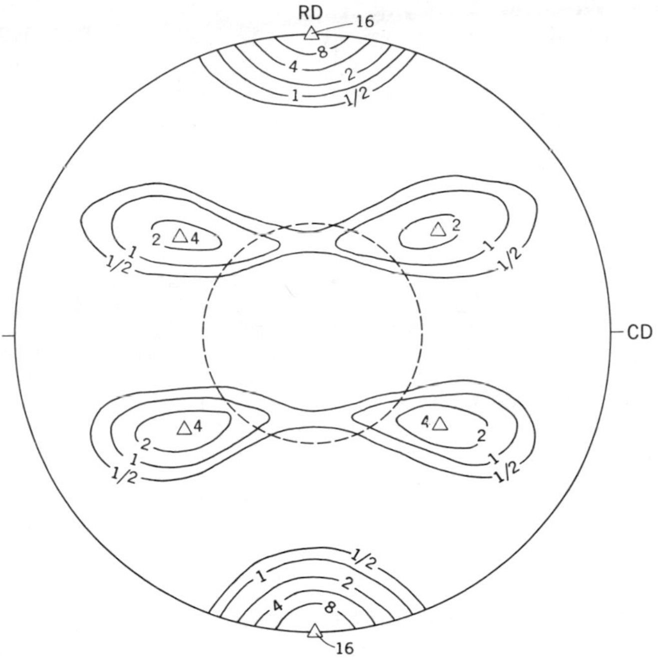

Fig. 20-18 (10$\bar{1}$0) pole figure for zirconium sheet coldrolled 99.8 percent. Triangles indicate (10$\bar{1}$0) poles after ±40° rotation around RD. Diffractometer measurements. (J. H. Keeler and A. H. Geisler.)

Fig. 20–17 shows the texture observed in unalloyed titanium[1]; various Ti-rich substitutional solid solutions are similar,[2] with the exception of Ti–3.8 percent aluminum alloy. The textures of beryllium, hafnium, and zirconium are rather similar[3] (see Fig. 20–18 for zirconium).

The texture of cold-rolled alpha-uranium (orthorhombic) is complex. One pair of components observed by several investigators is (102)[010] + (012)[021],[4] and additional "ideal orientations" are also needed.[5]

[1] D. N. Williams and D. S. Eppelsheimer, *Trans. AIME*, vol. 197, p. 1378, 1953. J. H. Keeler and A. H. Geisler, *Trans. AIME*, vol. 206, p. 80, 1956.

[2] C. J. McHargue, S. E. Adair, and J. P. Hammond, *Trans. AIME*, vol. 197, p. 1199, 1953.

[3] J. H. Keeler, *Trans. AIME*, vol. 212, p. 781, 1958 (Be, Hf). D. S. Eppelsheimer and D. S. Gould, *J. Inst. Metals*, vol. 85, p. 158, 1956 (Hf). J. H. Keeler and A. H. Geisler, *Trans. AIME*, vol. 203, p. 395, 1955 (Zr).

[4] W. Seymour, *Trans. AIME*, vol. 200, p. 999, 1954. C. M. Mitchell and J. F. Rowland, *Acta Met.*, vol. 2, p. 559, 1954. J. Adam and J. Stephenson, *J. Inst. Metals*, vol. 82, p. 561, 1953–1954.

[5] M. H. Mueller, H. W. Knott, and P. A. Beck, *Trans. AIME*, vol. 203, p. 1214, 1955. For cold-rolled alpha-uranium four components have been listed: I, (4, $\bar{17}$, 26) [410]; II, probably the most intense, (9, 0, 25) [010], or approximately (102) [010]; III, (4, 14, 45) [$\bar{5}$52]; IV, (0$\bar{3}$8) [031]; III and IV are perhaps merely a spread in orientation.

Theories of rolling textures A full list of attempts to develop a theory of rolling textures would be a long one.[1] No fully rigorous theory has been developed, and no measurements have been made of the parameters that would be needed in such a theory, for example, the stresses for different modes of slip and twinning after large strains. Many simplifying assumptions have been made in various theories, and it is well to remember that the success of a theory based on many assumptions does not prove the validity of all the assumptions.

Early theories were concerned with which *end orientations* would be expected to be stable under the rotations from symmetrically disposed slip systems, and which ones provide slip directions that lie nearest to the rolling direction. Explanations of the differences between textures of different f.c.c. metals in terms of differences in the tendencies of single crystals of these metals to "overshoot" were advanced,[2] but were soon attacked as inadequate.[3] It would seem that rotation directions should not be appreciably altered by overshoot after a considerable amount of flow has occurred, yet the *stability* of certain orientations may nevertheless be determined by the tendency to overshoot.

Dillamore and Roberts[4] assumed the following: (1) At least in large-grained metal, multiple slip systems were required to operate only in a small fraction of the volume of a grain along its boundaries, and the main body of the grain rotated approximately as a single crystal of the same orientation would. (2) The rotation was that expected of a crystal slipping on the slip system subjected to the highest resolved shear stress under forces applied normal to the sheet (compressive) and also forces applied parallel to the rolling direction (tensile), with rotation occurring until a position is reached that is stable. (3) A stable position was one that is not merely symmetrical, but is stable in the sense that if fluctuations take a grain away from the position, it returns—a criterion proposed by Tucker.[5] There is also a need, Tucker had emphasized, for considering stresses as a *triaxial* system in theories of textures, rather than making the assumption[6] that rotations would necessarily be those that simultaneously and independently satisfy uniaxial tension and uniaxial compression modes of deformation. (The Dillamore-Roberts theory did this, with the principal stress along the transverse direction assumed to be zero.) (4) The f.c.c. metals of high stacking-fault energy γ rotated not only in accord with the demands of primary and conjugate slip of single crystals, but also in response to *cross slip*, which would occur in relative amounts dependent upon the value of γ.

A complex theory by Liu[7] gave attention to strong attractive interactions between the dislocations on slip planes most effective in elongating the sheet, and introduced a minimum work criterion into the complex flow process. None of these theories involved twinning.

Hexagonal metal textures appear to be somewhat easier to understand in a general way than f.c.c. textures. The Calnan and Clews theory assumed that for magnesium ($c/a < \sqrt{3}$) the compression normal to the sheet twins all grains having (0001) poles more

[1] Early theories were discussed in the second edition of this book.

[2] J. F. W. Bishop, *J. Mech. Phys. Solids*, vol. 3, p. 130, 1954. E. A. Calnan, *Acta Met.*, vol. 2, p. 865, 1954.

[3] E. A. Calnan, *J. Inst. Metals*, Discussion, vol. 84, p. 503, 1955–1956. R. F. Braybrook and E. A. Calnan, *J. Inst. Metals*, vol. 85, p. 11, 1956–1957. R. E. Smallman and D. Green, *Acta Met.*, vol. 12, p. 145, 1964.

[4] I. L. Dillamore and W. T. Roberts, *Acta Met.*, vol. 12, p. 281, 1964.

[5] G. E. C. Tucker, *J. Inst. Metals*, vol. 82, p. 655, 1953–1954.

[6] As in some earlier theories, e.g., E. A. Calnan and C. J. B. Clews, *Phil. Mag.*, vol. 41, p. 1085, 1950; vol. 42, p. 616, 1951; vol. 42, p. 919, 1951.

[7] Y. C. Liu, *Trans. AIME*, vol. 230, p. 656, 1963.

than about 68° from the center of the pole figure and brings these (0001) poles near to the center, to which point the basal slip also rotates the grains, as in tension textures. The twinning mechanism assumed was {10$\bar{1}$2}. Slip also aligns [11$\bar{2}$0] with the rolling direction. Zinc, with $c/a > \sqrt{3}$, on the other hand, deforms differently. The theory proposed that grains rotate by slip until (0001) poles are within about 25° of the center of the pole figure, and that compression twins then move the grains out to positions near the circumference of the pole figure, followed by a renewal of rotation by slip. Williams and Eppelsheimer showed that the theory could be applied to cold-rolled titanium, and that with suitable assumptions regarding stress requirements for the known modes of deformation, the observed texture could be accounted for.[1] Slip was assumed on (10$\bar{1}$0) [11$\bar{2}$0] together with (10$\bar{1}$1)[11$\bar{2}$0] and (0001)[11$\bar{2}$0], and twinning was assumed on {10$\bar{1}$2} and {11$\bar{2}$2}. The rolling texture was seen as a combination of tension and compression textures such that tension in the rolling direction keeps one set of [10$\bar{1}$0] poles in the rolling direction while the remaining poles are free to rotate around that direction as an axis in response to the competing mechanisms of slip and twinning. Similar competition might be assumed for other metals of low c/a ratio (zirconium, hafnium, and beryllium).

We can anticipate that the attempts made thus far to understand deformation textures in the various metals will be superseded by still others, perhaps by some even before the reader sees this brief survey. Accordingly we have presented all theories in the past tense.[2]

Torsion textures The textures produced in a bar by twisting can be shown appropriately by a pole figure plotted on a plane tangent to the surface of the bar. Backofen has determined such plots for brass rods and for iron rods.[3] Three ideal orientations describe the *brass* texture in terms of the crystallographic plane parallel to the tangential plane of the rod and the direction parallel to the longitudinal direction: (11$\bar{1}$)[112], (112)[11$\bar{1}$], and (110)[001]. Each of these orientations provides a [110] slip direction along the transverse direction of maximum shear stress in the rod. The iron texture can be stated as a strong alignment of [110] directions longitudinally in the rod, which is limited in its range of spread around the rod axis, and with a (110)[1$\bar{1}$2] component superimposed. The textures are not the same as those found in cold-drawn wires and are neither well described by ideal textures nor understood in terms of fundamental principles. The textures do not appear to be altered if the direction of twisting is reserved.

Cold-drawn tubes Norton and Hiller[4] showed that in the cold reduction of seamless steel tubing the texture is determined by the relative reductions

[1] D. S. Williams and D. S. Eppelsheimer, *J. Inst. Metals*, vol. 81, p. 553, 1952–1953.

[2] The ASM symposium (see Ref. 2, p. 543) presents a new theory and data which corroborate it: Hu, Cline, and Goodman suggest that f.c.c. rolling textures are copper type if deformation faulting is absent or limited, but are brass type if faulting contributes substantially to the deformation, as it should when γ is small. The preferred orientation of active slip directions would be expected to cause (110) [112] crystals to widen more on rolling when γ is small than when it is large, if deformation faulting makes a significant contribution to the deformation when γ is small; such extra widening has been observed.

[3] W. A. Backofen, *Trans. AIME*, vol. 188, p. 1454, 1950. W. A. Backofen and B. B. Hundy, *Trans. AIME*, vol. 197, p. 61, 1953.

[4] J. T. Norton and R. E. Hiller, *Trans. AIME*, vol. 99, p. 190, 1932.

of wall thickness and circumference—thus by the magnitude of the principal strains. When only the wall thickness is reduced, the structure is identical to that of a rolled sheet, while when the wall thickness and circumference are reduced equally, the texture is that of a wire. Other variables in the commercial drawing operation are without effect on the principal orientations of the tubes.

Textures in deep drawing Textures in deep-drawn articles can be predicted, in general, from the magnitude of the three principal strains. Hermann and Sachs[1] found this to be the case in drawn-brass cups, where a pole-figure analysis showed that the texture varied from point to point in accordance with the varying nature of the deformation. At the center of the bottom of the cup the texture was identical with an ordinary compression texture and was caused by the thinning of the sheet and the radial flow outward in all directions, just as in a compression test. At the upper rim of the cup, on the other hand, there was a compression texture with the compression axis tangential. This texture was caused by the shortening of the circumference and the thickening of the sheet at the rim during the drawing operation.

An analysis of the plastic flow and orientations in a steel water pail[2] disclosed different conditions of strain but the same underlying correlation between the magnitudes of the principal strains and the nature of the textures. At the rim of the pail the circumference was reduced, and the steel was elongated vertically so that the deformation resembled that in a rolled sheet, provided the rolling direction was assumed to be vertical in the wall and the rolling plane to lie radially in the pail.

Textures after hot working Deformation at varying temperatures causes texture transitions in some f.c.c. metals, as mentioned earlier in this chapter. Texture changes of this kind are not always seen in b.c.c. metals, however, and would not be expected unless deformation modes changed with temperature. Low-carbon steel rolled 85 percent at 780°C was found to have a texture nearly the same as that of cold-rolled steel,[3] although rolling near 910°C produces a more nearly random texture. With almost any metal a change in texture can be expected if appreciable recrystallization occurs.

Beryllium,[4] zirconium,[5] and some titanium alloys[6] have been rolled at temperatures of 790°C without important changes in textures, although some change their textures when rolled at about 790°C and some at about 870°C. Titanium rolled at 565°C (1050°F) and at 790°C (1450°F) has the

[1] L. Hermann and G. Sachs, *Metallwirtschaft*, vol. 13, p. 745, 1934.

[2] Unpublished research by H. C. Arnold and C. S. Barrett.

[3] M. Gensamer and P. A. Vukmanic, *Trans. AIME*, vol. 125, p. 507, 1937.

[4] A. Smigelskas and C. S. Barrett, *Trans. AIME*, vol. 185, p. 149, 1949.

[5] R. McGeary and B. Lustman, *Trans. AIME*, vol. 191, p. 995, 1951.

[6] C. J. McHargue, J. R. Holland, and J. P. Hammond, *Trans. AIME*, vol. 206, p. 113, 1956.

usual (0001)[10$\bar{1}$0] texture but with considerable spread in both rolling and transverse directions[1]; the spread increases with increasing temperature.

Rolled rods of uranium have unchanged textures with rolling temperatures up to 300°C, provided reductions exceed 70 percent; the texture has a strong concentration of (010) and approximately (011) planes normal to the rod axis.[2]

[1] C. J. McHargue and J. P. Hammond, *Trans. AIME*, vol. 197, p. 57, 1953. C. J. McHargue, J. R. Holland, and J. P. Hammond, *Trans. AIME*, vol. 206, p. 113, 1956.

[2] G. B. Harris, *Phil. Mag.*, vol. 43, p. 113, 1952. M. H. Mueller et al., *Trans. AIME*, vol. 212, p. 793, 1958.

21

Preferred orientations after annealing

Textures resulting from annealing have been studied extensively both because of their influence on the directionality of properties in the finished products and because of their scientific interest. The literature on the subject has become voluminous. Although it was possible in a chapter of moderate size in the second edition of this book to summarize as of 1950 a major part of the significant research on annealing textures and to refer to the original papers in each instance, this is now impossible and has become unnecessary. Comprehensive summaries of the subject are now available. The large book by Wassermann and Grewen[1] includes more than 700 references to annealing textures and theories of these, and reproduces a majority of the available pole figures. A smaller book by Underwood also discusses many references.[2] The significance of experimental data in terms of theory has also been discussed in a comprehensive review by Beck[3] which includes many references and a vigorous defense of theories that stress the orientation dependence of growth rather than of nucleation.

It is our intent in the present chapter to introduce typical annealing textures, to highlight general principles that seem to have been established empirically, and to summarize the fundamentals of our (limited) understanding of the textures. The original papers to which we refer will be found to contain bibliographies of most of the prior work.

Variables and mechanisms affecting textures Some annealing textures are simple and sharply defined, and some are complex, with many components or with a nearly random distribution of grains. Many variables can have an influence on the nature of a texture: these include alloy composition, the mechanical and thermal history prior to the final annealing, the thermal history and sometimes even the atmosphere during the annealing process, the grain sizes at various stages of the working and annealing schedule, and the presence of nucleating agents.

The development of an annealing texture involves several fundamental mechanisms. An annealing texture may result from recovery without recrystallization (in which case it would be expected to duplicate the texture present before annealing), from primary recrystallization, or from grain growth subsequent to recrystallization. Grain growth may be "normal" ("continuous") with the grain-size distribution remaining normal through-

[1] G. Wassermann and J. Grewen, "Texturen metallischer Werkstoffe," Springer, Berlin, 1962.

[2] F. A. Underwood, "Textures in Metal Sheets," Macdonald, London, 1961.

[3] P. A. Beck, *Phil. Mag. Suppl.*, vol. 3, p. 245, 1954. See also Ref. 2, p. 543.

out the process, or a few grains may grow very large while the rest remain approximately unchanged until devoured by the large ones. The latter type of grain growth has been called "discontinuous," "abnormal," or "exaggerated," and the process has also been called "coarsening," "germination," or "secondary recrystallization." It occurs when normal grain growth is inhibited by particles of a dispersed phase or by a strongly developed texture. In general, an annealing texture may be said to result from a competition among recovery, primary recrystallization, and grain growth (normal or abnormal), and sometimes more than one of these processes influences the final result.

In an attempt to understand annealing textures, much attention has been directed toward finding the nature, location, and orientation of the nuclei of the recrystallizing grains, and to the factors that govern their growth into the cold-worked or recovered surrounding material and also their growth into other recrystallized grains if this occurs.

With so many factors and their interactions being involved to a different extent in different experiments, it is not surprising that alternative interpretations have been proposed throughout the history of research in this field, and that theories frequently rest on unproven or controversial assumptions, or seem to apply only under strictly limited conditions.

Recrystallized wires and compression specimens Many experiments on the alteration of fiber textures by annealing have failed to distinguish between recrystallization textures and textures that result from grain growth subsequent to annealing. Recrystallization temperatures are strongly influenced by small amounts of certain impurities; consequently, the mechanisms operating at a given temperature in a metal may depend upon the purity of the sample.

Face-centered cubic wires generally retain their deformation texture, [111] + [100], when recrystallized. Higher annealing temperatures generally tend to produce new textures and in some cases to increase randomness, particularly with aluminum wires of lower purity (98.7 to 99.6 percent) annealed above about 550°. Extruded rods of commercial aluminum and 99.95 percent aluminum, with [111] + [100] deformation texture, recrystallize at 350 to 550°C to grains forming a minor [111] component and a major component ranging from [103] to [113].[1] Secondary recrystallization occurs on annealing at 600°C in the purer aluminum and produces large grains belonging to the [103] to [113] component.

Copper develops various wire textures depending on the amount of reduction and the annealing temperature[2]; the recrystallization texture may be [100], [112], or random after a 400°C anneal, and the texture after grain growth at 950°C may be [112], [111], mixed, or random, depending upon the deformation texture. Similar complexities are reported in f.c.c. wires of other metals and alloys.

[1] K. V. Gow and R. W. Cahn, *Acta Met.*, vol. 1, p. 238, 1953. K. V. Gow, *Acta Met.*, vol. 2, p. 394, 1954.

[2] H. J. Wallbaum, *Z. Metallk.*, vol. 42, p. 207, 1951. G. Bassi, *Trans. AIME*, vol. 191, p. 533, 1951; vol. 194, pp. 515, 753, 1952.

Body-centered cubic wires, which have [110] deformation textures, retain these on primary recrystallization, but grains of other orientations are found after secondary recrystallization. Thus, tungsten may have secondary grains with orientations near [320], [321], [531], or [421] according to tests by different observers, and depending upon composition.[1] Iron, steel, and vanadium recrystallize with retention of the [110].

Close-packed hexagonal beryllium[2] retains its $[10\bar{1}0]$ texture in wires on recrystallization, but titanium[3] changes its $[10\bar{1}0]$ texture to one near $[11\bar{2}0]$ on annealing, as does zirconium.[4]

Uranium (orthorhombic, α), in the form of round rods rolled at 300°C and recrystallized, consists of a fiber texture with the rod axis near the (140) plane together with a component having the rod axis near (010).[5] Heating into the β range produces near-randomness.

The texture of aluminum after compression and after compression rolling is retained as a whole on recrystallization, but individual areas nevertheless alter their orientation.[6] Secondary recrystallization after compression rolling was found to produce a [111] texture in aluminum and a [110] texture in silver.[7]

Compression specimens of iron (b.c.c.) retain their major component, [111], but appear to lose the minor component, [100], on recrystallization at 580°C and 850°C.[8]

The cube texture in f.c.c. sheets The most remarkable recrystallization texture known is the *cube texture*, in which a cube plane lies parallel to the plane of the sheet and a cube edge is parallel to the rolling direction, i.e., (100)[001]. It is an extremely sharp texture when fully developed, as is seen from the pole figures of copper in Fig. 21–1. It resembles a single crystal with subgrains, but may contain a minor amount of material in twin relationship to the principal orientation.

This texture has been obtained in many f.c.c. materials, including the following (compositions in weight percent):

Cu	Cu–Al (0 to 0.2 percent Al)
Au	Cu–Cd (0 to 0.1 percent Cd)
Al	Cu-O (0 to 0.1 percent O_2)
Ni	Ni–Mn (0 to 1 percent Mn)
Cu–Zn (see below)	Fe–Ni–Cu (see below)
Th	Al–0.23 percent Fe–0.51 percent Si

[1] G. D. Rieck, *Acta Met.*, vol. 6, p. 360, 1958; vol. 9, p. 825, 1961.

[2] A. R. Kaufmann, P. Gordon, and D. W. Lillie, *Trans. Am. Soc. Metals*, vol. 42, p. 785, 1950. C. S. Barrett and A. Smigelskas, *Trans. AIME*, vol. 185, p. 149, 1949.

[3] C. J. McHargue and J. P. Hammond, *Trans. AIME*, vol. 197, p. 57, 1953.

[4] W. G. Burgers, J. D. Fast, and F. M. Jacobs, *Z. Metallk.*, vol. 29, p. 419, 1938.

[5] M. H. Mueller et al., *Trans. AIME*, vol. 212, p. 793, 1958.

[6] C. S. Barrett, *Trans. AIME*, vol. 137, p. 128, 1940.

[7] J. S. Bowles and W. Boas, *J. Inst. Metals*, vol. 74, p. 501, 1948.

[8] C. S. Barrett, "Structure of Metals," 2d ed., p. 490, McGraw-Hill, New York, 1952.

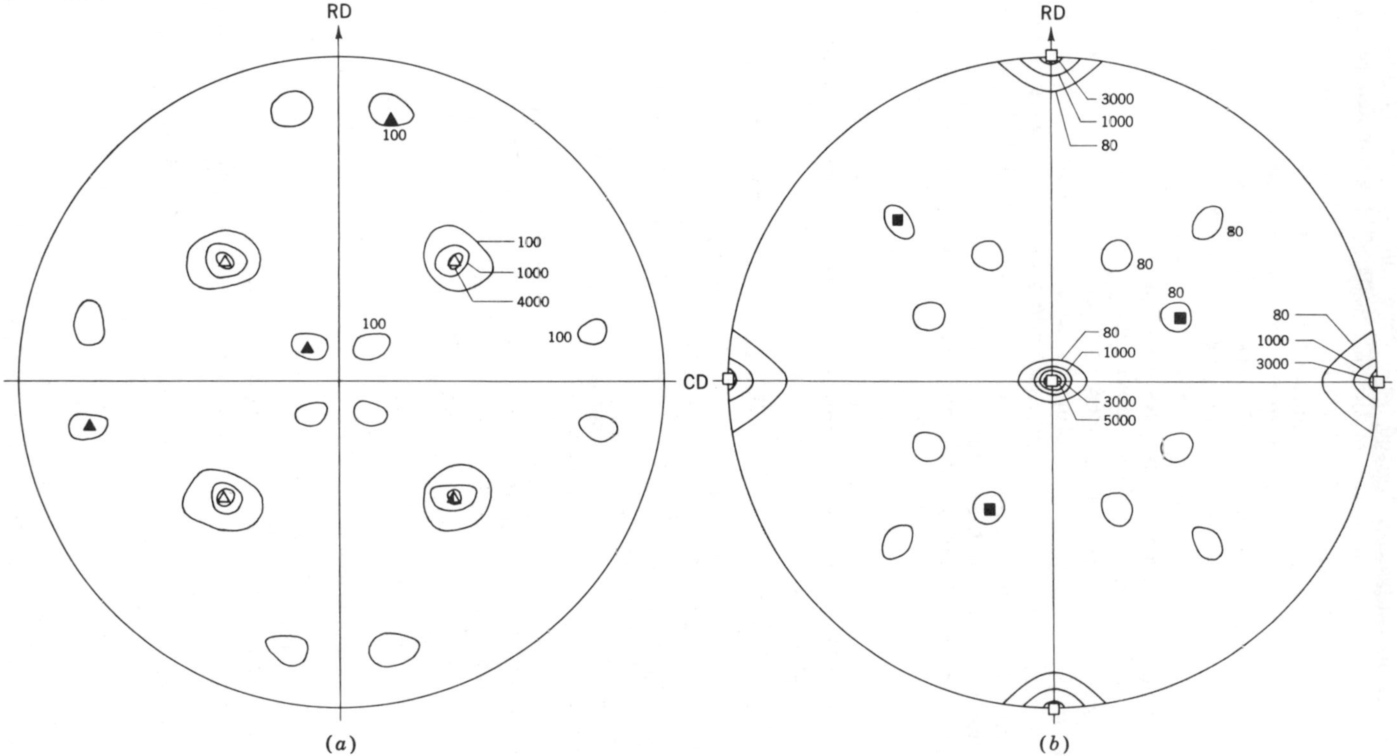

Fig. 21-1 (*a*) (111) pole figure for the annealing texture of tough pitch copper strip. Rolled as in Figs. 20-12 and 20-13 (p. 556), annealed 5 min at 200°C. Specimen 0.002 in. thick, from center portion of 0.020-in.-thick strip. △ (100)[001] "cube" orientation. ▲ (122)[21$\bar{2}$] "twin" orientation. The (111) pole common to both of these two textures is indicated by ◭. (*b*) (200) pole figure for the annealing texture of tough pitch copper strip. Same specimen as in *a*. □ (100)[001] ■ (122)[21$\bar{2}$]. (Courtesy P. A. Beck and H. Hu.)

The cube texture appears admixed with other orientations, including randomly oriented grains, in many of the materials of this list as well as others; and the percentage in (100)[001] orientation varies, in general, with processing variables as well as composition. A well-developed deformation texture must be obtained before recrystallization will give a well-developed cube texture; usually 80 to 95 percent cold reduction is required. The percentage in cube orientation in copper increases with annealing temperature and time[1] and is favored by having a high reduction prior to the final anneal, a small penultimate grain size, and a small thickness of the cold-rolled grains just prior to the final anneal.[2] To minimize the cube texture and the directionality in properties that it causes, intermediate anneals and small reductions are helpful, and also large penultimate grain size, low final annealing temperature, and addition of various alloying elements, particularly those that are known to change the rolling texture from the copper type to the brass type.[3]

Remarkably small additions of some alloying elements prevent or largely suppress the cube texture without causing major changes in the deformation textures[4]—for example, the following additions (in weight percent) suppress the cube texture: 5 percent Zn, 1 percent Sn, 4 percent Al, 0.5 percent Be, and 0.5 percent Cd, as well as additions (in atomic percent) of 0.0025 percent P, 0.3 percent Sb, 1.5 percent Mg, 4.2 percent Ni, 0.18 percent Cd, and 0.047 percent As. In general these elements differ markedly from copper in size, and cause strong solution-hardening. Müller[5] has found cube textures in Fe-Ni-Cu alloys at certain compositions.

Conditions that lead to maximizing the cube texture in aluminum are different from those in copper, and are dependent on impurity or alloying content. In commercial aluminum, increasing the rolling reduction tends to decrease the cube texture[6] and to cause retention of the rolling texture. The ratio of iron to silicon contents is found to be a critical variable.[7] Apparently, recovery is easier in aluminum than in copper and may completely replace recrystallization—especially on annealing after high reductions or with high iron contents.

Other f.c.c. annealing textures in sheets Annealing can produce various textures in f.c.c. metals: (1) the cube texture mentioned above, (2) a

[1] O. Dahl and F. Pawlek, *Z. Metallk.*, vol. 28, p. 266, 1936. P. A. Beck, *Trans. AIME*, vol. 191, p. 474, 1951.

[2] M. Cook and T. L. Richards, *J. Inst. Metals*, vol. 70, p. 159, 1944. W. H. Baldwin, *Trans. AIME*, vol. 166, p. 591, 1946.

[3] M. Cook and T. L. Richards, *J. Inst. Metals*, vol. 70, p. 159, 1944. H. Widmann, *Z. Physik*, vol. 45, p. 200, 1927. A. Merlini and P. A. Beck, *Trans. AIME*, vol. 203, pp. 385, 1267, 1955.

[4] O. Dahl and F. Pawlek, *Z. Metallk.*, vol. 28, p. 266, 1936. R. M. Brick, D. L. Martin, and R. P. Angier, *Trans. ASM*, vol. 31, p. 671, 1943. V. A. Phillips and A. Phillips, *J. Inst. Metals*, vol. 81, p. 185, 1952–1953.

[5] H. G. Müller, *Z. Metallk.*, vol. 31, p. 322, 1939.

[6] L. Frommer, *J. Inst. Metals*, vol. 66, p. 264, 1940.

[7] W. Bunk and P. Esslinger, *Z. Metallk.*, vol. 50, p. 278, 1959; *Metall*, vol. 13, p. 198, 1959.

texture approximately the same as the rolling texture, (3) one or more new types of textures, or (4) approximate randomness. The rolling texture of silver, often listed as $(110)[1\bar{1}2]$ and consisting of two twin-related components, changes into a new texture at low temperatures but becomes random with annealing above 800°C.[1] The new texture, sometimes designated as $(113)[21\bar{1}]$[2] and perhaps better as $(225)[73\bar{4}]$,[3] has been obtained also in various alloys with weight percentages as follows: Cu–33 percent Zn, Cu–5 percent Sn, Ag–1 percent Zn, Ag–30 percent Au. The primary recrystallization texture of silver is replaced upon long annealing at 433 to 533°C by secondary grains having the orientations of the deformation texture.[4]

Various observers[5] have noticed that a rotation of 30° around a [111] direction converts the rolling texture into the recrystallization texture and a 30° [111] rotation in the opposite direction converts the primary recrystallization texture into the secondary. It has been suggested, also,[6] that a $(358)[3\bar{5}2]$ component of the aluminum deformation texture converts to the cube-texture component and perhaps recrystallizes at a lower temperature than the $(110)[1\bar{1}2]$ component; and that alloying which suppresses the cube texture raises the recrystallization temperature of this $(358)[3\bar{5}2]$ component.

The recrystallization textures of brass,[7] Cu-Ge,[8] and Cu-Sn[8] gradually change with composition; pole figures for 70–30 brass are reproduced in Fig. 21–2.

Rolled aluminum partially retains its rolling texture on recrystallization, in addition to developing a cube texture. The retained component is due to "recrystallization in situ" (without orientation change) in some grains; other grains reorient, but they change to orientations already present in the texture as a whole.[9]

Recrystallization textures in b.c.c. sheets Recrystallized iron and steel sheet has a texture after large reductions and annealing between about 540 and 840°C as shown in Fig. 21–3 and described by Kurdjumov

[1] R. Glocker et al., *Z. Physik*, vol. 31, p. 386, 1925; *Z. Metallk.*, vol. 16, p. 377, 1925; vol. 17, p. 353, 1927.

[2] F. v. Göler and G. Sachs, *Z. Physik*, vol. 56, p. 485, 1929.

[3] P. A. Beck and H. Hu, *Trans. AIME*, vol. 194, p. 83, 1952.

[4] F. D. Rosi, B. H. Alexander, and C. A. Dube, *Trans. AIME*, vol. 194, p. 189, 1952.

[5] Ibid. M. Cook and T. L. Richards, *J. Inst. Metals*, vol. 69, p. 351, 1943. P. A. Beck, *Trans. AIME*, vol. 191, p. 474, 1951. M. L. Kronberg and F. H. Wilson, *Trans. AIME*, vol. 185, p. 501, 1949. Beck and Hu, loc. cit.

[6] Y. C. Liu and W. R. Hibbard, *Trans. AIME*, vol. 203, pp. 381, 1270, 1955.

[7] A. Merlini and P. A. Beck, *Trans. AIME*, vol. 203, p. 385, 1955. P. A. Beck and H. Hu, *Trans. AIME*, vol. 194, pp. 83, 1267, 1952 (15 texture components of five types are used to describe these textures).

[8] Y. C. Liu and R. H. Richman, *Trans. AIME*, vol. 218, p. 688, 1960.

[9] P. A. Beck and H. Hu, *Trans. AIME*, vol. 194, pp. 83, 1267, 1952.

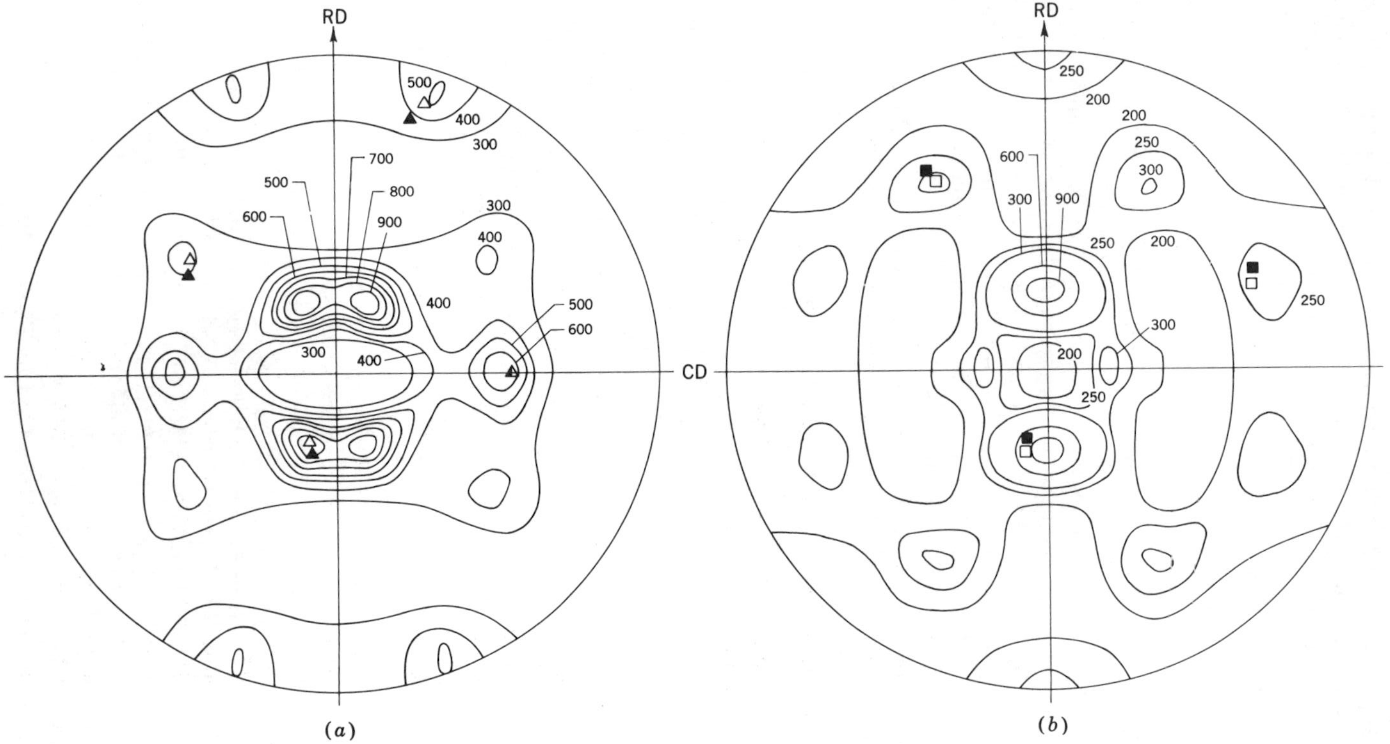

Fig. 21-2 (*a*) (111) pole figure for the annealing texture of strip of rolled commercial 70–30 brass. Rolled as in Figs. 20-14 and 20-15 (p. 558), annealed 5 min. at 340°C. Specimen 0.002 in. thick, from center portion of 0.024-in.-thick strip. Geiger-counter intensity values in arbitrary units indicated. △ (225)[73$\bar{4}$]. ▲ (113)[$\bar{2}$11]. (*b*) (200) pole figure for the annealing texture of rolled commercial 70–30 brass strip. Same specimen as in *a*. □ (225)[73$\bar{4}$]. ■ (113)[$\bar{2}$11]. (Courtesy P. A. Beck and H. Hu.)

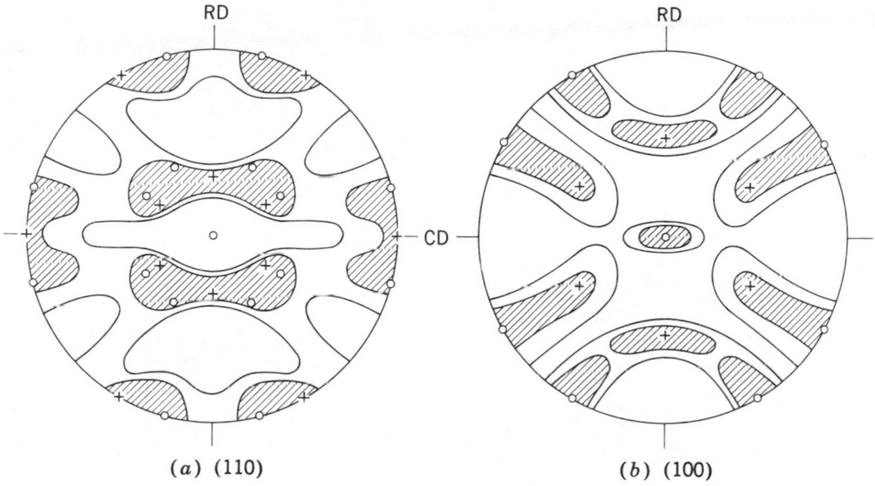

Fig. 21-3 Recrystallization texture of rolled iron. (○), position near (100)[011]; (+), (111)[11$\bar{2}$] position. (G. Kurdjumov and G. Sachs.)

and Sachs[1] in terms of three principal orientations:

1. Rolling plane (111), rolling direction [$\bar{2}$11]
2. Rolling plane (001), rolling direction 15° from [$\bar{1}$10]
3. Rolling plane (112), rolling direction 15° from [$\bar{1}$10]

The third is weak. A diffractometer study of carbonyl iron recrystallized at 540 to 600°C showed a progressively stronger texture as reductions were increased from 60 to 97.5 percent,[2] but with [110] clustered at the rolling direction (as in the rolling texture) rather than 15° each side of the rolling direction under most of the conditions used. Heating into the austenite range and cooling again tends to randomize the orientations. Both the recrystallization texture and the deformation texture of iron are very similar to those of vanadium,[3] iron-silicon alloys,[4] and a Zr-Cb alloy of b.c.c. structure.[5] Molybdenum,[6] on the other hand, was found to retain its deformation texture (with somewhat reduced spread). Tantalum[7] has a recrystallization texture represented by the single orientation 1 listed above, and tungsten annealed above 1800°C recrystallizes to orientation 2.[8]

[1] G. Kurdjumov and G. Sachs, *Z. Physik*, vol. 62, p. 592, 1930.

[2] F. Haessner and H. Weik, *Arch. Eisenhüttenw.*, vol. 27, p. 153, 1956.

[3] C. J. McHargue and J. P. Hammond, *Trans. AIME*, vol. 194, p. 745, 1952. Component 3 of the iron list, however, was not listed for vanadium.

[4] C. S. Barrett, G. Ansel, and R. F. Mehl, *Trans. AIME*, vol. 125, p. 516, 1937.

[5] J. H. Keeler, *Trans. AIME*, vol. 206, p. 122, 1956.

[6] M. Semchyshen and G. A. Timmons, *Trans. AIME*, vol. 194, p. 279, 1952.

[7] J. W. Pugh and W. R. Hibbard, *Trans. ASM*, vol. 48, p. 526, 1956.

[8] J. W. Pugh, *Trans. AIME*, vol. 212, p. 637, 1958.

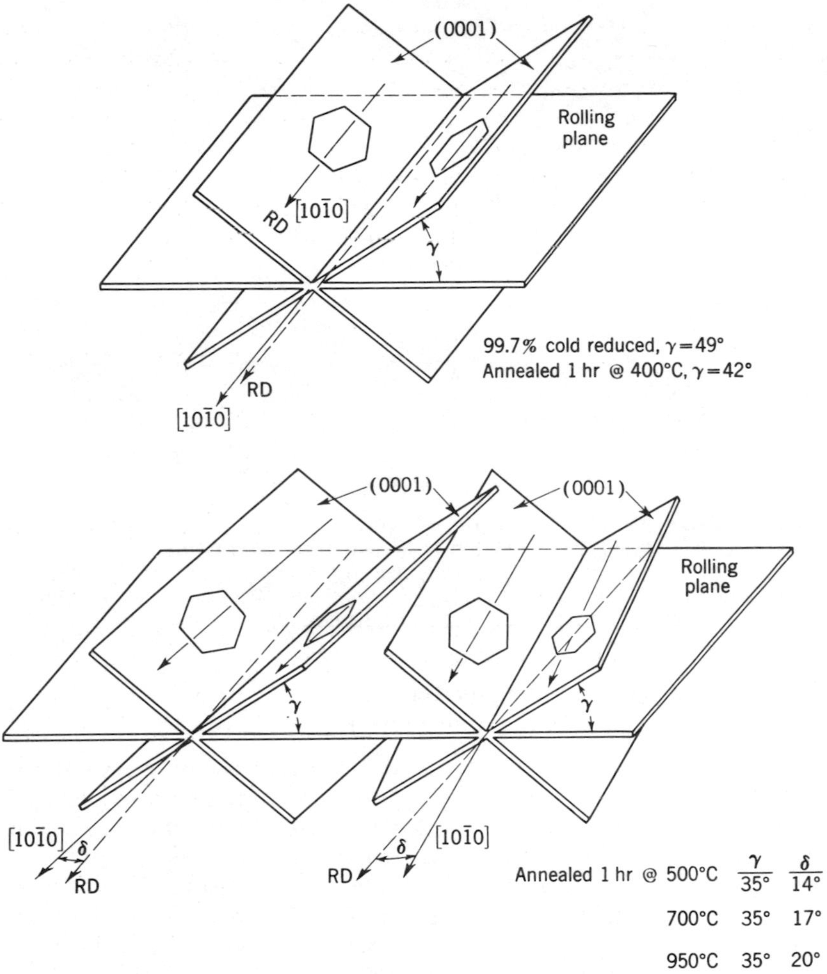

Fig. 21-4 Titanium sheet recrystallization orientations according to Keeler and Geisler. The [10$\bar{1}$0] direction was found to be rotated 14° from RD after annealing 1 hour at 500°C; a 20° rotation occurred after annealing 1 hour at 950°C.

Cross-rolled molybdenum retains the deformation texture on full recrystallization at 2400°C with a spread from the mean orientation of only 5°; compression-rolled molybdenum sheet also retains its deformation texture.[1]

Recrystallization textures of c.p.h. sheets and alpha-uranium Zinc and magnesium[2] retain their rolling texture on recrystallization, and the same is true of beryllium annealed at 700°C[3] and titanium annealed

[1] Semchyshen and Timmons, loc. cit.

[2] V. Caglioti and G. Sachs, *Metallwirtschaft*, vol. 11, p. 1, 1932.

[3] A. Smigelskas and C. S. Barrett, *Trans. AIME*, vol. 185, p. 145, 1949.

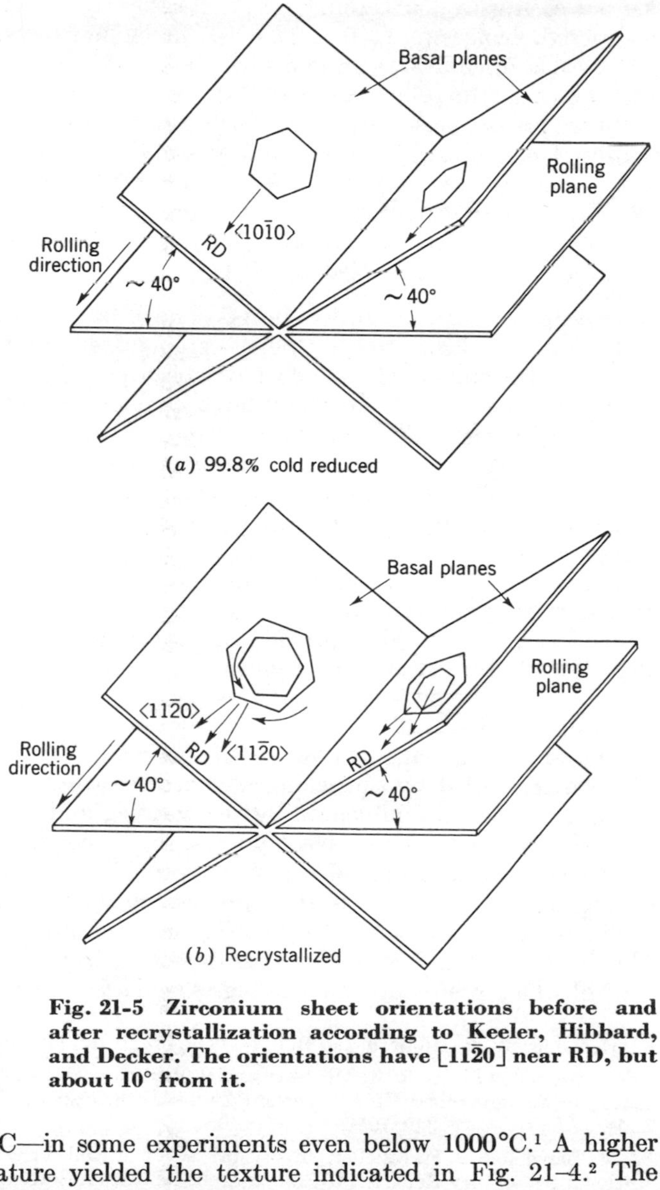

Fig. 21-5 Zirconium sheet orientations before and after recrystallization according to Keeler, Hibbard, and Decker. The orientations have [11$\bar{2}$0] near RD, but about 10° from it.

below about 500°C—in some experiments even below 1000°C.[1] A higher annealing temperature yielded the texture indicated in Fig. 21-4.[2] The texture after heating into the β range depends upon the temperatures reached, additional components being produced at 1000 to 1050°C that

[1] H. T. Clark, *Trans. AIME*, vol. 188, p. 1154, 1950. C. J. McHargue and J. P. Hammond, *Trans. AIME*, vol. 197, p. 57, 1953. J. H. Keeler and A. H. Geisler, *Trans. AIME*, vol. 206, p. 80, 1956.

[2] Three components are listed by D. N. Williams and D. S. Eppelsheimer, *Z. Metallk.*, vol. 44, p. 360, 1953: (1) (0001) [10$\bar{1}$0] rotated 30° around RD, (2) (0001) [11$\bar{2}$0] rotated 30° around RD, (3) (0001) [10$\bar{1}$0] rotated 30° around RD and then 20° around ND.

are not seen with 950°C annealing, and secondary recrystallization causes additional changes at 1150 to 1200°C. Zirconium, recrystallized at 400 to 600°C, has orientations as indicated in Fig. 21–5[1]; [11$\bar{2}$0] is rotated to positions near the rolling direction.[2]

When precautions were taken in rolling alpha-uranium to develop a uniform small grain size prior to a heavy rolling reduction at room temperature, the textures observed after complete recrystallization at 380 and 400°C were found to be similar to the rolled texture, but some alteration in texture occurred upon annealing at 450°C; secondary recrystallization was found in specimens annealed at 600°C.[3]

Sheets for magnetic applications The [100] direction in crystals of alpha-iron or silicon-steel is the direction of easiest magnetization, i.e., highest permeability. Great effort has been expended in developing steel for transformer cores that would have as many as possible of the grains with [100] in the rolling direction or (100) in the rolling plane.

Goss[4] in 1935 developed a method of producing a desirable (110)[001] texture in steel for transformers by using cycles of cold rolling and annealing. Processing variables are critical for getting the maximum number of grains in "cube-on-edge" orientation (110)[001], and the minimum number in undesirable orientations such as the major component of the ordinary primary recrystallization texture, the "cube-on-point" orientation (111)[$\bar{2}$11].[5] A desirable primary recrystallization structure is needed, which should consist of small grains. This is obtained by having a suitable distribution of second-phase particles such as manganese sulfide, silica, or vanadium nitride, which serve as grain-growth inhibitors (or perhaps by having grain-boundary grooves in the surface). The particle distribution may be controlled by quenching and heat-treating, and the orientations of the primaries are influenced by the working and annealing cycles. The desired texture is then accomplished by secondary recrystallization, (110)[001] grains growing to a size 10 to 100 times the sheet thickness during annealing in suitable atmospheres and temperature ranges.

The textures at various stages for various compositions are discussed by May and Turnbull.[6] Surface energy is one of the important factors that control which grains become the big secondaries.[7] With certain annealing

[1] J. H. Keeler, W. R. Hibbard, and B. F. Decker, *Trans. AIME*, vol. 197, p. 932, 1953.

[2] For cross-rolled Zr, see R. K. McGeary and B. Lustman, *Trans. AIME*, vol. 191, p. 995, 1951. For straight-rolled Hf + 3 percent Zr, see D. S. Eppelsheimer and D. S. Gould, *Z. Metallk.*, vol. 48, p. 349, 1957.

[3] L. T. Lloyd and M. H. Mueller, "Recrystallization in Rolled Uranium Sheet," Argonne National Laboratory Rept. ANL–6327, March, 1962. Quantitative pole figures are given

[4] N. P. Goss, *Trans. ASM*, vol. 23, p. 511, 1935; United States Patent 1,965,559.

[5] C. G. Dunn, in "Cold Working of Metals," p. 113, American Society for Metals Cleveland, Ohio, 1949. Some United States patents dealing with this are 2,112,084 2,287,466; 2,158,065.

[6] J. E. May and D. Turnbull, *Trans. AIME*, vol. 212, p. 769, 1958.

[7] Ibid. K. Detert, *Acta Met.*, vol. 7, p. 589, 1959. D. Kohler, *J. Appl. Phys.*, vol. 31, p 408S, 1960.

atmospheres (e.g., the presence of some oxygen) a grain with a (110) plane in the plane of the surface appears to have lower energy than one with (100) in the surface,[1] and with another atmosphere (for example, vacuum) this relationship is reversed so that (100) grains are favored. Since atmosphere compositions are critical, it has been proposed that adsorbed layers on the surface (for example, sulfur) can play a role in controlling the relative surface energies,[2] but not all workers accept this view. Although the processing requirements are rigid, it is possible to produce sheet commercially with more than 90 percent of the grains in proper orientation.

A highly oriented (100)[001] cube-texture sheet would also be very desirable, since it would have desirable magnetic qualities even when the magnetic flux was not unidirectional in the sheet. Although Sixtus[3] identified the (100)[001] component as early as 1935, not until more than twenty years later was a material with over 90 percent cube texture announced.[4] Information has now been obtained on how to favor the growth of (100)[001] secondaries,[5] but the best results appear to be limited to sheets a few thousandths of an inch thick, and the processing is expensive.

Theories of annealing textures Two principal types of mechanisms have been proposed to account for annealing textures. One involves *oriented nucleation*, and the other involves *oriented growth*. The oriented-nucleation theory rested on the hypothesis that the orientation of the recrystallized grains is determined entirely by the orientations of the nuclei for recrystallization.[6] However, a general principle for predicting which orientations of the many that are present will act as nuclei has not been obvious either from theory or experiment.

It was proposed[7] that theories might be based on the principle that certain orientations grow into the deformation-texture material more rapidly than others, which means that the shifting of atoms from the strained matrix to the recrystallized grain proceeds only slowly when the new grain has nearly the same orientation as the cold-worked matrix, and fastest when its orientation differs in certain ways from the orientation of the matrix. Beck[8] and coworkers have adopted this oriented-growth point of view for all textures except those produced in certain experiments in-

[1] C. G. Dunn and J. L. Walter, *Acta Met.*, vol. 7, p. 648, 1959; *Trans. AIME*, vol. 224, p. 518, 1962; vol. 227, p. 185, 1963. J. L. Walter, *J. Appl. Phys.*, vol. 36, p. 1213, 1965.

[2] Kohler, loc. cit.

[3] K. J. Sixtus, *Physica*, vol. 6, p. 105, 1935.

[4] F. Assmus, K. Detert, and G. Ibe, *Z. Metallk.*, vol. 48, p. 344, 1957. G. Wiener et al., *J. Appl. Phys.*, vol. 29, p. 366, 1958.

[5] R. G. Aspden, *Trans. AIME*, vol. 227, p. 905, 1963.

[6] W. G. Burgers and P. C. Louwerse, *Z. Physik*, vol. 67, p. 605, 1931. W. G. Burgers, "L'état solide," p. 73, Institute International de Physique, Brussels, 1951; *Acta Met.*, vol. 1, p. 234, 1953.

[7] C. S. Barrett, *Trans. AIME*, vol. 137, p. 128, 1940.

[8] P. A. Beck, *Phil. Mag.*, Suppl. (Advan. Phys.), vol. 3, p. 245, 1954.

volving deformed single crystals, in which there is a scarcity of nuclei in orientations favorable for growth. Much experimental evidence can be interpreted by an oriented-growth type of theory. A number of fundamental experiments have indicated that grain-boundary mobility is very small when the boundary is a twin boundary, or when a grain is growing into a fine-grained matrix with essentially a single-crystal texture of orientation similar to the growing grain. In general, low-energy boundaries between a recrystallized grain and the adjacent cold-worked metal have low mobility, and high-energy boundaries have high mobility.

In f.c.c. metals, low interface energies occur at 0 and 60° rotations around [111]; high energies occur at intermediate rotations; a 30 to 40° rotation around [111] is found to favor high boundary mobility.[1]

If it is assumed that the rate-determining process in boundary migration resembles self-diffusion along the boundary,[2] a correlation with grain-boundary energy may be accounted for, since boundary self-diffusion is faster in high-energy boundaries than in those of low energy.

Aust and Rutter have found that the relative rates of grain-boundary migration under a given (small) driving force depend strongly on orientation relationships, with certain "special boundaries" moving orders of magnitude faster than "random boundaries"—provided solute concentrations are within certain limits with certain solutes (e.g., some tens of parts per million of tin in zone-refined lead).[3] They suggest that the lack of exceptionally high mobilities of special boundaries with other solutes may be related to the magnitude of the interaction between solute and boundary, which varies from one alloy system to another. However, various other factors also are involved,[4] and the significance of special boundaries for a *general* theory is uncertain.

When a deformed metal has a fairly sharp single-orientation texture, the oriented-growth theory in its simplest form would predict the formation of an annealing texture consisting of all crystallographically equivalent orientations corresponding to maximum boundary mobility. In actuality, the complete set of equivalent orientations is not generally observed. Therefore, the oriented-growth theory is forced to rely on *ad hoc* assumptions regarding what orientations will occur and for what reasons.[5] This

[1] M. L. Kronberg and F. H. Wilson, *Trans. AIME*, vol. 185, p. 501, 1949. P. A. Beck and H. Hu, *Trans. AIME*, vol. 185, p. 627, 1949. G. W. Rathenau and J. F. H. Custers, *Philips Res. Rept.*, vol. 4, p. 241, 1949. P. A. Beck, P. R. Sperry, and H. Hu, *J. Appl. Phys.*, vol. 21, p. 420, 1950. See also Ref. 2, p. 543.

[2] S. Kohara, M. N. Parthasarathi, and P. A. Beck, *J. Appl. Phys.*, vol. 29, p. 1125, 1958.

[3] K. T. Aust and J. W. Rutter, *Trans. AIME*, vol. 218, pp. 50, 686, 1960. Some special boundaries were studied by M. L. Kronberg and F. H. Wilson, *Trans. AIME*, vol. 185, p. 501, 1949; their movement would require only minor atom shifts as the boundary moves past them.

[4] B. Liebmann and K. Lücke, *Trans. AIME*, vol. 206, p. 1413, 1956.

[5] J. E. Burke and D. Turnbull in "Progress in Metal Physics," vol. 3, p. 220, Interscience, New York, 1952.

and other criticisms have been reviewed and discussed by Beck[1] and by Burgers and Tiedema.[2,3]

From the researches of Dunn on silicon-iron and of Crussard on recrystallization in situ and from other studies on recovery and polygonization, it is clear that texture theory must also recognize the *retention* of deformation texture in some specimens or in certain grains. Electron microscopy shows that the mechanism for this involves polygonization and subgrain coalescence without the nucleation of newly oriented grains. There is also, however, the possibility of retaining some textures almost unchanged during recrystallization even though *individual* areas alter their orientation.[4]

When annealing textures are a function of the annealing temperature, it is often true that the change is due to grain growth becoming important at the higher temperatures. There is also the possibility that different texture components recrystallize at different temperatures; thus, an annealing texture may be a combination of several components, some of which are recrystallized and some not, for it is known that individual crystals can have wide differences in recrystallization tendency after a given amount of deformation.[5] Transmission electron microscopy, together with selected-area electron diffraction, has shown the importance of deformation bands and the narrow subdivisions of these bands, *microbands* or *transition-band regions*, which become nuclei for recrystallization when the boundary of one of these microbands migrates or when adjacent microbands coalesce in a way that develops high-angle, highly mobile boundaries.[6] In general, crystals that have a sharp, single-orientation deformation texture tend to recover rather than to recrystallize, and therefore to retain their deformation texture, at least until they are consumed by differently oriented grains. A low annealing temperature favors retention of the cold-rolled texture,[7] at least in iron, and the precipitation of copper in cold-rolled iron after cold-rolling and before the final annealing causes a retention of the deformation texture.[8] Probably the effects of alloying additions in various other cases also should be ascribed to the presence of precipitates rather than to the solid-solution effects.

In general, we may conclude that various factors may be important in

[1] P. A. Beck, *Phil. Mag.*, Suppl. (Advan. Phys.), vol. 3, p. 245, 1954; *Acta Met.*, vol. 1, p. 230, 1953.

[2] W. G. Burgers and T. J. Tiedema, *Acta Met.*, vol. 1, p. 234, 1953.

[3] For recent critical discussion see Ref. 2, p. 543.

[4] C. S. Barrett, *Trans. AIME*, vol. 137, p. 128, 1940. P. A. Beck and H. Hu, *Trans. AIME*, vol. 185, p. 627, 1949.

[5] B. F. Decker and D. Harker, *J. Appl. Phys.*, vol. 22, p. 900, 1951. P. K. Koh and C. G. Dunn, *Trans. AIME*, vol. 203, p. 401, 1955; vol. 206, p. 1017, 1956. H. Hu, *Trans. AIME*, vol. 209, p. 1164, 1957; vol. 215, p. 320, 1959. H. Hu in "Recovery and Recrystallization of Metals," p. 311, Interscience, New York, 1964.

[6] H. Hu, in "Recovery and Recrystallization of Metals," *loc. cit.*, p. 364.

[7] F. Haessner and H. Weik, *Arch. Eisenhüttenw.*, vol. 27, p. 1953, 1956.

[8] W. C. Leslie, *Trans. AIME*, vol. 221, p. 752, 1961.

any specific instance, including the orientation dependence of grain-boundary mobility and of surface energy, and the tendencies of grains with certain orientations to deform with deformation bands or microbands, or to polygonize rather than recrystallize when annealed.

The *driving forces* for boundary movement are (1) a reduction in strain energy in the material swept over by a boundary, (2) a reduction in boundary energy from a reduction in boundary area, from a reduction in energy per unit area, or from both, and (3) in some cases a reduction in energy per unit area exposed at the surface of the material. Factors that *inhibit* boundary movement and that may control its rate include (1) diffusion rates at the moving boundary, (2) diffusion rates for solute atmospheres that may be dragged along by a (slow-moving) boundary, (3) grain-growth interference caused by particles of a second phase, and (4) grain-growth interference because of the formation of grooves on the surface.

A theory of the cube texture in f.c.c. metals It should be easier to understand the development of the sharp, single-orientation cube texture than to understand almost any other more complicated one. A theory that rests on a simple basis has, in fact, been proposed.[1] It is that conditions are favorable for the formation of the cube texture in a rolled sheet if the recrystallized grains at an early stage of the annealing process reach a size at which they come into effective contact with several (or all four) principal orientations of the deformation texture. Since these grains of the cube-texture orientation are related to each of these four deformation texture components by a [111] rotation suited to high boundary mobility, they can grow faster than any grains with other orientations, and can grow into all four of these components.

The increase in the amount of cube-oriented material with increasing annealing temperature, which is concurrent with grain growth after recrystallization is complete, can be ascribed to a similar favorable orientation of the cube-oriented grains with respect to the recrystallization-texture components of the type $(123)[\bar{4}\bar{1}2]$. The driving force for growth is ascribed to the cube-oriented grains being larger than the others and therefore having boundaries that are curved to favor continued growth at the expense of the others.

Oriented-nucleation theories of the cube texture are not supported by the selected-area electron-diffraction tests of Haessner et al.[2] Copper was rolled to a reduction of 95 percent, thin specimens were prepared, and orientations were surveyed in 600 regularly spaced small areas. Although the individual orientations described the complete cold-rolling texture, not one pattern showed the cube orientation.

Theories of textures in silicon-steel A very important effect of varying the annealing temperature is found in silicon-iron and commercial

[1] P. A. Beck, *Trans. AIME*, vol. 191, p. 474, 1951. T. J. Koppenaal, M. N. Parthasarathi and P. A. Beck, *Trans. AIME*, vol. 218, p. 98, 1960.

[2] F. Haessner, U. Jakubowski, and M. Wilkens, *Phys. Stat. Sol.*, vol. 7, p. 701, 1964, and private communication.

silicon-steel. The highly oriented (110)[001] texture (cube-on-edge texture), desirable for its magnetic properties, is a result of secondary recrystallization when it occurs in a certain temperature range and with a grain-growth inhibitor present.[1] The primary recrystallization texture is much more complex, as is also a special texture obtained by rapid heating to 1250°C. The simple (110)[001] secondary recrystallization texture is obtained if heating is slow and grain growth is inhibited. It results from the abnormal growth of a limited number of grains at temperatures near 900°C.

Dunn[2] has proposed that secondary recrystallization textures are best understood on the basis of an "oriented-nucleation, growth-selectivity" theory. He concluded that large primary recrystallized grains of a minor component of the recrystallization texture are favored in growth and consume the major component. The mechanism for this in silicon-steel, as proposed by May and Turnbull,[3] is as follows: The larger of the (110)[001] grains of the primary recrystallization texture grows at the expense of the other primaries, which remain small until consumed, owing to the presence of second-phase inclusions that inhibit their growth. The (110)[001] grains not only grow faster than the grains of other orientations, but they also are the first to surpass the critical size that makes them act as nuclei for secondary recrystallization. (Their boundaries at this critical size develop enough contacts with the surrounding grains and become curved enough for the secondaries to grow at the expense of the surrounding primaries.) These (110)[001] grains are presumed to have lower free energy than grains of other orientations (a lower total for the sum of surface energy and grain-boundary energy as well as residual strain energy[4]).

Some tests indicate that *surface energy* is a controlling factor.[5] On the other hand, Philip and Lenhart[6] found evidence that in (110)[001] commercial sheet the nuclei of the secondary grains are usually located two or three primary-grain diameters *below the surface*, which suggests that at an early stage it is not principally the surface free energy that favors the (110)[001] grains. And certain studies of growth rates of (100) grains in sheets of different thicknesses have led to the conclusion that the driving force for growth is chiefly *grain-boundary* energy rather than surface energy.[7]

[1] C. G. Dunn, in "Cold Working of Metals," p. 113, American Society for Metals Cleveland, Ohio, 1949. J. E. May and D. Turnbull, *Trans. AIME*, vol. 212, p. 769, 1958 P. K. Koh and C. G. Dunn, *Trans. AIME*, vol. 218, p. 65, 1960. H. C. Fielder, *Trans AIME*, vol. 227, p. 777, 1963.

[2] C. G. Dunn, *Acta Met.*, vol. 2, p. 173, 1954.

[3] J. E. May and D. Turnbull, *Trans. AIME*, vol. 212, p. 769, 1958.

[4] C. G. Dunn and E. F. Koch, *Acta Met.*, vol. 5, p. 548, 1957.

[5] J. L. Walter, *Acta Met.*, vol. 7, p. 425, 1959. J. J. Kramer and K. Foster, private communication, 1964.

[6] T. V. Philip and R. E. Lenhart, *Trans. AIME*, vol. 221, p. 439, 1961.

[7] K. Foster, J. J. Kramer, and G. W. Wiener, *Trans. AIME*, vol. 227, p. 185, 1963.

22

Electron diffraction

The use of electrons for diffraction purposes dates from 1927 when Davisson and Germer[1] first showed that it was possible, after de Broglie had predicted, in 1924, that material particles should act as waves.[2] A few months after Davisson and Germer's preliminary report of diffraction of low-velocity electrons from the surface of a nickel crystal, G. P. Thomson and A. Reid reported the diffraction of high-velocity electrons by transmission through thin films, and quantitatively confirmed de Broglie's predictions.[3] Electron diffraction quickly proved its worth as a practical diffraction method not only with thin transmission specimens but also with polycrystalline bulk specimens.[4] The voltage range most convenient and useful proved to be the range 30 to 100 kv, but some work is done at higher voltages, and recently the range of voltages (near 100 volts) originally used by Davisson and Germer is again being used, but the electrons are now accelerated after they are reflected.

After a number of years in which electron diffraction cameras were modified but not changed radically in design, electron microscopes were developed which could be operated as diffraction cameras. These permit an investigator to obtain a diffraction pattern of the same minute crystallite that is seen at high magnification in the electron microscope image of the same specimen. The possibility these instruments offer for determining the identity and orientation of a crystallite in the submicron range, and also of producing diffraction patterns of polycrystalline samples and powders, made electron diffraction a tool of great value in many fields of research.[5]

[1] C. J. Davisson and L. H. Germer, *Nature*, vol. 119, p. 558, 1927; *Phys. Rev.*, vol. 30, p. 705, 1927.

[2] L. de Broglie, *Phil. Mag.*, vol. 47, p. 446, 1924.

[3] G. P. Thomson and A. Reid, *Nature*, vol. 119, p. 890, 1927. G. P. Thomson, *Proc. Roy. Soc. (London)*, vol. A117, p. 600, 1928.

[4] K. Matukawa and K. Shinohara, *Proc. Phys. Math. Soc. Japan*, vol. 12, p. 171, 1930. G. P. Thomson, *Proc. Roy. Soc. (London)*, vol. A128, p. 649, 1930.

[5] G. P. Thomson and W. Cochrane, "Theory and Practice of Electron Diffraction," Macmillan, London, 1939. H. Raether, "Electroneninterferenzen und Ihre Anwendung" in "Ergebnisse der Exacten Naturwissenschaften," vol. 24, p. 54, Springer, Berlin, 1951. C. F. Meyer, "The Diffraction of Light, X-Rays and Material Particles," Edwards, Ann Arbor, Mich., 1949. Z. G. Pinsker, "Electron Diffraction," Butterworth, London, 1953 (includes structure analysis of crystalline and amorphous solids and the structure of molecules). H. J. Yearian in "Methods of Experimental Physics," vol. 6, Solid State Physics, Part A, Academic, New York, 1959 (a convenient 32-page treatment of funda-

There is an intimate relationship between electron diffraction and the origin of contrast in transmission electron micrographs, as mentioned in Chap. 15; however, to account for this relationship in full detail it is necessary to use the dynamical theory, as is explained in Heidenreich's book and elsewhere. A simpler treatment of diffraction, which is based on kinematic theory, will serve our purpose here.

Electron waves The wave nature of electrons, introduced in Chap. 12 (page 311), accounts for their diffraction. Electrons interact with the periodic field in a crystal to produce diffracted beams as if they had a wavelength λ specified by the relation $\lambda = h/mv$ where h is Planck's constant and mv is the momentum (the mass times the velocity of the electron; v is the group velocity of the wave packet associated with the electron). The kinetic energy of an electron when it is accelerated by the potential applied to a vacuum tube in which it moves is $mv^2/2 = eV/300$ where V is the accelerating potential in volts and e is the charge in electrostatic units. From those relationships it follows that $\lambda = \sqrt{150}/\sqrt{V}$. However, this formula requires correction because the mass of an electron is not a constant but depends on the velocity of the electron, v, according to the relation $m = m_0/(1 - v^2/c^2)^{1/2}$, where m_0 is the mass of the electron at rest and c is the velocity of light. Relativity theory gives the correct expression for the wavelength: $\lambda = h/[2m_0eV(1 + eV/2m_0c^2)]^{1/2}$. The relativistically correct λ differs from the λ of the simpler formula by 5 percent at 100 kv. Some corrected values of electron wavelengths are as follows[6]:

V in volts	λ in Angstroms	V in volts	λ in Angstroms
10	3.878	50,000	0.05355
100	1.226	70,000	0.04485
500	0.5479	100,000	0.03702
1,000	0.3877	300,000	0.01968
10,000	0.1220	1,000,000	0.008720

It will be seen that in the range of voltages most used in electron diffraction, 50 to 100 kv, the wavelengths are much smaller than those used in x-ray diffraction. Diffraction of atoms, molecules, and neutrons is also possible because these and all material particles have the property of acting as waves, but of these only the diffraction of neutrons is of practical value (see Chap. 23).

mentals and techniques). K. J. Marsh and A. G. Quarrell in B. Chalmers and A. G. Quarrell (eds.), "The Physical Examination of Metals," 2d ed., Arnold, London, 1960 (chiefly metallurgical applications). R. D. Heidenreich, "Fundamentals of Transmission Electron Microscopy," Interscience, New York, 1964 (a thorough treatment of diffraction as it relates to microscopy).

[6] A larger table is given in the book by Heidenreich, op. cit., p. 393.

Apparatus Electron diffraction requires a highly evacuated container in which a collimated stream of electrons is accelerated through a constant potential and strikes a specimen. Some early cameras used a gas discharge as an electron source, but these have been superseded by cameras that use an electron gun containing a hot filament. The diffraction pattern is recorded photographically (except in certain low-velocity experiments), and may be viewed on a fluorescent screen.

In the earlier types of instruments the diffracted beams traveled in straight paths of 20 to 60 cm from specimen to photographic plate, and the apparatus was analogous to an x-ray camera with a flat film. In electron microscopes, however, the beams pass through a series of magnetic lenses which can vary the size of the diffraction pattern. To read the angle of diffraction from the position of any spot on the photographic plate therefore requires calibration of the magnification of the pattern for the lens settings used in recording the pattern. In a simple camera without lenses, if a diffracted beam at an angle $2\theta°$ from the incident beam strikes a plate at a distance L mm from the specimen, it will produce a spot R mm from the spot made by the direct beam, and the diffraction angle will be given by the relation $\tan 2\theta = R/L$. Since the short wavelengths in high-voltage diffraction make θ very small, it will be accurate to within about $\frac{1}{2}$ percent to set $2\theta = R/L$ and $\sin \theta = \theta$. The Bragg relation for interplanar spacing d_{hkl} may then be put in the form

$$d_{hkl} = \frac{\lambda L}{R_{hkl}} \tag{22–1}$$

The effective camera constant L is best determined by calibrating it with a standard specimen, regardless of the type of camera. The linear relation between d and R does not hold as accurately in cameras containing lenses, and the effective L does not remain as constant in these as it does in cameras without lenses.

The diffraction pattern can be formed by transmission through a thin specimen, or by reflection from the surface of a thick sample. The reflection method gives relatively higher background intensities, often gives broader diffracted beams, involves greater uncertainties in the effective value of L, and requires a surface that is rough on a very fine scale so that electrons can penetrate through the small projections on the surface. The specimen surface must be inclined at a very small angle to the primary beam to avoid blocking out too much of the pattern by absorption, an adjustment more easily done in a simple camera than in an electron microscope.

When the diffraction is done in an electron microscope of the usual design (Fig. 15–8, page 431), the focal length of the intermediate lens is adjusted differently than for microscopy. The intermediate lens is focused so as to form a sharp diffraction pattern at the intermediate image plane. The pattern is then enlarged by the projection lens. The selected-area aperture is placed in the image plane of the objective lens, where it passes the rays forming the portion of the image that is wanted for the diffraction pattern.

Some diffraction may come from peripheral areas outside the selected-area aperture (because of spherical aberration of the objective lens); this should be kept in mind in planning exposures. Various details concerned with adjustments for selected-area diffraction have been presented by Agar and Phillips.[1]

The diffraction pattern *Polycrystalline samples* of very small grain size produce concentric rings analogous to the Debye rings of the powder method of x-ray diffraction, and the same principles are used in assigning indices. Equation (22–1) permits very rapid indexing. For a cubic crystal the relations given in Chap. 4 for d combined with Eq. (22–1) give the relation $\sqrt{h^2 + k^2 + l^2} = a_0 R / L\lambda$, and analogous relations are easily derived for the noncubic lattices. The absences discussed in earlier chapters for f.c.c., b.c.c., and other lattices also hold, although the intensities of reflections are different than with x-rays.

Reflection patterns from magnetic materials are distorted by the magnetic field of the sample. To obtain suitable patterns from these samples, it is necessary to demagnetize them. The presence of a finely divided ferromagnetic precipitate in an otherwise nonmagnetic alloy can also affect the diffraction pattern.[2] Distortion can be introduced also by the electrostatic field around a sample if the sample is charged up by the electron beam. To avoid this an electron gun can be used to spray 400- to 1200-volt electrons at the specimen and to discharge the surface by secondary emission.

Single-crystal patterns are best interpreted with the aid of the reciprocal lattice (introduced in Chap. 4). With the short wavelengths used in high-voltage electron diffraction, the radius of the reflection sphere is very large. The sphere therefore deviates very little from a plane, and the pattern rather closely approximates a plane section of the reciprocal lattice. The plane-section appearance is enhanced by the fact that more reflections occur than would be expected from the intersection of a true sphere with the geometrical points of a reciprocal lattice, because the electron beam is not strictly monochromatic and the lattice points spread out, as discussed in Chap. 16, into rods of small volume as a result of an orientation range, crystal imperfections, thermal motion, or particle-size broadening in one or more directions. Particle-size broadening causes each reciprocal lattice point to be elongated in any direction for which a crystal lacks resolving power, as discussed on page 450, each point becoming a body whose dimension in any direction is reciprocal to the actual dimension in the crystal.

Indices are assigned to the spots and the lattice orientation is determined by inspection (see page 432, Chap. 15), by comparing the pattern with a model of the reciprocal lattice, or by comparing the d spacings obtained from Eq. (22–1) with a list of computed spacings for the specimen.

[1] A. W. Agar and R. Phillips, *Brit. J. Appl. Phys.*, vol. 11, pp. 185, 504, 1960.

[2] R. D. Heidenreich and E. A. Nesbit, *J. Appl. Phys.*, vol. 23, p. 325, 1952. Magnetic fields in a crystal can be mapped out by an electron beam: M. Blackman and N. D. Lisgarten, *Phil. Mag.*, vol. 3, p. 1609, 1958.

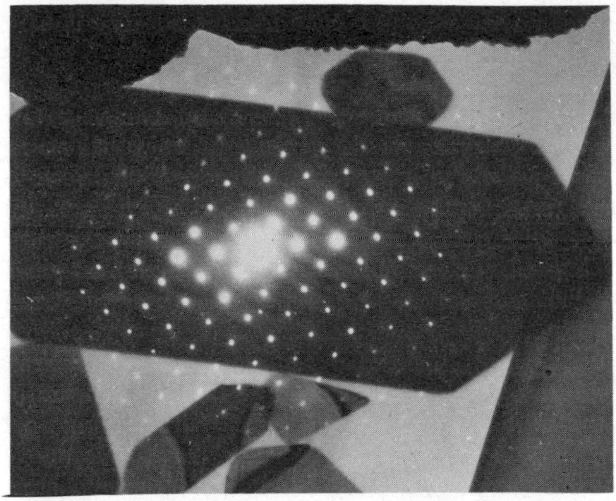

Fig. 22-1 Selected-area 100-kv diffraction pattern of MoO_3 crystallite superimposed on the electron microscope image of the same crystallite (×15,000) for the purpose of calibrating the rotation of the image. (Courtesy G. Thomas.)

To relate the *orientation* of a diffraction pattern to the actual orientation of a crystal, it is necessary to correct for the rotation introduced by the magnetic lenses. The relative azimuthal angle between a selected-area diffraction pattern and the electron microscope image of the same area in the specimen varies with the lens currents. The angle for given settings of the instrument is calibrated by using a crystal with some crystallographic feature, such as an edge of the crystal (Fig. 22–1) or an annealing-twin boundary in a f.c.c. metal crystal [a (111) plane]. Not only does crystallographic direction in the image rotate by the minimum apparent angle between the given direction in the image and the corresponding direction in the pattern, but also there is an inversion through the central spot so that the *sense* of a direction is reversed (in the cases where the sense can be distinguished from its opposite).[1]

Specimens A metallic film so thin as to be translucent or transparent will diffract electrons very strongly; a layer of oxide thin enough to give temper colors likewise produces a good pattern, and in many cases even invisible films can be detected.

It is somewhat difficult to prepare a film so thin that electrons can be transmitted through it, for this requires thicknesses of a few hundred angstroms or less (about a millionth of an inch). Nevertheless, thin trans-

[1] G. W. Groves and M. J. Whelan, *Phil. Mag.*, vol. 7, p. 1603, 1962. R. D. Heidenreich, "Fundamentals of Transmission Electron Microscopy," pp. 187, 377, Interscience, New York, 1964.

mission specimens are prepared in the following ways:

1. Etching a metal foil (usually by floating it on the etching bath).
2. Depositing a substance from the vapor state on a thin cellulose film (or equivalent).
3. Sputtering metal in a gas-discharge tube upon a cellulose film.
4. Precipitating a colloid on a film.
5. Electrodepositing on a substrate which is subsequently etched off.
6. Skimming an oxide layer from molten metal.
7. Oxidizing a thin film or etching an oxidized metal so as to leave the oxide layer.
8. Beating the specimen.
9. Electropolishing, with specimen as anode.
10. Thinning by ion bombardment.
11. Cutting very thin slices with special microtomes.
12. Fast cooling from the liquid state onto cold metal, a method sometimes called *splat cooling* or *crushing* (see Chap. 7, p. 161).
13. Cleaving a crystal.

There is frequently some uncertainty about whether a film has undergone some change during the process of removal.

Reflection specimens are more readily prepared and more widely applicable to metallurgical problems. A reflection specimen, for best results, should have a surface that is slightly rough on a submicroscopic scale so that the electrons can penetrate small projections. The surface must be as clean as the experimenter can make it. A light etch followed by washes in water, alcohol, and benzene serves for metals; oxides are often sufficiently rough when formed on polished surfaces. Electrodeposits and condensed or sputtered films are usually satisfactory without further treatment.

Identification of polycrystalline materials Most applications of electron diffraction involve the identification of reaction products, particularly those formed on metal surfaces. In such studies the patterns are analogous to x-ray powder patterns, and consist of concentric Debye rings that are interpreted just as x-ray patterns are. Typical reflection patterns are reproduced in Fig. 22–2.

The patterns provide the same kind of information that x-ray diffraction

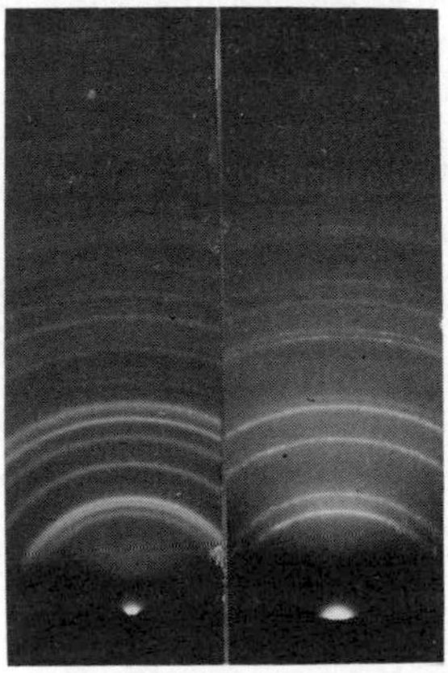

Fig. 22-2 Electron diffraction patterns formed by reflection from surface of a specimen. All electron-diffraction patterns reproduced here are direct prints of the original plates, in conformity with the usual practice. (*a*) Zinc oxide, (*b*) Al-Zn. (M. L. Fuller.)

provides for thicker layers, viz., the identification of the phases in the layer, the crystal structure of the material in the layer (provided this is not masked by anomalies in the diffraction pattern), the approximate grain size in the layer, and the orientations present.

The peculiar advantage of the use of electrons for work with surface layers lies in the fact that they are diffracted strongly by the outermost atoms of the specimen and convey information to the photographic plate concerning these atoms only, whereas x-rays penetrate more deeply and do not register the surface layers unless their thickness is measured in thousands of angstroms.

Identification is carried out by comparison with known materials or by computation of spacings; the ASTM-index system[1] developed for x-rays is applicable here, although some difficulties may be encountered from the lower precision of the spacing measurements and the somewhat altered line intensities.

Extra rings The extraordinary sensitivity of the electron diffraction camera to thin layers of material can be a considerable nuisance when it comes to carrying out an experiment. It is a standard complaint that almost anything, if given a chance, will collect on the specimen in sufficient quantities to cause trouble. The thinnest invisible deposits of grease, for example, give excellent diffraction patterns, as many investigators have found to their sorrow, and a little mercury deposited from vapor that has worked back into the camera from the vacuum pump will quickly affect the patterns and introduce "extra rings." If a specimen is rubbed lightly with rubber, say during etching, the spectrum of rubber appears. With some metals it is almost impossible to avoid obtaining the spectrum of an oxide film that forms while the specimen is being prepared and loaded into the camera—a few minutes at room temperature will produce a detectable oxide film on copper. One annoying source of extra rings was found to consist of an oriented deposit of $Ni(OH)_2$ dissolved from the surface of the specimen holder, which was a nickel alloy.[2] Some extra rings from thin deposited films have been identified with an altered crystal structure in the deposit.

Crystal-structure determination by electron diffraction is much less reliable than by x-rays, not only because of dangers of contamination of the sample, but also because anomalous intensities may be encountered. Anomalous intensities may arise, when rings are broad, from the extremely small size of diffracting crystals[3]; also, with sufficiently imperfect crystals of any size, anomalous intensities may result from a strong first-order reflection being again reflected in the first order by a portion of the same crystal. The doubly reflected beam has the appearance of a second-order

[1] See p. 151.

[2] L. H. Germer, *Z. Krist.*, vol. A100, p. 277, 1938.

[3] L. H. Germer and A. White, *Phys. Rev.*, vol. 60, p. 447, 1941.

reflection and adds to the intensity of any normal second-order reflection that is present.[1]

Double reflection is possible with electron diffraction because the diffraction angles are so small that they may be of the same order of magnitude as the range of orientation in the sample (less than 2°). Multiple reflections of various types, in fact, may be observed, leading to diffracted beams that are apparently incompatible with the crystal structure.[2] If a beam reflected by the $h_1k_1l_1$ plane is again reflected by the $h_2k_2l_2$ plane of the same crystal, it appears on the photograph as reflections having indices $h_1 \pm h_2$, $k_1 \pm k_2$, $l_1 \pm l_2$. Double reflections are often troublesome when two phases are present, such as an oxide layer on a metal crystal. They are prominent, also, when stacking faults are present, as in cobalt, where the forbidden 00·1 reflection, for example, results from the successive $h0\cdot1$ and $\bar{h}0\cdot0$ reflections.[3]

Powder-diffraction patterns with high resolution sometimes reveal fine structure in the lines that would not have been seen in x-ray patterns of the same powders.[4,5,6] Sturkey and Frevel,[5] for instance, found that *hhh* reflections from MgO and CdO smoke particles appeared to be doubled and various other lines were broadened. They suggested that the effect could be accounted for by refraction of the rays at the faces of regularly shaped particles (cubes). Cowley and Rees[7] found examples of one, two, three, and four components in various lines, and from a more thorough treatment of the refraction problem showed that cube-shaped particles could give up to six components (a cluster of six spots surrounding a normal spot position, with the normal spot being absent). The deviations for CdO and MgO indicated inner potentials of about 15 volts in these crystals. Powder particles of irregular shape or of varying inner potential will give varying refraction effects such that the lines will be widened from this cause as well as from other causes. The broadening due to refraction will be of the same magnitude as that due to 250-A crystallite size, Cowley and Rees point out, if the inner potential is 15 volts and the accelerating voltage of the electron beam is 50 kv. Refraction effects have also been seen in patterns made by reflection from etched metal surfaces[8] where pyramids having {111} faces projected from the surface and were penetrated by the beam.

Diffraction by very thin crystals When a crystal is extremely thin in one dimension, its reciprocal lattice points extend out into rods normal to the plane of the plate, i.e., one of the Laue conditions (see page 75) is relaxed; a diffraction pattern made with an electron beam normal to the plate then is a two-dimensional grating pattern (*cross-grating pattern*). An example is shown in Fig. 22–1. Somewhat thicker crystals will give the pattern expected of a three-dimensional lattice with one Laue condition relaxed. A beam normal to the major face of such a crystal will have a cross-grating array of spots, but their intensities will be altered in such a

[1] L. H. Germer, *Phys. Rev.*, vol. 61, p. 309, 1942. O. Hasse and R. D. Heidenreich, *J. Appl. Phys.*, vol. 32, p. 1840, 1961.

[2] H. Raether, *Z. Physik*, vol. 78, p. 527, 1932. G. I. Finch and H. Wilman, *Ergeb. Exakt. Naturw.*, vol. 16, p. 353, 1937. L. G. Schulz, *Phys. Rev.*, vol. 78, p. 316, 1950.

[3] S. Fujime, D. Watanabe, and S. Ogawa, *J. Phys. Soc. Japan*, vol. 19, p. 711, 1964.

[4] J. Hillier and R. F. Baker, *Phys. Rev.*, vol. 68, p. 98, 1945.

[5] L. Sturkey and L. K. Frevel, *Phys. Rev.*, vol. 68, p. 56, 1945; vol. 73, p. 183, 1948.

[6] J. M. Cowley and A. L. G. Rees, *Proc. Phys. Soc.* (*London*), vol. 59, p. 287, 1947; *Nature*, vol. 158, p. 550, 1946.

[7] Ibid.

[8] D. W. Pashley, *Metal Ind.*, vol. 45, pp. 27, 557, 1949.

way that one or more bands are abnormally strong, the bands being circular and concentric around the direct beam. Powdered mica and other layer structures yield a pattern in which there are bands with a sharp limit on their inside edges, corresponding to the *hk*0 diffractions in the case of mica, as expected from extremely thin crystal flakes.[1]

It has also been suggested that extra rings corresponding to submultiples of the normal reflections (*fractional orders*) may be obtained with very thin crystals, as a result of the subsidiary maxima of the interference function.[2] (The analogous effect with single crystals and divergent beams has been clearly seen.[3])

Penetration The *depth of penetration* of electrons may be estimated by depositing layers of various thicknesses on substrates of a different metal. Nickel deposited on copper begins to give a recognizable pattern by reflection when the average thickness reaches about 10 A; a 200-A layer diffracts strongly, and a 400-A layer obliterates the effect of the base.[4] These figures are a function of the roughness of the surface—on some surfaces a monomolecular layer or two or three atomic layers can be detected. The upper limit is imposed by the inelastic impacts of the electrons, which destroy their ability to cooperate in building the diffraction pattern. With heavy metals a film must be less than 10^{-6} cm (100 A) if the background scattering from inelastic impacts is to be small, and patterns cannot be expected if films are uniformly thicker than 10^{-5} cm. (For studies of film thickness up to several thousand angstrom, increasing use is being made of ellipsometry—see the references in the following section.)

Oxide layers There have been many investigations of oxidation with electron diffraction.[5] The various layers in the oxide scale that forms when iron is heated in air can be dissected and the various surfaces studied separately. For example, when the film has just increased in thickness through the color stage and has turned black, it consists of hexagonal α-Fe_2O_3 with the pattern illustrated in Fig. 22–3. When the scale has grown thick enough to be chipped off, the underside shows the spectrum of Fe_3O_4.* Numerous studies of the oxides of aluminum have shown the value of this method for identifying the various structures, their grain

[1] G. I. Finch, A. G. Quarrell, and H. Wilman, *Trans. Faraday Soc.*, vol. 31, p. 1051, 1935. A. Steinheil, *Z. Physik*, vol. 89, p. 50, 1934. G. I. Finch and H. Wilman, *Proc. Roy. Soc. (London)*, vol. A155, p. 345, 1936; *Trans. Faraday Soc.*, vol. 32, p. 1539, 1936.

[2] A discussion of this is given by G. I. Finch and H. Wilman, *Ergeb. Exakt. Naturw.*, vol. 16, p. 353, 1937.

[3] W. Kossel and G. Möllenstedt, *Naturwissenschaften*, vol. 26, p. 660, 1938; *Ann. Phys.*, vol. 36, p. 113, 1939. G. Möllenstedt, *Ann. Phys.*, vol. 40, p. 39, 1941. H. Raether, *Z. Physik*, vol. 126, p. 185, 1949.

[4] W. Cochrane, *Proc. Phys. Soc. (London)*, vol. 48, p. 723, 1936.

[5] Extensive reviews and bibliographies are given by G. P. Thomson and W. Cochrane, "Theory and Practice of Electron Diffraction," Macmillan, London, 1939.

* H. R. Nelson, *J. Appl. Phys.*, vol. 9, p. 623, 1938.

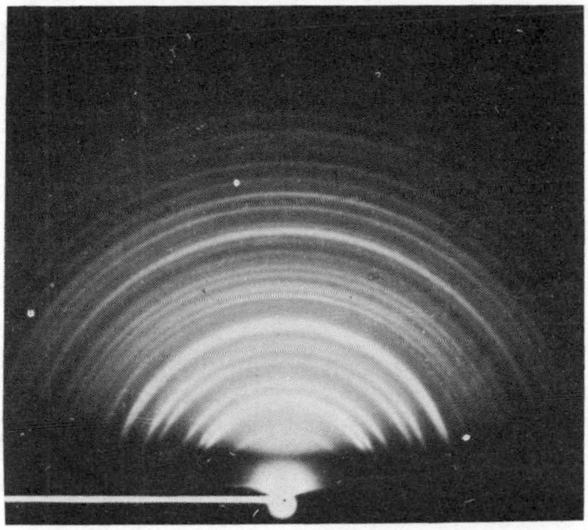

Fig. 22-3 Surface-reflection pattern of hexagonal α-Fe_2O_3. (H. R. Nelson.)

size, and their orientation.[1] Similar data have been obtained as a function of temperature and composition for Fe, Co, Ni, and their alloys by Gulbransen and Hickman.[2]

Preferred orientations It is usual for the thin deposits to have a preferred orientation of some sort. This is revealed in the patterns by a lack of uniformity in the intensity of each ring on the photographic plate or, in certain cases, by abnormal intensities of certain rings.

The interpretation of the texture patterns is exactly as with x-rays, except that here the angle of incidence is so small that one can consider $\theta = 0$ and the reflection circle is a great circle. An oriented electrodeposit of iron is illustrated in the reflection photograph of Fig. 22-4. Deposited films normally have a *fiber texture* with the fiber axis normal to the substrate, but deposits from an inclined stream of vapor may have a tilted fiber axis. (See also page 540.) The texture can usually be judged even from transmission patterns with the beam nearly normal to the film if attention is paid to the intensities of the different *hkl* rings, keeping in mind the intersections of the reflection sphere and the reciprocal lattice (or the reflection circle on a pole figure).

A special type of preferred orientation is often found in a filed surface of a single crystal: the crystallites near the surface have a range of orientations with a common axis parallel to the surface and normal to the direction of motion of the file. Surface reflection patterns of these contain *arcs* when

[1] Ellipsometers are used for *thickness* measurements of anodic oxide films on aluminum: M. A. Barrett and A. B. Winterbottom, in "First International Congress on Metallic Corrosion," Butterworths, London, 1961, p. 657. A. B. Winterbottom in "Ellipsometry in the Measurement of Surfaces and Thin Films," U.S. Department of Commerce, National Bureau of Standards, Washington D.C. (Misc. Pub. 256), 1954, p. 91; M. A. Barrett, Misc. Pub. 256, loc. cit.

[2] E. A. Gulbransen and J. W. Hickman, *Trans. AIME*, vol. 171, pp. 306, 344, 1947; vol. 180, pp. 519, 534, 1949. A summary of the data will be found in E. Gulbransen, *Trans. Electrochem. Soc.*, vol. 91, p. 573, 1947.

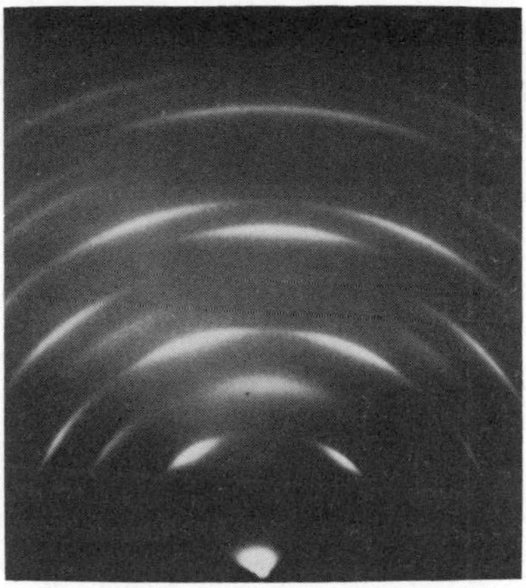

Fig. 22-4 Reflection pattern showing preferred orientation in electrodeposited iron. (Courtesy G. I. Finch.)

the beam is normal to the direction of file motion, but they contain spots when the beam is nearly parallel to the file motion, because the reciprocal lattice contains arcs of circles that are concentric with the axis of rotation of the crystallites. Among the crystals showing this effect are galena (PbS),[1] NaCl,[2] Cu, ZnO, and zinc blende (ZnS).[3] A potential source of error in judging preferred orientations has been found in polycrystalline specimens with *etched surfaces*.[3] Grains of certain orientations may develop etched-surface contours more favorable for diffraction than others, and these selected grains may predominate in forming the pattern, giving a misleading sampling of the various orientations present.

Polished surfaces Polished surfaces of crystalline materials are frequently found to give a very diffuse reflection pattern. Some diffuseness in reflection patterns can be caused by refraction, when there is a varying angle of incidence and emergence at a wavy surface of a specimen.[4] Broadening of the rings due to refraction may be wrongly interpreted as particle-size broadening. To judge whether some particle-size broadening is also present, transmission patterns are much more reliable, since the refraction effects are much smaller when the electron beam enters and leaves surfaces nearly perpendicularly—but precautions must be taken against the formation of oxides when a thin sheet is being prepared and cold-worked.

A layer of gold was evaporated onto a substrate of nickel, polished, isolated by etching the nickel away, and examined in transmission by Cochrane.[5] The very diffuse rings obtained showed that the coherently diffracting domains had dimensions of only a few angstroms and the metal was practically amorphous, as proposed by Sir George Bielby

[1] L. H. Germer, *Phys. Rev.*, vol. 50, p. 659, 1936.

[2] D. M. Evans and H. Wilman, *Proc. Phys. Soc.* (*London*), vol. 63, p. 298, 1950.

[3] R. P. Johnson and W. R. Grams, *Phys. Rev.*, vol. 62, p. 77, 1942.

[4] F. Kirchner, *Trans. Faraday Soc.*, vol. 31, p. 1114, 1935. L. H. Germer, *Phys. Rev.*, vo. 43, p. 724, 1933; vol. 49, p. 163, 1936. W. Kranert, K. H. Leise, and H. Raether, *Z. Physik*, vol. 122, p. 248, 1944.

[5] W. Cochrane, *Proc. Roy. Soc.* (*London*), vol. A166, p. 228, 1938.

at the beginning of the century. Crystallite sizes reduced to the order of 50 A have also been observed in metals after other kinds of severe cold-working operations.[1] Many divergent results and conclusions have been obtained in this field of research, however, and it is probable that added to difficulties arising from refraction there have been varying amounts of room-temperature recovery in samples used by different observers, and possibly in some experiments there were appreciable amounts of minute oxide inclusions which might act to stabilize the *Bielby layer*.[2]

A polished layer is most severely distorted at the surface; the effective grain size increases as one goes to a depth of 100, 1000, or 10,000 A below the surface. A preferred orientation has been found at the surface of polished copper and gold, with (110) planes lying parallel to the surface,[3] and it has been suggested[4] that the melting which occurs at the surface of metals during polishing (Bowden and Ridler[5]) may produce recrystallization of the surface layers which affects the orientations.

Wulff and his collaborators have applied electron diffraction to the study of phase transformation induced in stainless steel (18 percent Ni, 8 percent Cr) by polishing, grinding, sanding, superfinishing, and cold-rolling.[6] The surface layers resulting from dry or wet grinding are austenitic, indicating that they have been heated above 200°C. Below this austenitic layer is a layer that has been reduced to ferrite by cold-work; the ferritic layer is about 6×10^{-4} cm below the surface in samples that have been ground. Samples that have received any type of polishing contain a ferritic layer at the surface 1.5×10^{-5} (1500 A) or less in thickness. The ferrite layer in sheet reduced 50 percent by cold-rolling extends about 1×10^{-4} cm below the surface.

Studies of surface conditions are of particular importance in connection with problems of friction, adhesion, wear, lubrication, and chemical reactivity of surfaces; a brief summary of research in these fields, with references, has been assembled by a committee.[7] More detailed reviews are also available.[8]

Thin deposits Electron diffraction is extensively used in studies of the thinnest deposited films, particularly those too thin for x-ray diffraction.[9] Deposits on cleavage surfaces of single crystals are commonly found to be

[1] W. Kranert and H. Raether, *Ann. Phys.*, vol. 43, p. 520, 1943; *Z. Naturforsch.*, vol. 1, p. 512, 1946.

[2] F. P. Bowden and D. Tabor, "Symposium on Properties of Metallic Surfaces," p. 197, Institute of Metals, London, 1952.

[3] H. G. Hopkins, *Trans. Faraday Soc.*, vol. 31, p. 1095, 1935. C. S. Lees, *Trans. Faraday Soc.*, vol. 31, p. 1102, 1935.

[4] G. I. Finch, *Sci. Progr.*, vol. 31, p. 609, 1937.

[5] F. P. Bowden and K. E. W. Ridler, *Proc. Roy. Soc. (London)*, vol. A154, p. 640, 1936.

[6] J. Wulff, *Trans. AIME*, vol. 145, p. 295, 1941. J. T. Burwell and J. Wulff, *Trans. AIME*, vol. 135, p. 486, 1939.

[7] L. Himmel, J. J. Harwood, and W. J. Harris, Jr. (eds.), "Perspectives in Research," ACR-61, Surveys of Naval Science, No. 10, February, 1963, p. 498, U. S. Government Printing Office, Washington, D.C.

[8] F. P. Bowden and D. Tabor, "Friction and Lubrication of Solids," Clarendon, Oxford, 1954. R. Gomer and C. S. Smith (eds.), "Structures and Properties of Solid Surfaces," University of Chicago, Chicago, 1953 (see G. P. Thomson, p. 185, and F. P. Bowden, p. 203).

[9] Summaries of this field: G. I. Finch and H. Wilman, *Ergeb. Exakt. Naturw.*, vol. 16, p. 353, 1937 (in English). W. Cochrane, "Theory and Practice of Electron Diffraction," Macmillan, London, 1939. H. Richter, *Physik Z.*, vol. 44, p. 406, 1943. R. Gomer and C. S. Smith (eds.), "Structure and Properties of Solid Surfaces," University of Chicago, Chicago, 1953. P. W. Selwood in "Advances in Catalysis," Academic, New York (various volumes).

oriented (*epitaxial*), with the degree and type of orientation dependent upon the nature and temperature of the substrate as well as upon the nature of the deposited metal.[1] Oriented overgrowths of alkali halides on cleaved surfaces of other alkali halides and on mica are advantageously studied by such techniques, and have yielded information on orientation vs. degree of misfit and thickness of the deposit, mechanism of crystal growth, effect of contamination, and polymorphism induced by depositing.[2] Schulz has found, for example, that six salts (CsCl, CsBr, CsI, TlCl, TlBr, and TlI), which normally have the CsCl type of crystal structure, take the NaCl type of structure when grown from the vapor on suitable substrates.[3] Matching of the atom pattern at the interface is a factor in this polymorphism (but close matching is not necessary). Theory has shown that the free energies of the two forms are similar, and it is known that polymorphism can also be induced in some of these compounds by subjecting them to high pressures.

Deposits of alloys on single-crystal surfaces have provided, after removal, excellent epitaxial single-crystal films for electron diffraction studies of superlattices[4]; specimens prepared in this way have been particularly effective in studies of long-period superlattices. An example is reproduced in Fig. 11–16 (page 299).

The *grain size* in a thin-film transmission specimen is indicated by the breadth of the diffraction lines, as with x-rays, though this is recommended only with transmission specimens, not reflection, and for crystallites of about 3 to 100 unit cells in diameter. An electrodeposit of arsenic with exceedingly small crystals gives the pattern reproduced in Fig. 22–5,[5] in which the line broadening is evident. The usual Scherrer formula for computing crystallite size from line width may be employed; but with electrons this reduces to $b = \lambda/L$, where b is the line width (in radians) at half maximum intensity attributable to crystallite-size widening, λ is the wavelength, and L is the mean crystal size normal to the beam.[6] For particles so small that diffraction lines overlap, it is possible to judge crystal size by comparison with computed curves of intensity vs. angle; Germer and White plotted these by using the theoretical formula for the

[1] G. P. Thomson, *Proc. Roy. Soc.* (*London*), vol. A128, p. 649, 1930. H. Lassen, *Physik Z.*, vol. 35, p. 172, 1934. H. Lassen and L. Bruck, *Ann. Phys.*, vol. 22, p. 65, 1935. L. Bruck, *Ann. Phys.*, vol. 26, p. 233, 1936. M. Kubo and S. Miyake, *J. Phys. Soc.* (*Japan*), vol. 3, p. 114, 1948.

[2] L. G. Schulz, *Phys. Rev.*, vol. 77, p. 750, 1950; vol. 78, p. 638, 1950; *J. Chem. Phys.*, vol. 18, p. 896, 1950; *J. Appl. Phys.*, vol. 9, p. 942, 1950; *Acta Cryst.*, vol. 4, p. 487, 1951.

[3] L. G. Schulz, *Acta Cryst.*, vol. 4, p. 487, 1951.

[4] Papers by H. Sato and R. S. Toth, by S. Ogawa and coworkers, and by others in this field are mentioned in Chap. 11.

[5] G. I. Finch, A. G. Quarrell, and H. Wilman, *Trans. Faraday Soc.*, vol. 31, p. 1051, 1935.

[6] Small particles yield much sharper lines in electron diffraction patterns than in x-ray patterns because the electron wavelengths are shorter than the x-ray wavelengths.

Fig. 22-5 Transmission pattern for electro-deposited arsenic. Very small crystals, substantially amorphous. (G. I. Finch, A. G. Quarrell, and H. Wilman.)

intensity of coherent scattering from gas molecules.[1] Curves for clusters of atoms in f.c.c. array containing 13 atoms or less did not show the 111 ring as resolved from the 200, or 220 resolved from 311 and 222; clusters of 55 atoms resolved 220 from its neighbors, but not 111; clusters of 379 atoms resolved 111 from 200, and 220 from the unresolved doublet 311 + 222. Many metals deposited on glass or on amorphous substrates yield diffraction patterns with such broad halos and unresolved lines in electron diffraction transmission patterns that the number of atoms per grain cannot exceed 100 or 200, and the structure is commonly called amorphous. With grain sizes this small, much of the material is in the grain boundaries and the structure would be best described in a statistical manner by radial distribution plots of atomic density.

Germer found that vaporization of Cu, Ni, Au, and Pd on amorphous substrates produced crystallites—in the thinnest films—somewhat larger than the average thickness of the films; for example, a 3-A film of copper produces an electron-diffraction pattern characteristic of crystallites containing two or three hundred atoms.[2] Films of ionic compounds contain crystallites 50 to 100 A or more on a side in films of roughly 10 A average thickness.[2,3] It is therefore clear that there is considerable migration of atoms and molecules over the surface of the substrate, as has been concluded by resistivity, thermionic, photoelectric, and optical measurements,[1] and by electron microscope observations.[4] Schulz points out[3] that this mobility is limited and does not approach that of a liquid state, for a

[1] L. H. Germer and A. W. White, *Phys. Rev.*, vol. 60, p. 447, 1941.

[2] L. H. Germer, *Phys. Rev.*, vol. 56, p. 58, 1939.

[3] L. G. Schulz, *J. Chem. Phys.*, vol. 17, p. 1153, 1949.

[4] R. C. Pickard and O. S. Duffendack, *J. Appl. Phys.*, vol. 14, p. 291, 1943.

liquid state would not account for the tipping of the fiber axis toward or away from the obliquely incident stream of vapor, or for the uniformly small size of the crystallites.

Intensity of scattering The intensity of an electron beam scattered by a crystal is proportional to the efficiency of scattering of individual atoms. With x-ray scattering we use an atomic scattering factor f to take account of this efficiency; similarly, with electrons we can use a factor E for the purpose. Both E and f are functions of the angle θ. The value of f depends upon the number of electrons in the scattering atom, Z, but E depends on the quantity $(Z - f)$ because of scattering of electrons by the nucleus as well as by the outer electrons in the atom. The factor E has the value

$$E = (Z - f)\left(\frac{\lambda}{\sin\theta}\right)^2 \cdot c$$

where λ is expressed in angstroms, and c is a constant involving the electronic charge and mass and Planck's constant ($c = e^2m/2h^2$). The value of E varies with angle as f does, though somewhat more rapidly, but the numerical value of E at small angles is about 10^4 times greater than f; hence, electrons scatter much more intensely than x-rays. This is the reason electron diffraction can detect such minutely thin films of material. To compute the relative intensities of reflections from a crystal, the usual x-ray formulas for $F(hkl)$ are used, with E substituted for f. (See, however, page 437 regarding anomalous intensities.)

Tables of E vs. θ have been published which are based on various assumed wave functions.[1] These are subject to change as improved wave functions come into use; most available tabulations are subject to considerable uncertainty in the range of $(\sin\theta)/\lambda$ less than 0.05. With slow electrons, also, account should be taken of the polarization of the electron cloud by the bombarding electron,[2] a correction that has not been made in the tables cited. The atomic scattering function is actually a complex quantity, as it is with x-rays, and while the imaginary part is normally neglected in work with solids, it should not be neglected in scattering from gases containing both heavy and light atoms.

Structure determination Crystal structures have been determined by electron diffraction. Fourier methods are applicable, as with x-ray and neutron diffraction. But the method is generally inferior in practice to other diffraction methods. As mentioned above, there are uncertainties in the atomic scattering amplitudes, particularly in the range of greatest importance, near $(\sin\theta)/\lambda = 0$. In addition, multiple reflections are more troublesome than with x-rays, which means that the dynamical theory

[1] "International Tables for X-ray Crystallography," vol. 3, Tables 3.3.3. A(1) and A(2), Kynoch, Birmingham, England, 1962. G. Thomas, "Transmission Electron Microscopy of Metals," pp. 287, 290, Wiley, New York, 1962. B. Dawson, *Acta Cryst.*, vol. 14, p. 1120, 1961. R. D. Heidenreich, "Fundamentals of Transmission Electron Microscopy," p. 395 (introduced on p. 55), Interscience, New York, 1964.

[2] H. Browne and E. Bauer, in "Fifth International Congress for Electron Microscopy," vol. 1, p. AA-14, Academic, New York, 1962.

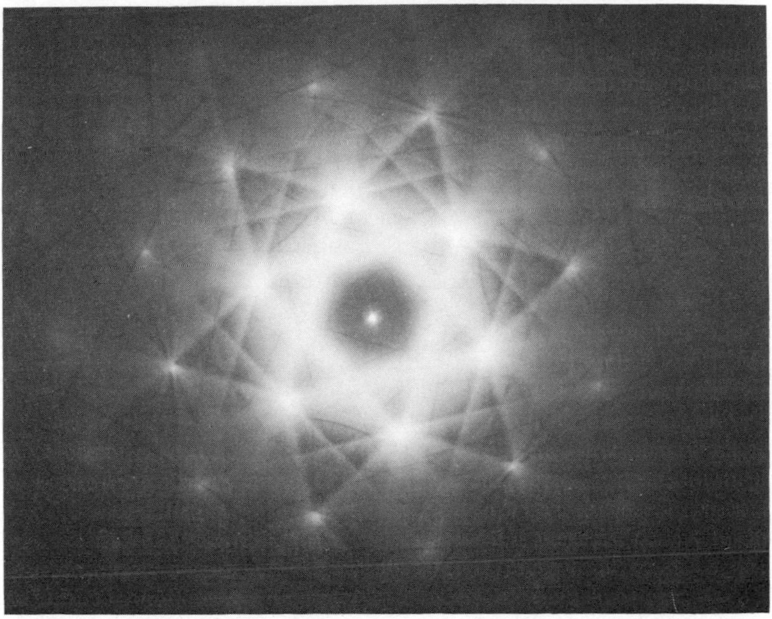

Fig. 22-6 Diffraction pattern containing black and white Kikuchi lines from a silicon crystal. Electron beam along [111]. Because both +g and −g vectors are operating, the symmetry is six-fold rather than three-fold. (**Courtesy G. Thomas.**)

rather than the simpler kinematic theory is needed. The intensity of a reflection *hkl* may have a significant contribution from a doubly diffracted or a multiply diffracted beam, as is becoming increasingly evident also with x-rays and neutrons. "Forbidden" reflections therefore often appear with apparent *hkl* values for which the structure factor is really zero (although this does not occur with f.c.c. or b.c.c. crystals). Intensity measurements using photographic films further increase the intensity uncertainties.

Kikuchi lines A number of black and white lines appear on a diffraction pattern when the diffracting crystal is not too thin and when it is not so imperfect that the lines are too blurred to be seen. The Kikuchi lines,[1] as they are called, are illustrated in Fig. 22–6.[2] They are formed by electrons that have been diffusely scattered with some loss of energy.[3] Some of these diffusely scattered electrons find themselves going in exactly the right directions to reflect from some plane in the crystal. In Fig. 22–7, AB and $A'B'$ represent a set of parallel reflecting planes in a crystal. Diffusely scattered electrons that would have followed the path OP are reflected to

[1] S. Kikuchi, *Proc. Imp. Acad.* (*Tokyo*), vol. 4, pp. 271, 354, 1928; *Japan. J. Phys.*, vol. 5, p. 83, 1928.

[2] G. I. Finch and H. Wilman, *Proc. Roy. Soc.* (*London*), vol. A155, p. 345, 1936.

[3] W. Hartl and H. Raether, *Z. Physik*, vol. 161, p. 238, 1961.

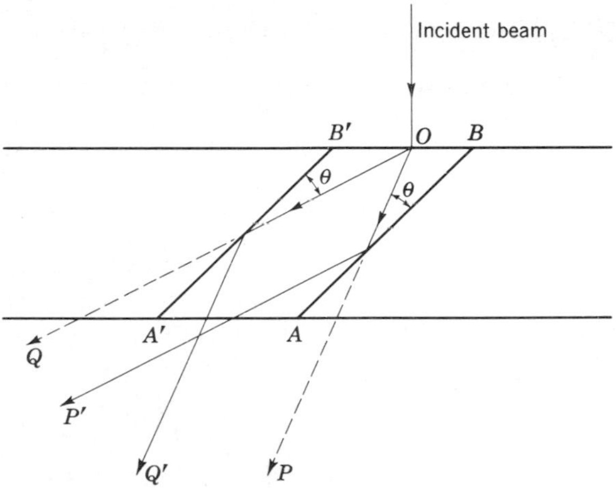

Fig. 22-7 Origin of Kikuchi lines.

P', while the scattered ray OQ is reflected to Q'. Now the intensity of the scattered ray in the direction OP is greater than that of the ray in the direction OQ, and so the energy robbed from the ray P is not fully returned by the reflected ray Q' and there is a net loss in the direction OP; similarly, there is a net gain in the direction OQ. Thus, there will appear on the plate a white (weakened) line and a black (enhanced) line parallel to the projection of each crystal plane and equidistant from it. Sometimes there are so many lines at slight angles to each other, tangent along a curve, that the curved envelope of the lines becomes a prominent feature of the pattern.

Kikuchi lines are closely related to the Kossel lines seen with x-rays[1]; they are seen only with crystals of considerable perfection and disappear when a crystal is distorted. They may be seen with both transmission and reflection experiments. They are also seen when crystals in the micron range of size are irradiated with an electron beam focused to a cross-section diameter of 100 to 200 A.[2] With this focused beam a pattern is obtained in which each spot is an enlarged image of the crystal; striations are obtained in the spots that agree with the predictions of the dynamical theory of diffraction as computed by MacGillavry.[3] Detailed theoretical treatments of Kikuchi lines, based on the dynamical theory of diffraction, have been published[4]; these are combined with a dynamical-theory treatment of

[1] Excellent reproductions of Kikuchi lines and Kossel lines, together with discussions of their geometry and indexing, will be found in *Ergeb. Exakt. Naturw.*, vol. 16, pp. 296, 353, 1937.

[2] N. Davidson and J. Hillier, *J. Appl. Phys.*, vol. 18, p. 499, 1947.

[3] C. H. MacGillavry, *Physica*, vol. 7, p. 329, 1940.

[4] K. Shinohara, *Sci. Papers Phys. Chem. Res. (Tokyo)*, vol. 18, p. 223, 1932; vol. 18, p. 39, 1932; *Phys. Rev.*, vol. 47, p. 730, 1935. S. Kikuchi and S. Nakagawa, *Sci. Papers Inst. Phys. Chem. Res. (Tokyo)*, vol. 21, p. 256, 1933. M. von Laue, *Physik Z.*, vol. 37, p. 544, 1936.

special features of single-crystal patterns that have been observed by Kossel and Möllenstedt[1] in the book by von Laue.[2]

Diffraction of atoms and molecules Beams of atoms and molecules can also be diffracted by crystals, for the de Broglie relation discussed in connection with electron diffraction, $\lambda = h/(mv)$, holds for any moving particle. The techniques, which are different from those used in other diffraction work, have been summarized by Estermann.[3] The source is an oven, which emits particles with a "continuous spectrum" of velocities. Monochromatization is possible either by sorting out definite velocities mechanically by the use of a rotating sector, or by a crystal monochromator. Beams of H, H_2, and He, when diffracted from cleaved ionic crystals, produce the diffracted beams that would be expected of diffraction from a crossed grating.[4] Since penetration into the diffracting crystal is negligible, the beam is directed by the pattern of atoms in the *surface.*

Low-energy electron diffraction Ordinary electron diffraction in the 50- to 100-kv range is not sensitive to a single layer of atoms on a crystal surface, yet such sensitivity should be of great value in surface physics and surface chemistry. (Field-emission microscopy, using electrons emitted from a point, and field-ion microscopy, using a field-emission tube containing helium or other gas, are making contributions of importance in this field.[5]) Diffraction with electrons of less than 1-kv energy and even less than 100 volts is capable of revealing single atomic layers and has come into more widespread use in the 1960s. Even the first electron diffraction experiments of Davisson and Germer in 1927 showed that low-energy electron diffraction was possible, and throughout the intervening years H. E. Farnsworth has done research in the field,[6] but the difficulties of experimentation and interpretation have tended to limit activity in the field during the period 1927 to 1963. It is now more widely realized that the diffracted beams can be accelerated so as to become visible on a fluorescent screen, and that low-energy diffraction can provide information that cannot be obtained by x-ray diffraction. Early conclusions from the renewed activity in this field are that adsorption of foreign atoms upon many crystal surfaces occurs with an accompanying drastic rearrangement of the atoms of the metal or semiconductor surface. Facts of this type and others have importance in understanding catalysis, and imply that increased activity in the field will continue.

[1] W. Kossel and G. Möllenstedt, *Naturwissenschaften,* vol. 26, p. 660, 1938; *Ann. Phys.,* vol. 36, p. 113, 1939; vol. 42, p. 287, 1942. W. Kossel, I. Ackerman, and G. Möllenstedt, *Z. Physik,* vol. 120, p. 553, 1943.

[2] M. von Laue, "Materiewellen und ihre Interferenzen," Akademische Verlagsgesellschaft m.b.H., Leipzig, 1944, and Edwards, Ann Arbor, Mich.

[3] I. Estermann, *Rev. Modern Phys.,* vol. 18, p. 300, 1946.

[4] I. Estermann and O. Stern, *Z. Physik,* vol. 61, p. 95, 1930. I. Estermann, R. Frisch, and O. Stern, *Z. Physik,* vol. 73, p. 348, 1931. T. H. Johnson, *J. Franklin Inst.,* vol. 210, p. 135, 1930.

[5] For a discussion and references see L. Himmel, J. J. Harwood, and W. J. Harris (eds.), "Perspectives in Materials Research," U.S. Government Printing Office, Washington, D.C., 1964.

[6] For a review article, see H. E. Farnsworth and R. F. Woodcock, in "Advances in Catalysis," vol. 9, p. 123, Academic, New York, N.Y., 1957. An interesting history of the field is included in an article by L. H. Germer, *Phys. Today,* vol. 17, p. 19, July, 1964.

23

Neutron diffraction and magnetic structures

The characteristic of neutrons that enables them to be diffracted as waves has permitted important advances in our knowledge of certain crystals, for neutron diffraction can supply information that cannot be obtained with x-rays or electrons.

Equipment and methods The high-energy neutrons resulting from nuclear fission are "thermalized" within a reactor by collisions in a "moderator" substance, such as graphite or heavy water; that is, they are slowed down to thermal equilibrium with the moderator. There is then a broad band of energies present, and since the wavelength of a neutron, λ, is related to its momentum mv by the equation $\lambda = h/mv$, where h is Planck's constant, a broad band of wavelengths is available. The peak intensity of a beam of "thermal neutrons" from a reactor operating near 100°C is at a wavelength suitable for diffraction from crystals. Many wavelengths are present, and it is possible to produce Laue patterns of crystals, with the aid of suitable intensifying screens.[1] The Laue method finds little use, however, for much better data are obtained by selecting from the thermal neutron beam a narrow band of wavelengths by means of a crystal monochromator and using this monochromatized beam in a diffractometer.[2] The diffraction pattern from the sample is detected with a proportional counter. The attached circuits serve to discriminate against the gamma rays. The usual counter is one that is filled with boron trifluoride gas, preferably enriched in the boron isotope B^{10}, and is thoroughly shielded to reduce the background count. The diffracted beams that must be measured are very weak, usually giving only 1000 to 10,000 counts per min above background with powder specimens, even though the collimated beams, monochromator, specimen, and counter are made many times as large as those used with x-rays. Automatic registration of the counts and automatic resetting of the counter angle are customary because of the long runs required in most experiments. Since the runs extend over many hours, frequently even several days, a monitor counter is usually installed between the mono-

[1] E. O. Wollan, C. G. Shull, and M. C. Marney, *Phys. Rev.*, vol. 73, p. 527, 1948. S. P. Wang, C. G. Shull, and W. C. Phillips, *Rev. Sci. Instr.*, vol. 33, p. 126, 1962.

[2] Crystals of lead or copper have been much used for this purpose, and tests (C. S. Barrett, M. H. Mueller, and L. Heaton, *Rev. Sci. Instr.*, vol. 34, p. 847, 1963) have shown that warm-worked germanium crystals are efficient. The germanium 111 reflection has the advantage that the half-wavelength component is essentially eliminated since the 222 reflection is forbidden.

chromator and the specimen. Details concerning construction and operation of neutron diffractometers will be found in the literature.[1] Auxiliary equipment such as cryostats[2] and furnaces[3] has been used with neutron diffractometers to study substances at temperatures throughout the range from 1.25°K to 1000°C.

Neutron scattering from atomic nuclei Because neutrons have a magnetic moment they interact with the magnetic moments of atoms. The scattering from atomic nuclei will be considered first, and the additional scattering from any magnetic moments outside of the nucleus will be treated later. Unpaired electrons of the 3*d* shell or the 4*f* shell are responsible for this magnetic extranuclear scattering (also, some molecules lacking 3*d* or 4*f* electrons are paramagnetic, e.g., O_2). Other electrons do not scatter neutrons appreciably since neutrons carry no electric charge.

A nucleus scatters neutrons both elastically and inelastically. When neutrons are *inelastically* scattered by the nuclei in a crystal, they excite vibrations in the crystal and thereby lose some of their energy, or they gain energy from the thermal vibrations in the crystal. In either event they are altered in wavelength and do not contribute to the coherent scattering which builds up the diffraction pattern.

The *elastically* scattered neutrons which scatter coherently build up the diffraction pattern. The total scattering power of an element is expressed in terms of a *total scattering cross section* σ_S, defined as the outgoing current of elastically and inelastically scattered neutrons divided by the incident flux of neutrons. The value of σ_S for a nucleus in a crystal (a bound nucleus) is different from that for the nucleus of a free atom; it is the value for a bound nucleus that applies in diffraction from a solid.

The *coherent scattering* from an element is also expressed as a cross section σ_{coh} and, like σ_S, is tabulated in units of 10^{-24} cm^2, a unit called a *barn*. A diffracted beam is made up of contributions, with different phases and amplitudes, from the various atoms of a unit cell, just as with x-rays. The coherent scattering amplitude for neutrons, which corresponds to f for x-rays, is termed b. For elements consisting of more than one isotope, it is the average b that is used in the diffraction formulas.

An atomic nucleus, being many orders of magnitude smaller than the wavelength of thermal neutrons, acts as a point source of scattering for neutrons. With x-rays, there is destructive interference between the waves from different parts of an atom, which results in a decreasing value of the

[1] W. H. Zinn, *Phys. Rev.*, vol. 71, p. 752, 1947. E. O. Wollan and C. G. Shull, *Phys. Rev.*, vol. 73, p. 830, 1948. G. E. Bacon, J. A. G. Smith, and C. D. Whitehead, *J. Sci. Instr.*, vol. 27, p. 330, 1950. G. E. Bacon and R. F. Dyer, *J. Sci. Instr.*, vol. 32, p. 256, 1955. G. R. Ringo in S. Flügge (ed.), "Handbuch der Physik," vol. XXXII, p. 552, Springer-Verlag, Berlin, 1957. G. E. Bacon, "Neutron Diffraction," 2d ed., Clarendon, Oxford, 1962.

[2] R. A. Erickson, *Phys. Rev.*, vol. 90, p. 779, 1953. E. O. Wollan, W. C. Koehler, and M. K. Wilkinson, *Phys. Rev.*, vol. 110, p. 638, 1958.

[3] S. S. Sidhu, L. Heaton, and M. H. Mueller, *J. Appl. Phys.*, vol. 30, p. 1323, 1959. M. K. Wilkinson and C. G. Shull, *Phys. Rev.*, vol. 103, p. 516, 1956.

x-ray atomic scattering amplitude f with increasing scattering angle; but the nuclear scattering amplitude for neutrons shows no variation with scattering angle. The scattering amplitude from the nucleus depends on atomic number in a complex and irregular way very different from the linear dependence of f on atomic number. Many of the light atoms scatter as strongly as most of the atoms of high atomic number. Deuterium, for instance, and even hydrogen do not have the very weak scattering amplitude in comparison to most metal atoms that they have when x-rays are used; the same is true of C, O, and N. Accordingly, neutron diffraction is particularly valuable for determining the atomic positions of light atoms in crystals containing heavy atoms.

Another feature in which neutron b values differ from x-ray f values is that adjacent elements in the Periodic Table may differ greatly in b, whereas they differ a minimum amount in f. The distribution of Fe and Co atoms in the structure of an alloy can be seen clearly with neutrons, for example, but is nearly invisible to x-ray diffraction.

A third striking feature is the presence of a few scattering amplitudes with a negative sign. These are associated with atoms that scatter 180° out of phase with the majority of the elements. The presence of negative signs must not be overlooked in computing structure factors for crystals containing elements with negative b values. The structure amplitude for neutron diffraction is

$$F_{hkl} = \Sigma b_j \exp 2\pi i(hu_j + kv_j + lw_j) \tag{23–1}$$

where the sum includes terms for each of the atoms of the unit cell, and the individual term for the jth atom has a positive or negative sign depending on the sign of b_j (in contrast to the analogous formula for x-rays in which the atomic scattering amplitudes f are all positive). A curious consequence of negative values of b is that alloys can be made which, although crystalline, yield no crystalline diffraction pattern. These "null-matrix" alloys are random solid solutions of two or more kinds of atoms or isotopes in which the relative proportions of atoms with positive and negative scattering amplitudes are such as to make $F_{hkl} = 0$ for all reflections.[1]

The absorption of a neutron beam is almost always much less than with x-rays. The intensity of the incoherent scattering, which forms the background for the diffraction peaks, can result from several diffuse scattering processes[2]: (1) multiple scattering, (2) thermal vibrations of the atoms, (3) randomness in the distribution of isotopes of differing scattering amplitude, and (4) randomness in the distribution of parallel and antiparallel spins of the nuclei.

Diffraction from powders is the usual practice, but diffraction from single crystals is helpful in solving some of the more difficult crystallo-

[1] S. S. Sidhu et al., *J. Appl. Phys.*, vol. 27, p. 1040, 1956.

[2] E. O. Wollan and C. G. Shull, *Phys. Rev.*, vol. 73, p. 830, 1948. G. E. Bacon, "Neutron Diffraction," 2d ed., Clarendon, Oxford, 1962.

graphic problems.[1] Resolution of the peaks is better with single crystals, and intensities are higher; more specific information about the distribution of magnetic moments can be obtained. Fourier synthesis is used, as with x-ray data, for the more difficult problems, and for certain problems there is advantage in using polarized neutrons and in diffracting from single crystals placed in a magnetic field. Multiple Bragg reflections can seriously distort the measurements of intensities, but can usually be eliminated by varying the azimuthal orientation of the reflecting plane.

Structure determinations with nonmagnetic substances Neutron diffraction has been of great value in determining the positions of light atoms in crystals where their positions could not be seen with x-rays or electrons. Most of these are in nonmetallic materials.[2] Many hydrides of the metals have been investigated, including hydrides of Pd,[3] Hf, Ti,[4] U, Th, Zr,[5] and the null-matrix alloy 62 atomic percent Ti–38 atomic percent Zr.[6] Neutron data have also been used to locate the light atoms of Be, C, N, O, and F in compounds containing heavy elements.

The ease with which neutrons can distinguish between atoms of nearly identical atomic number has been of advantage in various structure investigations, e.g., in verifying superlattice formation in FeCo, Ni_3Mn,[7] and the structure of $MgAl_2O_4$.[8] The ordering of the atoms in various σ-phase alloys such as the σ phases in the systems Ni-V, Fe-V, and Mn-Cr,[9] and the absence of ordered structures in some other σ phases can be shown by neutrons, but only with difficulty, if at all, with x-rays.

The study of short-range order in beta-brass by means of measurements of the diffuse scattering at temperatures above the critical ordering temperature has only become possible by using neutrons.[10] Radial density distributions in liquids are determined as with x-rays, by the Zernike and Prins method,[11] and the results are usually somewhat better than with x-rays.

[1] G. E. Bacon and R. S. Pease, *Proc. Roy. Soc. (London)*, vol. A220, p. 397, 1953. S. W. Peterson and H. A. Levy, *J. Chem. Phys.*, vol. 19, p. 1416, 1951.

[2] G. E. Bacon, *Rev. Modern Phys.*, vol. 30, p. 94, 1958; "Neutron Diffraction," Clarendon, Oxford, 1962. R. Pepinsky, *Rev. Modern Phys.*, vol. 30, p. 100, 1958. G. R. Ringo in S. Flügge (ed.), "Handbuch der Physik," vol. XXXII, p. 552, Springer-Verlag, Berlin, 1957. M. K. Wilkinson, E. O. Wollan, and W. C. Koehler, *Ann. Rev. Nucl. Sci.*, vol. 11, p. 303, 1961.

[3] J. E. Worsham, M. K. Wilkinson, and C. G. Shull, *Phys. Chem. Solids*, vol. 3, p. 303, 1957. J. Bergsma and J. A. Goedkoop, *Physica*, vol. 26, p. 744, 1960.

[4] S. S. Sidhu, L. Heaton, and D. D. Zauberis, *Acta Cryst.*, vol. 9, p. 607, 1956.

[5] R. E. Rundle, C. G. Shull, and E. O. Wollan, *Acta Cryst.*, vol. 5, p. 22, 1952.

[6] S. S. Sidhu, L. Heaton, and M. H. Mueller, *J. Appl. Phys.*, vol. 30, p. 1323, 1959.

[7] C. G. Shull and S. Siegel, *Phys. Rev.*, vol. 75, p. 1008, 1949.

[8] G. E. Bacon, *Acta Cryst.*, vol. 5, p. 684, 1952.

[9] J. S. Kasper and R. M. Waterstrat, *Acta Cryst.*, vol. 9, p. 289, 1956.

[10] C. B. Walker and D. T. Keating, *Phys. Rev.*, vol. 130, p. 1726, 1963.

[11] F. Zirnike and J. Prins, *Z. Physik*, vol. 41, p. 184, 1927.

Among the liquids studied by neutrons are Hg,[1] Pb, Bi,[2] Pb-Bi alloys,[3] and the alkali metals.[4] Various liquefied gases have also been investigated, including the rare gases.[5] Some molten salts have been studied effectively by neutrons, and in some cases a comparison of the radial distribution curves obtained by neutrons and by x-rays gives added information. Similarly, amorphous substances (such as amorphous silica) yield more diffraction maxima with neutrons than with x-rays.

Neutron scattering from magnetic materials Every atom in which the net magnetic moment is not zero contributes magnetic scattering and nuclear scattering to the total neutron scattered intensity. In the transition metals it is the unpaired $3d$ electrons that are responsible for the magnetic scattering; in the rare earths it is the $4f$ electrons. The magnitude of this magnetic component of the scattering depends upon the *orientation* and *magnitude* of the unbalanced electron spins with respect to the moments of the incoming neutrons and also on the *spatial distribution* of the magnetic moments within the scattering material. The moments may be localized in the immediate vicinity of the individual atoms or ions, or they may be distributed in some more uniform manner throughout space; they may be arranged on a space lattice that is identical to that of the atoms, or on a multiple of this, or even with a periodicity that is incommensurate with the periodicity of the lattice of the atoms. Even the most highly localized electron magnetic moments are distributed at the radial distances from the nucleus of the $3d$ or $4f$ electron shells; as a consequence, there is an angle dependence of the magnetic scattering amplitude. Analysis of neutron diffraction patterns in terms of the magnetic structure of crystals has been the most significant contribution of neutron diffraction to solid-state science.

We consider first a paramagnetic substance, in which the atomic magnetic moments are randomly oriented and exhibit no periodicity. The magnetic scattering from these is incoherent and appears as a general diffuse background.[6] The intensity of the magnetic scattering is measured by $d\sigma_{pm}$, a differential cross section which determines the intensity of scattering per unit solid angle around the scatterer, and which is a function of the magnetic amplitude form factor for the magnetic scattering f_m, the spin quantum number of the atom, and the magnetic moment of the neutron γ_n. The factor f_m has the maximum value of unity when the scattering angle is zero and it decreases as the angle increases. Some crystals contain ions

[1] G. H. Vineyard, *J. Chem. Phys.*, vol. 22, p. 1665, 1954.

[2] O. Chamberlin, *Phys. Rev.*, vol. 77, p. 305, 1950. P. C. Sharrah and G. P. Smith, *J. Chem. Phys.*, vol. 21, p. 228, 1953.

[3] P. C. Sharrah, J. I. Pertz, and R. F. Kruh, *J. Chem. Phys.*, vol. 32, p. 241, 1960.

[4] N. S. Gingrich and L. Heaton, *J. Chem. Phys.*, vol. 34, p. 873, 1961.

[5] D. G. Henshaw and D. G. Hurst, *Phys. Rev.*, vol. 91, p. 122, 1953. D. G. Henshaw, D. G. Hurst, and N. K. Pope, *Phys. Rev.*, vol. 92, p. 1229, 1953. G. T. Clayton and L. Heaton, *Phys. Rev.*, vol. 121, p. 649, 1961.

[6] O. Halpern and M. H. Johnson, *Phys. Rev.*, vol. 55, p. 898, 1939.

that have not only electron spin but also orbital angular momenta contributing to their magnetism.[1]

Consider next a *ferromagnetic* metal in which an applied magnetic field has aligned all the atomic moments in the sample so that they are all parallel and pointing in the same direction. The magnetic scattering from all the atoms will then be coherent and will add vectorially. If the magnetic moments of the atoms lie in a plane which is made to be the reflecting plane, then each atom will scatter with the maximum possible magnetic scattering amplitude p, but with other reflecting planes p will be reduced to

$$p_{eff} = p \sin \alpha$$

where α is the angle between the normal to the reflecting plane and the direction of magnetization. Reflections for which $\sin \alpha = 0$ receive no contribution from magnetic scattering, a fact which can be used in determining how much of the total scattering is magnetic scattering. The actual differential scattering cross section for magnetic materials is a vectorial combination of the nuclear and magnetic contributions, with proper regard for the state of polarization of the neutron beam and the relative orientation of the moments of the neutrons and the atoms.

The coherent diffraction pattern for neutrons diffracting from a ferromagnetic substance consists of reflected beams with intensities proportional to a structure factor F^2 which is the sum of the nuclear and the magnetic contributions with the magnetic contributions from all the atoms being in phase.

We now consider *antiferromagnetic* materials. In the simplest types of these the moments at some atomic sites are antiparallel to those at other sites and form an ordered array. The net moment of the crystal is zero. In these simple types of crystals, the phase angle between the scattered waves from different atoms will be zero for parallel spin sites and π for the antiparallel sites. The scattering power will then vary from one site to another in the ordered fashion that is characteristic of a superlattice, resulting in magnetic superlattice reflections which would not occur in the nuclear scattering or in a ferromagnetic array of spins on the same atomic sites.

Another class of antiferromagnetic crystals has now been discovered in which spins are all parallel or antiparallel to a given direction but in which the effective moment varies sinusoidally in space, as indicated by Fig. 23–1*a*. The periodicity of this static spin-density modulation can be described by a vector **R** in reciprocal space.[2] If one moves through the crystal in the direction of the vector **R**, observing the atoms on successive planes normal to **R**, one finds a certain degree of magnetic alignment on each

[1] For a discussion of the quenching of orbital angular momenta and the magnetic scattering amplitude, see G. E. Bacon, "Neutron Diffraction," Clarendon, Oxford, 1962; or W. C. Koehler and E. O. Wollan, *Rev. Mod. Phys.*, vol. 25, p. 128, 1953; *Phys. Rev.*, vol. 92, p. 1380, 1953.

[2] This sinusoidal structure has sometimes been spoken of as a "static spin density wave" or a "linear static spin wave."

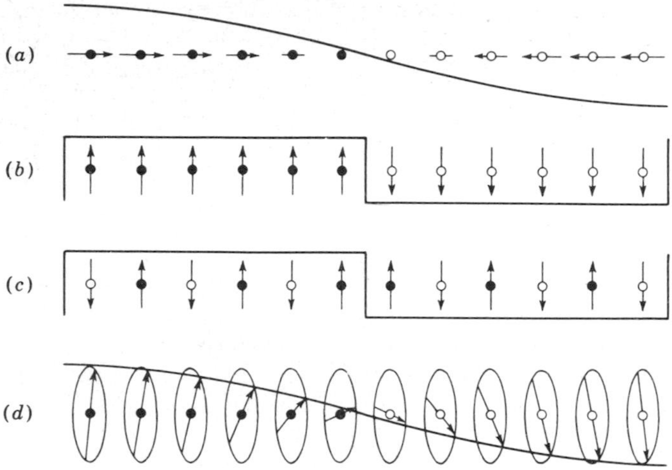

Fig. 23-1 Basic types of long-wavelength modulation in antiferromagnetic crystals: (*a*) Sinusoidal, (*b*) square-wave pattern of domains, (*c*) out-of-step domains, (*d*) helical.

plane, which varies from zero on some planes to a maximum on others in a sinusoidal fashion, but the direction of the magnetic moment on every plane is parallel or antiparallel to **R**. This periodic structure will generate satellites in the neutron diffraction pattern, as will another type of structure consisting of a layer structure of domains as indicated in Fig. 23–1*b* or *c*.

Some antiferromagnetic materials, for example holmium, have a *helical* arrangement of moments. In these, the moments in a certain plane are all parallel, but in the adjacent plane they are turned through an angle with respect to the first plane; in the next plane they are again turned through the same angle with respect to the preceding one, and so on. The successive turns give complete rotation in a certain distance, as indicated in Fig. 23–1*d*. In actual crystals the moments need not lie in the planes normal to the helical axes, but may be tipped out of these planes to some degree. Analysis shows that the magnetic scattering from this type of structure is also characterized by satellites around each nuclear reflection.[1] A careful study of diffraction patterns, particularly the satellites, can lead in favorable cases to an exact description of the type of magnetic structure.

Fourier methods can be employed with neutrons as well as with x-rays. For example, suitable single-crystal magnetic scattering data can give, by Fourier methods, the distribution of magnetic moments throughout the unit cell. Fundamental information on the orbital contribution to the magnetic moments is also derived from the observed magnetic scattering.

Polarized beams of neutrons can be produced by reflecting from a suitable ferromagnetic crystal placed in a magnetic field that saturates the magnetization, causing all moments in the crystal to be aligned in a suitable di-

[1] W. C. Koehler, *Acta Cryst.*, vol. 14, p. 535, 1961.

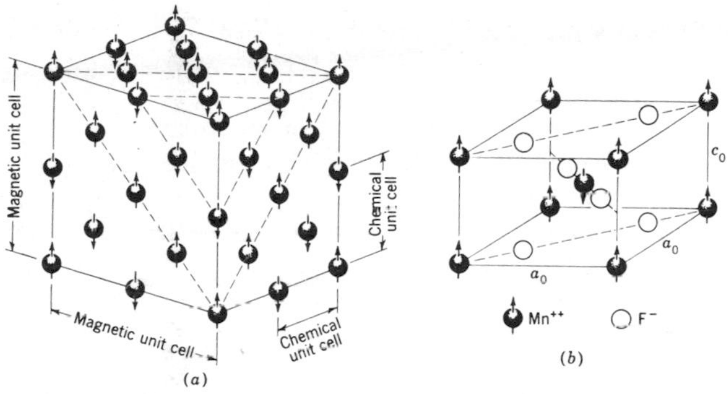

Fig. 23-2 Simple antiferromagnetic structures, with spin orientations indicated by arrows. (*a*) Arrangement of the Mn^{2+} ions in MnO (O atoms not shown). (After C. G. Shull, W. A. Strauser, and E. O. Wollan.) (*b*) Structure of MnF_2. (After R. A. Erickson, *Phys. Rev.*, vol. 90, p. 779, 1953.)

rection.[1] Data obtainable with polarized neutrons are of value when magnetic scattering is weak compared with nuclear scattering, or when certain ambiguities in the magnetic structure of a crystal can be resolved by the use of a polarized beam.

Summary of types of magnetic configurations Many antiferromagnetic crystals consist of two or more sublattices with the moments on some directed in opposition to the moments on the others and with the overall magnetization equal to zero. For example, in MnO (Fig. 23-2), (111) planes alternate with spins up and down.[2] The lattice distortion from the cubic cell that might be expected from such an arrangement is actually observed, although it is small. Different planes of the form {111} are favored in different domains of a polydomain single crystal.

Because of the different phase angles for the scattering amplitudes from differently directed spins, the presence of these sublattices causes superlattice reflections that are prohibited in the diffraction patterns of the nuclei alone. The unit cell for the magnetic diffraction pattern may be the same as or larger than the chemical unit cell, and the magnetic reflections may superimpose on nuclear reflections or may occur where there are no nuclear reflections.

The term *ferrimagnetic* has been applied to materials in which the antiferromagnetic arrangement does not cancel the magnetization completely, so that a net ferromagnetic component is left. Magnetite, Fe_3O_4, is an example.

Magnetic structures of metals and alloys Neutron diffraction has demonstrated the presence of antiferromagnetic structures in Cr, Mn,

[1] C. G. Shull, E. O. Wollan, and W. C. Koehler, *Phys. Rev.*, vol. 81, p. 527, 1951. R. Nathans et al., *J. Phys. Chem. Solids*, vol. 10, p. 138, 1959.

[2] C. G. Shull, W. A. Strauser, and E. O. Wollan, *Phys. Rev.*, vol. 83, p. 333, 1951.

α-U, many rare earths, and many alloys and compounds. The reader should consult a recent tabulation or a review article for a complete list since the number is growing rapidly.

The perfection of alignment of an antiferromagnetic material tends to be destroyed as temperature is increased, and may be destroyed completely at what is called the Néel temperature (corresponding to the Curie temperature of a ferromagnetic), may alter its characteristic lattice period, or may take up a new magnetic configuration. Externally applied magnetic fields also can transform the magnetic ordering from one pattern to another in some metals, or cause some domains to disappear.

When chromium, which is b.c.c., is cooled below 310°K,[1] it has diffraction characteristics that appear to indicate a sinusoidal structure, which in some samples persists to temperatures as high as 475°K. Upon cooling through 115°K the direction of the moments is altered by 90°. The structure of gamma-manganese, which is f.c.t., has moments parallel and antiparallel to [001] in alternate (001) layers.[2] Micron-size crystals of iron precipitated from solid solution in copper are f.c.c. in structure and are antiferromagnetic below a Néel temperature of roughly 8°K; a structure that accounts for the diffraction data consists of moments in alternate (001) planes nearly parallel or antiparallel to [001] as in γ-Mn, but with the spin vectors inclined about 18° from [001].[3]

The rare earths all become magnetically ordered at low temperatures. Gadolinium goes from a paramagnetic state to ferromagnetic on cooling, and the others go from paramagnetic to antiferromagnetic. On cooling further there is an additional magnetic transition in some of them. For example, terbium and dysprosium change from helical antiferromagnetic to ferromagnetic; holmium changes from one type of helical to another; erbium has a ferromagnetic component added to a sinusoidal configuration having a temperature-dependent period. The ferromagnetic moments are in the basal plane in dysprosium and along the *c* axis in erbium. Below 20°K, holmium and erbium have spins tipped appreciably out of the basal plane.

The interlayer turn angle of the helical arrangement in dysprosium decreases from 43.2 to 26.5° on cooling from the Néel temperature of 179 to 85°K, where the transition to ferromagnetism starts.[4] Terbium transforms on cooling to an ordered magnetic state at about 230°K and undergoes a transformation to a different magnetic structure at 220°K; in the narrow antiferromagnetic region between these temperatures, the structure is helical with an interlayer turn angle decreasing from 20.5° per layer at

[1] V. N. Bykov et al., *Soviet Phys. Dokl.*, vol. 4, p. 1070, 1960. G. Shirane and W. J. Takei *J. Phys. Soc. Japan*, Suppl. on Magnetism, 1962.

[2] G. E. Bacon et al., *Proc. Roy. Soc. (London)*, vol. A241, p. 233, 1957. D. Meneghett and S. S. Sidhu, *Phys. Rev.*, vol. 105, p. 130, 1957.

[3] S. C. Abrahams, L. Guttman, and J. S. Kasper, private communication, to be published.

[4] M. K. Wilkinson et al., *J. Appl. Phys.*, vol. 32, p. 48S, 1961.

the Néel point to 18.5° per layer at the 220°K transition. Below 220°K, terbium is ferromagnetic.[1]

Thulium has a simple oscillating z-component type of antiferromagnetic structure, similar to erbium, between 56 and 38°K.[2] Below 38°K the single-crystal diffraction pattern suggests that each atom has a magnetic moment along the c axis, alternating in sense with the sequence

$$++++---++++---,$$

i.e., with a period of seven layers.

Much more research with neutron diffraction can be expected on antiferromagnetic structures in alloys; studies have already been made of the magnetic structure of CrSb, MnAs, Mn_2Sb, MnBi, $MnAu_2$, MnAu, MnNi, and solid solutions of Mn in Cu, for example, and of Tb in Y.

Magnetic structures in compounds Helical structures also occur in compounds; the helix is parallel to the unique crystal axis in MnO_2,[3] $MnAu_2$,[4] and MnI_2,[5] for example.

The reviews of experimental data[6] list many antiferromagnetic compounds; a few important types are as follows: (1) Hexagonal layer structures (NiAs type) exist in MnTe, $FeCl_2$, $CoCl_2$, $NiCl_2$, FeS, and CrSb. (2) In MnO, MnS, FeO, CoO, and NiO the NaCl-type structures are variously distorted by interactions via intermediate oxygen atoms. The pattern in MnO (Fig. 23–2) produces rhombohedral distortion. The MnO-type antiferromagnetic structure is very common among the nitrides, phosphides, antimonides, and arsenides of the rare earths. (3) Tetragonal structures isomorphous with rutile (TiO_2) contain metal ions on a b.c.t. lattice which is divided into two sublattices with opposing moments (as in Fig. 23–2b), in MnF_2, FeF_2, and CoF_2, and similarly but with somewhat different spin orientations in NiF_2 and MnO_2. (4) Rhombohedral structures of space group $R\bar{3}C$ are characteristic of the trifluorides of Cr, Fe, and Co. In these each magnetic ion is coupled antiferromagnetically, via the intermediate anions, to all six of its nearest magnetic ions.[7] This type of coupling is called *superexchange*.

The ferrimagnetic compounds have as their most important class the spinels. Among these the ferrites have become of practical interest as magnetic materials for high-frequency applications because of their very

[1] W. C. Koehler, *J. Appl. Phys.*, vol. 36, p. 1078, 1965.

[2] W. C. Koehler et al., *J. Appl. Phys.*, vol. 33, p. 1124S, 1962.

[3] A. Yoshimori, *J. Phys. Soc. Japan*, vol. 14, p. 807, 1959.

[4] A. Herpin, P. Merial, and J. Villain, *Compt. Rend.*, vol. 249, p. 1334, 1949.

[5] J. W. Cable et al., *Phys. Rev.*, vol. 125, p. 1860, 1962.

[6] T. Nagamiya, K. Yosida and R. Kubo, *Advan. Phys.*, vol. 4, p. 1, 1955. A. B. Lidiard, *Rept. Progr. Phys.*, vol. 17, p. 201, 1954. C. Kittel, "Introduction to Solid State Physics," 2d ed., Wiley, New York, 1956. A. Arrott, in G. T. Rado and H. Suhl (eds.), "Magnetism," vol. 2B, Academic, New York, 1966.

[7] E. O. Wollan et al., *Phys. Rev.*, vol. 112, p. 1132, 1958.

high resistivities. Many of these have the formula $MOFe_2O_3$ where M is a divalent metal ion of Mn, Co, Ni, Cu, Mg, Zn, Cd, Fe, or a mixture of these, and have a cubic structure isomorphous with spinel ($MgAl_2O_4$). The ions with magnetic moments are distributed among the interstices between the oxygen atoms, which are larger than the metal ions and are arranged in cubic (or distorted) close packing. The list of *spinels* is large, for the general formula is PQ_2X_4, where $X = O^{2-}$, S^{2-}, or Se^{2-} and the divalent ions P are usually either Mn^{2+}, Fe^{2+}, Ni^{2+}, Cu^{2+}, Zn^{2+}, or Mg^{2+}; Q is a trivalent ion such as Mn^{3+}, Fe^{3+}, Co^{3+}, Al^{3+}, or Ga^{3+}. Many compounds exist with mixtures of these ions, as may be seen from published summaries,[1] and there are many types of distributions of the divalent and trivalent ions in the different spinels.

Ferrimagnetic structures are also found among the garnets, which belong to the cubic space group O_h^{10}—$Ia3d$. There are eight molecules $P_3Q_2R_3O_{12}$ per unit cell (P, Q, and R are metal ions that are usually but not always different from each other). In these, one set of equivalent lattice sites is not always occupied exclusively by one kind of ion. Compounds with the magnetoplumbite structure include useful permanent magnet materials; the formula for these is $AO\cdot 6B_2O_3$ where $A = Ba^{2+}$, Sr^{2+}, or Pb^{2+}; and $B = Fe^{3+}$, Al^{3+}, Ga^{3+}, or Cr^{3+}.

Inelastic scattering of neutrons A field of neutron scattering that is becoming of increasing importance in solid-state research is the measurement of inelastic scattering that involves gain or loss of energy by the neutron in its interaction with the scattering body. With very high flux reactors it is possible to scatter a monochromatic beam from a sample and then analyze the energy distribution in the scattered beam with a single-crystal spectrometer (or a neutron velocity selector).[2] The energy analysis of the scattered neutrons can be interpreted in terms of the vibration spectrum of the solid—and without the large corrections necessary when x-rays are used for this purpose. With magnetic solids, the analysis of magnetic-inelastic scattering yields information about the magnetic energy levels in the atoms, and affords tests of magnetic spin-wave theory.[3]

[1] E. W. Gorter, *Philips Res. Rept.*, vol. 9, p. 295, 1954. W. P. Wolf, *Rept. Progr. Phys.* vol. 24, p. 212, 1961.

[2] B. N. Brockhouse and A. T. Stewart, *Phys. Rev.*, vol. 100, p. 756, 1955; *Rev. Modern Phys.*, vol. 30, pp. 236, 250, 1958.

[3] For reviews of these and other inelastic scattering methods, see: M. K. Wilkinson E. O. Wollan, and W. C. Koehler, *Ann. Rev. Nucl. Sci.*, vol. 11, p. 303, 1961. G. R Ringo in S. Flügge (ed.), "Handbuch der Physik," vol. XXXII, p. 622, Springer-Verlag Berlin, 1957 (in English).

Appendix

Table A-1 Physical constants[1] and numerical factors

N, Avogardo's number $(6.02257 \pm 0.00009) \times 10^{23}$ (g mole)$^{-1}$ (chemical scale, $^{12}C = 12$ for use with International Atomic Weights)
h, Planck's constant $(6.62554 \pm 0.00015) \times 10^{-27}$ erg-sec
$\hbar$, $= h/2\pi = (1.05443 \pm 0.00004) \times 10^{-27}$ erg-sec
m, electron mass $(9.10904 \pm 0.00013) \times 10^{-28}$ g
e, electronic charge $(4.80296 \pm 0.00006) \times 10^{-10}$ esu
$(1.60206 \pm 0.00003) \times 10^{-20}$ emu $= e/c$
F, Faraday constant, $= ne = (2.89366 \pm 0.00003)$ (g mole)$^{-1}$
F', Faraday constant, $= Ne/c = (9652.19 \pm 0.11)$ emu (g mole)$^{-1}$
λ_0, x-ray wavelength associated with 1 ev $(12398.04 \pm 0.12) \times 10^{-8}$ ev cm
e/m, $(1.78590 \pm 0.00002) \times 10^{7}$ emu g^{-1}
$(5.27305 \pm 0.00007) \times 10^{17}$ esu g^{-1}
h/e, $(1.37942 \pm 0.00002) \times 10^{-17}$ erg-sec (esu)$^{-1}$
c, velocity of light, $299{,}792.5 \pm 0.2$ km sec^{-1}
n_0, Loschmidt's number $(2.68702 \pm 0.00008) \times 10^{19}$ cm^{-3}
k, Boltzmann's constant $(1.38053 \pm 0.00006) \times 10^{-16}$ erg deg^{-1}
R_0, gas constant per mole $(8.31696 \pm 0.00034) \times 10^{7}$ erg mole^{-1} deg^{-1}
V_0, standard volume of perfect gas $(22.4146 \pm 0.0006) \times 10^{3}$ cm^3 mole^{-1}
M_1, mass of atom of unit atomic weight $= 1/N = (1.66042 \pm 0.00003) \times 10^{-24}$ g
λ_g/λ_s ratio, grating wavelengths to Siegbahn wavelengths of x-rays, 1.002076 ± 0.000005 (Note: By international agreement of 1947 Siegbahn's values are converted to angstroms by multiplying by 1.00202.)
μ_B, Bohr magneton, $he/4\pi mc = (0.927314 \pm 0.000021) \times 10^{-20}$ erg gauss^{-1}
Ice point, $T_0 = 273.16$°K
1 electron volt $= (1.602095 \times 10^{-12} \pm 0.000022 \times 10^{-12})$ ergs
1 electron volt per molecule = 23.05 kcal per mole
1 kcal $= 4.186 \times 10^{10}$ ergs
1 kb (kilobar) $= 10^9$ dynes per cm^2 = 986.92 atmospheres = 14,504 psi
1 cal (15°C) = 4.186 joules
1 volt $= \frac{1}{300}$ esu $= 10^8$ emu
1 radian = 57.29578 deg
1 in. = 2.54001 cm
1 A $= 10^{-8}$ cm $\approx$ A* (see Table A-5, p. 622)
1 micron (μ) $= 10^{-3}$ mm $= 10^4$ A
1 psi = 0.070307 kg per cm^2

Table A-2 Crystal geometry

Any formula for the cubic system is obtained from the corresponding formula for the orthorhombic system by setting $a = b = c$; any formula for the tetragonal system is obtained from the corresponding formula for the orthorhombic system by setting $a = b$.

[1] E. R. Cohen et al., *Rev. Mod. Phys.*, vol. 27, p. 363, 1955. Adjusted to 1962 by E. R. Cohen, J. W. DuMond, and A. G. McNish in "International Tables for X-ray Crystallography," vol. 3, p. 40, Kynoch, Birmingham, England, 1962. See also Appendix Table A-5 for λ_g/λ_s.

Unit cells in real and reciprocal space The unit cell in a crystal (i.e. in real space) is specified by cell edges and angles a, b, c, α, β, γ; the cell edges are the vectors **a**, **b**, **c** and the cell volume is $V = \mathbf{a} \cdot \mathbf{b} \times \mathbf{c}$. The corresponding reciprocal lattice, defined in Chap. 4, has axes $\mathbf{a}^*$, $\mathbf{b}^*$, $\mathbf{c}^*$ and angles α^*, β^*, γ^*, where

$$\mathbf{a}^* = \frac{\mathbf{b} \times \mathbf{c}}{V} \qquad \mathbf{b}^* = \frac{\mathbf{a} \times \mathbf{c}}{V} \qquad \mathbf{c}^* = \frac{\mathbf{a} \times \mathbf{b}}{V}$$

$$\cos \alpha^* = \frac{\cos \beta \cos \gamma - \cos \alpha}{\sin \alpha \sin \beta} \qquad \cos \beta^* = \frac{\cos \alpha \cos \gamma - \cos \beta}{\sin \alpha \sin \gamma}$$

$$\cos \gamma^* = \frac{\cos \alpha \cos \beta - \cos \gamma}{\sin \alpha \sin \beta}$$

The formulas for the volume of a unit cell in crystals reduce to the following:

Triclinic and all crystal systems:

$$V = abc\sqrt{1 - \cos^2 \alpha - \cos^2 \beta - \cos^2 \gamma + 2 \cos \alpha \cos \beta \cos \gamma}$$

Monoclinic: $V = abc \sin \beta$

Orthorhombic: $V = abc$

Hexagonal: $V = \frac{\sqrt{3}}{2} a^2 c = 0.866a^2c$

Rhombohedral: $V = a^3\sqrt{1 - 3\cos^2 \alpha + 2\cos^3 \alpha}$

Interplanar spacings in crystals Spacings of (hkl) planes in the simple space lattices, d_{hkl}, are given by the lengths of vectors $\mathbf{r}_{hkl}$ in the reciprocal lattice, with $d_{hkl} = 1/|\mathbf{r}_{hkl}|$. Formulas for d are conveniently put in terms of $1/d^2$, a quantity referred to in some books as Q_{hkl}. Diffraction angles θ then are obtained from Bragg's law in the form $\sin^2 \theta = (\lambda^2/4)(1/d^2)$ as explained in Chap. 4.

Triclinic and all crystal systems:

$$\frac{1}{d^2} = h^2a^{*2} + k^2b^{*2} + l^2c^{*2} + 2hka^*b^* \cos \gamma^* + 2klb^*c^* \cos \alpha^* + 2lhc^*a^* \cos \beta^*$$

Monoclinic: set $\alpha^* = \gamma^* = 90°$ in the above.

Rhombohedral: set $\alpha^* = \beta^* = \gamma^*$ and $a^* = b^* = c^*$.

Hexagonal: Since $a^* = b^* = 2/(a\sqrt{3})$ and $c^* = 1/c$,

$$\frac{1}{d^2} = \frac{4}{3}\frac{h^2 + hk + k^2}{a^2} + \frac{l^2}{c^2}$$

Orthorhombic: $\frac{1}{d^2} = \frac{h^2}{a^2} + \frac{k^2}{b^2} + \frac{l^2}{c^2}$

Angles between planes in a crystal The angle ϕ between planes $(h_1k_1l_1)$ and $(h_2k_2l_2)$ of spacings d_1 and d_2 is equal to the angle between the reciprocal lattice vectors r_1^* and r_2^* that are normal to these planes; therefore $\mathbf{r}_1^* \cdot \mathbf{r}_2^* = r_1^* r_2^* \cos \phi$, from which one obtains:

Triclinic and all crystal systems:

$$\cos \phi = \{h_1h_2a^* + k_1k_2b^* + l_1l_2c^* + (h_1k_2 + h_2k_1)a^*b^* \cos \gamma^* + (k_1l_2 + k_2l_1)b^*c^* \cos \alpha^* + (l_1h_2 + l_2h_1)c^*a^* \cos \beta^*\}d_1d_2$$

For monoclinic and rhombohedral, set axes and angles as in the preceding case.

Hexagonal:

$$\cos \phi = \frac{h_1h_2 + k_1k_2 + \frac{1}{2}(h_1k_2 + h_2k_1) + \frac{3}{4}\frac{a^2}{c^2}l_1l_2}{\sqrt{\left(h_1^2 + k_1^2 + h_1k_1 + \frac{3}{4}\frac{a^2}{c^2}l_1^2\right)\left(h_2^2 + k_2^2 + h_2k_2 + \frac{3}{4}\frac{a^2}{c^2}l_2^2\right)}}$$

Orthorhombic:

$$\cos \phi = \frac{\frac{h_1h_2}{a^2} + \frac{k_1k_2}{b^2} + \frac{l_1l_2}{c^2}}{\sqrt{\left(\frac{h_1^2}{a^2} + \frac{k_1^2}{b^2} + \frac{l_1^2}{c^2}\right)\left(\frac{h_2^2}{a^2} + \frac{k_2^2}{b^2} + \frac{l_2^2}{c^2}\right)}}$$

Cubic: For tabulated values see Chap. 2, page 40.

Reciprocal lattice axes and angles (See Chap. 4.)

Triclinic and all crystal systems:

$$a^* = (bc \sin \alpha)/V, \quad b^* = (ca \sin \beta)/V, \quad c^* = (ab \sin \gamma)/V$$

$$\cos \alpha^* = (\cos \beta \cos \gamma - \cos \alpha)/(\sin \beta \sin \gamma)$$

$$\cos \beta^* = (\cos \gamma \cos \alpha - \cos \beta)/(\sin \gamma \sin \alpha)$$

$$\cos \gamma^* = (\cos \alpha \cos \beta - \cos \gamma)/(\sin \alpha \sin \beta)$$

$$V^* = a^*b^*c^* \sin \alpha \sin \beta^* \sin \gamma^*$$

$$= a^*b^*c^* \sin \alpha^* \sin \beta \sin \gamma^*$$

$$= a^*b^*c^* \sin \alpha^* \sin \beta^* \sin \gamma$$

Monoclinic: $a^* = 1/(a \sin \beta), \quad b^* = 1/b, \quad c^* = 1/c \sin \beta$

$$\alpha^* = \gamma^* = 90°, \quad \beta^* = 180° - \beta$$

Orthorhombic: $a^* = 1/a, \quad b^* = 1/b, \quad c^* = 1/c, \quad \alpha^* = \beta^* = \gamma^* = 90°$

Hexagonal: $a^* = b^* = 2/(a\sqrt{3}), \quad c^* = 1/c, \quad \alpha^* = \beta^* = 90°, \quad \gamma^* = 60$

Rhombohedral: $a^* = 1/(a \sin \alpha \sin \alpha^*), \quad \cos (\alpha^*/2) = 1/[2 \cos (\alpha/2)]$

Zone laws Crystal planes parallel to a given direction $[uvw]$ are planes of the zone $[uvw]$. Since the reciprocal lattice vector $\mathbf{r}_{hkl}$ is normal to the plane (hkl), the condition that (hkl) belongs to zone $[uvw]$ is that $\mathbf{r}^* \cdot \mathbf{r} = 0$ where r is the crystal lattice vector defined by $\mathbf{r} = u\mathbf{a} + v\mathbf{b} + w\mathbf{c}$; the condition is therefore

$$hu + kv + lw = 0$$

Planes $(h_1k_1l_1)$ and $(h_2k_2l_2)$ define a zone; its indices $[uvw]$ are obtained from two simultaneous equations like the above; therefore,

$$u:v:w = (k_1l_2 - k_2l_1):(l_1h_2 - l_2h_1):(h_1k_2 - h_2k_1)$$

If $(h_1k_1l_1)$ and $(h_2k_2l_2)$ belong to a zone, then the planes with indices $(h_1 \pm h_2, k_1 \pm k_2, l_1 \pm l_2)$ also belong to this zone, and also the planes $(mh_1 \pm nh_2, mk_1 \pm nk_2, ml_1 \pm nl_2)$.

Lattice points and planes The condition for the lattice point with coordinates uvw, at the position specified by vector $\mathbf{r}_{uvw} = u\mathbf{a} + v\mathbf{b} + w\mathbf{c}$, to lie on the lattice plane (hkl), is as follows: With the origin at a lattice point, the (hkl) planes of any simple lattice will lie at distances from the origin given by $nd_{hkl} = n/|\mathbf{r}^*|$, where n takes integral values. Therefore, the point uvw will lie on the nth plane of this set from the origin if the projection of the vector $\mathbf{r}$ on $\mathbf{r}^*$ is $nd_{hkl} = n/|\mathbf{r}^*|$, therefore if $\mathbf{r} \cdot \mathbf{r}^*/|\mathbf{r}^*| = n/|\mathbf{r}^*|$ or

$$hu + kv + lw = n$$

In the reciprocal lattice, points satisfying this relationship lie on a plane and form the nth layer line of a rotation photograph when a crystal is rotated about crystal axis $[uvw]$.

Identity distances The shortest distance between identical points along a row in the direction $[uvw]$ in a simple space lattice, i.e., the *identity distance* I, is given by:

Triclinic and all other crystal systems:

$$I = \sqrt{a^2u^2 + b^2v^2 + c^2w^2 + 2bcvw\cos\alpha + 2cawu\cos\beta + 2abuv\cos\gamma}$$

Hexagonal: $I = a\sqrt{u^2 + v^2 + \dfrac{w^2c^2}{a^2} - uv}$

Orthorhombic: $I = \sqrt{u^2a^2 + v^2b^2 + w^2c^2}$

Angles between directions The angle ρ between lattice directions $[u_1v_1w_1]$ and $[u_2v_2w_2]$ along which the identity distances are I_1 and I_2, respectively, is given by:

Triclinic and all other crystal systems:

$$\cos\rho = [a^2u_1u_2 + b^2v_1v_2 + c^2w_1w_2 + bc(v_1w_2 + v_2w_1)\cos\alpha + ac(w_1u_2 + w_2u_1)\cos\beta + ab(u_1v_2 + u_2v_1)\cos\gamma]/I_1I_2$$

Hexagonal: $\cos\rho = \left[u_1u_2 + v_1v_2 - \frac{1}{2}(u_1v_2 + u_2v_1) + w_1w_2\left(\dfrac{c}{a}\right)^2\right]/I_1I_2$

$$\textit{Orthorhombic: } \cos\rho = \frac{a^2u_1u_2 + b^2v_1v_2 + c^2w_1w_2}{\sqrt{a^2u_1^2 + b^2v_1^2 + c^2w_1^2}\sqrt{a^2u_2^2 + b^2v_2^2 + c^2w_2^2}}$$

Cubic: The angles are identical with the angles between planes, which are tabulated in Chap. 2, and in references given there.

Perpendicularity of line and plane The direction $[uvw]$ is necessarily perpendicular to the plane having exactly the same indices *only* in the cubic system. The general condition[1] for perpendicularity between direction $[uvw]$ and plane (hkl) reduces to the following special cases:

$$\textit{Monoclinic: } \frac{a}{N}(au + cw\cos\beta) = \frac{b^2}{k}v = \frac{c}{l}(au\cos\beta + cw)$$

$$\text{or } \frac{ua}{\dfrac{h}{a} - \dfrac{l}{c}\cos\beta} = \frac{vb}{\dfrac{k}{b}\sin^2\beta} = \frac{wc}{\dfrac{l}{c} - \dfrac{h}{a}\cos\beta}$$

$$\textit{Orthorhombic: } \frac{a^2}{h}u = \frac{b^2}{k}v = \frac{c^2}{l}w$$

$$\textit{Tetragonal: } \frac{a^2}{h}u = \frac{a^2}{k}v = \frac{c^2}{l}w$$

Therefore $(001) \perp [001]$ and $(hk0) \perp [hk0]$.

Hexagonal: $[uvw]$ (in 3-axes system) is perpendicular to $(hk\cdot l)$ if

$$\frac{1}{h}\left(u - \frac{v}{2}\right) = \frac{1}{k}\left(v - \frac{u}{2}\right) = \frac{1}{l}w\frac{c^2}{a^2}$$

or if

$$\frac{1}{h}\left(h + \frac{k}{2}\right) = \frac{1}{v}\left(k + \frac{h}{2}\right) = \frac{1}{w}l\frac{3}{4}\frac{a^2}{c^2}$$

Therefore, $(0001) \perp [001]$ and $(hk\cdot l) \perp [2h + k, h + 2k, 0]$.

Rhombohedral:

$$\frac{1}{h}[u + (v + w)\cos\alpha] = \frac{1}{k}[v + (w + u)\cos\alpha] = \frac{1}{l}[w + (u + v)\cos\alpha]$$

or

$$\frac{1}{u}[h(1 + \cos\alpha) - (k + l)\cos\alpha] = \frac{1}{v}[k(1 + \cos\alpha) - (l + h)\cos\alpha]$$

$$= \frac{1}{w}[l(1 + \cos\alpha) - (h + k)\cos\alpha]$$

Therefore, $(111) \perp [111]$ and every plane containing $[111]$ is perpendicular to some row in (111).

[1] J. D. H. Donnay and G. Donnay in "International Tables for X-Ray Crystallography," vol. 2, p. 106, Kynoch, Birmingham, England, 1959.

Table A-3 Quadratic forms*

	hkl	hk0			hkl	hk0	
N	Cubic h,k,l	Tetrag. h,k	Hex. k,l	N	Cubic h,k,l	Tetrag. h,k	Hex. k,l
1	1, 0, 0	1, 0	1, 0	29	5, 2, 0; 4, 3, 2	5, 2	
2	1, 1, 0	1, 1		30	5, 2, 1		
3	1, 1, 1; *F*		1, 1	31			5, 1
4	2, 0, 0; *F*	2, 0	2, 0	32	4, 4, 0; *F*	4, 4	
5	2, 1, 0	2, 1		33	5, 2, 2; 4, 4, 1		
6	2, 1, 1			34	5, 3, 0; 4, 3, 3	5, 3	
7			2, 1	35	5, 3, 1; *F*		
8	2, 2, 0; *F*	2, 2		36	6, 0, 0; 4, 4, 2; *F*	6, 0	6, 0
9	3, 0, 0; 2, 2, 1	3, 0	3, 0	37	6, 1, 0	6, 1	4, 3
10	3, 1, 0	3, 1		38	6, 1, 1; 5, 3, 2		
11	3, 1, 1; *F*			39			5, 2
12	2, 2, 2; *F*		2, 2	40	6, 2, 0; *F*	6, 2	
13	3, 2, 0	3, 2	3, 1	41	6, 2, 1; 5, 4, 0; 4, 4, 3	5, 4	
14	3, 2, 1			42	5, 4, 1		
15				43	5, 3, 3; *F*		6, 1
16	4, 0, 0; *F*	4, 0	4, 0	44	6, 2, 2; *F*		
17	4, 1, 0; 3, 2, 2	4, 1		45	6, 3, 0; 5, 4, 2	6, 3	
18	4, 1, 1; 3, 3, 0	3, 3		46	6, 3, 1		
19	3, 3, 1; *F*		3, 2	47			
20	4, 2, 0; *F*	4, 2		48	4, 4, 4; *F*		4, 4
21	4, 2, 1		4, 1	49	7, 0, 0; 6, 3, 2	7, 0	7, 0; 5, 3
22	3, 3, 2			50	7, 1, 0; 5, 5, 0; 5, 4, 3	7, 1; 5, 5	
23				51	7, 1, 1; 5, 5, 1; *F*		
24	4, 2, 2; *F*			52	6, 4, 0; *F*	6, 4	6, 2
25	5, 0, 0; 4, 3, 0	5, 0	5, 0	53	7, 2, 0; 6, 4, 1	7, 2	
26	5, 1, 0; 4, 3, 1	5, 1		54	7, 2, 1; 6, 3, 3; 5, 5, 2		
27	5, 1, 1; 3, 3, 3; *F*		3, 3	55			
28			4, 2	56	6, 4, 2; *F*		

Table A-3 Quadratic forms* (Cont.)

	hkl	*hk*0			*hkl*	*hk*0	
N	Cubic *h,k,l*	Tetrag. *h,k*	Hex. *k,l*	*N*	Cubic *h,k,l*	Tetrag. *h,k*	Hex. *k,l*
57	7,2,2; 5,4,4		7, 1	85	9,2,0; 7,6,0	9, 2; 7, 6	
58	7,3,0	7, 3		86	9,2,1; 7,6,1; 6,5,5		
59	7,3,1; 5,5,3; *F*			87			
60				88	6,6,4; *F*		
61	6,5,0; 6,4,3	6, 5	5, 4	89	9,2,2; 8,5,0; 8,4,3; 7,6,2	8, 5	
62	7,3,2; 6,5,1			90	9,3,0; 8,5,1; 7,5,4	9, 3	
63			6, 3	91	9,3,1; *F*		6, 5; 9, 1
64	8,0,0; *F*	8, 0	8, 0	92			
65	8,1,0; 7,4,0; 6,5,2	8, 1; 7, 4		93	8,5,2		7, 4
66	8,1,1; 7,4,1; 5,5,4			94	9,3,2; 7,6,3		
67	7,3,3; *F*		7, 2	95			
68	8,2,0; 6,4,4; *F*	8, 2		96	8,4,4; *F*		
69	8,2,1; 7,4,2			97	9,4,0; 6,6,5	9, 4	8, 3
70	6,5,3			98	9,4,1; 8,5,3; 7,7,0	7, 7	
71				99	9,3,3; 7,7,1; 7,5,5; *F*		
72	8,2,2; 6,6,0; *F*	6, 6		100	10,0,0; 8,6,0; *F*	10, 0; 8, 6	10, 0
73	8,3,0; 6,6,1	8, 3	8, 1	101	10,1,0; 9,4,2; 8,6,1; 7,6,4	10, 1	
74	8,3,1; 7,5,0; 7,4,3	7, 5		102	10,1,1; 7,7,2		
75	7,5,1; 5,5,5; *F*		5, 5	103			9, 2
76	6,6,2; *F*		6, 4	104	10,2,0; 8,6,2; *F*	10, 2	
77	8,3,2; 6,5,4			105	10,2,1; 8,5,4		
78	7,5,2			106	9,5,0; 9,4,3	9, 5	
79			7, 3	107	9,5,1; 7,7,3; *F*		
80	8,4,0; *F*	8, 4		108	10,2,2; 6,6,6; *F*		6, 6
81	9,0,0; 8,4,1; 7,4,4; 6,6,3	9, 0	9, 0	109	10,3,0; 8,6,3	10, 3	7, 5
82	9,1,0; 8,3,3	9, 1		110	10,3,1; 9,5,2; 7,6,5		
83	9,1,1; 7,5,3; *F*			111			10, 1
84	8,4,2; *F*		8, 2	112			8, 4

Table A-3 Quadratic forms* (Cont.)

N	hkl: Cubic h,k,l	hk0: Tetrag. h,k	hk0: Hex. k,l
113	10,3,2; 9,4,4; 8,7,0	8, 7	
114	8,7,1; 8,5,5; 7,7,4		
115	9,5,3; *F*		
116	10,4,0; 8,6,4; *F*	10, 4	
117	10,4,1; 9,6,0; 8,7,2	9, 6	9, 3
118	10,3,3; 9,6,1		
119			
120	10,4,2; *F*		
121	11,0,0; 9,6,2; 7,6,6	11, 0	11, 0
122	11,1,0; 9,5,4; 8,7,3	11, 1	10, 2
123	11,1,1; 7,7,5; *F*		
124			
125	11,2,0; 10,5,0; 10,4,3; 8,6,5	11, 2; 10, 5	
126	11,2,1; 10,5,1; 9,6,3		
127			7, 6
128	8,8,0; *F*	8, 8	
129	11,2,2; 10,5,2; 8,8,1; 8,7,4		8, 5
130	11,3,0; 9,7,0	11, 3; 9, 7	
131	11,3,1; 9,7,1; 9,5,5; *F*		
132	10,4,4; 8,8,2; *F*		
133	9,6,4		11, 1; 9, 4
134	11,3,2; 10,5,3; 9,7,2; 7,7,6		
135			
136	10,6,0; 8,6,6; *F*	10, 6	

N	hkl: Cubic h,k,l	hk0: Tetrag. h,k	hk0: Hex. k,l
137	11,4,0; 10,6,1; 8,8,3	11, 4	
138	11,4,1; 8,7,5		
139	11,3,3; 9,7,3; *F*		10, 3
140	10,6,2; *F*		
141	11,4,2; 10,5,4		
142	9,6,5		
143			
144	12,0,0; 8,8,4; *F*	12, 0; 12, 1; 9, 8	12, 0
145	12,1,0; 10,6,3; 9,8,0		
146	12,1,1; 11,5,0; 11,4,3; 9,8,1; 9,7,4	11, 5	
147	11,5,1; 7,7,7; *F*		11, 2; 7, 7
148	12,2,0; *F*	12, 2	8, 6
149	12,2,1; 10,7,0; 9,8,2; 8,7,6	10, 7	
150	11,5,2; 10,7,1; 10,5,5		
151			9, 5
152	12,2,2; 10,6,4; *F*		
153	12,3,0; 11,4,4; 10,7,2; 9,6,6; 8,8,5	12, 3	
154	12,3,1; 9,8,3		
155	11,5,3; 9,7,5; *F*		
156			10, 4
157	12,3,2; 11,6,0	11, 6	12, 1
158	11,6,1; 10,7,3		
159			

* Reflections from f.c.c. crystals are possible only for entries marked *F* and from b.c.c. only if N is even ($N = h^2 + k^2 + l^2$). Values of N for $hk0$ reflections of tetragonal and hexagonal crystals are also given.

Table A-4 Mass absorption coefficients (μ/ρ) for x-rays*

Radiation λ in A		Mo $K\alpha$ 0.7107	Cu $K\alpha$ 1.5418	Co $K\alpha$ 1.7902	Radiation λ in A		Mo $K\alpha$ 0.7107	Cu $K\alpha$ 1.5418	Co $K\alpha$ 1.7902
Absorber		Values			Absorber		Values		
H	1	*0.380*	*0.432*	*0.464*	Tc	43	19.7	172	257
He	2	*0.207*	*0.383*	*0.491*	Ru	44	21.1	183	272
Li	3	*0.217*	*0.716*	*1.03*	Rh	45	22.6	194	288
Be	4	*0.298*	*1.50*	*2.25*	Pd	46	24.1	206	304
B	5	*0.392*	*2.39*	*3.63*	Ag	47	25.8	218	321
C	6	*0.625*	*4.60*	7.07	Cd	48	27.5	231	338
N	7	*0.916*	7.52	11.6	In	49	29.3	243	356
O	8	*1.31*	11.5	17.8	Sn	50	31.1	256	373
F	9	*1.80*	16.4	25.4	Sb	51	33.1	270	391
Ne	10	*2.47*	22.9	35.4	Te	52	35.0	282	407
Na	11	*3.21*	30.1	46.5	I	53	37.1	294	422
Mg	12	*4.11*	38.6	59.5	Xe	54	39.2	306	436
Al	13	5.16	48.6	74.8	Cs	55	41.3	318	450
Si	14	6.44	60.6	93.3	Ba	56	43.5	330	463
P	15	7.89	74.1	114	La	57	45.8	341	475
S	16	9.55	89.1	136	Ce	58	48.2	352	486
Cl	17	11.4	106	161	Pr	59	50.7	363	497
A	18	13.5	123	187	Nd	60	53.2	374	*543*
K	19	15.8	143	215	Pm	61	55.9	386	*327*
Ca	20	18.3	162	243	Sm	62	58.6	397	*344*
Sc	21	21.1	184	273	Eu	63	61.5	*425*	156
Ti	22	24.2	208	308	Gd	64	64.4	*439*	165
V	23	27.5	233	343	Tb	65	67.5	*273*	173
Cr	24	31.1	260	381	Dy	66	70.6	*286*	182
Mn	25	34.7	285	414	Ho	67	73.9	128	191
Fe	26	38.5	308	52.8	Er	68	77.3	134	199
Co	27	42.5	313	61.1	Tm	69	80.8	140	208
Ni	28	46.6	45.7	70.5	Yb	70	84.5	146	217
Cu	29	50.9	52.9	81.6	Lu	71	88.2	153	226
Zn	30	55.4	60.3	93.0	Hf	72	91.7	159	235
Ga	31	60.1	67.9	105	Ta	73	95.4	166	244
Ge	32	64.8	75.6	116	W	74	99.1	172	253
As	33	69.7	83.4	128	Re	75	103	179	262
Se	34	74.7	91.4	140	Os	76	106	186	272
Br	35	79.8	99.6	152	Ir	77	110	193	282
Kr	36	84.9	108	165	Pt	78	113	200	291
Rb	37	90.0	117	177	Au	79	115	208	302
Sr	38	95.0	125	190	Hg	80	117	216	312
Y	39	100	134	203	Tl	81	119	224	323
Zr	40	15.9	143	216	Pb	82	120	232	334
Nb	41	17.1	153	230	Bi	83	120	240	346
Mo	42	18.4	162	243					

* Values given in italics are of low accuracy. Units are cm per g. The values are for $(\mu/\rho)_{total}$, consisting of photoelectric absorption plus scattering (both modified and unmodified); they are from "International Tables for X-ray Crystallography," vol. 3, p. 162, Kynoch, Birmingham, England, 1964.

Table A–5 X-ray emission and absorption wavelengths in angstroms

The strongest emission lines of the K series and the K absorption edges are listed in absolute angstroms; the weaker diagram lines, β_4, β_5, β_x, $O_{\mathrm{II,III}}$, and L_{I}, and nondiagram lines are omitted from this table. All wavelengths are from Bearden.[1] Although they are given in A, they are actually given in terms of the recently introduced A* unit, where the A* is *defined* by the wavelength of a particularly suitable line; W $K\alpha_1$: $\lambda_{\mathrm{W}\,K\alpha_1} =$ 0.2090100. This definition is such that the A* unit is very nearly equal to the angstrom, since 1 A* = 1 A ± 5 parts per million (probable error) according to the best available information up to the present. Nevertheless, although indistinguishable from the angstrom in most crystallographic work, the A* unit is defined differently and is operationally distinct from the angstrom.[2] Probable errors vary from 1 to 9 in the last digit quoted (specific values of errors as well as wavelengths of the weaker lines are given in the book cited above).

[1] J. A. Bearden, "X-ray Wavelengths," Clearinghouse for Federal Scientific and Technical Information, National Bureau of Standards, U.S. Department of Commerce, Springfield, Va., 1964.

[2] The A* unit avoids some confusion arising from the fact that we have in effect in the past had two x-unit standards, one apparently based on Mo $K\alpha_1$ as 707.831 xu, and the other based on Cu $K\alpha_1$ as 1537.396 xu. These disagree by approximately 20 parts per million. J. A. Bearden, *Phys. Rev.*, vol. 137, p. B 181, 1965.

Table A–5 X-ray emission and absorption wavelengths in angstroms

Element	$K\alpha_2$	$K\alpha_1$	$K\beta_3$	$K\beta_1$	$K\beta_2^{II}$	$K\beta_2^{I}$	K absorption edge
Approximate intensity	50	100	15 to 30		2 to 5		
1 H							
2 He							
3 Li		228					226.5
4 Be		114					111
5 B		67.6					
6 C		44.7					43.68
7 N		31.60					30.99
8 O		23.62					23.32
9 F		18.32					
10 Ne		14.610	14.452				14.3018
11 Na		11.9101	11.575				11.569
12 Mg		9.8900	9.5207				9.5122
13 Al	8.34173	8.33934	7.9605				7.94813
14 Si	7.12791	7.12542	6.7530				6.738
15 P	6.1598	6.1568	5.7960				5.784
16 S	5.37496	5.37216		5.0316			5.0185
17 Cl	4.7307	4.7278	4.4034				4.3971
18 A	4.19474	4.19180	3.8860				3.87090
19 K	3.7445	3.7414	3.4539				3.4365
20 Ca	3.36166	3.35839	3.0897				3.0703
21 Sc	3.0342	3.0309	2.7796				2.762
22 Ti	2.75216	2.74851	2.51391				2.49734
23 V	2.50738	2.50356	2.28440				2.2691
24 Cr	2.293606	2.28970	2.08487				2.07020
25 Mn	2.10578	2.101820	1.91021				1.89643
26 Fe	1.939980	1.936042	1.75661				1.74346
27 Co	1.792850	1.788965	1.62079				1.60815
28 Ni	1.661747	1.657910	1.500135				1.48807
29 Cu	1.544398	1.540562	1.392218			1.38109	1.38059
30 Zn	1.439000	1.435155	1.29525			1.28372	1.2834
31 Ga	1.34399	1.340083	1.20835	1.20789		1.19600	1.1958
32 Ge	1.258011	1.254054	1.12936	1.12894		1.11686	1.11658
33 As	1.17987	1.17588	1.05783	1.05730		1.04500	1.0450
34 Se	1.10882	1.10477	0.99268	0.99218		0.97992	0.97974
35 Br	1.04382	1.03974	0.93327	0.93279		0.92046	0.9204
36 Kr	0.9841	0.9801	0.8790	0.8785		0.8661	0.86552
37 Rb	0.92969	0.925553	0.82921	0.82868		0.81645	0.81554
38 Sr	0.87943	0.87526	0.78345	0.78292		0.77081	0.76973
39 Y	0.83305	0.82884	0.74126	0.74072		0.72864	0.72766
40 Zr	0.79015	0.78593	0.70228	0.70173		0.68993	0.68883

Table A–5 X-ray emission and absorption wavelengths in angstroms (*Cont.*)

Element	$K\alpha_2$	$K\alpha_1$	$K\beta_3$	$K\beta_1$	$K\beta_2^{II}$	$K\beta_2^{I}$	K absorption edge
Approximate intensity	50	100	15 to 30		2 to 5		
41 Nb	0.75044	0.74620	0.66634	0.66576	0.65416		0.65298
42 Mo	0.713590	0.709300	0.63286	0.632288	0.62099		0.61978
43 Tc	0.67932	0.67502	0.60188	0.60130	0.59024		0.58906
44 Ru	0.647408	0.643083	0.573067	0.572482	0.56166		0.56051
45 Rh	0.617630	0.613279	0.546200	0.545605	0.53503		0.53395
46 Pd	0.589821	0.585448	0.521123	0.520520	0.510228		0.50920
47 Ag	0.563798	0.5594075	0.497685	0.497069	0.487032		0.48589
48 Cd	0.539422	0.535010	0.475730	0.475105	0.465328		0.46407
49 In	0.516544	0.512113	0.455181	0.454545	0.44500		0.44371
50 Sn	0.495053	0.490599	0.435877	0.435236	0.425915		0.42467
51 Sb	0.474827	0.470354	0.417737	0.417085	0.407973		0.40668
52 Te	0.455784	0.451295	0.400659	0.399995	0.391102		0.38974
53 I	0.437829	0.433318	0.384564	0.383905	0.37523		0.37381
54 Xe	0.42087	0.41634	0.36941	0.36872	0.36026		0.3584
55 Cs	0.404835	0.400290	0.355050	0.354364	0.34611		0.34451
56 Ba	0.389668	0.385111	0.341507	0.340811	0.33277		0.33104
57 La	0.375313	0.370737	0.328686	0.327983	0.320117		0.31844
58 Ce	0.361683	0.357092	0.316520	0.315816	0.30816		0.30648
59 Pr	0.348749	0.344140	0.304975	0.304261	0.29679		0.29518
60 Nd	0.336472	0.331846	0.294027	0.293299	0.2862		0.28453
61 Pm	0.324803	0.320160	0.28363	0.28290	0.2760		0.27431
62 Sm	0.313698	0.309040	0.27376	0.27301	0.2663		0.26464
63 Eu	0.303118	0.298446	0.264332	0.263577	0.25715	0.256823	0.25553
64 Gd	0.293038	0.288353	0.25534	0.25460	0.24816		0.24681
65 Tb	0.283423	0.278724	0.24683	0.24608	0.2399		0.23841
66 Dy	0.274247	0.269533	0.23862	0.23788	0.2319		0.23048
67 Ho	0.265486	0.260756	0.23083	0.23012	0.2243		0.22291
68 Er	0.257110	0.252365	0.22341	0.22266	0.2170		0.21567
69 Tm	0.249095	0.244338	0.21636	0.21556	0.2101		0.20880
70 Yb	0.241424	0.236655	0.2096	0.20884	0.2036		0.20224
71 Lu	0.234081	0.229298	0.20309	0.20231	0.1973		0.19585
72 Hf	0.227024	0.222227	0.19686	0.19607	0.1912		0.18982
73 Ta	0.220305	0.215497	0.190890	0.190089	0.185188	0.185011	0.18394
74 W	0.213828	0.2090100	0.185181	0.184374	0.17960	0.179421	0.17837
75 Re	0.207611	0.202781	0.179697	0.178880	0.17425	0.174054	0.17302
76 Os	0.201639	0.196794	0.174431	0.173611	0.16910	0.168906	0.16787
77 Ir	0.195904	0.191047	0.169367	0.168542	0.16415	0.163956	0.16292
78 Pt	0.190381	0.185511	0.164501	0.163675	0.15939	0.15920	0.15818
79 Au	0.185075	0.180195	0.154800	0.158982	0.15483	0.154618	0.153593
80 Hg	0.179958	0.175068	0.154920	0.154487	0.15040	0.15020	0.14918

Table A-5 X-ray emission and absorption wavelengths in angstroms (*Cont.*)

Element	$K\alpha_2$	$K\alpha_1$	$K\beta_3$	$K\beta_1$	$K\beta_2^{II}$	$K\beta_2^{I}$	K absorption edge
Approximate intensity	50	100	15 to 30		2 to 5		
81 Tl	0.175036	0.170136	0.150980	0.150142	0.14614	0.14595	0.14495
82 Pb	0.170294	0.165376	0.146810	0.145970	0.14212	0.14191	0.140880
83 Bi	0.165717	0.160789	0.142779	0.141948	0.13817	0.13797	0.13694
84 Po	0.16130	0.15636	0.13892	0.13807	0.13438	0.13418	
85 At	0.15705	0.15210	0.13517	0.13432	0.13072	0.13052	
86 Rn	0.15294	0.14798	0.13155	0.13069	0.12719	0.12698	
87 Fr	0.14896	0.14399	0.12807	0.12719	0.12379	0.12358	
88 Ra	0.14512	0.14014	0.12469	0.12382	0.12050	0.12029	
89 Ac	0.14141	0.136417	0.12143	0.12055	0.11732	0.11711	
90 Th	0.137829	0.132813	0.118268	0.117396	0.11426	0.114040	0.11307
91 Pa	0.134343	0.129325	0.11523	0.114345	0.11129	0.11107	
92 U	0.130968	0.125947	0.112296	0.111394	0.10837	0.10818	0.10723

Table A-6 The crystal structure of the elements*

Element: form (transformation temp. °C)	Temp. °C	Structure	Type	Lattice constants: a in A (space group;	b in A parameters)	c in A (α or β)	Inter-atomic dist., A	Density g per cm³	Atomic weight (1959)**	Volume per atom in A³
Actinium (Ac)	R.T.?	F.c.c.	*A*1	5.311	*Fm*3*m*		3.755	10.07	227	37.48
Aluminum (Al)	25	F.c.c.	*A*1	4.0496	*Fm*3*m*		2.863	2.698	26.98	16.60
Americium (Am)	R.T.	Hex.		3.642	*ABAC*···‡	11.76	3.614	11.87	(243)	33.77
Antimony (Sb)	25	Rhomb.	*A*7	4.5067	$R\bar{3}m$	57°6′27″	2.907	6.692	121.76	30.20
	26	Rhomb.	*A*7	4.307	(Hex. axes)	11.273				
				z = 0.2335						
	4°K	Rhomb.	*A*7	4.3007	z = 0.2336	11.222	2.902			
Argon (Ar)	83.8	F.c.c.	*A*1	5.467	*Fm*3*m*		3.866	1.623	39.944	40.00
	4°K	F.c.c.	*A*1	5.3108	*Fm*3*m*		3.756	1.770		
(metastable)	4°K	C.p.h.	*A*3	3.761	$P6_3/mmc$	6.143				
Arsenic (As)†		Rhomb.	*A*7	4.131	$R\bar{3}m$	α = 54°10′	3.16	5.766	74.92	21.54
		Orthorhomb.			*Cmca*···					
		Cubic								
Barium (Ba)	R.T.	B.c.c.	*A*2	5.019	*Im*3*m*		4.347	3.594	137.36	62.59
Beryllium, α† (Be)	20	C.p.h.	*A*3	2.2856	$P6_3/mmc$	3.5832	2.225	1.846	9.013	8.11
β? >1250	1250	B.c.c.	*A*2	2.55			2.21			8.29
Bismuth (Bi)	25	Rhomb.	*A*7	4.546	$R\bar{3}m$	11.862	3.071	9.803	208.99	35.38
				(hex. axes)	z = 0.2339					
	78°K	Rhomb.	*A*7	4.535	z = 0.2341	11.814	3.064			
	4°K	Rhomb.	*A*7		z = 0.2340	11.862				
Boron (B) α	R.T.	Rhomb.						2.466	10.82	7.67
1100 to 1300		Tetrag.						2.33		
B > 1300		Rhomb.						2.356		
Bromine (Br)	−150	Orthorhomb.		4.48	6.67	8.72		4.05	79.916	42.50
Cadmium (Cd)	21	C.p.h.	*A*3	2.9788	$P6_3/mmc$	5.6167	2.979	8.647	112.41	21.58

Calcium, α† (Ca)	18	F.c.c.	*A*1	5.582	*Fm*3*m*		3.947	1.530	40.08	43.48
β = unconfirmed			(*A*3)							
γ 464 to m.p.	~500	B.c.c.	*A*2	4.477	*Im*3*m*		3.877	1.483		
Carbon, diamond†	20	Cubic	*A*4	3.5670	*Fd*3*m*		1.544	3.516	12.011	5.68
Graphite, α†	20	Hex.	*A*9	2.4612	$P6_3/mmc$	6.7078	1.421	2.266		
Graphite, β†		Rhomb.		2.4612		10.0618	1.421	2.266		
Cerium (Ce) −10 to 730	20	F.c.c.	*A*1	5.1604	*Fm*3*m*		3.649	6.771	140.13	34.37
−150 to −10		C.p.h.	*A*3	3.68	*ABAC*···‡	11.92	3.66	6.844		
Below −150		F.c.c.	*A*1	4.85	*Fm*3*m*		3.429	8.32		
730 to m.p.		B.c.c.	*A*2	4.12	*Im*3*m*		3.57	6.67		
Cesium (Cs)	−10	B.c.c.	*A*2	6.14	*Im*3*m*		5.32	1.910	132.91	115.17
	5°K	B.c.c.	*A*2	6.045	*Im*3*m*		5.235			
Chlorine, α (Cl)	−160	Orthorhomb.		6.24	4.48	8.26	2.02	2.03	35.457	
Chromium† (Cr)	20	B.c.c.	*A*2	2.8846	*Im*3*m*		2.498	7.194	52.01	12.00
β (?) above 1850	R.T.	F.c.c.	*A*1	3.68	*Fm*3*m*		2.60	7.92		
Cobalt, α† (Co)	18	C.p.h.	*A*3	2.506		4.069	2.497	8.8	58.94	11.13
β stable ~ 450 to m.p.	18	F.c.c.	*A*1	3.544			2.506	8.7		
Copper (Cu)	20	F.c.c.	*A*1	3.6147			2.556	8.932	63.54	11.81
	0°	F.c.c.	*A*1	3.6029						
Dysprosium† to 950 (Dy)	R.T.	C.p.h.	*A*3	3.5923		5.6545	3.506	8.531	162.51	31.52
950 to m.p.		?								
Erbium (Er)	20	C.p.h.	*A*3	3.5590		5.592	3.470	9.044	167.27	30.64
Europium (Eu)	20	B.c.c.	*A*2	4.578			3.965	5.248	152.0	48.86
Fluorine (F)									19.00	
Gadolinium† (Gd)	20	C.p.h.	*A*3	3.6315		5.777	3.583	7.895	157.26	33.10
1262 to m.p.		B.c.c.?		4.06			3.52	7.8		
Gallium (Ga)	20	Orthorhomb.		4.5258	4.5198	7.6602	2.44	5.908	69.72	19.59
				Cmca	y = 0.1525					
	4°K	Orthorhomb.		4.516	4.490	7.633				
Germanium (Ge)	25	Cubic	*A*4	5.6576	*Fd*3*m*		2.450	5.324	72.60	22.64
Gold (Au)	25	F.c.c.	*A*1	4.0788			2.884	19.281	197.0	16.96
Hafnium α† (Hf)	24	C.p.h.	*A*3	3.1946		5.0511	3.127	13.28	178.50	22.16
β ~ 1950 to m.p.	R.T.	B.c.c.	*A*2	~3.50	(extrap.)		3.03			
Holmium† (Ho) to 966	R.T.	C.p.h.	*A*3	3.5761		5.6174	3.486	8.797	164.94	31.12
966 to m.p.		?								

Table A–6 The crystal structures of the elements* (Cont.)

Element: form (transformation temp.°C)	Temp. °C	Structure	Type	Lattice constants: a in A (space group;	b in A parameters)	c in A (α or β)	Inter-atomic dist., A	Density g per cm³	Atomic weight (1959)**	Volume per atom in A³
Indium (In)	R.T.	Tetrag.	*A*6	4.5979	(f.c. cell)	4.9467	3.251	7.290	114.82	26.15
	R.T.	Tetrag.	*A*6	3.2512	(b.c. cell)	4.9467	3.251	7.290	114.82	26.15
	4°K	Tetrag.	*A*6	4.5557		4.9342				
Iodine (I)	26	Orthorhomb.		4.79	7.25	9.78	3.54	4.953	126.91	42.66
Iridium (Ir)	26	F.c.c.	*A*1	3.8389			2.714	22.55	192.2	14.14
Iron, α† (Fe)	20	B.c.c.	*A*2	2.8664			2.482	7.873	55.85	11.77
γ 911 to 1392	916	F.c.c.	*A*1	3.6468			2.579	7.646		
δ 1392 to m.p.	1394	B.c.c.	*A*2	2.9322			2.539	7.356		
Krypton (Kr)	4°K	F.c.c.	*A*1	5.644				3.093	83.80	46.05
Lanthanum, α† (La)	20	Hex.		3.770	*ABAC*···‡	12.131	3.733	6.174	138.92	37.12
β 310 to 868	500	F.c.c.	*A*1	5.303			3.750	6.126		
γ 868 to m.p.		B.c.c.	*A*2	4.26			3.69	5.98		
Lead (Pb)	25	F.c.c.	*A*1	4.9502			3.500	11.341	207.21	30.33
Lithium (Li)†	20	B.c.c.	*A*2	3.5092			3.039	0.533	6.940	21.61
Below ~ 72°K	78°K	C.p.h.	*A*3	3.111		5.093	3.11			
Deformed below ~ 110°K	78°K	F.c.c.	*A*1	~4.40			3.11			
Lutetium (Lu)†	R.T.	C.p.h.	*A*3	3.5050		5.5486	3.434	9.842	174.99	29.50
1400 to m.p.		?								
Magnesium (Mg)	25	C.p.h.	*A*3	3.2094		5.2105	3.197	1.737	24.32	23.23
Manganese, α† (Mn)	25	Cubic	*A*12	8.9139	($I\bar{4}3m$)		2.24	7.473	54.94	12.21
β 742 to 1095	25	Cubic	*A*13	6.315	($P\bar{4}3m$)		2.37	7.24		
γ 1095 to 1133	1095	F.c.c.	*A*1	3.862			2.731	6.33		
δ 1133 to m.p.	1134	B.c.c.	*A*2	3.081			2.668	6.23		
Mercury (Hg)	−46	Rhomb.	*A*10	3.005		α = 70°31′	3.005	14.26	200.61	23.42
<79°K after pressure	77°K	B.c.t.		3.995		2.825	2.825	14.77		
Molybdenum (Mo)	20	B.c.c.	*A*2	3.1468			2.725	10.22	95.95	15.58

Neodymium, α† (Nd)	R.T.	Hex.		3.6582	$P6_3mmc$	11.802	3.658	7.004	144.27	34.18
868 to m.p.		B.c.c.	*A*2	4.13			3.58	6.80		
Neon (Ne)	4°K	F.c.c.	*A*1	4.4636			3.155	1.508	20.183	22.93
Neptunium, α† (Np)	20	Orthorhomb.	*Pnma*	4.723	4.887	6.663	2.60–2.64	20.45	(237)	19.22
β 280 to 577	313	Tetrag.	$P42_1$	4.897	u = 0.375	3.388	2.75	19.36		
γ 577 to m.p.	600	B.c.c.	(?)	3.52				18.12		
Nickel† (Ni)	18	F.c.c.	*A*1	3.5236			2.492	8.907	58.71	10.94
Niobium (Nb)	20	B.c.c.	*A*2	3.3007			2.858	8.578	92.91	17.98
(Columbium)										
Nitrogen, α (N_2)	20°K	Cubic		5.661	*Pa*3			1.026	14.008	
β 35.6–63.1°K	50°K	Hex.		4.036	*P*6/*mmc*	6.630		0.987		
Osmium (Os)	20	C.p.h.	*A*3	2.7353		4.3191	2.675	22.58	190.2	13.99
Oxygen, α (O_2) <23°K	−252	?						1.46	16.000	
β 23–44°K	27°K	Rhomb.		4.210	$R\bar{3}m$	α = 46°16′				
γ 44°K − m.p.	48°K	Cubic		6.83	*Pm*3*m*			1.30		
Palladium (Pd)	22	F.c.c.	*A*1	3.8907			2.751	11.995	106.4	14.72
Phosphorus, white (P)	−35	Cubic		7.17				2.22		
Complex form	−186	?							30.975	
Black† (*Cmca*)	R.T.	Orthorhomb.		3.32	10.52	4.39	2.18	2.69		16.59
Red	R.T.	Cubic		11.31				2.35		
Yellow, above −70	R.T.	Cubic		18.8	$I\bar{4}3m$			1.80		
Platinum (Pt)	20	F.c.c.	*A*1	3.9239			2.775	21.47	195.09	15.10
Plutonium, α† (Pu)	21	Monoclin.		6.1835	4.8244	10.973	3.1–3.3	19.814	(242)	23.4
				$P2_1/m$	β = 101.81°					
β 122 to 206	190	Monoclin.		9.284	10.463	7.859	2.97–	17.70		
				*I*2/*m*		β = 92.13°	3.10			
γ 206 to 319	235	Orthorhomb.		3.159	5.768	10.162	3.026	17.14		
				Fddd						
δ 319 to 451	320	F.c.c.	*A*1	4.637			3.279	15.92		
δ′ 451 to 485	477	Tetrag.		3.339		4.446	3.27	16.01		
ε 476 to m.p.	490	B.c.c.	*A*2	3.636			3.149	16.51		
Polonium, α† (Po)	∼10	Simple cubic		3.345	(1 atom/cell)		3.338	9.31	210	37.43
β	∼75	Rhomb.		3.359	$R\bar{3}m$	α = 98°13′		9.47		
Potassium (K)	78°K	B.c.c.	*A*2	5.247			4.544	0.899	39.100	75.31
	5°K	B.c.c.	*A*2	5.225			4.525	0.909		

Table A–6 The crystal structures of the elements* (Cont.)

Element: form (transformation temp.°C)	Temp. °C	Structure	Type	Lattice constants: a in A (space group;	b in A parameters)	c in A (α or β)	Inter-atomic dist., A	Density g per cm³	Atomic weight (1959)**	Volume per atom in A³
Praseodynium, α† (Pr)	R.T.	Hex.		3.6702	*ABAC*····‡	11.828	3.633	6.779	140.92	34.15
β 798 to m.p.		B.c.c.	*A*2	4.13			3.58	6.64		
Protactinium (Pa)	R.T.	B.c. tetrag.		3.925		3.238	3.210	15.37	231	24.94
Radium (Ra)									226	
Radon (Rn)									222	
Rhenium (Re)	26	C.p.h.	*A*3	2.760		4.458	2.741	21.03	186.22	14.70
Rhodium (Rh)	20	F.c.c.	*A*1	3.8044			2.690	12.42	102.91	13.77
Rubidium (Rb)	−196	B.c.c.	*A*2	5.610			4.858	1.607	85.48	92.67
	5°K	B.c.c.	*A*2	5.585			4.837	1.629		
Ruthenium (Ru)	25°K	C.p.h.	*A*3	2.7057		4.2816	2.650	12.36	101.1	13.57
Samarium† (Sm)	20	Rhomb.		8.996 *ABABCBCAC*···‡	$R\bar{3}m$	23°13′	3.588	7.536	150.35	33.01
917 to m.p.		?								
Scandium, α† (Sc)	R.T.	C.p.h.	*A*3	3.3080		5.267	3.252	2.992	44.96	23.41
Selenium† (Se)	25	Rhomb.	*A*8	4.3656	(Hex. axes)	4.9590	2.32	4.808	78.96	27.27
α (red metastable)	R.T.	Monoclin.	$P2_1/n$	9.05	9.07	11.61 β = 90°46′				
β (red metastable)	R.T.	Monoclin.	$P2_1/c$	12.85	8.07	9.31 β = 93°8′				
Silicon (Si)	20	Cubic	*A*4	5.4305	*Fd*3*m*		2.351	2.329	28.09	20.02
Silver (Ag)	25	F.c.c.	*A*1	4.0857			2.889	10.50	107.873	17.06
Sodium (Na)	20	B.c.c.	*A*2	4.2906			3.716	0.9660	22.991	39.50
(metastable)	5°K	B.c.c.	*A*2	4.225			3.659			
Below ~36°K	5°K	C.p.h.	*A*3	3.767		6.154	3.767	1.009		
Strontium† (Sr)	25	F.c.c.	*A*1	6.0849			4.302	2.583	87.63	56.32
~225 to 570 (?)	300	C.p.h.	*A*3	4.32		7.056	4.32			
~570 to m.p.	614	B.c.c.	*A*2	4.87			4.23			
Sulfur α, yellow† (S)	R.T.	Orthorhomb.		10.414	10.845	24.369		2.086	32.066	25.52
β, Monoclinic	R.T.	Monoclin.	$P2_1/c$	10.92	10.98	11.04 β = 83°16′		2.063		

[illegible]	[illegible]	[illegible]	[illegible]	2.743		4.400	2.71	11.497	(99)	14.21
Tellurium (Te)	25	Hex.	*A*8	4.4566	$P3_121$ or $P3_221$	5.9268	2.864	6.247	127.61	33.98
Terbium† (Tb)	25	C.p.h.	*A*3	3.5990		5.696	3.525	8.272	158.93	31.14
1310 to m.p.		?								
Thallium, α† (Tl)	18	C.p.h.	*A*3	3.4566		5.5248	3.457	11.87	204.39	28.58
β 230 to m.p.	262	B.c.c.	*A*2	3.882			3.362	11.60		
Thorium, α† (Th)	~25	F.c.c.	*A*3	5.0843			3.595	11.72	232	32.86
β 1400 to m.p.	1450	B.c.c.	*A*2	4.11			3.56	11.10		
Thulium (Tm)	20	C.p.h.	*A*3	3.5372		5.5619	3.537	9.325	168.94	30.10
Tin, α, gray (Sn)	20	Cubic	*A*4	6.4892			2.810	5.765	118.70	27.65
β, white† 13.2 to m.p.	25	Tetrag.	*A*5	5.8315	$I4_1/amd$	3.1814	3.022	7.285		
Titanium, α† (Ti)	25	C.p.h.	*A*3	2.9506		4.6788	2.890	4.508	47.90	17.65
β ~882 to m.p.	900	B.c.c.	*A*2	3.3065			2.863	4.400		
Tungsten (wolfram) (W)	21	B.c.c.	*A*2	3.1650			2.741	19.253	183.86	15.85
Uranium, α† (U) *Cmcm*	25	Orthorhomb.	*A*20	2.8536	5.8699	4.9555	2.754	19.05	238.07	20.81
				$y = 0.10245$						
	4°K	Orthorhomb.	*A*20	2.8444	5.8689	4.9316				
				$y = 0.10242$						
β 662 to 774	720	Tetrag.		10.759	$P4/mnm$	5.656	2.7–3.3	18.11		
γ 774 to m.p.	800	B.c.c.	*A*2	3.534			3.061	18.06		
Vanadium (V)	30	B.c.c.	*A*2	3.0282			2.622	6.09	50.95	13.88
Xenon (Xe)	4°K	F.c.c.	*A*1	6.131			4.335	3.784	131.30	61.04
Ytterbium† (Yb)	R.T.	F.c.c.	*A*1	5.481			3.876	6.97	173.04	33.02
798 to m.p.		B.c.c.	*A*2	4.44			3.85	6.54		
Yttrium (Y)	20	C.p.h.	*A*3	3.6474		5.7306	3.554	4.475	88.91	33.01
1460 to m.p.		B.c.c.	*A*2	4.11			3.56			
Zinc (Zn)	25	C.p.h.	*A*3	2.6649		4.9468	2.665	7.134	65.38	15.24
Zirconium, α† (Zr)	25	C.p.h.	*A*3	3.2312		5.1477	3.172	6.507	91.22	23.27
β 862 to m.p.	862	B.c.c.	*A*2	3.6090			3.125	6.443		

* Density is computed from x-ray data (with a few exceptions). Unit cell dimensions in angstroms (values in kx were multiplied by 1.00202). Data are from original sources and from "International Tables for X-ray Crystallography", vol. 3, Kynoch, Birmingham, England, 1962; also from W. B. Pearson, "Handbook of Lattice Spacings of Metals and Alloys," Pergamon, New York, 1958.

† The form judged to be the stable one at ordinary temperatures and pressures. See p. 233 for high-pressure forms.

‡ The sequence of letters indicates the sequence of close-packed atomic layers.

** Atomic weights of 1959 were based on the $O^{16} = 16$ scale; to convert to the new international $C^{12} = 12$ scale, divide these by 1.000043. However, newer values of greater precision appear frequently.

Name Index*

* Omitted from this index: editors of symposia and contributors to the data of the tables on pages 234, 404, and 415.

Subject Index

Other Titles in the Series

ASHBY & JONES
Engineering Materials. An Introduction to their Properties and Applications

BARRETT & MASSALSKI
Structure of Metals, 3rd Edition

BISWAS & DAVENPORT
Extractive Metallurgy of Copper, 2nd Edition

CHRISTIAN
The Theory of Transformations in Metals and Alloys, Part 1, 2nd Edition

COUDURIER, HOPKINS & WILKOMIRSKY
Fundamentals of Metallurgical Processes

DOWSON & HIGGINSON
Elasto-Hydrodynamic Lubrication, SI Edition

GABE
Principles of Metal Surface Treatment and Protection, 2nd Edition

GILCHRIST
Extraction Metallurgy, 2nd Edition
Fuels, Furnaces and Refractories

HARRIS
Mechanical Working of Metals

HEARN
Mechanics of Materials

HULL
Introduction to Dislocations, 2nd Edition

KUBASCHEWSKI & ALCOCK
Metallurgical Thermochemistry, 5th Edition

MARSCHALL & MARINGER
Dimensional Stability — An Introduction

MASUBUCHI
Analysis of Welded Structures, Residual Stresses,
Distortion and their Consequences

PARKER
An Introduction to Chemical Metallurgy, 2nd Edition

PEACEY & DAVENPORT
The Iron Blast Furnace: Theory & Problems

SARKAR
Wear of Metals

SCULLY
Fundamentals of Corrosion, 2nd Edition

UPADHYAYA & DUBE
Problems in Metallurgical Thermodynamics and Kinetics

WILLS
Mineral Processing Technology, 2nd Edition